An Overview of Nuclear & Radioactive Accidents

Edited by Paul F. Kisak

Contents

Chapter 1

Nuclear and radiation accidents and incidents

See also: Lists of nuclear disasters and radioactive incidents

A **nuclear and radiation accident** is defined by the

Following the 2011 Japanese Fukushima nuclear disaster, authorities shut down the nation's 54 nuclear power plants. As of 2013, the Fukushima site remains highly radioactive, with some 160,000 evacuees still living in temporary housing, and some land will be unfarmable for centuries. The difficult cleanup job will take 40 or more years, and cost tens of billions of dollars.[1]*[2]

The Kashiwazaki-Kariwa Nuclear Power Plant, a Japanese nuclear plant with seven units, the largest single nuclear power station in the world, was completely shut down for 21 months following an earthquake in 2007.[3]

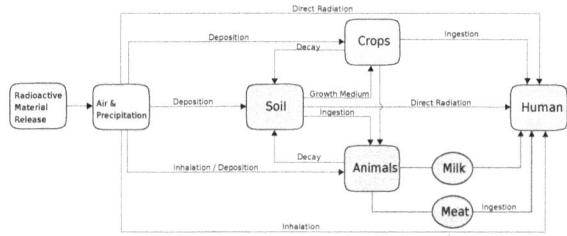

Pathways from airborne radioactive contamination to human

International Atomic Energy Agency (IAEA) as "an event that has led to significant consequences to people, the environment or the facility." Examples include lethal effects to individuals, large radioactivity release to the environment, or reactor core melt." *[4] The prime example of a "major nuclear accident" is one in which a reactor core is damaged

and significant amounts of radioactivity are released, such as in the Chernobyl disaster in 1986.*[5]

The impact of nuclear accidents has been a topic of debate since the first nuclear reactors were constructed in 1954, and has been a key factor in public concern about nuclear facilities.*[6] Technical measures to reduce the risk of accidents or to minimize the amount of radioactivity released to the environment have been adopted, however human error remains, and "there have been many accidents with varying impacts as well near misses and incidents".*[6]*[7] As of 2014, there have been more than 100 serious nuclear accidents and incidents from the use of nuclear power. Fifty-seven accidents have occurred since the Chernobyl disaster, and about 60% of all nuclear-related accidents have occurred in the USA.*[8] Serious nuclear power plant accidents include the Fukushima Daiichi nuclear disaster (2011), Chernobyl disaster (1986), Three Mile Island accident (1979), and the SL-1 accident (1961).*[9] Nuclear power accidents can involve loss of life and large monetary costs for remediation work.[*][10]

Nuclear-powered submarine core meltdown and other mishaps include the K-19 (1961), K-11 (1965), K-27 (1968), K-140 (1968), K-429 (1970), K-222 (1980), and K-431 (1985).[*][9][*][11][*][12] Serious radiation accidents include the Kyshtym disaster, Windscale fire, radiotherapy accident in Costa Rica,[*][13] radiotherapy accident in Zaragoza,[*][14] radiation accident in Morocco,[*][15] Goiania accident,[*][16] radiation accident in Mexico City, radiotherapy unit accident in Thailand,[*][17] and the Mayapuri radiological accident in India.[*][17]

The IAEA maintains a website reporting recent accidents.[*][18]

1.1 Nuclear power plant accidents

The abandoned city of Prypiat, Ukraine, following the Chernobyl disaster. The Chernobyl nuclear power plant is in the background.

See also: Nuclear reactor accidents in the United States, List of nuclear power accidents by country, and List of nuclear and radiation fatalities by country

One of the worst nuclear accidents to date was the Chernobyl disaster which occurred in 1986 in Ukraine. The accident killed 31 people directly and damaged approximately $7 billion of property. A study published in 2005 estimates that there will eventually be up to 4,000 additional cancer deaths related to the accident among those exposed to significant radiation levels.[*][19] Radioactive fallout from the accident was concentrated in areas of Belarus, Ukraine and Russia. Other studies have estimated as many as over a million eventual cancer deaths from Chernobyl.[*][20] [*][21] Estimates of eventual deaths from cancer are highly contested. Industry, UN and DOE agencies claim low num-

bers of legally provable cancer deaths will be traceable to the disaster. The UN, DOE and industry agencies all use the limits of the epidemiological resolvable deaths as the cutoff below which they cannot be legally proven to come from the disaster. Independent studies statistically calculate fatal cancers from dose and population, even though the number of additional cancers will be below the epidemiological threshold of measurement of around 1%. These are two very different concepts and lead to the huge variations in estimates. Both are reasonable projections with different meanings. Approximately 350,000 people were forcibly resettled away from these areas soon after the accident.[*][19]

Social scientist and energy policy expert, Benjamin K. Sovacool has reported that worldwide there have been 99 accidents at nuclear power plants from 1952 to 2009 (defined as incidents that either resulted in the loss of human life or more than US$50,000 of property damage, the amount the US federal government uses to define major energy accidents that must be reported), totaling US$20.5 billion in property damages.[*][8] Fifty-seven accidents have occurred since the Chernobyl disaster, and almost two-thirds (56 out of 99) of all nuclear-related accidents have occurred in the US. There have been comparatively few fatalities associated with nuclear power plant accidents.[*][8]

1.2 Nuclear reactor attacks

Main article: Vulnerability of nuclear plants to attack
See also: Nuclear terrorism

The vulnerability of nuclear plants to deliberate attack is of concern in the area of nuclear safety and security.[*][29] Nuclear power plants, civilian research reactors, certain naval fuel facilities, uranium enrichment plants, fuel fabrication plants, and even potentially uranium mines are vulnerable to attacks which could lead to widespread radioactive contamination. The attack threat is of several general types: commando-like ground-based attacks on equipment which if disabled could lead to a reactor core meltdown or widespread dispersal of radioactivity; and external attacks such as an aircraft crash into a reactor complex, or cyber attacks.[*][30]

The United States 9/11 Commission has said that nuclear power plants were potential targets originally considered for the September 11, 2001 attacks. If terrorist groups could sufficiently damage safety systems to cause a core meltdown at a nuclear power plant, and/or sufficiently damage spent fuel pools, such an attack could lead to widespread radioactive contamination. The Federation of American Scientists have said that if nuclear power use is to expand significantly, nuclear facilities will have to be made extremely safe from attacks that could release massive quantities of radioactivity into the community. New reactor designs have features of passive nuclear safety, which may help. In the United States, the NRC carries out "Force on Force" (FOF) exercises at all Nuclear Power Plant (NPP) sites at least once every three years.*[30]

Nuclear reactors become preferred targets during military conflict and, over the past three decades, have been repeatedly attacked during military air strikes, occupations, invasions and campaigns.*[31] Various acts of civil disobedience since 1980 by the peace group Plowshares have shown how nuclear weapons facilities can be penetrated, and the group's actions represent extraordinary breaches of security at nuclear weapons plants in the United States. The National Nuclear Security Administration has acknowledged the seriousness of the 2012 Plowshares action. Non-proliferation policy experts have questioned "the use of private contractors to provide security at facilities that manufacture and store the government's most dangerous military material".*[32] Nuclear weapons materials on the black market are a global concern,*[33]*[34] and there is concern about the possible detonation of a small, crude nuclear weapon or dirty bomb by a militant group in a major city, causing significant loss of life and property.*[35]*[36]

The number and sophistication of cyber attacks is on the rise. *Stuxnet* is a computer worm discovered in June 2010 that is believed to have been created by the United States and Israel to attack Iran's nuclear facilities. It switched off safety devices, causing centrifuges to spin out of control.*[37] The computers of South Korea's nuclear plant operator (KHNP) were hacked in December 2014. The cyber attacks involved thousands of phishing emails containing malicious codes, and information was stolen.*[38]

1.3 Radiation and other accidents and incidents

See also: List of civilian radiation accidents and List of nuclear weapons tests of the United States

Serious radiation and other accidents and incidents include:

1940s

- May 1945: Albert Stevens was one of several subjects of a human radiation experiment, and was injected with plutonium without his knowledge or informed consent. Although Stevens was the person who received the highest dose of radiation during the plutonium experiments, he was neither the first nor the last subject to be studied. Eighteen people aged 4 to

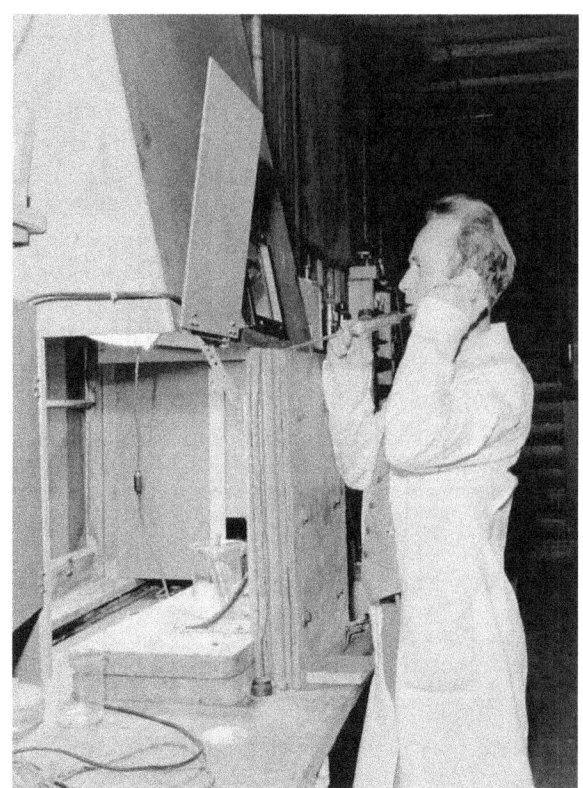

Dr. Joseph G. Hamilton was the primary researcher for the human plutonium experiments done at U.C. San Francisco from 1944 to 1947.[39] Hamilton wrote a memo in 1950 discouraging further human experiments because the AEC would be left open "to considerable criticism," since the experiments as proposed had "a little of the Buchenwald touch."* [40]*

69 were injected with plutonium. Subjects who were chosen for the experiment had been diagnosed with a terminal disease. They lived from 6 days up to 44 years past the time of their injection.*[39] Eight of the 18 died within two years of the injection.*[39] All died from their preexisting terminal illness, or cardiac illnesses. None died from the plutonium itself. Patients from Rochester, Chicago, and Oak Ridge were also injected with plutonium in the Manhattan Project human experiments.*[39]*[43]*[44]

- 6–9 August 1945: On the orders of President Harry S. Truman, a uranium-gun design bomb, Little Boy, was used against the city of Hiroshima, Japan. Fat Man, a plutonium implosion-design bomb was used against the city of Nagasaki. The two weapons killed approximately 120,000 to 140,000 civilians and military personnel instantly and thousands more have died over the years from radiation sickness and related cancers.

- August 1945: Criticality accident at US Los Alamos National Laboratory. Harry Daghlian dies.*[45]

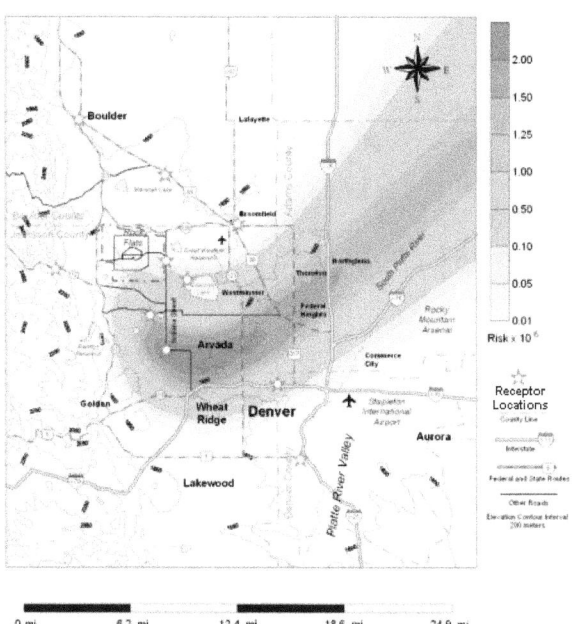

One of four example estimates of the plutonium (Pu-239) plume from the 1957 fire at the Rocky Flats Nuclear Weapons Plant. Public protests and a combined Federal Bureau of Investigation and United States Environmental Protection Agency raid in 1989 stopped production at the plant.

- May 1946: Criticality accident at Los Alamos National Laboratory. Louis Slotin dies.*[45]

1950s

- February 13, 1950: a Convair B-36B crashed in northern British Columbia after jettisoning a Mark IV atomic bomb. This was the first such nuclear weapon loss in history.

- December 12, 1952: NRX AECL Chalk River Laboratories, Chalk River, Ontario, Canada. Partial meltdown, about 10,000 Curies released.*[46] Approximately 1202 people were involved in the two year cleanup.*[47] Future president Jimmy Carter was one of the many people that helped clean up the accident.*[48]

- 15/03/1953 – Mayak, Former Soviet Union. Criticality accident. Contamination of plant personnel occurred.*[45]

- 1954: The 15 Mt Castle Bravo shot of 1954 which spread considerable nuclear fallout on many Pacific islands, including several which were inhabited, and some that had not been evacuated.*[49]

- March 1, 1954: Daigo Fukuryū Maru, 1 fatality.

Corroded and leaking 55-gallon drum, for storing radioactive waste at the Rocky Flats Plant, tipped on its side so the bottom is showing.

The Hanford site represents two-thirds of USA's high-level radioactive waste by volume. Nuclear reactors line the riverbank at the Hanford Site along the Columbia River in January 1960.

- September 1957: a plutonium fire occurred at the Rocky Flats Plant, which resulted in the contamination of Building 71 and the release of plutonium into the atmosphere, causing US $818,600 in damage.

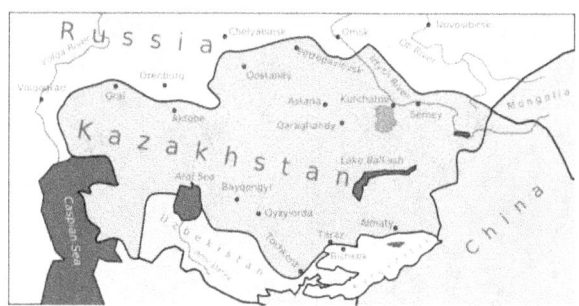

The 18,000 km^2 expanse of the Semipalatinsk Test Site (indicated in red), which covers an area the size of Wales. The Soviet Union conducted 456 nuclear tests at Semipalatinsk from 1949 until 1989 with little regard for their effect on the local people or environment. The full impact of radiation exposure was hidden for many years by Soviet authorities and has only come to light since the test site closed in 1991.[42]*

On Feb. 14, 2014, at the WIPP, radioactive materials leaked from a damaged storage drum (see photo). Analysis of several accidents, by DOE, have shown lack of a "safety culture" at the facility.[41]*

2007 ISO radioactivity danger symbol. The red background is intended to convey urgent danger, and the sign is intended to be used in places or on equipment where exceptionally intense radiation fields could be encountered or created through misuse or tampering. The intention is that a normal user will never see such a sign, however after partly dismantling the equipment the sign will be exposed warning that the person should stop work and leave the scene

- 21/04/1957 - Mayak, Former Soviet Union. Criticality accident in the factory number 20 in the collection oxalate decantate after filtering sediment oxalate enriched uranium. Six people received doses of 300 to 1,000 rem (four women and two men), one woman died.*[45]

- September 1957: Kyshtym disaster: Nuclear waste storage tank explosion at Chelyabinsk, Russia. 200+ fatalities, believed to be a conservative estimate; 270,000 people were exposed to dangerous radiation levels. Over thirty small communities were removed from Soviet maps between 1958 and 1991.*[50] (INES level 6)*[23]

- October 1957: Windscale fire, UK. Fire ignites plutonium piles and contaminates surrounding

dairy farms.*[8]*[51] An estimated 33 cancer deaths.*[8]*[51]

- 1957-1964: Rocketdyne located at the Santa Susanna Field Lab, 30 miles north of Los Angeles, California operated ten experimental nuclear reactors. Numerous accidents occurred including a core meltdown. Experimental reactors of that era were not required to have the same type of containment structures that shield modern nuclear reactors. During the Cold War time in which the accidents that occurred at Rocketdyne, these events were not publicly reported by the Department of Energy.*[52]

- 1958: Fuel rupture and fire at the National Research Universal reactor (NRU), Chalk River, Canada.

- 10/02/1958 - Mayak, Former Soviet Union. Criticality accident in SCR plant. Conducted experiments to determine the critical mass of enriched uranium in a cylindrical container with different concentrations of uranium in solution. Staff broke the rules and instructions for working with YADM (nuclear fissile material). When SCR personnel received doses from 7600 to 13,000 rem. Three people died, one man got radiation sickness and went blind.*[45]

- December 30, 1958: Cecil Kelley criticality accident at Los Alamos National Laboratory.*[45]*[53]

- March 1959: Santa Susana Field Laboratory, Los Angeles, California. Fire in a fuel processing facility.

- July 1959: Santa Susana Field Laboratory, Los Angeles, California. Partial meltdown.

1960s

- 7 June 1960: the 1960 Fort Dix IM-99 accident destroyed a CIM-10 Bomarc nuclear missile and shelter and contaminated the BOMARC Missile Accident Site in New Jersey.

- 24 January 1961: the 1961 Goldsboro B-52 crash occurred near Goldsboro, North Carolina. A B-52 Stratofortress carrying two Mark 39 nuclear bombs broke up in mid-air, dropping its nuclear payload in the process.*[54]*[55]

- July 1961: soviet submarine K-19 accident. Eight fatalities and more than 30 people were over-exposed to radiation.*[56]

- March, 21 -August 1962: radiation accident in Mexico City, four fatalities.

- May 1962: The Cuban Missile Crisis was a 13-day confrontation in October 1962 between the Soviet Union and Cuba on one side and the United States on the other side. The crisis is generally regarded as the moment in which the Cold War came closest to turning into a nuclear conflict*[57] and is also the first documented instance of mutual assured destruction (MAD) being discussed as a determining factor in a major international arms agreement.*[58]*[59]

- 23 July, 1964: Wood River Junction criticality accident. Resulted in 1 fatality

- 1964, 1969: Santa Susana Field Laboratory, Los Angeles, California. Partial meltdowns.

- 1965 Philippine Sea A-4 crash, where a Skyhawk attack aircraft with a nuclear weapon fell into the sea.*[60] The pilot, the aircraft, and the B43 nuclear bomb were never recovered.*[61] It was not until the 1980s that the Pentagon revealed the loss of the one-megaton bomb.*[62]

- October 1965: US CIA-led expedition abandons a nuclear-powered telemetry relay listening device on Nanda Devi *[63]

- January 17, 1966: the 1966 Palomares B-52 crash occurred when a B-52G bomber of the USAF collided with a KC-135 tanker during mid-air refuelling off the coast of Spain. The KC-135 was completely destroyed when its fuel load ignited, killing all four crew members. The B-52G broke apart, killing three of the seven crew members aboard.*[64] Of the four Mk28 type hydrogen bombs the B-52G carried,*[65] three were found on land near Almería, Spain. The non-nuclear explosives in two of the weapons detonated upon impact with the ground, resulting in the contamination of a 2-square-kilometer (490-acre) (0.78 square mile) area by radioactive plutonium.*[66] The fourth, which fell into the Mediterranean Sea, was recovered intact after a 2½-month-long search.*[67]

- January 21, 1968: the 1968 Thule Air Base B-52 crash involved a United States Air Force (USAF) B-52 bomber. The aircraft was carrying four hydrogen bombs when a cabin fire forced the crew to abandon the aircraft. Six crew members ejected safely, but one who did not have an ejection seat was killed while trying to bail out. The bomber crashed onto sea ice in Greenland, causing the nuclear payload to rupture and disperse, which resulted in widespread radioactive contamination.

- May 1968: Soviet submarine K-27 reactor near meltdown. 9 people died, 83 people were injured.*[12] In August 1968, the Project 667 A - Yankee class nuclear submarine K-140 was in the naval yard at Severodvinsk for repairs. On August 27, an uncontrolled increase of the reactor's power occurred following work to upgrade the vessel. One of the reactors started up automatically when the control rods were raised to a higher position. Power increased to 18 times its normal amount, while pressure and temperature levels in the reactor increased to four times the normal amount. The automatic start-up of the reactor was caused by the incorrect installation of the control rod electrical cables and by operator error. Radiation levels aboard the vessel deteriorated.

- 10/12/1968 - Mayak, Former Soviet Union. Criticality accident. Plutonium solution was poured into a cylindrical container with dangerous geometry. One person died, another took a high dose of radiation and radiation sickness, after which he had two legs and his right arm amputated.[*][45]

- January 1969: Lucens reactor in Switzerland undergoes partial core meltdown leading to massive radioactive contamination of a cavern.

1970s

- 1974–1976: Columbus radiotherapy accident, 10 fatalities, 88 injuries from cobalt-60 source.[*][12][*][68]

- July 1978: Anatoli Bugorski was working on U-70, the largest Soviet particle accelerator, when he accidentally exposed his head directly to the proton beam. He survived, despite suffering some long-term damage.

- July 1979: Church Rock Uranium Mill Spill in New Mexico, USA, when United Nuclear Corporation's uranium mill tailings disposal pond breached its dam. Over 1,000 tons of radioactive mill waste and millions of gallons of mine effluent flowed into the Puerco River, and contaminants traveled downstream.[*][69]

1980s

- 1980 to 1989: The Kramatorsk radiological accident happened in Kramatorsk, Ukrainian SSR. In 1989, a small capsule containing highly radioactive caesium-137 was found inside the concrete wall of an apartment building. 6 residents of the building died from leukemia and 17 more received varying radiation doses. The accident was detected only after the residents called in a health physicist.

- 1980: Houston radiotherapy accident, 7 fatalities.[*][12][*][68]

- October 5, 1982: Lost radiation source, Baku, Azerbaijan, USSR. 5 fatalities, 13 injuries.[*][12]

- March 1984: Radiation accident in Morocco, eight fatalities from overexposure to radiation from a lost iridium-192 source.[*][15]

- 1984: Fernald Feed Materials Production Center gained notoriety when it was learned that the plant was releasing millions of pounds of uranium dust into the atmosphere, causing major radioactive contamination of the surrounding areas. That same year, employee Dave Bocks, a 39-year-old pipefitter, disappeared during the facility's graveyard shift and was later reported

missing. Eventually, his remains were discovered inside a uranium processing furnace located in Plant 6.[*][70]

- August 1985: Soviet submarine K-431 accident. Ten fatalities and 49 other people suffered radiation injuries.[*][9]

- January 4, 1986: an overloaded tank at Sequoyah Fuels Corporation ruptured and released 14.5 tons of uranium hexafluoride gas (UF6), causing the death of a worker, the hospitalization of 37 other workers, and approximately 100 downwinders.[*][71] [*][72][*][73]

- October 1986: Soviet submarine K-219 reactor almost had a meltdown. Sergei Preminin died after he manually lowered the control rods, and stopped the explosion. The submarine sank three days later.

- September 1987: Goiania accident. Four fatalities, and following radiological screening of more than 100,000 people, it was ascertained that 249 people received serious radiation contamination from exposure to caesium-137.[*][16][*][74] In the cleanup operation, topsoil had to be removed from several sites, and several houses were demolished. All the objects from within those houses were removed and examined. *Time* magazine has identified the accident as one of the world's "worst nuclear disasters" and the International Atomic Energy Agency called it "one of the world's worst radiological incidents" .[*][74][*][75]

- 1989: San Salvador, El Salvador; one fatality due to violation of safety rules at cobalt-60 irradiation facility.[*][76]

1990s

- 1990: Soreq, Israel; one fatality due to violation of safety rules at cobalt-60 irradiation facility.[*][76]

- December 16 - 1990: radiotherapy accident in Zaragoza. Eleven fatalities and 27 other patients were injured.[*][56]

- 1991: Neswizh, Belarus; one fatality due to violation of safety rules at cobalt-60 irradiation facility.[*][76]

- 1992: Jilin, China; three fatalities at cobalt-60 irradiation facility.[*][76]

- 1992: USA; one fatality.[*][76]

- April 1993: accident at the Tomsk-7 Reprocessing Complex, when a tank exploded while being cleaned with nitric acid. The explosion released a cloud of radioactive gas. (INES level 4).[*][23]

- 1994: Tammiku, Estonia; one fatality from disposed caesium-137 source.*[76]

- August —December 1996: Radiotherapy accident in Costa Rica. Thirteen fatalities and 114 other patients received an overdose of radiation.*[13]

- 1996: an accident at Pelindaba research facility in South Africa results in the exposure of workers to radiation. Harold Daniels and several others die from cancers and radiation burns related to the exposure.*[77]

- June 1997: Sarov, Russia; one fatality due to violation of safety rules.*[76]

- May 1998: The Acerinox accident was an incident of radioactive contamination in Southern Spain. A caesium-137 source managed to pass through the monitoring equipment in an Acerinox scrap metal reprocessing plant. When melted, the caesium-137 caused the release of a radioactive cloud.

- September 1999: two fatalities at criticality accident at Tokaimura nuclear accident (Japan)

2000s

- January–February 2000: Samut Prakan radiation accident: three deaths and ten injuries resulted in Samut Prakan when a cobalt-60 radiation-therapy unit was dismantled.*[17]

- May 2000: Meet Halfa, Egypt; two fatalities due to radiography accident.*[76]

- August 2000 – March 2001: Instituto Oncologico Nacional of Panama, 17 fatalities. Patients receiving treatment for prostate cancer and cancer of the cervix receive lethal doses of radiation.*[12]*[78]

- August 9, 2004: Mihama Nuclear Power Plant accident, 4 fatalities. Hot water and steam leaked from a broken pipe (not actually a radiation accident).*[79]

- 9 May 2005: it was announced that Thermal Oxide Reprocessing Plant in the UK suffered a large leak of a highly radioactive solution, which first started in July 2004.*[80]

- April 2010: Mayapuri radiological accident, India, one fatality after a cobalt-60 research irradiator was sold to a scrap metal dealer and dismantled.*[17]

2010s

- March 2011: Fukushima I nuclear accidents, Japan and the radioactive discharge at the Fukushima Daiichi Power Station.*[81]

- January 17, 2014: At the Rössing Uranium Mine, Namibia, a catastrophic structural failure of a leach tank resulted in a major spill.*[82] The France-based laboratory, CRIIRAD, reported elevated levels of radioactive materials in the area surrounding the mine.*[83]*[84] Workers were not informed of the dangers of working with radioactive materials and the health effects thereof.*[85]*[86]*[87]

- February 1, 2014: Designed to last ten thousand years, the Waste Isolation Pilot Plant (WIPP) site had its first leak of airborne radioactive materials.*[88]*[89] 140 employees working underground at the time were sheltered indoors. 13 of these tested positive for internal radioactive contamination. Internal exposure to radioactive isotopes is more serious than external exposure, as these particles lodge in the body for decades, irradiating the surrounding tissues, thus increasing the risk of future cancers and other health effects. A second leak at the plant occurred shortly after the first, releasing plutonium and other radiotoxins, causing concern for communities living near the repository.*[90]

1.4 Worldwide nuclear testing summary

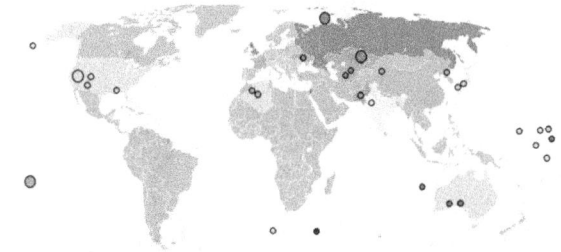

Over 2,000 nuclear tests have been conducted, in over a dozen different sites around the world. Red Russia/Soviet Union, blue France, light blue United States, violet Britain, black Israel, orange China, yellow India, brown Pakistan, green North Korea and light green (territories exposed to nuclear bombs)

Between 16 July 1945 and 23 September 1992, the United States maintained a program of vigorous nuclear testing, with the exception of a moratorium between November 1958 and September 1961. By official count, a total of 1,054 nuclear tests and two nuclear attacks were conducted, with over 100 of them taking place at sites in the Pacific Ocean, over 900 of them at the Nevada Test Site, and ten on miscellaneous sites in the United States (Alaska, Colorado, Mississippi, and New Mexico).*[91] Until November 1962, the vast majority of the U.S. tests were atmospheric (that is, above-ground); after the acceptance of the Partial Test Ban

Operation Crossroads Test Able, *a 23-kiloton air-deployed nuclear weapon detonated on July 1, 1946. This bomb used, and consumed, the infamous Demon core that took the lives of two scientists in two separate criticality accidents.*

Radioactive materials were accidentally released from the 1970 Baneberry Nuclear Test at the Nevada Test Site.

Treaty all testing was regulated underground, in order to prevent the dispersion of nuclear fallout.

The U.S. program of atmospheric nuclear testing exposed a number of the population to the hazards of fallout. Estimating exact numbers, and the exact consequences, of people exposed has been medically very difficult, with the exception of the high exposures of Marshall Islanders and Japanese fishers in the case of the Castle Bravo incident in 1954. A number of groups of U.S. citizens —especially farmers and inhabitants of cities downwind of the Nevada

Test Site and U.S. military workers at various tests —have sued for compensation and recognition of their exposure, many successfully. The passage of the Radiation Exposure Compensation Act of 1990 allowed for a systematic filing of compensation claims in relation to testing as well as those employed at nuclear weapons facilities. As of June 2009 over $1.4 billion total has been given in compensation, with over $660 million going to "downwinders".[*][92]

This view of downtown Las Vegas shows a mushroom cloud in the background. Scenes such as this were typical during the 1950s. From 1951 to 1962 the government conducted 100 atmospheric tests at the nearby Nevada Test Site.

[1] Including salvo tests counted as a single test.

[2] Detonations include zero-yield detonations in safety tests and failed full yield tests, but not those in the accident category listed above.

[3] As declared so by the nation testing; some may have been dual use.

[4] Defined as these classes of tests: atmospheric, surface, barge, cratering, space, and underwater tests.

[5] Including five tests in which the devices were destroyed before detonation, and the combat bombs dropped on Japan in World War II

[6] Includes both application tests and research tests at NTS.

[7] When the yield reads "< 20 kt" this total assumes the yield was half the maximum, i.e., 10 kt.

[8] Includes the test left behind in Semipalatinsk and 13 apparent failures not in the official list.

[9] 124 applications tests and 32 research tests which helped design better PNE charges.

[10] Includes the 31 *Vixen* tests, which were safety tests.

WARNING

January 11, 1951

From this day forward the U. S. Atomic Energy Commission has been authorized to use part of the Las Vegas Bombing and Gunnery Range for test work necessary to the atomic weapons development program.

Test activities will include experimental nuclear detonations for the development of atomic bombs – so-called "A-Bombs" – carried out under controlled conditions.

Tests will be conducted on a routine basis for an indefinite period.

NO PUBLIC ANNOUNCEMENT OF THE TIME OF ANY TEST WILL BE MADE

Unauthorized persons who pass inside the limits of the Las Vegas Bombing and Gunnery Range may be subject to injury from or as a result of the AEC test activities.

Health and safety authorities have determined that no danger from or as a result of AEC test activities may be expected outside the limits of the Las Vegas Bombing and Gunnery Range. All necessary precautions, including radiological surveys and patrolling of the surrounding territory, will be undertaken to insure that safety conditions are maintained.

Full security restrictions of the Atomic Energy Act will apply to the work in this area.

RALPH P. JOHNSON, Project Manager
Las Vegas Project Office
U. S. Atomic Energy Commission

This handbill was distributed 16 days before the first nuclear device was detonated at the Nevada Test Site.

[11] Including two possible safety tests in 1978, which don't appear on other lists.

[12] Four of the tests at In Ekker were the focus of attention by APEX (Application pacifique des expérimentations nucléaires). They even gave them different names, causing confusion.

[13] Includes one bomb destroyed before detonation by a failed parachute.

[14] Indira Gandhi, in her capacity as India's Minister of Atomic Energy at the time, declared the *Smiling Buddha* test to have been a test for the peaceful uses of atomic power.

[15] There is some uncertainty as to exactly how many bombs were exploded in each of Pakistan's tests. It could be as low as three altogether or as high as six.

1.5 Trafficking and thefts

See also: Vulnerability of nuclear plants to attack

The International Atomic Energy Agency says there is "a persistent problem with the illicit trafficking in nuclear and

other radioactive materials, thefts, losses and other unauthorized activities" .*[96] The IAEA Illicit Nuclear Trafficking Database notes 1,266 incidents reported by 99 countries over the last 12 years, including 18 incidents involving HEU or plutonium trafficking:*[97]*[74]*[98]

- Security specialist Shaun Gregory argued in an article that terrorists have attacked Pakistani nuclear facilities three times in the recent past; twice in 2007 and once in 2008.*[99]

- In November 2007, burglars with unknown intentions infiltrated the Pelindaba nuclear research facility near Pretoria, South Africa. The burglars escaped without acquiring any of the uranium held at the facility.*[100]*[101]

- In June 2007, the Federal Bureau of Investigation released to the press the name of Adnan Gulshair el Shukrijumah, allegedly the operations leader for developing tactical plans for detonating nuclear bombs in several American cities simultaneously.*[102]

- In November 2006, MI5 warned that al-Qaida were planning on using nuclear weapons against cities in the United Kingdom by obtaining the bombs via clandestine means.*[103]

- In February 2006, Oleg Khinsagov of Russia was arrested in Georgia, along with three Georgian accomplices, with 79.5 grams of 89 percent enriched HEU.*[104]

- The Alexander Litvinenko poisoning with radioactive polonium "represents an ominous landmark: the beginning of an era of nuclear terrorism," according to Andrew J. Patterson.*[105]

- In June 2002, U.S. citizen José Padilla was arrested for allegedly planning a radiological attack on the city of Chicago; however, he was never charged with such conduct. He was instead convicted of charges that he conspired to "murder, kidnap and maim" people overseas.

1.6 Accident categories

For a list of many of the most important accidents see the International Atomic Energy Agency site.*[106]

1.6.1 Nuclear meltdown

Main articles: Nuclear meltdown and Design basis accident

A nuclear meltdown is a severe nuclear reactor accident that results in reactor core damage from overheating. It has been defined as the accidental melting of the core of a nuclear reactor, and refers to the core's either complete or partial collapse.[107][108] A core melt accident occurs when the heat generated by a nuclear reactor exceeds the heat removed by the cooling systems to the point where at least one nuclear fuel element exceeds its melting point. This differs from a fuel element failure, which is not caused by high temperatures. A meltdown may be caused by a loss of coolant, loss of coolant pressure, or low coolant flow rate or be the result of a criticality excursion in which the reactor is operated at a power level that exceeds its design limits. Alternately, in a reactor plant such as the RBMK-1000, an external fire may endanger the core, leading to a meltdown.

Large-scale nuclear meltdowns at civilian nuclear power plants include:[11][45]

- the Lucens reactor, Switzerland, in 1969.

- the Three Mile Island accident in Pennsylvania, United States, in 1979.

- the Chernobyl disaster at Chernobyl Nuclear Power Plant, Ukraine, USSR, in 1986.

- the Fukushima Daiichi nuclear disaster following the earthquake and tsunami in Japan, March 2011.

Other core meltdowns have occurred at:[45]

- NRX (military), Ontario, Canada, in 1952

- BORAX-I (experimental), Idaho, U.S.A., in 1954

- EBR-I, Idaho, U.S.A., in 1955

- Windscale (military), Sellafield, England, in 1957 (see Windscale fire)

- Sodium Reactor Experiment, (civilian), California, U.S.A., in 1959

- Fermi 1 (civilian), Michigan, U.S.A., in 1966

- Chapelcross nuclear power station (civilian), Scotland, in 1967

- Saint-Laurent Nuclear Power Plant (civilian), France, in 1969

- A1 plant, (civilian) at Jaslovské Bohunice, Czechoslovakia, in 1977

- Saint-Laurent Nuclear Power Plant (civilian), France, in 1980

Eight Soviet Navy nuclear submarines have had nuclear core meltdowns or radiation incidents: K-19 (1961), K-11(1965), K-27 (1968), K-140 (1968), K-429 (1970), K-222 (1980), K-314 (1985), and K-431 (1985).[11]

1.6.2 Criticality accidents

A criticality accident (also sometimes referred to as an "excursion" or "power excursion") occurs when a nuclear chain reaction is accidentally allowed to occur in fissile material, such as enriched uranium or plutonium. The Chernobyl accident is an example of a criticality accident. This accident destroyed a reactor at the plant and left a large geographic area uninhabitable. In a smaller scale accident at Sarov a technician working with highly enriched uranium was irradiated while preparing an experiment involving a sphere of fissile material. The Sarov accident is interesting because the system remained critical for many days before it could be stopped, though safely located in a shielded experimental hall.[109] This is an example of a limited scope accident where only a few people can be harmed, while no release of radioactivity into the environment occurred. A criticality accident with limited off site release of both radiation (gamma and neutron) and a very small release of radioactivity occurred at Tokaimura in 1999 during the production of enriched uranium fuel.[110] Two workers died, a third was permanently injured, and 350 citizens were exposed to radiation.

1.6.3 Decay heat

Decay heat accidents are where the heat generated by the radioactive decay causes harm. In a large nuclear reactor, a loss of coolant accident can damage the core: for example, at Three Mile Island a recently shutdown (SCRAMed) PWR reactor was left for a length of time without cooling water. As a result, the nuclear fuel was damaged, and the core partially melted. The removal of the decay heat is a significant reactor safety concern, especially shortly after shutdown. Failure to remove decay heat may cause the reactor core temperature to rise to dangerous levels and has caused nuclear accidents. The heat removal is usually achieved through several redundant and diverse systems, and the heat is often dissipated to an 'ultimate heat sink' which has a large capacity and requires no active power, though this method is typically used after decay heat has reduced to a very small value. The main cause of release of radioactivity in the Three Mile Island accident was a pilot-operated relief valve on the primary loop which stuck in the open position. This caused the overflow tank into which it drained to rupture and release large amounts of radioactive cooling water into the containment building.

In 2011, an earthquake and tsunami caused a loss of power to two plants in Fukushima, Japan, crippling the reactor as decay heat caused 90% of the fuel rods in the core of the Daiichi Unit 3 reactor to become uncovered.*[111] As of May 30, 2011, the removal of decay heat is still a cause for concern.

1.6.4 Transport

Transport accidents can cause a release of radioactivity resulting in contamination or shielding to be damaged resulting in direct irradiation. In Cochabamba a defective gamma radiography set was transported in a passenger bus as cargo. The gamma source was outside the shielding, and it irradiated some bus passengers.

In the United Kingdom, it was revealed in a court case that in March 2002 a radiotherapy source was transported from Leeds to Sellafield with defective shielding. The shielding had a gap on the underside. It is thought that no human has been seriously harmed by the escaping radiation.*[112]

1.6.5 Equipment failure

Equipment failure is one possible type of accident. In Białystok, Poland, in 2001 the electronics associated with a particle accelerator used for the treatment of cancer suffered a malfunction.*[113] This then led to the overexposure of at least one patient. While the initial failure was the simple failure of a semiconductor diode, it set in motion a series of events which led to a radiation injury.

A related cause of accidents is failure of control software, as in the cases involving the Therac-25 medical radiotherapy equipment: the elimination of a hardware safety interlock in a new design model exposed a previously undetected bug in the control software, which could have led to patients receiving massive overdoses under a specific set of conditions.

1.6.6 Human error

Many of the major nuclear accidents have been directly attributable to operator or human error. This was obviously the case in the analysis of both the Chernobyl and TMI-2 accidents. At Chernobyl, a test procedure was being conducted prior to the accident. The leaders of the test permitted operators to disable and ignore key protection circuits and warnings that would have normally shut the reactor down. At TMI-2, operators permitted thousands of gallons of water to escape from the reactor plant before observing that the coolant pumps were behaving abnormally. The coolant pumps were thus turned off to protect the pumps,

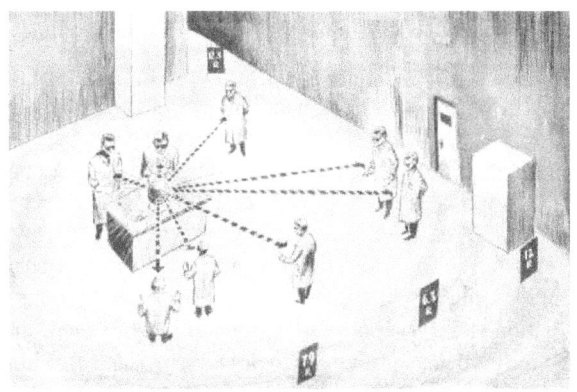

A sketch used by doctors to determine the amount of radiation to which each person had been exposed during the Slotin excursion

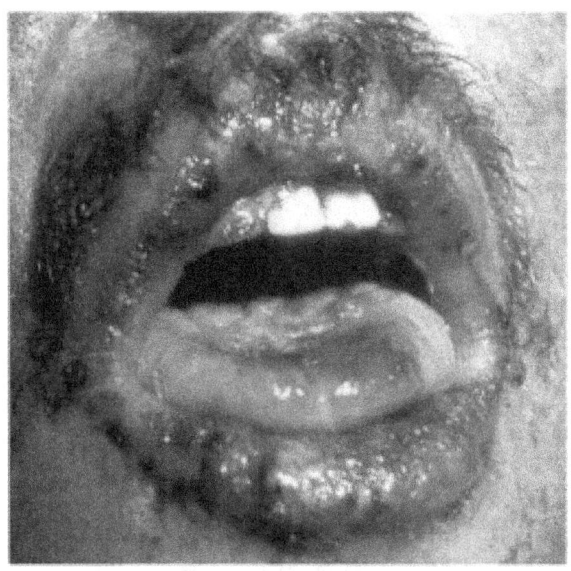

Part of a photo from an IAEA report on a radiation accident which occurred in Israel (Medical products treatment plant where the operator entered the irradiation room).[114]*

which in turn led to the destruction of the reactor itself as cooling was completely lost within the core.

A detailed investigation into SL-1 determined that one operator (perhaps inadvertently) manually pulled the 84-pound (38 kg) central control rod out about 26 inches rather than the maintenance procedure's intention of about 4 inches.*[115]

An assessment conducted by the Commissariat à l' Énergie Atomique (CEA) in France concluded that no amount of technical innovation can eliminate the risk of human-induced errors associated with the operation of nuclear power plants. Two types of mistakes were deemed most serious: errors committed during field operations, such as maintenance and testing, that can cause an accident; and human errors made during small accidents that cascade to complete failure.*[8]

In 1946 Canadian Manhattan Project physicist Louis Slotin performed a risky experiment known as "tickling the dragon's tail" *[116] which involved two hemispheres of neutron-reflective beryllium being brought together around a plutonium core to bring it to criticality. Against operating procedures, the hemispheres were separated only by a screwdriver. The screwdriver slipped and set off a chain reaction criticality accident filling the room with harmful radiation and a flash of blue light (caused by excited, ionized air particles returning to their unexcited states). Slotin reflexively separated the hemispheres in reaction to the heat flash and blue light, preventing further irradiation of several co-workers present in the room. However, Slotin absorbed a lethal dose of the radiation and died nine days later. The infamous plutonium mass used in the experiment was referred to as the demon core.

1.6.7 Lost source

Lost source accidents,*[117]*[118] also referred to as orphan sources, are incidents in which a radioactive source is lost, stolen or abandoned. The source then might cause harm to humans. One case occurred at Yanango where a radiography source was lost, also at Samut Prakarn a phosphorus teletherapy source was lost*[119] and at Gilan in Iran a radiography source harmed a welder.*[120] The best known example of this type of event is the Goiânia accident in Brazil.

The International Atomic Energy Agency has provided guides for scrap metal collectors on what a sealed source might look like.*[121]*[122] The scrap metal industry is the one where lost sources are most likely to be found.*[123]

1.7 Comparisons

Comparing the historical safety record of civilian nuclear energy with other forms of electrical generation, Ball, Roberts, and Simpson, the IAEA, and the Paul Scherrer Institute found in separate studies that during the period from 1970 to 1992, there were just 39 on-the-job deaths of nuclear power plant workers worldwide, while during the same time period, there were 6,400 on-the-job deaths of coal power plant workers, 1,200 on-the-job deaths of natural gas power plant workers and members of the general public caused by natural gas power plants, and 4,000 deaths of members of the general public caused by hydroelectric power plants.*[124]*[125]*[126] In particular, coal power plants are estimated to kill 24,000 Americans per year due to lung disease*[127] as well as causing 40,000 heart attacks per year*[128] in the United States. According to *Scientific American*, the average coal power plant emits 100 times more radiation per year than a comparatively sized nuclear power plant in the form of toxic coal waste known as fly ash.*[129]

Journalist Stephanie Cooke says that it is not very useful to make accident comparisons just in terms of number of immediate deaths, as the way people's lives are disrupted is also relevant, as in the case of the 2011 Japanese nuclear accidents, where 80,000 residents were forced to evacuate from neighborhoods around the Fukushima plant:*[130]

> You have people in Japan right now that are facing either not returning to their homes forever, or if they do return to their homes, living in a contaminated area... And knowing that whatever food they eat, it might be contaminated and always living with this sort of shadow of fear over them that they will die early because of cancer... It doesn't just kill now, it kills later, and it could kill centuries later... I'm not a great fan of coal-burning. I don't think any of these great big massive plants that spew pollution into the air are good. But I don't think it's really helpful to make these comparisons just in terms of number of deaths.*[131]

Physicist Amory Lovins has said: "Nuclear power is the only energy source where mishap or malice can destroy so much value or kill many faraway people; the only one whose materials, technologies, and skills can help make and hide nuclear weapons; the only proposed climate solution that substitutes proliferation, major accidents, and radioactive-waste dangers" .*[132]

In terms of energy accidents, hydroelectric plants were responsible for the most fatalities, but nuclear power plant accidents rank first in terms of their economic cost, accounting for 41 percent of all property damage. Oil and hydroelectric follow at around 25 percent each, followed by natural gas at 9 percent and coal at 2 percent.*[19] Excluding Chernobyl and the Shimantan Dam, the three other most expensive accidents involved the Exxon Valdez oil spill (Alaska), the Prestige oil spill (Spain), and the Three Mile Island nuclear accident (Pennsylvania).*[19]

1.8 Nuclear safety

Main article: Nuclear safety

Nuclear safety covers the actions taken to prevent nuclear and radiation accidents or to limit their consequences. This covers nuclear power plants as well as all other nuclear facilities, the transportation of nuclear materials, and the use and storage of nuclear materials for medical, power, industry, and military uses.

The nuclear power industry has improved the safety and performance of reactors, and has proposed new safer (but generally untested) reactor designs but there is no guarantee that the reactors will be designed, built and operated correctly.*[133] Mistakes do occur and the designers of reactors at Fukushima in Japan did not anticipate that a tsunami generated by an earthquake would disable the backup systems that were supposed to stabilize the reactor after the earthquake.*[134]*[135] According to UBS AG, the Fukushima I nuclear accidents have cast doubt on whether even an advanced economy like Japan can master nuclear safety.*[136] Catastrophic scenarios involving terrorist attacks are also conceivable.*[133]

In his book, *Normal accidents*, Charles Perrow says that multiple and unexpected failures are built into society's complex and tightly-coupled nuclear reactor systems. Nuclear power plants cannot be operated without some major accidents. Such accidents are unavoidable and cannot be designed around.*[137] An interdisciplinary team from MIT have estimated that given the expected growth of nuclear power from 2005 – 2055, at least four serious nuclear accidents would be expected in that period.*[138]*[139] To date, there have been five serious accidents (core damage) in the world since 1970 (one at Three Mile Island in 1979; one at Chernobyl in 1986; and three at Fukushima-Daiichi in 2011), corresponding to the beginning of the operation of generation II reactors. This leads to on average one serious accident happening every eight years worldwide.*[135]

In the 2003 book, *Brittle Power*, Amory Lovins talks about the need for a resilient, secure, energy system:

> The foundation of a secure energy system is to need less energy in the first place, then to get it from sources that are inherently invulnerable because they're diverse, dispersed, renewable, and mainly local. They're secure not because they're American but because of their design. Any highly centralised energy system -- pipelines, nuclear plants, refineries -- invite devastating attack. But invulnerable alternatives don't, and can't, fail on a large scale.*[140]

1.9   See also

1.10   References

[1] Richard Schiffman (12 March 2013). "Two years on, America hasn't learned lessons of Fukushima nuclear disaster". *The Guardian.*

[2] Martin Fackler (June 1, 2011). "Report Finds Japan Underestimated Tsunami Danger". *New York Times.*

[3] The European Parliament's Greens-EFA Group - The World Nuclear Industry Status Report 2007 p. 23. Archived June 25, 2008, at the Wayback Machine.

[4] Staff, IAEA, AEN/NEA. *International Nuclear and Radiological Events Scale Users' Manual, 2008 Edition* (PDF). Vienna, Austria: International Atomic Energy Agency. p. 184. Archived from the original (PDF) on May 15, 2011. Retrieved 2010-07-26.

[5] Yablokov, Alexey V.; Nesterenko, Vassily B.; Nesterenko, Alexey; Sherman-Nevinger, consulting editor, Jannette D. (2009). *Chernobyl: Consequences of the Catastrophe for People and the Environment*. Boston, MA: Blackwell Publishing for the Annals of the New York Academy of Sciences. ISBN 978-1-57331-757-3. Retrieved 11 June 2016.

[6] M.V. Ramana. Nuclear Power: Economic, Safety, Health, and Environmental Issues of Near-Term Technologies, *Annual Review of Environment and Resources*, 2009, 34, p. 136.

[7] Matthew Wald (February 29, 2012). "The Nuclear Ups and Downs of 2011". *New York Times.*

[8] Benjamin K. Sovacool. A Critical Evaluation of Nuclear Power and Renewable Electricity in Asia *Journal of Contemporary Asia*, Vol. 40, No. 3, August 2010, pp. 393–400.

[9] "The Worst Nuclear Disasters". *TIME.com.* 25 March 2009.

[10] Gralla, Fabienne, Abson, David J., and Muller, Anders, P. *et al.* "Nuclear accidents call for transidsciplinary energy research", *Sustainability Science*, January 2015.

[11] Kristin Shrader-Frechette (October 2011). "Fukushima, Flawed Epistemology, and Black-Swan Events" (PDF). *Ethics, Policy and Environment, Vol. 14, No. 3.*

[12] Johnston, Robert (September 23, 2007). "Deadliest radiation accidents and other events causing radiation casualties". Database of Radiological Incidents and Related Events.

[13] Gusev, Igor; Guskova, Angelina; Mettler, Fred A. (2001-03-28). *Medical Management of Radiation Accidents, Second Edition*. CRC Press. ISBN 9781420037197.

[14] Strengthening the Safety of Radiation Sources p. 15.

[15] "NRC: Information Notice No. 85-57: Lost Iridium-192 Source Resulting in the Death of Eight Persons in Morocco".

[16] The Radiological Accident in Goiania p. 2.

[17] Pallava Bagla. "Radiation Accident a 'Wake-Up Call' For India's Scientific Community" *Science*, Vol. 328, 7 May 2010, p. 679.

[18] "IAEA Scientific and Technical Publications of Special Interest". *www-pub.iaea.org*. Retrieved 2016-04-07.

[19] Benjamin K. Sovacool. A preliminary assessment of major energy accidents, 1907–2007, *Energy Policy* 36 (2008), pp. 1802-1820.

[20] "Predicting the global health consequences of the Chernobyl accident Methodology of the European Committee on Radiation Risk" (PDF).

[21] "Chernobyl Consequences of the Catastrophe for People and the Environment" (PDF).

[22] Benjamin K. Sovacool (2009). The Accidental Century - Prominent Energy Accidents in the Last 100 Years

[23] Timeline: Nuclear plant accidents *BBC News*, 11 July 2006.

[24] "Nuclear Accidents".

[25] cs:Havárie elektrárny Jaslovské Bohunice A-1

[26] anita.brunader. "UNSCEAR assessments of the Chernobyl accident". *www.unscear.org*. Retrieved 2016-10-19.

[27] "Worker dies at damaged Fukushima nuclear plant". *CBS News*. 2011-05-14.

[28] "Fukushima Nuclear Accident Update Log".

[29] Julia Mareike Neles, Christoph Pistner (Hrsg.), *Kernenergie. Eine Technik für die Zukunft?*, Berlin – Heidelberg 2012, S. 114 f.

[30] Charles D. Ferguson & Frank A. Settle (2012). "The Future of Nuclear Power in the United States" (PDF). *Federation of American Scientists*.

[31] Benjamin K. Sovacool (2011). *Contesting the Future of Nuclear Power: A Critical Global Assessment of Atomic Energy*, World Scientific, p. 192.

[32] Kennette Benedict (9 August 2012). "Civil disobedience". *Bulletin of the Atomic Scientists*.

[33] Jay Davis. After A Nuclear 9/11 *The Washington Post*, March 25, 2008.

[34] Brian Michael Jenkins. A Nuclear 9/11? *CNN.com*, September 11, 2008.

[35] Orde Kittrie. Averting Catastrophe: Why the Nuclear Nonproliferation Treaty is Losing its Deterrence Capacity and How to Restore It May 22, 2007, p. 338.

[36] Nicholas D. Kristof. A Nuclear 9/11 *The New York Times*, March 10, 2004.

[37] "Legal Experts: Stuxnet Attack on Iran Was Illegal 'Act of Force'". Wired. 25 March 2013.

[38] Penny Hitchin, "Cyber attacks on the nuclear industry", *Nuclear Engineering International*, 15 September 2015.

[39] Moss, William; Eckhardt, Roger (1995). "The Human Plutonium Injection Experiments" (PDF). *Los Alamos Science*. Radiation Protection and the Human Radiation Experiments (23): 177–223. Retrieved 13 November 2012.

[40] "The Media & Me: *[The Radiation Story No One Would Touch]*", *Geoffrey Sea*, Columbia Journalism Review, *March/April 1994.*

[41] Cameron L. Tracy, Megan K. Dustin & Rodney C. Ewing, Policy: Reassess New Mexico's nuclear-waste repository, *Nature*, 13 January 2016.

[42] Togzhan Kassenova (28 September 2009). "The lasting toll of Semipalatinsk's nuclear testing". *Bulletin of the Atomic Scientists*.

[43] Welsome, Eileen (1999). *The plutonium files* (PDF). New York, N.Y: Delacorte Press. p. 184. ISBN 0385314027.

[44] Final Report, Advisory Committee on Human Radiation Experiments, 1985

[45] "Annex C: Radiation exposures in accidents". *Sources and Effects of Ionizing Radiation – 2008 Report to the General Assembly* (pdf). *United Nations Scientific Committee on the Effects of Atomic Radiation.* II Scientific Annexes C, D, and E. 2011.

[46] "The Canadian Nuclear FAQ - Section D: Safety and Liability". *www.nuclearfaq.ca*. Retrieved 2016-04-07.

[47] "The NRX Incident".

[48] "Jimmy Carter's exposure to nuclear danger". Archived from the original on October 28, 2012.

[49] The evacuation of Rongelap Archived February 13, 2007, at the Wayback Machine.

[50] Newtan, Samuel Upton (2007-06-01). *Nuclear War I and Other Major Nuclear Disasters of the 20th Century*. AuthorHouse. ISBN 9781425985127.

[51] "Perhaps the Worst, Not the First". *Time*. May 12, 1986.

[52] Laramee, Eve Andree. "Tracking Our Nuclear Legacy". *WEAD*.

[53] McInroy, James F. (1995), "A true measure of plutonium exposure: the human tissue analysis program at Los Alamos" (PDF), *Los Alamos Science*, **23**: 235–255

[54] Barry Schneider (May 1975). "Big Bangs from Little Bombs". *Bulletin of Atomic Scientists*: 28. Retrieved 2009-07-13.

[55] James C. Oskins; Michael H. Maggelet (2008). *Broken Arrow —The Declassified History of U.S. Nuclear Weapons Accidents*. lulu.com. ISBN 1-4357-0361-8. Retrieved 2008-12-29.

[56] Strengthening the Safety of Radiation Sources p. 14.

[57] Marfleet, B. Gregory. "The Operational Code of John F. Kennedy During the Cuban Missile Crisis: A Comparison of Public and Private Rhetoric". *Political Psychology*. **21** (3): 545. doi:10.1111/0162-895x.00203.

[58] "Briefing Room". *Fourteen Days in October: The Cuban Missile Crisis*. ThinkQuest. 1997. Retrieved December 30, 2010.

[59] "Letters between Khrushchev and Kennedy". 2010. Retrieved December 30, 2010. Archive of correspondence between Kennedy and Khrushchev during Cuban Missile Crisis.

[60] "Ticonderoga Cruise Reports" (Navy.mil weblist of Aug 2003 compilation from cruise reports). Retrieved 2012-04-20. The National Archives hold*[s]* deck logs for aircraft carriers for the Vietnam Conflict.

[61] Broken Arrows at www.atomicarchive.com. Accessed Aug 24, 2007.

[62] "U.S. Confirms '65 Loss of H-Bomb Near Japanese Islands". *The Washington Post*. Reuters. May 9, 1989. p. A–27.

[63] Vinod K. Jose (1 December 2010). "River Deep Mountain High". Caravan Magazine. Retrieved 20 May 2013.

[64] Hayes, Ron (January 17, 2007). "H-bomb incident crippled pilot's career". Palm Beach Post. Archived from the original on 2011-06-16. Retrieved 2006-05-24.

[65] Maydew, Randall C. (1997). *America's Lost H-Bomb: Palomares, Spain, 1966*. Sunflower University Press. ISBN 978-0-89745-214-4.

[66] Phillips, Dave (June 19, 2016). "Decades Later, Sickness Among Airmen After a Hydrogen Bomb Accident". The New York TImes. Retrieved 20 June 2016.

[67] Long, Tony (January 17, 2008). "Jan. 17, 1966: H-Bombs Rain Down on a Spanish Fishing Village". WIRED. Archived from the original on December 3, 2008. Retrieved 2008-02-16.

[68] Ricks, Robert C.; et al. (2000). "REAC/TS Radiation Accident Registry: Update of Accidents in the United States" (PDF). International Radiation Protection Association. p. 6.

[69] Second Five-Year Review Report for the. United Nuclear Corporation. Ground Water Operable Unit EPA, September 2003

[70] https://www.youtube.com/watch?v=DGemW-pKCZw&feature=relmfu%5B%5D

[71] Shum, Edward Y. "Accidental Release of UF6 at Sequoyah Fuels Corporation Facility at Gore, Oklahoma, U.S.A." (PDF). Nuclear Regulatory Commission. Retrieved 12 February 2017.

[72] Brugge, Doug; deLemos, Jamie L.; Bui, Cat. "The Sequoyah Corporation Fuels Release and the Church Rock Spill: Unpublicized Nuclear Releases in American Indian Communities". *American Journal of Public Health*. **97** (9): 1595–1600. doi:10.2105/ajph.2006.103044. PMC 1963288. PMID 17666688.

[73] Kennedy, J. Michael (January 8, 1986). "Oklahoma Town Ponders Impact of Nuclear Fuel Plant's Fatal Accident". The Los Angeles Times. Retrieved 12 February 2017.

[74] Yukiya Amano (March 26, 2012). "Time to better secure radioactive materials". *Washington Post*.

[75] "The Worst Nuclear Disasters". *TIME.com*. 25 March 2009.

[76] Turai, István; Veress, Katalin (2001). "Radiation Accidents: Occurrence, Types, Consequences, Medical Management, and the Lessons to be Learned". *CEJOEM*.

[77] http://www.pmg.org.za/mp3/2007/070620pcenviro1.mp3

[78] Investigation of an accidental Exposure of radiotherapy patients in Panama - International Atomic Energy Agency

[79] Facts and Details on Nuclear energy in Japan Archived September 11, 2013, at the Wayback Machine.

[80]

[81] "TEPCO : Press Release - Plant Status of Fukushima Daini Nuclear Power Station (as of 2:00am March 13th)".

[82] WISE Uranium Project. "Issues at Rössing Uranium Mine, Namibia". World Information Service on Energy, Uranium Project. Retrieved 7 April 2014.

[83] Commission de Recherche et d' Information Indépendantes sur la Radioactivité. "Preliminary results of CRIIRAD radiation monitoring near uranium mines in Namibia" (PDF). *April 11, 2012*. CRIIRAD. Retrieved 7 April 2014.

[84] Commission de Recherche et d' Information Indépendantes sur la Radioactivité. "CRIIRAD Preliminary Report No. 12-32b Preliminary results of radiation monitoring near uranium mines in Namibia" (PDF). *April 5, 2012*. CRIIRAD EJOLT Project. Retrieved 7 April 2014.

[85] Labor Resource and Research Institute. "Namibian workers in times of uncertainty: The Labour Movement 20 years after independence". *2009*. LaRRI. Retrieved 7 April 2014.

[86] LaRRI. "Our Work: Labour Resource and Research Institute". *April 25, 2013*. LaRII. Retrieved 7 April 2014.

[87] Shinbdondola-Mote, Hilma (January 2009). "Uranium mining in Namibia: The mystery behind 'low level radiation'". Labor Resource and Research Institute (LaRRI). Retrieved 7 April 2014.

[88] Fleck, John (March 8, 2013). "WIPP radiation leak was never supposed to happen". *Albuquerque Journal*. Retrieved 28 March 2014.

[89] "What Happened at WIPP in February 2014". U.S. Department of Energy. Retrieved 28 March 2014.

[90] Jamail, Dahr. "Radiation Leak at New Mexico Nuclear Waste Storage Site Highlights Problems". Truth-Out.org. Retrieved 28 March 2014.

[91] "Gallery of U.S. Nuclear Tests". *The Nuclear Weapon Archive*. 6 August 2001.

[92] "Radiation Exposure Compensation System Claims to Date Summary of Claims Received by 08/15/2013 All Claims" (pdf). *United States Department of Justice*. 16 August 2013. – updated regularly

[93] "United States Nuclear Tests: July 1945 through September 1992 (Revision 15)" (PDF). Department of Energy, Nevada Operations Office. December 2000. Archived from the original (PDF) on 2010-06-15. Retrieved 2013-10-26. Generally regarded as the "official" list of American tests.

[94] "USSR Nuclear Weapons Tests and Peaceful Nuclear Explosions 1949 through 1990". Sarov, Russia: RFNC-VNIIEF. 1996. Unfortunately is no longer accessible over the internet.

[95] Yang, Xiaoping; North, Robert; Romney, Carl; Richards, Paul G. (August 2000). "Worldwide Nuclear Explosions" (PDF). Retrieved 2013-12-31.

[96] IAEA Illicit Trafficking Database (ITDB) p. 3.

[97] "IAEA Report". *In Focus: Chernobyl*. Retrieved 2008-05-31.

[98] Bunn, Matthew. "Securing the Bomb 2010: Securing All Nuclear Materials in Four Years" (PDF). President and Fellows of Harvard College. Retrieved 28 January 2013.

[99] Rhys Blakeley, "Terrorists 'have attacked Pakistan nuclear sites three times'," *Times Online* (August 11, 2009).

[100] "IOL | Pretoria News | IOL". *IOL*. Retrieved 2016-04-07.

[101] Washington Post, December 20, 2007, Op-Ed by Micah Zenko

[102] "Feds Hoped to Snag Bin Laden Nuke Expert in JFK Bomb Plot". *Fox News*. June 4, 2007.

[103] Dodd, Vikram (2006-11-13). "Al-Qaida plotting nuclear attack on UK, officials warn". *The Guardian*. ISSN 0261-3077. Retrieved 2016-04-07.

[104] Bunn, Matthew & Col-Gen. E.P. Maslin (2010). "All Stocks of Weapons-Usable Nuclear Materials Worldwide Must be Protected Against Global Terrorist Threats" (PDF). Belfer Center for Science and International Affairs, Harvard University. Retrieved July 26, 2012.

[105] "Ushering in the era of nuclear terrorism," by Patterson, Andrew J. MD, PhD, *Critical Care Medicine*, v. 35, p.953-954, 2007.

[106] "WebCite query result" (PDF). *www.webcitation.org*. Archived from the original (PDF) on June 8, 2009. Retrieved 2016-04-07.

[107] *Reactor Safety Study*.

[108] "Meltdown - Definition and More from the Free Merriam-Webster Dictionary".

[109] "The Criticality Accident in Sarov" (PDF). International Atomic Energy Agency. February 2001. Retrieved 12 February 2012.

[110] http://www-pub.iaea.org/MTCD/publications/PDF/TOAC_web.pdf

[111] Analysis: Seawater helps but Japan nuclear crisis is not over by Scott DiSavino and Fredrik Dahl, March 13, 2011.

[112] "Road container 'leaked radiation'". *BBC News*. February 17, 2006.

[113] "Accidental Overexposure of Radiotherapy Patients in Bialystok" (PDF). International Atomic Energy Agency. February 2004. Retrieved 12 February 2012.

[114] http://www-pub.iaea.org/MTCD/publications/PDF/Pub925_web.pdf

[115] Tucker, Todd (2009). *Atomic America: How a Deadly Explosion and a Feared Admiral Changed the Course of Nuclear History*. New York: Free Press. ISBN 978-1-4165-4433-3. See summary:

[116] Jungk, Robert. Brighter than a Thousand Suns. 1956. p.194

[117] "WebCite query result" (PDF). Archived from the original (PDF) on July 30, 2011.

[118] "WebCite query result" (PDF). *www.webcitation.org*. Archived from the original (PDF) on July 30, 2011. Retrieved 2016-04-07.

[119] "The Radiological Accident in Samut Prakarn" (PDF). International Atomic Energy Agency. 2002.

[120] http://www-pub.iaea.org/MTCD/publications/PDF/Pub1123_scr.pdf

[121] "IAEA Topical Booklets and Overviews" (PDF).

[122] "IAEA Topical Booklets and Overviews" (PDF).

[123] http://web.archive.org/web/20090304080024/http://www.srp-uk.org/srpcdrom/p8-5.doc

[124] Ball, Roberts, Simpson; et al. (1994). *Research Report #20. Center for Environmental & Risk Management*. United Kingdom: University of East Anglia.

[125] Hirschberg et al, Paul Scherrer Institut, 1996; in: IAEA, Sustainable Development and Nuclear Power, 1997

[126] Severe Accidents in the Energy Sector, Paul Scherrer Institut, 2001.

[127] "Senator Reid tells America coal makes them sick". 2008-07-10. Retrieved 2009-05-18.

[128] "Deadly power plants? Study fuels debate". 2004-06-09. Retrieved 2009-05-18.

[129] Scientific American, December 13, 2007 "Coal Ash Is More Radioactive than Nuclear Waste". 2009-05-18. Retrieved 2009-05-18.

[130] "Japan says it was unprepared for post-quake nuclear disaster". *Los Angeles Times*. June 8, 2011. Archived from the original on 2011-06-08.

[131] Annabelle Quince (30 March 2011). "The history of nuclear power". *ABC Radio National*.

[132] Amory Lovins (2011). "Soft Energy Paths for the 21st Century".

[133] Jacobson, Mark Z. & Delucchi, Mark A. (2010). "Providing all Global Energy with Wind, Water, and Solar Power, Part I: Technologies, Energy Resources, Quantities and Areas of Infrastructure, and Materials" (PDF). *Energy Policy*. p. 6.

[134] Hugh Gusterson (16 March 2011). "The lessons of Fukushima". *Bulletin of the Atomic Scientists*.

[135] Diaz Maurin, François (26 March 2011). "Fukushima: Consequences of Systemic Problems in Nuclear Plant Design". *Economic & Political Weekly*. **46** (13): 10–12.

[136] James Paton (April 4, 2011). "Fukushima Crisis Worse for Atomic Power Than Chernobyl, UBS Says". *Bloomberg Businessweek*.

[137] Daniel E Whitney (2003). "Normal Accidents by Charles Perrow" (PDF). *Massachusetts Institute of Technology*.

[138] Benjamin K. Sovacool (January 2011). "Second Thoughts About Nuclear Power" (PDF). National University of Singapore. p. 8.

[139] Massachusetts Institute of Technology (2003). "The Future of Nuclear Power" (PDF). p. 48.

[140] Amory B. Lovins and L. Hunter Lovins. "Terrorism and Brittle Technology" in *Technology and the Future* by Albert H. Teich, Ninth edition, Thomson, 2003, p. 169.

1.11   Further reading

- *Chernobyl: Consequences of the Catastrophe for People and the Environment (2009)*
- *Chernobyl. Vengeance of peaceful atom. (2006)*
- *Conservation Fallout: Nuclear Protest at Diablo Canyon (2006)*
- *Contesting the Future of Nuclear Power (2011)*
- *Essence of Decision: Explaining the Cuban Missile Crisis (1971)*
- *Fallout: An American Nuclear Tragedy (2004)*
- *Fallout Protection (1961)*
- *Fukushima: Japan's Tsunami and the Inside Story of the Nuclear Meltdowns (2013)*
- *Full Body Burden: Growing Up in the Nuclear Shadow of Rocky Flats (2012)*
- *Hiroshima (1946)*
- *Killing Our Own: The Disaster of America's Experience with Atomic Radiation (1982)*
- *In Mortal Hands: A Cautionary History of the Nuclear Age (2009)*
- *Making a Real Killing: Rocky Flats and the Nuclear West (1999)*
- *Maralinga: Australia's Nuclear Waste Cover-up (2007)*
- *Non-Nuclear Futures: The Case for an Ethical Energy Strategy (1975)*
- *Normal Accidents: Living with High-Risk Technologies (1984)*
- *Nuclear or Not? Does Nuclear Power Have a Place in a Sustainable Energy Future? (2007)*
- *Nuclear Politics in America (1997)*
- *Nuclear Power and the Environment (1976)*
- *Nuclear Terrorism: The Ultimate Preventable Catastrophe (2004)*
- *Nuclear War Survival Skills (1979)*
- *Nuclear Weapons: The Road to Zero (1998)*
- *Nukespeak: Nuclear Language, Visions and Mindset (1982)*
- *On Nuclear Terrorism (2007)*
- *Plutopia (2013)*

1.12 External links

- U.S. Nuclear Accidents (lutins.org) most comprehensive online list of incidents involving U.S. nuclear facilities and vessels, 1950–present

- US Nuclear Regulatory Commission (NRC) website with search function and electronic public reading room

- International Atomic Energy Agency website with extensive online library

- Plutopia: Nuclear Families, Atomic Cities, and the Great Soviet and American Plutonium Disasters

- Concerned Citizens for Nuclear Safety Detailed articles on nuclear watchdog activities in the US

- World Nuclear Association: Radiation Doses Background on ionizing radiation and doses

- Radiological Incidents Database Extensive, well-referenced list of radiological incidents

- "A Review of Criticality Accidents". Archived from the original on 2004-12-09. Retrieved 2004-12-09.

- Nuclear Files.org List of nuclear accidents

- Annotated bibliography for civilian nuclear accidents from the Alsos Digital Library for Nuclear Issues

- Critical Hour: Three Mile Island, The Nuclear Legacy, And National Security. Albert J. Fritsch, Arthur H. Purcell, and Mary Byrd Davis (2005).Updated edition, June 2006

- Nuclear Emergency and Radiation Resources Literature review: what to do in the event of a nuclear accident

- Radioactivity.eu.com Radiation accidents

Chapter 2

Crimes involving radioactive substances

*The radiation warning symbol (*trefoil*)*

This is a list of criminal (or arguably, allegedly, or potentially criminal) acts intentionally involving radioactive substances. Inclusion in this list does not necessarily imply that anyone involved was guilty of a crime. For accidents or crimes that involved radioactive substances unbeknownst to those involved, see the list of radiation accidents.

2.1 Murder/attempted murder

2.1.1 The Karlsruhe plutonium affair

An unnamed man was convicted of attempting to poison his ex-wife in 2001 with plutonium stolen from WAK (Wiederaufbereitungsanlage Karlsruhe), a small scale reprocessing plant where he worked. He did not steal a large amount of plutonium, only rags used for wiping surfaces and a small amount of liquid waste.[1][2] At least two people (besides the criminal) were contaminated by the plutonium.[3] Two

flats in Landau in the Rhineland-Palatinate were contaminated, and had to be cleaned at a cost of two million euro.[4] Photographs of the case and details of other nuclear crimes have been presented by a worker at the Institute for Transuranium Elements.[5]

2.1.2 The Litvinenko assassination

Alexander Litvinenko died from polonium-210 poisoning in 2006. British officials said investigators had concluded the murder of Litvinenko was "a 'state-sponsored' assassination orchestrated by Russian security services." [6] On 20 January 2007 British police announced that they had "identified the man they believe poisoned Alexander Litvinenko," Andrei Lugovoi.[7]

On the 21st of September 2012 a story was posted in various UK newspapers suggesting the existence of an ongoing cover-up by the British Government over the material facts of the case. The report suggests that many aspects of the case may "never see the light of day" due to the significant risk to UK/Russian relations and the implications of the declaration that an act of nuclear terrorism took place on British soil.

2.1.3 Roman Tsepov homicide

Roman Tsepov, a politically influential Russian who provided security to Vladimir Putin and others, fell sick on September 11, 2004 after a trip to Moscow, and died on September 24. A postmortem investigation found a poisoning by an unspecified radioactive material. He had symptoms similar to Aleksandr Litvinenko.[8][9][10]

2.1.4 Zheleznodorozhny criminal radiological act

An unnamed truck driver was killed by 5 months of radiation exposure to a 1.3 curies (48 GBq) cesium-137 source that had been put into the door of his truck around February 1995. He died of radiation-induced leukemia on 27 April 1997.*[11]

2.1.5 Vladimir Kaplun radiation homicide

In 1993, director of the Kartontara packing company Vladimir Kaplun was killed by radioactive material (probably cesium-137) placed in his chair. He died of radiation sickness after a month of hospitalization. The source of the radiation was found after his death.*[12]

2.1.6 Karen Silkwood poisoning allegations

In November 5, 1974, Kerr-McGee worker and labor union activist Karen Silkwood found herself exposed to plutonium-239 after working to grind and polish plutonium pellets by way of a glovebox to be used in nuclear fuel rods at the Cimarron Fuel Fabrication Site in Oklahoma. Inspection of the gloves would yield no evidence of external leakage of radioactive contaminant from the glovebox, despite the fact that plutonium had been found on those surfaces of the gloves that had contact with Silkwood's hands, and no other source for a plutonium leak could be ascertained despite thorough inspections of the air vents and surrounding surfaces. However, on Nov. 6th, despite Silkwood's prior decontamination and self-inspection, a detection on her part the next day yielded more signs of alpha activity on her hands, while health physics staff at the plant subsequently detected further alpha activity on her right forearm, neck and face. On November 7, Silkwood tested positive for very significant levels of alpha activity, and an inspection of her apartment showed high levels of radioactive contamination. Thereafter, Silkwood and two other coworkers personally associated with her (roommate Sherri Ellis and boyfriend Drew Stephens) would be tested at Los Alamos National Laboratory; while the latter two only tested positive for insignificant amounts of plutonium exposure, Silkwood was found to have 6–7 nanocuries (220–260 Bq) of plutonium-239 in her lungs, though researcher Dr. George Voelz insisted that this amount was still negligible and non-harmful. Later posthumous measurements taken after Silkwood's death under separate but likewise controversial circumstances were shown to be roughly consistent with the initial findings described by Dr. Voelz, but also indicated that she had somehow ingested plutonium prior to her demise. It would also be found that Karen Silkwood would not have had access to plutonium-239 for months after her transfer from metallography to rod assembly.

Allegations that Silkwood's exposure to plutonium-239 was a deliberate act of radiation poisoning are fueled by the fact that she was in possession of potentially compromising evidence that linked Kerr-McGee with egregious safety violations, encompassing unsafe workplace conditions at the plant, faulty manufacture of fuel rod components that posed a potential public safety risk, and even substantial missing plutonium supplies that were unaccounted for; Silkwood also contended that she had evidence that photographs containing evidence of hairline cracks in the fuel rods may have been doctored by company personnel as a cover-up. As Silkwood was coordinating with fellow union members and had been *en route* to meet a journalist to present and discuss evidence she had found of Kerr-McGee's actions at the time of her death, unverified assertions that Kerr-McGee or other associated parties may have been involved in her radiation exposure and later fatal car accident have circulated in the following period.*[13]

2.2 Intentional or attempted theft of radioactive material

For accidental theft or attempted theft of radioactive materials, see the list of radiation accidents.

2.2.1 Grozny cobalt theft or attempted theft

On 13 September 1999, six people attempted to steal radioactive cobalt-60 rods from a chemical plant in the city of Grozny in the Chechen Republic.*[14] During the theft, the suspects opened the radioactive material container and handled it, resulting in the deaths of three of the suspects and injury of the remaining three. The suspect who held the material directly in his hands died of radiation exposure 30 minutes later. This incident is described as an attempted theft, but some of the rods are reportedly still missing.*[15]

2.3 Official use of X-ray equipment and other radiation technology by secret police

Some former East German dissidents claim that the Stasi used X-ray equipment to induce cancer in political prisoners.*[16]

Similarly, some anti-Castro activists claim that the Cuban secret police sometimes used radioactive isotopes to induce cancer in "adversaries they wished to destroy with as little notice as possible".*[17] In 1997, the Cuban expatriate columnist Carlos Alberto Montaner called this method "the Bulgarian Treatment", after its alleged use by the Bulgarian secret police.*[18]

2.4   Illicit, fraudulent, and patent medicine

In the early 20th century a series of products claiming medicinal properties, which contained radioactive elements were marketed to the general public. This does not include certain medications that contain radioactive isotopes (e.g. iodine-131 for its oncological uses) but pertains to elixirs and other medications that made preposterous claims (see below) that were neither scientific nor verifiable.

Radithor, a well known patent medicine/snake oil, is possibly the best known example of radioactive quackery. It consisted of triple distilled water containing at a minimum 1 microcurie (37 kBq) each of the radium−226 and radium-228 isotopes.[*][19]

Radithor was manufactured from 1918 - 1928 by the Bailey Radium Laboratories, Inc., of East Orange, New Jersey. The head of the laboratories was listed as Dr. William J. A. Bailey, not a medical doctor.[*][20] It was advertised as "A Cure for the Living Dead"[*][21] as well as "Perpetual Sunshine".

These radium elixirs were marketed similar to the way opiates were peddled to the masses with laudanum an age earlier, and electrical cure-alls during the same time period such as the Prostate Warmer.[*][22]

The eventual death of the socialite Eben Byers from Radithor consumption and the associated radiation poisoning led to the strengthening of the Food and Drug Administration's powers and the demise of most radiation based patent medication.

2.4.1   Associated links

- Scientific American; August 1993; The Great Radium Scandal; by Roger Macklis

- Theodore Gray's Periodic Table of Elements

2.5   See also

- Lists of nuclear disasters and radioactive incidents

- List of attacks on nuclear plants

- Radioactive scrap metal

- Radioactive waste

2.6   External links

- Johnston's Archive: Criminal acts causing radiation casualties

2.7   References

[1] "Welcome". World Information Service on Energy. Retrieved 2006-12-05.

[2] "Germany: Plutonium soup as a murder weapon?". World Information Service on Energy. October 5, 2001. Retrieved 2006-12-05.

[3] "Radiation Victim Demands Compensatory Damages". *German News* (English ed.). 24 February 2005. Archived from the original on 2007-06-24. Retrieved 2006-12-05.

[4] "Clean-up of a GIGA-BQ-PU contamination of two apartments" (PDF). Hagen Hoefer. Archived from the original (pdf) on January 5, 2005. Retrieved 2006-12-05.

[5] Ray, Ian. "Nuclear Forensic Science and Illicit Trafficking" (PDF). Karlsruhe (Germany): Institute for Transuranium Elements. Archived from the original (PDF) on 2006-07-25. Retrieved 2006-12-05.

[6] "Murder in a Teapot". "The Blotter" on ABCNews.com. 26 January 2007. Retrieved 26 January 2006.

[7] McGrory, Daniel; Halpin, Tony (20 January 2007). "Police match image of Litvinenko's real assassin with his death-bed description". London: Times Online. Retrieved 22 January 2006.

[8] The Putin bodyguard riddle, The Sunday Times, December 3, 2006

[9] Полоний и три Владимира

[10] Расследование отравления радиоактивным изотопом Романа Цепова, бывшего телохранителя Анатолия Собчака и Владимира Путина

[11] Wm. Robert Johnston (24 September 2007). "Zheleznodorozhny criminal radiological act, 1995". Retrieved 2 June 2012.

[12] Wm. Robert Johnston (22 September 2007). "Moscow radiological homicide, 1993". Retrieved 2 June 2012.

[13] Tricia Romano. "The Life and Mysterious Death of Karen Silkwood". Archived from the original on 6 June 2011. Retrieved 8 October 2013.

[14] Wm. Robert Johnston (8 April 2005). "Gronzy orphaned source, 1999". Retrieved 27 February 2013.

[15] "Criminal Dies Stealing Radioactive Material". James Martin Center for Nonproliferation Studies at the Monterey Institute of International Studies. 14 September 1999. Retrieved 2 June 2012.

[16] "Dissidents say Stasi gave them cancer". BBC. 25 May 1999. Retrieved 2006-12-07.

[17] Stride, Jonathan T. (30 December 1997). "Castro said to be using cancer instigating weapons for warfare". Florida International University. Archived from the original on September 18, 2006. Retrieved 2006-12-07.

[18] Montaner, Carlos Alberto (28 December 1997). "The Bulgarian Treatment". Firmas Press. Archived from the original on December 5, 2006. Retrieved 2006-12-07.

[19] "Radithor (ca. 1918).". Oak Ridge Associated Universities. 15 September 2004. Retrieved 2006-12-07.

[20] "U/A". *Literary Digest.* 16 April 1932.

[21] "Radium Cures". Museum of Questionable Medical Devices (Science Museum of Minnesota). 4 February 2000. Archived from the original on 2006-11-10. Retrieved 2006-12-07.

[22] "Prostate Cures". Museum of Questionable Medical Devices (Science Museum of Minnesota). 18 April 1999. Archived from the original on 2006-12-06. Retrieved 2006-12-07.

Chapter 3

Vulnerability of nuclear plants to attack

On September 9, 1980, Daniel Berrigan (above), his brother Philip, and six others (the "Plowshares Eight") began the Plowshares Movement. They illegally trespassed onto the General Electric Nuclear Missile facility in King of Prussia, Pennsylvania, where they damaged nuclear warhead nose cones and poured blood onto documents and files. They were arrested and charged with over ten different felony and misdemeanor counts.[1]

The **vulnerability of nuclear plants to deliberate attack** is of concern in the area of nuclear safety and security. Nuclear power plants, civilian research reactors, certain naval fuel facilities, uranium enrichment plants, fuel fabrication plants, and even potentially uranium mines are vulnerable to attacks which could lead to widespread radioactive contamination. The attack threat is of several general types: commando-like ground-based attacks on equipment which if disabled could lead to a reactor core meltdown or widespread dispersal of radioactivity; and external attacks such as an aircraft crash into a reactor complex, or cyber attacks.*[2]

The United States 9/11 Commission has said that nuclear power plants were potential targets originally considered for the September 11, 2001 attacks. If terrorist groups could sufficiently damage safety systems to cause a core meltdown at a nuclear power plant, and/or sufficiently damage spent fuel pools, such an attack could lead to widespread radioactive contamination. The Federation of American Scientists have said that if nuclear power use is to expand significantly, nuclear facilities will have to be made extremely safe from attacks that could release massive quantities of radioactivity into the community. New reactor designs have features of passive nuclear safety, which may help. In the United States, the NRC carries out "Force on Force" exercises at all nuclear power plant sites at least once every three years.*[2]

Nuclear reactors become preferred targets during military conflict and, over the past three decades, have been repeatedly attacked during military air strikes, occupations, invasions and campaigns.*[3] Various acts of civil disobedience since 1980 by the peace group Plowshares have shown how nuclear weapons facilities can be penetrated, and the group's actions represent extraordinary breaches of security at nuclear weapons plants in the United States. The National Nuclear Security Administration has acknowledged the seriousness of the 2012 Plowshares action. Non-proliferation policy experts have questioned "the use of private contractors to provide security at facilities that manufacture and store the government's most dangerous military material".*[4] Nuclear weapons materials on the black market are a global concern,*[5]*[6] and there is concern about the possible detonation of a dirty bomb by a militant group in a major city.*[7]*[8]

The number and sophistication of cyber attacks is on the rise. *Stuxnet* is a computer worm discovered in June 2010 that is believed to have been created by the United States and Israel to attack Iran's nuclear facilities. It switched off safety devices, causing centrifuges to spin out of control.*[9] The computers of South Korea's nuclear plant operator (KHNP) were hacked in December 2014. The cyber attacks involved thousands of phishing emails containing malicious code, and information was stolen.*[10]

3.1 Attacks on nuclear power plants

Terrorists could target nuclear power plants in an attempt to release radioactive contamination into the community. The United States 9/11 Commission has said that nuclear power plants were potential targets originally considered for the September 11, 2001 attacks. If terrorist groups could sufficiently damage safety systems to cause a core meltdown at a nuclear power plant, and/or sufficiently damage spent fuel pools, such an attack could lead to a widespread radioactive contamination. According to a 2004 report by the U.S. Congressional Budget Office, "The human, environmental, and economic costs from a successful attack on a nuclear power plant that results in the release of substantial quantities of radioactive material to the environment could be great." *[11] An attack on a reactor' s spent fuel pool could also be serious, as these pools are less protected than the reactor core. The release of radioactivity could lead to thousands of near-term deaths and greater numbers of long-term fatalities.*[2]

If nuclear power use is to expand significantly, nuclear facilities will have to be made extremely safe from attacks that could release massive quantities of radioactivity into the community. New reactor designs have features of passive safety, such as the flooding of the reactor core without active intervention by reactor operators. But these safety measures have generally been developed and studied with respect to accidents, not to the deliberate reactor attack by a terrorist group. However, the US Nuclear Regulatory Commission does now also require new reactor license applications to consider security during the design stage.*[2]

In the United States, the NRC carries out "Force on Force" (FOF) exercises at all Nuclear Power Plant (NPP) sites at least once every three years. The FOF exercise, which is typically conducted over 3 weeks, "includes both tabletop drills and exercises that simulate combat between a mock adversary force and the licensee' s security force. At an NPP, the adversary force attempts to reach and simulate damage to key safety systems and components, defined as "target sets" that protect the reactor' s core or the spent fuel pool, which could potentially cause a radioactive release to the environment. The licensee' s security force, in turn,

interposes itself to prevent the adversaries from reaching target sets and thus causing such a release" .*[2]

In the U.S., plants are surrounded by a double row of tall fences which are electronically monitored. The plant grounds are patrolled by a sizeable force of armed guards.*[12]

3.2 Military attacks

Nuclear reactors become preferred targets during military conflict and, over the past three decades, have been repeatedly attacked during military air strikes, occupations, invasions and campaigns:*[3]

- In September 1980, Iran bombed the Al Tuwaitha nuclear complex in Iraq, in Operation Scorch Sword, which was a surprise IRIAF (Islamic Republic of Iran Air Force) airstrike carried out on 30 September 1980, that damaged an almost complete nuclear reactor 17 km south-east of Baghdad, Iraq.*[13]

- In June 1981, an Israeli air strike completely destroyed Iraq's Osirak nuclear research facility.

- Between 1984 and 1987, Iraq bombed Iran's Bushehr nuclear plant six times.

- In 1991, the U.S. bombed three nuclear reactors and an enrichment pilot facility in Iraq.

- In 1991, Iraq launched Scud missiles at Israel's Dimona nuclear power plant.

- In September 2007, Israel bombed a Syrian reactor under construction.*[3]

After several incidents in Pakistan in which terrorists attacked three of its military nuclear facilities, it became clear that there emerged a serious danger that they would gain access to the country' s nuclear arsenal, according to a journal published by the US Military Academy at West Point.*[14] In January 2010, it was revealed that the US army was training a specialised unit "to seal off and snatch back" Pakistani nuclear weapons in the event that militants would obtain a nuclear device or materials that could make one. Pakistan supposedly possesses about 80 nuclear warheads. US officials refused to speak on the record about the American safety plans.*[15]

3.3 Nuclear terrorism

Main article: Nuclear terrorism

Amory Lovins says that the United States has for decades been running on energy that is "brittle" (easily shattered by accident or malice) and that this poses a grave and growing threat to national security, life, and liberty.*[16] Lovins' claims that these vulnerabilities are increasingly being exploited. His book *Brittle Power* documents many significant assaults on energy facilities, other than during a war, in forty countries and within the United States, in some twenty-four states.*[17]

Lovins further claims that in 1966, twenty natural uranium fuel rods were stolen from the Bradwell nuclear power station in England, and in 1971, five more were stolen at the Wylfa Nuclear Power Station. In 1971, an intruder wounded a night watchman at the Vermont Yankee reactor in the USA. The New York University reactor building was broken into in 1972, as was the Oconee Nuclear Station's fuel storage building in 1973. In 1975, the Kerr McGee plutonium plant had thousands of dollars worth of platinum stolen and taken home by workers. In 1975, at the Biblis Nuclear Power Plant in Germany, a Member of Parliament demonstrated the lack of security by carrying a bazooka into the plant under his coat.*[18]

Nuclear plants were designed to withstand earthquakes, hurricanes, and other extreme natural events. But deliberate attacks involving large airliners loaded with fuel, such as those that crashed into the World Trade Center and Pentagon, were not considered when design requirements for today's fleet of reactors were determined. It was in 1972 when three hijackers took control of a domestic passenger flight along the east coast of the U.S. and threatened to crash the plane into a U.S. nuclear weapons plant in Oak Ridge, Tennessee. The plane got as close as 8,000 feet above the site before the hijackers' demands were met.*[19]*[20]

In February 1993, a man drove his car past a check point the Three Mile Island Nuclear plant, then broke through an entry gate. He eventually crashed the car through a secure door and entered the Unit 1 reactor turbine building. The intruder, who had a history of mental illness, hid in a building and was not apprehended for four hours. Stephanie Cooke asks: "What if he'd been a terrorist armed with a ticking bomb?"*[21]

Fissile material may be stolen from nuclear plants and this may promote the spread of nuclear weapons. Many terrorist groups are eager to acquire the fissile material needed to make a crude nuclear device, or a dirty bomb. Nuclear weapons materials on the black market are a global concern,*[5]*[6] and there is concern about the possible detonation of a small, crude nuclear weapon by a militant group in a major city, with significant loss of life and property.*[7]*[8] It is feared that a terrorist group could detonate a radiological or "dirty bomb", composed of any radioactive source and a conventional explosive. The ra-

dioactive material is dispersed by the detonation of the explosive. Detonation of such a weapon is not as powerful as a nuclear blast, but can produce considerable radioactive fallout. Alternatively, a terrorist group may position some of its members, or sympathisers, within the plant to sabotage it from inside.*[22]

The IAEA Illicit Nuclear Trafficking Database notes 1,266 incidents reported by 99 countries over the last 12 years, including 18 incidents involving HEU or plutonium trafficking:*[23]

- There have been 18 incidences of theft or loss of highly enriched uranium (HEU) and plutonium confirmed by the IAEA.*[24]

- Security specialist Shaun Gregory argued in an article that terrorists have attacked Pakistani nuclear facilities three times in the recent past; twice in 2007 and once in 2008.*[25]

- In November 2007, burglars with unknown intentions infiltrated the Pelindaba nuclear research facility near Pretoria, South Africa. The burglars escaped without acquiring any of the uranium held at the facility.*[26]*[27]

- In June 2007, the Federal Bureau of Investigation released to the press the name of Adnan Gulshair el Shukrijumah, allegedly the operations leader for developing tactical plans for detonating nuclear bombs in several American cities simultaneously.*[28]

- In November 2006, MI5 warned that al-Qaida were planning on using nuclear weapons against cities in the United Kingdom by obtaining the bombs via clandestine means.*[29]

- In February 2006, Oleg Khinsagov of Russia was arrested in Georgia, along with three Georgian accomplices, with 79.5 grams of 89 percent enriched HEU.*[24]

- The Alexander Litvinenko poisoning with radioactive polonium "represents an ominous landmark: the beginning of an era of nuclear terrorism," according to Andrew J. Patterson.*[30]

- In June 2002, U.S. citizen José Padilla was arrested for allegedly planning a radiological attack on the city of Chicago; however, he was never charged with such conduct. He was instead convicted of charges that he conspired to "murder, kidnap and maim" people overseas.

3.4 Sabotage by insiders

Insider sabotage regularly occurs, because insiders can observe and work around security measures. In a study of insider crimes, the authors repeatedly said that successful insider crimes depended on the perpetrators' observation and knowledge of security vulnerabilities. Since the atomic age began, the U.S. Department of Energy's nuclear laboratories have been known for widespread violations of security rules. During the Manhattan Project, physicist Richard Feynman was barred from entering certain nuclear facilities; he would crack safes and violate other rules as pranks to reveal deficiencies in security. A better understanding of the reality of the threat will help to overcome complacency and is critical to getting countries to take stronger preventive measures.*[31]

A fire caused 5–10 million dollars worth of damage to New York's Indian Point Energy Center in 1971. The arsonist turned out to be a plant maintenance worker. Sabotage by workers has been reported at many other reactors in the United States: at Zion Nuclear Power Station (1974), Quad Cities Nuclear Generating Station, Peach Bottom Nuclear Generating Station, Fort St. Vrain Generating Station, Trojan Nuclear Power Plant (1974), Browns Ferry Nuclear Power Plant (1980), and Beaver Valley Nuclear Generating Station (1981). Many reactors overseas have also reported sabotage by workers. Suspected arson has occurred in the USA and overseas.*[18]

On 8 January 1982, the 70th anniversary of the formation of the African National Congress, Umkhonto we Sizwe, the armed wing of the ANC attacked Koeberg Nuclear Power Station while it was still under construction.*[32] Damage was estimated at R 500 million and the commissioning of the plant was put back by 18 months.*[33] In 1998 a group of workers at one of Russia's largest nuclear weapons facilities attempted to steal 18.5 kilograms of HEU—enough for a bomb.*[18]

3.5 Civil disobedience

Various acts of civil disobedience since 1980 by the peace group Plowshares have shown how nuclear weapons facilities can be penetrated, and the group's actions represent extraordinary breaches of security at nuclear weapons plants in the United States. On July 28, 2012, three members of Plowshares cut through fences at the Y-12 National Security Complex in Oak Ridge, Tennessee, which manufactures US nuclear weapons and stockpiles highly enriched uranium. The group spray-painted protest messages, hung banners, and splashed blood.*[4]

The National Nuclear Security Administration has acknowledged the seriousness of the 2012 Plowshares action, which involved the protesters walking into a high-security zone of the plant, calling the security breach "unprecedented." Independent security contractor, WSI, has since had a weeklong "security stand-down," a halt to weapons production, and mandatory refresher training for all security staff.*[4]

Non-proliferation policy experts are concerned about the relative ease with which these unarmed, unsophisticated protesters could cut through a fence and walk into the center of the facility. This is further evidence that nuclear security —the securing of highly enriched uranium and plutonium— should be a top priority to prevent terrorist groups from acquiring nuclear bomb-making material. These experts have questioned "the use of private contractors to provide security at facilities that manufacture and store the government's most dangerous military material" .*[4]

In 2010, there was a security breach at a Belgian air force base which possessed U.S. nuclear warheads. The incident involved six anti-nuclear activists entering Kleine Brogel Air Base. The activists stayed in the snow-covered base for about 20 minutes, before being arrested. A similar event occurred in 2009.*[34]

On December 5, 2011, two anti-nuclear campaigners breached the perimeter of the Cruas Nuclear Power Plant, escaping detection for more than 14 hours, while posting videos of their sit-in on the internet.*[35]

3.6 Cyber attacks

Stuxnet is a computer worm discovered in June 2010 that is believed to have been created by the United States and Israel to attack Iran's nuclear facilities.*[9] It switched off safety devices, causing centrifuges to spin out of control. Stuxnet initially spreads via Microsoft Windows, and targets Siemens industrial control systems. While it is not the first time that hackers have targeted industrial systems,*[36] it is the first discovered malware that spies on and subverts industrial systems,*[37] and the first to include a programmable logic controller (PLC) rootkit.*[38]*[39]

Different variants of Stuxnet targeted five Iranian organizations,*[40] with the probable target widely suspected to be uranium enrichment infrastructure in Iran;*[41]*[42] Symantec noted in August 2010 that 60% of the infected computers worldwide were in Iran.*[43] Siemens stated that the worm has not caused any damage to its customers,*[44] but the Iran nuclear program, which uses embargoed Siemens equipment procured secretly, has been damaged by Stuxnet.*[45]*[46] Kaspersky Lab concluded that the sophisticated attack could only have been conducted "with nation-state support" .*[47]

Idaho National Laboratory ran the Aurora Experiment in 2007 to demonstrate how a cyber attack could destroy physical components of the electric grid.*[48] The experiment used a computer program to rapidly open and close a diesel generator's circuit breakers out of phase from the rest of the grid and explode. This vulnerability is referred to as the *Aurora Vulnerability*.

The number and sophistication of cyber attacks is on the rise. The computers of South Korea's nuclear plant operator (KHNP) were hacked in December 2014. The cyber attacks involved thousands of phishing emails containing malicious code, and information was stolen.*[10]

3.7   Population surrounding plants

Population density is one critical lens through which risks have to be assessed, says Laurent Stricker, a nuclear engineer and chairman of the World Association of Nuclear Operators:*[49]

> The KANUPP plant in Karachi, Pakistan, has the most people—8.2 million—living within 30 kilometres of a nuclear plants have populations larger than 3 million within that radius.*[49] plant, although it has just one relatively small reactor with an output of 125 megawatts. Next in the league, however, are much larger plants —Taiwan's 1,933-megawatt Kuosheng plant with 5.5 million people within a 30-kilometre radius and the 1,208-megawatt Chin Shan plant with 4.7 million; both zones include the capital city of Taipei.*[49]

172,000 people living within a 30 kilometre radius of the Fukushima Daiichi nuclear power plant, have been forced or advised to evacuate the area. More generally, a 2011 analysis by *Nature* and Columbia University, New York, shows that some 21 nuclear plants have populations larger than 1 million within a 30-km radius, and six plants have populations larger than 3 million within that radius.*[49]

3.8   Implications

In his book, *Normal accidents*, Charles Perrow says that multiple and unexpected failures are built into society's complex and tightly-coupled nuclear reactor systems. Such accidents are unavoidable and cannot be designed around.*[50]

In the 2003 book, *Brittle Power*, Amory Lovins talks about the need for a resilient, secure, energy system:

The foundation of a secure energy system is to need less energy in the first place, then to get it from sources that are inherently invulnerable because they're diverse, dispersed, renewable, and mainly local. They're secure not because they're American but because of their design. Any highly centralised energy system—pipelines, nuclear plants, refineries—invite devastating attack. But invulnerable alternatives don't, and can't, fail on a large scale.*[51]

3.9   See also

- Lists of nuclear disasters and radioactive incidents

- Atomic spies

- Crimes involving radioactive substances

- Design basis accident

- Environmental impact of nuclear power

- International Nuclear Events Scale

- Nuclear and radiation accidents

- Nuclear close calls

- Nuclear criticality safety

- Nuclear fuel response to reactor accidents

- Nuclear power debate

- Nuclear power plant emergency response team

- Nuclear whistleblowers

- Nuclear weapon

- Micro nuclear reactor

- Passive nuclear safety

- Safety code (nuclear reactor)

- World Association of Nuclear Operators

- Category:Victims of radiological poisoning

3.10 Further reading

- Allison, Graham (9 August 2004). *Nuclear Terrorism: The Ultimate Preventable Catastrophe*. New York, New York: Times Books. ISBN 978-0-8050-7651-6.

- Byrne, John and Steven M. Hoffman (1996). *Governing the Atom: The Politics of Risk*, Transaction Publishers.

- Cooke, Stephanie (2009). *In Mortal Hands: A Cautionary History of the Nuclear Age*, Black Inc.

- Ferguson, Charles D., and William C. Potter, with Amy Sands, Leonard S. Spector and Fred L. Wehling (2004). *The Four Faces of Nuclear Terrorism*. Monterey, California: Center for Nonproliferation Studies. ISBN 1-885350-09-0.

- Jones, Ishmael (2010) [2008]. *The Human Factor: Inside the CIA's Dysfunctional Intelligence Culture*. Encounter Books. ISBN 978-1-59403-382-7.

- Levi, Michael (2007). *On Nuclear Terrorism*. Cambridge, Massachusetts: Harvard University Press. ISBN 978-0-674-02649-0.

- Lovins, Amory B. and John H. Price (1975). *Non-Nuclear Futures: The Case for an Ethical Energy Strategy*, Ballinger Publishing Company, 1975, ISBN 0-88410-602-0

- Schell, Jonathan (2007). *The Seventh Decade: The New Shape of Nuclear Danger*. New York, New York: Metropolitan Books.

- Kuperman, Alan J (2013). *Nuclear Terrorism and Global Security: The Challenge of Phasing out Highly Enriched Uranium*. New York, New York: Routledge.

3.11 References

[1] *Commonwealth v. Berrigan*, 501 A.2d 226, 509 Pa. 118 (1985)

[2] Charles D. Ferguson & Frank A. Settle (2012). "The Future of Nuclear Power in the United States" (PDF). *Federation of American Scientists*.

[3] Benjamin K. Sovacool (2011). *Contesting the Future of Nuclear Power: A Critical Global Assessment of Atomic Energy*, World Scientific, p. 192.

[4] Kennette Benedict (9 August 2012). "Civil disobedience". *Bulletin of the Atomic Scientists*.

[5] Jay Davis. After A Nuclear 9/11 *The Washington Post*, March 25, 2008.

[6] Brian Michael Jenkins. A Nuclear 9/11? *CNN.com*, September 11, 2008.

[7] Orde Kittrie. Averting Catastrophe: Why the Nuclear Nonproliferation Treaty is Losing its Deterrence Capacity and How to Restore It May 22, 2007, p. 338.

[8] Nicholas D. Kristof. A Nuclear 9/11 *The New York Times*, March 10, 2004.

[9] "Legal Experts: Stuxnet Attack on Iran Was Illegal 'Act of Force'". Wired. 25 March 2013.

[10] Penny Hitchin, "Cyber attacks on the nuclear industry", *Nuclear Engineering International*, 15 September 2015.

[11] "Congressional Budget Office Vulnerabilities from Attacks on Power Reactors and Spent Material".

[12] U.S. NRC: "Nuclear Security – Five Years After 9/11". Accessed 23 July 2007

[13] When Iran Bombed Iraq's Nuclear Reactor, Iraq's Osirak Destruction.

[14] Blakely, Rhys (August 11, 2009), "Terrorists 'have attacked Pakistan nuclear sites three times'", *Times Online*, London

[15] "Elite US troops ready to combat Pakistani nuclear hijacks", *Times*

[16] Brittle Power, Chapter 1, p. 1.

[17] *Brittle Power*, Chapter 1, p. 2.

[18] Amory Lovins (2001). *Brittle Power* (PDF). pp. 145–146.

[19] Threat Assessment: U.S. Nuclear Plants Near Airports May Be at Risk of Airplane Attack (Link Defunct), Global Security Newswire, June 11, 2003.

[20] Newtan, Samuel Upton (2007). *Nuclear War 1 and Other Major Nuclear Disasters of the 20th Century*, AuthorHouse, p.146.

[21] Stephanie Cooke (March 19, 2011). "Nuclear power is on trial". CNN.

[22] Frank Barnaby (2007). "Consequences of a Nuclear Renaissance" (PDF). *International Symposium*.

[23] Bunn, Matthew. "Securing the Bomb 2010: Securing All Nuclear Materials in Four Years" (PDF). President and Fellows of Harvard College. Retrieved 28 January 2013.

[24] Bunn, Matthew & Col-Gen. E.P. Maslin (2010). "All Stocks of Weapons-Usable Nuclear Materials Worldwide Must be Protected Against Global Terrorist Threats" (PDF). Belfer Center for Science and International Affairs, Harvard University. Retrieved July 26, 2012.

[25] Rhys Blakeley, "Terrorists 'have attacked Pakistan nuclear sites three times'," *Times Online* (August 11, 2009).

[26] http://www.pretorianews.co.za/?fSectionId=&fArticleId= vn20071109061218448C528585

[27] *Washington Post*, December 20, 2007, Op-Ed by Micah Zenko

[28] "Feds Hoped to Snag Bin Laden Nuke Expert in JFK Bomb Plot". Fox News. June 4, 2007.

[29] http://politics.guardian.co.uk/terrorism/story/0,,1947295, 00.html

[30] "Ushering in the era of nuclear terrorism," by Patterson, Andrew J. MD, PhD, *Critical Care Medicine*, v. 35, p.953–954, 2007.

[31] Matthew Bunn and Scott Sagan (2014). "A Worst Practices Guide to Insider Threats: Lessons from Past Mistakes". The American Academy of Arts & Sciences.

[32] "History of MK". African National Congress. Archived from the original on 4 April 2007. Retrieved 14 May 2007.

[33] Helen Bamford (11 March 2006). "Koeberg: SA's ill-starred nuclear power plant". *Cape Argus*. Retrieved 14 May 2007.

[34] Kevin Dougherty (February 6, 2010). "Belgian base breach sparks nuclear worries". *Stars and Stripes*.

[35] Tara Patel (December 16, 2011). "Breaches at N-plants heighten France's debate over reactors". *Seattle Times*.

[36] "Building a Cyber Secure Plant". Siemens. 30 September 2010. Retrieved 5 December 2010.

[37] Robert McMillan (16 September 2010). "Siemens: Stuxnet worm hit industrial systems". Computerworld. Retrieved 16 September 2010.

[38] "Last-minute paper: An indepth look into Stuxnet". Virus Bulletin.

[39] "Stuxnet worm hits Iran nuclear plant staff computers". BBC News. 26 September 2010.

[40] "Stuxnet Virus Targets and Spread Revealed". BBC News. 15 February 2011. Retrieved 17 February 2011.

[41] Steven Cherry; with Ralph Langner (13 October 2010). "How Stuxnet Is Rewriting the Cyberterrorism Playbook". IEEE Spectrum.

[42] Beaumont, Claudine (23 September 2010). "Stuxnet virus: worm 'could be aimed at high-profile Iranian targets'". London: The Daily Telegraph. Retrieved 28 September 2010.

[43] MacLean, William (24 September 2010). "UPDATE 2-Cyber attack appears to target Iran-tech firms". *Reuters*.

[44] ComputerWorld (14 September 2010). "Siemens: Stuxnet worm hit industrial systems". Computerworld. Retrieved 3 October 2010.

[45] "Iran Confirms Stuxnet Worm Halted Centrifuges". *CBS News*. 29 November 2010.

[46] Ethan Bronner & William J. Broad (29 September 2010). "In a Computer Worm, a Possible Biblical Clue". *NYTimes*. Retrieved 2 October 2010. "Software smart bomb fired at Iranian nuclear plant: Experts". Economictimes.indiatimes.com. 24 September 2010. Retrieved 28 September 2010.

[47] "Kaspersky Lab provides its insights on Stuxnet worm". *Kaspersky*. Russia. 24 September 2010.

[48] "Mouse click could plunge city into darkness, experts say", *CNN*, September 27, 2007. Source: http://www.cnn.com/ 2007/US/09/27/power.at.risk/index.html

[49] Declan Butler (21 April 2011). "Reactors, residents and risk". *Nature*.

[50] Daniel E Whitney (2003). "Normal Accidents by Charles Perrow" (PDF). *Massachusetts Institute of Technology*.

[51] Amory B. Lovins and L. Hunter Lovins. "Terrorism and Brittle Technology" in *Technology and the Future* by Albert H. Teich, Ninth edition, Thomson, 2003, p. 169.

3.12   External links

- Annotated bibliography, Alsos Digital Library for Nuclear Issues

- Fallout: After a Nuclear Attack - slideshow by *Life magazine*

- Nuclear Emergency and Radiation Resources

- What if the terrorists go nuclear?, Center for Defense Information

- Preventing Catastrophic Nuclear Terrorism, Council on Foreign Relations

- Use of nuclear and radiological weapons by terrorists?, International Review of the Red Cross

- Nuclear-free future award

Chapter 4

Criticality accident

A **criticality accident** is an uncontrolled nuclear chain reaction. It is sometimes referred to as a **critical excursion** or a **critical power excursion** and represents the unintentional assembly of a critical mass of a given fissile material, such as enriched uranium or plutonium, in an unprotected environment. A critical or supercritical fission reaction (one that is sustained in power or increasing in power) generally only occurs inside reactor cores and occasionally within test environments; a criticality accident occurs when the same reaction is achieved unintentionally and in an unsafe environment. Though dangerous and frequently lethal to humans within the immediate area, the critical mass formed is still incapable of producing a nuclear detonation of the type seen in fission bombs, as the reaction lacks the many engineering elements that are necessary to induce explosive supercriticality. The heat released by the nuclear reaction will typically cause the fissile material to expand, so that the nuclear reaction becomes subcritical again within a few seconds.

In the history of atomic power development, 60 criticality accidents have occurred, including 22 in collections of fissile materials located in process environments outside of a nuclear reactor or critical experiments assembly. Although process accidents occurring outside of reactors are characterized by a large release of radiation, the release is localized and has caused fatal radiation exposure only to persons very near to the event (less than 1 metre), resulting in 14 fatalities. No criticality accidents have resulted in nuclear explosions.[*][1]

4.1 Cause

Criticality occurs when sufficient fissile material (a "critical mass") is in one place such that each fission of an atom of the material, on average, produces a neutron that in turn strikes another atom causing another fission; this causes the chain reaction to become self-sustaining within the mass of material. Criticality can be achieved by using metallic uranium or plutonium or by mixing compounds or liquid solutions of these elements. The chain reaction is influenced by parameters noted by the acronym MAGIC MERV - for Mass, Absorption, Geometry, Interaction, Concentration, Moderation, Enrichment, Reflection and Volume.

The calculations that predict the likelihood of a material going into a critical state can be complex, so both civil and military installations that handle fissile materials employ specially trained personnel to monitor operations and prevent criticality accidents. The calculations that predict the excursion characteristics can also be complex, as this requires knowledge of the likely process upset conditions.

The assembly of a critical mass establishes a nuclear chain reaction, resulting in an exponential rate of change in the neutron population over space and time leading to neutron radiation and a neutron flux. This radiation contains both a neutron and gamma ray component and is extremely dangerous to any unprotected nearby life-form. The rate of change of neutron population depends on the neutron generation time, which is characteristic of the neutron population, the state of "criticality", and the fissile medium.

A nuclear fission creates approximately 2.5 neutrons per fission event on average.[*][2] For every 1000 neutrons released by fission, 7 are delayed neutrons which are emitted from the fission product precursors, called *delayed neutron emitters*. This delayed neutron fraction, on the order of 0.007 for uranium, is crucial for the control of the neutron chain reaction in reactors. It is called one dollar of reactivity. The lifetime of delayed neutrons ranges from fractions of seconds to almost 100 seconds after fission. The neutrons are usually classified in 6 delayed neutron groups.[*][2] The average neutron lifetime considering delayed neutrons is approximately 0.1 sec, which makes the chain reaction relatively easy to control over time. The remaining 993 prompt neutrons are released very quickly, approximately 1 μs after the fission event.

Nuclear reactors operate at exact criticality. When at least one dollar of reactivity is added above the exact critical point (the point where neutrons produced is balanced by neutrons lost per generation) then the chain reaction does not rely on delayed neutrons, and the rate of change of neutron population increases exponentially as the time constant is the prompt neutron lifetime. Thus there is a very large increase in neutron population over a very short time frame. Since each fission event contributes approximately 200 MeV per fission, this results in a very large energy burst as a "prompt critical spike". This spike can be easily detected by radiation dosimetry instrumentation and "criticality accident alarm system" detectors that are properly deployed.

4.2 Accident types

Criticality accidents are divided into one of two categories:

- *Process accidents*, where controls in place to prevent any criticality are breached;

- *Reactor accidents*, where deliberately achieved criticality in a nuclear reactor becomes uncontrollable.

Excursion types can be classified into four categories depicting the nature of the evolution over time:

1. Prompt criticality excursion

2. Transient criticality excursion

3. Exponential excursion

4. Steady state excursion

The prompt critical excursion is characterized by a power history with an initial prompt critical spike as previously noted, that either self terminates or continues for an extended period as a tail region that decreases over time. The former was only 1 of the 22 process accidents, the latter is noted for reactors and critical assemblies. The transient critical excursion is characterized by a continuing or repeating spike pattern after the initial prompt critical excursion. The longest of the 22 process accidents lasted 37 hours. The 1997 Tokaimura nuclear accident lasted 18 hours. The exponential excursion is characterized by a reactivity of less than one dollar added, where the neutron population rises as an exponential over time, but not reaching prompt critical. The exponential excursion can reach a peak power level, then decrease over time, or reach a steady state power level, where the critical state is exactly achieved for a "steady state" excursion.

The steady state excursion is also a state which the heat generated by fission is balanced by the heat losses to the ambient environment. This excursion has been characterized by the Oklo natural reactor that was naturally produced within uranium deposits in Gabon, Africa about 1.7 billion years ago.

4.3 Recorded incidents

At least sixty criticality accidents have been recorded since 1945. These have caused at least twenty-one deaths: seven in the United States, ten in the Soviet Union, two in Japan, one in Argentina, and one in Yugoslavia. Nine have been due to process accidents, and the others from research reactor accidents.[*][1]

Criticality accidents have occurred both in the context of nuclear weapons and nuclear reactors.

- The sphere of plutonium surrounded by neutron-reflecting tungsten carbide blocks in a re-enactment of Harry Daghlian's 1945 experiment.[*][31]

- A re-creation of the Slotin incident. The inside hemisphere with the thumb-hole next to the hand is beryllium (replacing the uranium tamper in a Fat Man bomb), with an external larger metal sphere under it, of aluminium. The 3.5-inch-diameter (89 mm) plutonium "demon core" (the same as in the Daghlian incident) was inside at the time of the accident, and would not be visible. However, its dimensions are comparable with the two small half-spheres shown resting nearby.

- Image of the Lady Godiva assembly in the scrammed (safe) configuration.[*][32]

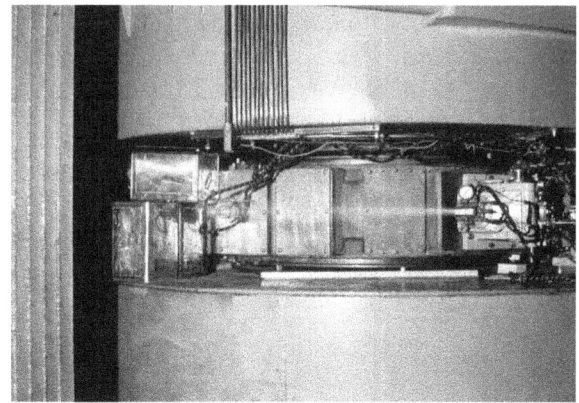

• Image of the Lady Godiva assembly, showing the damage caused to the supporting rods after the excursion of February 1954. Note the images are of different assemblies.*[32]

Image of a 60-inch cyclotron, circa 1939, showing an external beam of accelerated ions (perhaps protons or deuterons) ionizing the surrounding air and causing an ionized-air glow. Due to the similar mechanism of production, the blue glow is thought to resemble the "blue flash" seen by Harry Daghlian and other witnesses of criticality accidents.

There was speculation although not confirmed within criticality accident experts, that Fukushima 3 suffered a criticality accident. Based on incomplete information about the 2011 Fukushima I nuclear accidents, Dr. Ferenc Dalnoki-Veress speculates that transient criticalities may have occurred there.*[33] Noting that limited, uncontrolled chain reactions might occur at Fukushima I, a spokesman for the International Atomic Energy Agency (IAEA) "emphasized that the nuclear reactors won' t explode." *[34] By March 23, 2011, neutron beams had already been observed 13 times at the crippled Fukushima nuclear power plant. While a criticality accident was not believed to account for these beams, the beams could indicate nuclear fission is occurring.*[35] On April 15, TEPCO reported that nuclear fuel had melted and fallen to the lower containment sections of three of the Fukushima I reactors, including reactor three. The melted material was not expected to breach one of the lower containers, which could cause a massive radioactivity release. Instead, the melted fuel is thought to have dispersed uniformly across the lower portions of the containers of reactors No. 1, No. 2 and No. 3, making the resumption of the fission process, known as a "recriticality" , most unlikely.*[36]

4.4 Observed effects

4.4.1 Blue glow

See also: Ionized-air glow

Many criticality accidents have been observed to emit a blue flash of light.

The blue glow of a criticality accident results from the fluorescence of the excited ions, atoms and molecules of air (mostly oxygen and nitrogen) falling back to unexcited states, which produces an abundance of blue light. This is also the reason electrical sparks in air, including lightning, appear electric blue. The smell of ozone was said to be a sign of high ambient radioactivity by Chernobyl liquidators.

This blue flash or "blue glow" is often incorrectly attributed to Cherenkov radiation. It is a coincidence that the color of Cherenkov light and light emitted by ionized air are a very similar blue as their methods of production are different. Cherenkov radiation does occur in air for high energy particles (such as particle showers from cosmic rays)*[37] but not for the lower energy charged particles emitted from nuclear decay. In a nuclear setting, Cherenkov radiation is instead seen in dense media such as water or in a solution such as uranyl nitrate in a reprocessing plant. Cherenkov radiation could also be responsible for the "blue flash" experienced in an excursion due to the intersection of particles with the vitreous humour within the eyeballs of those in the presence of the criticality. This would also explain the absence of any record of blue light in video surveillance of the more recent incidents.

4.4.2 Heat effects

Some people reported feeling a "heat wave" during a criticality event.*[38]*[39] It is not known whether this may be a psychosomatic reaction to the terrifying realization of what has just occurred, or if it is a physical effect of heating (or nonthermal stimulation of heat sensing nerves in the skin) due to energy emitted by the criticality event.

A review of all of the criticality accidents with eyewitness accounts indicates that the heat waves were only observed when the fluorescent blue glow (the non-Cherenkov light, see above) was also observed. This would suggest a possible relationship between the two, and indeed, one can be readily identified. In dense air, over 30% of the emissions lines from nitrogen and oxygen are in the ultraviolet range, and about 45% are in the infrared range. Only about 25% are in the visible range. Since the skin feels infrared light directly as heat, and ultraviolet light is a cause of sunburn, it is likely that this phenomenon can explain the heat wave observations.[*][40]

4.5 See also

4.5.1 Related terms and concepts

- Criticality (status)
- Nuclear and radiation accidents
- Nuclear criticality safety

4.5.2 In popular culture

- *The Beginning or the End*, a 1947 MGM movie that was the first Hollywood film to depict a person (played by actor Tom Drake) killed in an accident similar to the real-life Slotin criticality event.

- *Edge of Darkness*, a 1985 British television drama where a character dying from radiation poisoning deliberately induces a criticality event as proof that he is in possession of plutonium, and to expose an enemy to a fatal radiation dose.

- *Fat Man and Little Boy*, a 1989 Paramount picture, portrays a fictional composite of Harry K. Daghlian and Louis Slotin who dies of exposure when two hemispheres, which are separated by a wedge, connect accidentally.

- "Meridian", an episode of *Stargate SG-1*, where a criticality accident similar to the Slotin incident occurs during research into using the fictional superheavy element "Naquadriah" in atomic bombs.

- *Infinity*, a 1996 story of Richard Feynman played and directed by Matthew Broderick. There was a sub story of a death due to a criticality accident.

- *Day One*, (TV 1989) A history of the A-bomb development.

- *1000 Ways to Die*, a program airing on Spike TV related a docufiction story titled "Gone Fission", depicting terrorists involved in a criticality event similar to the Daghlian accident.

- List of films about nuclear issues

4.6 Notes

[1] McLaughlin, Thomas P.; et al. (2000). *A Review of Criticality Accidents* (PDF). Los Alamos: Los Alamos National Laboratory. LA-13638. Retrieved 5 November 2012.

[2] Lewis, Elmer E. *Fundamentals of Nuclear Reactor Physics.* Elsevier, 2008, p. 123. ISBN 978-0-12-370631-7

[3] Diana Preston *Before the Fall-Out - From Marie Curie to Hiroshima* - Transworld - 2005 - ISBN 0-385-60438-6 p, 278

[4] McLaughlin et al. pages 78, 80-83

[5] McLaughlin et al. page 93, "In this excursion, three people received radiation doses in the amounts of 66, 66, and 7.4 rep.", LA Appendix A: "rep: An obsolete term for absorbed dose in human tissue, replaced by rad. Originally derived from roentgen equivalent, physical."

[6] Dion, Arnold S., *Harry Daghlian: America's first peacetime atom bomb fatality*, retrieved April 13, 2010

[7] McLaughlin et al. pages 74-76, "His dose was estimated as 510 rem"

[8] "The blue flash". *Restricted Data: The Nuclear Secrecy Blog*. Retrieved 2016-06-29.

[9] Declassified report See pg. 23 for dimensions of beryllium hand-controlled sphere.

[10] McLaughlin et al. pages 74-76, "The eight people in the room received doses of about 2100, 360, 250, 160, 110, 65, 47, and 37 rem."

[11] Y-12's 1958 nuclear criticality accident and increased safety

[12] Criticality accident at the Y-12 plant. Diagnosis and treatment of acute radiation injury, 1961, Geneva, World Health Organization, pp. 27-48.

[13] McLaughlin et al. page 96, "Radiation doses were intense, being estimated at 205, 320, 410, 415, 422, and 433 rem. Of the six persons present, one died shortly afterward, and the other five recovered after severe cases of radiation sickness."

[14] "1958-01-01". Retrieved 2011-01-02.

[15] Vinca reactor accident, 1958, compiled by Wm. Robert Johnston

[16] Nuove esplosioni a Fukushima: danni al nocciolo. Ue: "In Giappone l' apocalisse", 14 marzo 2011

[17] The Cecil Kelley Criticality Accident

[18] Stacy, Susan M. (2000). "Chapter 15: The SL-1 Incident" (PDF). *Proving the Principle: A History of The Idaho National Engineering and Environmental Laboratory, 1949-1999* (PDF). U.S. Department of Energy, Idaho Operations Office. pp. 138–149. ISBN 0-16-059185-6.

[19] McLaughlin et al. pages 33-34

[20] Johnstone

[21] "Wood River criticality accident, 1964". Retrieved 7 December 2016.

[22] McLaughlin et al. pages 40-43

[23] McLaughlin et al. page 103

[24] "NRC: Information Notice No. 83-66, Supplement 1: Fatality at Argentine Critical Facility". Retrieved 7 December 2016.

[25] Johnston, Wm. Robert. "Arzamas-16 criticality accident, 19". Retrieved July 8, 2013.

[26] Kudrik, Igor (June 23, 1997). "Arzamas-16 researcher died on June 20". Retrieved July 8, 2013.

[27] The criticality accident in Sarov, IAEA, 2001.

[28] McLaughlin et al. pages 53-56

[29] http://www.nrc.gov/reading-rm/doc-collections/commission/secys/2000/secy2000-0085/2000-0085scy.pdf

[30] http://www.nrc.gov/reading-rm/doc-collections/commission/secys/2000/secy2000-0085/attachment3.pdf

[31] McLaughlin et al. pages 74-75

[32] McLaughlin et al. pages 81-82

[33] "Has Fukushima's Reactor No. 1 Gone Critical?". *Ecocentric - TIME.com*. 2011-03-30. Retrieved 2011-04-01.

[34] Jonathan Tirone, Sachiko Sakamaki and Yuriy Humber (March 31, 2011). "Fukushima Workers Threatened by Heat Bursts; Sea Radiation Rises".

[35] **Neutron beam observed 13 times at crippled Fukushima nuke plant**. These "neutron beams" as explained in the popular media, do not explain or prove a criticality excursion, as the requisite signature (combined neutron/gamma ratio of approximately 1:3 was not confirmed). A more credible explanation is the presence of neutrons from continued fissions from the decay process. It is highly unlikely that a recriticality occurred in Fukushima 3 since workers near the reactor were not exposed to a high neutron dose in a very short time (milliseconds), and plant radiation instruments would have captured any "repeating spikes" that are characteristic of a continuing moderated criticality accident. TOKYO, March 23, Kyodo News http://web.archive.org/web/20110323214235/http://english.kyodonews.jp/news/2011/03/80539.html

[36] **Japan Plant Fuel Melted Partway Through Reactors: Report** Because there was no large radiation release in the proximity of the reactor, and available dosimetry did not indicate an abnormal neutron dose or neutron/gamma dose ratio, there is no evidence of a criticality accident at Fukushima. Friday, April 15, 2011 "Archived copy". Archived from the original on 2 December 2011. Retrieved 24 April 2011.

[37] "Science". Retrieved 7 December 2016.

[38] McLaughlin et al. page 42, "the operator saw a flash of light and felt a pulse of heat."

[39] McLaughlin et al. page 88, "There was a flash, a shock, a stream of heat in our faces."

[40] Minnema, "Criticality Accidents and the Blue Glow," American Nuclear Society Winter Meeting, 2007.

4.7 References

- Johnstone. List of radiation accidents, Wm. Robert Johnston

- Johnstone. Wood River criticality accident, 1964, Wm. Robert Johnston

- McLaughlin et al. "A Review of Criticality Accidents" by Los Alamos National Laboratory (Report LA-13638), May 2000. Coverage includes United States, Russia, United Kingdom, and Japan. Also available at this page, which also tries to track down documents referenced in the report.

4.8 External links

- Press release on a report on criticality accidents from Los Alamos National Laboratory

- U.S. report from 1971 on criticality accidents to date

Chapter 5

Nuclear meltdown

Three of the reactors at Fukushima I overheated because the cooling systems failed after a tsunami flooded the power station, causing core meltdowns. This was compounded by hydrogen gas explosions and the venting of contaminated steam which released large amounts of radioactive material into the air.[1]

Three Mile Island Nuclear Generating Station consisted of two pressurized water reactors manufactured by Babcock & Wilcox, each inside its own containment building and connected cooling towers. Unit 2, which suffered a partial core melt, is in the background.

A **nuclear meltdown** (**core melt accident** or **partial core melt**[2]) is a severe nuclear reactor accident that results in core damage from overheating. The term *nuclear meltdown* is not officially defined by the International Atomic Energy Agency[3] or by the Nuclear Regulatory Commission.[4] However, it has been defined to mean the accidental melting of the core of a nuclear reactor,[5] and is in common usage a reference to the core's either complete or partial collapse.

A core melt accident occurs when the heat generated by a nuclear reactor exceeds the heat removed by the cooling systems to the point where at least one nuclear fuel element exceeds its melting point. This differs from a fuel element failure, which is not caused by high temperatures. A meltdown may be caused by a loss of coolant, loss of coolant pressure, or low coolant flow rate or be the result of a criticality excursion in which the reactor is operated at a power level that exceeds its design limits. Alternately, in a reactor plant such as the RBMK-1000, an external fire may endanger the core, leading to a meltdown.

Once the fuel elements of a reactor begin to melt, the fuel cladding has been breached, and the nuclear fuel (such as uranium, plutonium, or thorium[n 1]) and fission products (such as cesium-137, krypton-85, or iodine-131) within the fuel elements can leach out into the coolant. Subsequent failures can permit these radioisotopes to breach further layers of containment. Superheated steam and hot metal inside the core can lead to fuel-coolant interactions, hydrogen explosions, or water hammer, any of which could destroy parts of the containment. A meltdown is considered very serious because of the potential for radioactive materials to breach all containment and escape (or be released) into the environment, resulting in radioactive contamination and fallout, and potentially leading to radiation poisoning of people and animals nearby.

5.1 Causes

Nuclear power plants generate electricity by heating fluid via a nuclear reaction to run a generator. If the heat from that reaction is not removed adequately, the fuel assemblies in a reactor core can melt. A core damage incident can occur even after a reactor is shut down because the fuel continues to produce decay heat.

A core damage accident is caused by the loss of sufficient cooling for the nuclear fuel within the reactor core. The reason may be one of several factors, including a loss-of-pressure-control accident, a loss-of-coolant accident (LOCA), an uncontrolled power excursion or, in reactors without a pressure vessel, a fire within the reactor core. Failures in control systems may cause a series of events resulting in loss of cooling. Contemporary safety principles of defense in depth ensure that multiple layers of safety systems are always present to make such accidents unlikely.

The containment building is the last of several safeguards that prevent the release of radioactivity to the environment. Many commercial reactors are contained within a 1.2-to-2.4-metre (3.9 to 7.9 ft) thick pre-stressed, steel-reinforced, air-tight concrete structure that can withstand hurricane-force winds and severe earthquakes.

- In a loss-of-coolant accident, either the physical loss of coolant (which is typically deionized water, an inert gas, NaK, or liquid sodium) or the loss of a method to ensure a sufficient flow rate of the coolant occurs. A loss-of-coolant accident and a loss-of-pressure-control accident are closely related in some reactors. In a pressurized water reactor, a LOCA can also cause a "steam bubble" to form in the core due to excessive heating of stalled coolant or by the subsequent loss-of-pressure-control accident caused by a rapid loss of coolant. In a loss-of-forced-circulation accident, a gas cooled reactor's circulators (generally motor or steam driven turbines) fail to circulate the gas coolant within the core, and heat transfer is impeded by this loss of forced circulation, though natural circulation through convection will keep the fuel cool as long as the reactor is not depressurized.*[6]

- In a loss-of-pressure-control accident, the pressure of the confined coolant falls below specification without the means to restore it. In some cases this may reduce the heat transfer efficiency (when using an inert gas as a coolant) and in others may form an insulating "bubble" of steam surrounding the fuel assemblies (for pressurized water reactors). In the latter case, due to localized heating of the "steam bubble" due to decay heat, the pressure required to collapse the "steam bubble" may exceed reactor design specifications until the reactor has had time to cool down. (This event is less likely to occur in boiling water reactors, where the core may be deliberately depressurized so that the Emergency Core Cooling System may be turned on). In a depressurization fault, a gas-cooled reactor loses gas pressure within the core, reducing heat transfer efficiency and posing a challenge to the cooling of fuel;

however, as long as at least one gas circulator is available, the fuel will be kept cool.*[6]

- In an uncontrolled power excursion accident, a sudden power spike in the reactor exceeds reactor design specifications due to a sudden increase in reactor reactivity. An uncontrolled power excursion occurs due to significantly altering a parameter that affects the neutron multiplication rate of a chain reaction (examples include ejecting a control rod or significantly altering the nuclear characteristics of the moderator, such as by rapid cooling). In extreme cases the reactor may proceed to a condition known as prompt critical. This is especially a problem in reactors that have a positive void coefficient of reactivity, a positive temperature coefficient, are overmoderated, or can trap excess quantities of deleterious fission products within their fuel or moderators. Many of these characteristics are present in the RBMK design, and the Chernobyl disaster was caused by such deficiencies as well as by severe operator negligence. Western light water reactors are not subject to very large uncontrolled power excursions because loss of coolant decreases, rather than increases, core reactivity (a negative void coefficient of reactivity); "transients," as the minor power fluctuations within Western light water reactors are called, are limited to momentary increases in reactivity that will rapidly decrease with time (approximately 200% - 250% of maximum neutronic power for a few seconds in the event of a complete rapid shutdown failure combined with a transient).

- Core-based fires endanger the core and can cause the fuel assemblies to melt. A fire may be caused by air entering a graphite moderated reactor, or a liquid-sodium cooled reactor. Graphite is also subject to accumulation of Wigner energy, which can overheat the graphite (as happened at the Windscale fire). Light water reactors do not have flammable cores or moderators and are not subject to core fires. Gas-cooled civilian reactors, such as the Magnox, UNGG, and AGCR type reactors, keep their cores blanketed with non reactive carbon dioxide gas, which cannot support a fire. Modern gas-cooled civilian reactors use helium, which cannot burn, and have fuel that can withstand high temperatures without melting (such as the High Temperature Gas Cooled Reactor and the Pebble Bed Modular Reactor).

- Byzantine faults and cascading failures within instrumentation and control systems may cause severe problems in reactor operation, potentially leading to core damage if not mitigated. For example, the Browns Ferry fire damaged control cables and required the plant operators to manually activate cooling systems.

The Three Mile Island accident was caused by a stuck-open pilot-operated pressure relief valve combined with a deceptive water level gauge that misled reactor operators, which resulted in core damage.

5.2   Light water reactors (LWRs)

Before the core of a light water nuclear reactor can be damaged, two precursor events must have already occurred:

- A limiting fault (or a set of compounded emergency conditions) that leads to the failure of heat removal within the core (the loss of cooling). Low water level uncovers the core, allowing it to heat up.

- Failure of the Emergency Core Cooling System (ECCS). The ECCS is designed to rapidly cool the core and make it safe in the event of the maximum fault (the design basis accident) that nuclear regulators and plant engineers could imagine. There are at least two copies of the ECCS built for every reactor. Each division (copy) of the ECCS is capable, by itself, of responding to the design basis accident. The latest reactors have as many as four divisions of the ECCS. This is the principle of redundancy, or duplication. As long as at least one ECCS division functions, no core damage can occur. Each of the several divisions of the ECCS has several internal "trains" of components. Thus the ECCS divisions themselves have internal redundancy – and can withstand failures of components within them.

The Three Mile Island accident was a compounded group of emergencies that led to core damage. What led to this was an erroneous decision by operators to shut down the ECCS during an emergency condition due to gauge readings that were either incorrect or misinterpreted; this caused another emergency condition that, several hours after the fact, led to core exposure and a core damage incident. If the ECCS had been allowed to function, it would have prevented both exposure and core damage. During the Fukushima incident the emergency cooling system had also been manually shut down several minutes after it started.*[7]

If such a limiting fault were to occur, and a complete failure of all ECCS divisions were to occur, both Kuan, *et al* and Haskin, *et al* describe six stages between the start of the limiting fault (the loss of cooling) and the potential escape of molten corium into the containment (a so-called "full meltdown"):*[8]*[9]

1. **Uncovering of the Core** – In the event of a transient, upset, emergency, or limiting fault, LWRs are

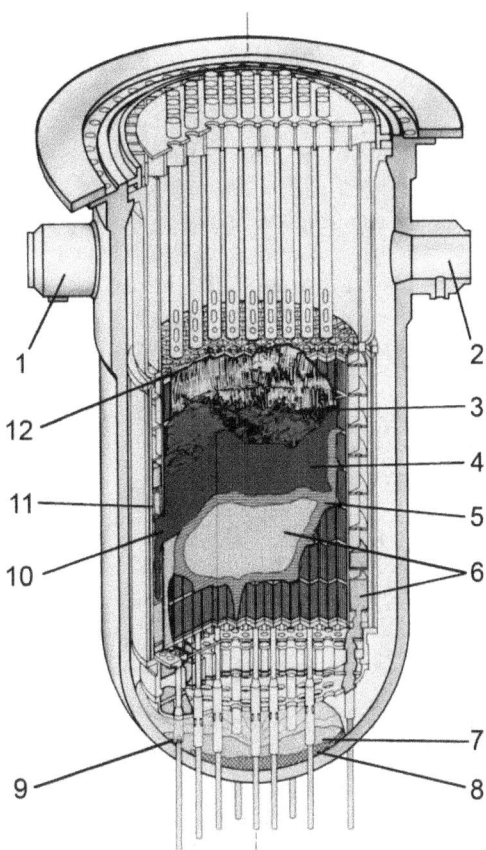

The Three Mile Island reactor 2 after the meltdown.

1. *Inlet 2B*
2. *Inlet 1A*
3. *Cavity*
4. *Loose core debris*
5. *Crust*
6. *Previously molten material*
7. *Lower plenum debris*
8. *Possible region depleted in uranium*
9. *Ablated incore instrument guide*
10. *Hole in baffle plate*
11. *Coating of previously-molten material on bypass region interior surfaces*
12. *Upper grid damage*

designed to automatically SCRAM (a SCRAM being the immediate and full insertion of all control rods) and spin up the ECCS. This greatly reduces reactor thermal power (but does not remove it completely); this delays core becoming uncovered, which is defined as the point when the fuel rods are no longer covered by coolant and can begin to heat up. As Kuan states: "In a small-break LOCA with no emergency core coolant injection, core uncovery [sic] generally begins approximately an hour after the initiation of the break. If the reactor coolant pumps are not running, the upper part of the core will be exposed to a steam environment and heatup of the core will begin. However, if the coolant pumps are running, the core will be cooled by a two-phase mixture of steam and water, and heatup of the fuel rods will be delayed until almost all of the water in the two-phase mixture is vaporized. The TMI-2 accident showed that operation of reactor coolant pumps may be sustained for up to approximately two hours to deliver a two phase mixture that can prevent core heatup." *[8]

2. **Pre-damage heat up** – "In the absence of a two-phase mixture going through the core or of water addition to the core to compensate water boiloff, the fuel rods in a steam environment will heat up at a rate between 0.3 °C/s (0.5 °F/s) and 1 °C/s (1.8 °F/s) (3)." *[8]

3. **Fuel ballooning and bursting** – "In less than half an hour, the peak core temperature would reach 1,100 K (830 °C). At this temperature the zircaloy cladding of the fuel rods may balloon and burst. This is the first stage of core damage. Cladding ballooning may block a substantial portion of the flow area of the core and restrict the flow of coolant. However, complete blockage of the core is unlikely because not all fuel rods balloon at the same axial location. In this case, sufficient water addition can cool the core and stop core damage progression." *[8]

4. **Rapid oxidation** – "The next stage of core damage, beginning at approximately 1,500 K (1,230 °C), is the rapid oxidation of the Zircaloy by steam. In the oxidation process, hydrogen is produced and a large amount of heat is released. Above 1,500 K (1,230 °C), the power from oxidation exceeds that from decay heat (4,5) unless the oxidation rate is limited by the supply of either zircaloy or steam." *[8]

5. **Debris bed formation** – "When the temperature in the core reaches about 1,700 K (1,430 °C), molten control materials (1,6) will flow to and solidify in the space between the lower parts of the fuel rods where the temperature is comparatively low. Above 1,700 K (1,430 °C), the core temperature may escalate in a few minutes to the melting point of zircaloy [2,150 K (1,880 °C)] due to increased oxidation rate. When the oxidized cladding breaks, the molten zircaloy, along with dissolved UO$_2$ (1,7) would flow downward and freeze in the cooler, lower region of the core. Together with solidified control materials from earlier down-

flows, the relocated zircaloy and UO$_2$ would form the lower crust of a developing cohesive debris bed." *[8]

6. **(Corium) Relocation to the lower plenum** – "In scenarios of small-break LOCAs, there is generally a pool of water in the lower plenum of the vessel at the time of core relocation. Release of molten core materials into water always generates large amounts of steam. If the molten stream of core materials breaks up rapidly in water, there is also a possibility of a steam explosion. During relocation, any unoxidized zirconium in the molten material may also be oxidized by steam, and in the process hydrogen is produced. Recriticality also may be a concern if the control materials are left behind in the core and the relocated material breaks up in unborated water in the lower plenum." *[8]

At the point at which the corium relocates to the lower plenum, Haskin, *et al* relate that the possibility exists for an incident called a *fuel-coolant interaction (FCI)* to substantially stress or breach the primary pressure boundary when the corium relocates to the lower plenum of the reactor pressure vessel ("RPV").*[10] This is because the lower plenum of the RPV may have a substantial quantity of water - the reactor coolant - in it, and, assuming the primary system has not been depressurized, the water will likely be in the liquid phase, and consequently dense, and at a vastly lower temperature than the corium. Since corium is a liquid metal-ceramic eutectic at temperatures of 2,200 to 3,200 K (1,930 to 2,930 °C), its fall into liquid water at 550 to 600 K (277 to 327 °C) may cause an extremely rapid evolution of steam that could cause a sudden extreme overpressure and consequent gross structural failure of the primary system or RPV.*[10] Though most modern studies hold that it is physically infeasible, or at least extraordinarily unlikely, Haskin, *et al* state that there exists a remote possibility of an extremely violent FCI leading to something referred to as an *alpha-mode failure*, or the gross failure of the RPV itself, and subsequent ejection of the upper plenum of the RPV as a missile against the inside of the containment, which would likely lead to the failure of the containment and release of the fission products of the core to the outside environment without any substantial decay having taken place.*[11]

The American Nuclear Society has commented on the TMI-2 accident, that despite melting of about one-third of the fuel, the reactor vessel itself maintained its integrity and contained the damaged fuel.*[12]

5.2.1 Breach of the Primary Pressure Boundary

There are several possibilities as to how the primary pressure boundary could be breached by corium.

- Steam Explosion

As previously described, FCI could lead to an overpressure event leading to RPV fail, and thus, primary pressure boundary fail. Haskin, *et al.* report that in the event of a steam explosion, failure of the lower plenum is far more likely than ejection of the upper plenum in the alpha-mode. In the event of lower plenum failure, debris at varied temperatures can be expected to be projected into the cavity below the core. The containment may be subject to overpressure, though this is not likely to fail the containment. The alpha-mode failure will lead to the consequences previously discussed.

- Pressurized Melt Ejection (PME)

It is quite possible, especially in pressurized water reactors, that the primary loop will remain pressurized following corium relocation to the lower plenum. As such, pressure stresses on the RPV will be present in addition to the weight stress that the molten corium places on the lower plenum of the RPV; when the metal of the RPV weakens sufficiently due to the heat of the molten corium, it is likely that the liquid corium will be discharged under pressure out of the bottom of the RPV in a pressurized stream, together with entrained gases. This mode of corium ejection may lead to direct containment heating (DCH).

5.2.2 Severe Accident Ex-Vessel Interactions and Challenges to Containment

Haskin, *et al* identify six modes by which the containment could be credibly challenged; some of these modes are not applicable to core melt accidents.

1. Overpressure

2. Dynamic pressure (shockwaves)

3. Internal missiles

4. External missiles (not applicable to core melt accidents)

5. Meltthrough

6. Bypass

5.2.3 Standard failure modes

If the melted core penetrates the pressure vessel, there are theories and speculations as to what may then occur.

In modern Russian plants, there is a "core catching device" in the bottom of the containment building. The melted core is supposed to hit a thick layer of a "sacrificial metal" which would melt, dilute the core and increase the heat conductivity, and finally the diluted core can be cooled down by water circulating in the floor. However, there has never been any full-scale testing of this device.*[13]

In Western plants there is an airtight containment building. Though radiation would be at a high level within the containment, doses outside of it would be lower. Containment buildings are designed for the orderly release of pressure without releasing radionuclides, through a pressure release valve and filters. Hydrogen/oxygen recombiners also are installed within the containment to prevent gas explosions.

In a melting event, one spot or area on the RPV will become hotter than other areas, and will eventually melt. When it melts, corium will pour into the cavity under the reactor. Though the cavity is designed to remain dry, several NUREG-class documents advise operators to flood the cavity in the event of a fuel melt incident. This water will become steam and pressurize the containment. Automatic water sprays will pump large quantities of water into the steamy environment to keep the pressure down. Catalytic recombiners will rapidly convert the hydrogen and oxygen back into water. One positive effect of the corium falling into water is that it is cooled and returns to a solid state.

Extensive water spray systems within the containment along with the ECCS, when it is reactivated, will allow operators to spray water within the containment to cool the core on the floor and reduce it to a low temperature.

These procedures are intended to prevent release of radioactivity. In the Three Mile Island event in 1979, a theoretical person standing at the plant property line during the entire event would have received a dose of approximately 2 millisieverts (200 millirem), between a chest X-ray's and a CT scan's worth of radiation. This was due to outgassing by an uncontrolled system that, today, would have been backfitted with activated carbon and HEPA filters to prevent radionuclide release.

However in case of Fukushima incident this design failed: Despite the efforts of the operators at the Fukushima Daiichi nuclear power plant to maintain control, the reactor cores in units 1-3 overheated, the nuclear fuel melted and the three containment vessels were breached. Hydrogen was released from the reactor pressure vessels, leading to explosions inside the reactor buildings in units 1, 3 and 4 that damaged structures and equipment and injured personnel. Radionuclides were released from the plant to the atmosphere and were deposited on land and on the ocean. There were also direct releases into the sea.*[14]

Cooling will take quite a while, until the natural decay heat of the corium reduces to the point where natural convection and conduction of heat to the containment walls and re-radiation of heat from the containment allows for water spray systems to be shut down and the reactor put into safe storage. The containment can be sealed with release of extremely limited offsite radioactivity and release of pressure within the containment. After a number of years for fission products to decay - probably around a decade - the containment can be reopened for decontamination and demolition.

5.2.4 Unexpected failure modes

Another scenario sees a buildup of hydrogen, which may lead to a detonation event, as happened for three reactors during Fukushima incident. Catalytic hydrogen recombiners located within containment are designed to prevent this from occurring; however, in Fukushima recombiners did not work due the absence of power and hydrogen detonation breached the containment. During the 1979 Three Mile Island accident a hydrogen bubble formed in the pressure vessel dome. There were initial concerns that this hydrogen bubble might ignite and damage the pressure vessel or even damage the containment building; but it was soon realized that a lack of oxygen precluded a burnable or explosive mixture from forming inside the pressure vessel.[15]

5.2.5 Speculative failure modes

One scenario consists of the reactor pressure vessel failing all at once, with the entire mass of corium dropping into a pool of water (for example, coolant or moderator) and causing extremely rapid generation of steam. The pressure rise within the containment could threaten integrity if rupture disks could not relieve the stress. Exposed flammable substances could burn, but there are few, if any, flammable substances within the containment.

Another theory called an 'alpha mode' failure by the 1975 Rasmussen (WASH-1400) study asserted steam could produce enough pressure to blow the head off the reactor pressure vessel (RPV). The containment could be threatened if the RPV head collided with it. (The WASH-1400 report was replaced by better-based newer studies, and now the Nuclear Regulatory Commission has disavowed them all and is preparing the overarching State-of-the-Art Reactor Consequence Analyses [SOARCA] study - see the Disclaimer in NUREG-1150.)

It has not been determined to what extent a molten mass can melt through a structure (although that was tested in the Loss-of-Fluid-Test Reactor described in Test Area North's fact sheet[16]). The Three Mile Island accident provided some real-life experience, with an actual molten core within an actual structure; the molten corium failed to melt through

the Reactor Pressure Vessel after over six hours of exposure, due to dilution of the melt by the control rods and other reactor internals, validating the emphasis on defense in depth against core damage incidents. According to some, a molten reactor core could penetrate the reactor pressure vessel and containment structure and burn downwards to the level of the groundwater.[17]

By 1970, there were doubts about the ability of the emergency cooling systems of a nuclear reactor to prevent a loss of coolant accident and the consequent meltdown of the fuel core; the subject proved popular in the technical and the popular presses.[18] In 1971, in the article *Thoughts on Nuclear Plumbing*, former Manhattan Project nuclear physicist Ralph Lapp used the term "China syndrome" to describe a possible burn-through of the containment structures, and the subsequent escape of radioactive material(s) into the atmosphere and environment. The hypothesis derived from a 1967 report by a group of nuclear physicists, headed by W. K. Ergen.[19]

5.3 Other reactor types

Other types of reactors have different capabilities and safety profiles than the LWR does. Advanced varieties of several of these reactors have the potential to be inherently safe.

5.3.1 CANDU reactors

CANDU reactors, Canadian-invented deuterium-uranium design, are designed with at least one, and generally two, large low-temperature and low-pressure water reservoirs around their fuel/coolant channels. The first is the bulk heavy-water moderator (a separate system from the coolant), and the second is the light-water-filled shield tank(or calandria vault). These backup heat sinks are sufficient to prevent either the fuel meltdown in the first place (using the moderator heat sink), or the breaching of the core vessel should the moderator eventually boil off (using the shield tank heat sink).[20] Other failure modes aside from fuel melt will probably occur in a CANDU rather than a meltdown, such as deformation of the calandria into a non-critical configuration. All CANDU reactors are located within standard Western containments as well.

5.3.2 Gas-cooled reactors

One type of Western reactor, known as the advanced gas-cooled reactor (or AGCR), built by the United Kingdom, is not very vulnerable to loss-of-cooling accidents or to core damage except in the most extreme of circumstances. By virtue of the relatively inert coolant (carbon dioxide), the large volume and high pressure of the coolant, and the relatively high heat transfer efficiency of the reactor, the time frame for core damage in the event of a limiting fault is measured in days. Restoration of some means of coolant flow will prevent core damage from occurring.

Other types of highly advanced gas cooled reactors, generally known as high-temperature gas-cooled reactors (HTGRs) such as the Japanese High Temperature Test Reactor and the United States' Very High Temperature Reactor, are inherently safe, meaning that meltdown or other forms of core damage are physically impossible, due to the structure of the core, which consists of hexagonal prismatic blocks of silicon carbide reinforced graphite infused with TRISO or QUADRISO pellets of uranium, thorium, or mixed oxide buried underground in a helium-filled steel pressure vessel within a concrete containment. Though this type of reactor is not susceptible to meltdown, additional capabilities of heat removal are provided by using regular atmospheric airflow as a means of backup heat removal, by having it pass through a heat exchanger and rising into the atmosphere due to convection, achieving full residual heat removal. The VHTR is scheduled to be prototyped and tested at Idaho National Laboratory within the next decade (as of 2009) as the design selected for the Next Generation Nuclear Plant by the US Department of Energy. This reactor will use a gas as a coolant, which can then be used for process heat (such as in hydrogen production) or for the driving of gas turbines and the generation of electricity.

A similar highly advanced gas cooled reactor originally designed by West Germany (the AVR reactor) and now developed by South Africa is known as the Pebble Bed Modular Reactor. It is an inherently safe design, meaning that core damage is physically impossible, due to the design of the fuel (spherical graphite "pebbles" arranged in a bed within a metal RPV and filled with TRISO (or QUADRISO) pellets of uranium, thorium, or mixed oxide within). A prototype of a very similar type of reactor has been built by the Chinese, HTR-10, and has worked beyond researchers' expectations, leading the Chinese to announce plans to build a pair of follow-on, full-scale 250 MWe, inherently safe, power production reactors based on the same concept. (See Nuclear power in the People's Republic of China for more information.)

5.3.3 Lead and Lead-Bismuth-cooled reactors

Recently it was identified a special phenomenology for heavy liquid metal-cooled fast reactors -HLM, as lead and lead-bismuth-cooled reactors.[*][21] Because of the similar densities of the fuel and the HLM, an inherent passive safety self-removal feedback mechanism due to buoyancy forces

is developed, which propels the packed bed away from the wall when certain threshold of temperature is attained and the bed becomes lighter than the surrounding coolant, thus preventing temperatures that can jeopardize the vessel's structural integrity and also reducing the recriticality potential by limiting the allowable bed depth.

5.3.4 Experimental or conceptual designs

Some design concepts for nuclear reactors emphasize resistance to meltdown and operating safety.

The PIUS (process inherent ultimate safety) designs, originally engineered by the Swedes in the late 1970s and early 1980s, are LWRs that by virtue of their design are resistant to core damage. No units have ever been built.

Power reactors, including the Deployable Electrical Energy Reactor, a larger-scale mobile version of the TRIGA for power generation in disaster areas and on military missions, and the TRIGA Power System, a small power plant and heat source for small and remote community use, have been put forward by interested engineers, and share the safety characteristics of the TRIGA due to the uranium zirconium hydride fuel used.

The Hydrogen Moderated Self-regulating Nuclear Power Module, a reactor that uses uranium hydride as a moderator and fuel, similar in chemistry and safety to the TRIGA, also possesses these extreme safety and stability characteristics, and has attracted a good deal of interest in recent times.

The liquid fluoride thorium reactor is designed to naturally have its core in a molten state, as a eutectic mix of thorium and fluorine salts. As such, a molten core is reflective of the normal and safe state of operation of this reactor type. In the event the core overheats, a metal plug will melt, and the molten salt core will drain into tanks where it will cool in a non-critical configuration. Since the core is liquid, and already melted, it cannot be damaged.

Advanced liquid metal reactors, such as the U.S. Integral Fast Reactor and the Russian BN-350, BN-600, and BN-800, all have a coolant with very high heat capacity, sodium metal. As such, they can withstand a loss of cooling without SCRAM and a loss of heat sink without SCRAM, qualifying them as inherently safe.

5.4 Soviet Union-designed reactors

5.4.1 RBMKs

Soviet-designed RBMKs, found only in Russia and the CIS and now shut down everywhere except Russia, do not have containment buildings, are naturally unstable (tending to dangerous power fluctuations), and also have ECCS systems that are considered grossly inadequate by Western safety standards. The reactor from the Chernobyl Disaster was an RBMK reactor.

RBMK ECCS systems only have one division and have less than sufficient redundancy within that division. Though the large core size of the RBMK makes it less energy-dense than the Western LWR core, it makes it harder to cool. The RBMK is moderated by graphite. In the presence of both steam and oxygen, at high temperatures, graphite forms synthesis gas and with the water gas shift reaction the resultant hydrogen burns explosively. If oxygen contacts hot graphite, it will burn. The RBMK tends towards dangerous power fluctuations. Control rods used to be tipped with graphite, a material that slows neutrons and thus speeds up the chain reaction. Water is used as a coolant, but not a moderator. If the water boils away, cooling is lost, but moderation continues. This is termed a positive void coefficient of reactivity.

Control rods can become stuck if the reactor suddenly heats up and they are moving. Xenon-135, a neutron absorbent fission product, has a tendency to build up in the core and burn off unpredictably in the event of low power operation. This can lead to inaccurate neutronic and thermal power ratings.

The RBMK does not have any containment above the core. The only substantial solid barrier above the fuel is the upper part of the core, called the upper biological shield, which is a piece of concrete interpenetrated with control rods and with access holes for refueling while online. Other parts of the RBMK were shielded better than the core itself. Rapid shutdown (SCRAM) takes 10 to 15 seconds. Western reactors take 1 - 2.5 seconds.

Western aid has been given to provide certain real-time safety monitoring capacities to the human staff. Whether this extends to automatic initiation of emergency cooling is not known. Training has been provided in safety assessment from Western sources, and Russian reactors have evolved in result to the weaknesses that were in the RBMK. However, numerous RBMKs still operate.

It is safe to say that it might be possible to stop a loss-of-coolant event prior to core damage occurring, but that any core damage incidents will probably assure massive release of radioactive materials. Further, dangerous power fluctuations are natural to the design.

Lithuania joined the EU recently, and upon acceding, it has been required to shut the two RBMKs that it has at Ignalina NPP, as such reactors are totally incompatible with the nuclear safety standards of Europe. It will be replacing them with some safer form of reactor.

5.4.2 MKER

The MKER is a modern Russian-engineered channel type reactor that is a distant descendant of the RBMK. It approaches the concept from a different and superior direction, optimizing the benefits, and fixing the flaws of the original RBMK design.

There are several unique features of the MKER's design that make it a credible and interesting option: One unique benefit of the MKER's design is that in the event of a challenge to cooling within the core - a pipe break of a channel, the channel can be isolated from the plenums supplying water, decreasing the potential for common-mode failures.

The lower power density of the core greatly enhances thermal regulation. Graphite moderation enhances neutronic characteristics beyond light water ranges. The passive emergency cooling system provides a high level of protection by using natural phenomena to cool the core rather than depending on motor-driven pumps. The containment structure is modern and designed to withstand a very high level of punishment.

Refueling is accomplished while online, ensuring that outages are for maintenance only and are very few and far between. 97-99% uptime is a definite possibility. Lower enrichment fuels can be used, and high burnup can be achieved due to the moderator design. Neutronics characteristics have been revamped to optimize for purely civilian fuel fertilization and recycling.

Due to the enhanced quality control of parts, advanced computer controls, comprehensive passive emergency core cooling system, and very strong containment structure, along with a negative void coefficient and a fast acting rapid shutdown system, the MKER's safety can generally be regarded as being in the range of the Western Generation III reactors, and the unique benefits of the design may enhance its competitiveness in countries considering full fuel-cycle options for nuclear development.

5.4.3 VVER

The VVER is a pressurized light water reactor that is far more stable and safe than the RBMK. This is because it uses light water as a moderator (rather than graphite), has well understood operating characteristics, and has a negative void coefficient of reactivity. In addition, some have been built with more than marginal containments, some have quality ECCS systems, and some have been upgraded to international standards of control and instrumentation. Present generations of VVERs (the VVER-1000) are built to Western-equivalent levels of instrumentation, control, and containment systems.

However, even with these positive developments, certain older VVER models raise a high level of concern, especially the VVER-440 V230.*[22]

The VVER-440 V230 has no containment building, but only has a structure capable of confining steam surrounding the RPV. This is a volume of thin steel, perhaps an inch or two in thickness, grossly insufficient by Western standards.

- Has no ECCS. Can survive at most one 4 inch pipe break (there are many pipes greater than 4 inches within the design).

- Has six steam generator loops, adding unnecessary complexity.

  - However, apparently steam generator loops can be isolated, in the event that a break occurs in one of these loops. The plant can remain operating with one isolated loop - a feature found in few Western reactors.

The interior of the pressure vessel is plain alloy steel, exposed to water. This can lead to rust, if the reactor is exposed to water. One point of distinction in which the VVER surpasses the West is the reactor water cleanup facility - built, no doubt, to deal with the enormous volume of rust within the primary coolant loop - the product of the slow corrosion of the RPV. This model is viewed as having inadequate process control systems.

Bulgaria had a number of VVER-440 V230 models, but they opted to shut them down upon joining the EU rather than backfit them, and are instead building new VVER-1000 models. Many non-EU states maintain V230 models, including Russia and the CIS. Many of these states - rather than abandoning the reactors entirely - have opted to install an ECCS, develop standard procedures, and install proper instrumentation and control systems. Though confinements cannot be transformed into containments, the risk of a limiting fault resulting in core damage can be greatly reduced.

The VVER-440 V213 model was built to the first set of Soviet nuclear safety standards. It possesses a modest containment building, and the ECCS systems, though not completely to Western standards, are reasonably comprehensive. Many VVER-440 V213 models operated by former Soviet bloc countries have been upgraded to fully automated Western-style instrumentation and control systems, improving safety to Western levels for accident prevention - but not for accident containment, which is of a modest level compared to Western plants. These reactors are regarded as "safe enough" by Western standards to continue operation without major modifications, though most owners have performed major modifications to bring them up to generally equivalent levels of nuclear safety.

During the 1970s, Finland built two VVER-440 V213 models to Western standards with a large-volume full containment and world-class instrumentation, control standards and an ECCS with multiply redundant and diversified components. In addition, passive safety features such as 900-tonne ice condensers have been installed, making these two units safety-wise the most advanced VVER-440's in the world.

The VVER-1000 type has a definitely adequate Western-style containment, the ECCS is sufficient by Western standards, and instrumentation and control has been markedly improved to Western 1970s-era levels.

5.4.4 Chernobyl disaster

Main article: Chernobyl disaster

In the Chernobyl disaster the fuel became non-critical when it melted and flowed away from the graphite moderator - however, it took considerable time to cool. The molten core of Chernobyl (that part that was not blown outside the reactor or did not vaporize in the fire) flowed in a channel created by the structure of its reactor building and froze in place before a core-concrete interaction could happen. In the basement of the reactor at Chernobyl, a large "elephant's foot" of congealed core material was found, one example of the freely-flowing corium. Time delay, and prevention of direct emission to the atmosphere (i.e., containment), would have reduced the radiological release. If the basement of the reactor building had been penetrated, the groundwater would be severely contaminated, and its flow could carry the contamination far afield.

The Chernobyl reactor was a RBMK type. The disaster was caused by a power excursion that led to a steam explosion, meltdown and extensive offsite consequences. Operator error and a faulty shutdown system led to a sudden, massive spike in the neutron multiplication rate, a sudden decrease in the neutron period, and a consequent increase in neutron population; thus, core heat flux increased rapidly beyond the design limits of the reactor. This caused the water coolant to flash to steam, causing a sudden overpressure within the reactor pressure vessel (RPV), leading to granulation of the upper portion of the core and the ejection of the upper plenum of said pressure vessel along with core debris from the reactor building in a widely dispersed pattern. The lower portion of the reactor remained somewhat intact; the graphite neutron moderator was exposed to oxygen-containing air; heat from the power excursion in addition to residual heat flux from the remaining fuel rods left without coolant induced oxidation in the moderator and in the opened fuel rods; this in turn evolved more heat and contributed to the melting of more of the fuel rods and the outgassing of the fission products contained therein. The liquefied remains of the melted fuel rods, pulverized concrete and any other objects in the path flowed through a drainage pipe into the basement of the reactor building and solidified in a mass, though the primary threat to the public safety was the dispersed core ejecta, vaporized and gaseous fission products and fuel, and the gasses evolved from the oxidation of the moderator.

Although the Chernobyl accident had dire off-site effects, much of the radioactivity remained within the building. If the building were to fail and dust was to be released into the environment then the release of a given mass of fission products which have aged for almost thirty years would have a smaller effect than the release of the same mass of fission products (in the same chemical and physical form) which had only undergone a short cooling time (such as one hour) after the nuclear reaction has been terminated. However, if a nuclear reaction was to occur again within the Chernobyl plant (for instance if rainwater was to collect and act as a moderator) then the new fission products would have a higher specific activity and thus pose a greater threat if they were released. To prevent a post-accident nuclear reaction, steps have been taken, such as adding neutron poisons to key parts of the basement.

5.5 Effects

The effects of a nuclear meltdown depend on the safety features designed into a reactor. A modern reactor is designed both to make a meltdown unlikely, and to contain one should it occur.

In a modern reactor, a nuclear meltdown, whether partial or total, should be contained inside the reactor's containment structure. Thus (assuming that no other major disasters occur) while the meltdown will severely damage the reactor itself, possibly contaminating the whole structure with highly radioactive material, a meltdown alone should not lead to significant radioactivity release or danger to the public.[*][23]

In practice, however, a nuclear meltdown is often part of a larger chain of disasters (although there have been so few meltdowns in the history of nuclear power that there is not a large pool of statistical information from which to draw a credible conclusion as to what "often" happens in such circumstances). For example, in the Chernobyl accident, by the time the core melted, there had already been a large steam explosion and graphite fire and major release of radioactive contamination (as with almost all Soviet reactors, there was no containment structure at Chernobyl). Also, before a possible meltdown occurs, pressure can already be rising in the reactor, and to prevent a meltdown by restor-

ing the cooling of the core, operators are allowed to reduce the pressure in the reactor by releasing (radioactive) steam into the environment. This enables them to inject additional cooling water into the reactor again.

5.6 Reactor design

Although pressurized water reactors are more susceptible to nuclear meltdown in the absence of active safety measures, this is not a universal feature of civilian nuclear reactors. Much of the research in civilian nuclear reactors is for designs with passive nuclear safety features that may be less susceptible to meltdown, even if all emergency systems failed. For example, pebble bed reactors are designed so that complete loss of coolant for an indefinite period does not result in the reactor overheating. The General Electric ESBWR and Westinghouse AP1000 have passively activated safety systems. The CANDU reactor has two low-temperature and low-pressure water systems surrounding the fuel (i.e. moderator and shield tank) that act as back-up heat sinks and preclude meltdowns and core-breaching scenarios.[*][20] Liquid fueled reactors can be stopped by draining the fuel into tankage which not only prevents further fission but draws decay heat away statically, and by drawing off the fission products (which are the source of post-shutdown heating) incrementally. The ideal is to have reactors that fail-safe through physics rather than through redundant safety systems or human intervention.

Fast breeder reactors are more susceptible to meltdown than other reactor types, due to the larger quantity of fissile material and the higher neutron flux inside the reactor core, which makes it more difficult to control the reaction. This is not true of the Integral Fast Reactor model EBR II,[*][24] which was explicitly designed to be meltdown-immune. It was tested in April 1986, just before the Chernobyl failure, to simulate loss of coolant pumping power, by switching off the power to the primary pumps. As designed, it shut itself down, in about 300 seconds, as soon as the temperature rose to a point designed as higher than proper operation would require. This was well below the boiling point of the unpressurised liquid metal coolant, which had entirely sufficient cooling ability to deal with the heat of fission product radioactivity, by simple convection. The second test, deliberate shut-off of the secondary coolant loop that supplies the generators, caused the primary circuit to undergo the same safe shutdown. This test simulated the case of a water-cooled reactor losing its steam turbine circuit, perhaps by a leak.

Accidental fires are widely acknowledged to be risk factors that can contribute to a nuclear meltdown.

5.7   Nuclear meltdown events

This is a list of the major reactor failures in which meltdown played a role:[*][25]

5.7.1   United States

SL-1 core damage after a nuclear excursion.

- BORAX-I was a test reactor designed to explore criticality excursions and observe if a reactor would self limit. In the final test, it was deliberately destroyed and revealed that the reactor reached much higher temperatures than were predicted at the time.[*][26]

- The reactor at EBR-I suffered a partial meltdown during a coolant flow test on 29 November 1955.

- The Sodium Reactor Experiment in Santa Susana Field Laboratory was an experimental nuclear reactor which operated from 1957 to 1964 and was the first commercial power plant in the world to experience a core meltdown in July 1959.

- Stationary Low-Power Reactor Number One (SL-1) was a United States Army experimental nuclear power reactor which underwent a criticality excursion, a steam explosion, and a meltdown on 3 January 1961, killing three operators.

- The SNAP8ER reactor at the Santa Susana Field Laboratory experienced damage to 80% of its fuel in an accident in 1964.

- The partial meltdown at the Fermi 1 experimental fast breeder reactor, in 1966, required the reactor to be repaired, though it never achieved full operation afterward.

- The SNAP8DR reactor at the Santa Susana Field Laboratory experienced damage to approximately a third of its fuel in an accident in 1969.

- The Three Mile Island accident, in 1979, referred to in the press as a "partial core melt"[*][27] led to the total dismantlement and the permanent shutdown of that reactor. Unit-1 still continues to operate at TMI.

5.7.2   Soviet Union

- In the most serious example, the Chernobyl disaster, design flaws and operator negligence led to a power excursion that subsequently caused a meltdown. According to a report released by the Chernobyl Forum (consisting of numerous United Nations agencies, including the International Atomic Energy Agency and the World Health Organization; the World Bank; and the Governments of Ukraine, Belarus, and Russia) the disaster killed twenty-eight people due to acute radiation syndrome,[*][28] could possibly result in up to four thousand fatal cancers at an unknown time in the future[*][29] and required the permanent evacuation of an exclusion zone around the reactor.

- A number of Soviet Navy nuclear submarines experienced nuclear meltdowns, including K-27, K-140, and K-431.

5.7.3   Japan

- During the Fukushima Daiichi nuclear disaster following the earthquake and tsunami in March 2011, three of the power plant's six reactors suffered meltdowns. Most of the fuel in the reactor No. 1 Nuclear Power Plant melted.[*][30][*][31]

5.7.4   Switzerland

- The Lucens reactor, Switzerland, in 1969.

5.7.5 Canada

- NRX (military), Ontario, Canada, in 1952

5.7.6 United Kingdom

- Windscale (military), Sellafield, England, in 1957 (see Windscale fire)

- Chapelcross nuclear power station (civilian), Scotland, in 1967

5.7.7 France

- Saint-Laurent Nuclear Power Plant (civilian), France, in 1969

- Saint-Laurent Nuclear Power Plant (civilian), France, in 1980

5.7.8 Czechoslovakia

- A1 plant, (civilian) at Jaslovské Bohunice, Czechoslovakia, in 1977

5.8 China syndrome

For the 1979 film, see The China Syndrome. For *The King of Queens* episode, see China Syndrome (The King of Queens).
See also: Core catcher

The **China syndrome** (loss-of-coolant accident) is a hypothetical nuclear reactor operations accident characterized by the severe meltdown of the core components of the reactor, which then burn through the containment vessel and the housing building, then (figuratively) through the crust and body of the Earth until reaching the opposite side (which, in the United States, is colloquially referred to as China).[32][33] The phrasing is metaphorical; there is no way a core could penetrate the several-kilometer thickness of the Earth's crust, and even if it did melt to the center of the Earth, it would not travel back upwards against the pull of gravity. Moreover, any tunnel behind the material would be closed by immense lithostatic pressure. Furthermore, China does not contain the antipode of any landmass in North America.

In reality, under a complete loss of coolant scenario, the fast erosion phase of the concrete basement lasts for about an hour and progresses into about one meter depth, then slows to several centimeters per hour, and stops completely when the corium melt cools below the decomposition temperature of concrete (about 1100 °C). Complete melt-through can occur in several days, even through several meters of concrete; the corium then penetrates several meters into the underlying soil, spreads around, cools, and solidifies.[34] It is also possible that there is already a harmless dense natural concentration of radioactive material in the Earth's core (primarily uranium-238, thorium-232 and potassium-40, which have half-lives of 4.47 billion years, 14.05 billion years and 1.25 billion years respectively.)[35][36][37]

The real scare, however, came from a quote in the 1979 film *The China Syndrome*, which stated, "It melts right down through the bottom of the plant—theoretically to China, but of course, as soon as it hits ground water, it blasts into the atmosphere and sends out clouds of radioactivity. The number of people killed would depend on which way the wind was blowing, rendering an area the size of Pennsylvania permanently uninhabitable." The actual threat of this was tested just 12 days after the release of the film when a meltdown at Pennsylvania's Three Mile Island Plant 2 (TMI-2) created a molten core that moved 15 millimeters toward "China" before the core froze at the bottom of the reactor pressure vessel.[38] Thus, the TMI-2 reactor fuel and fission products breached the fuel plates, but the melted core itself did not break the containment of the reactor vessel.[39] Hours after the meltdown, concern about hydrogen build-up led operators to release some radioactive gasses into the atmosphere, including gaseous fission products. Release of the fission products led to a temporary evacuation of the surrounding area, but no injuries.

A similar situation occurred during the Chernobyl disaster: after the reactor was destroyed and began to burn, a liquid corium mass from the melting core began to breach the concrete floor of the reactor vessel, which was situated above the bubbler pool (a large water reservoir for emergency pumps, also designed to safely contain steam pipe ruptures). The RBMK had no allowance or planning for core meltdowns, and the imminent interaction of the core mass with the bubbler pool would have produced a massive steam explosion, likely destroying the entire plant (vastly increasing the spread and magnitude of the radioactive plume). It was therefore necessary to drain the bubbler pool before the corium reached it. However, the initial explosion had broken the control circuitry which allowed the pool to be emptied. Three volunteer divers were required to manually operate the valves necessary to drain this pool, and later images of the corium mass in the pipes of the bubbler pool's basement reinforced the heroic necessity of their actions.[40][41][42]

5.8.1 History

The system design of the nuclear power plants built in the late 1960s raised questions of operational safety, and raised the concern that a severe reactor accident could release large quantities of radioactive materials into the atmosphere and environment. By 1970, there were doubts about the ability of the emergency core cooling system of a nuclear reactor to prevent a loss of coolant accident and the consequent meltdown of the fuel core; the subject proved popular in the technical and the popular presses.*[18] In 1971, in the article *Thoughts on Nuclear Plumbing*, former Manhattan Project (1942–1946) nuclear physicist Ralph Lapp used the term "China syndrome" to describe a possible burn-through, after a loss of coolant accident, of the nuclear fuel rods and core components melting the containment structures, and the subsequent escape of radioactive material(s) into the atmosphere and environment; the hypothesis derived from a 1967 report by a group of nuclear physicists, headed by W. K. Ergen.*[19] In the event, Lapp's hypothetical nuclear accident was cinematically adapted as *The China Syndrome* (1979).

5.9   See also

- Behavior of nuclear fuel during a reactor accident
- Chernobyl compared to other radioactivity releases
- Chernobyl disaster effects
- High-level radioactive waste management
- International Nuclear Event Scale
- List of civilian nuclear accidents
- Lists of nuclear disasters and radioactive incidents
- Nuclear fuel response to reactor accidents
- Nuclear safety
- Nuclear power
- Nuclear power debate
- Scram or SCRAM, an emergency shutdown of a nuclear reactor

5.10   Notes

[1] Note that use of thorium fuel in a liquid fluoride thorium reactor (LFTR) carries no risk of core meltdown because a LFTR does not develop an overheated core.

5.11   References

[1] Martin Fackler (1 June 2011). "Report Finds Japan Underestimated Tsunami Danger". *New York Times*.

[2] Reactor safety study: an assessment of accident risks in U.S. commercial nuclear power plants, Volume 1

[3] International Atomic Energy Agency (IAEA) (2007). *IAEA Safety Glossary: Terminology Used in Nuclear Safety and Radiation Protection* (PDF). Vienna, Austria: International Atomic Energy Agency. ISBN 92-0-100707-8. Retrieved 17 August 2009.

[4] United States Nuclear Regulatory Commission (NRC) (14 September 2009). "Glossary". *Website*. Rockville, Maryland, USA: Federal Government of the United States. pp. See Entries for Letter M and Entries for Letter N. Retrieved 3 October 2009.

[5] http://www.merriam-webster.com/dictionary/meltdown

[6] Hewitt, Geoffrey Frederick; Collier, John Gordon (2000). "4.6.1 Design Basis Accident for the AGR: Depressurization Fault". *Introduction to nuclear power*. London, UK: Taylor & Francis. p. 133. ISBN 978-1-56032-454-6. Retrieved 5 June 2010.

[7] "Earthquake Report No. 91" (PDF). JAIF. 25 May 2011. Retrieved 25 May 2011.

[8] Kuan, P.; Hanson, D. J.; Odar, F. (1991). *Managing water addition to a degraded core*. Retrieved 22 November 2010.

[9] Haskin, F.E.; Camp, A.L. (1994). *Perspectives on Reactor Safety (NUREG/CR-6042) (Reactor Safety Course R-800), 1st Edition*. Beltsville, MD: U.S. Nuclear Regulatory Commission. p. 3.1–5. Retrieved 23 November 2010.

[10] Haskin, F.E.; Camp, A.L. (1994). *Perspectives on Reactor Safety (NUREG/CR-6042) (Reactor Safety Course R-800), 1st Edition*. Beltsville, MD: U.S. Nuclear Regulatory Commission. pp. 3.5–1 to 3.5–4. Retrieved 24 December 2010.

[11] Haskin, F.E.; Camp, A.L. (1994). *Perspectives on Reactor Safety (NUREG/CR-6042) (Reactor Safety Course R-800), 1st Edition*. Beltsville, MD: U.S. Nuclear Regulatory Commission. pp. 3.5–4 to 3.5–5. Retrieved 24 December 2010.

[12] ANS : Public Information : Resources : Special Topics : History at Three Mile Island : What Happened and What Didn't in the TMI-2 Accident

[13] Nuclear Industry in Russia Sells Safety, Taught by Chernobyl

[14] world nuclear org fukushima-accident http://www.world-nuclear.org/information-library/safety-and-security/safety-of-plants/fukushima-accident.aspx

[15] "Backgrounder on the Three Mile Island Accident". United States Nuclear Regulatory Commission. Retrieved 1 December 2013.

[16] Test Area North

[17] Terra Pitta (5 August 2015). *Catastrophe: A Guide to World's Worst Industrial Disasters*. Vij Books India Pvt Ltd. pp. 25–. ISBN 978-93-85505-17-1.

[18] Walker, J. Samuel (2004). *Three Mile Island: A Nuclear Crisis in Historical Perspective* (Berkeley: University of California Press), p. 11.

[19] Lapp, Ralph E. "Thoughts on nuclear plumbing." *The New York Times*, 12 December 1971, pg. E11.

[20] Allen, P.J.; J.Q. Howieson; H.S. Shapiro; J.T. Rogers; P. Mostert; R.W. van Otterloo (April–June 1990). "Summary of CANDU 6 Probabilistic Safety Assessment Study Results". *Nuclear Safety*. **31** (2): 202–214.

[21] F.J Arias. The Phenomenology of Packed Beds in Heavy Liquid Metal Fast Reactors During Postaccident Heat Removal: The Self-Removal Feedback Mechanism. Nuclear Science and Engineering / Volume 178 / Number 2 / October 2014 / Pages 240-249

[22] INL VVER Sourcebook

[23] Partial Fuel Meltdown Events

[24] Integral fast reactor

[25] "Sources and Effects of Ionizing Radiation – 2008 Report to the General Assembly" (PDF). *United Nations Scientific Committee on the Effects of Atomic Radiation*. 2011. |chapter= ignored (help)

[26] ANL-W Reactor History: BORAX I

[27] Wald, Matthew L. (11 March 2011). "Japan Expands Evacuation Around Nuclear Plant". *The New York Times*.

[28] The Chernobyl Forum: 2003-2005 (April 2006). "Chernobyl's Legacy: Health, Environmental and Socio-economic Impacts" (PDF). International Atomic Energy Agency. p. 14. Retrieved 26 January 2011.

[29] The Chernobyl Forum: 2003-2005 (April 2006). "Chernobyl's Legacy: Health, Environmental and Socio-Economic Impacts" (PDF). International Atomic Energy Agency. p. 16. Retrieved 26 January 2011.

[30]

[31] Hiroko Tabuchi (24 May 2011). "Company Believes 3 Reactors Melted Down in Japan". *The New York Times*. Retrieved 25 May 2011.

[32] "China Syndrome". Merriam-Webster. Retrieved 11 December 2012.

[33] Presenter: Martha Raddatz (15 March 2011). *ABC World News*. ABC.

[34] Jacques Libmann (1996). *Elements of nuclear safety*. L'Editeur : EDP Sciences. p. 194. ISBN 2-86883-286-5.

[35] Scientific American (6 October 1997): *Why is the earth's core so hot? And how do scientists measure its temperature?*

[36] Phys.org (30 March 2006): *Probing Question: What heats the earth's core?*

[37] Popular Mechanics (15 June 2012): *How Do We Know What's in the Earth's Core? PM Explains*

[38]

[39] Gianni Petrangeli (2006). *Nuclear safety*. Butterworth-Heinemann. p. 37. ISBN 0-7506-6723-0.

[40] Chernobyl: The End of the Nuclear Dream, 1986, p.178, by Nigel Hawkes et al., ISBN 0-330-29743-0

[41] "Stephen McGinty: Lead coffins and a nation's thanks for the Chernobyl suicide squad". scotsman.com. 16 March 2011.

[42] "Soviets Report Heroic Acts at Chernobyl Reactor With AM Chernobyl Nuclear Bjt". Associated Press. 1986-05-15. Retrieved 2014-04-26. This story is based on a TASS release (in Russian) Чернобыль: адрес мужества.

5.12 External links

- Annotated bibliography on civilian nuclear accidents from the Alsos Digital Library for Nuclear Issues

- Partial Fuel Meltdown Events

- "The world's worst nuclear power disasters". *Power Technology*. 7 October 2013.

Chapter 6

Three Mile Island

Three Mile Island is the Three Mile Island Nuclear Generating Station in eastern Pennsylvania.

Three Mile Island also may refer to:

- Matters related to the Three Mile Island plant:

  - Three Mile Island accident, a partial core meltdown occurring in 1979
  - Three Mile Island accident health effects
  - Books:
    - *Three Mile Island: Thirty Minutes to Meltdown* (1982)
    - *Three Mile Island: A Nuclear Crisis in Historical Perspective* (2004)

- Three Mile Island (Lake Winnipesaukee), an island in New Hampshire, United States

Chapter 7

Nagasaki

This article is about the Japanese city. For the popular 1920s song, see Nagasaki (song).
Warning: Page using Template:Infobox settlement with unknown parameter "official_" (this message is shown only in preview).

Nagasaki (長崎市 *Nagasaki-shi*, Japanese: [nàgá↓sàkì]) (◀)) listen) is the capital and the largest city of Nagasaki Prefecture on the island of Kyushu in Japan. It became a centre of Portuguese and Dutch influence in the 16th through 19th centuries, and the Churches and Christian Sites in Nagasaki have been proposed for inscription on the UNESCO World Heritage List. Part of Nagasaki was home to a major Imperial Japanese Navy base during the First Sino-Japanese War and Russo-Japanese War. Its name means "long cape".

During World War II, the American atomic bombings of Hiroshima and Nagasaki made Nagasaki the second and, to date, last city in the world to experience a nuclear attack.[*][1]

As of 1 January 2009, the city has an estimated population of 446,007 and a population density of 1,100 persons per km^2. The total area is 406.35 km^2.

7.1 History

See also: Timeline of Nagasaki

7.1.1 Medieval and early modern history

A small fishing village set in a secluded harbor, Nagasaki had little historical significance until contact with Portuguese explorers in 1543. An early visitor was Fernão Mendes Pinto, who came on a Portuguese ship which landed nearby in Tanegashima.

Soon after, Portuguese ships started sailing to Japan as reg-

Portuguese (green) *and Spanish* (yellow) *trade routes to Macao and Nagasaki*

A Japanese Nanban byōbu *detail depicting a Portuguese carrack arriving at Nagasaki, c. 1571*

ular trade freighters, thus increasing the contact and trade relations between Japan and the rest of the world, and particularly with mainland China, with whom Japan had previously severed its commercial and political ties, mainly due to a number of incidents involving Wokou piracy in the South China Sea, with the Portuguese now serving as intermediaries between the two Asian countries.

Despite the mutual advantages derived from these trading contacts, which would soon be acknowledged by all parties involved, the lack of a proper seaport in Kyūshū for the purpose of harboring foreign ships posed a major problem for both merchants and the Kyushu *daimyōs* (feudal lords) who expected to collect great advantages from the trade with the Portuguese.

In the meantime, Navarrese Jesuit missionary St. Francis Xavier arrived in Kagoshima, South Kyūshū, in 1549, and soon initiated a thorough campaign of evangelization throughout Japan, but left for China in 1552 and died soon afterwards. His followers who remained behind converted a number of *daimyōs*. The most notable among them was Ōmura Sumitada. In 1569, Ōmura granted a permit for the establishment of a port with the purpose of harboring Portuguese ships in Nagasaki, which was finally set up in 1571, under the supervision of the Jesuit missionary Gaspar Vilela and Portuguese Captain-Major Tristão Vaz de Veiga, with Ōmura's personal assistance.[2]

The little harbor village quickly grew into a diverse port city , and Portuguese products imported through Nagasaki (such as tobacco, bread, textiles and a Portuguese sponge-cake called *castellas*) were assimilated into popular Japanese culture. Tempura derived from a popular Portuguese recipe originally known as *peixinho-da-horta*, and takes its name from the Portuguese word, 'tempero' another example of the enduring effects of this cultural exchange. The Portuguese also brought with them many goods from China.

Due to the instability during the Sengoku period, Sumitada and Jesuit leader Alexandro Valignano conceived a plan to pass administrative control over to the Society of Jesus rather than see the Catholic city taken over by a non-Catholic *daimyō*. Thus, for a brief period after 1580, the city of Nagasaki was a Jesuit colony, under their administrative and military control. It was administered by the captain of the Portuguese "black ship" , the highest representative of the Portuguese Crown, when present. It became a refuge for Christians escaping maltreatment in other regions of Japan.[3] In 1587, however, Toyotomi Hideyoshi's campaign to unify the country arrived in Kyūshū. Concerned with the large Christian influence in southern Japan, as well as the active and what was perceived as the arrogant role the Jesuits were playing in the Japanese political arena, Hideyoshi ordered the expulsion of all missionaries, and placed the city under his direct control. However, the expulsion order went largely unenforced, and the fact remained that most of Nagasaki's population remained openly practicing Catholic.

In 1596, the Spanish ship *San Felipe* was wrecked off the coast of Shikoku, and Hideyoshi learned from its pilot[4] that the Spanish Franciscans were the vanguard of an Iberian invasion of Japan. In response, Hideyoshi or-

dered the crucifixions of twenty-six Catholics in Nagasaki on February 5 of that year (i.e. the "Twenty-six Martyrs of Japan"). Portuguese traders were not ostracized, however, and so the city continued to thrive.

Some of Nagasaki's stone bridges over the Nakashima River in the 1870s

In 1602, Augustinian missionaries also arrived in Japan, and when Tokugawa Ieyasu took power in 1603, Catholicism was still tolerated. Many Catholic *daimyōs* had been critical allies at the Battle of Sekigahara, and the Tokugawa position was not strong enough to move against them. Once Osaka Castle had been taken and Toyotomi Hideyoshi's offspring killed, though, the Tokugawa dominance was assured. In addition, the Dutch and English presence allowed trade without religious strings attached. Thus, in 1614, Catholicism was officially banned and all missionaries ordered to leave. Most Catholic daimyo apostatized, and forced their subjects to do so, although a few would not renounce the religion and left the country for Macau, Luzon and Japantowns in Southeast Asia. A brutal campaign of persecution followed, with thousands of converts across Kyūshū and other parts of Japan killed, tortured, or forced to renounce their religion (see Martyrs of Japan).

Catholicism's last gasp as an open religion and the last major military action in Japan until the Meiji Restoration was the Shimabara Rebellion of 1637. While there is no evidence that Europeans directly incited the rebellion, Shimabara Domain had been a Christian *han* for several decades, and the rebels adopted many Portuguese motifs and Christian icons. Consequently, in Tokugawa society the word "Shimabara" solidified the connection between Christianity and disloyalty, constantly used again and again in Tokugawa propaganda. The Shimabara Rebellion also convinced many policy-makers that foreign influences were more trouble than they were worth, leading to the national isolation policy. The Portuguese, who had been previously living on a specially constructed island-prison in Nagasaki harbour called Dejima, were expelled from the archipelago altogether, and the Dutch were moved from their base at Hirado into the trading island.

The Great Fire of Nagasaki destroyed much of the city in 1663, including the Mazu shrine at the Kofukuji Temple patronized by the Chinese sailors and merchants visiting the port.*[5]

In 1720 the ban on Dutch books was lifted, causing hundreds of scholars to flood into Nagasaki to study European science and art. Consequently, Nagasaki became a major center of *rangaku*, or "Dutch Learning". During the Edo period, the Tokugawa shogunate governed the city, appointing a *hatamoto*, the *Nagasaki bugyō*, as its chief administrator.

Kameyama-ware jar with a Dutch trading ship, 19th Century

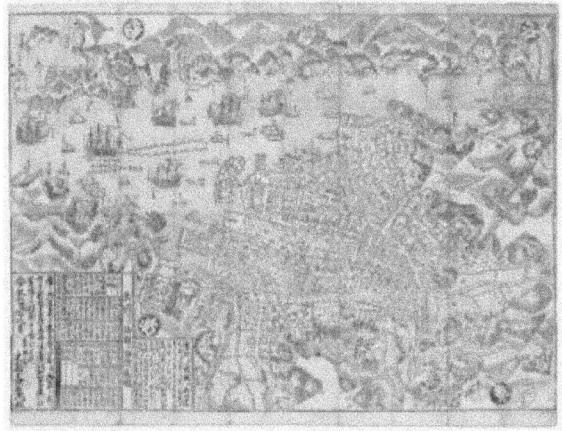

Plan of Nagasaki, Hizen province, 1778

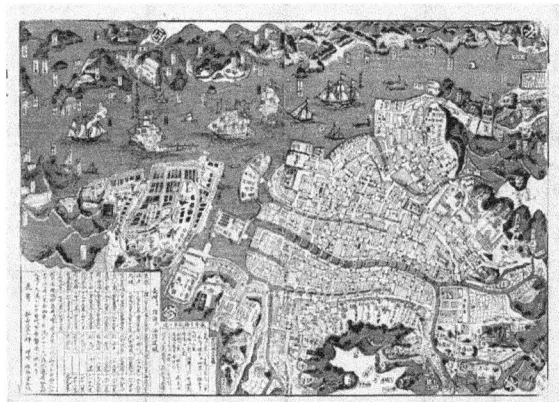

Plan of Nagasaki, 1860

Uchida Kuichi's image of Nagasaki c. 1870

Consensus among historians was once that Nagasaki was Japan's only window on the world during its time as a closed country in the Tokugawa era. However, nowadays it is generally accepted that this was not the case, since Japan interacted and traded with the Ryūkyū Kingdom, Korea and Russia through Satsuma, Tsushima and Matsumae respec-

tively. Nevertheless, Nagasaki was depicted in contemporary art and literature as a cosmopolitan port brimming with exotic curiosities from the Western World.*[6]

In 1808, during the Napoleonic Wars the Royal Navy frigate HMS *Phaeton* entered Nagasaki Harbor in search of Dutch

Ōura Church built in 1864, a national treasure of Japan

Nagasaki Prefect Office, Meiji period

Nagasaki City Office, Taishō period

of merchant and artist such as 18th century Yi Hai. It is believed that as much as one-third of the population of Nagasaki at this time may have been Chinese.*[7]

7.1.2 Modern history

With the Meiji Restoration, Japan opened its doors once again to foreign trade and diplomatic relations. Nagasaki became a free port in 1859 and modernization began in earnest in 1868. Nagasaki was officially proclaimed a city on April 1, 1889. With Christianity legalized and the Kakure Kirishitan coming out of hiding, Nagasaki regained its earlier role as a center for Roman Catholicism in Japan.

During the Meiji period, Nagasaki became a center of heavy industry. Its main industry was ship-building, with the dockyards under control of Mitsubishi Heavy Industries becoming one of the prime contractors for the Imperial Japanese Navy, and with Nagasaki harbor used as an anchorage under the control of nearby Sasebo Naval District. During World War II, at the time of the nuclear attack, Nagasaki was an important industrial city, containing both plants of the Mitsubishi Steel and Arms Works, the Akunoura Engine Works, Mitsubishi Arms Plant, Mitsubishi Electric Shipyards, Mitsubishi Steel and Arms Works, Mitsubishi-Urakami Ordnance Works, several other small factories, and most of the ports storage and trans-shipment facilities, which employed about 90% of the city's labor force, and accounted for 90% of the city's industry. These connections with the Japanese war effort made Nagasaki a major target for strategic bombing by the Allies during the war.*[8]*[9]

7.1.3 Atomic bombing of Nagasaki during World War II

Main article: Atomic bombings of Hiroshima and Nagasaki
For 12 months prior to the nuclear attack, Nagasaki had ex-

trading ships. The local magistrate was unable to resist the British demand for food, fuel, and water, later committing *seppuku* as a result. Laws were passed in the wake of this incident strengthening coastal defenses, threatening death to intruding foreigners, and prompting the training of English and Russian translators.

The *Tōjinyashiki* (唐人屋敷) or Chinese Factory in Nagasaki was also an important conduit for Chinese goods and information for the Japanese market. Various colourful Chinese merchants and artists sailed between the Chinese mainland and Nagasaki. Some actually combined the roles

Mushroom cloud from the atomic explosion over Nagasaki at 11:02 a.m., August 9, 1945

perienced five small-scale air attacks by an aggregate of 136 U.S. planes which dropped a total of 270 tons of high explosive, 53 tons of incendiary, and 20 tons of fragmentation bombs. Of these, a raid of August 1, 1945, was most effective, with a few of the bombs hitting the shipyards and dock areas in the southwest portion of the city, several hitting the Mitsubishi Steel and Arms Works, and six bombs landing at the Nagasaki Medical School and Hospital, with three direct hits on buildings there. While the damage from these few bombs was relatively small, it created considerable concern in Nagasaki and a number of people, principally school children, were evacuated to rural areas for safety, thus reducing the population in the city at the time of the atomic attack.[8][10][11][12]

On the day of the nuclear strike (August 9, 1945) the population in Nagasaki was estimated to be 263,000, which consisted of 240,000 Japanese residents, 10,000 Korean residents, 2,500 conscripted Korean workers, 9,000 Japanese soldiers, 600 conscripted Chinese workers, and 400 Allied POWs.[12] That day, the Boeing B-29 Superfortress *Bockscar*, commanded by Major Charles Sweeney, departed from Tinian's North Field just before dawn, this time carrying a plutonium bomb, code named "Fat Man". The primary target for the bomb was Kokura, with the secondary target, Nagasaki, if the primary target was too cloudy to make a visual sighting. When the plane reached Kokura at 9:44 a.m. (10:44 a.m. Tinian Time), the city was

obscured by clouds and smoke, as the nearby city of Yawata had been firebombed on the previous day. Unable to make a bombing attack on visual due to the clouds and smoke and with limited fuel, the plane left the city at 10:30 a.m. for the secondary target. After 20 minutes, the plane arrived at 10:50 a.m. over Nagasaki, but the city was also concealed by clouds. Desperately short of fuel and after making a couple of bombing runs without obtaining any visual target, the crew was forced to use radar in order to drop the bomb. At the last minute, the opening of the clouds allowed them to make visual contact with a racetrack in Nagasaki, and they dropped the bomb on the city's Urakami Valley midway between the Mitsubishi Steel and Arms Works in the south, and the Mitsubishi-Urakami Ordnance Works in the north.[13] After 53 seconds of its release, the bomb exploded at 11:02 a.m. at an approximate altitude of 1,800 feet.[14]

Within less than a second after the detonation, the north of the city was destroyed and 35,000 people were killed.[15] Among the deaths were 6,200 out of the 7,500 employees of the Mitsubishi Munitions plant, and 24,000 others (including 2,000 Koreans) who worked in other war plants and factories in the city, as well as 150 Japanese soldiers. The industrial damage in Nagasaki was high, leaving 68–80% of the non-dock industrial production destroyed. It was the second and, to date, the last use of a nuclear weapon in combat, and also the second detonation of a plutonium bomb. The first combat use of a nuclear weapon was the "Little Boy" bomb, which was dropped on the Japanese city of Hiroshima on August 6, 1945, and the first plutonium bomb was tested in central New Mexico, United States, on July 16, 1945. The Fat Man bomb was somewhat more powerful than the one dropped over Hiroshima, but because of Nagasaki's more uneven terrain, there was less damage.[16][17][18][19]

7.1.4 After the war

The city was rebuilt after the war, albeit dramatically changed. The pace of reconstruction was slow. The first simple emergency dwellings were not provided until 1946. The focus on redevelopment was the replacement of war industries with foreign trade, shipbuilding and fishing. This was formally declared when the Nagasaki International Culture City Reconstruction Law was passed in May 1949.[20] New temples were built, as well as new churches owing to an increase in the presence of Christianity.[21] Some of the rubble was left as a memorial, such as a one-legged *torii* at Sannō Shrine and an arch near ground zero. New structures were also raised as memorials, such as the Atomic Bomb Museum. Nagasaki remains first and foremost a port city, supporting a rich ship building industry and setting a strong example of perseverance and peace.

One-legged torii, *Sannō Shrine, Nagasaki, Japan. The other half was toppled in the explosion of the nuclear bomb.*

Torii, *Nagasaki, Japan. One-legged torii in the background*

On January 4, 2005, the towns of Iōjima, Kōyagi, Nomozaki, Sanwa, Sotome and Takashima (all from Nishisonogi District) were officially merged into Nagasaki.

7.2   Geography and climate

Nagasaki and Nishisonogi Peninsulas are located within the city limits. The city is surrounded by the cities of Isahaya and Saikai, and the towns of Togitsu and Nagayo in Nishisonogi District.

Nagasaki lies at the head of a long bay which forms the best natural harbor on the island of Kyūshū. The main commercial and residential area of the city lies on a small plain near the end of the bay. Two rivers divided by a mountain spur form the two main valleys in which the city lies. The heavily built-up area of the city is confined by the terrain to less than 4 square miles (10 km^2).

Nagasaki has the typical humid subtropical climate of Kyūshū and Honshū, characterized by mild winters and long, hot, and humid summers. Apart from Kanazawa and Shizuoka it is the wettest sizeable city in Japan and indeed all of temperate Eurasia. In the summer, the combination of persistent heat and high humidity results in unpleasant conditions, with wet-bulb temperatures sometimes reaching 26 °C (79 °F). In the winter, however, Nagasaki is drier and sunnier than Gotō to the west, and temperatures are slightly milder than further inland in Kyūshū. Since records began in 1878 the wettest month has been July 1982 with 1,178 millimetres (46 in) including 555 millimetres (21.9 in) in a single day, whilst the driest month has been September 1967 with 1.8 millimetres (0.07 in). Precipitation occurs year-round, though winter is the driest season; rainfall peaks sharply in June & July. August is the warmest month of the year. On January 24, 2016, a snowfall of 17 centimetres (6.7 in) was recorded.*[22]

7.3   Education

7.3.1   Universities

- Nagasaki University

- Nagasaki Institute of Applied Science

- Nagasaki University of Foreign Studies*[25]

- Kwassui Women's College

- Nagasaki Junshin University

- Siebold University of Nagasaki

7.3.2   Junior colleges

- Nagasaki Junior College

- Nagasaki Junshin Women's Junior College

- Tamaki Women's Junior College (玉木女子短期大学)

- Nagasaki Women's Junior College (長崎女子短期大学)

7.4 Transportation

A busy street in Nagasaki

The nearest airport is Nagasaki Airport in the nearby city of Ōmura. The Kyushu Railway Company (JR Kyushu) provides rail transportation on the Nagasaki Main Line, whose terminal is at Nagasaki Station. In addition, the Nagasaki Electric Tramway operates five routes in the city. The Nagasaki Expressway serves vehicular traffic with interchanges at Nagasaki and Susukizuka. In addition, six national highways crisscross the city: Routes 34, 202, 251, 324, and 499.

7.5 Sports

Nagasaki is represented in the J. League of football with its local club, V-Varen Nagasaki.

7.6 Main sites

- Confucius Shrine

- Dejima Museum of History

- Former residence of Shuhan Takashima

- Former site of Latin Seminario

- Former site of the British Consulate in Nagasaki

- Former site of Hong Kong and Shanghai Banking Corporation Nagasaki Branch

- Glover Garden
  - Former Glover Residence
  - Former Alt Residence
  - Former Ringer Residence
  - Former Walker Residence

- Fukusai-ji

- Gunkanjima

- Higashi-Yamate Juniban Mansion

- Kazagashira Park

- Kofukuji

- Megane Bridge

- Mount Inasa

- Nagasaki Atomic Bomb Museum[*][26] (Located next to the Peace Park)

- Nagasaki Museum of History and Culture[*][27]

- Nagasaki National Peace Memorial Hall for the Atomic Bomb Victims

- Nagasaki Peace Park
  - Atomic Bomb Hypocenter (Located near the Peace Park)

- Nagasaki Peace Pagoda

- Nagasaki Penguin Aquarium[*][28]

- Nagasaki Chinatown

- Nagasaki Science Museum[*][28]

- Nagasaki Subtropical Botanical Garden

- Nyoko-do Hermitage

- Ōura Church

- Sannō Shrine - One-legged stone *torii*, sometimes called an arch or gateway

- Sakamoto International Cemetery

- Shōfuku-ji

- Siebold Memorial Museum

- Sōfuku-ji - Daiyūhōden and Daiippomon are national treasures of Japan.

- Suwa Shrine

- Syusaku Endo Literature Museum

- Tateyama Park

- Twenty-Six Martyrs Museum and Monument

- Nagasaki Prefectural Art Museum

- Urakami Cathedral

- Miyo-Ken, a temple where the white snake is worshiped*[29]

-  Monument at the atomic bomb hypocenter in Nagasaki

-  Nagasaki National Peace Memorial Hall for the Atomic Bomb Victims

Nagasaki Lantern Festival

each November.

Kunchi, the most famous festival in Nagasaki, is held from 7–9 October.

The Nagasaki Lantern Festival,*[30] celebrating the Chinese New Year, is celebrated from February 18 to March 4.

- Sōfuku-ji (National treasure of Japan)

of Nagasaki

7.7 Events

The Prince Takamatsu Cup Nishinippon Round-Kyūshū Ekiden, the world's longest relay race, begins in Nagasaki

7.9 Twin towns

See also: List of twin towns and sister cities in Japan

The city of Nagasaki maintains sister cities or friendship relations with other cities worldwide.*[31]

- ● Hiroshima, Japan

- 🇺🇸 Saint Paul, Minnesota, United States, since 1955*[31]

- Dupnitsa, Bulgaria

- Santos, Brazil, since 1972*[31]

- Fuzhou, China, since 1980*[31]

- Middelburg, Netherlands, since 1978*[31]

- Porto, Portugal, since 1978*[31]*[32]

- Vaux-sur-Aure, France, since 2005; sister city of Sotome since 1978*[31]

7.10   See also

- Cultural treatments of the atomic bombings of Hiroshima and Nagasaki

- Foreign cemeteries in Japan

- Hashima Island (Gunkanjima)

7.11   References

[1] Hakim, Joy (1995). *A History of Us: War, Peace and all that Jazz*. New York: Oxford University Press. ISBN 0-19-509514-6.

[2] Boxer, *The Christian Century In Japan 1549–1650*, p. 100–101

[3] Diego Paccheco, Monumenta Nipponica, 1970

[4] so says the Jesuit account

[5] "Cultural Properties", Official site, Nagasaki: Thomeizan Kofukuji, retrieved 23 December 2016.

[6] Cambridge Encyclopedia of Japan, Richard Bowring and Haruko Laurie

[7] Screech, Timon. *The Western Scientific Gaze and Popular Imagery in Later Edo Japan: The Lens Within the Heart*. Cambridge: Cambridge University Press, 1996. p15.

[8] "Chapter II The Effects of the Atomic Bombings". United States Strategic Bombing Survey.

[9] *How Effective is Strategic Bombing?: Lessons Learned From World War II to Kosovo (World of War)*. NYU Press. December 1, 2000. pp. 86–87.

[10] "Avalon Project - The Atomic Bombings of Hiroshima and Nagasaki".

[11] Bradley, F.J. (1999). *No Strategic Targets Left*. Turner Publishing Company. p. 103. ISBN 1-5631-1483-6.

[12] Skylark, Tom (2002). *Final Months of the Pacific War*. Georgetown University Press. p. 178.

[13] Bruce Cameron Reed (October 16, 2013). *The History and Science of the Manhattan Project*. Springer Publishing. p. 400. ISBN 3-6424-0296-8.

[14] "BBC - WW2 People's War - Timeline".

[15] Robert Hull (October 11, 2011). *Welcome To Planet Earth - 2050 - Population Zero*. AuthorHouse. p. 215. ISBN 1-4634-2604-6.

[16] *Nuke-Rebuke: Writers & Artists Against Nuclear Energy & Weapons (The Contemporary anthology series)*. The Spirit That Moves Us Press. May 1, 1984. pp. 22–29.

[17] Groves 1962, pp. 343–346.

[18] Hoddeson et al., pp. 396-397

[19] Hoddeson et al. 1993, pp. 396–397

[20] "AtomicBombMuseum.org - After the Bomb".

[21] "Nagasaki History Facts and Timeline".

[22] あすにかけ全国的に厳しい冷え込み続く 気象庁

[23] "平年値（年・月ごとの値）". Japan Meteorological Agency. Retrieved 2011-12-02.

[24] "観測史上１～10位の値（年間を通じての値）". Japan Meteorological Agency. Retrieved 2011-12-02.

[25] "長崎外国語大学 - Nagasaki University of Foreign Studies". Nagasaki-gaigo.ac.jp. Retrieved 2013-03-12.

[26] "お知らせ長崎市平和・原爆のホームページが変わりました。". .city.nagasaki.nagasaki.jp. Retrieved 2011-06-01.

[27] "長崎歴史文化博物館". Nmhc.jp. Retrieved 2011-06-01.

[28] "移転のお知らせ". .city.nagasaki.nagasaki.jp. Retrieved 2011-06-01.

[29] *The Encircled Serpent: A Study of Serpent Symbolism in All Countries and Ages - M. Oldfield Howey - Google Books*. Books.google.com. 2005-03-31. Retrieved 2013-03-12.

[30] "長崎ランタンフェスティバル". Nagasaki-lantern.com. Retrieved 2011-06-01.

[31] "Sister Cities of Nagasaki City". © 2008-2009 International Affairs Section Nagasaki City Hall. Archived from the original on July 29, 2009. Retrieved 2009-07-10. External link in |publisher= (help)

[32] "International Relations of the City of Porto" (PDF). Municipal Directorateofthe Presidency Services International Relations Office. Retrieved 2009-07-10.

7.12 External links

- Nagasaki City official website (Japanese)

- Nagasaki City official website (English)

- Is Nagasaki still radioactive? - No. Includes explanation.

- Nagasaki after atomic bombing - interactive aerial map

- Nuclear Files.org Comprehensive information on the history, and political and social implications of the US atomic bombings of Hiroshima and Nagasaki

- Nagasaki Prefectural Tourism Federation

- Nagasaki Product Promotion Association

- Useful information for foreign residents, produced by Nagasaki International Association

Chapter 8

Hiroshima

For other uses, see Hiroshima (disambiguation).

    Hiroshima (広島市 *Hiroshima-shi*, Japanese:

Hiroshima Metropolitan Employment Area

[çiɾoɕimaɕi]) is the capital of Hiroshima Prefecture and the largest city in the Chūgoku region of western Honshu - the largest island of Japan. The city's name, 広島, means "Broad Island" in Japanese. Hiroshima gained city status on April 1, 1889. On April 1, 1980, Hiroshima became a designated city. As of August 2016, the city had an estimated population of 1,196,274. The GDP in Greater Hiroshima, Hiroshima Metropolitan Employment Area, is US$61.3 billion as of 2010.*[1]*[2] Kazumi Matsui has been the city's mayor since April 2011.

Hiroshima is perhaps best known as the first city in history to be targeted by a nuclear weapon when the United States Army Air Forces (USAAF) dropped an atomic bomb on the city (and later on Nagasaki) at 8:15 a.m. on August 6, 1945, near the end of World War II.*[3]

8.1 History

See also: Timeline of Hiroshima

8.1.1 Sengoku period (1589–1871)

![Hiroshima Castle]

Hiroshima Castle

Hiroshima was established on the river delta coastline of the Seto Inland Sea in 1589 by powerful warlord Mōri Terumoto, who made it his capital after leaving Kōriyama Castle in Aki Province.*[4]*[5] Hiroshima Castle was quickly built, and in 1593 Terumoto moved in. Terumoto was on the losing side at the Battle of Sekigahara. The winner of the battle, Tokugawa Ieyasu, deprived Mōri Terumoto of most of his fiefs, including Hiroshima and gave Aki Province to Masanori Fukushima, a *daimyō* who had supported Tokugawa.*[6]

8.1.2 Imperial period (1871–1939)

After the han was abolished in 1871, the city became the capital of Hiroshima Prefecture. Hiroshima became a major urban center during the imperial period, as the Japanese

Hiroshima Commercial Museum 1915

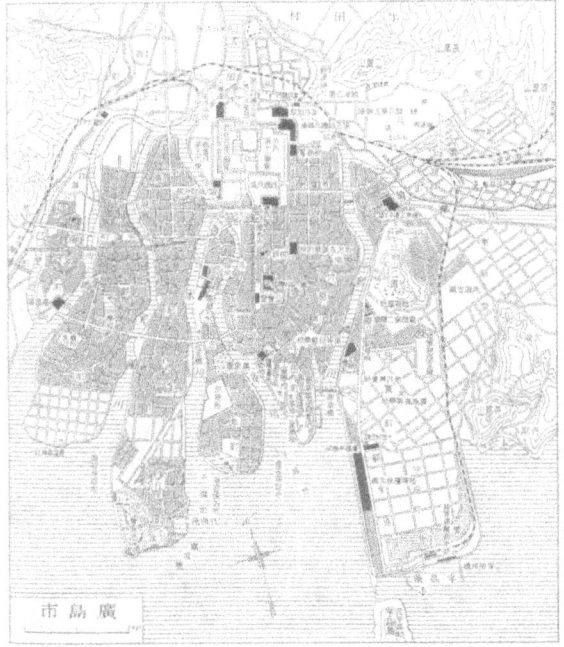

Map of Hiroshima City in the 1930s (Japanese edition)

economy shifted from primarily rural to urban industries. During the 1870s, one of the seven government-sponsored English language schools was established in Hiroshima.[7] Ujina Harbor was constructed through the efforts of Hiroshima Governor Sadaaki Senda in the 1880s, allowing Hiroshima to become an important port city.

The San'yō Railway was extended to Hiroshima in 1894, and a rail line from the main station to the harbor was constructed for military transportation during the First Sino-Japanese War.[8] During that war, the Japanese government moved temporarily to Hiroshima, and Emperor Meiji maintained his headquarters at Hiroshima Castle from September 15, 1894, to April 27, 1895.[8] The significance of Hiroshima for the Japanese government can be discerned from the fact that the first round of talks between Chinese and Japanese representatives to end the

Sino-Japanese War was held in Hiroshima, from February 1 to February 4, 1895.[9] New industrial plants, including cotton mills, were established in Hiroshima in the late 19th century.[10] Further industrialization in Hiroshima was stimulated during the Russo-Japanese War in 1904, which required development and production of military supplies. The Hiroshima Prefectural Commercial Exhibition Hall was constructed in 1915 as a center for trade and exhibition of new products. Later, its name was changed to Hiroshima Prefectural Product Exhibition Hall, and again to Hiroshima Prefectural Industrial Promotion Hall.[11]

During World War I, Hiroshima became a focal point of military activity, as the Japanese government entered the war on the Allied side. About 500 German prisoners of war were held in Ninoshima Island in Hiroshima Bay.[12] The growth of Hiroshima as a city continued after the First World War, as the city now attracted the attention of the Catholic Church, and on May 4, 1923, an Apostolic Vicar was appointed for that city.[13]

8.1.3   World War II and the atomic bombing (1939–1945)

Main article: Atomic bombings of Hiroshima and Nagasaki § Hiroshima

During World War II, the 2nd General Army and Chugoku Regional Army were headquartered in Hiroshima, and the Army Marine Headquarters was located at Ujina port. The city also had large depots of military supplies, and was a key center for shipping.[14]

The bombing of Tokyo and other cities in Japan during World War II caused widespread destruction and hundreds of thousands of civilian deaths.[15] There were no such air raids on Hiroshima. However, a real threat existed and was recognized. In order to protect against potential firebombings in Hiroshima, school children aged 11–14 years were mobilized to demolish houses and create firebreaks.[16]

On Monday, August 6, 1945, at 8:15 a.m., the nuclear weapon "Little Boy" was dropped on Hiroshima by an American B-29 bomber, the *Enola Gay*, flown by Colonel Paul Tibbets,[17] directly killing an estimated 70,000 people, including 20,000 Japanese combatants and 2,000 Korean slave laborers. By the end of the year, injury and radiation brought the total number of deaths to 90,000–166,000.[18] The population before the bombing was around 340,000 to 350,000. About 70% of the city's buildings were destroyed, and another 7% severely damaged.

The public release of film footage of the city following the attack, and some of the Atomic Bomb Casualty Commission research about the human effects of the attack, was restricted during the occupation of Japan, and much of this information was censored until the signing of the San Francisco Peace Treaty in 1951, restoring control to the Japanese.*[19]

As Ian Buruma observed, "News of the terrible consequences of the atom bomb attacks on Japan was deliberately withheld from the Japanese public by US military censors during the Allied occupation—even as they sought to teach the natives the virtues of a free press. Casualty statistics were suppressed. Film shot by Japanese cameramen in Hiroshima and Nagasaki after the bombings was confiscated. "Hiroshima", the account written by John Hersey for *The New Yorker*, had a huge impact in the US, but was banned in Japan. As [John] Dower says: 'In the localities themselves, suffering was compounded not merely by the unprecedented nature of the catastrophe ... but also by the fact that public struggle with this traumatic experience was not permitted.'" The US occupation authorities maintained a monopoly on scientific and medical information about the effects of the atomic bomb through the work of the Atomic Bomb Casualty Commission, which treated the data gathered in studies of hibakusha as privileged information rather than making the results available for the treatment of victims or providing financial or medical support to aid victims. The US also stood by official denial of the ravages associated with radiation. Finally, not only was the press tightly censored on atomic issues, but literature and the arts were also subject to rigorous control prior.*[20]

The book *Hiroshima* by John Hersey was originally featured in article form and published in the magazine *The New Yorker*,*[21] on 31 August 1946. It is reported to have reached Tokyo, in English, at least by January 1947 and the translated version was released in Japan in 1949.*[22] Despite the fact that the article was planned to be published over four issues, "Hiroshima" made up the entire contents of one issue of the magazine.*[23]*[24] *Hiroshima* narrates the stories of six bomb survivors immediately prior to and for months after the dropping of the Little Boy bomb.*[21]*[25]

The oleander is the official flower of the city of Hiroshima because it was the first to bloom again after the explosion of the atomic bomb in 1945.*[26]

● Hiroshima

after the bombing

● *Hiroshima
after the bombing

8.1.4 Postwar period (1945–present)

Folded paper cranes representing prayers for peace and Sadako Sasaki

On September 17, 1945, Hiroshima was struck by the Makurazaki Typhoon (Typhoon Ida). Hiroshima Prefecture suffered more than 3,000 deaths and injuries, about half the national total.*[27] More than half the bridges in the city were destroyed, along with heavy damage to roads and railroads, further devastating the city.*[28]

Hiroshima was rebuilt after the war, with help from the national government through the Hiroshima Peace Memorial City Construction Law passed in 1949. It provided financial assistance for reconstruction, along with land donated that was previously owned by the national government and used for military purposes.*[29]

In 1949, a design was selected for the Hiroshima Peace Memorial Park. Hiroshima Prefectural Industrial Promo-

Atomic Bomb Dome by Jan Letzel and modern Hiroshima

tion Hall, the closest surviving building to the location of the bomb's detonation, was designated the Genbaku Dome (原爆ドーム) or "Atomic Dome", a part of the Hiroshima Peace Memorial Park. The Hiroshima Peace Memorial Museum was opened in 1955 in the Peace Park.[*][30]

The peace park also contains a Peace Pagoda, built in 1966 by Nipponzan-Myōhōji. Uniquely, the pagoda is made of steel, rather than the usual stone.

Hiroshima was proclaimed a City of Peace by the Japanese parliament in 1949, at the initiative of its mayor, Shinzo Hamai (1905–1968). As a result, the city of Hiroshima received more international attention as a desirable location for holding international conferences on peace as well as social issues. As part of that effort, the Hiroshima Interpreters' and Guide's Association (HIGA) was established in 1992 in order to facilitate interpretation for conferences, and the Hiroshima Peace Institute was established in 1998 within the Hiroshima University. The city government continues to advocate the abolition of all nuclear weapons and the Mayor of Hiroshima is the president of Mayors for Peace, an international mayoral organization mobilizing cities and citizens worldwide to abolish and eliminate nuclear weapons by the year 2020.[*][31][*][32]

On May 27, 2016, Barack Obama visited Hiroshima, being the first sitting president of the United States to visit since the drop of the atomic bomb.[*][33]

8.2 Geography

Hiroshima is situated on the Ōta River delta, on Hiroshima Bay, facing the Seto Inland Sea on its south side. The river's six channels divide Hiroshima into several islets.

8.2.1 Climate

Hiroshima has a humid subtropical climate characterized by cool to mild winters and hot humid summers. Like much of the rest of Japan, Hiroshima experiences a seasonal temperature lag in summer, with August rather than July being the warmest month of the year. Precipitation occurs year-round, although winter is the driest season. Rainfall peaks in June and July, with August experiencing sunnier and drier conditions.

8.2.2 Wards

Hiroshima has eight wards (*ku*):

8.3 Demographics

Hondōri shopping arcade in Hiroshima

As of 2006, the city has an estimated population of 1,154,391, while the total population for the metropolitan area was estimated as 2,043,788 in 2000.[*][35] The total area of the city is 905.08 square kilometres (349.45 sq mi), with a population density of 1275.4 persons per km^2.[*][36]

The population around 1910 was 143,000.[*][37] Before World War II, Hiroshima's population had grown to 360,000, and peaked at 419,182 in 1942.[*][36] Following the atomic bombing in 1945, the population dropped to 137,197.[*][36] By 1955, the city's population had returned to pre-war levels.[*][38]

8.4 Transportation

8.4.1 Air

Hiroshima is served by Hiroshima Airport (IATA: **HIJ**, ICAO: **RJOA**), located 50 kilometres (31 mi) east of the city, with regular flights to Tokyo, Sapporo, Sendai, Okinawa, and also to China, Taiwan and South Korea.

8.4.2 Trains

- JR West

    - Sanyō Shinkansen, San'yō Main Line, Kure Line, Geibi Line, Kabe Line

- Hiroshima New Transit Line 1

- Hiroshima Short Distance Transit Seno Line

8.4.3 Streetcars

A modern tram in Hiroshima, May 2008

Hiroshima is notable, in Japan, for its light rail system, nicknamed *Hiroden,* and the "Moving Streetcar Museum." Streetcar service started in 1912,[39] was interrupted by the atomic bomb, and was restored as soon as was practical. (Service between Koi/Nishi Hiroshima and Tenma-cho was started up three days after the bombing.[40])

Streetcars and light rail vehicles are still rolling down Hiroshima's streets, including nuked streetcars 651 and 652, which are among the older streetcars in the system. When Kyoto and Fukuoka discontinued their trolley systems, Hiroshima bought them up at discounted prices, and, by 2011, the city had 298 streetcars, more than any other city in Japan.[40]

- Hiroden

- Main Line, Ujina Line, Eba Line, Hakushima Line, Hijiyama Line, Yokogawa Line, Miyajima Line

8.4.4 Automobiles

Hiroshima is served by Japan National Route 54, Hiroshima Prefectural Route 37 (Hiroshima-Miyoshi Route), Hiroshima Prefectural Route 70 (Hiroshima-Nakashima Route), Hiroshima Prefectural Route 84 (Higashi Kaita Hiroshima Route), Hiroshima Prefectural Route 164 (Hiroshima-Kaita Route), and Hiroshima Prefectural Route 264 (Nakayama-Onaga Route).

8.5 Events

Hiroshima Flower Festival 2011

- Hiroshima Flower Festival, May 3–5, Heiwa Odori, Hiroshima Peace Memorial Park

- Toukasan, first Friday to Sunday in June, Mikawa-cho, Chuo Dori

- Ebisu Festival, November 18–20, Ebisucho, Hacchobori, Chuo Dori

- Hiroshima Peace Memorial Ceremony, August 6, Hiroshima Peace Memorial Park

8.6 Culture

Hiroshima has a professional symphony orchestra, which has performed at Wel City Hiroshima since 1963.[41] There are also many museums in Hiroshima, including the Hiroshima Peace Memorial Museum, along with several art

Shukkei-en

A man prepares okonomiyaki in a restaurant in Hiroshima

museums. The Hiroshima Museum of Art, which has a large collection of French renaissance art, opened in 1978. The Hiroshima Prefectural Art Museum opened in 1968, and is located near Shukkei-en gardens. The Hiroshima City Museum of Contemporary Art, which opened in 1989, is located near Hijiyama Park. Festivals include Hiroshima Flower Festival and Hiroshima International Animation Festival.

Hiroshima Peace Memorial Park, which includes the Hiroshima Peace Memorial, draws many visitors from around the world, especially for the Hiroshima Peace Memorial Ceremony, an annual commemoration held on the date of the atomic bombing. The park also contains a large collection of monuments, including the Children's Peace Monument, the Hiroshima National Peace Memorial Hall for the Atomic Bomb Victims and many others.

Hiroshima's rebuilt castle (nicknamed *Rijō*, meaning *Koi Castle*) houses a museum of life in the Edo period. Hiroshima Gokoku Shrine is within the walls of the castle. Other attractions in Hiroshima include Shukkei-en, Fudōin, Mitaki-dera, and Hijiyama Park.

8.6.1   Cuisine

Hiroshima is known for okonomiyaki, a savory (umami) pancake cooked on a hot-plate, usually in front of the customer. It is cooked with various ingredients, which are layered rather than mixed together as done with the Osaka version of okonomiyaki. The layers are typically egg, cabbage, bean sprouts (moyashi), sliced pork/bacon with optional items (mayonnaise, fried squid, octopus, cheese, mochi, kimchi, etc.), and noodles (soba, udon) topped with another layer of egg and a generous dollop of okonomiyaki sauce (Carp and Otafuku are two popular brands). The amount of cabbage used is usually 3 to 4 times the amount used in the Osaka style, therefore arguably a healthier version. It starts out piled very high and is generally pushed down as

the cabbage cooks. The order of the layers may vary slightly depending on the chef's style and preference, and ingredients will vary depending on the preference of the customer.

8.6.2   Media

The Chugoku Shimbun is the local newspaper serving Hiroshima. It publishes both morning paper and evening editions. Television stations include Hiroshima Home Television, Hiroshima TV, TV Shinhiroshima, and the RCC Broadcasting Company. Radio stations include Hiroshima FM, Chugoku Communication Network, FM Fukuyama, FM Nanami, and Onomichi FM. Hiroshima is also served by NHK, Japan's public broadcaster, with television and radio broadcasting.

8.6.3   Education

Hiroshima University was established in 1949, as part of a national restructuring of the education system. One national university was set up in each prefecture, including Hiroshima University, which combined eight existing institutions (Hiroshima University of Literature and Science, Hiroshima School of Secondary Education, Hiroshima School of Education, Hiroshima Women's School of Sec-

Satake Memorial Hall at Hiroshima University (in Higashihiroshima City)

ondary Education, Hiroshima School of Education for Youth, Hiroshima Higher School, Hiroshima Higher Technical School, and Hiroshima Municipal Higher Technical School), with the Hiroshima Prefectural Medical College added in 1953. But, in 1972 the relocation of Hiroshima University was decided from urban areas of Hiroshima City to wider campus in Higashihiroshima City. By 1995 almost all campuses were relocated to Higashihiroshima. But, School of Medicine, School of Dentistry, School of Pharmaceutical Sciences and Graduate School in these fields in Kasumi Campus and Law School and Center for Research on Regional Economic System in Higashi-Senda Campus are still in Hiroshima City.[42]

8.6.4 Sport

Hiroshima has several professional sports clubs. The city's main football club are Sanfrecce Hiroshima, who play at the Hiroshima Big Arch. As Toyo Kogyo Soccer Club, they won the Japan Soccer League five times between 1965 and 1970 and the Emperor's Cup in 1965, 1967 and 1969. After adopting their current name in 1992, the club won the J. League in 2012 and 2013. The city's main women's football club is Angeviolet Hiroshima. Defunct clubs include Rijo Shukyu, who won the Emperor's Cup in 1924 and 1925, and Ẽfini Hiroshima.

Hiroshima Toyo Carp are the city's major baseball club, and play at the Mazda Stadium. Members of the Central League, the club won the Japan Series in 1979, 1980 and 1984. Other sports clubs include Hiroshima Dragonflies (basketball), Hiroshima Maple Reds (handball) and JT Thunders (volleyball).

The Woodone Open Hiroshima was part of the Japan Golf Tour between 1973 and 2007. The city also hosted the 1994

Asian Games, using the Big Arch stadium, which is now used for the annual Mikio Oda Memorial International Amateur Athletic Game.

8.7 Hospitals

- Hiroshima City Hospital
- Hiroshima City Asa Hospital
- Hiroshima Prefectural Hospital
- Hiroshima Red Cross Hospital & Atomic-bomb Survivors Hospital
- Hiroshima University Hospital

8.8 International relations

See also: List of twin towns and sister cities in Japan

8.8.1 Twin towns and sister cities

Hiroshima has six overseas sister cities:[43]

- Honolulu, United States (1959)
- Volgograd, Russia (1972)[44]
- Hanover, Germany (1983)[45]
- Chongqing, People's Republic of China (1986)
- Daegu, South Korea (1997)
- Montreal, Quebec, Canada (1998)

Within Japan, Hiroshima has a similar relationship with Nagasaki.

8.9 Further reading

- Pacific War Research Society, *Japan's Longest Day* (Kodansha, 2002, ISBN 4-7700-2887-3), the internal Japanese account of the surrender and how it was almost thwarted by fanatic soldiers who attempted a coup against the Emperor.
- Richard B. Frank, *Downfall: The End of the Imperial Japanese Empire* (Penguin, 2001 ISBN 0-14-100146-1)

- Robert Jungk, *Children of the Ashes*, 1st Eng. ed. 1961*[46]

- Gar Alperovitz, *The Decision to Use the Atomic Bomb*, ISBN 0-679-76285-X

- John Hersey, *Hiroshima*, ISBN 0-679-72103-7

- Michihiko Hachiya, *Hiroshima Diary: The Journal of a Japanese Physician*, August 6 - September 30, 1945 (Chapel Hill: University of North Carolina Press, 1955), since reprinted.

- Masuji Ibuse, *Black Rain*, ISBN 0-87011-364-X

- Tamiki Hara, *Summer Flowers* ISBN 0-691-00837-X

- Robert Jay Lifton *Death in life: The survivors of Hiroshima*, Weidenfeld & Nicolson 1st edition (1968) ISBN 0-297-76466-7

8.10   See also

- *Barefoot Gen*

- Cultural treatments of the atomic bombings of Hiroshima and Nagasaki

- Kokura

- Masaharu Morimoto

- Nagasaki

- Perfume, a pop group from Hiroshima

- Sadako Kurihara

- Sadako Sasaki (1943–1955)

- Town of Evening Calm, Country of Cherry Blossoms

- Yōko Ōta, author of several works of Atomic bomb literature

- Yoshito Matsushige

8.11   Notes

8.12   References

-

-

8.13   External links

- Hiroshima City official website

- Hiroshima City official website

- Hiroshima before and after atomic bombing - interactive aerial maps

- Hiroshima atomic bomb damage - interactive aerial map

- Is Hiroshima still radioactive? - No. Includes explanation.

-

- City Mayors article

- CBC Digital Archives - Shadows of Hiroshima

- Hiroshima Map - interactive with points of interest

- BBC World Service BBC Witness programme interviews a schoolgirl who survived the bomb

- Hope Elizabeth May, "Creating Peace through Law: the City of Hiroshima"

[1] Yoshitsugu Kanemoto. "Metropolitan Employment Area (MEA) Data". Center for Spatial Information Science, The University of Tokyo.

[2] Conversion rates - Exchange rates - OECD Data

[3] Hakim, Joy (1995). *A History of Us: War, Peace and all that Jazz.* New York: Oxford University Press. ISBN 0-19-509514-6.

[4] "The Origin of Hiroshima". Hiroshima Peace Culture Foundation. Archived from the original on 2008-01-30. Retrieved 2007-08-17.

[5] Scott O'Bryan (2009). "Hiroshima: History, City, Event". About Japan: A Teacher's Resource. Retrieved 2010-03-14.

[6] Kosaikai, Yoshiteru (2007). "History of Hiroshima". *Hiroshima Peace Reader.* Hiroshima Peace Culture Foundation.

[7] Bingham (US Legation in Tokyo) to Fish (US Department of State), September 20, 1876, in *Papers relating to the foreign relations of the United States, transmitted to congress, with the annual message of the president, December 4, 1876*, p. 384

[8] Kosakai, *Hiroshima Peace Reader*

[9] Dun (US Legation in Tokyo) to Gresham, February 4, 1895, in *Foreign relations of United States, 1894*, Appendix I, p. 97

[10] Jacobs, Norman (1958). *The Origin of Modern Capitalism and Eastern Asia.* Hong Kong University. p. 51.

[11] Sanko (1998). *Hiroshima Peace Memorial (Genbaku Dome)*. The City of Hiroshima and the Hiroshima Peace Culture Foundation.

[12]

[13] "Diocese of Hiroshima". *Catholic-Hierarchy.org*. David M. Cheney. Retrieved 21 January 2015.

[14] United States Strategic Bombing Survey (June 1946). "U. S. Strategic Bombing Survey: The Effects of the Atomic Bombings of Hiroshima and Nagasaki". nuclearfiles.org. Archived from the original on 2004-10-11. Retrieved 2009-07-26.

[15] Pape, Robert (1996). *Bombing to Win: Airpower and Coercion in War*. Cornell University Press. p. 129. ISBN 978-0-8014-8311-0.

[16] "Japan in the Modern Age and Hiroshima as a Military City". The Chugoku Shimbun. Retrieved 2007-08-19.

[17] The Atomic Bombing of Hiroshima, U.S. Department of Energy, Office of History and Heritage Resources.

[18] "Frequently Asked Questions - Radiation Effects Research Foundation". Rerf.or.jp. Retrieved 2011-07-29.

[19] Ishikawa and Swain (1981), p. 5

[20] Selden, Mark. "Bombs Bursting in Air: The US Firebombing and Atomic Bombing of Japan". Asia Pacific Journal. Retrieved March 27, 2015.

[21] Roger Angell, From the Archives, "HERSEY AND HISTORY", *The New Yorker*, July 31, 1995, p. 66.

[22] http://www.japantimes.co.jp/culture/2009/08/16/books/the-pure-horror-of-hiroshima/#.UdhVsfnVDTc The pure horror of Hiroshima, published in *The Japan Times* by Donald Richie.

[23] Sharp, "From Yellow Peril to Japanese Wasteland: John Hersey's 'Hiroshima'", Twentieth Century Literature 46 (2000): 434–452, accessed March 15, 2012.

[24] Jon Michaub, "EIGHTY-FIVE FROM THE ARCHIVE: JOHN HERSEY" *The New Yorker*, June 8, 2010, np.

[25] John Hersey, Hiroshima (New York: Random House, 1989).

[26] "広島市市の木・市の花". Retrieved 2012-07-15.

[27] "Excite エキサイト".

[28] Ishikawa and also Swain (1981), p. 6

[29] "Peace Memorial City, Hiroshima". Hiroshima Peace Culture Foundation. Archived from the original on 2008-02-06. Retrieved 2007-08-14.

[30] "Fifty Years for the Peace Memorial Museum". Hiroshima Peace Memorial Museum. Retrieved 2007-08-17.

[31] "Surviving the Atomic Attack on Hiroshima, 1944". Eyewitnesstohistory.com. 1945-08-06. Retrieved 2009-07-17.

[32] "Library: Media Gallery: Video Files: Rare film documents devastation at Hiroshima". Nuclear Files. Retrieved 2009-07-17.

[33] "President Obama Visits Hiroshima". The New York Times. Archived from the original on 2016-05-29. Retrieved 2016-05-31.

[34] "気象庁 / 平年値（年・月ごとの値）". Japan Meteorological Agency.

[35] "Population of Japan, Table 92". Statistics Bureau. Retrieved 2007-08-14.

[36] "2006 Statistical Profile". The City of Hiroshima. Archived from the original on 2008-02-06. Retrieved 2007-08-14.

[37] Terry, Thomas Philip (1914). *Terry's Japanese Empire*. Houghton Mifflin Co. p. 640.

[38] de Rham-Azimi, Nassrine, Matt Fuller, and Hiroko Nakayama (2003). *Post-conflict Reconstruction in Japan, Republic of Korea, Vietnam, Cambodia, East Timor*. United Nations Publications. p. 69.

[39] "広島市交通科学館／ Hiroshima City Transportation Museum".

[40] "Peace Newspaper produced by Japanese teenagers: Peace Seeds:feature story".

[41] "Wel City Hiroshima". Wel-hknk.com. Retrieved 2011-06-13.

[42] "History of Hiroshima University". Hiroshima University. Retrieved 2007-06-25.

[43] "Introduction to our Sister and Friendship Cities". City.hiroshima.jp. Retrieved 2010-05-10.

[44] "Friendly relationship at Official website of Volgograd". Volgadmin.ru. 1994-12-01. Retrieved 2011-06-13.

[45] "Twinnings of the City of Hannover". *Hanover.de - Offizielles Portal der Landeshauptstadt und der Region Hannover* (in German). Presse- und Öffentlichkeitsarbeit der Landeshauptstadt Hannover. Retrieved 2014-10-13. External link in |website= (help)

[46] Gyanpedia.in

Chapter 9

Pollution of Lake Karachay

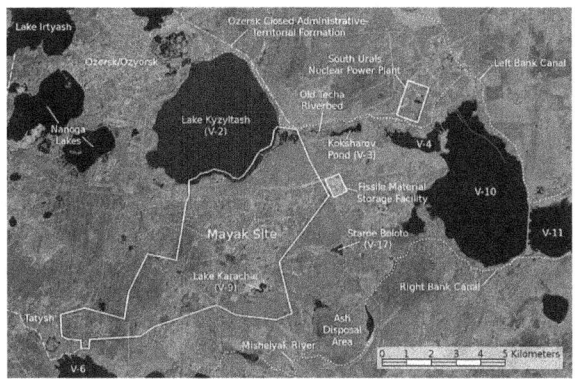

A satellite view of Lake Karachay.

Lake Karachay, located in the southern Ural Mountains in eastern Russia, was a dumping ground for the Soviet Union's nuclear weapon facilities. It was also affected by a string of accidents and disasters causing the surrounding areas to be highly contaminated with radioactive waste. Washington, D.C.-based Worldwatch Institute has described it as the "most polluted spot on Earth".*[1]*[2]

9.1 History

Built in the late 1940s Mayak was one of Russia's most prominent nuclear weapons factories. Mayak was kept secret by the government until 1990. When Russian president Boris Yeltsin signed a 1992 decree opening the area, Western scientists were able to gain access. The sediment of the lake bed is estimated to be composed almost entirely of high level radioactive waste deposits to a depth of roughly 11 feet (3.4 m).

In 1994, a report revealed that 5 million cubic meters of polluted water had migrated from Lake Karachay, and was spreading to the south and north at 80 meters per year, "threatening to enter water intakes and rivers".*[3] The authors acknowledged that "theoretical hazards developed into actual events".

In November 1994, officials from the Russian Ministry of Atomic Energy stated that Soviet officials initiated a process following the 1957 Kyshtym disaster resulting in the transfer of 3 billion curies of high level nuclear waste into deep wells at three other sites.*[4]

The Techa river, which provides water to nearby areas, was contaminated, and about 65% of local residents fell ill with radiation sickness. Doctors called it the "special disease" because they were not allowed to note radiation in their diagnoses as long as the facility was secret. In the village of Metlino, it was found that 65% of residents were suffering from chronic radiation sickness. Workers at the plutonium plant were also affected.

9.2 Causes

The pollution of Lake Karachay is connected to the disposal of nuclear materials from Mayak. Among workers, cancer mortality remains an issue.*[5] By the time Mayak's existence was officially recognized, there had been a 21% rise in cancer cases, a 25% rise in birth defects, and a 41% rise in leukemia in the surrounding region of Chelyabinsk.*[6] By one estimate, the river contains 120 million curies of radioactive waste.*[7]

9.3 Prevalence of pollution

Nuclear waste, either from civilian or military nuclear projects, remains a serious threat to the environment of Russia.*[8] Reports suggest that there are few or no road signs warning about the polluted areas surrounding Lake Karachay.*[9]

Some parts of the lake are extremely radioactive (600 röntgens/hour) and one could receive a lethal dose of radiation in 30 minutes (300 röntgens).

9.4 See also

- Lake Karachay

- Water pollution

- Plutopia

- Ozyorsk, Chelyabinsk Oblast

- Semipalatinsk Test Site

- Soviet atomic bomb project

9.5 References

[1] Lenssen, "Nuclear Waste: The Problem that Won't Go Away", Worldwatch Institute, Washington, D.C., 1991: 15.

[2] Andrea Pelleschi (2013). *Russia*. ABDO Publishing Company. ISBN 9781614808787.

[3] (Laverov, Omelianeako and Velichkin 1994, 3)

[4] "Critical Masses: Citizens, Nuclear Weapons Production, and Environmental Destruction in the United States and Russia", by Russell J. Dalton, 1999, p.79

[5] Cancer mortality risk among workers at the Mayak nuclear complex., June 2003

[6] Is this the most polluted place on Earth? The Russian lake where an hour on the beach would kill you 9 October 2012

[7] "Russia", p. 121, publisher = Lonely Planet

[8] "The Politics of Environmental Policy in Russia", p. 34, by David Lewis Feldman, Ivan Blokov

[9] "The Burning Lake: A Volk Thriller", by Brent Ghelfi, p. 101

9.6 External links

- Lake Karachay

Chapter 10

Techa River

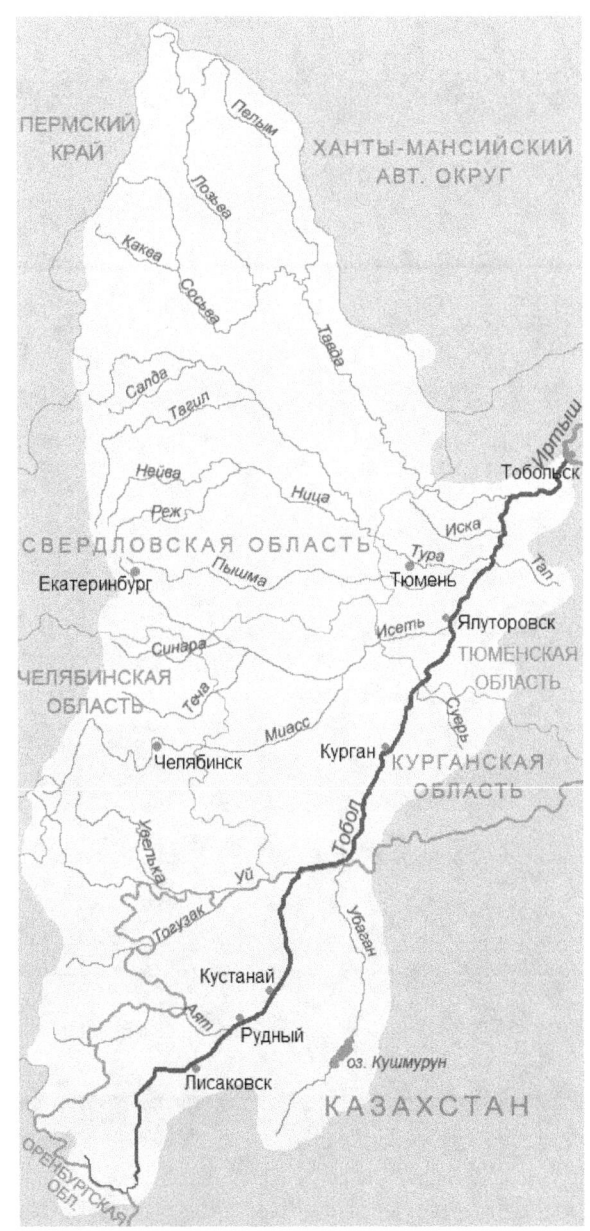

Tobol basin

The **Techa River** is a river on the eastern flank of the southern Ural Mountains noted for its nuclear contamination. It is about 240 kilometres (150 mi) long, and its basin covers 7,500 square kilometres (2,900 sq mi). It begins at the formerly secret nuclear-processing town of Ozyorsk, Chelyabinsk Oblast about 80 kilometres (50 mi) northwest of Chelyabinsk and flows northeast to Dalmatovo on the Iset River, a tributary of the Tobol River. Its basin is enclosed on the southeast by that of the Miass River, another river that flows northeast into the Iset.

10.1 Water pollution

From 1949 to 1956 the Mayak complex[*][1] dumped an estimated 76 million cubic metres (2.7×10^9 cu ft) of radioactive waste water into the Techa River,[*][2] a cumulative dispersal of 2.75 MCi (102 PBq) of radioactivity.[*][3]

As many as forty villages, with a combined population of about 28,000 residents, lined the river at the time.[*][4] For 24 of them, the Techa was a major source of water; 23 of them were eventually evacuated.[*][5] In the past 45 years, about half a million people in the region have been irradiated in one or more of the incidents,[*][4][*][6] exposing them to as much as 20 times the radiation suffered by the Chernobyl disaster victims.[*][2] Coordinates: 55°46′07″N 60°44′02″E / 55.7686°N 60.7339°E

10.2 See also

- Pollution of the Lake Karachay

- Water pollution

- Plutopia

- Ozyorsk, Chelyabinsk Oblast

- Semipalatinsk Test Site

10.3 References

[1] Techa River Archived 11 September 2007 at the Wayback Machine.

[2] CHELYABINSK "The Most Contaminated Spot on the Planet" - a documentary film by Slawomir Grunberg - Log In Productions - distributed by LogTV LTD

[3] Pike, John. "Chelyabinsk-65 / Ozersk Combine 817 / Production Association Mayak". GlobalSecurity.org. Retrieved 29 September 2010.

[4] Radioactive Contamination of the Techa River and its Effects

[5] Clay, Rebecca (April 2001). "Cold War, Hot Nukes: Legacy of an Era". *Environmental Health Perspectives.* National Institute of Environmental Health Sciences. **109** (4): a162–a169. doi:10.1289/ehp.109-a162. Archived from the original on 2 June 2010. Retrieved 29 September 2010.

[6] Zaitchik, Alexander (8 Oct 2007). "Inside the Zone". The Exile. Retrieved 29 September 2010.

Chapter 11

Rocky Flats Plant

See also: Radioactive contamination from the Rocky Flats Plant

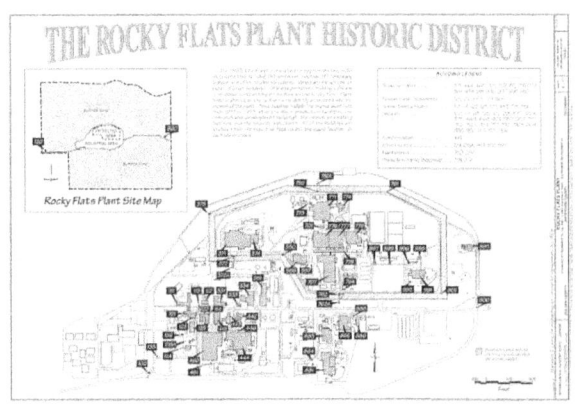

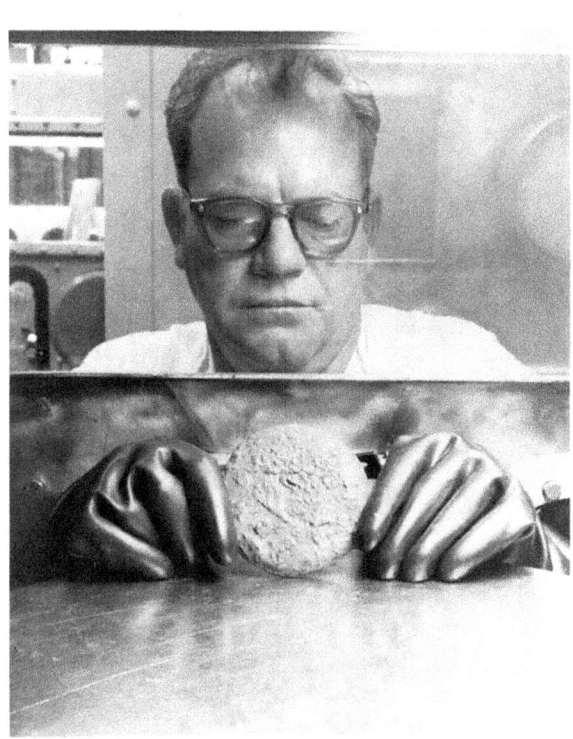

Worker holding plutonium "button" in glove box

The **Rocky Flats Plant** was a former nuclear weapons production facility in the western United States, near Denver, Colorado.[*][2] Operated from 1952 to 1992, it was under the control of the U.S. Atomic Energy Commission (AEC), succeeded by the Department of Energy (DOE) in 1977.

Weapons production was temporarily halted in 1989 after EPA and FBI agents raided the facility.[*][3] Operators of the plant later pleaded guilty to criminal violations of environmental law.[*][4] At the time, the fine was one of the largest penalties ever in an environmental law case.[*][5]

Cleanup began in the early 1990s,[*][6][*][7][*][8] and the site achieved regulatory closure in 2006.[*][9] The cleanup effort decommissioned and demolished over 800 structures; removed over 21 tons of weapons-grade material; removed over 1.3 million cubic meters of waste; and, treated more than 16 million gallons of water. Four groundwater treatment systems were also constructed.[*][10] Today, the Rocky Flats Plant is gone. The site of the former facility consists of two distinct areas: (1) the "Central Operable Unit" (including the former industrial area), which remains off-limits to the public as a CERCLA "Superfund" site, owned and managed by the U.S. Department of Energy,[*][11] and

(2) the Rocky Flats National Wildlife Refuge, owned and managed by the U.S. Fish and Wildlife Service.[*][12] The Refuge (also known as the "Peripheral Operable Unit") was determined to be suitable for unrestricted use. Every five years, the U.S. Department of Energy, U.S. Environmental Protection Agency, and Colorado Department of Public Health and Environment review environmental data to assess whether the remedy is functioning as intended. The last five-year review concluded the remedy is effective.[*][13] The next five-year review will be produced in 2017. [*][14]

11.1   History

Precision plutonium foundry mold, 1959

*Room damaged by
1969 Rocky Flats Fire*

Control panel, Critical Mass Laboratory, 1970

11.1.1   1950s

Following World War II, the United States increased production of nuclear weapons. The AEC chose the Dow Chemical Company to manage the bomb production facility. A 4-square-mile (10 km^2) site about 15 miles (25 km) northwest of Denver on a windy plateau called Rocky Flats was chosen for the facility. On July 10, 1951, ground was broken on the first building in the facility. Contemporary news reports stated that the site would not be used to produce nuclear bombs.[*][15]

In 1953, the plant began production of bomb components, manufacturing plutonium triggers, or "pits", which were used at the Pantex plant in Amarillo, Texas to assemble nuclear weapons. By 1957, the plant had expanded to 27 buildings.

On September 11, 1957, a plutonium fire occurred in one of the gloveboxes used to handle radioactive materials, igniting the combustible rubber gloves and plexiglas windows of the box. Metallic plutonium is a fire hazard and pyrophoric; under the right conditions it may ignite in air at room temperature. The accident resulted in the contamination of Building 771, the release of plutonium into the atmosphere, and caused $818,600 in damage. An incinerator for plutonium-contaminated waste was installed in Building 771 in 1958.

Barrels of radioactive waste were found to be leaking into an open field in 1959. This was not made publicly known until 1970 when wind-borne particles were detected in Denver.

11.1.2   1960s

Throughout the 1960s, the plant continued to enlarge and add buildings. The 1960s also brought more contamination to the site. In 1967, 3,500 barrels (560 m^3) of plutonium-contaminated lubricants and solvents were stored on Pad 903. A large number of them were found to be leaking, and low-level contaminated soil was becoming wind-borne from this area. This pad was covered with gravel and paved over with asphalt in 1969.

May 11, 1969 saw a major fire in a glovebox in Building 776/777.*[16] This was the costliest industrial accident ever to occur in the United States up to that time. Cleanup from the accident took two years and led to safety upgrades on the site, including fire sprinkler systems and firewalls.

11.1.3   1970s

In order to reduce the danger of public contamination and to create a security area around the plant following protests, the United States Congress authorized the purchase of a 4,600-acre (7.2 sq mi; 19 km^2) buffer zone around the plant in 1972. In 1973, nearby Walnut Creek and the Great Western Reservoir were found to have elevated tritium levels. The tritium was determined to have been released from contaminated materials shipped to Rocky Flats from the Lawrence Livermore Laboratory. Discovery of the contamination by the Colorado Department of Health led to investigations by the AEC and United States Environmental Protection Agency (EPA). As a result of the investigation, several mitigation efforts were put in place to prevent further contamination. Some of the elements included channeling of wastewater runoff to three dams for testing before release into the water system and construction of a reverse osmosis facility to clean up wastewater.

The next year, elevated plutonium levels were found in the topsoil near the now covered Pad 903. An additional 4,500 acres (18 km^2) of buffer zone were purchased.

1975 saw Rockwell International replacing Dow Chemical as the contractor for the site. This year also saw local landowners suing for property contamination caused by the plant.

In 1978, 60 protesters belonging to the Rocky Flats Truth Force, or Satyagraha Affinity Group, based in Boulder, Colorado, were arrested for trespassing at Rocky Flats, and were brought to trial before Judge Kim Goldberger. Dr. John Candler Cobb, Professor of Preventive Medicine at the University of Colorado Medical Center, testified that the most significant danger of radioactive contamination came from the 1967 incident in which oil barrels containing plutonium leaked 5,000 US gallons (19,000 L) of oil into sand under the barrels, which was then blown by strong winds as far away as Denver. Radioactivity of the sand at the spill site was measured at 30 million disintegrations per minute per gram (about 220 ppm plutonium in the sand by weight), 15 million times higher than the state standard of

two disintegrations per minute.*[17]

Dr. Carl Johnson, Jefferson County health director from 1973 to 1981, directed numerous studies on contamination levels and health risks the plant posed to public health. Based on his conclusions, Johnson opposed housing development near Rocky Flats. He was fired. Later studies confirmed many of his findings.*[18]

On April 28, 1979, a few weeks after the Three Mile Island accident, a crowd of close to 15,000 protesters assembled at a nearby site.*[19]*[20] Singers Jackson Browne and Bonnie Raitt took the stage along with various speakers. The following day, 286 protesters including Daniel Ellsberg were arrested for civil disobedience/trespassing on the Rocky Flats facility.*[21]*[22]

11.1.4   1980s

See also: Rocky Flats Truth Force

Dark Circle is a 1982 American documentary film that focuses on the Rocky Flats Plant and its plutonium contamination of the area's environment. The film won the Grand Prize for documentary at the Sundance Film Festival and received a national Emmy Award for "Outstanding individual achievement in news and documentary" .*[23]

Rocky Flats became a focus of protest by peace activists throughout the 1980s. In 1983, a demonstration was organized that brought together 17,000 people who joined hands in an encirclement around the 17-mile (27 km) perimeter of the plant.*[24]*[25]

A perimeter security zone was installed around the facility in 1983 and was upgraded with remote detection abilities in 1985. Also in 1983, the first radioactive waste was processed through the aqueous recovery system, creating a plutonium button.

A celebration of 250,000 continuous safe hours by the employees at Rocky Flats happened in 1985. The same year, Rockwell received Industrial Research Magazine's IR-100 award for a process to remove actinide contamination from wastewater at the plant. The next year, the site received a National Safety Council Award of Honor for outstanding safety performance.

On August 10, 1987, 320 demonstrators were arrested after they tried to force a one-day shutdown of the Rocky Flats nuclear weapons plant.*[26]

In 1988, a Department of Energy (DOE) safety evaluation resulted in a report that was critical of safety measures at the plant. The EPA fined the plant for polychlorinated biphenyl (PCB) leaks from a transformer. A solid waste form, called pondcrete, was found not to have cured properly and was leaking from containers. A boxcar of transuranic waste from the site was refused entry into Idaho and returned to the plant. Plans to potentially close the plant were released.

In 1989 an employee left a faucet running, resulting in chromic acid being released into the sanitary water system. The Colorado Department of Health and the EPA both posted full-time personnel at the plant to monitor safety. Plutonium production was suspended due to safety violations.

In August 1989, an estimated 3,500 people turned out for a demonstration at Rocky Flats.[*][25]

Investigation by FBI and EPA

Insiders at the plant started to covertly inform the Environmental Protection Agency (EPA) and the Federal Bureau of Investigation (FBI) about the unsafe conditions in 1987. Late that year the FBI commenced clandestine flights of light aircraft over the area and noticed that the incinerator was apparently being used late into the night. After several months of collecting evidence both from workers and via direct measurement, the FBI informed the DOE on June 6, 1989 that they wanted to meet to discuss a potential terrorist threat.

Dubbed "Operation Desert Glow", the raid, sponsored by the United States Department of Justice (DOJ), began at 9 a.m. on June 6, 1989.[*][27] After the FBI got past the DOE's heavily armed, authorized to shoot-to-kill security—whose armament included surface-to-air missiles—they served the search warrant to Dominick Sanchini,[*][28] Rockwell International's manager of Rocky Flats. (Ironically, Sanchini died the next year in Boulder of cancer.[*][29][*][30]). The FBI discovered numerous violations of federal anti-pollution laws, including limited[*][16] contamination of water and soil. In 1992, Rockwell International was charged with environmental crimes including violations of the Resource Conservation and Recovery Act (RCRA), and the Clean Water Act. Rockwell pleaded guilty and paid an $18.5 million fine. This was the largest fine for an environmental crime to that date.

After the June 1989 FBI raid, federal authorities used the subsequent grand jury investigation to gather evidence of wrongdoing and then sealed the record. In October 2006, DOE announced completion of the Rocky Flats cleanup without this information being available.[*][31]

The FBI raid led to the formation of Colorado's first special grand jury in 1989, the juried testimony of 110 witnesses, reviews of 2,000 exhibits, and ultimately a 1992 plea agreement in which Rockwell admitted to 10 federal environmental crimes and agreed to pay $18.5 million in fines out of its own funds. This amount was less than the com-

pany had been paid in bonuses for running the plant as determined by the Government Accounting Office (GAO), and yet was also by far the highest hazardous-waste fine ever; four times larger than the previous record.[*][32] Due to indemnification of nuclear contractors, without some form of settlement being arrived at between the U.S. Justice Department and Rockwell, the cost of paying any civil penalties would ultimately have been borne by U.S. taxpayers. While any criminal penalties allotted to Rockwell would not have been covered, for its part Rockwell claimed that the Department of Energy had specifically exempted them from most environmental laws, including hazardous waste.[*][33][*][34][*][35][*][36][*][37][*][38][*][39]

Regardless, and as forewarned by the prosecuting U.S. Attorney, Ken Fimberg/Scott,[*][40] the Department of Justice's stated findings and plea agreement with Rockwell were heavily contested by its own, 23-member special grand jury. Press leaks on both sides—members of the DOJ and the grand jury—occurred in violation of secrecy regarding grand jury information, a federal crime punishable by a prison sentence.[*][41] The public contest led to U.S. Congressional oversight committee hearings chaired by Congressman Howard Wolpe, which issued subpoenas to DOJ principals despite several instances of DOJ's refusal to comply. The hearings, whose findings include that the Justice Department had "bargained away the truth",[*][42] ultimately still did not fully reveal to the public the special grand jury's report, which remains sealed by the DOJ courts.[*][37][*][43]

The special grand jury report[*][44] was nonetheless leaked to *Westword*. According to its subsequent publications, the Rocky Flats special grand jury had compiled indictments charging three DOE officials and five Rockwell employees with environmental crimes. The grand jury also wrote a report, intended for the public's consumption per their charter, lambasting the conduct of DOE and Rocky Flats contractors for "engaging in a continuing campaign of distraction, deception and dishonesty" and noted that Rocky Flats, for many years, had discharged pollutants, hazardous materials and radioactive matter into nearby creeks and Broomfield's and Westminster's water supplies.[*][45]

The DOE itself, in a study released in December of the year prior to the FBI raid, had called Rocky Flats' ground water the single greatest environmental hazard at any of its nuclear facilities.[*][46]

Withheld records

Court records from the grand jury proceeding on Rocky Flats have been sealed for a number of years. The Federal Rules of Criminal Procedure, which govern federal grand jury proceedings, explicitly require grand jury proceedings to be kept secret unless otherwise provided by the Rules.*[47] Rocky Flats' secret grand jury proceedings were not unique.

However, some activists dispute the reasons for records confidentiality:*[48] Dr. Leroy Brown, a Boulder scientist, retired FBI Special Agent Jon Lipsky,*[45] who led the FBI's raid of the Rocky Flats plant to investigate illegal plutonium burning and other environmental crimes, and Wes McKinley, who was the foreman of the grand jury investigation into the operations at Rocky Flats and is today a Colorado State Representative.*[33]*[49]*[50]

Former grand jury foreman McKinley chronicles his experiences in the 2004 book he co-authored with attorney Caron Balkany, *The Ambushed Grand Jury*, which begins with an open letter to the U.S. Congress from Special Agent Lipsky:

> I am an FBI agent. My superiors have ordered me to lie about a criminal investigation I headed in 1989. We were investigating the US Department of Energy, but the US Justice Department covered up the truth.
>
> I have refused to follow the orders to lie about what really happened during that criminal investigation at Rocky Flats Nuclear Weapons Plant. Instead, I have told the author of this book the truth. Her promise to me if I told her what really happened was that she would put it in a book to tell Congress and the American people.
>
> Some dangerous decisions are now being made based on that government cover-up. Please read this book. I believe you know what needs to happen.*[51]

11.1.5   1990s

Rockwell International was replaced by EG&G as primary contractor for the Rocky Flats plant.*[52] EG&G began an aggressive work safety and cleanup plan for the site that included construction of a system to remove contamination from the groundwater of the site. The *Sierra Club vs. Rockwell* case was decided in favor of the Sierra Club. The ruling directed Rocky Flats to manage plutonium residues as hazardous waste.

In 1991, an interagency agreement between DOE, the Colorado Department of Health, and the EPA outlined multiyear schedules for environmental restoration studies and remediation activities. DOE released a report that advocated downsizing the plant's production into a more streamlined facility. Due to the fall of the Soviet Union, production of most of the systems at Rocky Flats was no longer needed, leaving only the W88 warhead triggers.

In 1992, due to an order by President G.H.W. Bush, production of submarine-based missiles using the W88 trigger was discontinued, leading to the layoff of 4,500 employees at the plant; 4,000 others were retained for long-term cleanup of the facility. The Rocky Flats Plant Transition Plan outlined the environmental restoration process. The DOE announced that 61 pounds (28 kg) of plutonium lined the exhaust ductwork in six buildings on the site.

Starting in 1993, weapons-grade plutonium began to be shipped to the Oak Ridge National Laboratory, Los Alamos National Laboratory, and the Savannah River Site.

In 1994 the site was renamed the Rocky Flats Environmental Technology Site, reflecting the changed nature of the site from weapon production to environmental cleanup and restoration. The cleanup effort was contracted to the Kaiser-Hill Company, which proposed the release of 4,100 acres (6.4 sq mi; 16.6 km^2) of the buffer zone for public access.

In 1998, the Colorado Department of Public Health and Environment's Cancer Registry conducted an independent study of cancer rates in areas around the Rocky Flats Site. Data showed no pattern of increased cancers tied to Rocky Flats.*[53]

Throughout the remainder of the 1990s and into the 2000s, cleanup of contaminated sites and dismantling of contaminated buildings continued with the waste materials being shipped to the Nevada Test Site, the Waste Isolation Pilot Plant in New Mexico, and the Envirocare company facility in Utah,*[8] which is now EnergySolutions.

11.1.6   2000s

In 2001, Congress passed the **Rocky Flats National Wildlife Refuge Act**.*[54] In July 2007, the U.S. Department of Energy transferred nearly 4,000 acres (6 sq mi; 16 km^2) of land on the Rocky Flats site to the U.S. Fish and Wildlife Service to establish the Rocky Flats National Wildlife Refuge.*[55] Surveys of the site reveal 630 species of vascular plants, 76% of which are native.*[56] Herds of elk are commonly seen on the site. However, the DOE retained the central area of the site, the Central Operable Unit.

The last contaminated building was removed and the last weapons-grade plutonium was shipped out in 2003, ending the cleanup based on a modified cleanup agreement. The modified agreement required a higher level of cleanup in the first 3 feet (0.9 m) of soil in exchange for not having to remove any contamination below that point unless it posed a chance of migrating to the surface or contaminating the groundwater.*[57] About half of the 800 buildings previously existing on the site had been dismantled by early December 2004.

The site is contaminated with residual plutonium due to several industrial fires that occurred on the site and other inadvertent releases caused by wind at a waste storage area. The other major contaminant is carbon tetrachloride (CCl_4). Both of these substances affected areas adjacent to the site. In addition, there were small releases of beryllium and tritium, as well as dioxin from incineration.[58][59]

Cleanup was declared complete on October 13, 2005.[9] About 1,300 acres (2 sq mi; 5 km^2) of the original site, the former industrial area, remains under U.S. DOE Office of Legacy Management control for ongoing environmental monitoring and remediation. On March 14, 2007, DOE, EPA, and CDPHE entered into the Rocky Flats Legacy Management Agreement (RFLMA). The agreement establishes the regulatory framework for implementing the final remedy for the Rocky Flats site and ensuring the protection of human health and the environment.

In September 2010, after a 20-year legal battle, the Tenth Circuit Court of Appeals reversed a $926 million award in a class-action lawsuit against Dow Chemical and Rockwell International.[60] The three-judge panel said that the jury reached its decision on faulty instructions that incorrectly stated the law. The appeals court tossed the jury verdict and sent the case back to the District Court. According to the Appellate Court, the owners of 12,000 properties in the class-action area had not proved their properties were damaged or that they suffered bodily injury from plutonium that blew onto their properties.[60][61]

In response to historic and ongoing reports of health issues by people who live and lived near Rocky Flats, an online health survey was launched in May 2016 by Metropolitan State University, Rocky Flats Downwinders,[62] and other local universities and health agencies to test thousands of Coloradans who lived east of the Rocky Flats plant while it was operational.[63]

On May 19, 2016, a $375 million settlement was reached over claims by more than 15,000 nearby homeowners that plutonium releases from the plant risked their health and devalued their property. This settlement ended a 26-year legal battle between residents and the two corporations that ran the Rocky Flats Plant, Dow Chemical and Rockwell International, for the Department of Energy.[64]

11.2 See also

- Atomic Energy Act

- Cold War

- Kristen Iversen, author of *Full Body Burden: Growing Up in the Nuclear Shadow of Rocky Flats*

- *Making a Real Killing: Rocky Flats and the Nuclear West*

- Manhattan Project

- Price-Anderson Act

- Timeline of nuclear weapons development

11.3 Notes

[1] National Park Service (2010-07-09). "National Register Information System". *National Register of Historic Places*. National Park Service.

[2] Colorado Department of Public Health and Environment, Rocky Flats Site, https://www.colorado.gov/pacific/cdphe/rocky-flats-faq

[3] "Rocky Flats cover-ups alleged". *Pittsburgh Press*. wire services. June 10, 1989. p. A4.

[4] "Ex-Rocky Flats operator pleads guilty; agrees to $18.5 million fine". *Prescott Courier*. (Arizona). Associated Press. March 27, 1992. p. 7A.

[5] https://www.washingtonpost.com/archive/politics/1992/03/27/rockwell-accepts-185-million-fine/8da37f73-8580-4136-b6a1-5b4a072badd3/

[6] "Agencies sign Rocky Flats cleanup pact". *Eugene Register-Guard*. (Oregon). Associated Press. January 23, 1991. p. 8A.

[7] "Watkins says Rocky Flats wont' reopen". *Spokesman-Review*. (Spokane, Washington). Associated Press. January 30, 1992. p. A3.

[8] "Radioactive soil heads for Utah". *Deseret News*. (Salt Lake City, Utah). Scripps Howard News Service. March 29, 1995. p. B4.

[9] Elliott, Dan (October 14, 2005). "Nuclear cleanup done at Rocky Flats, firms says". *Spokesman-Review*. (Spokane, Washington). Associated Press. p. A4.

[10] Colorado Department of Public Health and Environment, Rocky Flats, Site History, https://www.colorado.gov/pacific/cdphe/rocky-flats-site-history

[11] U.S. Department of Energy, Rocky Flats, http://www.lm.doe.gov/Rocky_Flats/Sites.aspx

[12] U.S. Fish and Wildlife Service, Rocky Flats Wildlife Refuge, https://www.fws.gov/refuge/rocky_flats/

[13] U.S. Environmental Protection Agency, https://cumulis.epa.gov/supercpad/cursites/dsp_ssppSiteData1.cfm?id=0800360#Why

[14] U.S. Department of Energy, Legacy Management, Update on Rocky Flats Five-Year Review, https://energy.gov/lm/articles/update-rocky-flats-cercla-five-year-review

[15] "New Atom Plant to be Built". *The Pittsburgh Press*. March 23, 1951. Retrieved 2 July 2013.

[16] Hobbs, Farrel (2010). *An Insider's View of Rocky Flats: Urban Myths Debunked*. CreateSpace. ISBN 978-1460911471.

[17] Bob Reuteman, "Vindication at last for all who feared Rocky Flats", *The Rocky Mountain News*, February 18, 2006.

[18] Kristen Iversen (March 10, 2012). "Nuclear Fallout". *New York Times*.

[19] Nonviolent Social Movements p. 295.

[20] Headline: Rocky Flats Nuclear Plant / Protest

[21] "280 arrested in protest at nuclear weapons plant". *Sumter Daily Item*. (South Carolina). Associated Press. April 30, 1979. p. 6A.

[22] "Ellsberg fined for protest". *Wilmington Morning Star*. (North Carolina). wire services. July 15, 1979.

[23] *Dark Circle*, DVD release date March 27, 2007, Directors: Judy Irving, Chris Beaver, Ruth Landy. ISBN 0-7670-9304-6.

[24] Headline: Colorado / Anti-Nuclear Demonstration

[25] Activists fail to encircle Rocky Flats/ Too few join hands in symbolic protest

[26] "570 Arrested in A-Bomb Protests". *The New York Times*. August 10, 1987.

[27] Siegel, Barry (August 8, 1993). "Showdown at Rocky Flats : When Federal Agents Take On a Government Nuclear-Bomb Plant, Lines of Law and Politics Blur, and Moral Responsibility Is Tested". *Los Angeles Times*.

[28] "Dominick J. Sanchini". *Memorial Tributes: National Academy of Engineering,*. **6**. The National Academies Press. 1993. doi:10.17226/2231. Retrieved 2016-01-31.

[29] Siegel, Barry (August 8, 1993). "Showdown at Rocky Flats : When Federal Agents Take On a Government Nuclear-Bomb Plant, Lines of Law and Politics Blur, and Moral Responsibility Is Tested". *Los Angeles Times*.

[30] "Dominick Sanchini". Orlando Sentinel. 1990-11-22. Retrieved 2016-01-31.

[31] Moore, LeRoy (2012). "Democracy and Public Health at Rocky Flats: The Examples of Edward A. Martell and Carl J. Johnson" (PDF). In Quigley, Dianne; Lowman, Amy; Wing, Steve. *Tortured Science: Health Studies, Ethics, and Nuclear Weapons in the United States*. Amityville, New York: Baywood Publishing Company. pp. 69–98. Archived (PDF) from the original on March 31, 2012. Retrieved 2011-09-01.

[32] Siegel, Barry (August 15, 1993). "Showdown at Rocky Flats : The Justice Department had negotiated a Rocky Flats settlement, but the grand jury could not keep quiet about what happened there". *Los Angeles Times*.

[33] Hardesty, Greg (March 29, 2006). "Retired FBI agent helped close nuclear-weapons site". *The Orange County Register*. Retrieved September 17, 2011.

[34] Calhoun, Patricia (2004-08-19). "True Lies". Westword. Retrieved 2016-01-31.

[35] Schneider, Keith (March 27, 1992). "U.S. Shares Blame in Abuses at A-Plant". *The New York Times*.

[36] Siegel, Barry (August 8, 1993). "Showdown at Rocky Flats : When Federal Agents Take On a Government Nuclear-Bomb Plant, Lines of Law and Politics Blur, and Moral Responsibility Is Tested". *Los Angeles Times*.

[37] Siegel, Barry (August 15, 1993). "Showdown at Rocky Flats : The Justice Department had negotiated a Rocky Flats settlement, but the grand jury could not keep quiet about what happened there". *Los Angeles Times*.

[38] Wald, Matthew L. (September 23, 1989). "Rockwell Is Giving Up Rocky Flats Plant". *The New York Times*.

[39] "U.S. Shares Blame in Abuses at A-Plant". *The New York Times*. March 27, 1992.

[40] Prosecuting U.S. attorney Fimberg changed his last name to Scott after the Rocky Flats deliberations were finalized; see *The Ambushed Grand Jury*, page 118.

[41] Rule 6(e)

[42] *The Ambushed Grand Jury*, Chapter 6, page 98.

[43] *The Ambushed Grand Jury*, Chapter 6, Note 54.

[44] The special grand jury report

[45] Warner, Joel (2005-01-06). "Servant of the people". *Boulder Weekly*. Archived from the original on August 27, 2012. Retrieved 2011-10-10.

[46] Siegel, Barry (August 15, 1993). "Showdown at Rocky Flats : The Justice Department had negotiated a Rocky Flats settlement, but the grand jury could not keep quiet about what happened there". *Los Angeles Times*.

[47] See Federal Rules of Criminal Procedure, Rule 6, https://www.law.cornell.edu/rules/frcrmp/rule_6#

[48] 5280 Magazine, Rogue Agent, http://www.5280.com/news/magazine/2016/04/rogue-agent?page=full

[49] "Rocky Flats Nuclear Site Too Hot for Public Access, Citizens Warn". Environment News Service. August 5, 2010. Retrieved September 17, 2011.

[50] Wald, Matthew L. (March 13, 2004). "Book Says U.S. Aides Lied In Nuclear-Arms Plant Case". *The New York Times*. Retrieved September 17, 2011.

[51] McKinley, Wes; Balkany, Caron (2004). *The Ambushed Grand Jury: How the Justice Department Covered up Government Nuclear Crimes and How We Caught Them Red Handed*. New York: Apex Press. ISBN 978-1-891843-28-0.

[52] "Rockwell to abandon Rocky Flats plant". *Bend Bulletin*. (Oregon). Associated Press. p. A-14.

[53] Colorado Department of Public Health and Environment, Cancer Registry, "Ratios of Cancer Incidence in Ten Areas Around Rocky Flats, Colorado" (1998), https://www.colorado.gov/pacific/cdphe/cdphe-rocky-flats-cancer-study

[54] U.S. Fish and Wildlife Service (January 31, 2007). "Final Rocky Flats Sign Text" (PDF). Retrieved April 19, 2007.

[55] "Rocky Flats National Wildlife Refuge". U.S. Fish and Wildlife Service. Retrieved 2016-02-10.

[56] Nelson, J. K. (2010). Vascular flora of the Rocky Flats Area, Jefferson County, Colorado, USA. *Phytologia* 92:2 121.

[57] "Glovebox removal heralds new Flats era". The Denver Post. December 9, 2004. Archived from the original on January 16, 2005.

[58] "Key questions addressed by the research" (PDF). *Rocky Flats, Historical Public Exposures Studies*. Colorado Department of Public Health and Environment. Archived from the original on May 17, 2006. Retrieved 2011-09-03.

[59] "Contaminants Released to Surface Water from Rocky Flats" (PDF). Department of Public Health and Environment. Archived (PDF) from the original on February 10, 2016. Retrieved September 3, 2011.

[60] "Appeals court tosses jury award in Rocky Flats case". The Denver Post. September 4, 2010.

[61] *Cook, et al. v. Rockwell International Corp., et al.*, Nos. 08-1224, 08-1226 and 08-1239 (U.S. Court of Appeals, 10th Circuit September 3, 2010).

[62] "ROCKY FLATS DOWNWINDERS | A community organization advocating for residents living downwind from Rocky Flats.". *rockyflatsdownwinders.com*. Retrieved 2016-05-22.

[63] "Health survey of Rocky Flats neighbors to launch Tuesday". *www.denverpost.com*. Retrieved 2016-05-22.

[64] "Deal Reached Between Homeowners, Rocky Flats Operators". Retrieved 2016-05-22.

11.4 External links

- U.S. Department of Energy, Legacy Management, Rocky Flats

- U.S. Environmental Protection Agency, Rocky Flats

- U.S. Fish & Wildlife Service, Rocky Flats Wildlife Refuge

- Colorado Department of Public Health & Environment (CDPHE), Rocky Flats

- Kristen Iversen, author of *Full Body Burden*

- Rocky Flats History

- Bomb Production at Rocky Flats: Death Downwind

- Survey number HAER CO-83

- A Technically Useful History of the Critical Mass Laboratory at Rocky Flats

- Photography: a year of disobedience

- Rocky Flats before (1995) and after (2005) cleanup.

- An Insider's View of Rocky Flats: Urban Myths Debunked By Farrel D. Hobbs

- Annotated bibliography on Rocky Flats from the Alsos Digital Library for Nuclear Issues

Chapter 12

Radioactive contamination from the Rocky Flats Plant

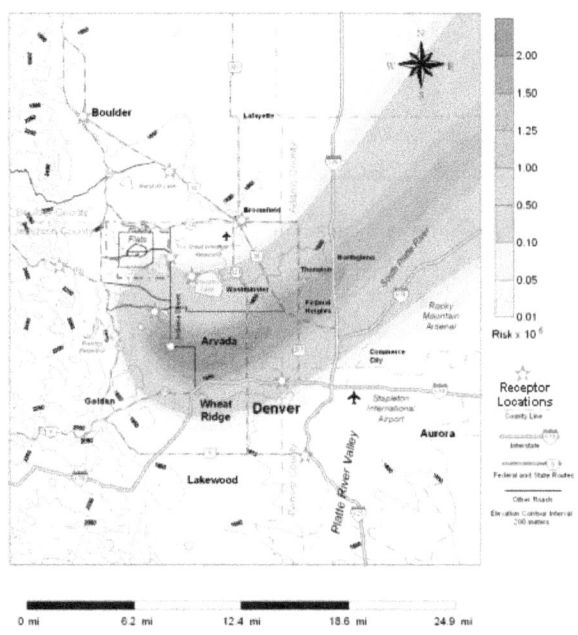

One of four example estimates of the plutonium (Pu-239) plume from the 1957 fire at the Rocky Flats nuclear weapons plant.

The Rocky Flats Plant, a former U.S. nuclear weapons production facility about 15 miles northwest of Denver, Colorado, caused radioactive contamination, primarily plutonium, americium, and uranium, within and outside its boundaries.*[1]*[2] The facility was dismantled and removed, and the central industrial area of the property became a Superfund site, surrounded by the newly created Rocky Flats National Wildlife Refuge.*[3]

The contamination primarily resulted from two major plutonium fires in 1957 and 1969 (plutonium is pyrophoric and shavings can spontaneously combust) and from wind-blown plutonium that leaked from barrels of radioactive waste. Much lower concentrations of radioactive isotopes were released throughout the operational life of the plant from 1952 to 1992, from smaller accidents and from nor-

mal operational releases of plutonium particles too small to be filtered. Prevailing winds from the plant swept airborne contamination south and east, into populated areas northwest of Denver.

The contamination of the Denver area by plutonium from the fires and other sources was not publicly reported until the 1970s. According to a 1972 study coauthored by Edward Martell, "In the more densely populated areas of Denver, the Pu contamination level in surface soils is several times fallout", and the plutonium contamination "just east of the Rocky Flats plant ranges up to hundreds of times that from nuclear tests." *[4] As noted by Carl Johnson in Ambio, "Exposures of a large population in the Denver area to plutonium and other radionuclides in the exhaust plumes from the plant date back to 1953." *[5]

In the 1990s, a series of Historical Public Exposure Studies were conducted to assess past releases and public exposures.*[6] For example, the figure at right illustrates the calculated lifetime cancer risk to a laborer from the 1957 Rocky Flats fire. The key shows the cancer risk due to exposure during the 1957 event per million persons. This figure means that an outdoor laborer in the reddest part of Arvada would have roughly a two-in-a-million risk of contracting cancer (0.000002) from being outside during the 1957 fire. The original, complete figure - as well as many others showing the risks of past exposures - can be viewed in the Summary of Findings from the Historical Public Exposure Studies.*[7]

Weapons production at the plant was halted after a combined FBI and EPA raid in 1989. Due to the end of the Cold War, changes in nuclear weapons policy, and years of protests, the Rocky Flats Plant was shut down, with its buildings demolished and completely removed from the site. The site's mission then changed to cleanup. The Rocky Flats Plant was declared a CERCLASuperfund site in 1989 and began its transformation to a cleanup site in February 1992. Removal of the plant and surface contamination was largely completed in the late 1990s and early 2000s, with federal and state agency support and stakeholder input. The cleanup effort decommissioned and demolished over 800 structures; removed over 21 tons of weapons-grade material; removed over 1.3 million cubic meters of waste; and, treated more than 16 million gallons of water.[8] Four groundwater treatment systems were also constructed. The site achieved regulatory closure in 2006.

Today, the site consists of two areas. The "Central Operable Unit" encompasses the former industrial/plant area of the site. This area is still a CERCLA "Superfund" site, retained and managed by the U.S. Department of Energy.[9] Environmental monitoring and sampling are conducted by the Department of Energy here on a regular basis. Remediation efforts are also ongoing in the Central Operable Unit. Four groundwater treatment systems are currently installed and operating in the Central Operable Unit. Every five years, the DOE, EPA, and Colorado Department of Public Health and Environment review environmental data to assess whether the remedy is functioning as intended.[10] The last five-year review concluded the site remedy is effective. However, this area remains off-limits to the public due to residual contamination, and to protect site treatment systems and the integrity of remedial efforts. The public comment period for the next five-year review is open until December 31, 2016.

The outer "Peripheral Operable Unit" is now the Rocky Flats National Wildlife Refuge. This area was the site's former security buffer zone and did not require remediation. In 2001, the U.S. Congress passed the Rocky Flats National Wildlife Refuge Act, dedicating this buffer zone to conservation. Accordingly, the DOE transferred land ownership of this area to the U.S. Fish and Wildlife Service, which currently exercises jurisdiction over the Refuge. It is anticipated the Refuge will open to the public in 2018.

While the U.S. Department of Energy continues to monitor and collect samples from the Central Operable Unit, a some groups and citizens remain concerned about the extent and long-term public health consequences of the contamination.[11][12][13][14] Estimates of the public health risk caused by the contamination vary. Activist groups are concerned about the potential risks posed by residual contamination, which exists on-Site.[15] However, the Comprehensive Risk Assessment for the site found the post-cleanup risks posed by the site to be very low and within EPA guidelines. A 1998 independent study by the Colorado Department of Public Health and Environment on cancer rates in communities surrounding Rocky Flats also found no pattern of increased cancers tied to Rocky Flats.[16]

12.1 Background

Main article: Rocky Flats Plant
The Rocky Flats Plant was located south of Boulder, Col-

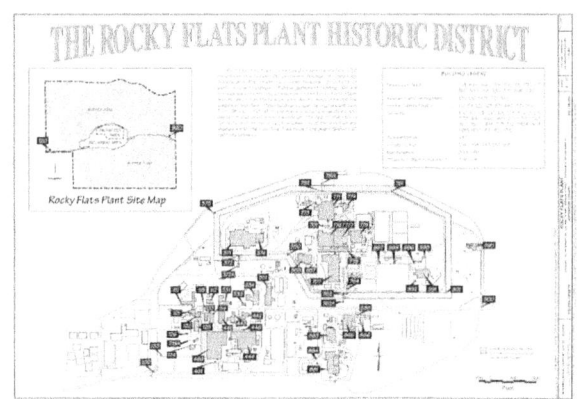

A map of the Rocky Flats Plant prior to its decommissioning. All buildings have since been demolished from the site.

orado and northwest of Denver. Originally under management of the Dow Chemical Company, management was transferred to Rockwell in 1975.[17]:13 Initially having an area of 4 sq mi (10 km^2), the site was expanded with a 4,600 acres (19 km^2) buffer zone in 1972.[17]:12

Construction of the first buildings was started on the site on July 10, 1951. Production of parts for nuclear weapons began in 1953. At the time, the precise nature of the work at Rocky Flats was a closely guarded secret. The plant produced fission cores for nuclear weapons, used to "ignite" fusion and fissionable fuel in all modern nuclear weapons.[18] Fission cores resemble miniaturized versions of the Fat Man nuclear bomb detonated above Nagasaki. They are often referred to as "triggers" in official and news documents to obfuscate their function.[19][20]:190 For much of its operational lifetime, Rocky Flats was the sole mass-producer of plutonium components for America's nuclear stockpile.[21]

Management of the site passed to EG&G in 1990, which did not reapply for the contract in 1994.[22] Management of the site then passed to the Kaiser-Hill Company as of July 1, 1995.[23] The Department of Energy now manages the central portion of the site, where production buildings were once located, while the Fish and Wildlife Service has taken over management of the Peripheral Outer Unit.[24]

12.2 Sources of contamination

Most of the radioactive contamination from Rocky Flats came from three sources: a catastrophic fire in 1957,[11] leaking barrels in an outdoor storage area in 1964-1968, and another less severe fire in 1969.[25] Plutonium, used to construct the weapons' fissile components, can spontaneously combust at room temperatures in air. Additional sources of actinide contamination include inadequate pondcrete vitrification attempts and routine releases during the decades of plant operations.

12.2.1   1957 fire

The glove box where the 1957 fire started.

HEPA filter banks meant to remove microscopic particles of plutonium from the glove box exhaust streams were destroyed by the fire, allowing radioactive smoke to escape the building.

On the evening of September 11, 1957, plutonium shavings in a glove box located in building 771 (the Plutonium Recovery and Fabrication Facility) spontaneously ignited. The fire spread to the flammable glove box materials, including plexiglas windows and rubber gloves. The fire rapidly spread through the interconnected glove boxes and ignited the large bank of High-efficiency particulate air (HEPA) filters located in a plenum downstream. Within minutes the first filters had burned out, allowing plutonium particles to escape from the building exhaust stacks. The building exhaust fans stopped operating due to fire damage at 10:40 PM, which ended the majority of the plutonium release. Fire fighters initially used carbon dioxide fire extinguishers because water can act as a moderator and cause plutonium to go critical. They resorted to water hoses when the dry fire extinguishers proved ineffective.[26]:27

The 1957 fire released 11-36 Ci (160–510 grams or 0.35–1.12 pounds) of plutonium, much of which contaminated off-site areas as microscopic particles entrained in smoke from the fire.[27]:22–29 Isopleth diagrams from studies show portions of the city of Denver included in the area where surface sampling detected plutonium.[2]:87–89 The fact that the fire had resulted in significant plutonium contamination of surrounding populated areas remained secret. News reports at the time reported, per the

Atomic Energy Commission's briefing, that there was slight risk of light contamination and that no fire fighters had been contaminated.[28][29] No abnormal radioactivity was reported by the Colorado Public Health Service.[30]

12.2.2   Pad 903 leakage

Plutonium milling operations produced large quantities of toxic cutting fluid contaminated with particles of plutonium and uranium. Thousands of 55-gallon drums of the waste were stored outside in an unprotected earthen area called the 903 pad storage area,[26]:28 where they corroded and leaked radionuclides over years into the soil and water.[31][32] An estimated 5,000 gallons of plutonium-contaminated oil leached into the soil between 1964 and 1967.[33] Portions of this waste, mixed with dust that composed Pad 903, became airborne in the heavy winds of the Front Range and contaminated offsite areas to the south and east.[25][34][35][36][37]

Leaking storage barrels at Pad 903 released 1.4-15 Ci (19–208 grams or 0.042–0.459 pounds) of plutonium as airborne dust during the storage and subsequent attempts at cleanup.[27]:29 Much more remains interred under the Pad 903 area, which has been paved over with asphalt.[32]

Corroded waste storage barrel at Pad 903.

A room in building 776 damaged by the 1969 fire.

12.2.3 1969 fire

Another major fire occurred on May 11, 1969 in building 776/777 (the Plutonium Processing Facility), again starting due to spontaneous combustion of plutonium shavings in a glove box. Fire fighters again resorted to fighting the fire with water after dry extinguishers proved ineffective. Despite recommendations after the 1957 fire, suppression systems were not built into the glove boxes.[38]

While the fire bore marked similarities to the 1957 fire,[38] the level of contamination was less severe because the HEPA filters in the exhaust system did not burn through[27]:25 (After the 1957 fire, the filter material was changed from cellulose to nonflammable fiberglass).[38] Had the filters failed or the roof (which sustained heavy fire damage) been breached, the release could have been more severe than the 1957 fire. About 1,400 kilograms (3,100 lb) of plutonium was in the storage area where the fire occurred, and about 3,400 kilograms (7,500 lb) total plutonium was in building 776/777.[38]

The 1969 fire released 13-62 mCi (140–900 milligrams or 0.00031–0.00198 pounds) of plutonium,[27]:25 about 1000th as much as was released in the 1957 fire. The

1969 fire, however, led local health officials to perform independent tests of the area surrounding Rocky Flats to determine the extent of the contamination. This resulted in the first releases of information to the public that populated areas southeast of Rocky Flats had been contaminated.[2]:3[25]

12.2.4 Other sources

Rockwell workers mixed hazardous and other wastes with concrete to create one-ton solid blocks called pondcrete. These were stored in the open under tarps on asphalt pads. The pondcrete turned out to be weak storage, an outcome that had been predicted by Rockwell's own engineers.[39] Relatively unprotected from the elements, the blocks began to leak and sag.[40] Nitrates, cadmium and low-level radioactive waste began to leach into the ground and run downhill toward Walnut Creek and Woman Creek.[33]

Most of the plutonium from Rocky Flats was oxidized plutonium, which does not readily dissolve in water. A large portion of the plutonium released into the creeks sank to the bottom and is now found in the streambeds of Walnut and Woman Creeks, and on the bottom of local public reservoirs just outside Rocky Flats: Great Western Reservoir, (no longer used for city of Broomfield drinking water consumption as of 1997 but still used for irrigation),*[41] and Standley Lake, a drinking water supply for the cities of Westminster, Thornton, Northglenn and some residents of Federal Heights.*[42] As one of several forms of remediation and once the extent of the lapses at Rocky Flats became public knowledge, several streams that were formed by drainage through the contaminated areas of the Rocky Flats Plant were diverted such that they would no longer flow directly into some of the local reservoirs, such as Mower Reservoir and Standley Lake.*[43] Also, a surface water control system was built to allow runoff from contaminated creeks to collect in holding ponds and thus reduce or prevent direct runoff into Standley Lake.*[44] Proposals to remove or breach some of these dams to reduce the cost of maintenance have been protested by the cities downstream.*[45]

12.3   Reporting of contamination

No radioactivity warning, advisement or cleanup was provided to the public in the 1957 fire, the worse of the two major fires. At the time of the 1957 fire, AEC officials told the *Denver Post* that the fire "resulted in no spread of radioactive contamination of any consequence." *[46] The public was not informed of substantial contamination from the 1957 plutonium fire until after the highly visible 1969 fire,*[2]*:3 when civilian monitoring teams confronted government officials with measurements made outside the plant of radioactive contamination suspected to be from the 1969 fire, which consumed hundreds of pounds of plutonium (850 kg).*[38]

The 1969 fire raised public awareness of potential hazards posed by the plant and led to years of increasing citizen protests and demands for plant closure. Releases from previous years had not been reported publicly prior to the fire;*[47] airborne-become-groundborne radioactive contamination extending well beyond the Rocky Flats plant was not publicly reported until the 1970s.*[2]*[5]*:page 177 and table 3

In 2002, the U.S. Fish and Wildlife Service surveyed tissues harvested from deer that lived at Rocky Flats for plutonium and other actinides. Isotopes of plutonium, americium, and uranium were detected, with the highest measured activity being 0.0125 pCi/g (2360 seconds per disintegration) for uranium-233 or uranium-234. The increased cancer risk, as reported by the study, to an individual who ate 28 kilograms (62 lb) of Rocky Flats deer meat per year over a 70-year lifetime was estimated to be as high as 1 in 210,000. This is near the conservative end of the EPA's acceptable risk range.*[48]

12.4   Contamination and health studies

See also: Plutonium in the environment

Plutonium-239 and 240 emit ionizing radiation in the form of alpha particles. Inhalation is the primary pathway by which plutonium enters the body, though plutonium can also enter the body through a wound.*[49] Once inhaled, plutonium increases the risk of lung cancer, liver cancer, bone cancer, and leukemia.*[27] Once absorbed into the body, the biological half life of plutonium is about 200 years.*[50]

Following the public 1969 fire, surveys were taken of the land outside the boundaries of Rocky Flats to quantify the amount of plutonium contamination. Researchers noted that plutonium contamination from the plant was present, but did not match the wind conditions of the 1969 fire. The 1957 fire and leaking barrels on Pad 903 have since been confirmed to be the main sources of plutonium contamination. Authors Krey and Hardy estimated the total quantity of plutonium contamination outside of Rocky Flats's boundaries to be 2.6 Ci (36 grams or 0.079 pounds),*[51] while Poet and Martell estimated the value to be 6.6 Ci (92 grams or 0.203 pounds). The study also noted that plutonium levels just outside the boundaries of the plant were hundreds of times higher than the background level caused by global fallout from nuclear testing, and that contamination to the north of the plant was probably caused by normal operations rather than accidental releases.*[4]

From September 1947 to April 1969 there were 5 or more accidental surface water releases of tritium. Tritium, a radioactive element which was found in scrap material from Rocky Flats, was therefore directed in to the Great Western Reservoir. This was uncovered in 1973 and following this, urine samples were taken from people living or working near Broomfield who could have drank water from the reservoir. The findings of the samples showed that those who were exposed to contaminated water had tritium concentrations near seven times higher than normal (4,300 picocuries per liter versus 600 picocuries per liter). However, when the same group underwent urine sampling three years later, their tritium concentrations had returned to the standard.*[52]

In a 1981 study by Dr. Carl Johnson, health director for Jefferson County, showed a 45 percent increase in congenital birth defects in Denver suburbs downwind of Rocky Flats compared to the rest of Colorado. Moreover, he found a 16% increase in cancer rates for those living closest to the plant as compared to those on the outer perimeter of the area, and he estimated 491 excess cancer cases whereas the DOE estimated one. A 1987 study by Crump and others did not find the cancer rates in the northwestern portion of Denver to be significantly higher than other parts of the city and attributed variance in cancer rates to the population density of urban areas.[*][53] Crump's conclusions were contested by Johnson in a letter to the journal editor.[*][54] In a 1992 survey of radiation risk analysis, the authors concluded, "Johnson failed to describe an effective and complete model for the cause of the cancers and its relationship to other knowledge as Crump et al. have done. Therefore, Crump et al.'s explanation must be preferred." [*][55][*]:137

In 1983, Colorado University Medical School professor John C. Cobb and the EPA reported plutonium concentrations from about 500 persons who had died in Colorado. A comparison study was done of those who lived near Rocky Flats with those who lived far from this nuclear weapons production site. The ratio of Pu-240 to Pu-239 was "minutely lower" for persons who lived within 50 km of Rocky Flats, but was more strongly correlated to age, gender, and smoking habits than proximity to the plant.[*][56]

In 1991, the Department of Energy's public affairs group published a pamphlet stating that the inhalation of sediments that become resuspended in the air is considered the most significant pathway that could expose human beings to plutonium from the contaminated local reservoirs, but also stated that the airborne plutonium concentrations as measured by downwind air monitors remained below the DOE standard.[*][44] In a 1999 analysis, it was found that "the major event contributing the highest individual risk from plutonium released from Rocky Flats was the 1957 fire," with wind distribution of plutonium from the 903 Pad Storage Area being the next greatest source of health risk. In this analysis, health risk estimates for off-site humans had a variance of four orders of magnitude, from "between $2.0 \times 10^{*}-4$ (95th percentile) and $2.2 \times 10^{*}-8$ (5th percentile), with a median risk estimate of $2.3 \times 10^{*}-6$." [*][27] The DOE maintains a list of Rocky Flats epidemiological studies.[*][57]

In 1995, a report over 8,000 pages long was released by the Plutonium Working Group Report on Environmental, Safety and Health Vulnerabilities Associated with the Department's Plutonium Storage. This report listed Rocky Flats as having 5 of the 14 most vulnerable facilities based on plutonium environmental, safety, and health vulnerability at all Department Of Energy facilities.[*][58]

During the early 1990s, an independent Health Advisory Panel - appointed by then-Governor Roy Romer - oversaw a series of reports, called the Historical Public Exposure Studies. The 12-member Health Advisory Panel included multiple medical doctors, scientists, PhDs, and local officials. The Rocky Flats Historical Public Exposure Studies involved nine years of research. The Studies had three main

objectives: (1) create a public record of plant operations and accidents that contributed to contaminant releases from the Rocky Flats Plant between 1952 and 1989; (2) assess public exposures to contaminants and potential risks from past releases; and, (3) determine the need for future studies. The Studies' research included identification and assessment of chemicals and radioactive materials from past releases; estimates of risk to residents living or working in surrounding communities during the Plant's operation from 1952 to 1989; an evaluation of possible exposure pathways; and, dose assessments for historical releases.[*][59]

In 2003, Dr. James Ruttenber led a study on the health effects of plutonium. Conducted by the University of Colorado Health Sciences Center and the Colorado Department of Public Health and Environment, the study concluded that lung cancer is linked to plutonium inhalation. "We have supporting evidence from other studies that, along with our findings, support the hypothesis that plutonium exposure causes lung cancer," Ruttenber said. His group's findings were part of a broader study that tracked 16,303 people who worked at the Rocky Flats plant between 1952 and 1989. Their research also found that these workers were 2.5 times more likely to develop brain tumors than other people.[*][60]

Many findings linking workers and other cancer development are muddled due to the "strong healthy worker effect" (that workers tend to have lower overall death rates than general population because those that are ill or disabled are restricted from working).[*][61] Also the standard mortality rates for cancers of stomach and rectum were found to be much higher than other studies of nuclear workers which indicates the necessity for further study since inhalation of plutonium can distribute to these areas.[*][62]

In February 2006, the Rocky Flats Stewardship Council was formed to address post-closure management of Rocky Flats. The council includes elected officials from nine municipal governments neighboring Rocky Flats and four skilled/experienced organizations and/or individuals. Information about the council is available on their website.[*][63]

In 2016, the Colorado Department of Public Health and Environment announced its Cancer Registry was preparing a follow up cancer study to its original 1998 report on cancer incidence in the vicinity of the former Rocky Flats Plant.[*][64] The original report found no pattern of increased cancers in communities around Rocky Flats.

12.5 Legal actions

Subsequent to reports of environmental crimes being committed at Rocky Flats, the United States Department of Justice sponsored an FBI raid dubbed "Operation Desert

Glow," which began at 9 a.m. on June 6, 1989.*[33] The FBI entered the premises under the ruse of providing a terrorist threat briefing, and served its search warrant to Dominick Sanchini, Rockwell International's manager of Rocky Flats.*[33]

The FBI raid led to the formation of Colorado's first special grand jury, the juried testimony of 110 witnesses, reviews of 2,000 exhibits and ultimately a 1992 plea agreement in which Rockwell admitted to 10 federal environmental crimes and agreed to pay $18.5 million in fines out of its own funds.   This amount was less than the company had been paid in bonuses for running the plant as determined by the GAO, and yet was also by far the highest hazardous-waste fine ever; four times larger than the previous record.*[65] Due to DOE indemnification of its contractors, without some form of settlement being arrived at between the U.S. Justice Department and Rockwell the cost of paying any civil penalties would ultimately have been borne by U.S. taxpayers. While any criminal penalties allotted to Rockwell would not have been covered by U.S. taxpayers, Rockwell claimed that the Department of Energy had specifically exempted them from most environmental laws, including hazardous waste.*[33]*[65]*[66]*[67]*[68]*[69]

As forewarned by the prosecuting U.S. Attorney, Ken Fimberg (later Ken Scott),*[70]*:118 the Department of Justice's stated findings and plea agreement with Rockwell were heavily contested by its own, 23-member special grand jury. Press leaks by both members of the DOJ and the grand jury occurred in violation of secrecy Rule 6(e) regarding Grand Jury information. The public contest led to U.S. Congressional oversight committee hearings chaired by Congressman Howard Wolpe, which issued subpoenas to DOJ principals despite several instances of the DOJ's refusal to comply. The hearings, whose findings include that the Justice Department had "bargained away the truth," *[70]*:98 ultimately still did not fully reveal the special grand jury's report to the public, which remains sealed by the DOJ courts.*[65]*[70]*:Ch 6, note 54

The special grand jury report was nonetheless leaked to *Westword.* According to its subsequent publications, the Rocky Flats special grand jury had compiled indictments charging three DOE officials and five Rockwell employees with environmental crimes. The grand jury also wrote a report, intended for the public's consumption per their charter, lambasting the conduct of DOE and Rocky Flats contractors for "engaging in a continuing campaign of distraction, deception and dishonesty" and noted that Rocky Flats, for many years, had discharged pollutants, hazardous materials and radioactive matter into nearby creeks and Broomfield's and Westminster's water supplies.*[71]

The DOE itself, in a study released in December of the year

prior to the FBI raid, called Rocky Flats' ground water the single greatest environmental hazard at any of its nuclear facilities.*[65] From the grand jury's report: "The DOE reached this conclusion because the groundwater contamination was so extensive, toxic, and migrating toward the drinking water supplies for the Cities of Broomfield and Westminster, Colorado." *[72]

A class action lawsuit, *Cook v.   Rockwell International Corp.*, was filed in January 1990 against Rockwell and Dow Chemical (due to the indemnity of nuclear contractors, the award would have been paid by the federal government). Sixteen years later, the plaintiffs were awarded $926 million in economic damages, punitive damages, and interest. The 10th U.S. Circuit Court of Appeals subsequently threw out the verdict and ordered a retrial. A further appeal was rejected without comment by the United States Supreme Court in June 2012.*[73]

In May 2016, U.S. District Judge John L. Kane gave preliminary approval for a $375 million settlement against the Rockwell International Corp. and Dow Chemical Co. Nearly 26 years later, approximately 13,000 to 15,000 eligible property owners could receive monetary payments for damages and decreased property values. Property and homeowners who owned property on June 7, 1989, the day the FBI raided the plant, are eligible to file a claim for property devaluation. The deadline to file a claim is June 1, 2017.*[74]

Carl Johnson sued Jefferson County for unlawful termination, after he was forced to resign from his position as Director of the Jefferson County Health Department. He alleged that his termination was due to concerns by the board members that his reports of contamination would lower property values. The suit was settled out of court for $150,000.*[2]*:106–107*[75]

12.6   Legacy

Denver's automotive beltway does not include a component in the northwest sector, partly due to concerns over unremediated plutonium contamination.*[76]*[77]*[78]

According to the "Rocky Flats National Wildlife Refuge Act of 2001", the land transferred from DOE to the US Fish and Wildlife Service was to be used as a wildlife refuge once Rocky Flats cleanup was complete.*[79] In order to help guide the future of Rocky Flats care and management, the Rocky Flats Stewardship Council was formed in 2006 after the US Congress, DOE and previous organization created the new council. Cleanup of Rocky Flats was finished in 2005 and verified by the EPA and CDPHE in 2007 after ten years and almost $7 billion.*[80]

In 2006, according to DOE, "The selected remedy/corrective action for the Peripheral OU is no action. The RI/FS report (RCRA Facility Investigation-Remedial Investigation/Corrective Measures Study- Feasibility Study) concludes that the Peripheral OU is already in a state protective of human health and the environment." *[81]

In 2007, the "Peripheral Operable Unit" (Peripheral OU) land area of Rocky Flats was redesignated as the Rocky Flats National Wildlife Refuge and fell under U.S. Fish and Wildlife Service (USFWS) stewardship in 2007 following the EPA's determination that final corrective actions had been completed. According to the USFWS, "the refuge has remained closed to the public due to a lack of appropriations for refuge management operations" .*[24] The U.S. Government's efforts to make the area surrounding the former plant into the Rocky Flats National Wildlife Refuge have been controversial due to the contamination, much of which is underground and not remediated.*[13]*[82] The substantially contaminated "Central Operable Unit" (COU) land area of Rocky Flats remains under DOE control, and is now surrounded by the refuge.

Plutonium 239, with a 24,000 year half life, will persist in the environment hundreds of thousands of years. The DOE's assessment of the Central Operating Unit indicates that the long-term risk to citizens living outside the boundaries of Rocky Flats is negligible,*[81]*:30 but citizen organizations state that the remediation of the site was inadequate.*[12]*[13]

In March 2006, the Rocky Flats Stewardship Council was formed to address post-closure management of Rocky Flats and provide a forum for public discussion. This organization was the successor organization to the Rocky Flats Coalition of Local Governments, which advocated for stakeholders during the site cleanup. The Council includes elected officials from nine municipal governments neighboring Rocky Flats and four skilled/experienced organizations and/or individuals. Information and Council meetings minutes and reports are available on its website.*[80] Members of the public are welcome to attend Council meetings.

In 2014, the U.S. Fish and Wildlife Services proposed a controlled burn on 701 acres of the Wildlife Refuge. In 2015, they reported that they will postpone those burns until 2017. In 2015, there was a "soft opening" of the Rocky Flats Wildlife Refuge where small groups of people could reserve space on a three-mile guided nature walk. The official opening of the Refuge in now planned for 2017.*[83]

In 2015 Rocky Mountain Downwinders was founded to study health effects in people who lived east of the facility while it was operational. The group set up an online health survey conducted by Metropolitan State University of Denver.*[84]*[85]

12.7 Public opposition

Main article: Anti-nuclear movement

On the weekend of April 28, 1979, more than 15,000 people demonstrated against the Rocky Flats Nuclear Weapons Plant. The protest was coordinated with other anti-nuclear demonstrations across the country. Daniel Ellsberg and Allen Ginsberg were among the 284 people who were arrested. The demonstration followed more than six months of continuous protests that included an attempted blockade of the railroad tracks leading to the site.*[86]*[87]*[88] Large pro-nuclear counter demonstrations were also staged that year.*[89]

In 1983 the Rocky Mountain Peace and Justice Center was founded with a goal of closing the Rocky Flats plant.*[90]*[91] The Center has since set goals of keeping the Rocky Flats National Wildlife Refuge closed to the public, preventing construction of highways in or near the site of the former plant, and preventing new housing construction in the area.*[92] The Center is a 501(c)(3) non-profit corporation an, as of 2014, has one full-time employee.

On October 15, 1983, about 10,000 demonstrators turned out for protest at the Rocky Flats Nuclear Weapons Plant (well short of the 21,000 hoped for by protest organizers). No arrests were made.*[93]*[94] On August 10, 1987 (the 42nd anniversary of the atomic bombing of Nagasaki), 320 demonstrators were arrested after they tried to force a one-day shutdown of the plant.*[95]*[96] A similar protest with a turnout of about 3,500 was staged on August 6, 1989 (the anniversary of the nuclear bombing of Hiroshima).*[97]

Though public demonstrations against plant operations ceased with the decommissioning of the plant,*[98] activists continue to protest disposal of nuclear waste from the site*[99] and the scale and scope of cleanup operations.*[2] Since 2013, opposition has focused on the Candelas development located along the southern border of the former Plant site.

With the establishment of the Rocky Flats National Wildlife Refuge Act in 2001, a 300 ft strip on the eastern edge of the refuge was allocated to Jefferson County for construction of the Jefferson County Parkway. In May 2008, the Jefferson Parkway Public Highway Authority was established to complete this last portion of the Denver metro beltway.*[100] Opponents of the parkway are concerned about the disruption of plutonium laden soil from excavation of the area to build the parkway.*[101] In April 2015, the WestConnect Corridor Coalition was formed with the hopes of bringing about the end of a decades long dispute to the completion of the Jefferson County Parkway.*[102] However, by October 2015, the WestConnect Corridor had withdrawn its support from the parkway, determining that the decision to build the parkway should be made outside of the coalition's process. As of 2015, the Jefferson Parkway Public Authority was still searching for private and public funding to complete the beltway.

A group named Candelas Glows is opposed to a large housing and commercial development planned in the area, which the group calls a "plutonium dust bowl." The Department of Energy responded by saying that studies show more risk from naturally occurring radioactive elements than from very low-level amounts of plutonium remaining around the former plant.*[103] Candelas Glows argued that a July 2015 radiation report from the Rocky Flats Stewardship Council shows plutonium levels at 1.02 pCi/L, compared to the regulatory standard of 0.15 pCi/L.*[104]

12.8   See also

- U.S. Department of Energy
- U.S. Environmental Protection Agency
- Hazardous waste management
- Risk assessment
- Environmental law
- CERCLA
- Superfund
- Price-Anderson Act
- Dark Circle (film)
- Making a Real Killing: Rocky Flats and the Nuclear West
- Rocky Flats Truth Force
- Nuclear and radiation accidents by country
- Candelas, Arvada, CO
- Cold War

12.9   Notes

[1] U.S. Department of Energy, Historical Release Report, http://www.lm.doe.gov/Rocky_Flats/Regulations.aspx

[2] Moore, LeRoy (2007). "Democracy and Public Health at Rocky Flats: The Examples of Edward Martell and Carl J. Johnson". In Quigley, Dianne; Lowman, Amy; Wing, Steve. *Ethics of Research on Health Impacts of Nuclear Weapons Activities in the United States* (PDF). Collaborative Initiative

for Research Ethics and Environmental Health (CIREEH) at Syracuse University. pp. 55–97. Retrieved September 17, 2011.

[3] Colorado Department of Public Health and Environment (CDPHE), Rocky Flats - Site History

[4] Poet, SE; Martell, EA (October 1972). "Plutonium-239 and americium-241 contamination in the Denver area." . *Health Physics*. **23** (4): 537–48. doi:10.1097/00004032-197210000-00012. PMID 4634934. Retrieved 12 June 2013.

[5] Johnson, CJ (October 1981). "Cancer Incidence in an area contaminated with radionuclides near a nuclear installation" . *Ambio*. **10** (4): 176–182. JSTOR 4312671. Reprinted in "Cancer Incidence in an area contaminated with radionuclides near a nuclear installation." . *Colo Med*. **78**: 385–92. Oct 1981. PMID 7348208.

[6] CDPHE, Rocky Flats, Historical Public Exposure Studies, https://www.colorado.gov/pacific/cdphe/rocky-flats-historical-public-exposure-studies

[7] CDPHE, Summary of Findings: Historical Public Exposure Studies (1999), p. 19, https://www.colorado.gov/pacific/sites/default/files/HM_sf-rocky-flats-smry-indings-hist-pub-exposure-studies-bklt-1999. pdf

[8] CDPHE, https://www.colorado.gov/pacific/cdphe/rocky-flats-facts-glance

[9] U.S. DOE, Legacy Management, Rocky Flats, http://www.lm.doe.gov/Rocky_Flats/Sites.aspx

[10] U.S. Environmental Protection Agency (EPA), https://cumulis.epa.gov/supercpad/cursites/csitinfo.cfm?id=0800360

[11] "The September 1957 Rocky Flats fire: A guide to records series of the Department of Energy" . United States Department of Energy. Retrieved September 3, 2011.

[12] "Rocky Flats Nuclear Site Too Hot for Public Access, Citizens Warn" . Environment News Service. August 5, 2010. Retrieved September 17, 2011.

[13] Hooper, Troy (August 4, 2011). "Invasive weeds raise nuclear concerns at Rocky Flats" . The Colorado Independent. Retrieved September 17, 2011.

[14] "1969 Fire Page 7" . Colorado.edu. Retrieved 2011-10-27.

[15] "Atomic Obsession: Nuclear Alarmism from Hiroshima to Al-Qaeda" . RockyFlatsFacts.com. Retrieved 4 July 2013.

[16] Colorado Department of Public Health and Environment, Cancer Registry, "Ratios of Cancer Incidence in Ten Areas Around Rocky Flats, Colorado" (1998), https://www.colorado.gov/pacific/cdphe/cdphe-rocky-flats-cancer-study

[17] Buffer, Patricia (2003). *Rocky Flats History*. Department of Energy.

[18] Carey Sublette (July 3, 2007). "Nuclear Weapons FAQ Section 4.4.1.4 The Teller–Ulam Design". *Nuclear Weapons FAQ*. Retrieved 17 July 2011.

[19] Wolf, Naomi (February 21, 2012). "From Rocky Flats to Fukushima: this nuclear folly". guardian.co.uk. Retrieved July 7, 2013.

[20] Ackland, Len (2002). *Making a real killing : Rocky Flats and the nuclear West* ([Updated ed.]--cover ; [1st ed.]--verso. ed.). Albuquerque [N.M.]: University of New Mexico Press. ISBN 0826327982.

[21] "Rocky Flats Cover-ups Alleged". *Pittsburgh Press*. June 9, 1989. Retrieved July 7, 2013.

[22] "EG&G Won't Try for New Contract at Rocky Flats". *Orlando Sentinel*. August 18, 1994. Retrieved July 7, 2013.

[23] "Property Management Has Improved at DOE's Rocky Flats Site". *GAO/RCED-96-39*. Department of Energy. December 28, 1995. Retrieved July 7, 2013.

[24] "Rocky Flats National Wildlife Refuge". U.S. Fish & Wildlife Service. Retrieved 2 July 2013.

[25] "Rocky Flats Virtual Museum". University of Colorado at Boulder. Retrieved September 3, 2011.

[26] Cochran, Thomas B. (November 21, 1996). "Overview of Rocky Flats Operations" (PDF). Retrieved July 7, 2013.

[27] Till, John E. (Principal Investigator); et al. (September 1999). "Technical Summary Report for the Historical Public Exposures Studies for Rocky Flats Phase II" (PDF). *Radiological Assessments Corporation*. Retrieved 13 June 2013.

[28] "Atomic Plant Hit by $50,000 Fire". *St. Joseph Gazette*. Sep 12, 1957. Retrieved 2 July 2013.

[29] "$50,000 Damage at Atomic Plant". *Youngstown Vindicator*. Sep 13, 1957. Retrieved 2 July 2013.

[30] "The September 1957 Rocky Flats Fire: A Guide To Records Series Of The Department Of Energy". *Department of Energy: Health and Safety*. Retrieved 2 July 2013.

[31] Iversen, Kristen (June 12, 2012). "Under The 'Nuclear Shadow' Of Colorado's Rocky Flats". NPR. Retrieved June 14, 2013.

[32] "Pad 903 Area" (PDF). Colorado.gov. Retrieved July 3, 2013.

[33] Siegel, Barry (1993-08-08). "Showdown at Rocky Flats : When Federal Agents Take On a Government Nuclear-Bomb Plant, Lines of Law and Politics Blur, and Moral Responsibility Is Tested". *Los Angeles Times*.

[34] Moore, LeRoy (May 1992). *1957: Fateful Year for the Nuclear Weapons Industry* (PDF). Environmental Consequences of Producing Nuclear Weapons. Chelyabinsk, Russia. Retrieved September 17, 2011.

[35] "Summary of Findings: Rocky Flats Public Exposure Studies: Key questions addressed by the research". Colorado Department of Public Health and Environment. Retrieved September 17, 2011.

[36] "Rocky Flats Virtual Museum: The Fire Was Inevitable". University of Colorado at Boulder. Retrieved September 17, 2011.

[37] "Operable Units: Environmental Restoration Areas at the Rocky Flats Plant" (PDF) (Press release). DOE Public Affairs. c. 1992. Retrieved 2013-06-14.

[38] Campbell, Bruce. "Past DOE Fires/ 1969 Rocky Flats Fire" (PDF). Hughes Associates, Inc. Retrieved 15 June 2013.

[39] Calhoun, Patricia (Apr 12, 2007). "Carved in Stone". *Westword*. Retrieved 8 July 2013.

[40] "Problems Continue for Rocky Flats Solar Pond Cleanup Program". *RCED-92-18*. GAO.gov. October 17, 1991. Retrieved 8 July 2013.

[41] Rubino, Joe (2011-05-26). "Broomfield officials concerned new monitoring stations at Rocky Flats not far enough downstream". *Broomfield Enterprise*. Retrieved 3 July 2013.

[42] "Contaminants Released Surface Water from Rocky Flats". Cdphe.state.co.us. Retrieved 2011-10-27.

[43] Associated Press (June 13, 1989). "City digging ditch to divert stream". *The Spokesman-Review*. Retrieved 13 June 2013.

[44] Paukert, Jill (1991-08-21). "Reservoirs" (PDF). *Operable Unit 3: Offsite Releases*. Department of Energy. Retrieved 13 June 2013.

[45] Davidson, Michael (2010-08-21). "Broomfield fighting plan to remove Rocky Flats dams". *Daily Camera*. Retrieved 3 July 2013.

[46] "Atomic Plant Fire Causes $50,000 Loss," Denver Post, 12 September 1957.

[47] Brooke, James (1996-12-11). "Plutonium Stockpile Fosters Fears of 'a Disaster Waiting to Happen' - New York Times". Nytimes.com. Retrieved 2011-10-27.

[48] Todd, Andrew S; R. Mark Sattelberg (November 2005). "Actinides in Deer Tissues at the Rocky Flats Environmental Technology Site". *Integrated Environmental Assessment and Management*. **1** (4): 391–396. doi:10.1002/ieam.5630010408. PMID 16639905. Retrieved 9 June 2013.

[49] Falk, RB; Daugherty, NM; Aldrich, JM; Furman, FJ; Hilmas, DE (August 2006). "Application of multi-compartment wound models to plutonium-contaminated wounds incurred by former workers at rocky flats". *Health physics.* **91** (2): 128–43. doi:10.1097/01.HP.0000203314.17612.63. PMID 16832194.

[50] Dixon, Jack A. Tuszynski, John M. (2002). *Biomedical applications of introductory physics.* New York: Wiley. p. 323. ISBN 0471412953.

[51] Krey, P.W.; Hardy, E.P. (August 1, 1970). "Plutonium in Soil around the Rocky Flats Plant". *United States Atomic Energy Commission Health and Safety Laboratory.* HASL-235.

[52] "Contaminants Released to Surface Water from Rocky Flats". Retrieved October 26, 2016.

[53] Crump, KS; Ng, TH; Cuddihy, RG (July 1987). "Cancer incidence patterns in the Denver metropolitan area in relation to the Rocky Flats plant.". *American Journal of Epidemiology.* **126** (1): 127–35. PMID 3591777. Retrieved 1 July 2013.

[54] Johnson, Carl (July 1987). "Re: "Cancer incidence patterns in the Denver metropolitan area in relation to the Rocky Flats plant".". *American Journal of Epidemiology.* **126** (1): 153–5. PMID 3591781. Retrieved 1 July 2013.

[55] Shihab-Eldin, Adnan; Alexander Shlyakhter; Richard Wilson (1992). "Is there a large risk of radiation? A critical review of pessimistic claims" (PDF). *Environment International.* **18**: 117–151. doi:10.1016/0160-4120(92)90002-l. Retrieved 1 July 2013.

[56] Cobb et al., "Plutonium Burdens in People Living Around the Rocky Flats Plant," March 1983, EPA-600/4-82-069, Springfield, VA: National Technical Information Service

[57] "The DOE's Rocky Flats Plant: A Guide to Record Series Useful for Health-Related Research". Hss.doe.gov. Retrieved June 14, 2013.

[58] "DOE - Office of Legacy Management-Rocky Flats History". Retrieved October 26, 2016.

[59] See CDPHE, Rocky Flats Historical Public Exposure Studies, https://www.colorado.gov/pacific/cdphe/rocky-flats-historical-public-exposure-studies

[60] Associated Press (April 19, 2003). "Plutonium exposure increases lung cancer risk | Lubbock Online | Lubbock Avalanche-Journal". Lubbock Online. Retrieved June 4, 2013.

[61] Shah, Divying (2009). "Healthy Worker Effect Phenomenon". *Indian Journal of Occupational and Environmental Medicine.* **77** (13.2).

[62] Ruttenburg Ph.D, MD, A.J.; Schoenbeck, M.; Brown Ph.D., S; Wells D.V.M. MPH, T; McClure M.S., D; McCrea, J; Popken, D (March 3, 2003). "Report of Epidemiological Analyses Performed for Rocky Flats Production Workers Employed Between 1952-1989" (PDF).

[63] "Rocky Flats Stewardship Council :: Home". Retrieved October 26, 2016.

[64] Denver Post, http://www.denverpost.com/2016/12/02/rocky-flats-cancer-study/

[65] Siegel, Barry (1993-08-15). "Showdown at Rocky Flats : The Justice Department had negotiated a Rocky Flats settlement, but the grand jury could not keep quiet about what happened there". *Los Angeles Times.*

[66] Schneider, Keith (1992-03-27). "U.S. Shares Blame in Abuses at A-Plant". *The New York Times.*

[67] Hardesty, Greg (March 29, 2006). "Retired FBI agent helped close nuclear-weapons site". The Orange County Register. Retrieved September 17, 2011.

[68] Calhoun, Patricia (Aug 19, 2004). "True Lies: The FBI agent who raided Rocky Flats finally sounds off". *Westword.* Retrieved 14 June 2013.

[69] Wald, Matthew L. (1989-09-23). "Rockwell Is Giving Up Rocky Flats Plant". *The New York Times.*

[70] McKinley, Wes; Balkany, Caron (2004). *The Ambushed Grand Jury: How the Justice Department Covered up Government Nuclear Crimes and how We Caught Them Red Handed.* New York: Apex Press. ISBN 9781891843297.

[71] Wald, Matthew L. (March 13, 2004). "Book Says U.S. Aides Lied in Nuclear-Arms Plant Case" (PDF). *The New York Times on the web.* Retrieved 3 July 2013.

[72] "Rocky Flats Grand Jury Report". Constitution.org. January 24, 1992. Retrieved 2013-06-14.

[73] Drummond, Bob (June 25, 2012). "Rocky Flats Suit Back to Square One in 22-Year Dispute". *Business Week.* Retrieved 4 July 2013.

[74] Aguilar, J (August 8, 2016). "Payouts to property owners in long-running Rocky Flats suit should start in 2017". The Denver Post. Retrieved October 26, 2016.

[75] Weyler, Rex (2004). *Greenpeace: How a group of journalists, ecologists and visionaries changed the world.* Emmaus, PA: Rodale Books. ISBN 978-1-59486-106-2.

[76] "Metro Denver's Northwest Quadrant Transportation Solution". Go The Betterway. Retrieved 2011-10-27.

[77] Schneider, Keith (1990-02-15). "Weapons Plant Pressed for Accounting of Toll on Environment and Health - New York Times". Nytimes.com. Retrieved 2011-10-27.

[78] Aguilar, John (2012-04-13). "'Beltway completion authority' bill enrages Jefferson Parkway opponents". *Daily Camera*. Retrieved 3 July 2013.

[79] "U.S. Fish and Wildlife Service Establishes Rocky Flats National Wildlife Refuge". U.S. Fish and Wildlife Service. Retrieved October 26, 2016.

[80] "Rocky Flats Stewardship Council :: Home". *Rocky Flats Stewardship Council*. Retrieved October 26, 2016.

[81] "Corrective Action Decision/Record of Decision for Rocky Flats Plant (USDOE) POU and COU". September 2006. Retrieved July 3, 2013.

[82] Salazar, Quibian (July 21, 2011). "Plutonium parkway". Boulderweekly.com. Retrieved October 27, 2011.

[83] "Rocky Flats Nuclear Guardianship | About". *Rocky Flats Nuclear Guardianship*. Retrieved October 26, 2016.

[84] "Health survey of Rocky Flats neighbors to launch Tuesday" Denver Post. Retrieved November 9, 2016.

[85] "Health Survey". *Rocky Flats Downwinders*. Retrieved October 26, 2016.

[86] Nonviolent Social Movements p. 295.

[87] Headline: Rocky Flats Nuclear Plant / Protest

[88] "Nearly 300 Protesters Arrested at Rocky Flats". *St. Joseph Gazette*. April 30, 1979. Retrieved July 7, 2013.

[89] "Pro-nuclear rally is held". *Lodi News-Sentinel*. August 27, 1979. Retrieved July 7, 2013.

[90] "Peace and justice endures" Boulder Weekly. Retrieved November 9, 2016

[91] "Nuclear Guardianship Collective". *RMPJC*. Retrieved November 9, 2016.

[92] "Rocky Flats Nuclear Guardianship". *Rocky Flats Nuclear Guardianship*. Retrieved November 9, 2016.

[93] Headline: Colorado / Anti-Nuclear Demonstration

[94] "AROUND THE NATION: 10,000 Demonstrate Against Nuclear Arms". *The New York Times*. October 16, 1983. Retrieved July 7, 2013.

[95] "570 Arrested in A-Bomb Protests". *New York Times*. August 10, 1987.

[96] "Nuclear Arms Protests Mark Nagasaki Bombing". *Gainesville Sun*. August 10, 1987. Retrieved July 7, 2013.

[97] "Hiroshima Marks Atomic Bomb Attack". *Sarasota Herald-Tribune*. August 7, 1989. Retrieved July 7, 2013.

[98] "Rocky Flats Protest Era Ends". *TimesDaily*. Feb 5, 1992. Retrieved July 7, 2013.

[99] Kelly, Sean (June 16, 1999). "First barrels roll out of Flats: protesters, supporters take opposite sides of roadway". *Denver Post*. Retrieved July 7, 2013.

[100] *Jefferson County Government Official Website*. Jefferson County, Colorado http://jeffco.us. Retrieved October 26, 2016. Missing or empty |title= (help)

[101] Durham, N (January 3, 2013). "Toxic Suburbia: Fantastic Rocky Flats vistas, plutonium breezes". The Colorado Independent. Retrieved October 26, 2016.

[102] Zitzman, A (April 5, 2015). "New coalition plans to study completion of 470 loop around metro Denver-7News". 7News Denver The Denver Channel. Retrieved October 26, 2016.

[103] "Former Rocky Flats site stirs concerns for some neighbors". The Denver Post. Retrieved October 26, 2016.

[104] "Colorado's Green Housing Development...Adjacent to a Plutonium Burial Site". *Candelas Glows*. Retrieved October 26, 2016.

12.10 External links

- Department of Energy - Rocky Flats Legacy Management

- *New York Times* Rocky Flats news archive

- Rocky Flats on Colorado.gov

- Rocky Flats Plant on the EPA web site

- Rocky Flats nuclear guardianship

- Rocky Flats virtual museum

- Department of Energy Health Assessment for Rocky Flats

Chapter 13

Hanford Site

Coordinates: 46°38′51″N 119°35′55″W / 46.64750°N 119.59861°W

Nuclear reactors line the riverbank at the Hanford Site along the Columbia River in January 1960. The N Reactor is in the foreground, with the twin KE and KW Reactors in the immediate background. The historic B Reactor, the world's first plutonium production reactor, is visible in the distance.

The **Hanford Site** is a mostly decommissioned nuclear production complex operated by the United States federal government on the Columbia River in the U.S. state of Washington. The site has been known by many names, including: **Hanford Project**, **Hanford Works**, **Hanford Engineer Works** and **Hanford Nuclear Reservation**. Established in 1943 as part of the Manhattan Project in Hanford, south-central Washington, the site was home to the B Reactor, the first full-scale plutonium production reactor in the world.[1] Plutonium manufactured at the site was used in the first nuclear bomb, tested at the Trinity site, and in Fat Man, the bomb detonated over Nagasaki, Japan.

During the Cold War, the project expanded to include nine nuclear reactors and five large plutonium processing complexes, which produced plutonium for most of the more than 60,000 weapons in the U.S. nuclear arsenal.[2][3] Nuclear technology developed rapidly during this period, and Hanford scientists produced major tech-

nological achievements. Many early safety procedures and waste disposal practices were inadequate, and government documents have confirmed that Hanford's operations released significant amounts of radioactive materials into the air and the Columbia River.

The weapons production reactors were decommissioned at the end of the Cold War, and decades of manufacturing left behind 53 million US gallons (200,000 m^3) of high-level radioactive waste[4] stored within 177 storage tanks, an additional 25 million cubic feet (710,000 m^3) of solid radioactive waste, and 200 square miles (520 km^2) of contaminated groundwater beneath the site.[5] In 2011, DOE emptied 149 single-shell tanks by pumping nearly all of the liquid waste out into 28 newer double-shell tanks. DOE later found water intruding into at least 14 single-shell tanks and that one of them had been leaking about 640 US gallons (2,400 l; 530 imp gal) per year into the ground since about 2010. In 2012, DOE discovered a leak also from a double-shell tank caused by construction flaws and corrosion in the bottom, and that 12 double-shell tanks have similar construction flaws. Since then, DOE changed to monitoring single-shell tanks monthly and double-shell tanks every 3 years, and also changed monitoring methods. In March 2014, DOE announced further delays in the construction of the Waste Treatment Plant, which will affect the schedule for removing waste from the tanks.[6] Intermittent discoveries of undocumented contamination have slowed the pace and raised the cost of cleanup.[7]

In 2007, the Hanford site represented two-thirds of the nation's high-level radioactive waste by volume.[8] Hanford is currently the most contaminated nuclear site in the United States[9][10] and is the focus of the nation's largest environmental cleanup.[2] Besides the cleanup project, Hanford also hosts a commercial nuclear power plant, the Columbia Generating Station, and various centers for scientific research and development, such as the Pacific Northwest National Laboratory and the LIGO Hanford Observatory.

On November 10, 2015, it was designated as part of the Manhattan Project National Historical Park alongside other sites in Oak Ridge and Los Alamos.[*][11]

13.1 Geography

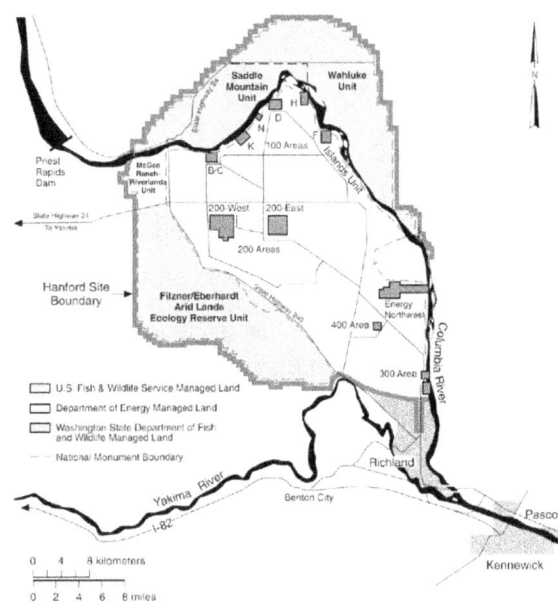

A map shows the main areas of the Hanford Site, as well as the buffer zone that was turned over to the Hanford Reach National Monument in 2000

The Hanford Site occupies 586 square miles (1,518 km^2) —roughly equivalent to half of the total area of Rhode Island—within Benton County, Washington.[*][2] This land is closed to the general public. It is a desert environment receiving under 10 inches of annual precipitation, covered mostly by shrub-steppe vegetation. The Columbia River flows along the site for approximately 50 miles (80 km), forming its northern and eastern boundary.[*][12] The original site was 670 square miles (1,740 km^2) and included buffer areas across the river in Grant and Franklin counties.[*][13] Some of this land has been returned to private use and is now covered with orchards and irrigated fields. In 2000, large portions of the site were turned over to the Hanford Reach National Monument.[*][14] The site is divided by function into three main areas. The nuclear reactors were located along the river in an area designated as the 100 Area; the chemical separations complexes were located inland in the Central Plateau, designated as the 200 Area; and various support facilities were located in the southeast corner of the site, designated as the 300 Area.[*][15]

The site is bordered on the southeast by the Tri-Cities, a metropolitan area composed of Richland, Kennewick, Pasco, and smaller communities, and home to over 230,000 residents. Hanford is a primary economic base for these

cities.[*][16]

13.2 Climate

13.3 Early history

The confluence of the Yakima, Snake, and Columbia rivers has been a meeting place for native peoples for centuries. The archaeological record of Native American habitation of this area stretches back over ten thousand years. Tribes and nations including the Yakama, Nez Perce, and Umatilla used the area for hunting, fishing, and gathering plant foods.[*][18] Hanford archaeologists have identified numerous Native American sites, including "pit house villages, open campsites, fishing sites, hunting/kill sites, game drive complexes, quarries, and spirit quest sites",[*][13] and two archaeological sites were listed on the National Register of Historic Places in 1976.[*][19] Native American use of the area continued into the 20th century, even as the tribes were relocated to reservations. The Wanapum people were never forced onto a reservation, and they lived along the Columbia River in the Priest Rapids Valley until 1943.[*][13] Euro-Americans began to settle the region in the 1860s, initially along the Columbia River south of Priest Rapids. They established farms and orchards supported by small-scale irrigation projects and railroad transportation, with small town centers at Hanford, White Bluffs, and Richland.[*][20]

13.4 Manhattan Project

Main article: Manhattan Project

During World War II, the S-1 Section of the federal Office of Scientific Research and Development (OSRD) sponsored an intensive research project on plutonium. The research contract was awarded to scientists at the University of Chicago Metallurgical Laboratory (Met Lab). At the time, plutonium was a rare element that had only recently been isolated in a University of California laboratory. The Met Lab researchers worked on producing chain-reacting "piles" of uranium to convert it to plutonium and finding ways to separate plutonium from uranium. The program was accelerated in 1942, as the United States government became concerned that scientists in Nazi Germany were developing a nuclear weapons program.[*][21]

Hanford High School, shown before residents were displaced by the creation of the Hanford Site

Hanford High after abandonment

13.4.1   Site selection

In September 1942, the Army Corps of Engineers placed the newly formed Manhattan Project under the command of General Leslie R. Groves, charging him with the construction of industrial-size plants for manufacturing plutonium and uranium.[13] Groves recruited the DuPont Company to be the prime contractor for the construction of the plutonium production complex. DuPont recommended that it be located far away from the existing uranium production facility at Oak Ridge, Tennessee. The ideal site was described by these criteria:[22]

- A large and remote tract of land

- A "hazardous manufacturing area" of at least 12 by 16 miles (19 by 26 km)

- Space for laboratory facilities at least 8 miles (13 km) from the nearest reactor or separations plant

- No towns of more than 1,000 people closer than 20 miles (32 km) from the hazardous rectangle

- No main highway, railway, or employee village closer than 10 miles (16 km) from the hazardous rectangle

- A clean and abundant water supply

- A large electric power supply

- Ground that could bear heavy loads.

In December 1942, Groves dispatched his assistant Colonel Franklin T. Matthias and DuPont engineers to scout potential sites. Matthias reported that Hanford was "ideal in virtually all respects," except for the farming towns of White Bluffs and Hanford.[23] General Groves visited the site in January and established the Hanford Engineer Works, codenamed "Site W". The federal government quickly acquired the land under its war powers authority[24] and relocated some 1,500 residents of Hanford, White Bluffs, and nearby settlements, as well as the Wanapum people, Confederated Tribes and Bands of the Yakama Nation, the Confederated Tribes of the Umatilla Indian Reservation, and the Nez Perce Tribe.[25][26]

13.4.2   Construction begins

B Reactor construction (1944)

The Hanford Engineer Works (HEW) broke ground in March 1943 and immediately launched a massive and technically challenging construction project.[27] DuPont advertised for workers in newspapers for an unspecified "war construction project" in southeastern Washington, offering "attractive scale of wages" and living facilities.[28]

The construction workers (who reached a peak of 44,900 in June 1944) lived in a construction camp near the old Hanford townsite. The administrators and engineers lived in the government town established at Richland Village, which eventually had accommodation in 4,300 family units and 25 dormitories. [29] [30]

Construction of the nuclear facilities proceeded rapidly. Before the end of the war in August 1945, the HEW built

554 buildings at Hanford, including three nuclear reactors (105-B, 105-D, and 105-F) and three plutonium processing canyons (221-T, 221-B, and 221-U), each 250 meters (820 ft) long.

To receive the radioactive wastes from the chemical separations process, the HEW built "tank farms" consisting of 64 single-shell underground waste tanks (241-B, 241-C, 241-T, and 241-U).[31] The project required 386 miles (621 km) of roads, 158 miles (254 km) of railway, and four electrical substations. The HEW used 780,000 cubic yards (600,000 m^3) of concrete and 40,000 short tons (36,000 t) of structural steel and consumed \$230 million between 1943 and 1946.[32]

13.4.3 Plutonium production

Further information: B Reactor

The B Reactor (105-B) at Hanford was the first large-scale plutonium production reactor in the world. It was designed and built by DuPont based on an experimental design by Enrico Fermi, and originally operated at 250 megawatts (thermal). The reactor was graphite moderated and water cooled. It consisted of a 28-by-36-foot (8.5 by 11.0 m), 1,200-short-ton (1,100 t) graphite cylinder lying on its side, penetrated through its entire length horizontally by 2,004 aluminium tubes.[33] Two hundred short tons (180 t) of uranium slugs, 1.625 inches (4.13 cm) diameter by 8 inches (20 cm) long, sealed in aluminium cans went into the tubes.[34] Cooling water was pumped through the aluminium tubes around the uranium slugs at the rate of 30,000 US gallons (110,000 L) per minute.[33]

The B Reactor during construction

Construction on B Reactor began in August 1943 and was completed on September 13, 1944. The reactor went

critical in late September and, after overcoming nuclear poisoning, produced its first plutonium on November 6, 1944.[35] Plutonium was produced in the Hanford reactors when a uranium-238 atom in a fuel slug absorbed a neutron to form uranium-239. U-239 rapidly undergoes beta decay to form neptunium-239, which rapidly undergoes a second beta decay to form plutonium-239. The irradiated fuel slugs were transported by rail to three huge remotely operated chemical separation plants called "canyons" that were about 10 miles (16 km) away. A series of chemical processing steps separated the small amount of plutonium that was produced from the remaining uranium and the fission waste products. This first batch of plutonium was refined in the 221-T plant from December 26, 1944, to February 2, 1945, and delivered to the Los Alamos laboratory in New Mexico on February 5, 1945.[36]

Two identical reactors, D Reactor and F reactor, came online in December 1944 and February 1945, respectively. By April 1945, shipments of plutonium were headed to Los Alamos every five days, and Hanford soon provided enough material for the bombs tested at Trinity and dropped over Nagasaki.[37] Throughout this period, the Manhattan Project maintained a top secret classification. Until news arrived of the bomb dropped on Hiroshima, fewer than one percent of Hanford's workers knew they were working on a nuclear weapons project.[38] General Groves noted in his memoirs that "We made certain that each member of the project thoroughly understood his part in the total effort; that, and nothing more." [39]

Initially six reactors or "piles" were proposed, when the plutonium was to be used in the gun-type Thin Man bomb. In mid-1944 a simple gun-type bomb was found to be impractical for plutonium, and the more advanced Fat Man bomb required less plutonium. The number of piles was reduced to four and then three; and the number of chemical separation plants from four to three.[40]

13.4.4 Technological innovations

In the short time frame of the Manhattan Project, Hanford engineers produced many significant technological advances. As no one had ever built an industrial-scale nuclear reactor before, scientists were unsure how much heat would be generated by fission during normal operations. Seeking the greatest possible production while maintaining an adequate safety margin, DuPont engineers installed ammonia-based refrigeration systems with the D and F reactors to further chill the river water before its use as reactor coolant.[41]

Another difficulty the engineers struggled with was how to deal with radioactive contamination. Once the canyons began processing irradiated slugs, the machinery would become so radioactive that it would be unsafe for humans ever to come in contact with it. The engineers therefore had to devise methods to allow for the replacement of any component via remote control. They came up with a modular cell concept, which allowed major components to be removed and replaced by an operator sitting in a heavily shielded overhead crane. This method required early practical application of two technologies that later gained widespread use: Teflon, used as a gasket material, and closed-circuit television, used to give the crane operator a better view of the process.[*][42]

13.5   Cold War expansion

Decommissioning D Reactor

In September 1946, the General Electric Company assumed management of the Hanford Works under the supervision of the newly created Atomic Energy Commission. As the Cold War began, the United States faced a new strategic threat in the rise of the Soviet nuclear weapons program. In August 1947, the Hanford Works announced funding for the construction of two new weapons reactors and research leading to the development of a new chemical separations process. With this announcement, Hanford entered a new phase of expansion.[*][43]

By 1963, the Hanford Site was home to nine nuclear reactors along the Columbia River, five reprocessing plants on the central plateau, and more than 900 support buildings and radiological laboratories around the site.[*][2] Extensive modifications and upgrades were made to the original three World War II reactors, and a total of 177 un-

derground waste tanks were built.[*][2] Hanford was at its peak production from 1956 to 1965. Over the entire 40 years of operations, the site produced about 63 short tons (57 t) of plutonium, supplying the majority of the 60,000 weapons in the U.S. arsenal.[*][2][*][3] Uranium-233 was also produced.[*][44][*][45][*][46][*][47]

13.5.1   Decommissioning

Most of the reactors were shut down between 1964 and 1971, with an average individual life span of 22 years. The last reactor, N Reactor, continued to operate as a dual-purpose reactor, being both a power reactor used to feed the civilian electrical grid via the Washington Public Power Supply System (WPPSS) and a plutonium production reactor for nuclear weapons. N Reactor operated until 1987. Since then, most of the Hanford reactors have been entombed ("cocooned") to allow the radioactive materials to decay, and the surrounding structures have been removed and buried.[*][48] The B-Reactor has not been cocooned and is accessible to the public on occasional guided tours. It was listed on the National Register of Historic Places in 1992,[*][49] and some historians advocate converting it into a museum.[*][50][*][51] B reactor was designated a National Historic Landmark by the National Park Service on August 19, 2008.[*][52][*][53]

13.6   Later operations

Highway sign on a road entering the Hanford Site

The United States Department of Energy assumed control of the Hanford Site in 1977. Although uranium enrichment and plutonium breeding were slowly phased out, the nuclear legacy left an indelible mark on the Tri-Cities. Since World War II, the area had developed from a small farming community to a booming "Atomic Frontier" to a powerhouse of the nuclear-industrial complex.[*][55] Decades of federal investment created a community of highly skilled scientists and engineers. As a result of this concentration of specialized skills, the Hanford Site was able to diversify its operations to include scientific research, test facilities, and commercial nuclear power production.

As of 2013, operational facilities located at the Hanford Site include:

- The Pacific Northwest National Laboratory, owned by the Department of Energy and operated by Battelle Memorial Institute

- The Fast Flux Test Facility (FFTF), a national research facility in operation from 1980 to 1992 (its last fuel was removed in 2008[*56])

- LIGO's Hanford Observatory, an interferometer searching for gravitational waves[*57][*58][*59][*60]

- Columbia Generating Station, a commercial nuclear power plant operated by Energy Northwest.

- A US Navy nuclear submarine reactor dry storage site contains sealed reactor sections of 114 US Navy submarines (as of 2008).[*61]

The Department of Energy and its contractors offer tours of the site. Sixty public tours, each five hours long, were planned for 2009. The tours are free, require advance reservation via the department's web site, and are limited to U.S. citizens at least 18 years of age.[*62]

13.7 Environmental concerns

The Hanford Reach of the Columbia River, where radioactivity was released from 1944 to 1971

A huge volume of water from the Columbia River was required to dissipate the heat produced by Hanford's nuclear reactors. From 1944 to 1971, pump systems drew cooling water from the river and, after treating this water for use by the reactors, returned it to the river. Before its release into the river, the used water was held in large tanks known as retention basins for up to six hours. Longer-lived isotopes were not affected by this retention, and several terabecquerels entered the river every day. The federal government kept knowledge about these radioactive releases secret.[*63] Radiation was later measured 200 miles downstream as far west as the Washington and Oregon coasts.[*64]

The plutonium separation process resulted in the release of radioactive isotopes into the air, which were carried by the wind throughout southeastern Washington and into parts of Idaho, Montana, Oregon, and British Columbia.[*63] Downwinders were exposed to radionuclides, particularly iodine-131, with the heaviest releases during the period from 1945 to 1951. These radionuclides entered the food chain via dairy cows grazing on contaminated fields; hazardous fallout was ingested by communities who consumed radioactive food and milk. Most of these airborne releases were a part of Hanford's routine operations, while a few of the larger releases occurred in isolated incidents. In 1949, an intentional release known as the "Green Run" released 8,000 curies of iodine-131 over two days.[*65] Another source of contaminated food came from Columbia River fish, an impact felt disproportionately by Native American communities who depended on the river for their customary diets.[*63] A U.S. government report released in 1992 estimated that 685,000 curies of radioactive iodine-131 had been released into the river and air from the Hanford site between 1944 and 1947.[*66]

Salmon spawning in the Hanford Reach near the H-Reactor

Beginning in the 1960s, scientists with the U.S. Public Health Service published reports about radioactivity released from Hanford, and there were protests from the health departments of Oregon and Washington. In response to an article in the Spokane Spokesman Review in September 1985, the Department of Energy announced to declassify environmental records and, in February 1986, released

19,000 pages of previously unavailable historical documents about Hanford's operations.[*][63] The Washington State Department of Health collaborated with the citizen-led Hanford Health Information Network (HHIN) to publicize data about the health effects of Hanford's operations. HHIN reports concluded that residents who lived downwind from Hanford or who used the Columbia River downstream were exposed to elevated doses of radiation that placed them at increased risk for various cancers and other diseases.[*][63] A mass tort lawsuit brought by two thousand Hanford downwinders against the federal government has been in the court system for many years.[*][67] Two of six plaintiffs who went to trial in 2005 were awarded $500,000 in damages.[*][68]

On February 15, 2013, Governor Jay Inslee announced that a tank storing radioactive waste at the site had been leaking liquids on average of 150 to 300 gallons per year. He said that the leak posed no immediate health risk to the public, but said that should not be an excuse for not doing anything.[*][69] On February 22, 2013, the Governor stated that "6 more tanks at Hanford site" were "leaking radioactive waste" [*][70] As of 2013, there are 177 tanks at Hanford, 149 of which have a single shell. Historically single shell tanks were used for storing radioactive liquid waste and designed to last 20 years. By 2005, some liquid waste was transferred from single shell tanks to (safer) double shell tanks. A substantial amount of residue remains in the older single shell tanks with one containing an estimated 447,000 gallons of radioactive sludge, for example. It is believed that up to six of these "empty" tanks are leaking. Two tanks are reportedly leaking at a rate of 300 gallons (1,136 liters) per year each, while the remaining four tanks are leaking at a rate of 15 gallons (57 liters) per year each.[*][71][*][72]

Since 2003, radioactive materials are known to be leaking from Hanford into the environment. "The highest tritium concentration detected in riverbank springs during 2002 was 58,000 pCi/L (2,100 Bq/L) at the Hanford Townsite. The highest iodine-129 concentration of 0.19 pCi/L (0.007 Bq/L) was also found in a Hanford Townsite spring. The WHO guidelines for radionuclides in drinking-water limits levels of iodine-129 at 1 Bq/L, and tritium at 10,000 Bq/L.[*][73] Concentrations of radionuclides including tritium, technetium-99, and iodine-129 in riverbank springs near the Hanford Townsite have generally been increasing since 1994. This is an area where a major groundwater plume from the 200 East Area intercepts the river ... Detected radionuclides include strontium-90, technetium-99, iodine-129, uranium-234, −235, and −238, and tritium. Other detected contaminants include arsenic, chromium, chloride, fluoride, nitrate, and sulfate." [*][74]

13.7.1 Occupational health concerns

Since 1987, workers have reported exposure to harmful vapors after working around underground nuclear storage tanks, with no solution found. More than 40 workers in 2014 alone reported smelling vapors and became ill with "nosebleeds, headaches, watery eyes, burning skin, contact dermatitis, increased heart rate, difficulty breathing, coughing, sore throats, expectorating, dizziness and nausea, ... Several of these workers have long-term disabilities." Doctors checked workers and cleared them to return to work. Monitors worn by tank workers have found no samples with chemicals close to the federal limit for occupational exposure.[*][75]

In August 2014, OSHA ordered the facility to rehire a contractor and pay $220,000 in back wages for firing them for whistleblowing on safety concerns at the site.[*][76]

On November 19, 2014, Washington Attorney General Bob Ferguson said the state planned to sue the DOE and its contractor to protect workers from hazardous vapors at Hanford. A 2014 report by the DOE Savannah River National Laboratory initiated by 'Washington River Protection Solutions' found that DOE's methods to study vapor releases were inadequate, particularly, that they did not account for short but intense vapor releases. They recommended "proactively sampling the air inside tanks to determine its chemical makeup; accelerating new practices to prevent worker exposures; and modifying medical evaluations to reflect how workers are exposed to vapors" .[*][75]

13.8   Cleanup era

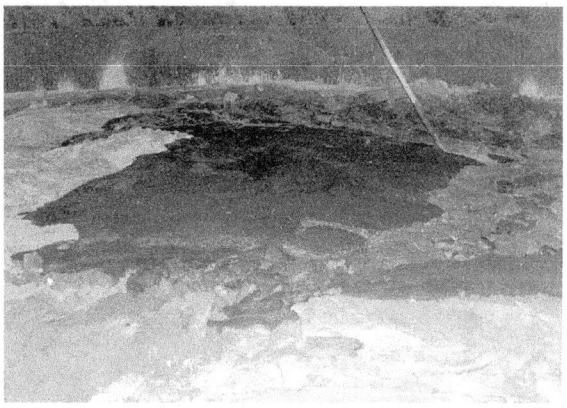

Image of the surface of waste found inside double-shell tank 101-SY at the Hanford Site, April 1989

On June 25, 1988, the Hanford site was divided into four areas and proposed for inclusion on the National Priorities List.[*][77] On May 15, 1989, the Washington Department of Ecology, the United States Environmental Protection Agency, and the Department of Energy entered into the Tri-Party Agreement, which provides a legal framework for environmental remediation at Hanford.[*][10] As of 2014 the agencies are engaged in the world's largest environmental cleanup, with many challenges to be resolved in the face of overlapping technical, political, regulatory, and cultural interests. The cleanup effort is focused on three outcomes: restoring the Columbia River corridor for other uses, converting the central plateau to long-term waste treatment and storage, and preparing for the future.[*][78] The cleanup effort is managed by the Department of Energy under the oversight of the two regulatory agencies. A citizen-led Hanford Advisory Board provides recommendations from community stakeholders, including local and state governments, regional environmental organizations, business interests, and Native American tribes.[*][79] Citing the 2014 Hanford Lifecycle Scope Schedule and Cost report, the 2014 estimated cost of the remaining Hanford clean up is $113.6 billion – more than $3 billion per year for the next six years, with a lower cost projection of approximately $2 billion per year until 2046.[*][80][*][81][*][82] About 11,000 workers are on site to consolidate, clean up, and mitigate waste, contaminated buildings, and contaminated soil.[*][4] Originally scheduled to be complete within thirty years, the cleanup was less than half finished by 2008.[*][82] Of the four areas that were formally listed as Superfund sites on October 4, 1989, only one has been removed from the list following cleanup.[*][83]

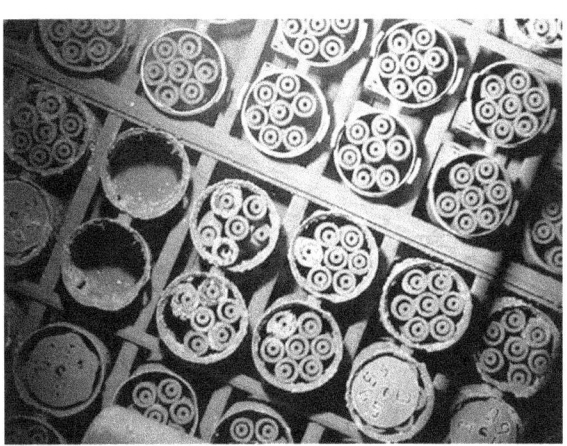

Spent nuclear fuel stored underwater and uncapped in Hanford's K-East Basin

While major releases of radioactive material ended with the reactor shutdown in the 1970s and many of the most dangerous wastes are contained, there are continued concerns about contaminated groundwater headed toward the Columbia River and about workers' health and safety.[*][82]

The most significant challenge at Hanford is stabilizing the 53,000,000 US gallons (200,000,000 l; 44,000,000

imp gal) of high-level radioactive waste stored in 177 underground tanks. By 1998, about a third of these tanks had leaked waste into the soil and groundwater.[*][84] As of 2008, most of the liquid waste has been transferred to more secure double-shelled tanks; however, 2,800,000 US gallons (11,000,000 l; 2,300,000 imp gal)of liquid waste, together with 27,000,000 US gallons (100,000,000 l; 22,000,000 imp gal) of salt cake and sludge, remains in the single-shelled tanks.[*][4] DOE lacks information about the extent to which the 27 double-shell tanks may be susceptible to corrosion. Without determining the extent to which the factors that contributed to the leak in AY-102 were similar to the other 27 double-shell tanks, DOE cannot be sure how long its double-shell tanks can safely store waste.[*][6] That waste was originally scheduled to be removed by 2018. As of 2008, the revised deadline was 2040.[*][82] Nearby aquifers contain an estimated 270,000,000,000 US gallons (1.0×10^{12} l; 2.2×10^{11} imp gal) of contaminated groundwater as a result of the leaks.[*][85] As of 2008, 1,000,000 US gallons (3,800,000 l; 830,000 imp gal) of radioactive waste is traveling through the groundwater toward the Columbia River. This waste is expected to reach the river in 12 to 50 years if cleanup does not proceed on schedule.[*][4] The site includes 25 million cubic feet (710,000 m^3) of solid radioactive waste.[*][85]

Grand opening of the Environmental Restoration Disposal Facility (ERDF)

Under the Tri-Party Agreement, lower-level hazardous wastes are buried in huge lined pits that will be sealed and monitored with sophisticated instruments for many years. Disposal of plutonium and other high-level wastes is a more difficult problem that continues to be a subject of intense debate. As an example, plutonium-239 has a half-life of 24,100 years, and a decay of ten half-lives is required before a sample is considered to cease its radioactivity.[*][86][*][87] In 2000, the Department of Energy awarded a $4.3 billion contract to Bechtel, a San Francisco-based construction and engineering firm, to build a vitrification plant to combine the dangerous wastes with glass to render them stable. Construction began in 2002. The plant was originally scheduled to be operational by 2011, with vitrification completed by 2028.*[82]*[88]*[89] As of 2012, according to a study by the General Accounting Office, there were a number of serious unresolved technical and managerial problems.*[90] As of 2013 estimated costs were $13.4 billion with commencement of operations estimated to be in 2022 and about 3 decades of operation.*[91]

In May 2007, state and federal officials began closed-door negotiations about the possibility of extending legal cleanup deadlines for waste vitrification in exchange for shifting the focus of the cleanup to urgent priorities, such as groundwater remediation. Those talks stalled in October 2007. In early 2008, a $600 million cut to the Hanford cleanup budget was proposed. Washington state officials expressed concern about the budget cuts, as well as missed deadlines and recent safety lapses at the site, and threatened to file a lawsuit alleging that the Department of Energy is in violation of environmental laws.*[82] They appeared to step back from that threat in April 2008 after another meeting of federal and state officials resulted in progress toward a tentative agreement.*[92]

During excavations from 2004 to 2007 a sample of purified plutonium was uncovered inside a safe in a waste trench, and has been dated to about the 1940s, making it the second-oldest sample of purified plutonium known to exist. Analyses published in 2009 concluded that the sample originated at Oak Ridge, and was one of several sent to Hanford for optimization tests of the T-Plant until Hanford could produce its own plutonium. Documents refer to such a sample, belonging to "Watt's group", which was disposed of in its safe when a radiation leak was suspected.*[93]*[94]

Some of the radioactive waste at Hanford was supposed to be stored in the planned Yucca Mountain nuclear waste repository, but after that project was cancelled due to the opposition of citizens of Nevada, Washington State sued. They were joined by South Carolina. Their first suit was dismissed, and second suits have been filed.

A potential radioactive leak was reported in 2013; the clean up was estimated to have cost $40 billion with $115 billion more required.*[95]

13.9   Hanford organizations

The Hanford site operations were initially directed by Colonel Franklin Matthias of the U.S. Army Corps of Engineers. Postwar the Atomic Energy Commission took over, and then the Energy Research and Development Administration. Hanford operations are currently directed by the U.S. Department of Energy. It has been operated under government contract by various private companies over the years – the table which follows summarizes the operating contractors through 2000.*[96]

13.10   Other divisions of the site (historical)

- Plutonium Finishing Plant (PFP) – made plutonium metal for use in weapons*[97]

- B Plant, S Plant, T Plant – processing, separation, and extraction of various chemicals and isotopes*[98]*[99]

- Health Instruments Section – an attempt to keep workers and the environment safe*[99]

- REDOX Plant / C Plant – recovered wasted uranium from World War II processes*[99]

- Experimental Animal Farm and Aquatic Biology Laboratory*[99]

- Technical Center – radiochemistry, physics, metallurgy, biophysics, radioactive sewer, neutralization, metal fab, fuels manufacturing*[99]

- Tank Farms – storage of liquid nuclear waste*[99]

- Metal Recovery Plant / U Plant – recover uranium from tank farms*[99]

- Uranium Trioxide Plant (aka Uranium Oxide Plant aka UO3 Plant) – took output from other plants (i.e. liquid uranyl nitrate hexahydrate from U plant and PUREX plant), made uranium trioxide powder*[98]*[99]*[100]*[101]*[102]*[103]*[104]

- Plutonium-Uranium Extraction Plant / PUREX Plant – extracted useful material from spent fuel waste (also see the PUREX article)*[98]*[99]

- Plutonium Recycle Test Reactor (PRTR) – experimented with alternative fuel mixtures*[97]*[99]*[105]*[106]

- Plutonium Fuels Pilot Plant (PFPP) – see PRTR

13.11 Historic photos

- Inside the PUREX facility

Cooling water retention basins at the F-Reactor

- View of the central plateau from Rattlesnake Mountain

- Underground tank farm with 12 of the site's 177 waste storage tanks

- The government town of Richland in the early days of the site

- Inside one of the waste storage tanks

- Hanford workers lining up for paychecks

• Hanford scientists feeding radioactive food to sheep

• Testing a sheep's thyroid for radiation

• Cold War-era billboard

• "Atomic Frontier Days" parade in Richland

• The Fast Flux Test Facility

13.12 See also

• Black Ops 2 (Green Run location)

• Fernald Feed Materials Production Center

• Harry Shearer's Le Show—a weekly radio show and podcast featuring "Clean, Safe, Too Cheap to Meter", a series of regular updates on the Hanford Site, and nuclear power news around the world.

• Idaho National Laboratory

• James Acord

• Lists of nuclear disasters and radioactive incidents

• Los Alamos National Laboratory

• Oak Ridge National Laboratory

• Paducah Gaseous Diffusion Plant

• Pantex Plant

• *Plutopia*

• Radioactive contamination from the Rocky Flats Plant

• Rocky Flats Plant

• *Safe As Mother's Milk*

• Sandia National Laboratories

• Savannah River Site

• Timeline of nuclear weapons development

13.13 References

[1] "B Reactor". United States Department of Energy. Archived from the original on February 2, 2010. Retrieved January 29, 2007.

[2] "Hanford Site: Hanford Overview". United States Department of Energy. Archived from the original on June 5, 2012. Retrieved February 13, 2012.

[3] "Science Watch: Growing Nuclear Arsenal". *The New York Times.* April 28, 1987. Retrieved January 29, 2007.

[4] "Hanford Quick Facts". Washington Department of Ecology. Archived from the original on June 24, 2008. Retrieved January 19, 2010.

[5] "Hanford Facts". psr.org. Retrieved February 7, 2015.

[6] GAO (November 25, 2014). "Condition of Tanks May Further Limit DOE's Ability to Respond to Leaks and Intrusions- Highlights". U.S. GAO. Retrieved December 22, 2014.

[7] Stang, John (December 21, 2010). "Spike in radioactivity a setback for Hanford cleanup". *Seattle Post-Intelligencer.*

[8] Harden, Blaine; Dan Morgan (June 2, 2007). "Debate Intensifies on Nuclear Waste". *Washington Post.* p. A02. Retrieved January 29, 2007.

[9] Dininny, Shannon (April 3, 2007). "U.S. to Assess the Harm from Hanford". *Seattle Post-Intelligencer.* Associated Press. Retrieved January 29, 2007.

[10] Schneider, Keith (February 28, 1989). "Agreement for a Cleanup at Nuclear Site". *The New York Times.* Retrieved January 30, 2008.

[11] Richard, Terry (November 10, 2015). "Washington's Hanford becomes part of national historical park". *The Oregonian.* Retrieved April 4, 2016.

[12] "The Columbia River at Risk: Why Hanford Cleanup is Vital to Oregon". oregon.gov. August 1, 2007. Archived from the original on June 2, 2010. Retrieved March 31, 2008.

[13] Hanford Cultural Resources Program, U.S. Department of Energy (2002). *Hanford Site Historic District: History of the Plutonium Production Facilities, 1943–1990.* Columbus, OH: Battelle Press. p. 1.12. ISBN 1-57477-133-7.

[14] Seelye, Katharine (June 10, 2000). "Gore Praises Move to Aid Salmon Run". *The New York Times.* Retrieved January 29, 2007.

[15] "Site Map Area and Description". Columbia Riverkeepers. Archived from the original on February 8, 2007. Retrieved January 29, 2007.

[16] Lewis, Mike (April 19, 2002). "In strange twist, Hanford cleanup creates latest boom". *Seattle Post-Intelligencer.* Retrieved January 29, 2007.

[17] "HANFORD A E C, WASHINGTON (453444)". Western Regional Climate Center. Retrieved April 27, 2016.

[18] "Hanford Reach National Monument". *HistoryLink.org: The Online Encyclopedia of Washington State History.* Retrieved January 29, 2007.

[19] Hanford Island Archaeological Site (NRHP #76001870) and Hanford North Archaeological District (NHRP #76001871). National Park Service (2007-01-23). "National Register Information System". *National Register of Historic Places.* National Park Service. (See also the commercial site National Register of Historic Places.)

[20] Gerber, Michele (2002). *On the Home Front: The Cold War Legacy of the Hanford Nuclear Site* (2nd ed.). Lincoln, NE: University of Nebraska Press. pp. 16–22. ISBN 0-8032-7101-8.

[21] Hanford Cultural Resources Program, U.S. Department of Energy (2002). *Hanford Site Historic District: History of the Plutonium Production Facilities, 1943–1990.* Columbus, OH: Battelle Press. p. 1.10. ISBN 1-57477-133-7.

[22] Gerber, Michele (1992). *Legend and Legacy: Fifty Years of Defense Production at the Hanford Site.* Richland, Washington: Westinghouse Hanford Company. p. 6.

[23] Franklin, Matthias (January 14, 1987). *Hanford Engineer Works, Manhattan Engineer District: Early History. Speech to the Technical Exchange Program.*

[24] Second War Powers Act 56 Stat. 176 (1942)

[25] Department of Energy: Hanford. "Department of Energy's Tribal Program: The DOE Tribal Program at Hanford". DOE Hanford.gov. Retrieved April 20, 2014.

[26] Brown, Kate (2013). *Plutopia : nuclear families, atomic cities, and the great Soviet and American plutonium disasters.* New York: Oxford University Press. pp. 33–36. ISBN 978-0-19-985576-6.

[27] Oldham, Kit (March 5, 2003). "Construction of massive plutonium production complex at Hanford begins in March 1943". *History Link.* Retrieved April 6, 2008.

[28] "Needed by E. I. duPont de Nemours & Company for Pacific Northwest (advertisement)". *Milwaukee Sentinel.* June 6, 1944. pp. 1–5. Retrieved March 25, 2013.

[29] Nichols, K. D. *The Road to Trinity* page 138 (1987, Morrow, New York) ISBN 0-688-06910-X

[30] Thayer, H. (1996). *Management of the Hanford Engineer Works in World War II.* New York, NY: American Society of Civil Engineers Press.

[31] Hanford Cultural Resources Program, U.S. Department of Energy (2002). *Hanford Site Historic District: History of the Plutonium Production Facilities, 1943–1990.* Columbus, OH: Battelle Press. p. 1.21–1.23. ISBN 1-57477-133-7.

[32] Gerber, Michele (2002). *On the Home Front: The Cold War Legacy of the Hanford Nuclear Site* (2nd ed.). Lincoln, NE: University of Nebraska Press. pp. 35–36. ISBN 0-8032-7101-8.

[33] Hanford Cultural Resources Program, U.S. Department of Energy (2002). *Hanford Site Historic District: History of the Plutonium Production Facilities, 1943–1990*. Columbus, OH: Battelle Press. p. 1.15, 1.30. ISBN 1-57477-133-7.

[34] Harvey, David; O'Conner, Georganne. "History of the Hanford Site 1943–1990" (PDF). Pacific Northwest National Laboratory. p. 11. Retrieved November 6, 2015.

[35] Hanford Cultural Resources Program, U.S. Department of Energy (2002). *Hanford Site Historic District: History of the Plutonium Production Facilities, 1943–1990*. Columbus, OH: Battelle Press. p. 1.22–1.27. ISBN 1-57477-133-7.

[36] Findlay, John; Bruce Hevly (1995). *Nuclear Technologies and Nuclear Communities: A History of Hanford and the Tri-Cities, 1943–1993*. Seattle, WA: Hanford History Project, Center for the Study of the Pacific Northwest, University of Washington. p. 50.

[37] Hanford Cultural Resources Program, U.S. Department of Energy (2002). *Hanford Site Historic District: History of the Plutonium Production Facilities, 1943–1990*. Columbus, OH: Battelle Press. p. 1.27. ISBN 1-57477-133-7.

[38] Hanford Cultural Resources Program, U.S. Department of Energy (2002). *Hanford Site Historic District: History of the Plutonium Production Facilities, 1943–1990*. Columbus, OH: Battelle Press. p. 1.22. ISBN 1-57477-133-7.

[39] Groves, Leslie (1983). *Now It Can Be Told: The Story of the Manhattan Project*. New York, NY: Da Capo Press. p. xv.

[40] Nichols, Kenneth (1987). *The Road to Trinity*. New York: William Morrow. ISBN 0-688-06910-X. p136

[41] Sanger, S. L. *Working on the Bomb: an Oral History of WWII Hanford*. Portland, Oregon: Continuing Education Press, Portland State University. p. 70.

[42] Sanger, S. L. *Working on the Bomb: an Oral History of WWII Hanford*. Portland, Oregon: Continuing Education Press, Portland State University. interview with Generaux.

[43] Hanford Cultural Resources Program, U.S. Department of Energy (2002). *Hanford Site Historic District: History of the Plutonium Production Facilities, 1943–1990*. Columbus, OH: Battelle Press. p. 1.42–45. ISBN 1-57477-133-7.

[44] "Historical use of thorium at Hanford" (PDF). hanford-challenge.org. Retrieved February 7, 2015.

[45] "Chronology of Important FOIA Documents: Hanford's Semi-Secret Thorium to U-233 Production Campaign" (PDF). hanfordchallenge.org. Retrieved February 7, 2015.

[46] "Questions and Answers on Uranium-233 at Hanford" (PDF). radioactivist.org. Retrieved February 7, 2015.

[47] "Hanford Radioactivity in Salmon Spawning Grounds" (PDF). clarku.edu. Retrieved February 7, 2015.

[48] "Cocooning Hanford Reactors". City of Richland. December 2, 2003. Archived from the original on June 11, 2008. Retrieved January 31, 2008.

[49] NRHP site #92000245. National Park Service (2007-01-23). "National Register Information System". *National Register of Historic Places*. National Park Service. (See also the commercial site National Register of Historic Places.)

[50] "B-Reactor Museum Association". B Reactor Museum Association. January 2008. Retrieved January 29, 2007.

[51] "Big Step Toward B Reactor Preservation". KNDO/KNDU News. March 12, 2008. Retrieved April 6, 2008.

[52] Chemical & Engineering News Vol. 86 No. 35, September 1, 2008, "Hanford's B Reactor gets LANDMARK Status", p. 37

[53] "National Historic Landmarks Program – B Reactor". National Park Service. August 19, 2007. Retrieved January 5, 2009.

[54] "Plutonium: the first 50 years: United States plutonium production, acquisition, and utilization from 1944 through 1994". U.S. Department of Energy. Retrieved January 29, 2007.

[55] Hevly, Bruce; John Findlay (1998). *The Atomic West*. Seattle, WA: University of Washington Press.

[56] Cary, Annette (June 3, 2009). "Fast Flux Test Facility shutdown completed at Hanford". Hanford News. Archived from the original on November 17, 2010.

[57] Twilley, Nicola. "Gravitational Waves Exist: The Inside Story of How Scientists Finally Found Them". *The New Yorker*. ISSN 0028-792X. Retrieved 2016-02-12.

[58] Abbott, B.P.; et al. (2016). "Observation of Gravitational Waves from a Binary Black Hole Merger". *Phys. Rev. Lett.* **116**: 061102. doi:10.1103/PhysRevLett.116.061102. PMID 26918975.

[59] Naeye, Robert (11 February 2016). "Gravitational Wave Detection Heralds New Era of Science". *Sky and Telescope*. Retrieved 12 February 2016.

[60] Castelvecchi, Davide; Witze, Alexandra (11 February 2016). "Einstein's gravitational waves found at last". *Nature News*. doi:10.1038/nature.2016.19361. Retrieved 11 February 2016.

[61] "Submarine". navy.memorieshop.com. Retrieved February 7, 2015.

[62] "Hanford Site Tours". United States Department of Energy. Archived from the original on April 2, 2012. Retrieved April 1, 2012.

[63] "An Overview of Hanford and Radiation Health Effects". Hanford Health Information Network. Archived from the original on January 6, 2010. Retrieved January 29, 2007.

[64] "Radiation Flowed 200 Miles to Sea, Study Finds". *The New York Times*. July 17, 1992. Retrieved January 29, 2007.

[65] Gerber, Michele (2002). *On the Home Front: The Cold War Legacy of the Hanford Nuclear Site* (2nd ed.). Lincoln, NE: University of Nebraska Press. pp. 78–80. ISBN 0-8032-7101-8.

[66] Martin, Hugo (August 13, 2008). "Nuclear site now a tourist hot spot". The Los Angeles Times.

[67] Hanford Downwinders Litigation Website. Downwinders.com. Retrieved on September 27, 2009.

[68] McClure, Robert (May 21, 2005). "Downwinders' court win seen as 'great victory'". *Seattle Post-Intelligencer*. Retrieved January 29, 2007.

[69] "Tank storing radioactive waste leaking in Washington". CNN. February 16, 2013. Retrieved February 15, 2013.

[70] "Washington Gov. Inslee's office: 6 more tanks at Hanford site are leaking radioactive waste". Breaking News. Retrieved February 22, 2013.

[71] "Gov: 6 underground Hanford nuclear tanks leaking | Inquirer News". Newsinfo.inquirer.net. March 23, 2004. Retrieved February 23, 2013.

[72] Johnson, Eric (February 1, 2013). "Radioactive waste leaking from six tanks at Washington state nuclear site". Reuters. Retrieved February 23, 2013.

[73] "Radiological Aspects (water sanitation)" (PDF). *www.who.int*. WHO.

[74] "Hanford Site National Environmental Policy Act (NEPA) Characterization" (PDF). *PNNL-6415 Rev. 17*. Pacific Northwest National Laboratory. September 2005.

[75] Nicholas K. Geranios (November 19, 2014). "Washington to sue over nuclear site's tank vapors". *Katu.com*. Associated Press. Retrieved December 19, 2014.

[76] "OSHA orders Hanford nuclear facility contractor to reinstate worker fired for raising environmental safety concerns". OSHA. August 20, 2014.

[77] "Hanford – Washington Superfund site". U.S. EPA. Retrieved February 3, 2010.

[78] "Hanford Site Tour Script" (PDF). United States Department of Energy. October 2007. Archived from the original (PDF) on February 27, 2008. Retrieved January 29, 2007.

[79] "Hanford Site: Hanford Advisory Board". United States Department of Energy. Retrieved February 14, 2012.

[80] Tri-Party Agreement: Department of Energy, Washington State Department of Ecology and the U.S. Environmental Protection Aency. "2014 Hanford Lifecycle Scope, Schedule and Cost Report" (PDF). *February, 2014*. DOE, WSDE, EPA. Retrieved April 20, 2014.

[81] Cary, Annette (February 21, 2014). "New Hanford clean up price tag is $113.6B". *Yakima Herald*. Retrieved April 20, 2014.

[82] Stiffler, Lisa (March 20, 2008). "Troubled Hanford cleanup has state mulling lawsuit". Seattle Post-Intelligencer.

[83] "Hanford 1100-Area (USDOE) Superfund site". U.S. EPA. Retrieved February 3, 2010.

[84] Wald, Matthew (January 16, 1998). "Panel Details Management Flaws at Hanford Nuclear Waste Site". *The New York Times*. Archived from the original on June 11, 2008. Retrieved January 29, 2007.

[85] Wolman, David (April 2007). "Fission Trip". *Wired Magazine*. p. 78.

[86] Hanson, Laura A. (November 2000). "Radioactive Waste Contamination of Soil and Groundwater at the Hanford Site" (PDF). University of Idaho. Retrieved January 31, 2008.

[87] Gephart, Roy (2003). *Hanford: A Conversation About Nuclear Waste and Cleanup*. Columbus, OH: Battelle Press. ISBN 1-57477-134-5.

[88] Dininny, Shannon (September 8, 2006). "Hanford plant now $12.2 billion". *Seattle Post-Intelligencer*. Retrieved January 29, 2007.

[89] *The Economist*, "Nuclear waste: From bombs to $800 handbags", March 19, 2011, p. 40.

[90] "Hanford Waste Treatment Plant: DOE Needs to Take Action to Resolve Technical and Management Challenges". *GAO-13-38*. General Accounting Office. December 19, 2012. Retrieved May 9, 2013.

[91] Valerie Brown (May 9, 2013). "Hanford Nuclear Waste Cleanup Plant May Be Too Dangerous: Safety issues make plans to clean up a mess left over from the construction of the U.S. nuclear arsenal uncertain". *Scientific American*. Retrieved May 9, 2013. The Vit Plant was supposed to start operating in 2007 and is now projected to begin in 2022. Its original budget was $4.3 billion and is now estimated at $13.4 billion.

[92] Stiffler, Lisa (April 3, 2008). "State steps back from brink of Hanford suit". Seattle Post-Intelligencer. Retrieved May 8, 2008.

[93] Chemical & Engineering News: Antique Plutonium: Manhattan Project-era plutonium is found in a glass jug during Hanford Site cleanup(subscription required)

[94] Annette Cary (January 25, 2009). "Historic plutonium found in safe at Hanford". *TRI-CITY HERALD*. Seattle Pi.

[95] "Possible radioactive leak into soil at Hanford". CBS News. June 21, 2013.

[96] Briggs, J.D. (March 22, 2001). "Historical Time Line and Information about the Hanford Site, Richland, Washington" (PDF). Pacific Northwest National Laboratory. Retrieved February 14, 2012.

[97] Lini, D.C.; L. H. Rodgers / SAIC / FH (March 2002). "Plutonium Finishing Plant Plutonium- Uranium Oxide" (PDF). US DOE. Retrieved October 1, 2009.

[98] "222-S Hanford Site". Advanced Technologies and Laboratories Intl. Retrieved October 1, 2009.

[99] Gerber, M.S. (February 2001). "History of Hanford Site Defense Production (Brief)" (PDF). Fluor Hanford / US DOE. Retrieved October 1, 2009.

[100] Freer, Brian; Charles Conway (June 2002). "History of the Plutonium Production Facilities at the Hanford Site Historic District, 1943–1990. Section 4 – Chemical Separations." (PDF). US DOE. doi:10.2172/807939.

[101] Brevick, Stroup, Funk; et al. (1997). "Supporting Document for the Historical Tank Content Estimate for SY-Tank Farm". US DOE. Retrieved October 5, 2009.

[102] Johnson; et al. (1994). "Historical records of radioactive contamination in biota at the 200 Areas of the Hanford Site". US DOE. Retrieved October 5, 2009.

[103] Carbaugh E.H., Bihl, D.E., and MacLellan, J.A. (January 1, 2003). "Methods and Models of the Hanford Internal Dosimetry Program, PNNL-MA-860" (PDF). Retrieved October 5, 2009.

[104] "Executive Summary Hanford Recycled Uranium Project" (PDF). US DOE. June 2000. doi:10.2172/803918. Retrieved October 5, 2009.

[105] Nuclear Engineering International (January 23, 2014). "Hanford removes plutonium test reactor". neimagazine.com. Retrieved January 24, 2014.

[106] - Department of Energy (January 22, 2014). "Massive Hanford Test Reactor Removed – Plutonium Recycle Test Reactor removed from Hanford's 300 Area". energy.gov. Retrieved January 24, 2014.

13.14 Further reading

- John M. Findlay and Bruce Hevly. *Atomic Frontier Days: Hanford and the American West* (University of Washington Press; 2011) 368 pages; explores the history of the Hanford nuclear reservation and the tricities of Richland, Pasco, and Kennewick, Washington

13.15 External links

- Official Hanford website Department of Energy.

- Washington Department of Ecology – Nuclear Waste Program State agency that regulates Hanford cleanup.

- U.S. Environmental Protection Agency Federal agency that regulates Hanford cleanup.

- Hanford Challenge Hanford watchdog group, based in Seattle.

- Hanford News Current news from the *Tri-City Herald*.

- Hanford Site Environmental Report Detailed annual report on radioactive concentrations measured at the Hanford Site.

- Atomic Heritage Foundation Historic Preservation of Manhattan Project Sites at Hanford.

- "Ranger in Your Pocket" Online tours of Hanford's Manhattan Project sites Atomic Heritage Foundation

- B Reactor Museum Association A collection of Hanford-related documents from a group fighting to preserve the B-100 Reactor at Hanford.

- Contaminated US site faces 'catastrophic' nuclear leak 2008 *New Scientist* report.

- Heart of America Northwest Hanford watchdog group, based in Seattle.

- The Alsos Digital Library for Nuclear Issues Annotated bibliography for the Hanford Site.

- A Review of Data Triples Plutonium Waste Figures Matthew L. Wald, *The New York Times*, July 10, 2010

- Washington River Protection Solutions Hanford environmental remediation contractor.

- Safe as Mother's Milk: The Hanford Project

- Hanford Documentary produced by Oregon Public Broadcasting

Chapter 14

Bikini Atoll

For the post-rock band, see Bikini Atoll (band).

Bikini

Location of Bikini Atoll in the Pacific Ocean

Bikini Atoll (pronounced /ˈbɪkɪˌniː/ or /bɪˈkiːni/; Marshallese: **Pikinni**, [pʲi͡ɯɡu͡ĩnʲːii̯], meaning *coconut place*)[2] is an atoll in the Marshall Islands which consists of 23 islands totaling 3.4 square miles (8.8 km²) surrounding a deep 229.4-square-mile (594.1 km²) central lagoon. It is at the northern end of the Ralik Chain, approximately 87 kilometres (54 mi) northwest of Ailinginae Atoll and 850 kilometres (530 mi) northwest of Majuro. Within Bikini Atoll, Bikini, Eneu, Nam, and Enidrik islands comprise just over 70% of the land area. Bikini and Eneu are the only islands of the atoll that hosted a permanent population. Bikini Island is the northeastern most and largest islet. Before World War II, the atoll was known by its German name, Eschscholtz Atoll.[3]

14.1 Etymology

The island's English name is derived from the German colonial name *Bikini* given to the atoll when it was part of German New Guinea. The German name is transliterated from the Marshallese name for the island, *Pikinni*, ([pʲi͡ɯɡu͡ĩnʲːii̯]), *Pik"* meaning "surface" and "*Ni*" meaning "coconut", or *surface of coconuts.*[2]

14.2 Culture

Main article: Marshallese culture

Before the advent of Western influence, the Bikini is-

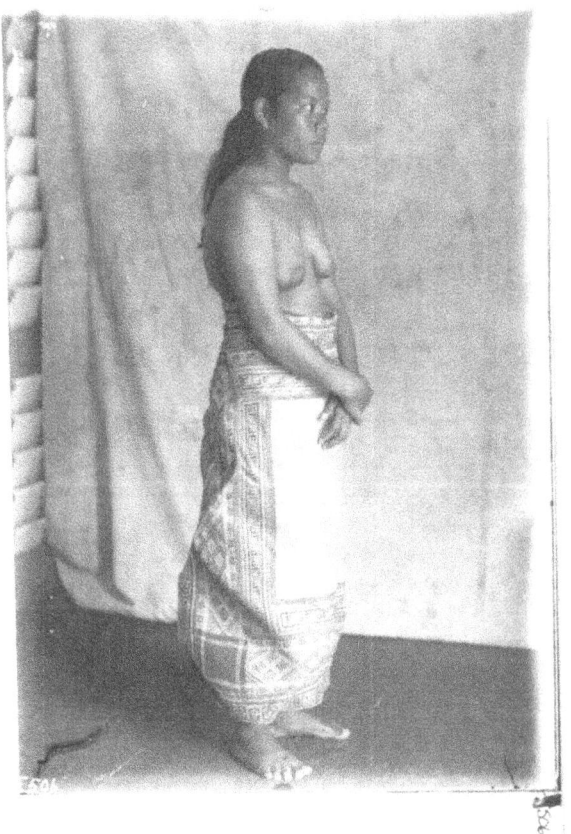

A woman named Liijabor from Likiep Island, Likiep Atoll in the Marshall Islands, wears a traditional nieded *or clothing mat circa 1918*

landers' sustenance-based lifestyle was based on cultivating native plants and eating shellfish and fish. They were skilled boat builders and navigators, sailing the two-hulled proa to and from islets around the Bikini and other atolls in the Marshall Islands.[4] The islanders were relatively isolated and had developed a well-integrated society bound by close extended family association and tradition.[4]

Every lagoon was led by a king and queen and a following of chieftains and chief women who comprised a ruling caste. Some of the leaders maintained Asian-style bungalows and maintained servants, including secretaries, maids, and valets. Poverty was non-existent. The islanders worked the copra plantations under the watchful eye of the Japanese, who took a portion of the sales. Chiefs could retain as much as $20,000 per year, and the remainder was distributed to the workers. The Marshall islanders were formerly aggressive, but the influence of the mission churches eliminated most conflict. They took pride in extending hospitality to one another, even distant relatives.[5]

14.2.1 Clothing and dress

The men wore a fringe skirt of native materials about 25 to 30 inches (60 to 80 cm) long. Women traditionally[6] wore two mats about a yard (metre) square each, made by weaving pandanus and hibiscus leaves together[4] and belted around the waist.[7] Children were usually naked.[4]

The Christian missionaries who began arriving in the late 19th century influenced the islanders' notions of modesty. In 1919, a visitor reported that Marshall Islands women "are perfect models of prudery. Not one would think of exposing her ankles." Women in the Marshall Islands today are still very modest. They believe that a woman's thighs[8] and shoulders should be covered.[9] Women generally wear cotton mu'umu'us or similar clothing that covers most of the body. Personal health is never discussed except within the family, and women are especially private about female-related health issues,[6] although they are willing to talk about their breasts.[6]

Marshall island women swim in mu'umu'us which are made of a fine polyester that quickly dries. In the capital of Majuro, revealing cocktail dresses are inappropriate for both islanders and guests.[9] With the increasing influence of Western media, the younger generation may wear shorts, though the older generation equates shorts with loose morals. T-shirts, jeans, skirts, and makeup are making their way via the media to the islands.[10]

14.2.2 Land-based wealth

The Bikini islanders continue to maintain land rights as the primary measure of wealth.[11]

> To all Marshallese, land is gold. If you were an owner of land, you would be held up as a very important figure in our society. Without land you would be viewed as a person of no consequence... But land here on Bikini is now poison land.[12]

Each family is part of a clan (*Bwij*), which owns all land. The clan owes allegiance to a chief (*Iroij*). The chiefs oversee the clan heads (*Alap*), who are supported by laborers (*Dri-jerbal*). The Iroij control land tenure, resource use and distribution, and settle disputes. The Alap supervise land maintenance and daily activities. The Dri-jerbal work the land including farming, cleaning, and construction.[4]

The Marshallese society is matrilineal and land is passed down from generation to generation through the mother. Land ownership ties families together into clans. Grandparents, parents, grandchildren, aunts, uncles, and cousins form extended, close-knit family groups. Gatherings tend to become big events. One of the most significant family events is the first birthday of a child (*kemem*), which relatives and friends celebrate with feasts and song.[4][13]

Payments made in the 20th century as reparations for damage to the Bikini Atoll and the islanders' way of life have elevated their income relative to other Marshall Island residents. It has caused some Bikini islanders to become economically dependent on the payments from the trust fund. This dependency has eroded individuals' interest in traditional economic pursuits like taro and copra production. The move also altered traditional patterns of social alliance and political organization. On Bikini, rights to land and land ownership were the major factor in social and political organization and leadership. After relocation and settlement on Kili, a dual system of land tenure evolved. Disbursements from the trust fund were based in part to land ownership on Bikini and based on current land tenure on Kili.[14]

Before the residents were relocated, they were led by a local chief and under the nominal control of the Paramount Chief of the Marshall Islands. Afterward, they had greater interaction with representatives of the trust fund and the U.S. government and began to look to them for support.[14]

14.2.3 Language

Most Marshallese speak both the Marshallese language and at least some English. Government agencies use Marshallese. One important word in Marshallese is "yokwe" which is similar to the Hawaiian "aloha" and means "hello", "goodbye" and "love" .[*][15]

14.3 Environment

Vegetation on the Bikini Atoll

The Bikini Atoll is part of the Ralik Chain (for "sunset chain") within the Marshall Islands.

14.3.1 Nuclear test site

Main article: Nuclear testing at Bikini Atoll

Between 1946 and 1958, 23 nuclear devices were detonated by the United States at seven test sites located on the reef, inside the atoll, in the air, or underwater.[*][16] They had a combined fission yield of 42.2 Mt. The testing began with the Operation Crossroads series in July 1946. Prior to nuclear testing, the residents initially accepted resettlement voluntarily to Rongerik Atoll, believing that they would be able to return home within a short time. Rongerik Atoll could not produce enough food and the islanders starved. When they could not return home, they were relocated to Kwajalein Atoll for six months before choosing to live on Kili Island, a small island one-sixth the size of their home island. Some were able to return to the Bikini Island in 1970 until further testing revealed dangerous levels of strontium-90. The islanders have been the beneficiary of several trust funds created by the United States government which as of 2013 covered medical treatment and other costs and paid about $550 annually to each individual.

14.3.2 Geography

Main article: Geography of the Marshall Islands

There are 23 islands in the Bikini Atoll; the islands of Bokonijien, Aerokojlol, and Nam were vaporized during the nuclear tests.[*][17] The islands are composed of low coral limestone and sand.[*][15] The average elevation is only about 7 feet (2.1 m) above low tide level. The total lagoon area is 229.4-square-mile (594.1 km^2). The primary home of the islanders was the most northeast and largest islet, Bikini Island, totaling 586 acres (237 ha) and 4 kilometres (2.5 mi) long.

14.3.3 Flora and fauna

The islanders cultivated native foods including coconut, pandanus, papaya, banana, arrowroot, taro, limes, breadfruit, and pumpkin. A wide variety of other trees and plants are also present on the islands.[*][17]

The islanders were skilled fishermen. They used fishing line made from coconut husk and hooks from sharpened sea shells. They used more than 25 methods of fishing.[*][4] The islanders raised ducks, pigs, and chickens for food and kept dogs and cats as pets. Animal life in the atoll was severely impacted by the atomic bomb testing. Existing land species include small lizards, hermit crabs, and coconut crabs. The islands are frequented by a wide variety of birds.[*][17]

To allow vessels with a larger draft to enter the lagoon and to prepare for the atomic bomb testing, the United States used explosives to cut a channel through the reef and to blow up large coral heads in the lagoon. The underwater nuclear explosions carved large holes in the bottom of the lagoon that were partially refilled by blast debris. The explosions distributed vast amounts of irradiated, pulverized coral and mud across wide expanses of the lagoon and surrounding islands. As of 2008, the atoll had recovered nearly 65% of the biodiversity that existed prior to radioactive contamination, but 28 species of coral appear to be locally extinct.[*][16]

14.3.4 Climate

The islands are hot and humid. The temperature on Bikini Atoll is a constant 80 to 85 °F (27 to 29 °C) year round. The water temperature is also 80 to 85 °F (27 to 29 °C) all year. The islands border the Pacific typhoon belt. The wet season is from May to December while the trade winds from January through May produce higher wave action.[*][17]

14.4 Resident and non-resident population

When the United States asked the islanders to relocate in 1946, 19 islanders lived elsewhere. The 167 residents comprising about 40 families[18] who lived on the atoll voluntarily moved to Rongerik Atoll, and then to Kwajalein Atoll, and once again in November 1948 to Kili Island, when the population numbered 184. They were later given public lands on Ejit and a few families initially moved there to grow copra. In 1970, about 160 Bikini islanders returned to live on the atoll after they were reassured that it was safe. They remained for about 10 years until scientists found an 11-fold increase in the caesium-137 body burdens and determined that the island wasn't safe after all. The 178 residents were evacuated in September 1978 once again.[11]

Since then a number of descendants have moved to Majuro (the Marshall Islands' capital), other Marshall Islands, and the United States. In 1999, there were 2600 total individuals; 1000 islanders living on Kiji, 700 in Majuro, 275 on Ejit, 175 on other Marshall Islands or atolls, and 450 in the United States. Of those, 81 were among those who left the atoll in 1946.[19] In 2001, the population of the dispersed islanders was 2800.[20]

As of February 2013, there were 4880 living Bikini islanders: 1250 islanders living on Kili, 2150 on Majuro, 280 on Ejit, 350 on other Marshall Islands, and 850 in the United States and other countries. Of that number, 31 lived on Bikini in 1946.[17] The resident population of the atoll is currently 4–6 caretakers,[1][20] including Edward Maddison. Maddison has lived on Bikini Island since 1985. His grandfather was one of the original residents relocated in 1947.[21] He also helps the U.S. Department of Energy with soil monitoring, testing cleanup methods, mapping the wrecks in the lagoon, and accompanying visitors on dives.[22] He is also the divemaster of Bikini Atoll Divers.[22]

14.5 Government

Main article: Politics of the Marshall Islands

The Bikini islanders were historically loyal to a king, or Irojj. After the Marshall Islands separated from the United States in the Compact of Free Association in 1986, its constitution established a bicameral parliament. The upper house is only a consultative body. It consists of traditional leaders (Iroijlaplap), known as the Council of Irooj, who advise the lower house on traditional, cultural issues.[23] As of 2013, there are four members of the Council.

The lower house or Nitijela consists of 33 senators elected by 24 electoral districts. Universal suffrage is available to all citizens 18 years of age and older. The 24 electoral districts correspond roughly to each Marshall Islands atoll. The lower house elects the president who, with the approval of the Nitijela, selects a cabinet from among members of the Nitijela.[24][25]

14.5.1 Local government

Four district centers in Majuro, Ebeye, Jaluit, and Wotje provide local government. Each district elects a council and mayor and may appoint local officials. The district centers are funded by the national government and by local revenues. There are two political parties. Elections are held every four years. In 2011 Nishma Jamore was elected mayor of the district representing the Bikini people. Council members are elected from two wards on Ejit Island (three seats) and Kili Island (12 seats).[24]

14.5.2 U.S. liaison

The local government works with a U.S. paid Liaison Officer for Bikini Atoll Local Government, Jack Niedenthal, who is acting Bikini/Kili/Majuro Projects Manager. He is also the Tourism Operations Manager and oversees Bikini Atoll Divers.

14.6 History

Human beings have inhabited the Bikini Atoll for about 3,600 years.[26] U.S. Army Corps of Engineers archaeologist Charles F. Streck, Jr., found bits of charcoal, fish bones, shells and other artifacts under 3 feet (1 metre) of sand. Carbon-dating placed the age of the artifacts at between 1960-1650, B.C. Other discoveries on Bikini and Eneu island were carbon-dated to between 1,000 B.C. and 1 B.C., and others between 400-1,400 A.D.[27]

The first recorded sighting by Europeans was in September 1529 by the Spanish navigator Álvaro de Saavedra on board his ship La Florida when trying to return to New Spain, and was charted as Buenos Jardines (Good Gardens in Spanish).[28] The Marshalls lacked the wealth to encourage exploitation or mapping. The British captain Samuel Wallis chanced upon Rongerik and Rongelap atolls while sailing from Tahiti to Tinian. The British naval captains John Marshall and Thomas Gilbert partially explored the Marshalls in 1788.[29]

The first Westerner to see the atoll in the mid-1820s was the Baltic German captain and explorer Otto von Kotzebue,

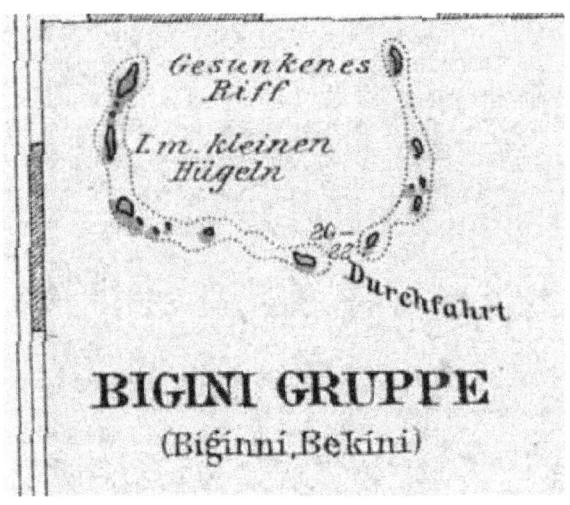

Map of Bikini Atoll, taken from the 1893 map Schutzgebiet der Marshall Inseln, *published in 1897.*

sailing in service of the Russian Empire. He visited three times during 1816 and 1817.[30] He named the atoll *Eschscholtz Atoll* after Johann Friedrich von Eschscholtz, the naturalist of von Kotzebue's ship. The Baltic Germans used the atoll to produce copra oil from coconuts, although contact with the native population was infrequent. The atoll's climate is dryer than the more fertile southern Marshall Islands which produced more copra. Bikini islanders were recruited into developing the copra trade during the German colonial period.[14]

14.6.1   Christian missionaries arrive

Protestant missionaries from the American Board of Commissioners for Foreign Missions arrived on Ebon, in the southern Ralik Chain, in 1857. They first introduced the islanders to Christianity in 1857, which gradually displaced their native religion.[31][32]

14.6.2   Spanish-German Treaty of 1899

Main article: German–Spanish Treaty (1899)

The explosion in Havana Harbor of the battleship USS Maine—sent by the U.S. to protect American commercial interests in Cuba—led to the Spanish–American War in 1898. It resulted in Spain's losing many of its remaining colonies, Cuba became independent while the United States took possession of Puerto Rico and Spain's Pacific colonies of the Philippines and Guam. This left Spain with the remainder of the Spanish East Indies in the Pacific, about 6000 islands that were tiny, sparsely populated. After the

loss of the administrative center of Manila, the minor islands became ungovernable and, after the entire loss of two Spanish fleets in 1898, indefensible. The year is still known in Spain as the "Year of the national disaster" or "the loss of the 400 years Empire".

The Spanish government sold the islands to Germany.[24] The treaty was signed on 12 February 1899, by Spanish Prime Minister Francisco Silvela and transferred the Caroline Islands, the Mariana Islands, Palau and other possessions to Germany. The islands were then placed under control of German New Guinea. The first American missionary arrived in 1908.

14.6.3   Japanese occupation

Bikini was captured along with the rest of the Marshall Islands by the Imperial Japanese Navy in 1914 during World War I and mandated to the Empire of Japan by the League of Nations in 1920. The Japanese administered the island under the South Pacific Mandate, but mostly left local affairs in the hands of traditional local leaders until the start of World War II. At the outset of the war, the Marshall Islands suddenly became a strategic outpost for the Japanese. They built and manned a watchtower on the island, an outpost for the Japanese headquarters on Kwajalein Atoll, to guard against an American invasion of the islands.[33]

14.6.4   World War II

Main article: Battle of Kwajalein

The islands remained relatively unscathed by the war until February 1944, when in a terrific bloody battle, the American forces captured Kwajalein Atoll. There were only five Japanese soldiers on Bikini and they committed suicide rather than allow themselves to be captured.[33]

For the U.S., the battle represented both the next step in its island-hopping march to Japan and a significant moral victory because it was the first time the Americans had penetrated the "outer ring" of the Japanese Pacific sphere. For the Japanese, the battle represented the failure of the beach-line defense. Japanese defenses became prepared in depth, and the battles of Peleliu, Guam, and the Marianas proved far more costly to the U.S.

14.7   Residents relocated

Main article: Nuclear testing at Bikini Atoll

After World War II, the United States was engaged in a

7 March 1946, 161 residents of Bikini Island board LST 1108 as they depart from Bikini Atoll

Bikini islanders arrive on Rongerik Atoll and unload pandanus for thatching the roofs of their new buildings.[34]

Cold War Nuclear arms race with the Soviet Union to build bigger and more destructive bombs.*[33]

The nuclear weapons testing at Bikini Atoll program was a series of 23 nuclear devices detonated by the United States between 1946 and 1958 at seven test sites. The test weapons were detonated on the reef itself, on the sea, in the air and underwater*[16] with a combined fission yield of 42.2 Mt. The testing began with the Operation Crossroads series in July 1946. Shortly after World War II ended, President Harry S. Truman directed Army and Navy officials to secure a site for testing nuclear weapons on American war-

ships. While the Army had seen the results of a land-based explosion, the Navy wanted to know the effect of a nuclear weapon on ships. They wanted to determine whether ships could be spaced at sea and in ports in a way that would make nuclear weapons ineffective against vessels.*[35]

Bikini was distant from both regular sea and air traffic, making it an ideal location. In February 1946, Navy Commodore Ben H. Wyatt, the military governor of the Marshall Islands, asked the 167 Micronesian inhabitants of the atoll to voluntarily and temporarily relocate so the United States government could begin testing atomic bombs for "the good of mankind and to end all world wars." After "confused and sorrowful deliberation" among the Bikinians, their leader, King Juda, agreed to the U.S. relocation request, announcing "We will go believing that everything is in the hands of God." *[33] Nine of the eleven family heads, or *alaps*, chose Rongerik as their new home.*[36]

In February, Navy Seabees helped them to disassemble their church and community house and prepare to relocate them to their new home. On 7 March 1946, the residents gathered their personal belongings and saved building supplies. They were transported 125 miles (201 km) eastward on U.S. Navy landing craft 1108 to the uninhabited Rongerik Atoll,*[36] which was one-sixth the size of Bikini Atoll.*[36] No one lived on Rongerik because it had an inadequate water and food supply and due to deep-rooted traditional beliefs that the island was haunted by the *Demon Girls of Ujae*. The Navy left them with a few weeks of food and water which soon proved to be inadequate.*[33]

14.7.1   Nuclear testing program

The weapons testing began with the Operation Crossroads series in July 1946. The *Baker* test's radioactive contamination of all the target ships was the first case of immediate, concentrated radioactive fallout from a nuclear explosion. Chemist Glenn T. Seaborg, the longest-serving chairman of the Atomic Energy Commission, called *Baker* "the world's first nuclear disaster." *[37] This was followed by a series of later tests that left the island contaminated with radioactivity, particularly caesium-137, and uninhabitable.

14.7.2   Strategic Trust Territory

Main article: Trust Territory of the Pacific Islands

In 1947, the United States convinced the United Nations to designate the islands of Micronesia a United Nations Strategic Trust Territory. This was the only trust ever granted by the U.N.*[38] The United States Navy controlled the Trust from a headquarters in Guam until 1951, when the United States Department of the Interior took over control, administering the territory from a base in Saipan.[*][39] The directive stated that the United States should "promote the economic advancement and self-sufficiency of the inhabitants, and to this end shall... protect the inhabitants against the loss of their lands and resources..." [*][33]

Despite the promise to "protect the inhabitants", from July 1946 through July 1947, the residents of Bikini Atoll were left alone on Rongerik Atoll and were starving for lack of food. A team of U.S. investigators concluded in late 1947 that the islanders must be moved immediately. Press from around the world harshly criticized the U.S. Navy for ignoring the people. Harold Ickes, a syndicated columnist, wrote "The natives are actually and literally starving to death." [*][33]

14.7.3 Move to Kili Island

For more details on this topic, see Kili Island.
In January 1948, Dr. Leonard Mason, an anthropologist

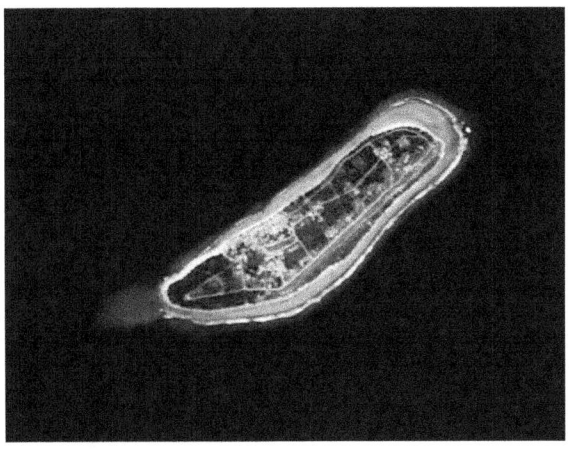

Kili Island is one of the smallest islands in the Marshall Islands.

from the University of Hawaii, visited Rongerik Atoll and was horrified at what he found. One resident of Rongerik commented,[*][12]

> We'd get a few fish, then the entire community would have to share this meager amount... The fish were not fit to eat there. They were poisonous because of what they ate on the reef. We got sick from them, like when your arms and legs fall asleep and you can't feel anything. We'd get up in the morning to go to our canoes and fall over because we were so ill... Then we started asking these men from America [to] bring us food... We were dying, but they didn't listen to us.

Mason requested that food be brought to the islanders on Rongerik immediately along with a medical officer. The Navy then selected Ujelang Atoll for their temporary home and some young men from the Bikini Atoll population went ahead to begin constructing living accommodations. But U.S. Trust Authorities changed their mind. They decided to use Enewetak Atoll as a second nuclear weapons test site and relocated that atoll's residents to Ujelang Atoll instead and to the homes built for the Bikini Islanders.[*][33]

In March 1948, 184 malnourished Bikini islanders were relocated again to Kwajalein Atoll. They were given tents on a strip of grass alongside the airport runway to live in.[*][38] In June 1948, the Bikini residents chose Kili Island as a long-term home.[*][33] The small, 200 acres (81 ha) (.36 square miles (0.93 km^2)) island was uninhabited and wasn't ruled by a paramount *iroij*, or king. In June, the Bikini community chose two dozen men to accompany eight Seabees to Kili to begin construction of a village. In November 1948, the residents, now totaling 184 individuals, moved to Kili Island,[*][33] at 0.93 square kilometres (0.36 sq mi), one of the smallest islands in the Marshall Island chain. They soon learned they could no longer fish the way they had on Bikini Atoll. Kili lacked the calm, protected, bountiful lagoon.[*][38] Living on Kili Island effectively destroyed their culture that had been based on fishing and island-hopping canoe voyages to various islets around the Bikini Atoll.[*][12] Kili does not provide enough food for the transplanted residents.

14.7.4 Failed resettlement

After their relocation to Kili, the Bikini residents continued to suffer from inadequate food supplies. Kili is a small island without a lagoon, and most of the year it is exposed to 10 to 20 feet (3.0 to 6.1 m) waves that make fishing and putting canoes out difficult. Starvation ensued. In 1949, the Trust Territory administration donated a 40 feet (12 m) ship for transporting copra between Kili and Jaluit Atoll, but the ship was wrecked in heavy surf while delivering copra and other fruit.[*][33] The U.S. Trust Authorities airdropped food onto Kili. The residents were forced to rely on imported USDA rice and canned goods and had to buy food with their supplemental income.[*][33]

During 1955 and 1956, ships dispatched by the U.S. Trust Territory continually experienced problems unloading food because of the rough seas around the island, leading to additional food shortages once again. The people once again suffered from starvation and food shortages increased in 1956. The U.S. suggested that some of the Bikini Islanders move to Jaluit where food was more readily available. A few people moved.[*][38]

The United States opened a satellite community for the

families on public land on Jaluit Atoll, 30 miles (48 km) north. Three families moved there to produce copra for sale and other families rotated living there later on.*[33] Their homes on both Kili and Jaluit were struck by typhoons during 1957 and 1958, sinking their supply ship and damaging crops.

14.7.5   Return to Bikini Island

In June 1968, based on scientific advice that the radiation levels were sufficiently reduced, President Lyndon B. Johnson promised the 540 Bikini Atoll family members living on Kili and other islands that they would be able to return to their home. The Atomic Energy Commission cleared radioactive debris from the island, and the U.S. Trust Territory was in charge of rebuilding structures and replanting crops on the atoll. But shortly afterward the Trust Territory ended regular air flights between Kwajalein Atoll and Bikini Atoll which seriously impeded progress. Coconut trees were finally replanted in 1972, but the AEC learned that the coconut crabs retained high levels of radioactivity and could not be eaten. The Bikini Council voted to delay a return the island as a result.*[33]

Three extended families, eventually totaling about 100 people, moved back to their home island in 1972 despite the risk. But 10 years later, a team of French scientists performed additional tests on the island and its inhabitants. They found some wells were too radioactive for use and determined that the pandanus and breadfruit were also dangerous for human consumption. Urine samples from the islanders on Bikini Atoll showed low levels of plutonium 239 and 240. As a result, the Bikini community filed a federal lawsuit seeking a complete scientific survey of Bikini and the northern Marshall Islands. Inter-departmental squabbling over responsibility for the costs delayed the work for three years.*[33] Then in May 1977 scientists found dangerously high levels of strontium-90 in the well water exceeding the U.S. maximum allowed limits.*[40] In June, the Department of Energy stated that "All living patterns involving Bikini Island exceed Federal [radiation] guidelines for thirty year population doses." Later that year scientists discovered an 11-fold increase in the caesium-137 body burdens in all of the people living on the atoll.*[33] In May 1978 officials from the U.S. Department of the Interior described the 75% increase in radioactive caesium 137 found as "incredible".*[11]

Women were experiencing miscarriages, stillbirths, and genetic abnormalities in their children.*[41]*[42] Researchers learned that the coral soil behaved differently from mainland soil because it contains very little potassium. Plants and trees readily absorb potassium as part of the normal biological process, but since caesium is part of the same

group on the periodic table, it is absorbed by plants in a very similar chemical process. The islanders who unknowingly consumed contaminated coconut milk were found to have abnormally high concentrations of caesium in their bodies. The Trust Territory decided that the islanders had to be evacuated from the atoll a second time.*[43]*[44]

The islanders received USD$75 million in damages in 1986 as part of a new Compact of Free Association with the U.S. and in 1988, another $90 million to be used specifically for radiological cleanup. In 1987, a few Bikini elders traveled to Eneu island to reestablish old property lines. Construction crews began building a hotel on Bikini, docks, roads, and installed generators, desalinators, and power lines. A packed coral and sand runway still exists on Eneu Island. The Bikini Atoll Divers was established to provide income. But in 1995, council learned that the US Environmental Protection Agency standard required reducing radiation levels to 15 millirems, substantially less than the US Department of Energy standard of 100 millirems. This discovery significantly increased the potential cost of cleanup and stalled the effort.*[20]

14.7.6   Relocation to Kili Island

As a result of the military use of the island and the failed resettlement, the islands are littered with abandoned concrete bunkers and tons of heavy equipment, vehicles, supplies, machines, and buildings.*[45] In September 1978, Trust Territory officials finally arrived to relocate the residents. The radiological survey of the northern Marshalls, compelled by the 1975 lawsuit, began only after the residents were removed*[33] and returned to Kili Island.*[33]

As of 2013, the tiny 0.93 square kilometres (0.36 sq mi) Kili Island supported about 600 residents who live in cinderblock houses. They must rely on contributions from a settlement trust fund to supplement what they produce locally. Each family receives 1-2 boxes of frozen chicken, 2-4 23 kilograms (51 lb) bags of flour, and 2-4 bags of rice 2-3 times per year. The islanders operate several small stores out of their homes to supply nonperishable food items like salt, Tabasco, candy, and canned items. A generator provides electricity.

Children attend elementary school on Kili through eighth grade. Toward the end of the eighth grade, students must pass a standardized test to gain admission to attend public high school in Jaluit or Majuro.

Beginning in 2011 the resettled residents of Kili Island began to experience periods of ocean flooding during king tides aggravated by what they believe are the effects of global warming. The highest point of Kili Island is only 3 metres (9.8 ft) above sea level. Ocean waves have covered portions of the island at least five times from 2011 to 2015, contaminating the wells on the island. The runway servicing the island is unusable during and after rains and ocean flooding because it becomes extremely muddy. In August 2015, the Bikini Council passed a resolution requesting assistance from US government to modify terms of the Resettlement Trust Fund for the People of Bikini to be used to relocate the population once again, this time outside of the Marshall Islands.*[20]*[46]

14.8   Trust funds and failed claims

In 1975, when the islanders who had returned to Bikini Atoll learned that it wasn't safe, they sued the United States for the first time, demanding a radiological study of the northern islands.*[47]

In 1975, the United States set up *The Hawaiian Trust Fund for the People of Bikini*, totaling $3 million. When the islanders were removed from the island in 1978, the U.S. added $3 million to the fund. The U.S. created a second trust fund, *The Resettlement Trust Fund for the People of Bikini*, containing $20 million in 1982. The U.S. added another $90 million to that fund to pay to clean up, reconstruct homes and facilities, and resettle the islanders on Bikini and Eneu islands.*[48]

In 1983, the U.S. and the Marshall islanders signed the Compact of Free Association, which gave the Marshall Islands independence. The Compact became effective in 1986 and was subsequently modified by the Amended Compact that became effective in 2004.*[49] It also established the Nuclear Claims Tribunal, which was given the task of adjudicating compensation for victims and families affected by the nuclear testing program. Section 177 of the compact provided for reparations to the Bikini islanders and other northern atolls for damages. It included $75 million to be paid over 15 years.*[48] On 5 March 2001, the Nuclear Claims Tribunal ruled against the United States for damages done to the islands and its people.*[33]

The payments began in 1987 with $2.4 million paid annually to entire Bikini population, while the remaining $2.6 million is paid into *The Bikini Claims Trust Fund*. This trust is intended to exist in perpetuity and to provide the islanders a 5% payment from the trust annually.*[48]

The United States provided $150 million in compensation for damage caused by the nuclear testing program and their displacement from their home island.*[50]

In 2001, the Nuclear Claims Tribunal awarded the islanders a total of $563,315,500 after deducting past awards. However, the U.S. Congress has failed to fund the settlement. The only recourse is for the Bikini people to petition the U.S. Congress to fund the payment and fulfill this award. The United States Supreme Court turned down the islanders' appeal of the United States Court of Appeals decision that refused to compel the government to fund their claim. By 2001, of the original 167 residents who were relocated, 70 were still alive, and the entire population has grown to 2800.*[12] Most of the islanders and their descendents lived on Kili, in Majuro, and in the United States. In 2012, only 37 still lived.*[20]

The Hawaiian Trust Fund for the People of Bikini was liquidated as required by law in December 2006. The value of *The Resettlement Trust Fund for the People of Bikini* as of 31 March 2013 was approximately $82 million and *The Bikini Claims Trust Fund* was worth approximately $60 million. Each member of the trust received about $550 a year, making them relatively well-off (compared to other Marshall Island residents).*[48] In 2012, the trusts produced about USD$6 to $8 million annually in investment income, and the trusts paid out about USD$15,000 per family each year in benefits.*[20]

Representatives for the Bikini people expect this process to take many years and do not know whether the United States will honor the terms of the Compact of Free Association.*[48]

14.9   World Heritage Site

Because the site bears direct tangible evidence of the nuclear tests conducted there amid the paradoxical tropical location, UNESCO determined that the atoll symbolizes the dawn of the nuclear age and named it a World Heritage Site on 3 August 2010.*[51]*[52]

> Bikini Atoll has conserved direct tangible evidence ... conveying the power of ... nuclear tests, i.e. the sunken ships sent to the bottom of the lagoon by the tests in 1946 and the gigantic Bravo crater. Equivalent to 7,000 times the force of the Hiroshima bomb, the tests had major consequences on the geology and natural environment of Bikini Atoll and on the health of those who were exposed to radiation. Through its history, the atoll symbolises the dawn of the nuclear age, despite its paradoxical image of peace and of earthly paradise.*[51]*[53]

14.10   Visitor access

Bikini Atoll is open to visitors aboard vessels that are completely self-sufficient if they obtain prior approval. They must also pay for a diver and two local government council representatives to accompany them. The local representation is required to verify that visitors don't remove artifacts from the wrecks in the lagoon.*[54]

14.10.1 Bikini Lagoon diving

In June 1996, the Bikini Council authorized diving operations as a means to generate income for Bikini islanders currently and upon their eventual return. The Bikini Council hired dive guide Edward Maddison who had lived on Bikini Island since 1985 and Fabio Amaral, a Brazilian citizen at the time, as head divemaster and resort manager.*[22] The tours are limited to fewer than a dozen experienced divers a week, cost more than US$5,000, and include detailed histories of the nuclear tests. The operation brought in more than $500,000 during the season from May to October during 2001.*[55]

14.10.2 On-shore facilities

To accommodate the dive program and anglers, the Bikini Council built new air-conditioned rooms with private bathrooms and showers. They included verandas overlooking the lagoon. There was a dining facility that served American-style meals. The head chef Mios Maddison also prepared Marshallese dishes featuring fresh seafood. Only 12 visitors were hosted at one time.*[22] Because of the lingering contamination, all fruits and vegetables used for the Bikini Atoll dive and sport fishing operation were imported.*[19] In September 2007, the last of Air Marshall Islands' commuter aircraft ceased operations when spare parts could not be located and the aircraft were no longer airworthy. A half dozen divers and a journalist were stranded for a week on Bikini Island.*[38] The Bikini islanders suspended land-based dive operations beginning in August 2008.

14.10.3 Live aboard diving program

In October 2010, a live-aboard, self-contained vessel successfully conducted dive operations. In 2011, the local government licensed the live-aboard operator as the sole provider of dive expeditions on the nuclear ghost fleet at Bikini Atoll. The dive season runs from May through October. As of 2013, the 12-day dive trip costs US$5,100 per person.Visitors are still able to land on the island for brief stays.*[17]

Because the lagoon has remained undisturbed for so long, it contains a larger amount of sea life than usual, including sharks, which increases divers' interest in the area.*[1]

Visibility depth is over 100 feet (30 m). The lagoon is immensely popular with divers and is regarded as among the top 10 diving locations in the world.*[22] As of 2016, the dive program was managed by Indie Traders.*[56]

Dive visitors receive a history lesson along with the dive experience, including movies and complete briefings about each of the ships, their respective histories, and a tour of the island and the atoll.*[55] Divers are able to visit the USS Saratoga.*[55]

As of 2016, Air Marshall Islands operates one Bombardier Dash 8 Q100 aircraft and one 19-seat Dornier Do 228.*[57] Only vessels that are fully self-contained who make prior arrangements can currently visit the atoll.*[58]

14.10.4 Sportfishing

Bikini Island authorities opened sport fishing to visitors along with diving. Although the atomic blasts obliterated three islands and contaminated much of the atoll, after 50 years the coral reefs have largely recovered. The reefs attract reef fish and their predators: 30 pounds (14 kg) dogtooth tuna, 20 pounds (9.1 kg) barracuda, and bluefin trevally as big as 50 pounds (23 kg). Given the long-term absence of humans, the Bikini lagoon offers sportsmen one of the most pristine fishing environments in the world.*[20]

14.10.5 Shipwrecks

Shipwrecks in the lagoon include the following:

- USS *Saratoga* (CV-3) - aircraft carrier

- USS *Arkansas* (BB-33) - battleship

- USS *Gilliam* (APA-57) - attack transport

- USS *Carlisle* (APA-69) - attack transport

- USS *Lamson* (DD-367) - destroyer

- USS *Anderson* (DD-411) - destroyer

- USS *Apogon* (SS-308) - submarine

- USS *Pilotfish* (SS-386) - submarine

- Japanese battleship *Nagato* - battleship

- Japanese cruiser *Sakawa* - light cruiser

14.11 Current habitable state

In 1998 an IAEA advisory group, formed in response to a request by the Government of the Marshall Islands for an independent international review of the radiological conditions at Bikini Atoll, recommended that Bikini Island should not be permanently resettled under the present radiological conditions.[59]

The potential to make the island habitable has substantially improved since then. A 2012 assessment from Lawrence Livermore National Laboratory found that caesium-137 levels are dropping considerably faster than anyone expected. Terry Hamilton, scientific director of Lawrence Livermore National Laboratory's Marshall Islands Dose Assessment and Radioecology Program, reported that "Conditions have really changed on Bikini. They are improving at an accelerated rate. By using the combined option of removing soil and adding potassium, we can get very close to the 15 millirem standard. That has been true for roughly the past 10 years. So now is the time when the Bikinians, if they desired, could go back." [20]

As of 2013, about 4,880 Bikini people live on Kili and other Marshall Islands, and some have emigrated to the United States. Bikini Island is currently visited by a few scientists and inhabited by 4–6 caretakers.[1][60] The islanders want the top soil removed, but the money is not there for the cleanup. The opportunity for some Bikini islanders to potentially relocate back to their home island creates a dilemma. While the island may be habitable in the near term, virtually all of the islanders alive today have never lived there. Most of the younger generation have never visited. As of 2013, unemployment in the Marshall Islands was at about 40 percent. The population is growing at a four-percent growth rate, so increasing numbers are taking advantage of terms in the Marshall Islands' Compact of Free Association that allow them to obtain jobs in the United States.[20]

After the islanders were relocated in 1946, while the Bikini islanders were experiencing starvation on Rongerik Atoll, Lore Kessibuki wrote an anthem for the island:[20]

> *No longer can I stay, it's true*
>
> *No longer can I live in peace and harmony*
>
> *No longer can I rest on my sleeping mat and pillow*
>
> *Because of my island and the life I once knew there*
>
> *The thought is overwhelming*
>
> *Rendering me helpless and in great despair.*

14.12 In popular culture

14.12.1 Television shows

The animated television series *SpongeBob SquarePants* primarily takes place in the city of Bikini Bottom, which is frequently considered to be located beneath Bikini Atoll.[61][62] Voice actor Tom Kenny, who portrays the series' main character, confirmed that the fictitious city was named after Bikini Atoll, but denied a fan theory that the series' characters held any relation to the real-life nuclear testing that occurred in the atoll. Also, not coincidentally, is shown in the theme song and in other parts of episodes like The Season 9 episode, Lame and Fortune.[63]

14.12.2 Swimsuit design

Main article: bikini

The atoll's name became well-known worldwide as a result of the nuclear tests. On 5 July 1946, five days after the first nuclear device (nicknamed *Able*) was detonated over the Bikini Atoll during Operation Crossroads,[64] Louis Réard introduced a new swimsuit design named the *bikini* after the atoll. Réard was a French mechanical engineer by training and manager of his mother's lingerie shop in Paris. He introduced the new garment to the media and public[65] on 5 July 1946[66] at Piscine Molitor, a public pool in Paris.[67]

He hired Micheline Bernardini, a 19-year-old nude dancer from the Casino de Paris,[68] to demonstrate his design. It featured a g-string back of 30 square inches (200 cm^2) of cloth with newspaper-type print and was an immediate sensation. Bernardini received 50,000 fan letters, many of them from men.[67][69] Réard hoped that his swimsuit's revealing style would create an "explosive commercial and cultural reaction" similar in intensity to the 1946 nuclear explosion at Bikini Atoll.[70][71][72][73][71][72][73] Fashion writer Diana Vreeland described the bikini as the "atom bomb of fashion".[74]

Ironically, the bikini's design violates the Marshall Islanders' modern customs of modesty because it exposes a woman's thighs and shoulders.[8][9] However, before contact with Western missionaries, Marshall Island women were traditionally topless and still do not sexually objectify female breasts as is common in much of Western society[6] which the bikini *does* cover. Marshall Island women swim in their muumuus which are made of a fine polyester that dries quickly.[9] Wearing a bikini in the Marshall Islands is mainly limited to restricted-access beaches and pools like those at private resorts or on United States government facilities on the Kwajalein Atoll within the Ronald Reagan Ballistic Missile Defense Test Site.[75][76]

14.13   See also

- Operation Castle
- Operation Ivy
- Radio Bikini

14.14   References

14.14.1   Notes

[1] Borrett, Lloyd (March 2013). "Diving the Nuclear Ghost Fleet at Bikini Atoll". Retrieved 20 August 2013.

[2] "Marshallese-English Dictionary-Place Name Index". Retrieved 8 August 2013.

[3] "The Marshall Islands: A Brief History". Retrieved 14 August 2013.

[4] "Introduction to Marshallese Culture". Retrieved 17 August 2013.

[5] McMahon, Thomas J. (November 1919). "The Land of the Model Husband". *Travel*. **34** (1).

[6] Briand, Greta; Peters, Ruth (2010). "Community Perspectives on Cultural Considerations for Breast and Cervical Cancer Education among Marshallese Women in Orange County, California" (PDF). *Californian Journal of Health Promotion* (8): 84–89. Retrieved 25 August 2013.

[7] Bliss, Edwin Munsell (1891). *The Encyclopedia of Missions*. **II**. New York: Funk & Wagnalls.

[8] "Customs". *Marshall Islands*. FIU College of Business Administration. Retrieved 25 August 2013.

[9] "Marshall Islands". Encyclopedai.com. Retrieved 25 August 2013.

[10] "Republic of the Marshall Islands" (PDF). *Culture Grams 2008*. Ann Arbor, Michigan: ProQuest. Retrieved 25 August 2013.

[11] "Bikini History". Archived from the original on 23 June 2007. Retrieved 4 December 2013.

[12] Guyer, Ruth Levy (September 2001). "Radioactivity and Rights". *American Journal of Public Health*. **91** (9, issue 9): 1371–1376. doi:10.2105/AJPH.91.9.1371. PMC 1446783. PMID 11527760.

[13] "Marshallese Culture". Retrieved 16 August 2013.

[14] "Bikini". Countries and their Cultures. Retrieved 12 August 2013.

[15] Destinations / Marshall Islands

[16] Zoe T. Richards, Maria Beger, Silvia Pinca, and Carden C. Wallace (2008). "Bikini Atoll coral biodiversity resilience five decades after nuclear testing" (PDF). *Marine Pollution Bulletin*. **56** (3): 503–515. doi:10.1016/j.marpolbul.2007.11.018. PMID 18187160. Retrieved 13 August 2013.

[17] "Bikini Atoll Reference Facts". Retrieved 12 August 2013.

[18] "Bikini Atoll". Retrieved 12 August 2013.

[19] "Bikini Facts". Retrieved 11 August 2013.

[20] Gwynne, S.C. (5 October 2012). "Paradise With an Asterisk". Outside Magazine. Retrieved 9 August 2013.

[21] Pena, Tony. "Return To Bikini Atoll". Retrieved 20 August 2013.

[22] Kattenburg, Dave. "After the bombs". New Internationalist Magazine. Retrieved 20 August 2013.

[23] "History of the Nitijela". Republic of the Marshall Islands. Retrieved 14 August 2013.

[24] "The Nitijela (Parliament)". Republic of the Marshall Islands. Retrieved 14 August 2013.

[25] "The Presidency and Cabinet". Republic of the Marshall Islands. Retrieved 14 August 2013.

[26] "The Natural History of Enewetak Atoll". 1987. p. 333.

[27] Taggart, Stewart. "Bikini Excavation Indicates Early Man in Micronesia". Associated Press. Retrieved 12 August 2013.

[28] Montana, Alberto *Descubrimientos, exploraciones y conquistas de españoles y portugueses en América y Oceania*, Miguel Salvatella, Barcelona, 1943, p.81

[29] "Marshall Islands". Encyclopedia Britannica. Retrieved 16 August 2013.

[30] "Bikini". Retrieved 16 August 2013.

[31] "Marshall Islands Story Project". *mistories.org*. Retrieved 7 January 2017.

[32] "Culture of Marshall Islands - history, people, traditions, women, beliefs, food, family, social, dress". *www.everyculture.com*. Retrieved 7 January 2017.

[33] Niedenthal, Jack. "A Short History of the People of Bikini Atoll". Retrieved 7 August 2013.

[34] "he Republic of the Marshall Islands and the United States: A Strategic Partnership". Embassy of the Marshall Islands of the United States. Retrieved 18 August 2013.

[35] "Bikini". Newsweek. 1 July 1946. Retrieved 13 August 2013.

[36] "Operation Crossroads - The Official Pictorial Record" (PDF). New York: W. H. Wise and Co. Inc. 1946. p. 21.

[37] Weisgall 1994, p. ix.

[38] Kattenburg, David (December 2012). "Stranded on Bikini". Green Planet Monitor. Retrieved 19 August 2013.

[39] "Trust Territory of the Pacific Islands". University of Hawaii.

[40] "A Short History of the People of Bikini Atoll". Archived from the original on 25 June 2007. Retrieved 27 June 2007.

[41] Hamilton, Chris (4 March 2012). "Survivors of nuke testing seek justice: Marshall Islanders on Maui rally to share nation's story". *Maui News.*

[42] "Victims of the Nuclear Age". Archived from the original on 9 August 2007. Retrieved 22 July 2007.

[43] "Operation Castle". *nuclearweaponarchive.org.* May 17, 2006. Retrieved 2016-05-20.

[44] *The Ghost Fleet of Bikini Atoll* at the Internet Movie Database

[45] "Cruising Bikini Atoll 60 Years after the bomb". July 2006. Retrieved 13 August 2013.

[46] Johnson, Giff (8 August 2015). "Exiled by nuclear testing, rising seas force Bikinians to flee again". RNZI. Retrieved 28 April 2016.

[47] "Despite High Court Denial, Battle Over Bikini Atoll Bombing Endures". Bikini Island Local Government. 26 April 2010. Retrieved 20 August 2013.

[48] "U.S. Reparations for Damages". Bikini Atoll. Retrieved 12 August 2013.

[49] "U.S. Relations With Marshall Island". U.S. Department of State. Retrieved 14 August 2013.

[50] "Marshall Islands Nuclear Claims Tribunal". Archived from the original on 13 June 2007. Retrieved 22 July 2007.

[51] "Bikini Atoll, Nuclear Tests Site" (PDF). *Thirty-fourth Session.* World Heritage Committee. 3 August 2010. Retrieved 6 June 2012.

[52] "Bikini Atoll Nuclear Test Site". UNESCO. Retrieved 7 August 2010.

[53] Bikini Atoll Nuclear Test Site, unesco.org

[54] "Indies Trader". Bikini Atoll Divers. Retrieved 13 August 2013.

[55] Niedenthal, Jack (6 August 2002). "Paradise Lost - 'For the Good of Mankind'". *The Guardian.* Retrieved 11 August 2013.

[56] "Bikini Dive Trip". *IndieTrader.com.* Retrieved 29 April 2016.

[57] Johnson, Giff (July 2013). "Air Marshall faces difficult future". Islands Business. Archived from the original on 22 March 2014. Retrieved 22 March 2014.

[58] "Bikini Atoll Dive Tourism Information". bikiniatoll.com. 2008-08-23. Retrieved 4 December 2013.

[59] "Conditions at Bikini Atoll". International Atomic Energy Agency. Retrieved 12 August 2013.

[60] "Scuba Diving in Bikini Lagoon". diveadventures.com.au. Retrieved 30 October 2009.

[61] "THE HYPE SOAKING IT UP' SPONGEBOB' ACTOR LOVES THE ATTENTION". *Daily News.* Los Angeles, CA. 8 March 2001. Retrieved 30 October 2013. – via HighBeam (subscription required)

[62] QSR Staff (7 June 2001). "Burger King SpongeBob SquarePants". *QSR Magazone.* Archived from the original on 21 October 2007. Retrieved August 19, 2010.

[63] Bradley, Bill (10 Feb 2015). "SpongeBob SquarePants Answers 7 Big Questions And Debunks 1 Popular Theory". The Huffington Post. Retrieved 10 July 2016.

[64] Campbell, Richard H. (2005). *The Silverplate Bombers: A History and Registry of the Enola Gay and Other B-29s Configured to Carry Atomic Bombs.* Jefferson, North Carolina: McFarland & Company. pp. 18, 186–189. ISBN 0-7864-2139-8. OCLC 58554961.

[65] Westcott, Kathryn (5 June 2006). "The Bikini: Not a brief affair". BBC News. Retrieved 17 September 2008.

[66] David Louis Gold (2009). *Studies in Etymology and Etiology: With Emphasis on Germanic, Jewish, Romance and Slavic Languages.* Universidad de Alicante. pp. 99–. ISBN 978-84-7908-517-9. Retrieved 9 March 2013.

[67] Bikini Introduced, This Day in History, *History Channel*

[68] Rosebush, Judson. "Michele Bernadini: The First Bikini". *Bikini Science.* Retrieved 19 September 2007.

[69] Hoover, Elizabeth D. (5 July 2006). "60 Years of Bikinis". American Heritage Inc. Archived from the original on 9 September 2007. Retrieved 13 November 2007.

[70] "The History of the Bikini". Time. 3 July 2009. Retrieved 20 August 2013.

[71] "Tiny Swimsuit That Rocked the World: A History of the Bikini". Randomhistory.com. 1 May 2007. Retrieved 3 December 2011.

[72] Brij V. Lal; Kate Fortune (2000). *The Pacific Islands: an Encyclopedia.* University of Hawaii Press. p. 259. ISBN 978-0-8248-2265-1. Retrieved 5 July 2011.

[73] Foster, Ruth (June 2007). *Nonfiction Reading Comprehension: Social Studies, Grade 5.* Teacher Created Resources. p. 130. ISBN 978-1-4206-8030-0. Retrieved 5 July 2011.

[74] Judson Rosebush, "1945–1950: The Very First Bikini".
     Bikini Science. Retrieved 25 November 2012.

[75] "Marshallese Culture". Safaritheglobe.com. Retrieved 18
     July 2013.

[76] "Marshall Islands Facts, information, pictures". Encyclo-
     pedia.com. Retrieved 18 July 2013.

14.14.2 Bibliography

- Niedenthal, Jack, *For the Good of Mankind: A History of the People of Bikini and their Islands*, Bravo Publishers, (November 2002), ISBN 982-9050-02-5

- Wiesgall, Jonathan M, *Operation Crossroads: Atomic Tests at Bikini Atoll*, Naval Institute Press (21 April 1994), ISBN 1-55750-919-0

14.15 External links

- A Short History of the People of Bikini Atoll

- What About Radiation on Bikini Atoll?

- Department of Energy Marshall Islands Program: Chronology of nuclear testing, relocation of islanders and results of radiation tests

- Annotated bibliography for Bikini Atoll from the Alsos Digital Library for Nuclear Issues

- Islanders Want The Truth About Bikini Nuclear Test

- Marshall Islands site

- Entry at Oceandots.com at the Wayback Machine (archived 23 December 2010)

- Everything Marshall Islands

- Lauren R. Donaldson Collection, served as a radiation monitor for Operation Crossroads; the codename for the first atomic bomb tests at Bikini Atoll. - University of Washington Digital Collection

Chapter 15

Desert Rock exercises

Aerial view of Camp Desert Rock.

Desert Rock was the code name of a series of exercises conducted by the US military in conjunction with atmospheric nuclear tests. They were carried out at the Nevada Proving Grounds between 1951 and 1957.

Their purpose was to train troops and gain knowledge of military maneuvers and operations on the nuclear battlefield. They included observer programs, tactical maneuvers, and damage effects tests.

Camp Desert Rock was established in 1951, 1.5 miles (2.4 kilometers) south of Camp Mercury. The site was used to billet troops and stage equipment. The camp was discontinued as an Army installation in 1964.

15.1   Summary

15.2   Desert Rock I, II, III

Observer programs were conducted at shots *Dog*, *Sugar*, and *Uncle*. Tactical maneuvers were conducted after shot *Dog*. Damage effects tests were conducted at shots *Dog*, *Sugar*, and *Uncle* to determine the effects of a nuclear detonation on military equipment and field fortifications.

Desert Rock I - Buster-Jangle Dog - November 1, 1951.

15.3   Desert Rock IV

Observer programs were conducted at shots *Charlie*, *Dog*, *Fox*, and *George*. Tactical maneuvers were conducted after shots *Charlie*, *Dog*, and *George*. Psychological tests were conducted at shots *Charlie*, *Fox*, and *George* to determine the troops' reactions to witnessing a nuclear detonation.

15.4   Desert Rock V

Exercise Desert Rock V included troop orientation and training, a volunteer officer observer program, tactical troop maneuvers, operational helicopter tests, and damage

Desert Rock IV - Tumbler-Snapper George - June 1, 1952.

effects evaluation.

15.5   Desert Rock VI

Observer programs were conducted at shots *Wasp*, *Moth*, *Tesla*, *Turk*, *Bee*, *Ess*, *Apple 1*, and *Apple 2*. Tactical maneuvers were conducted after shots *Bee* and *Apple 2*. Technical studies were conducted at shots *Wasp*, *Moth*, *Tesla*, *Turk*, *Bee*, *Ess*, *Apple 1*, *Wasp Prime*, *Met*, and *Apple 2*.

A test of an armored task force, RAZOR, was conducted at shot *Apple 2* to demonstrate the capability of a reinforced tank battalion to seize an objective immediately after a nuclear detonation.

15.6   Desert Rock VII, VIII

Tactical maneuvers were conducted after shots *Hood*, *Smoky*, and *Galileo*. At shot *Hood*, the Marine Corps conducted a maneuver involving the use of a helicopter airlift and tactical air support. At shot *Smoky*, Army troops conducted an airlift assault, and at shot *Galileo*, Army troops were tested to determine their psychological reactions to witnessing a nuclear detonation.

15.7   See also

Totskoye nuclear exercise of 1954, a somewhat comparable series of Soviet exercises.

15.8   Research

[1] Operation BUSTER-JANGLE Fact Sheet Defense Threat Reduction Agency

[2] Operation TUMBLER-SNAPPER Fact Sheet Defense Threat Reduction Agency

[3] Operation UPSHOT-KNOTHOLE Fact Sheet Defense Threat Reduction Agency

[4] Operation TEAPOT Fact Sheet Defense Threat Reduction Agency

[5] Operation PLUMBBOB Fact Sheet Defense Threat Reduction Agency

15.9   External links

- Camp Desert Rock Tests on YouTube

- The short film *Big Picture: Atomic Battlefield* is available for free download at the Internet Archive

- The short film *Exercise Desert Rock (1951)* is available for free download at the Internet Archive

Chapter 16

Totskoye nuclear exercise

Coordinates: 52°38.54′N 52°48.55′E / 52.64233°N 52.80917°E

The **Totskoye nuclear exercise** was a military exercise undertaken by the Soviet Army to explore defensive and offensive warfare during nuclear war. The exercise, under the code name "Snowball", involved an aerial detonation of a 40 kt*[1] RDS-4 nuclear bomb. The stated goal of the operation was military training for breaking through heavily fortified defensive lines of a military opponent using nuclear weapons.*[2]*[3] An army of 45,000 soldiers marched through the area around the epicenter soon after the nuclear blast. The exercise was conducted on September 14, 1954, at 9:33 a.m.,*[4]*[5] under the command of Marshal Georgy Zhukov to the north of Totskoye village in Orenburg Oblast, Russia, in the South Ural Military District.

16.1 History

In mid-September 1954, nuclear bombing tests were performed at the Totskoye proving ground during the training exercise *Snezhok* (Russian: Снежок, *Snowball* or *Light Snow*) with some 45,000 people, all Soviet soldiers and officers,*[3] who explored the explosion site of a bomb twice as powerful as the one dropped on Hiroshima nine years earlier. The participants were carefully selected from Soviet military servicemen, informed that they would take part in an exercise with the use of a new kind of weapons, sworn to secrecy and earned a salary for three months ahead.*[6] A delegation of high-ranking government officials and senior military officers arrived to the region on the eve of the exercise, which included First Secretary Nikita Khrushchev, Nikolai Bulganin, Generals Aleksandr Vasilevsky, Konstantin Rokossovsky, Ivan Konev and Rodion Malinovsky.*[4] The operation was commanded by Marshal of the Soviet Union Georgy Zhukov and initiated by the Soviet Ministry of Defense.*[7] At 9:33 a.m. on 14 September 1954, a Soviet Tu-4 bomber dropped a 40-kilotonne (170 TJ)*[3] atomic weapon -

RDS-4 bomb, which had been previously tested in 1951 at the Semipalatinsk Test Site,*[3]*[8] - from 8,000 metres (26,000 ft). The bomb exploded 350 metres (1,150 ft) above Totskoye range, 13 kilometres (8 mi) from Totskoye.*[3]

The exercise involved the 270th Rifle Division,*[9] 320 planes, 600 tanks and self-propelled guns, 600 armoured personnel carriers, 500 artillery pieces and mortars and 6,000 automobiles.*[3]

Following the explosion, a Li-2 airplane was put to use on a reconnaissance mission to report the movement of a radioactive cloud produced by the blast,*[10] and the most dangerous areas were explored and marked by special reconnaissance troops.*[11] After the reconnaissance was complete and the Soviet command gained enough information on the level of radiation, the army moved in. Much attention was paid to personal safety: the participants were provided with personal protective equipment, tinted glasses or lenses for gas masks and had individual radiation dosimeters.*[12] Gamma-roentgenometers measured the level of radiation exposure in the epicentre, dosimeters were used to estimate the radiation dose deposited in an individual wearing or a vehicle after the troops completed their task, and the troops received a 'chemical alert' signal if the radiation was too high.*[13] The soldiers wore gas masks, protective suits and respirators,*[11] special gloves and capes*[13] and moved around the territory in armoured personnel carriers, holding the distance of 400*[10]−600 metres from the hypocentre and avoiding the most dangerous areas of the explosion site.*[4] A relatively low level of radiation, strong wind and the extensive use of personal protective equipment allowed them to move 400-500 metres from the epicentre, whereas tanks and armoured personnel carriers could safely get even closer.*[10]

Deputy Defense Minister Georgy Zhukov witnessed the blast from an underground nuclear bunker. The planes were ordered to bomb the explosion site five minutes after the blast, and three hours later (after the demarcation of the radioactive zone) the armored vehicles were ordered to practice the taking of a hostile area after a nuclear attack.[*][3]

The residents of villages (Bogdanovka, Fyodorovka and others) that were situated around 6 km (4 mi) from the epicenter of the future explosion were offered temporary evacuation outside the 50 km (31 mi) radius and given instructions. They were evacuated by the military and temporarily accommodated in military tents. During the exercise, the residents received daily payment, while their property was insured. Those of them who decided not to return after the operation was complete, were provided with newly built four-room furnished houses near the Samarka river or obtained financial compensation.[*][4] The nearest villages were generally not affected by the blast,[*][4] except for a number of houses located less than 8 km (5.0 mi) from the explosion site that caught fire and burned down.[*][10] Their owners received new housing.[*][14]

A few days after the explosion, Soviet scientists received detailed reports on the test and began to study the impact of the nuclear blast on model houses, shelters, vehicles, vegetation and experimental animals that had been affected by the explosion.[*][15] On 17 September 1954, the Soviet newspaper Pravda published a report on the exercise: "In accordance with the plan of scientific and experimental works, a test of one of the types of nuclear weapons has been conducted in the Soviet Union in the last few days. The purpose of the test was to examine the effects of nuclear explosion. Valuable results have been obtained that will help Soviet scientists and engineers to successfully solve the task of protecting the country from nuclear attack".[*][16] These results were discussed at a large scientific conference at the Kuybyshev Military Academy in Moscow and for many years served as the basis for the Soviet program of defense against nuclear warfare.[*][17]

16.2   See also

- Desert Rock exercises, the United States's closest counterpart.

- Operation Plumbbob

16.3   References

[1] Memoirs of Colonel V. I. Levykin published in *Nuclear Exercises*, V. II, 2006, p. 19

[2] *Totskyoe exercise. Measures of safety (Russian) by Sergei Markov*

[3] *Nuclear Exercises*, V. II. 2006. P. 19

[4] Memoirs of Lieutenant-Colonel N. V. Danilenko published in *Nuclear Exercises*, V. II, 2006, p. 144

[5] Memoirs of Colonel V. I. Levykin published in *Nuclear Exercises*, V. II, 2006, p. 141

[6] Memoirs of M. A. Kutsenko, a participant of the operation Snowball, published in *Nuclear Exercises*, V. II, 2006, p. 122

[7] *Nuclear Exercises*, V. II. 2006. P. 18

[8] *Nuclear Exercises*, V. II, 2006, p. 11

[9] V.I. Feskov et al., "The Soviet Army in the Cold War 1945–90", Tomsk, 2004, p. 94

[10] *Nuclear Exercises*, V. II. 2006. P. 41

[11] Memoirs of Colonel V. I. Levykin published in *Nuclear Exercises*, V. II, 2006, p. 142

[12] Memoirs of Colonel Professor M. P. Arkhipov published in *Nuclear Exercises*, V. II, 2006, p. 132

[13] *Nuclear Exercises*, V. II. 2006. P. 68

[14] *Nuclear Exercises*, V. II, 2006, p. 65

[15] Memoirs of Colonel V. I. Levykin published in *Nuclear Exercises*, V. II, 2006, p. 142

[16] Pravda, 17 September 1954

[17] Memoirs of Colonel V. I. Levykin published in *Nuclear Exercises*, V. II, 2006, p. 143

- "Nuclear Testing in the USSR. Volume 2. Soviet Nuclear Testing Technologies. Environmental Effects. Safety Provisions. Nuclear Test Sites", Begell-House, Inc., New York, 1998

- A.A. Romanyukha, E.A. Ignatiev, D.V. Ivanov and A.G. Vasilyev, "The Distance Effect on the Individual Exposures Evaluated from the Soviet Nuclear Bomb Test in 1954 at Totskoye Test Site in 1954", Radiation Protection Dosimetry 86:53-58 (1999) online abstract

- Генерал-лейтенант С.А. Зеленцов. Тоцкое войсковое учение (научно-публицистическая монография)[Totskoye Military Exercise] (in Russian). Retrieved 2011-03-05.

- Wm. Robert Johnston (2005-05-05). "Totsk nuclear test, 1954". Retrieved 2011-03-05.

- In the zone of nuclear blast (Russian) by General of Aviation Ostroumov

- Truth about the testing site of death (Russian), a publication by Moskovskii Komsomolets

Chapter 17

Operation Plumbbob

Operation Plumbbob was a series of nuclear tests conducted between May 28 and October 7, 1957, at the Nevada Test Site, following *Project 57*, and preceding *Project 58/58A*.*[1] It was the biggest, longest, and most controversial test series in the continental United States.

17.1 Background

The operation consisted of 29 explosions, of which only two did not produce any nuclear yield. Twenty-one laboratories and government agencies were involved. While most *Operation Plumbbob* tests contributed to the development of warheads for intercontinental and intermediate range missiles, they also tested air defense and anti-submarine warheads with smaller yields. They included forty-three military effects tests on civil and military structures, radiation and bio-medical studies, and aircraft structural tests. *Operation Plumbbob* had the tallest tower tests to date in the U.S. nuclear testing program as well as high-altitude balloon tests. One nuclear test involved the largest troop maneuver ever associated with U.S. nuclear testing.

Approximately 18,000 members of the U.S. Air Force, Army, Navy and Marines participated in exercises Desert Rock VII and VIII during *Operation Plumbbob*. The military was interested in knowing how the average foot-soldier would stand up, physically and psychologically, to the rigors of the tactical nuclear battlefield.

Almost 1,200 pigs were subjected to bio-medical experiments and blast-effects studies during *Operation Plumbbob*. On shot *Priscilla* (37 kt), 719 pigs were used in various experiments on Frenchman Flat. Some pigs were placed in elevated cages and provided with suits made of different materials, to test which materials provided best protection from the thermal radiation. As shown and reported in the PBS documentary *Dark Circle*, the pigs survived, but with third-degree burns to 80% of their bodies.*[2] Other pigs were placed in pens behind large sheets of glass at measured distances from the hypocenter to test the effects of flying debris on living targets. Studies were conducted of radioactive contamination and fallout from a simulated accidental detonation of a weapon; and projects concerning earth motion, blast loading and neutron output were carried out.

Nuclear weapons safety experiments were conducted to study the possibility of a nuclear weapon detonation during an accident. On July 26, 1957, a safety experiment, *Pascal-A*, was detonated in an unstemmed hole at NTS, becoming the first underground shaft nuclear test. The knowledge gained here would provide data to prevent nuclear yields in case of accidental detonations–for example, in a plane crash.

The *John* shot on July 19, 1957 was the only test of the Air Force's AIR-2 Genie missile with a nuclear warhead.*[3] It was fired from an F-89 Scorpion fighter over Yucca Flats at the NNSS. On the ground, the Air Force carried out a public relations event by having five Air Force officers and a photographer stand under ground zero of the blast, which took place at between 18,500 and 20,000 feet altitude, with the idea of demonstrating the possibility of the use of the weapon over civilian populations without ill effects. In 2012 the photographer and the last survivor of the five met in a restaurant in Dallas to reminisce.*[4] The photographer, Akira "George" Yoshitake, died in October 2013,*[5] and the last of the six, Donald A. Luttrell, died December 2014.*[6]

The *Rainier* shot, conducted September 19, 1957, was the first fully contained underground nuclear test, meaning that no fission products were vented into the atmosphere. This test of 1.7 kt could be detected around the world by seismologists using ordinary seismic instruments. The *Rainier* test became the prototype for larger and more powerful underground tests.

Some images from *Upshot-Knothole Grable* were accidentally relabeled as belonging to the *Priscilla* shot from *Operation Plumbbob* in 1957. As a consequence many publications including official government documents have the photo mislabeled.*[7]

17.2 Radiological effects

A test blimp that was aloft five miles from Plumbbob/Franklin Prime, *deflated from the shockwave*

Main article: Downwinders

Plumbbob released 58,300 kilocuries (2.16 EBq) of radioiodine (I-131) into the atmosphere. This produced total civilian radiation exposures amounting to 120 million person-rads of thyroid tissue exposure (about 32% of all exposure due to continental nuclear tests).

Statistically, this level of exposure would be expected to eventually cause between 11,000 and 212,000 excess cases of thyroid cancer, leading to between 1,000 and 20,000 deaths.[*][8]

In addition to civilian exposure, troop exercises conducted near the ground near shot *Smoky* exposed over 3000 servicemen to relatively high levels of radiation. A survey of these servicemen in 1980 found significantly elevated rates of leukemia: ten cases, instead of the baseline expected four.

17.3 Propulsion of steel plate cap

During the **Pascal-B** nuclear test, a 900-kilogram (2,000 lb) steel plate cap (a piece of armor plate) was blasted off the top of a test shaft at a speed of more than 66 km/s (41 mi/s; 240,000 km/h; 150,000 mph). Before the test, experimental designer Dr. Brownlee had estimated that the nuclear explosion, combined with the specific design of the shaft, would accelerate the plate to approximately six times Earth's escape velocity.[*][9] The plate was never found, but Dr. Brownlee believes that the plate did not leave the atmosphere, as it may even have been vaporized by compression heating of the atmosphere due to its high speed. The calculated velocity was sufficiently interesting that the crew trained a high-speed camera on the plate, which unfortunately only appeared in one frame, but this nevertheless gave a very high lower bound for its speed. After the event, Dr. Robert R. Brownlee described the best estimate of the cover's speed from the photographic evidence as "going like a bat out of hell!"[*][9][*][10]

17.4 List of tests

See also: List of nuclear weapons tests of the United States

[1] The US, France and Great Britain have code-named their test events, while the USSR and China did not, and therefore have only test numbers (with some exceptions – Soviet peaceful explosions were named). Word translations into English in parentheses unless the name is a proper noun. A dash followed by a number indicates a member of a salvo event. The US also sometimes named the individual explosions in such a salvo test, which results in "name1 – 1(with name2)". If test is canceled or aborted, then the row data like date and location discloses the intended plans, where known.

[2] To convert the UT time into standard local, add the number of hours in parentheses to the UT time; for local daylight saving time, add one additional hour. If the result is earlier than 00:00, add 24 hours and subtract 1 from the day; if it is 24:00 or later, subtract 24 hours and add 1 to the day. All historical timezone data are derived from here:

[3] Rough place name and a latitude/longitude reference; for rocket-carried tests, the launch location is specified before the detonation location, if known. Some locations are extremely accurate; others (like airdrops and space blasts) may be quite inaccurate. "~" indicates a likely pro-forma rough location, shared with other tests in that same area.

[4] Elevation is the ground level at the point directly below the explosion relative to sea level; height is the additional distance added or subtracted by tower, balloon, shaft, tunnel, air drop or other contrivance. For rocket bursts the ground level is "N/A". In some cases it is not clear if the height is absolute or relative to ground, for example, *Plumbbob/John*. No number or units indicates the value is unknown, while "0" means zero. Sorting on this column is by elevation and height added together.

[5] Atmospheric, airdrop, balloon, gun, cruise missile, rocket, surface, tower, and barge are all disallowed by the Partial Nuclear Test Ban Treaty. Sealed shaft and tunnel are underground, and remained useful under the PTBT. Intentional cratering tests are borderline; they occurred under the treaty, were sometimes protested, and generally overlooked if the test was declared to be a peaceful use.

[6] Include weapons development, weapon effects, safety test, transport safety test, war, science, joint verification and industrial/peaceful, which may be further broken down.

[7] Designations for test items where known, "?" indicates some uncertainty about the preceding value, nicknames for particular devices in quotes. This category of information is often not officially disclosed.

[8] Estimated energy yield in tons, kilotons, and megatons. A ton of TNT equivalent is defined as 4.184 gigajoules (1 gigacalorie).

[9] Radioactive emission to the atmosphere aside from prompt neutrons, where known. The measured species is only iodine-131 if mentioned, otherwise it is all species. No entry means unknown, probably none if underground and "all" if not; otherwise notation for whether measured on the site only or off the site, where known, and the measured amount of radioactivity released.

17.5   See also

- Lists of nuclear disasters and radioactive incidents

- Totskoye nuclear exercise

17.6   References

[1] Yang, Xiaoping; North, Robert; Romney, Carl (August 2000), *CMR Nuclear Explosion Database (Revision 3)*, SMDC Monitoring Research

[2] Dark Circle, DVD release date March 27, 2007, Directors: Judy Irving, Chris Beaver, Ruth Landy. ISBN 0-7670-9304-6. http://www.pbs.org/pov/darkcircle/

[3] Robert Krulwich. "Five Men Agree To Stand Directly Under An Exploding Nuclear Bomb". NPR.

[4] Timothy Stenovec (2012-07-20). "George Yoshitake, Nuclear Test Photographer, Recalls Filming Nuclear Blast 55 Years Ago". Huffington Post.

[5] "Akira "George" Yoshitake (obituary)". *Lompoc Record.* Lompoc, California, US. 22 October 2013. Retrieved 17 May 2014.

[6] "Donald Allen Luttrell (obituary)". *Dallas Morning News.* 1 January 2015. Retrieved 15 February 2015.

[7] Carey Sublette, "Operation Plumbbob," Nuclear Weapon Archive, http://nuclearweaponarchive.org/Usa/Tests/Plumbob.html. (accessed December 27, 2006).

[8] Institute of Medicine (U.S.). Committee on Thyroid Screening Related to I-131 Exposure, National Research Council (U.S.). Committee on Exposure of the American People to I-131 from the Nevada Atomic Bomb Tests, ed. (1999). *Exposure of the American people to Iodine-131 from Nevada nuclear-bomb tests: review of the National Cancer Institute report and public health implications.* National Academies Press. pp. 113–114. ISBN 978-0-309-06175-9.

[9] Brownlee, Robert R. (June 2002). "Learning to Contain Underground Nuclear Explosions". Retrieved 2006-07-31.

[10] Pascal B test at the Nuclear Weapon Archive.

[11] "Timezone Historical Database". iana.com. Retrieved March 8, 2014.

[12] Harris, =P.S.; Lowery, C.; Nelson, A. (1981), *Plumbbob Series, 1957 Final* (PDF) (DNA6005F), Defense Nuclear Agency, retrieved 2014-01-06

[13] *Estimated exposures and thyroid doses received by the American people from Iodine-131 in fallout following Nevada atmospheric nuclear bomb tests, Chapter 2* (PDF), National Cancer Institute, 1997, retrieved 2014-01-05

[14] Sublette, Carey, *Nuclear Weapons Archive*, retrieved 2014-01-06

[15] Hansen, Chuck (1995), *The Swords of Armageddon, Vol. 8*, Sunnyvale, CA: Chukelea Publications, ISBN 978-0-9791915-1-0

[16] *United States Nuclear Tests: July 1945 through September 1992* (PDF) (DOE/NV-209 REV15), Las Vegas, NV: Department of Energy, Nevada Operations Office, 2000-12-01, retrieved 2013-12-18

[17] Norris, Robert Standish; Cochran, Thomas B. (1 February 1994), "United States nuclear tests, July 1945 to 31 December 1992 (NWD 94-1)" (PDF), *Nuclear Weapons Databook Working Paper*, Washington, DC: Natural Resources Defense Council, retrieved 2013-10-26

[18] *Official list of underground nuclear explosions*, Sandia National Laboratories, 1994-07-01, retrieved 2013-12-18

[19] *Shot Smoky: A Test of the Plumbbob Series, 31 August 1957 (DNA-6004F)*, Washington, DC: Defense Nuclear Agency, Department of Defense, 1981, retrieved 2013-10-28

17.7   External links

- Video clip: Historic color footage of shot "Owens" during Operation Plumbbob

- Plumbbob page on the Nuclear Weapons Archive (also refers to manhole cover issue mentioned above).

Chapter 18

Windscale fire

The Windscale Piles (centre and right) in 1985

18.1 Windscale Piles

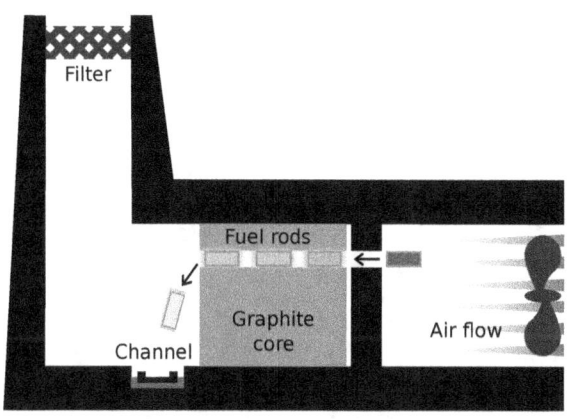

The design of Windscale Pile No. 1, with one of the many fuel channels illustrated.

The **Windscale fire** of 10 October 1957 was the worst nuclear accident in Great Britain's history, ranked in severity at level 5 out of a possible 7 on the International Nuclear Event Scale.[*][1] The fire took place in Unit 1 of the two-pile Windscale facility on the northwest coast of England in Cumberland (now Sellafield, Cumbria). The two graphite-moderated reactors, referred to at the time as "piles", had been built as part of the British atomic bomb project.[*][2] Windscale Pile No. 1 was operational in October 1950 followed by Pile No. 2 in June 1951.[*][3]

The fire burned for three days and there was a release of radioactive contamination that spread across the UK and Europe.[*][4] Of particular concern at the time was the radioactive isotope iodine-131, which may lead to cancer of the thyroid, and it has been estimated that the incident caused 240 additional cancer cases.[*][4] No one was evacuated from the surrounding area, but there was a worry that milk might be dangerously contaminated. Milk from about 500 km^2 of nearby countryside was diluted and destroyed for about a month. A 2010 study of workers directly involved in the cleanup found no significant long term health effects from their involvement.[*][5][*][6]

After the Second World War, the British government embarked on a programme to build nuclear weapons. Skipping the lower-performance uranium-based weapons in favour of those based on plutonium, a plutonium-breeding reactor system was designed to produce this material. The design was based on the graphite-moderated B Reactor built at the Hanford Site, which was known to British physicists who had been involved in the Manhattan Project during the war. The reactors were built in a short time near the village of Seascale, Cumberland. They were known as Windscale Pile 1 and Pile 2, housed in large concrete buildings a few hundred feet apart.

The core of the reactors consisted of a large block of graphite with horizontal channels drilled through it for the fuel cartridges. Each cartridge consisted of a uranium rod about 30 cm long encased in an aluminium canister to protect it from the air, as uranium becomes highly reactive when hot and can catch fire. The cartridge was finned, allowing heat exchange with the environment to cool the fuel rods while they were in the reactor. Rods were pushed in the front of the core, the "charge face", with new rods being added at a calculated rate. This pushed the other cartridges

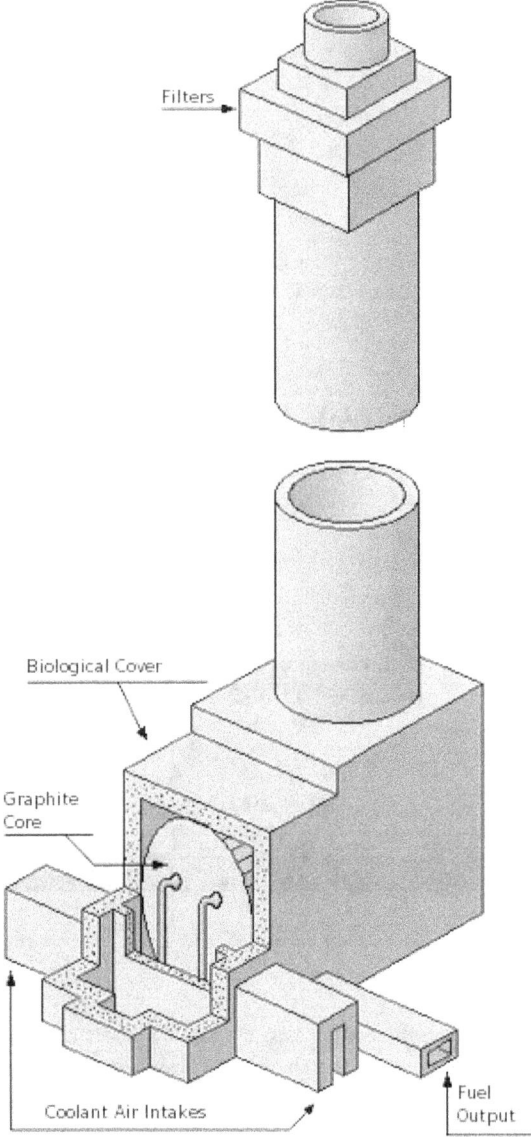

Cutaway diagram of Windscale reactor.

in the channel towards the rear of the reactor, eventually causing them to fall out the back, the "discharge face", into a water filled channel where they cooled and could be collected.[*][7] The chain reaction in the core converted the uranium into a variety of isotopes, including some plutonium, which was separated from the other materials using chemical processing. As this plutonium was intended for weapons purposes, the burnup of the fuel would have been kept low to reduce production of the heavier plutonium isotopes (^{240}Pu, ^{241}Pu etc.).

The design initially called for the core to be cooled like the B Reactor, which used a constant supply of water that poured through the channels in the graphite. There was considerable concern that such a system was subject to catas-

trophic failure in the event of a loss-of-coolant accident. This would cause the reactor to run out of control in seconds, potentially exploding. At Hanford, this possibility was dealt with by constructing a 30 miles (48 km) escape road to evacuate the staff were this to occur, abandoning the site.[*][8] Lacking any location where a 30 mile area could be abandoned if a similar event were to occur in the UK, the designers desired a passively safe cooling system. In place of water, they used air cooling driven by convection through a 400 feet (120 m) tall chimney, which could create enough airflow to cool the reactor under normal operating conditions. The chimney was arranged so it pulled air through the channels in the core, cooling the fuel via fins on the cartridges. For additional cooling, huge fans were positioned in front of the core, which could greatly increase the airflow rate.[*][9]

During construction, Terence Price, one of the many physicists working on the project, began to consider what would happen if one of the fuel cartridges being pushed out the back of the core were to break open. This could happen, for example, if a new cartridge being inserted was pushed too hard, causing the one at the back of the channel to fall past the relatively narrow water channel and strike the floor behind it. In that event, the hot uranium could catch fire, with the fine uranium oxide dust being blown up the chimney to escape.[*][10] When he raised the issue at a meeting and suggested that filters be added to the chimneys, the concern was dismissed as being too difficult to deal with and was not even recorded in the minutes. Sir John Cockcroft, leading the project team, was alarmed enough to order that filters be installed, which required them to be constructed on the ground while the chimneys were still being built, and then winched into position at the top once the chimney's concrete had set.[*][11] These became known as "Cockcroft's Folly" by workers and engineers.

In the end, Price's concerns came to pass. So many cartridges missed the water channel that it became routine for staff to walk through the chimney ductwork with shovels and scoop the cartridges back into the water.[*][12] On other occasions, fuel cartridges became stuck in the channels and burst open while still in the core.[*][13] In spite of these precautions and the stack filters, scientist Frank Leslie had discovered radioactivity around the site and the village, but this information was kept secret, even from the staff at the station.[*][14][*][15]

18.2 Wigner energy

Main article: Wigner effect

Once commissioned and settled into operations, Pile 2 experienced a mysterious rise in core temperature. Unlike the Americans and the Soviets, the British had little experience with the behaviour of graphite when exposed to neutrons. Hungarian-American physicist Eugene Wigner had discovered that graphite, when bombarded by neutrons, suffers dislocations in its crystalline structure, causing a build-up of potential energy. This energy, if allowed to accumulate, could escape spontaneously in a powerful rush of heat.

The sudden bursts of energy worried the operators, who turned to the only viable solution, heating the reactor core in a process known as annealing. When graphite is heated beyond 250° C it becomes plastic, and the Wigner dislocations can relax into their natural state. This process was gradual and caused a uniform release which spread throughout the core.[16]

18.3   Tritium production

Winston Churchill publicly committed the UK to building a hydrogen bomb, and gave the scientists a tight schedule in which to do so. This was then hastened after the US and USSR began working on a test ban and possible disarmament agreements which would begin to take effect in 1958. To meet this deadline there was no chance of building a new reactor to produce the required tritium, so the Windscale Pile 1 fuel loads were modified by adding enriched uranium and lithium-magnesium, the latter of which would produce tritium during neutron bombardment.[17] All of these materials were highly flammable, and a number of the Windscale staff raised the issue of the inherent dangers of the new fuel loads. These concerns were brushed aside.

When their first H-bomb test failed, the decision was made to build a large fusion-boosted-fission weapon instead. This required huge quantities of tritium, five times as much, and it had to be produced as rapidly as possible as the test deadlines approached. To boost the production rates, they used a trick that had been successful in increasing plutonium production in the past; by reducing the size of the cooling fins on the fuel cartridges, the temperature of the fuel loads increased, which caused a small but useful increase in neutron enrichment rates. This time they also took advantage of the smaller fins by building larger interiors in the cartridges, allowing more fuel in each one. These changes triggered further warnings from the technical staff, which were again brushed aside. Christopher Hinton, Windscale's director, left in frustration.[18]

After a first successful production run of tritium in Pile 1, the heat problem was presumed to be negligible and full-scale production began. But by raising the temperature of the reactor beyond the design specifications, the scientists had altered the normal distribution of heat in the core, caus-

ing hot spots to develop in Pile 1. These were not detected because the thermocouples used to measure the core temperatures were positioned based on the original heat distribution design, and were not measuring the parts of the reactor which became hottest.

18.4   Accident

18.4.1   Ignition

On 7 October 1957 operators of Pile 1 noticed that the reactor was heating up more than normal, and a Wigner release was ordered.[19] This had been carried out eight times in the past, and it was known that the cycle would cause the entire reactor core to heat up evenly. During this attempt the temperatures anomalously began falling across the reactor core, except in channel 2053, whose temperature was rising.[20] Concluding that 2053 was releasing energy but none of the others were, on the morning of 8 October the decision was made to try a second Wigner release. This attempt caused the temperature of the entire reactor to rise, indicating a successful release.[21]

Early in the morning of 10 October it was suspected that something unusual was going on. The temperature in the core was supposed to gradually fall as Wigner release ended, but the monitoring equipment showed something more ambiguous, and one thermocouple indicated that core temperature was instead rising. As this process continued, the temperature continued to rise and eventually reached 400 °C. In an effort to help cool the pile, the cooling fans were sped up and airflow was increased. Radiation detectors in the chimney then indicated a release, and it was assumed that a cartridge had burst. This was not a fatal problem, and had happened in the past. Unknown to the operators, the cartridge had not just burst, but caught fire, and this was the source of the anomalous heating in channel 2053, not a Wigner release.[22]

18.4.2   Fire

Speeding up the fans increased the airflow in the channel, fanning the flames. The fire spread to surrounding fuel channels, and soon the radioactivity in the chimney was rapidly increasing.[23] A foreman, arriving for work, noticed smoke coming out of the chimney. The core temperature continued to rise, and the operators began to suspect the core was on fire.[24]

Operators tried to examine the pile with a remote scanner but it had jammed. Tom Hughes, second in command to the Reactor Manager, suggested examining the reactor personally and so he and another operator went to the charge face of the reactor, clad in protective gear. A fuel channel inspection plug was taken out close to a thermocouple registering high temperatures and it was then that the operators saw that the fuel was red hot.

"An inspection plug was taken out," said Tom Hughes in a later interview, "and we saw, to our complete horror, four channels of fuel glowing bright cherry red."

There was now no doubt that the reactor was on fire, and had been for almost 48 hours. Reactor Manager Tom Tuohy[25] donned full protective equipment and breathing apparatus and scaled the 80-foot ladder to the top of the reactor building, where he stood atop the reactor lid to examine the rear of the reactor, the discharge face. Here he reported a dull red luminescence visible, lighting up the void between the back of the reactor and the rear containment. Red hot fuel cartridges were glowing in the fuel channels on the discharge face. He returned to the reactor upper containment several times throughout the incident, at the height of which a fierce conflagration was raging from the discharge face and playing on the back of the reinforced concrete containment —concrete whose specifications required that it be kept below a certain temperature to prevent its collapse.[26]

18.4.3   Initial fire fighting attempts

Operators were unsure what to do about the fire. First they tried to blow the flames out by running the fans at maximum speed, but this fed the flames. Tom Hughes and his colleague had already created a fire break by ejecting some undamaged fuel cartridges from around the blaze, and Tom Tuohy suggested trying to eject some from the heart of the fire by bludgeoning the melted cartridges through the reactor and into the cooling pond behind it with scaffolding poles. This proved impossible and the fuel rods refused to budge, no matter how much force was applied. The poles were withdrawn with their ends red hot; one returned dripping molten metal. Hughes knew this had to be molten irradiated uranium, causing serious radiation problems on the charge hoist itself.

"It [the exposed fuel channel] was white hot," said Hughes' colleague on the charge hoist with him, "it was just white hot. Nobody, I mean, nobody, can believe how hot it could possibly be."

18.4.4   Carbon dioxide

Next, the operators tried to extinguish the fire using carbon dioxide. The new gas-cooled Calder Hall reactors on the site had just received a delivery of 25 tonnes of liquid carbon dioxide and this was rigged up to the charge face of Windscale Pile 1, but there were problems getting it to the fire in useful quantities. The fire was so hot that it stripped the oxygen from what carbon dioxide could be applied.

"So we got this rigged up," Hughes recounted, "and we had this poor little tube of carbon dioxide and I had absolutely no hope it was going to work."

18.4.5   Use of water

On the morning of Friday 11 October, when the fire was at its worst, eleven tons of uranium were ablaze. Temperatures were becoming extreme (one thermocouple registered 1,300 °C) and the biological shield around the stricken reactor was now in severe danger of collapse. Faced with this crisis, Tuohy suggested using water. This was risky, as molten metal oxidises in contact with water, stripping oxygen from the water molecules and leaving free hydrogen, which could mix with incoming air and explode, tearing open the weakened containment. Faced with a lack of other options, the operators decided to go ahead with the plan.[27]

About a dozen fire hoses were hauled to the charge face of the reactor; their nozzles were cut off and the lines themselves connected to scaffolding poles and fed into fuel channels about a metre (roughly 3 feet) above the heart of the fire. Tuohy once again hauled himself onto the reactor shielding and ordered the water to be turned on, listening carefully at the inspection holes for any sign of a hydrogen reaction as the pressure was increased. The water was unsuccessful in extinguishing the fire, requiring further measures to be taken.

18.4.6   Shutting off air

Tuohy then ordered everyone out of the reactor building except himself and the Fire Chief in order to shut off all cooling and ventilating air entering the reactor. Tuohy then climbed up several times and reported watching the flames leaping from the discharge face slowly dying away. During one of the inspections, he found that the inspection plates—which were removed with a metal hook to facilitate viewing of the discharge face of the core—were stuck fast. This, he reported, was due to the fire trying to suck air in from wherever it could.

"I have no doubt it was even sucking air in through the chimney at this point to try and maintain itself," he remarked in an interview.

Finally he managed to pull the inspection plate away and was greeted with the sight of the fire dying away.

"First the flames went, then the flames reduced and the glow began to die down," he described, "I went up to check several times until I was satisfied that the fire was out. I did stand to one side, sort of hopefully," he went on to say, "but if you're staring straight at the core of a shut down reactor you're going to get quite a bit of radiation." (Tuohy lived to the age of 90, despite his exposure.)

Water was kept flowing through the pile for a further 24 hours until it was completely cold.

The reactor tank itself has remained sealed since the accident and still contains about 15 tons of uranium fuel. It was thought that the remaining fuel could still reignite if disturbed, due to the presence of pyrophoric uranium hydride formed in the original water dousing.*[28] Subsequent research, conducted as part of the decommissioning process, has ruled out this possibility.*[29] The pile is not scheduled for final decommissioning until 2037.

18.5   Aftermath

18.5.1   Radioactive release

There was a release of radioactive material that spread across the UK and Europe.*[4] The fire released an estimated 740 terabecquerels (20,000 curies) of iodine-131, as well as 22 TBq (594 curies) of caesium-137 and 12,000 TBq (324,000 curies) of xenon-133, among other radionuclides.*[30] Later reworking of contamination data has shown national and international contamination may have been higher than previously estimated.*[4] For comparison, the 1986 Chernobyl explosion released approximately 1,760,000 TBq of iodine-131; 79,500 TBq caesium-137; 6,500,000 TBq xenon-133; 80,000 TBq strontium-90; and 6100 TBq plutonium, along with about a dozen other radionuclides in large amounts.*[30] The Three Mile Island accident in 1979 released 25 times more xenon-135 than Windscale, but much less iodine, caesium and strontium.*[30] Estimates by the Norwegian Institute of Air Research indicate that atmospheric releases of xenon-133 by the Fukushima Daiichi nuclear disaster were broadly similar to those released at Chernobyl, and thus well above the Windscale fire releases.*[31]

The presence of the chimney scrubbers at Windscale was credited with maintaining partial containment and thus minimizing the radioactive content of the smoke that poured from the chimney during the fire. These scrubbers were installed at great expense on the insistence of John Cockcroft and were known as Cockcroft's folly, until the 1957 fire.*[32]

18.5.2   Health effects

Of particular concern at the time was the radioactive isotope iodine-131, which has a half-life of only 8 days but is taken up by the human body and stored in the thyroid. As a result, consumption of iodine-131 often leads to cancer of the thyroid. Estimates of additional cancer cases and mortality resulting from the radiological release have varied.*[33]

No one was evacuated from the surrounding area, but there was concern that milk might be dangerously contaminated. Milk from about 500 km² of nearby countryside was destroyed (diluted a thousandfold and dumped in the Irish Sea) for about a month. A 2010 study of workers directly involved in the cleanup—and thus expected to have seen the highest exposure rates—found no significant long term health effects from their involvement.*[5]*[6]

18.5.3   Salvage operations

The reactor was unsalvageable; where possible, the fuel rods were removed, and the reactor bioshield was sealed and left intact. Approximately 6,700 fire-damaged fuel elements and 1,700 fire-damaged isotope cartridges remain in the pile. The damaged reactor core was still slightly warm as a result of continuing nuclear reactions.*[34] Windscale Pile 2, though undamaged by the fire, was considered too unsafe for continued use. It was shut down shortly afterwards. No air-cooled reactors have been built since. The final removal of fuel from the damaged reactor was scheduled to begin in 2008 and continue for a further four years.*[29]

Inspections showed that there had not been a graphite fire, and the damage to the graphite was localised, caused by severely overheated uranium fuel assemblies nearby.*[29]

18.5.4   Board of Inquiry

The Board of Inquiry met under the chairmanship of Sir William Penney from 17 to 25 October 1957. Its report (the "Penney Report") was submitted to the Chairman of the United Kingdom Atomic Energy Authority and formed the basis of the Government White Paper submitted to Parliament in November 1957. The report itself was released at the Public Record Office in January 1988. In 1989 a revised transcript was released, following work to improve the transcription of the original recordings.*[35]*[36]

Penney reported on 26 October 1957, 16 days after the fire was extinguished*[37] and reached four conclusions:

- The primary cause of the accident had been the second nuclear heating on 8 October, applied too soon and too rapidly.

- Steps taken to deal with the accident, once discovered, were "prompt and efficient and displayed considerable devotion to duty on the part of all concerned".

- Measures taken to deal with the consequences of the accident were adequate and there had been "no immediate damage to health of any of the public or of the workers at Windscale". It was most unlikely that any harmful effects would develop. But the report was very critical of technical and organisational deficiencies.

- A more detailed technical assessment was needed, leading to organisational changes, clearer responsibilities for health and safety, and better definition of radiation dose limits.

Those who had been directly involved in the events were heartened by Penney's conclusion that the steps taken had been "prompt and efficient" and had "displayed considerable devotion to duty". Some considered that the determination and courage shown by Thomas Tuohy, and the critical role he played in the aversion of complete disaster, had not been properly recognised. Tuohy died on 12 March 2008; he had never received any kind of public recognition for his decisive actions.[*][25] The Board of Inquiry's report concluded officially that the fire had been caused by "an error of judgment" by the same people who then risked their lives to contain the blaze. It was later suggested by the grandson of Harold Macmillan, Prime Minister at the time of the fire, that the US Congress might have vetoed plans of Macmillan and US president Dwight Eisenhower for joint nuclear weapons development if they had known that it was due to reckless decisions by the UK government, and that Macmillan had covered up what really happened. Tuohy said of the officials who told the US that his staff had caused the fire that "they were a shower of bastards".[*][38]

The Windscale site was decontaminated and is still in use. Part of the site was later renamed Sellafield after being transferred to BNFL; the whole site is now owned by the Nuclear Decommissioning Authority.

18.6 Comparison with other accidents

The release of radiation by the Windscale fire was greatly exceeded by the Chernobyl disaster in 1986, but the fire has been described as the worst reactor accident until Three Mile Island in 1979. Epidemiological estimates put the number of additional cancers caused by the Three Mile Island accident at not more than one; only Chernobyl produced immediate casualties.[*][39]

Three Mile Island was a civilian reactor, and Chernobyl primarily so, both being used for electrical power production. In contrast Windscale was for purely military purposes.

The reactors at Three Mile Island, unlike those at Windscale and Chernobyl, were in buildings designed to contain radioactive materials released by a reactor accident.

Other military reactors have produced immediate, known casualties such as the 1961 incident at the SL-1 plant in Idaho which killed three operators.

The accident at Windscale was also contemporary to the Kyshtym disaster, a far more serious accident which happened on 29 September 1957 at the Mayak plant in the Soviet Union, when the failure of the cooling system for a tank storing tens of thousands of tons of dissolved nuclear waste resulted in a non-nuclear explosion.

18.7 Irish sea contamination

In 1968 a paper was submitted to the journal *Nature*, on a study of radioisotopes found in oysters from the Irish Sea, using gamma spectroscopy. The oysters were found to contain 141Ce, 144Ce, 103Ru, 106Ru, 137Cs, 95Zr and 95Nb. In addition a zinc activation product (65Zn) was found; this is thought to be due to the corrosion of magnox fuel cladding in cooling ponds.[*][40] A number of harder-to-detect pure alpha and beta decaying radionuclides were also present, such as Sr-90 and plutonium-239, but these do not appear in gamma spectroscopy as they do not generate any appreciable gamma rays as they decay.

18.8 Television documentaries

In 1999, the BBC produced an educational documentary film about the fire as a 30-minute episode of "Disaster" (Series 3) entitled *The Windscale Fire*. It subsequently was released on DVD.[*][41]

In 2007, the BBC produced another documentary about the accident entitled "Windscale: Britain's Biggest Nuclear Disaster",[*][35] which investigates the history of the first British nuclear facility and its role in the development of nuclear weapons. The documentary features interviews with key scientists and plant operators, such as Tom Tuohy, who was the deputy general manager of Windscale. The documentary suggests that the fire —the first fire in any nuclear facility —was caused by the relaxation of safety measures, as a result of pressure from the British government to quickly produce fissile materials for nuclear weapons.*[42]

18.9   Isotope cartridges

The following substances were placed inside metal cartridges and subjected to neutron irradiation to create radioisotopes. Both the target material and some of the product isotopes are listed below. Of these, the polonium-210 release made the most significant contribution to the collective dose on the general population.*[43]

- Lithium-magnesium alloy: tritium

- Aluminium nitride: carbon-14

- Potassium chloride: chlorine-36

- Cobalt: cobalt-60

- Thulium: thulium-170

- Thallium: thallium-204

- Bismuth oxide: polonium-210

- Thorium: uranium-233

18.10   See also

- Nuclear and radiation accidents

- Nuclear meltdown

18.11   References

[1] Richard Black (18 March 2011). "Fukushima - disaster or distraction?". BBC. Retrieved 7 April 2011.

[2] Gowing, M, Independence and Deterrence, Vol 2, p 386 ff.

[3] "Editorial". *J. Radiol. Prot.* **27**: 211–215. 2007. doi:10.1088/0952-4746/27/3/e02.

[4] Morelle, Rebecca (6 October 2007). "Windscale fallout underestimated". BBC News.

[5] McGeoghegan, D.; Whaley, S.; Binks, K.; Gillies, M.; Thompson, K.; McElvenny, D. M. (2010). "Mortality and cancer registration experience of the Sellafield workers known to have been involved in the 1957 Windscale accident: 50 year follow-up". *Journal of Radiological Protection.* **30** (3): 407–431. Bibcode:2010JRP....30..407M. doi:10.1088/0952-4746/30/3/001. PMID 20798473.

[6] McGeoghegan, D.; Binks, K. (2000). "Mortality and cancer registration experience of the Sellafield employees known to have been involved in the 1957 Windscale accident". *Journal of Radiological Protection.* **20** (3): 261–274. Bibcode:2000JRP....20..261M. doi:10.1088/0952-4746/20/3/301. PMID 11008931.

[7] Windscale, 19:15.

[8] Windscale, 19:50.

[9] Windscale, 20:40.

[10] Windscale, 22:15.

[11] Windscale, 22:30.

[12] Windscale, 42.35.

[13] Windscale, 41.10.

[14] Windscale, 41.45.

[15] https://www.wsws.org/en/articles/2008/04/nucl-a29.html

[16] W. BOTZEM, J. WÖRNER (NUKEM Nuklear GmbH, Alzenau, Germany) (2001-06-14). "INERT ANNEALING OF IRRADIATED GRAPHITE BY INDUCTIVE HEATING" (PDF).

[17] Windscale, 46.20.

[18] Windscale, 49:45.

[19] Windscale, 57:20.

[20] Windscale, 58:20.

[21] Windscale, 59:00.

[22] Windscale, 1:00:30.

[23] Windscale, 1:02:00.

[24] Windscale, 1:03:00.

[25] "Windscale Manager who doused the flames of 1957 fire - Obituary in The Independent 2008-03-26". London. 26 March 2008. Retrieved 2008-03-27.

[26] Arnold, L. (1992). *Windscale 1957: Anatomy of a Nuclear Accident.* Macmillan. p. 235. ISBN 0-333-65036-0.

[27] Windscale, 1:10:30.

[28] "Getting to the core issue", *The Engineer*, 14 May 2004.

[29] "Meeting of RG2 with Windscale Pile 1 Decommissioning Project Team" (PDF). Nuclear Safety Advisory Committee. 2005-09-29. NuSAC(2005)P 18. Retrieved 2008-11-26.

[30] John R. Cooper; Keith Randle; Ranjeet S. Sokhi (2003). *Radioactive releases in the environment: impact and assessment.* Wiley. p. 150. ISBN 978-0-471-89923-5.. *Citing:* M. J. Crick; G. S. Linsley (1984). *An assessment of the radiological impact of the Windscale reactor fire, October 1957.* National Emergency Training Center. ISBN 0-85951-182-0.

[31] Geoff Brumfiel (25 October 2011). "Fallout forensics hike radiation toll". *Nature*. **478**: 435–436. Bibcode:2011Natur.478..435B. doi:10.1038/478435a. Retrieved 8 February 2012.

[32] John Cockcroft#Cockcroft's folly

[33] "The view from outside Windscale in 1957". *BBC*. 2 October 2007. Retrieved 17 September 2013. *"No-one died in the fire but despite what the AEA said in 1957 about there being no risk to human health, it's now widely accepted that some deaths in the UK, and elsewhere in Europe, could have been caused by the release of the radioactivity. But the figures vary depending on which study you look at. Some have suggested 30, others around 100 and some well over that. Brian Wynne, professor of science studies at Lancaster University says the deaths are what are known as statistical deaths i.e not actual named people and it will be always difficult to prove whether any one person died as a direct result of an incident like the fire."*

[34] Details of the levels and nature of the radioactivity remaining in the core can be seen at (PDF) http://www.irpa.net/irpa10/cdrom/00322.pdf. Missing or empty |title= (help) (64.7 KiB).

[35] Paul Dwyer (5 October 2007). "Windscale: A nuclear disaster". BBC News.

[36] "Proceedings into the fire at Windscale Pile Number One (1989 revised transcript of the "Penney Report")" (PDF). UKAEA. 18 April 1989.

[37] When Windscale burned

[38] The Telegraph: Tom Tuohy, obituary

[39] Gerry Matlack (2007-05-07). "The Windscale Disaster".

[40] A. PRESTON, J. W. R. DUTTON & B. R. HARVEY (18 May 1968). "Detection, Estimation and Radiological Significance of Silver-110m in Oysters in the Irish Sea and the Blackwater Estuary" (PDF). *Nature*. pp. 689–690. doi:10.1038/218689a0.

[41] "Disaster - Series 3". bbcactivevideoforlearning.com. 1999.

[42] "BBC documentary reveals government reckless in drive for the production of nuclear weapons". WSWS. 2008-04-29.

[43] Crick, MJ; Linsley GS (November 1984). "An assessment of the radiological impact of the Windscale reactor fire, October 1957". *Int J Radiat Biol Relat Stud Phys Chem Med.* **46** (5): 479–506. doi:10.1080/09553008414551711. PMID 6335136.

18.12   Further reading

- Wakeford, R. (2007). "The Windscale reactor accident--50 years on" (PDF). *Journal of Radiological Protection*. **27** (3): 211–5. doi:10.1088/0952-4746/27/3/e02.

- "Windscale fallout blew right across Europe", Rob Edwards. *New Scientist*, October 6, 2007.

- *Windscale, 1957: Anatomy of a Nuclear Accident*, Lorna Arnold. New York : St. Martin's Press 1992

- "Chernobyl: worst but not first", Walter C. Patterson. *Bulletin of the Atomic Scientists*, August/September 1986.

- 'Secrets of the Windscale fire revealed', F. Pearce. New Scientist vol 99 29 September 1983 p. 911

- 'Windscale; increased cancer incidence alleged', T. Beardsley. Nature vol 306 Issue 5938 Nov 3 1983 p. 5

- "Accident at Windscale No.1 Pile on 10 October 1957". Cmnd. 302. (H.M.S.O., 1958).

- "Accident at Windscale: World's First Atomic Alarm", Hartley Howe. *Popular Science*, October 1958, Vol. 173, No. 4.

- "An Assessment of the Radiological Impact of the Windscale Reactor Fire", M.J. Crick, G.S. Linsley. NRPB Reports, Oct. 1957, Nov. 1982.

- *An airborne radiometric survey of the Windscale area, October 19–22nd, 1957*. A.E.R.E. reports, no. R2890. (Atomic Energy Research Establishment).

- *The deposition of strontium 89 and strontium 90 on agricultural land and their entry into milk after the reactor accident at Windscale in October, 1957*. A.H.S.B. (United Kingdom Atomic Energy Authority).

- 'Accident at Windscale' British Medical Journal 16 Nov 1957;2 (5054) pp 1166-8.

18.13   External links

- "Windscale". Nuclear Decommissioning Authority.

- "THE 1957 WINDSCALE FIRE". lakestay.co.uk. 2009-07-05.

- "Windscale Nuclear Incident". The Virtual Nuclear Tourist. 2005-12-22.

- "1957: Inquiry publishes cause of nuclear fire". BBC. 8 November 1957.

- Marsden, B.J.; Preston, S.D.; Wickham, A.J. (AEA Technology plc, Warrington (United Kingdom)); Tyson, A. (Process and Radwaste Chemistry Dept., AEA TEchnology plc, Windscale (United Kingdom)) (8–10 September 1997). "Evaluation of graphite safety issues for the British production piles at Windscale". IAEA.

- Paul Dodgson (8–9 October 2007). "Radio Plays - Energy Industry:WINDSCALE....2007". suttonelms.org.uk.

Coordinates: 54°25′29.50″N   3°30′00.00″W   /
54.4248611°N 3.5000000°W

Chapter 19

Kyshtym disaster

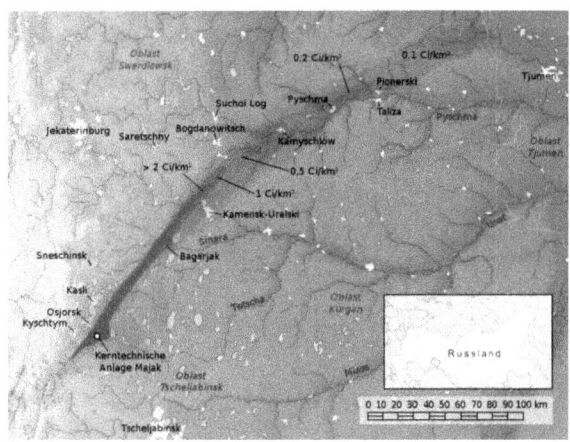

Ozyorsk today

Map of the East Urals Radioactive Trace (EURT): area contaminated by the Kyshtym disaster.

The **Kyshtym disaster** was a radioactive contamination accident that occurred on 29 September 1957 at Mayak, a plutonium production site for nuclear weapons and nuclear fuel reprocessing plant in the Soviet Union. It measured as a Level 6 disaster on the International Nuclear Event Scale (INES),[*][1] making it the third most serious nuclear accident ever recorded, behind the Fukushima Daiichi nuclear disaster and the Chernobyl disaster (both Level 7 on the INES). The event occurred in the town of Ozyorsk, Chelyabinsk Oblast, a closed city built around the Mayak plant. Since Ozyorsk/Mayak (named Chelyabinsk-40, then Chelyabinsk-65, until 1994) was not marked on maps, the disaster was named after Kyshtym, the nearest known town.

19.1 Background

After World War II, the Soviet Union lagged behind the US in development of nuclear weapons, so it started a rapid research and development program to produce a sufficient amount of weapons-grade uranium and plutonium. The Mayak plant was built in haste between 1945 and 1948. Gaps in Soviet physicists' knowledge about nuclear physics at the time made it difficult to judge the safety of many decisions. Environmental concerns were not taken seriously during the early development stage. Initially Mayak was dumping high-level radioactive waste into a nearby river, which was flowed to the river Ob, flowing further down to the Arctic Ocean. All six reactors were on Lake Kyzyltash and used an open cycle cooling system, discharging contaminated water directly back into the lake.[*][2] When Lake Kyzyltash quickly became contaminated, Lake Karachay was used for open-air storage, keeping the contamination a slight distance from the reactors but soon making Lake Karachay the "most polluted spot on Earth".[*][3][*][4][*][5]

A storage facility for liquid nuclear waste was added around 1953. It consisted of steel tanks mounted in a concrete base, 8.2 meters underground. Because of the high level of radioactivity, the waste was heating itself through decay heat (though a chain reaction was not possible). For that reason, a cooler was built around each bank containing 20 tanks. Facilities for monitoring operation of the coolers and the content of the tanks were inadequate.[*][6]

19.2   Explosion

In 1957 the cooling system in one of the tanks containing about 70–80 tons of liquid radioactive waste failed and was not repaired. The temperature in it started to rise, resulting in evaporation and a chemical explosion of the dried waste, consisting mainly of ammonium nitrate and acetates (see ammonium nitrate/fuel oil bomb). The explosion, on 29 September, 1957, estimated to have a force of about 70–100 tons of TNT, threw the 160-ton concrete lid into the air.[6] There were no immediate casualties as a result of the explosion, but it released an estimated 20 MCi (800 PBq) of radioactivity. Most of this contamination settled out near the site of the accident and contributed to the pollution of the Techa River, but a plume containing 2 MCi (80 PBq) of radionuclides spread out over hundreds of kilometers.[7] Previously contaminated areas within the affected area include the Techa river which had previously received 2.75 MCi (100 PBq) of deliberately dumped waste, and Lake Karachay which had received 120 MCi (4,000 PBq).[5]

In the next 10 to 11 hours, the radioactive cloud moved towards the north-east, reaching 300–350 kilometers from the accident. The fallout of the cloud resulted in a long-term contamination of an area of more than 800 to 20,000 square kilometers (depending on what contamination level is considered significant), primarily with caesium-137 and strontium-90.[5] This area is usually referred to as the East-Ural Radioactive Trace (EURT).[8]

Kyshtym Memorial

19.3   Evacuations

At least 22 villages were exposed to radiation from the disaster, with a total population of around 10,000 people evacuated. Some were evacuated after a week but it took almost 2 years for evacuations to occur at other sites.[9]

19.4   Aftermath

Because of the secrecy surrounding Mayak, the populations of affected areas were not initially informed of the accident. A week later (on 6 October) an operation for evacuating 10,000 people from the affected area started, still without giving an explanation of the reasons for evacuation.

Vague reports of a "catastrophic accident" causing "radioactive fallout over the Soviet and many neighboring states" began appearing in the western press between 13 and 14 April 1958, and the first details emerged in the Viennese paper *Die Presse* on 17 March 1959.[10][11] But it was only in 1976 that Zhores Medvedev made the nature

and extent of the disaster known to the world.[12][13] In the absence of verifiable information, exaggerated accounts of the disaster were given. People "grew hysterical with fear with the incidence of unknown 'mysterious' diseases breaking out. Victims were seen with skin 'sloughing off' their faces, hands and other exposed parts of their bodies." [14] Medvedev's description of the disaster in the *New Scientist* was initially derided by western nuclear industry sources, but the core of his story was soon confirmed by Professor Leo Tumerman, former head of the Biophysics Laboratory at the Engelhardt Institute of Molecular Biology in Moscow.[15]

The true number of fatalities remains uncertain because radiation-induced cancer is clinically indistinguishable from any other cancer, and its incidence rate can only be measured through epidemiological studies. One book claims that "in 1992, a study conducted by the Institute of Biophysics at the former Soviet Health Ministry in Chelyabinsk found that 8,015 people had died within the preceding 32 years as a result of the accident." [2] By contrast, only 6,000 death certificates have been found for residents of the Techa riverside between 1950 and 1982 from all causes of death,[16] though perhaps the Soviet study considered a larger geographic area affected by the airborne plume. The most commonly quoted estimate is 200 deaths due to cancer, but the origin of this number is not clear. More recent epidemiological studies suggest that around 49 to 55 cancer deaths among riverside residents can be associated to radiation exposure.[16] This would include the effects of all radioactive releases into the river, 98% of which happened long before the 1957 accident, but it would not include the effects of the airborne plume that was carried north-east.[17] The area closest to the accident produced 66 diagnosed cases of chronic radiation syndrome, providing the bulk of the data about this condition.[18]

To reduce the spread of radioactive contamination after the accident, contaminated soil was excavated and stockpiled in fenced enclosures that were called "graveyards of the earth".[19] The Soviet government in 1968 disguised the EURT area by creating the East Ural Nature Reserve, which prohibited any unauthorised access to the affected area.

According to Gyorgy,[20] who invoked the Freedom of Information Act to gain access to the relevant Central Intelligence Agency (CIA) files, the CIA knew of the 1957 Mayak accident since 1959, but kept it secret to prevent adverse consequences for the fledgling American nuclear industry.[21] Starting in 1989 the Soviet government gradually declassified documents pertaining to the disaster.[22][23]

19.5 Current situation

The level of radiation in Ozyorsk itself at about 0.1 mSv a year[24] is claimed to be safe for humans, but the area of the EURT is still heavily contaminated with radioactivity.[17]

19.6 See also

- Andreev Bay nuclear accident

19.7 References

[1] Lollino et al. 2014, p. 192

[2] Schlager 1994

[3] Lenssen, "Nuclear Waste: The Problem that Won't Go Away", Worldwatch Institute, Washington, D.C., 1991: 15.

[4] Andrea Pelleschi (2013). *Russia.* ABDO Publishing Company. ISBN 9781614808787.

[5] "Chelyabinsk-65".

[6] "Conclusions of government commission" (in Russian).

[7] Kabakchi & Putilov 1995, pp. 46–50

[8] Dicus 1997

[9] Kostyuchenko & Krestinina 1994, pp. 119–125

[10] Soran & Stillman 1982

[11] John Barry; E. Gene Frankland (25 February 2014). *International Encyclopedia of Environmental Politics.* Routledge. p. 297. ISBN 978-1-135-55396-8.

[12] Medvedev 1976, pp. 264-7

[13] Medvedev 1980

[14] Pollock 1978

[15] The Nuclear Disaster They Didn't Want To Tell You About/Andrews Cockburn/ Esquire Magazine/ 26 April 1978

[16] Standring 2009, pp. 174–199

[17] Kellerer 2002, pp. 307–316

[18] Gusev, Gus'kova & Mettler 2001, pp. 15–29

[19] Trabalka 1979

[20] Gyorgy 1979

[21] Newtan 2007, pp. 237–240

[22] "The decision of Nikipelov Commission" (in Russian).

[23] Smith 1989

[24] Suslova, KG; Romanov, SA; Efimov, AV; Sokolova, AB; Sneve, M; Smith, G. "Journal of Radiological Protection, December 2015, pp. 789-818". *J Radiol Prot.* **35**: 789–818. doi:10.1088/0952-4746/35/4/789. PMID 26485118.

- Lollino, Giorgio; Arattano, Massimo; Giardino, Marco; Oliveira, Ricardo; Silvia, Peppoloni, eds. (2014). *Engineering Geology for Society and Territory: Education, professional ethics and public recognition of engineering geology, Volume 7.* International Association for Engineering Geology and the Environment (IAEG). International Congress. Springer. ISBN 978-3-319-09303-1.

- Schlager, Neil (1994). *When Technology Fails.* Detroit: Gale Research. ISBN 0-8103-8908-8.

- Kabakchi, S. A.; Putilov, A. V. (January 1995). "Data Analysis and Physicochemical Modeling of the Radiation Accident in the Southern Urals in 1957". *Moscow ATOMNAYA ENERGIYA* (1).

- Dicus, Greta Joy (16 January 1997). "Joint American-Russian Radiation Health Effects Research". United States Nuclear Regulatory Commission. Retrieved 30 September 2010.

- Kostyuchenko, V.A.; Krestinina, L.Yu. (1994). "Long-term irradiation effects in the population evacuated from the East-Urals radioactive trace area". *Science of the Total Environment*. Elsevier. **142** (1–2): 119–25. doi:10.1016/0048-9697(94)90080-9. PMID 8178130. Retrieved 5 January 2012.

- Medvedev, Zhores A. (4 November 1976), *Two Decades of Dissidence*, New Scientist

- Medvedev, Zhores A. (1980), *Nuclear disaster in the Urals translated by George Saunders. 1st Vintage Books ed.*, New York: Vintage Books, ISBN 0-394-74445-4 (c1979)

- Soran, Diane M.; Stillman, Danny B. (1982). *An Analysis of the Alleged Kyshtym Disaster*. Los Alamos National Laboratory. doi:10.2172/5254763.

- Pollock, Richard (1978). "Soviets Experience Nuclear Accident". *Critical Mass Journal*.

- Standring, William J.F.; Dowdall, Mark; Strand, Per (2009). "Overview of Dose Assessment Developments and the Health of Riverside Residents Close to the "Mayak" PA Facilities, Russia". *International Journal of Environmental Research and Public Health*. **6** (1): 174–199. doi:10.3390/ijerph6010174. ISSN 1660-4601. PMC 2672329. PMID 19440276. Retrieved 11 June 2012.

- Kellerer, AM. (2002). "The Southern Urals radiation studies. A reappraisal of the current status." *Radiation and Environmental Biophysics*. **41** (4): 307–16. doi:10.1007/s00411-002-0168-1. ISSN 0301-634X. PMID 12541078.

- Gusev, Igor A.; Gus'kova, Angelina Konstantinovna; Mettler, Fred Albert (28 March 2001). *Medical Management of Radiation Accidents*. CRC Press. ISBN 978-0-8493-7004-5. Retrieved 11 June 2012.

- Trabalka, John R. (1979), *Russian Experience* (PDF) pp. 3–8 in *Environmental Decontamination: Proceedings of the Workshop, 4–5 December 1979, Oak Ridge, Tennessee*, Oak Ridge National Laboratory, CONF-791234

- Gyorgy, A. (1979). *No Nukes: Everyone's Guide to Nuclear Power*. ISBN 0-919618-95-2.

- Newtan, Samuel Upton (2007), *Nuclear War I and Other Major Nuclear Disasters of the 20th century*

- Smith, R. Jeffrey (10 July 1989). "Soviets Tell About Nuclear Plant Disaster; 1957 Reactor Mishap May Be Worst Ever". *The Washington Post*: A1.

- Rabl, Thomas (2012), *The Nuclear Disaster of Kyshtym 1957 and the Politics of the Cold War.*" Environment & Society Portal, Arcadia 2012, no. 20. Rachel Carson Center for Environment and Society.

19.8   External links

- An Analysis of the alleged Kyshtym Disaster

- Der nukleare Archipel (German)

- Official documents pertaining to the disaster (Russian)

Coordinates: 55°43′N 60°49′E / 55.717°N 60.817°E

Chapter 20

Santa Susana Field Laboratory

Aerial view of the Santa Susana Field Laboratory in the Simi Hills, with the San Fernando Valley and San Gabriel Mountains beyond to the east. The Energy Technology Engineering Center site is in the flat Area IV at the lower left, with the Rocket Test Field Laboratory sites in the hills at the center. (Spring 2005).

The **Santa Susana Field Laboratory** is a complex of industrial research and development facilities located on a 2,668-acre (1,080 ha)*[1] portion of the Southern California Simi Hills in Simi Valley, California. It was used mainly for the development and testing of liquid-propellant rocket engines for the United States space program from 1949 to 2006,*[1] nuclear reactors from 1953 to 1980 and the operation of a U.S. government-sponsored liquid metals research center from 1966 to 1998.*[2] The site is located approximately 7 miles (11 km) northwest from the community of Canoga Park and approximately 30 miles (48 km) northwest of Downtown Los Angeles. Sage Ranch Park is adjacent on part of the northern boundary and the community of Bell Canyon along the entire southern boundary.*[3]

Throughout the years, about ten low-power nuclear reactors operated at SSFL, in addition to several "critical facilities" which helped develop nuclear science and applications. At least four of the ten nuclear reactors had accidents during their operation. The reactors located on the grounds of SSFL were considered experimental, and therefore had

no containment structures.

20.1   Introduction

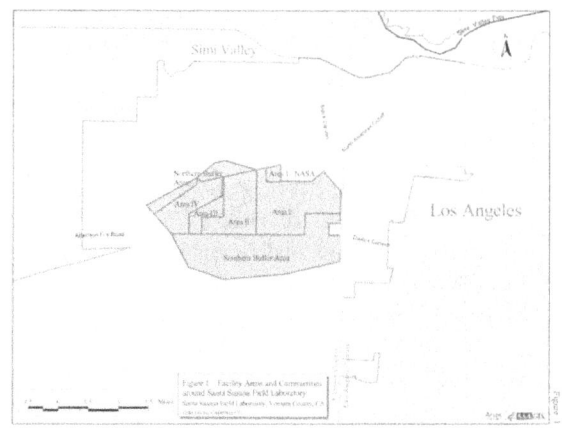

Santa Susana Field Laboratory administrative areas, and the surrounding communities.

Since 1947 the Santa Susana Field Laboratory location has been used by a number of companies and agencies. The first was Rocketdyne, originally a division of North American Aviation-NAA, which developed a variety of pioneering, successful and reliable liquid rocket engines.*[4] Some were those used in the Navaho cruise missile, the Redstone rocket, the Thor and Jupiter ballistic missiles, early versions of the Delta and Atlas rockets, the Saturn rocket family and the Space Shuttle Main Engine.*[5] The Atomics International division of North American Aviation utilized a separate and dedicated portion of the Santa Susana Field Laboratory to build and operate the first commercial nuclear power plant in the United States*[6] and for the testing and development of compact nuclear reactors including the first and only known nuclear reactor launched into Low Earth Orbit by the United States, the SNAP-10A.*[7] Atomics International also operated the Energy Technology

Engineering Center for the U.S. Department of Energy at the site. The Santa Susana Field Laboratory includes sites identified as historic by the American Institute of Aeronautics and Astronautics and by the American Nuclear Society. In 1996, The Boeing Company became the primary owner and operator of the Santa Susana Field Laboratory and later closed the site.

Three California state agencies and three federal agencies have been overseeing a detailed investigation of environmental impacts from historical site operations since at least 1990.[8] Concerns about the environmental impact of past disposal practices have inspired at least two lawsuits seeking payment from Boeing and several interest groups are actively involved with steering the ongoing environmental investigation.

The Santa Susana Field Laboratory is the focus of diverse interests. The National Register of Historic Places listed Burro Flats Painted Cave is located within the Santa Susana Field Laboratory, on a portion of the site owned by the U.S. government. The drawings within the cave have been termed "the best preserved Indian pictograph in Southern California." Several tributary streams to the Los Angeles River have headwater watersheds on the SSFL property, including Bell Creek (90% of SSFL drainage), Dayton Creek, Woolsey Canyon, and Runkle Creek.[9]

20.2   History

Aerial view looking north, of the Energy Technology Engineering Center in Area IV (1990).

SSFL was slated as a United States government facility dedicated to the development and testing of nuclear reactors, powerful rockets such as the Delta II, and the systems that powered the Apollo missions. The location of SSFL was chosen in 1947 for its remoteness in order to conduct work that was considered too dangerous and too noisy to be performed in more densely populated areas. In subsequent years however, the Southern California population grew, along with housing developments surrounding "The Hill." Today, more than 150,000 people live within 5 miles (8 km) of the facility, and at least half a million people live within 10 miles (16 km).

The site is divided into four production and two buffer areas, (Area I, II, III, and IV, and the northern and southern buffer zones). Areas I through III were used for rocket testing, missile testing, and munitions development. Area IV was used primarily for nuclear reactor experimentation and development. Laser research for the Strategic Defense Initiative (popularly known as "Star Wars"), was also conducted in Area IV.[10]

20.2.1   Rocket engine development

North American Aviation (NAA) began its development of liquid propellant rocket engines after the end of WWII. The Rocketdyne division of NAA, which came into being under its own name in the mid-1950s, designed and tested several rocket engines at the facility. They included engines for the Army's Redstone (an advanced short-range version of the German V-2), and the Army Jupiter intermediate range ballistic missile (IRBM) as well as the Air Force's counterpart IRBM, the Thor. Also included were engines for the Atlas Intercontinental Ballistic Missile (ICBM), as well as the twin combustion chamber alcohol/liquid oxygen booster engine for the Navaho, a large, intercontinental cruise missile that never became operational. Later, Rocketdyne designed and tested the huge F-1 engine that was eventually used as one of a cluster of engines powering the Apollo booster, as well as the J-2 liquid oxygen/hydrogen upper stage engine also used on the Project Apollo spacecraft.[11]

20.2.2   Nuclear and energy research and development

Main article: Atomics International

The Atomics International Division of North American Aviation utilized SSFL Area IV as the site of United States first commercial nuclear power plant[12] and the testing and development of the SNAP-10A, the first nuclear reactor launched into outer space by the United States.[13] Atomics International also operated the Energy Technology Engineering Center for the U.S. government. As overall interest in nuclear power declined, Atomics International transitioned to non-nuclear energy-related projects, such as coal gasification, and gradually ceased designing and testing nuclear reactors. Atomics International was eventually merged with the Rocketdyne division in 1978.[14]

Systems for Nuclear Auxiliary Power (SNAP):
• Atomics International (AI) program to develop space nuclear power systems.
• A system was launched from Vandenberg Air Force Base on April 3, 1965.
• Remains the only nuclear reactor placed in space by the U.S.

SSFL: the Atomics International Snap reactor.

Sodium reactor experiment

Main article: Sodium Reactor Experiment

The Sodium Reactor Experiment-SRE was an experimental nuclear reactor which operated from 1957 to 1964 and was the first commercial power plant in the world to experience a core meltdown. There was a decades-long cover-up by the US Department of Energy.*[15] The operation predated environmental regulation, so early disposal techniques are not recorded in detail.*[15] Thousands of pounds of sodium coolant from the time of the meltdown are not yet accounted for.*[16]*[17]

The reactor and support systems were removed in 1981 and the building torn down in 1999.

Energy Technology Engineering Center

Main article: Energy Technology Engineering Center

The Energy Technology Engineering Center-ETEC, was a government-owned, contractor-operated complex of industrial facilities located within Area IV of the Santa Susana Field Laboratory. The ETEC specialized in non-nuclear testing of components which were designed to transfer heat from a nuclear reactor using liquid metals instead of water or gas. The center operated from 1966 to 1998.*[18] The ETEC site has been closed and is now undergoing building removal and environmental remediation by the U.S. De-

partment of Energy.

20.3 Accidents and site contamination

20.3.1 Nuclear Reactors

Throughout the years, approximately ten low-power nuclear reactors operated at SSFL, in addition to several "critical facilities": a sodium burn pit in which sodium-coated objects were burned in an open pit; a plutonium fuel fabrication facility; a uranium carbide fuel fabrication facility; and the purportedly largest "Hot Lab" facility in the United States at the time.*[19] (A Hot Lab is a facility used for remotely cutting up irradiated nuclear fuel.) Irradiated nuclear fuel from other Atomic Energy Commission (AEC) and Department of Energy (DOE) facilities from around the country were shipped to SSFL to be decladded and examined.

The Hot Lab suffered a number of fires involving radioactive materials. For example, in 1957, a fire in the Hot Cell "got out of control and ... massive contamination" resulted. (see: NAA-SR-1941, Sodium Graphite Reactor, Quarterly Progress Report, January–March 1957).

At least four of the ten nuclear reactors suffered accidents: 1) The AE6 reactor experienced a release of fission gases in March 1959.*[20] 2) In July 1959, the site suffered a partial nuclear meltdown that has been named "the worst in U.S. history", releasing an undisclosed amount of radioactivity, but thought to be much more than the Three Mile Island disaster in 1979.*[21] Another radioactive fire occurred in 1971, involving combustible primary reactor coolant (NaK) contaminated with mixed fission products.*[22]*[23] The AE6 reactor experienced a release of fission gases in March 1959, the SRE experienced a power excursion and partial meltdown in July 1959; the SNAP8ER in 1964 experienced damage to 80% of its fuel; and the SNAP8DR in 1969 experienced similar damage to one-third of its fuel.*[24] 3) The SNAP8ER in 1964 experienced damage to 80% of its fuel.*[25] 4) The SNAP8DR in 1969 experienced similar damage to one-third of its fuel.*[26]

The reactors located on the grounds of SSFL were considered experimental, and therefore had no containment structures. Reactors and highly radioactive components were housed without the large concrete domes that surround modern power reactors.

Toxic substances burn and are released into the air.

A worker disposes of toxic chemicals by blowing up full barrels with a rifle shot (the reaction to the shot caused an explosion).

20.3.2   Sodium burn pits

The sodium burn pit, an open-air pit for cleaning sodium-contaminated components, was also contaminated when radioactively and chemically contaminated items were burned in it, in contravention of safety requirements. In an article in the *Ventura County Star*, James Palmer, a former SSFL worker was interviewed. The article notes that "of the 27 men on Palmer's crew, 22 died of cancers." On some nights Palmer returned home from work and kissed "his wife [hello], only to burn her lips with the chemicals he had breathed at work." The report also noted that "During their breaks, Palmer's crew would fish in one of three ponds ... The men would use a solution that was 90 percent hydrogen peroxide to neutralize the contamination. Sometimes, the water was so polluted it bubbled. The fish died off." Palmer's interview ended on a somber note: "They had seven wells up there, water wells, and every damn one of them was contaminated," Palmer said, "It was a horror story." *[27]

Other spills and releases occurred over the decades of operation as well. In 1989, a DOE investigation found widespread chemical and radioactive contamination on the property. Widely publicized in the local press, the revelations led to substantial concern among community members and elected officials, resulting in a challenge to and subsequent shutdown of continued nuclear activity at the site, and the filing of lawsuits. Cleanup commenced, and the United States Environmental Protection Agency (EPA) was brought in at the request of local legislators to provide oversight.

On December 11, 2002, a top Department of Energy (DOE) official, Mike Lopez, described typical clean-up procedures executed by Field Lab employees in the past. Workers would dispose of barrels filled with highly toxic

waste by shooting the barrels with rifles so that they would explode and release their contents into the air. It is unclear when this process ended, but for certain did end prior to the 1990s.*[28]

On July 26, 1994, two scientists, Otto K. Heiney, 52, of Canoga Park in Chatsworth and Larry A. Pugh, 51, of Thousand Oaks, were killed when the chemicals they were illegally burning in open pits exploded. After a grand jury investigation and FBI raid on the facility, three Rocketdyne officials pleaded guilty in June 2004 to illegally storing explosive materials. The jury deadlocked on the more serious charges related to illegal burning of hazardous waste.*[29]

At trial, a retired Rocketdyne mechanic testified as to what he witnessed at the time of the explosion: "I assumed we were burning waste," Lee Wells testified, comparing the process used on July 21 and 26, 1994, to that once used to legally dispose of leftover chemicals at the company's old burn pit. As Heiney poured the chemicals for what would have been the third burn of the day, the blast occurred, Wells said. "[The background noise] was so loud I didn't hear anything ... I felt the blast and I looked down and my shirt was coming apart."

When he realized what had occurred, Wells said, "I felt to see if I was all there ... I knew I was burned but I didn't know how bad." Wells suffered second- and third-degree burns to his face, arms and stomach.*[30]

20.3.3   Wildfires and contamination

In 2005, wildfires swept through northern Los Angeles County and parts of Ventura County. The fires consumed most of the dry brush throughout the Simi Hills where SSFL is located. The facility received substantial fire damage. Since the fire, allegations have emerged that vast quantities of on-site contamination were released into the air. Most recently, Los Angeles County firefighters who were assigned to SSFL during the fire have been sent for medical testing to see if any harmful doses were ingested or inhaled while protecting the facility.

While community members and firefighters have expressed concern about the amount of exposure, Boeing officials stand by their position that no contamination of the air resulted from the fire, and that any contamination that may have been consumed by the fire was negligible.

California Department of Toxic Substances Control also claims that no significant contamination occurred as a result of the fire. Although the Field Lab is under current criticism for violating almost 50 discharge permits, state agencies have been silent on the issue. Recently, lawyers disclosed to the California State Water Resources Control Board that over 80 exceedances of Boeing's discharge permits were found in the preceding year alone. In January 2006, the State Water Resources Control Board finally stepped in, and refused some requests by Boeing for even lighter standards.

20.3.4 Medical claims

Also in October 2005, plaintiff Margaret-Ann Galasso, in a suit against Boeing, criticized her attorneys, who, as she claimed, accepted a $30 million settlement with Boeing without her approval. The attorneys stand to collect $18 million, or 60% of the settlement amount after their costs and fees are subtracted. The plaintiff who disclosed the allegedly tainted deal, is splitting the rest of the settlement with other plaintiffs and will only receive around $30,000, a far cry from the amount she will need for extensive future medical treatments for diseases that were linked to contamination from the SSFL facility.

In October 2006, the Santa Susana Field Laboratory Advisory Panel, made up of independent scientists and researchers from around the United States, concluded that based on available data and computer models, contamination at the facility resulted in an estimated 260 cancer related deaths, with a 95% confidence interval of up to 1800 deaths. The report also concluded that the SRE meltdown caused the release of more than 458 times the amount of radioactivity released by the Three Mile Island accident. While the nuclear core of the SRE released 10 times less radiation than the TMI incident, the lack of proper containment such as concrete structures caused this radiation to be released into the surrounding environment. The radiation released by the core of the TMI was largely contained.*[31]

20.4 Cleanup

During its years of operation widespread use occurred of highly toxic chemical additives to power over 30,000 rocket engine tests and to clean the rocket test-stands afterwards, as well as considerable nuclear research and at least four nuclear accidents, which has resulted in the SSFL becoming a seriously contaminated site and offsite pollution source requiring a sophisticated multi-agency and corporate Cleanup Project.*[32] An ongoing process to determine the site contamination levels and locations, cleanup standards to meet, methods to use, costs, timelines and completion requirements - are still being debated, and litigated.*[33]

As of 2015 the site's owner is Boeing, with NASA and DOE liable for several parcels within that. On August 2, 2005, Pratt & Whitney purchased Boeing's Rocketdyne division, but declined to acquire SSFL as part of the sale.

20.4.1 Standards history

In 1989, the DOE found widespread chemical and radioactive contamination at their site, and a cleanup program commenced. In 1995 EPA and DOE announced that they had entered into a joint policy agreement to assure that all DOE sites would be cleaned up to standards consistent with the EPA's Comprehensive Environmental Response, Compensation, and Liability Act (CERCLA) standards.

However, in March 2003, the DOE reversed its position and announced that SSFL would not be cleaned up to EPA standards. While the DOE simultaneously claimed compliance with the 1995 joint policy agreement, the new plan included a cleanup of only 1% of the contaminated soil, and the release of SSFL for unrestricted residential use in as little as ten years. The EPA responded to this announcement by claiming that the DOE was not subject to EPA regulation due to the fact that the DOE existed as a separate entity under the executive branch of the federal government, and refused take steps to force DOE adherence to the 1995 agreement.

In August 2003, the Senate Appropriations Committee issued a report on Energy and Water Appropriations, urging the DOE to meet its commitments in the 1995 agreement and clean up SSFL to the EPA's CERCLA standards. The DOE responded to the Senate, claiming it was in fact consistent with both the agreement and EPA's CERCLA standards. In December 2003, soon after DOE's announcement that it was consistent with the 1995 agreement, EPA determined that the cleanup was not consistent with its CERCLA standards, and that sufficient contamination would remain at levels that would be dangerously inappropriate for unrestricted residential, and that the only safe use under DOE's revised cleanup standards would be restricted day hikes with limitations on picnicking.

Critics point out that if the DOE-Boeing cleanup plan was followed through and the site was released for unrestricted residential use, the property would likely become a Superfund site subject to EPA standards. After the sale, the site would no longer be a DOE facility, and thus, the exemption from CERCLA standards would no longer be in effect. The end result being that the site would only be brought into compliance with CERCLA cleanup standards after Boeing has sold the property, relieving the company of any burden of cleanup costs. The costs would likely be passed on to taxpayers, and not those responsible for the actual contamination.

In early May 2007, a Federal Court in San Francisco issued a major ruling which concluded that DOE has not been cleaning up the site to proper standards, and that the site would have to be cleaned up to higher standards if DOE ever wanted to release the site to Boeing, which in turn, would most likely release the land for unrestricted residential development.*[34] Judge "Conti's ruling requires DOE to prepare a more stringent review of the lab, which is on the border of Los Angeles County. Conti wrote that the department's decision to prepare a less-stringent environmental document prior to cleanup is in violation of the National Environmental Policy Act and noted that the lab 'is located only miles away from one of the largest population centers in the world.'"

Runoff issues

On July 26, 2007, staff at the Los Angeles Regional Water Quality Control Board recommended a $471,190 fine against Boeing for 79 violations of the California Water Code during an 18-month period. From October 2004 to January 2006, wastewater and storm water runoff coming from the lab had increased levels of chromium, dioxin, lead, mercury and other pollutants, the board said. The contaminated water flowed into Bell Creek and the Los Angeles River in violation of a July 1, 2004, permit that allowed release of wastewater and storm water runoff as long as it didn't contain high levels of pollutants.*[35]

Parkland

On October 15, 2007, Boeing announced that "In a landmark agreement between Boeing and California officials, nearly 2,400 acres (10 km^2) of land that is currently Boeing's Santa Susana Field Laboratory will become state parkland. According to the plan jointly announced by California Gov. Arnold Schwarzenegger, Boeing, and state Sen. Sheila Kuehl, the property will be donated and preserved as a vital undeveloped open-space link in the Simi Hills, above

the Simi Valley and the San Fernando Valley. The agreement will permanently restrict the land for nonresidential, noncommercial use." *[36]

20.4.2   New cleanup developments

SB 990

The California state senate bill SB 990, passed into law in 2007, set the standards for the site's cleanup.*[37] To achieve them, the R.P.s (responsible parties) consisting of Boeing, DOE, and NASA, need to sign agreements of acceptance and cleanup compliance.

Boeing

Boeing has contested the law, filing a lawsuit in September 2009 to release it from compliance, with a court date set for summer 2011. Boeing won the suit and claims it will clean up the site, although to levels far below those outlined in SB 990.*[38]

DOE and NASA

In September 2010 DOE and NASA agreed to meet the stringent cleanup standards set for the site in the state's SB 990 legislation, and to cover all costs for their cleanup's implementation. This agreement is significant progress in the SSFL cleanup sequence.*[39] In 2014, NASA issued a final environmental impact statement containing mitigation measures that would demolish all structures and remediate soil and groundwater contamination.*[40] NASA issued a report highlighting cleanup technology feasibility studies, soil and groundwater fieldwork, and additional archaeology surveys that would be performed in preparation for the demolition of the structures.*[41]

Demolition of the abandoned buildings, including a cafeteria, laboratory and offices for engineers and draftsmen built in the 1950s and 1960s, was scheduled to start in January or February 2015 after abatement of asbestos, lead paint and other regulated materials. The test stands will follow and are the most complex to tear down but all demolition should be completed in 2016. Because of their historical significance, one test stand and one control building will remain if the cleanup goals can still be met.*[42] The cleanup is projected to be completed in 2017.*[39]

20.4.3   Community involvement

PPG - Public Participation Group

The CA-DTSC: SSFL Project, the lead regulatory agency for the site cleanup, is forming a new [Sept. 2010] PPG - Public Participation Group, in response to their community 'Listening Sessions' held earlier in the year and the proposed Listening Session Response Plan.*[43] Applications from all the 'stakeholder' I.P.s - interested parties: the public, community groups, neighbors, local environmental and cultural groups, and others are being accepted currently [Sept. 2010].*[44]

SSFL Workgroup

Every quarter the SSFL Workgroup meetings regarding the cleanup are held that are open for public attendance. The SSFL Workgroup is the current version of the The Santa Susana Advisory Panel. The workgroup consists of representatives from the California Department of Toxic Substances Control, the U.S. EPA., public policy organizations, and community representatives. The Boeing Company, current owner of the SSFL site, the DOE are also invited. Other organizations and private companies also attend as part of the workgroup depending on the topic pending. The meetings are usually held at The Simi Valley Cultural Arts Center, and are posted on the DTSC-SSFL Calendar page of their website.*[45]

Community advisory group

A petition to form a "CAG" or community advisory group was denied in March 2010 by DTSC .*[46]*[47] In 2012, the current CAG's petition was approved, and their website is at ssflcag.net. The SSFL CAG recommends that all responsible parties execute a risk-based cleanup to EPA's suburban residential standard that will minimize excavation, soil removal and backfill and thus reduce danger to public health and functions of surrounding communities. However, the CAG has a clear conflict of interest, as it is funded in large part by a grant from the U.S. Department of Energy, and three of its members are former employees of Boeing or its parent company, North American Aviation.*[48] The CAG tried to keep the source of its funding, over $34,000, anonymous.

Physicians for Social Responsibility

The Los Angeles chapter of the Physicians for Social Responsibility has been working with the SSFL Work Group and Rocketdyne Cleanup Coalition.*[49] PSR expressed concern over conflict of interest involving Boeing, CAG, DTSC and others related to the cleanup that were revealed in a 55-page report, *Inside Job - How Boeing Fixers Captured Regulators and Derailed a Nuclear and Chemical Cleanup in LA's Backyard*, published in 2014 by Consumer Watchdog.*[50]

20.5 See also

- Nuclear labor issues

- Nuclear accidents

- Nuclear accidents in the United States

- Nuclear and radiation accidents and incidents

20.6 References

[1] Archeological Consultants, Inc.; Weitz Research (March 2009). "Historical resources survey and assessment of the NASA facility at the Santa Susana Field Laboratory, Simi Valley, California" (PDF). NASA. pp. 1–1. Retrieved January 25, 2010.

[2] Sapere and Boeing (May 2005). *Santa Susana Field Laboratory, Area IV Historical Site Assessment*. pp. 2–2. Retrieved January 25, 2010.

[3] Sage.Park

[4] http://www.boeing.com/aboutus/environment/santa_susana/history.html . accessed 8/30/2010

[5] American Institute of Aeronautics and Astronautics (2001). "Historic Aerospace Site: The Rocketdyne Santa Susana Field Laboratory, Canoga Park, California" (PDF). AIAA. Retrieved January 25, 2010.

[6] DuTemple, Octave. "American Nuclear Society Sodium Reactor Experiment Nuclear Historic Landmark awarded, February 21, 1986" (PDF). Retrieved January 25, 2010.

[7] Stokely, C. & Stansbury, E. (2008), "Identification of a debris cloud from the nuclear powered SNAP-SHOT satellite with Haystack radar measurements" , *Advances in Space Research*, **41** (7), pp. 1004–1009, doi:10.1016/j.asr.2007.03.046

[8] Charter.pdf "Santa Susana Field Laboratory Workgroup Charter" Check |url= value (help) (PDF). September 20, 1990. Retrieved January 1, 2010. |first1= missing |last1= in Authors list (help)

[9] http://www.enviroreporter.com/images/ESADA/2003-SSFL-surface%20water-map.jpg (accessed 4/10/2010) SSFL Watersheds Map

[10] "SITE SAFETY AND HEALTH PLAN AREA IV RADI-OLOGICAL STUDY SANTA SUSANA FIELD LABO-RATORY VENTURA COUNTY, CALIFORNIA" (PDF). *U.S. Environmental Protection Agency.* Retrieved 28 September 2016.

[11] The F-1 engine was so big that it could not be tested at the Rocketdyne Field Laboratory which was too close to populated San Fernando Valley areas, and tests on it were run out in the desert at the Edwards Air Force base. "Apollo Expeditions to the Moon, Chapter 3.2". NASA.

[12] U.S. Energy Information Agency. "California Nuclear Industry". Retrieved January 1, 2010.

[13] Voss, Susan (August 1984). *SNAP Reactor Overview.* U.S. Air Force Weapons Laboratory, Kirtland AFB, New Mexico. p. 57. AFWL-TN-84-14.

[14] Sapere and Boeing (May 2005). *Santa Susana Field Laboratory Area IV, Historical Site Assessment.* pp. 2–1. Retrieved January 1, 2010.

[15]

[16] Rockwell International Corporation, Energy Systems Group. "Sodium Reactor Experiment Decommissioning Final Report" (PDF). ESG-DOE-13403. Retrieved 17 February 2011. (see sections 2.1.7.4, 2.2.3, 4.4.2 and 9.3 for discrepancies concerning sodium amounts)

[17] Grover, Joel; Glasser, Matthew. "L.A.'s Nuclear Secret". *NBC.* National Broadcasting Company. Retrieved 31 October 2016.

[18] Sapere and Boeing (May 2005). *Santa Susana Field Laboratory, Area IV, Historical Site Assessment.* pp. 2–1. Retrieved January 20, 2010.

[19] Grover, Joel; Glasser, Matthew. "L.A.'s Nuclear Secret". *I-Team: 7-part NBC News special report.* NBC News. Retrieved 31 January 2017.

[20] "Report of the Santa Susana Field Laboratory Advisory Panel, October 2006" (PDF). Retrieved September 30, 2010.

[21] http://www.laweekly.com/2010-09-23/news/rocketdyne-cleanup-won-t-help-runkle-canyon/

[22] Rockwell International, Nuclear Operations at Rockwell's Santa Susana Field Laboratory —A Factual Perspective, September 6, 1991

[23] "Oak Ridge Associated Universities TEAM Dose Reconstruction Project for NIOSH, Document No. ORAUT-TKBS-0038-2, Rev. 0. page 24" (PDF). Retrieved September 30, 2010.

[24] "Report of the Santa Susana Field Laboratory Advisory Panel, October 2006" (PDF). Retrieved September 30, 2010.

[25] "Report of the Santa Susana Field Laboratory Advisory Panel, October 2006" (PDF). Retrieved September 30, 2010.

[26] "Report of the Santa Susana Field Laboratory Advisory Panel, October 2006" (PDF). Retrieved September 30, 2010.

[27] *The Cancer Effect*, October 30, 2006, The Ventura County Star

[28] "Rocketdyne, it's the pits", Ventura County Reporter, December 12, 2002; also see SB990, a bill before the California legislature relating this almost unbelievable procedure

[29] "Scientist Fined $100 in Lab Blast That Killed 2," Los Angeles Times, December 11, 2003 Thursday; also see "Executive Sentenced in '94 Blast; A former Rocketdyne official gets probation for violations linked to two scientists' deaths." Los Angeles Times, January 28, 2003 Tuesday

[30] "Ex-Rocketdyne Worker Describes Fatal 1994 Blast," Los Angeles Times, January 5, 2002

[31] http://www.ssflpanel.org/files/SSFLPanelReport.pdf

[32] http://www.boeing.com/aboutus/environment/santa_susana/index.html . accessed 8/30/2010

[33] Hiltzik, Michael (June 13, 2014). "Santa Susana toxic cleanup effort is a mess". The Los Angeles Times. Retrieved 31 January 2017.

[34] Griggs, Gregory W. (May 3, 2007). "Judge assails Rocketdyne cleanup". *Los Angeles Times.* Retrieved 23 December 2015.

[35] Griggs, Gregory W. (July 27, 2007). "Boeing faces fines over field lab runoff". *Los Angeles Times.* Retrieved 23 December 2015.

[36] http://www.boeing.com/aboutus/environment/santa_susana/future.html . accessed 8/3-/2010

[37] http://info.sen.ca.gov/pub/07-08/bill/sen/sb_0951-1000/sb_990_cfa_20070413_153535_sen_comm.html . accessed 8/30/2010

[38] Healy, Patrick and Lloyd, Jonathan (April 29, 2011) "Judge Sides With Boeing in Rocket Site Cleanup" NBC Southern California

[39] Sahagun, Louis (2010-09-04). "Nuclear cleanup at Santa Susana facility would finish by 2017 under settlement". *Los Angeles Times.* Retrieved 7 September 2010.

[40] Harris, Mike (2014-03-14). "NASA's Santa Susana cleanup could have significant impacts, report says". *Ventura County Star.* Retrieved 15 March 2014.

[41] Sullivan, Bartholomew (2014-05-01). "NASA plans to raze structures at Santa Susana Field Laboratory". *Ventura County Star.* Retrieved 3 May 2014.

[42] Harris, Mike (2014-12-06). "Historic NASA structures to be razed in Santa Susana cleanup" . *Ventura County Star.*

[43] http://www.dtscssfl. com/files/lib_pub_involve/other_docs/64651_LSRP.pdf . accessed 8/30/2010

[44] http://www.dtsc.ca.gov/SiteCleanup/Santa_Susana_Field_Lab/ . accessed 8/30/2010

[45] http://www.dtsc.ca.gov/SiteCleanup/Santa_Susana_Field_Lab/ssfl_calendar.cfm . accessed 8/30/2010

[46] http://cleanuprocketdyne.org/cleanuprocketdyne.org/Community_Advisory_Group/Community_Advisory_Group.html . accessed 8/30/2010

[47] http://www.dtsc.ca.gov/SiteCleanup/Projects/upload/CAG-Petition-Final-Response.pdf

[48] http://www.enviroreporter.com/2016/09/dept-of-energy-secretly-funding-front-group-to-sabotage-its-own-santa-susana-field-lab-cleanup/

[49] "Cleaning Up the Santa Susana Field Laboratory" . Physicians for Social Responsibility, Los Angeles. Retrieved 5 February 2017.

[50] Tucker, Liza. "Inside Job: How Boeing Fixers Captured Regulators and Derailed a Nuclear and Chemical Cleanup in LA's Backyard" (PDF). Consumer Watchdog. Retrieved 5 February 2017.

20.7 External links and sources

20.7.1 Agencies

- CA-DTSC-Santa Susana Field Laboratory website: site investigation and cleanup news, Listserv e-mail newsletter, calendar, documentation download links, contacts.

- "U.S. DOE ETEC Closure Project Website" . *DOE-sponsored project website provides historical ETEC technology development, site usage and current closure project information. Interactive graphic found in Regulation section explains the various involved regulatory agencies and their roles at the site. Large number of documents located in the Reading Room.* Retrieved September 12, 2008.

- "DTSC-Santa Susana Field Laboratory Site Investigation and Cleanup website" . *Hosted by the California State Department of Toxic Substances Control which oversees the investigation and cleanup of chemicals in the soil and groundwater at the SSFL. Project status documents, reports and public comment materials are available.* Retrieved September 14, 2006.

- "SSFL Ground Water Contamination: Preliminary Analysis" (PDF). Retrieved September 30, 2005. - PDF of a presentation given on August 19, 2003.

- "Santa Susana Field Laboratory (SSFL)". *The Decontamination and Decommissioning Science Consortium.* Retrieved September 30, 2005.

- "Discussed at FARK: Radioactive emissions from a nuclear meltdown in California 47 years ago are worse than anybody thought. In other news, there was a nuclear meltdown in the US back in 1959" . Retrieved October 6, 2006.

- The Santa Susana Advisory Panel

- The Rocketdyne Information Society Public Forum on SSFL cleanup.

- History Channel "Rocketdyne Meltdown" on YouTube

- lamountains Sage Ranch Park website.

- "Boeing: About Us - Santa Susana website" . *Brief website hosted by The Boeing Company, the largest landowner of the Santa Susana Field Laboratory. This site contains general information and a cleanup completion schedule for the soil and groundwater projects. Surface water discharge-related information for SSFL is posted in the Environmental Programs section.* Retrieved September 14, 2008.

- "Santa Susana Field Laboratory" . *U.S. DOE Office of Environmental Management.* Retrieved September 30, 2005.

- "Draft Preliminary Site Evaluation of Santa Susana Field Laboratory (SSFL)". *U.S. DOE ATSDR - Agency for Toxic Substances & Disease Registry.* Retrieved April 4, 2011.

20.7.2 Groups and info

- Rocketdyne Cleanup Coalition: The Rocketdyne Cleanup Coalition (RCC) is a community-based alliance dedicated to the cleanup of SSFL. The group has been active since 1989 when it helped stop all nuclear activities at the site. RCC's website contains news, resources, and information about community meetings and events.

- ACME - Aerospace Cancer Museum of Education website: SSFL information repository; documents, photos, maps, and history archives, numerous project links.

- Cleanup Rocketdyne.org website: SSFL news, information resources, documents [pdf], public participation and community advisory group links.

- RIS - Rocketdyne Information Society: public online forum on SSFL cleanup process.

- SSFL-CAG community advisory group forum: SSFL cleanup support group, public online meetings and resources.

- The Santa Susana Advisory Panel: 2006 report archives

- Sage Ranch Park website

- "RocketdyneWatch". *website dedicated to proper public disclosure of the activities at SSFL with: a historic news archive to the 1980s; a document archive; and other information relevant to this 'contentious site.'.* Retrieved January 1, 2007.

- "Environment Site Restoration Summary - Santa Susana Field Laboratory". *U.S. DOE Office of Environmental Management.* Retrieved September 30, 2005.

- "Energy Technology Engineering Center, Santa Susana Field Lab". *Center for Land Use Interpretation.* Retrieved September 30, 2005.

- "SSFL Ground Water Contamination: Preliminary Analysis" (PDF). Retrieved September 30, 2005. - PDF of a presentation given on August 19, 2003.

- "Santa Susana Field Laboratory (SSFL)". *The Decontamination and Decommissioning Science Consortium.* Retrieved September 30, 2005.

20.7.3   Media

- EnviroReporter.com: Investigative news website that has coverage of Rocketdyne issues since 1998, often in partnership with regional publications including the *LA Weekly* and *Ventura County Reporter* newspapers.

- History Channel History Channel: "Rocketdyne Meltdown" on YouTube

- "The Rockets' Red Glare: First Installment". *An upcoming documentary film recounts the horrors and hazards of the work done at Boeing's Santa Susana Field Laboratory. This first installment focuses on the workers and their every-day exposure to the hazardous environment provided by the owners and operators of this lab.* Retrieved July 27, 2007.

- "Discussed at FARK: Radioactive emissions from a nuclear meltdown in California 47 years ago are worse than anybody thought. In other news, there was a nuclear meltdown in the US back in 1959". Retrieved October 6, 2006.

- Joel Grover and Matthew Glasser LA'S Nuclear Secret, Part 1-5 NBC4, 21 September 2015, retrieved 23 December 2015.

20.7.4   Reactor accident sources

- "-NAA-SR-MEMO-3757" (PDF). Retrieved March 5, 2007., Release of Fission Gas from the AE-6 Reactor, hosted by RocketdyneWatch.org

- "-NAA-SR-5898" (PDF). Retrieved March 5, 2007., Analysis of SRE Power Excursion, hosted by RocketdyneWatch.org

- "-NAA-SR-4488" (PDF). Retrieved March 14, 2007., SRE Fuel Element Damage an Interim Report, hosted by RocketdyneWatch.org

- "-NAA-SR-4488-Suppl" (PDF). Retrieved March 14, 2007., SRE Fuel Element Damage Final Report, hosted by RocketdyneWatch.org

- "-NAA-SR-MEMO-12210" (PDF). Retrieved March 14, 2007., SNAP8 Experimental Reactor Fuel Element Behavior: Atomics International Task Force Review, hosted by RocketdyneWatch.org

- "-NAA-SR-12029" (PDF). Retrieved March 14, 2007., Postoperation Evaluation of Fuel Elements from the SNAP8 Experimental Reactor hosted by RocketdyneWatch.org

- "-AI-AEC-13003" (PDF). Retrieved March 19, 2007., Findings of the SNAP 8 Developmental Reactor (S8DR) Post-Test Examination, hosted by RocketdyneWatch.org

Coordinates: 34°13′51″N 118°41′47″W / 34.230822°N 118.696375°W

Chapter 21

Soviet submarine K-19

This article is about the Soviet submarine K-19. For the 2002 film dramatization of events involving the submarine, see K-19: The Widowmaker. For other uses, see K-19 (disambiguation).

Not to be confused with Soviet submarine K-219.

K-19 was one of the first two Soviet submarines of the **658** class (NATO reporting name Hotel-class submarine), the first generation nuclear submarine equipped with nuclear ballistic missiles, specifically the R-13 SLBM. The boat was hurriedly built by the Soviets, who were anxious to catch up with the United States' lead in nuclear submarines. Before it was launched, 10 civilian workers and a sailor died due to accidents and fires. After it was commissioned, it was plagued with breakdowns and accidents, several of which threatened to sink the sub.

On its initial voyage on 4 July 1961, it suffered a complete loss of coolant to its reactor. With no backup system, the captain ordered members of the engineering crew to find a solution to avoid a nuclear meltdown. Sacrificing their own lives, the engineering crew jury-rigged a secondary coolant system and kept the reactor from a meltdown. Twenty-two crew members died from radiation sickness during the following two years. The sub experienced several other accidents, including two fires and a collision. The series of accidents inspired crew members to nickname the sub *Hiroshima*.

21.1 Background

In the late 1950s, the leaders of the Soviet Union were determined to catch up with the United States and began to build a nuclear sub fleet, pushing subs through production and testing so rapidly that many Russian naval officers felt that the ships were not fit for combat.*[1] The crew aboard the first nuclear submarines of the Soviet fleet was provided with a very high quality standard of food including smoked fish, sausages, fine chocolates and cheeses, unlike the standard fare given the crews of other naval vessels.*[2]

21.1.1 Construction deaths

K-19 was ordered by the Soviet Navy on 16 October 1957.*[3] Her keel was laid on 17 October 1958 at the naval yard in Severodvinsk. Several workers died building the submarine: two workers were killed when a fire broke out, and later six women gluing rubber lining to a water cistern were killed by fumes.*[2] While missiles were being loaded, an electrician was crushed to death by a missile-tube cover, and an engineer fell between two compartments and died.*[1]

21.1.2 Gains unlucky reputation

The ship was launched and christened on 8 April 1959.*[1] Breaking with tradition, a man—Captain 3rd Rank V. V. Panov of the 5th Urgent Unit—instead of a woman, was chosen to smash the ceremonial champagne bottle across the ship's stern. The bottle failed to break, instead sliding along the screws and bouncing off the rubber-coated hull. This is traditionally viewed among sea crews as a sign that the ship is unlucky.*[4] Captain 1st Rank Nikolai Vladimirovich Zateyev was the first commander of the submarine.*[1]

21.1.3 Early problems

In January 1960, confusion among the crew during a watch change led to improper operation of the reactor and a reactor-control rod was bent. The damage required the reactor to be dismantled for repairs. The officers on duty were removed and Captain Panov was demoted.

The submarine's ensign was hoisted for the first time on 12 July 1960. It underwent sea trials from 13 through 17 July 1960 and again from 12 August through 8 November 1960, transiting 17,347 kilometres (10,779 mi). The ship was considered completed on 12 November 1960.*[1] After surfacing from a full-power run, the crew discovered that most of the hull's rubber coating had detached, and the entire surface of the boat had to be re-coated.

During a test dive to the maximum depth of 300 m (980 ft), flooding was reported in the reactor compartment, and Captain Zateyev ordered the sub to immediately surface, where it keeled over on its port side due to the water it had taken on. It was later determined that during construction the workers had failed to replace a gasket.*[1] In October, 1960, the galley crew disposed of wood from equipment crates through the galley's waste system, clogging it. This led to flooding of the ninth compartment, which filled one third full of water. In December 1960, a loss of coolant caused failure of the main circuit pump. Specialists called from Severodvinsk managed to repair it at sea within a week.

The boat was commissioned on 30 April 1961. The sub had a total of 139 men aboard, including missile men, reactor officers, torpedo men, doctors, cooks, stewards, and several observing officers who were not part of the standard crew.

21.2   Nuclear accident

Nikolai Vladimirovich Zateyev, commander of the submarine at the time of the nuclear accident

On 4 July 1961, under the command of Captain First Rank Nikolai Vladimirovich Zateyev, *K-19* was conducting exercises in the North Atlantic close to Southern Greenland when it developed a major leak in its reactor coolant system, causing the water pressure in the aft reactor to drop to zero and causing failure of the coolant pumps. A separate accident had disabled the long-range radio system, so they could not contact Moscow. Despite the control rods being inserted via a SCRAM mechanism, the reactor temperature rose uncontrollably due to decay heat from fission products created during normal operation, eventually reaching 800 °C (1,470 °F). The reactor continued to heat up as the required coolant was not available during shutdown because of design shortcomings which did not include a backup cooling system.

Making a drastic decision, Zateyev ordered the engineering section to fabricate a new coolant system by cutting off an air vent valve and welding a water-supplying pipe into it. This required the men to work in high radiation for extended periods. The accident released radioactive steam containing fission products which were drawn into the ship's ventilation system and spread to other compartments of the ship. The jury-rigged cooling water system successfully reduced the temperature in the reactor.

The incident irradiated the entire crew, most of the ship, and some of the ballistic missiles on board. All seven members of the engineering crew and their divisional officer died of radiation exposure within the next month. Fifteen more sailors died from the after-effects of radiation exposure within the next two years.*[5]

Instead of continuing on the mission's planned route, the captain decided to head south to meet diesel-powered submarines expected to be there. Worries about a potential crew mutiny prompted Zateyev to have all small arms thrown overboard except for five pistols distributed to his most trusted officers. A diesel submarine, *S-270*, picked up *K-19*'s low-power distress transmissions and joined up with it.

American warships nearby had also heard the transmission and offered to help, but Zateyev, afraid of giving away Soviet military secrets to the West, refused and sailed to meet the *S-270*. He evacuated the crew and had the boat towed to its home base. After its return to port, the vessel contaminated a zone within 700 m (2,300 ft). Over the next two years, repair crews removed and replaced the damaged reactors. The repair process contaminated the nearby environment and the repair crew. The Soviet Navy dumped the original radioactive compartment into the Kara Sea.*[6]

According to the government's official explanation of the disaster, the repair crews discovered that the catastrophe had been caused by a faulty welding incident during initial construction. They discovered that during installation of the primary cooling system piping, a welder had failed to cover exposed pipe surfaces with asbestos drop cloths (required to protect piping systems from accidental exposure to welding sparks), due to the cramped working space. A drop from a welding electrode fell on an unprotected surface, producing an invisible crack. This crack was subject to prolonged and intensive pressure (over 200 atmospheres), compromising the pipe's integrity and finally causing it to fail.*[7]

Others disputed this conclusion. Retired Rear-Admiral Nikolai Mormul asserted that when the reactor was first started ashore, the construction crew had not attached a pressure gauge to the primary cooling circuit. Before anyone realized there was a problem, the cooling pipes were subjected to a pressure of 400 atmospheres, double the acceptable limit. *K-19* returned to the fleet with the nickname "Hiroshima".*[7]

On 1 February 2006, former President of the Soviet Union Mikhail Gorbachev proposed in a letter to the Norwegian Nobel Committee that the crew of *K-19* be nominated for a Nobel Peace Prize for their actions on 4 July 1961.*[8]

21.2.1 Deceased crew members

Several crew members received fatal doses of radiation during repairs on the reserve coolant system of Reactor #8. All of them died between one and three weeks after the accident from severe radiation sickness. A person who receives a dose of 4 to 5 Sv (about 400-500 rem) has a 50% chance of dying.*[A 1]

[1] United States Nuclear Regulatory Commission "Lethal Dose" definition

[2] Convert from Roentgen (R) to rad or from Roentgen (R) to rem?

[3] СУБМАРИНА, СБЕРЕГШАЯ МИР ТРУд. 21 November 2002. (Russian)

Many other crew members also received doses of radiation exceeding permissible levels. They underwent medical treatment during the following year. The treatment was devised by Professor Z. Volynskiy and included bone marrow transplantation and blood transfusion. It saved, among others, Chief Lieutenant Mikhail Krasichkov and Captain 3rd class Vladimir Yenin, who had received doses of radiation that were otherwise considered deadly. For reasons of secrecy, the official diagnosis was not "radiation sickness" but "astheno-vegetative syndrome" which is a kind of mental disorder.*[9]

21.2.2 Crew members decorated

On 6 August 1961, 26 members of the crew were decorated for courage and valor shown during the accident.

21.3 Later operational history

On 14 December 1961, the boat was fully upgraded to the Hotel II (*658м*) variant, which included upgrading to R-21 missiles, which had twice the effective range of the earlier missiles.

21.3.1 Collision

At 07:13 on 15 November 1969, *K-19* collided with the attack submarine USS *Gato* in the Barents Sea at a depth of 60 m (200 ft). It was able to surface using an emergency main ballast tank blow. The impact completely destroyed the bow sonar systems and mangled the covers of the forward torpedo tubes. *K-19* was able to return to port where she was repaired and returned to the fleet. *Gato* was relatively undamaged and continued her patrol.*[10]

21.3.2 Fires

On 24 February 1972, a fire broke out while the submarine was at a depth of 120 m (390 ft), some 1,300 km (700 nmi; 810 mi) from Newfoundland, in Canada. The boat surfaced and the crew was evacuated to surface warships except for 12 men trapped in the aft torpedo room. Towing was delayed by a gale, and rescuers could not reach the aft torpedo room because of conditions in the engine room. The fire killed 28 sailors aboard the *K-19* and two others who died after they were transferred to rescue ships. Investigators determined that the fire was caused by hydraulic fluid leak onto a hot filter.

The boat was finally towed to Severomorsk on 4 April, and the men were rescued after surviving 24 days in the lightless, heatless torpedo room. The rescue operation lasted more than 40 days and involved over 30 ships. From 15 June through 5 November 1972, *K-19* was repaired and put back into service.

On 15 November 1972, another fire broke out in compartment 6, but it was put out by the chemical fire-extinguisher system and there were no casualties.

21.3.3 Reclassification

On 25 July 1977, *K-19* was reclassified as a *Large Submarine*, and on 26 July 1979, she was reclassified as a communications submarine and given the symbol KS-19 (КС−19). On 15 August 1982, an electrical short circuit resulted in severe burns to two sailors, and sailor V. A. Kravchuk died five days later.

On 28 November 1985, the ship was upgraded to the 658s (*658c*) variant.

21.4   Decommissioning

On 19 April 1990 the submarine was decommissioned, and was transferred in 1994 to the naval repair yard at Polyarny. In March 2002, it was towed to the Nerpa Shipyard, Snezhnogorsk, Murmansk, to be scrapped. Over its service life, it transited 332,396 miles (534,940 km) during 20,223 operational hours.

In August 2003, the crew visited the boat in the Polyarny shipyard for the last time. It was announced in October 2003 that scrapping would start soon. Only the sail was saved to be used as a burial site for fallen crew members. Due to the large number of accidents during its construction and service life, it gained the unofficial nickname "Hiroshima" among naval sailors and officers.*[11]

In 2006, a section of *K-19* was purchased by Vladimir Romanov, who once served on the sub as a conscript, with the intention of *"Turning it into a Moscow-based meeting place to build links between submarine veterans from Russia and other countries."* So far, the plans remain on hold, and many of *K-19*'s survivors have objected to them.*[12]

21.5   Theatre and film

In the year 1969 the poet Vasily Aksyonov created his version of the play about the nuclear incident. This critical version is one of the few known works of unofficial literature in the former Soviet Union.*[13]

The movie *K-19: The Widowmaker* (2002), starring Harrison Ford and Liam Neeson, is based on the story of the *K-19*'s first disaster.*[14] The production company attempted in March 2002 to secure access to the boat as a set for its production, but the Russian Navy declined. The nickname "The Widowmaker" referred to by the movie was fiction. The submarine did not gain a nickname until the nuclear accident on 4 July 1961, when she was called "Hiroshima".*[2]

21.6   References

[1] "1958-60: The Construction of K-19". *K-19: The History*. National Geographic. Retrieved 28 May 2014.

[2] Bivens, Matt (January 3, 1994). "Horror of Soviet Nuclear Sub 61' Tragedy Retold". The Los Angeles Times. Retrieved 22 September 2012.

[3] Historical overview (Russian)

[4] McNamara, Robert. "Ships, Champagne, and Superstition". About.com. Retrieved 26 May 2014.

[5] "Epilogue: Tragedy Upon Tragedy". *K-19: The History*. National Geographic. Retrieved May 5, 2011.

[6] Polmar, Norman (2003). *Cold War Submarines. The Design and Construction of U.S. and Soviet Submarines*. Potomac Books, Inc. p. 112. ISBN 1-57488-530-8.

[7] "K19 Widowmaker - A Nuclear Accident". Retrieved 24 February 2015.

[8] "Lenta.ru: Оружие: К−19 – достойная награда спустя 45 лет (Weapon: K-19 – distinguished award after 45 years)" (in Russian). Old.lenta.ru. 2003-09-13. Retrieved 2011-05-14.

[9] Bos, Carole. "K19 Widowmaker - SECRET HEROES". Retrieved 24 February 2015.

[10] Miller, David (2006). *Submarine disasters*. Guilford, Conn.: Lyons Press. p. 65. ISBN 978-1592288151. Retrieved 17 July 2015.

[11] «Own truth» (September 2003) (Russian)

[12] Haggerty, Anthony (20 July 2006). "Jambos chief Vlad splashes out on sub". *The Daily Record*. Archived from the original on 23 May 2007.

[13] "The poet Vasily Aksyonov" thesis of Herbert Gantschacher for obtaining the academic title "Master of Arts" at the Academy, today University of Music and Performing Arts, Graz, Re / 1653/1988, July 1988

[14] *Soviet submarine K-19* at the Internet Movie Database

21.7   External links

• Stewart, Will (12 May 2006). "Romanov invites Soviet sub heroes to cup final". *The Scotsman*. Scotsman Publications.

• K-19: the History (from National Geographic)

• Bratkov, Vitaly (July 26, 2002). "K-19, prototype of Hollywood thriller, to be cut up". *Pravda*.

Chapter 22

SL-1

This article is about The SL-1 Nuclear Reactor. For the Nortel Meridian SL1 PBX, see Nortel Meridian.

Coordinates: 43°31′06″N 112°49′25″W / 43.518233°N 112.823727°W

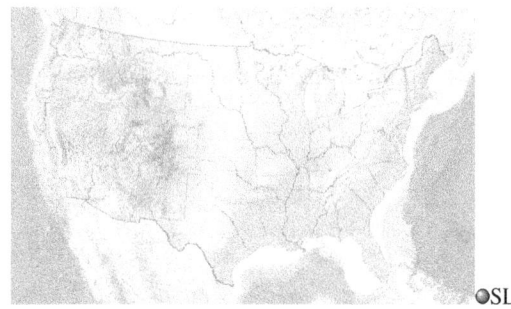

◉SL-1

Location in the United States

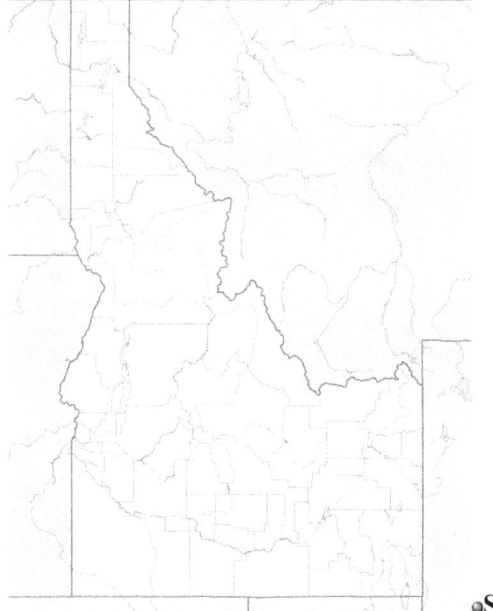

◉SL-1

Location in Idaho, west of Idaho Falls

The **SL-1**, or **Stationary Low-Power Reactor Number One**, was a United States Army experimental nuclear power reactor which underwent a steam explosion and meltdown on January 3, 1961, killing its three operators. The direct

November 29, 1961: The SL-1 reactor vessel being removed from the reactor building, which acted substantially like the containment building used in modern nuclear facilities. The 60-ton Manitowoc Model 3900 crane had a 5.25-inch (13.3 cm) steel shield with a 9-inch (23 cm) thick lead glass window to protect the operator.

cause was the improper withdrawal of the central control rod, responsible for absorbing neutrons in the reactor core. The event is the only reactor incident in the United States which resulted in immediate fatalities.*[1] The incident released about 80 curies (3.0 TBq) of iodine-131,*[2] which was not considered significant due to its location in the remote high desert of eastern Idaho. About 1,100 curies (41 TBq) of fission products were released into the atmosphere.*[3]

The facility, located at the National Reactor Testing Station (NRTS) approximately 40 miles (64 km) west of Idaho Falls, was part of the Army Nuclear Power Program and was known as the Argonne Low Power Reactor (ALPR) during its design and build phase. It was intended to provide electrical power and heat for small, remote military facilities, such as radar sites near the Arctic Circle, and those in the DEW Line.*[4] The design power was 3 MW (thermal), but some 4.7 MW tests were performed in the months prior to the accident. Operating power was 200 kW electrical and 400 kW thermal for space heating.

During the incident the core power level reached nearly 20 GW in just four milliseconds, precipitating the steam explosion.[*][5][*][6][*][7][*][8]

22.1   Design

The ALPR before the accident. The large cylindrical building holds the nuclear reactor embedded in gravel at the bottom, the main operating area or operating floor in the middle, and the condenser fan room near the top. Miscellaneous support and administration buildings surround it.

From 1954 to 1955, the U.S. Army evaluated their need for nuclear reactor plants that would be operable in remote regions of the Arctic. The reactors were to replace diesel generators and boilers that provided electricity and space heating for the Army's radar stations. The Army Reactors Branch formed the guidelines for the project and contracted with Argonne National Laboratory to design, build, and test a prototype reactor plant to be called the Argonne Low Power Reactor (ALPR).

Some of the more important criteria included:

- All components able to be transported by air
- Use of standard components
- Minimal on-site construction
- Simplicity and reliability
- 3-year fuel operating lifetime per core loading

The prototype was constructed at the NRTS site from July 1957 to July 1958. It was operational on October 24, 1958. The 3 MW (thermal) boiling water reactor (BWR) used 93.20% highly enriched uranium fuel. It operated with natural circulation, using light water as a coolant (vs. heavy water) and moderator. ANL used its experience from the BORAX experiments to design the BWR. The circulating water system operated at 300 pounds per square inch (2,100 kPa) flowing through fuel plates of uranium-aluminum alloy. The plant was turned over to the U.S. Army in December 1958 after extensive testing, with Combustion Engineering acting as the lead contractor beginning in February 1959. Members of the U.S. Army, called *cadre*, began training as plant operators, although some Navy personnel also trained with them.

The majority of the plant equipment was located in a cylindrical steel reactor building 38.5 feet (11.7 m) in diameter and an overall height of 48 feet (15 m). The reactor building known as ARA-602 was made of plate steel, most of which had a thickness of 1/4 inch (6 mm). Access to the building was provided by an ordinary door through an enclosed exterior stairwell from ARA-603, the Support Facilities Building. An emergency exit door was also included, with an exterior stairwell going to the ground level. The reactor building was not a pressure-type containment shell as would have been used for reactors located in populated areas. Nevertheless, the building was able to contain most of the radioactive particles released by the eventual explosion.

The reactor core structure was built for a capacity of 59 fuel assemblies, one source assembly, and 9 control rods. The core in use, however, had 40 fuel elements and was controlled by 5 cruciform rods. The 5 active rods were in the shape of a plus symbol (+) in cross section: 1 in the center (Rod Number 9), and 4 on the periphery of the active core (Rods 1, 3, 5, and 7). The control rods were made of 60 mils (1.5 mm) thick cadmium, clad with 80 mils (2.0 mm) of aluminum. They had an overall span of 14 inches (36 cm) and an effective length of 32 inches (81 cm). The 40 fuel assemblies were composed of 9 fuel plates each. The plates were 120 mils (3.0 mm) thick consisting of 50 mils (1.3 mm) of uranium-aluminium alloy "meat" covered by 35 mils (0.89 mm) of X-8001 aluminum cladding. The meat was 25.8 inches (66 cm) long and 3.5 inches (8.9 cm) wide. The water gap between fuel plates was 310 mils (7.9 mm). The initial loading of the 40 assembly core was highly enriched with 93.2% uranium-235 and contained 31 pounds (14 kg) of U-235.

The deliberate choice of a smaller fuel loading made the region near the center more active than it would have been with 59 fuel assemblies. The four outer control rods weren't even used in the smaller core. In the operating SL-1 core, Rods Number 2, 4, 6, and 8 were dummy rods, had cadmium shims, or were filled with test sensors, and were shaped like the capital letter T. The effort to minimize the size of the core gave the central rod an abnormally large reactivity worth.

22.2 Incident and response

On December 21, 1960, the reactor was shut down for maintenance, calibration of the instruments, installation of auxiliary instruments, and installation of 44 flux wires to monitor the neutron flux levels in the reactor core. The wires were made of aluminum, and contained slugs of aluminum–cobalt alloy.

On January 3, 1961, the reactor was being prepared for restart after a shutdown of eleven days over the holidays. Maintenance procedures required that the main central control rod be manually withdrawn a few inches to reconnect it to its drive mechanism. At 9:01 p.m., this rod was suddenly withdrawn too far, causing SL-1 to go prompt critical instantly. In four milliseconds, the heat generated by the resulting enormous power excursion caused water surrounding the core to begin to explosively vaporize. The water vapor caused a pressure wave to strike the top of the reactor vessel, causing water and steam to spray from the top of the vessel. This extreme form of water hammer propelled control rods, shield plugs, and the entire reactor vessel upward. A later investigation concluded that the 12,000-kilogram (26,000 lb) vessel had jumped 2.77 metres (9.1 ft) and the upper control rod drive mechanisms had struck the ceiling of the reactor building prior to settling back into its original location.*[6]*[9] The spray of water and steam knocked two operators onto the floor, killing one and severely injuring another. One of the shield plugs on top of the reactor vessel impaled the third man through his groin and exited his shoulder, pinning him to the ceiling.*[6] The victims were Army Specialists John A. Byrnes (age 27) and Richard Leroy McKinley (age 22), and Navy Seabee Construction Electrician First Class (CE1) Richard C. Legg (age 26).*[10]*[11]*[12] It was later established that Byrnes (the reactor operator) had lifted the rod and caused the excursion, Legg (the shift supervisor) was standing on top of the reactor vessel and was impaled and pinned to the ceiling, and McKinley, the trainee who stood nearby, was later found alive by rescuers.*[6] All three men succumbed to injuries from physical trauma;*[6] however, the radiation from the nuclear excursion would have given the men no chance of survival even if they had not been killed by the explosion stemming from the criticality accident.

22.2.1 Reactor principles and events

Several "kinetic" factors affect the rate at which the power (heat) produced in a nuclear reactor responds to changes in the position of a control rod. Other features of the design govern how rapidly heat is transferred from the reactor fuel to the coolant.

The nuclear chain reaction has a positive feedback component whenever a critical mass is created; specifically, excess neutrons are produced for every fission. Inside a nuclear reactor, these excess neutrons must be controlled as long as a critical mass exists. The most significant and effective control mechanism is the use of control rods to absorb the excess neutrons. Other controls include the size and shape of the reactor and the presence of neutron reflectors in and around a core. Changing the amount of absorption or reflection of neutrons will affect neutron flux, and therefore, the power of the reactor.

One kinetics factor is the tendency of most light-water-moderated reactor (LWR) designs to have negative moderator temperature and void coefficients of reactivity. (Due to the low density of steam, pockets of water vapor are known as "voids" in a LWR.) A negative reactivity coefficient means that as the water moderator heats up, molecules move farther apart (water expands and eventually boils) and neutrons are less likely to be slowed by collisions to energies favorable for inducing fission in the fuel. Because of these negative feedback mechanisms, most LWRs will naturally tend to decrease their rate of fissioning in response to additional heat produced within the reactor core. If enough heat is produced that water boils inside the core, fissions in that vicinity will drastically decrease.

However, when power output from the nuclear reaction increases rapidly, it may take longer for the water to heat up and boil than it does for the voids to cause the nuclear reactions to decrease. In such an event, reactor power can grow rapidly without any negative feedback from the expansion or boiling of the water, even if it is in a channel just 1 cm away. Dramatic heating will occur to the nuclear fuel, leading to melting and vaporization of the metals within the core. Rapid expansion, increases in pressure, and failure of core components may lead to the destruction of the nuclear reactor, as was the case with SL-1. As the energy of expansion and heat travel from the nuclear fuel to the water and the vessel, it becomes likely that the nuclear reaction will shut down, either from the lack of sufficient moderator or from the fuel expanding beyond the realm of a critical mass. In the post-accident analysis of SL-1, scientists determined that the two shutdown mechanisms were almost equally matched (see below).

Another relevant kinetics factor is the contribution of what are called delayed neutrons to the chain reaction in the core. Most neutrons (the *prompt* neutrons) are produced nearly instantaneously via fission. But a few —approximately 0.7 percent in a U-235-fueled reactor operating at steady-state —are produced through the relatively slow radioactive decay of certain fission products. (These fission products are trapped inside the fuel plates in close proximity to the uranium-235 fuel.) Delayed production of a fraction of the neutrons is what enables reactor power changes to be controllable on a time scale that is amenable to humans and machinery.*[13]

In the case of an ejected control assembly, it is possible for the reactor to become critical *on the prompt neutrons alone* (i.e. prompt critical). When the reactor is prompt critical, the time to double the power is on the order of 10 microseconds. The duration necessary for temperature to follow the power level depends on the design of the reactor core. Typically, the coolant temperature lags behind the power by 3 to 5 seconds in a conventional LWR. In the SL-1 design, it was about 6 milliseconds before steam formation started.

SL-1 was constructed with a main central control rod that was capable of producing a very large excess reactivity if it were completely removed. The extra rod worth was in part due to the decision to load only 40 of the 59 fuel assemblies with nuclear fuel, thus making the prototype reactor core more active in the center. In normal operation control rods are withdrawn only enough to cause sufficient reactivity for a sustained nuclear reaction and power generation. In this accident, however, the reactivity addition was sufficient to take the reactor prompt critical within a time estimated at 3.6 milliseconds. That was too fast for the heat from the fuel to get through the aluminum cladding and boil enough water to fully stop the power growth in all parts of the core via negative moderator temperature and void feedback.

Post-accident analysis concluded that the final control method (i.e., dissipation of the prompt critical state) occurred by means of catastrophic core disassembly: destructive melting, vaporization, and consequent conventional explosive expansion of the parts of the reactor core where the greatest amount of heat was being produced most quickly. It was estimated that this core heating and vaporization process happened in about 7.5 milliseconds, before enough steam had been formed to shut down the reaction, beating the steam shutdown by a few milliseconds. A key statistic makes it clear why the core blew apart: the reactor designed for a 3 MW power output operated momentarily at a peak of nearly 20 GW, a power density over 6,000 times higher than its safe operating limit.

22.2.2   Events after the power excursion

There were no other people at the reactor site. The ending of the nuclear reaction was caused solely by the design of the reactor and the basic physics of heated water and core elements melting, separating the core elements and removing the moderator.

Heat sensors above the reactor set off an alarm at the central test site security facility at 9:01 p.m. MST, the time of the incident. False alarms had occurred in the morning and afternoon that same day. The first response crew, of six firemen (Ken Dearden Asst Chief, Mel Hess Lt., Bob

Checking for radioactive contamination on nearby Highway 20

Archer, Carl Johnson, Egon Lamprecht, Gerald Stuart, & Vern Conlon), arrived nine minutes later, expecting another false alarm*[14] and initially noticed nothing unusual, with only a little steam rising from the building, normal for the cold 6 °F (−14 °C) night. The control building appeared normal. The firefighters entered the reactor building and noticed a radiation warning light. Their radiation detectors jumped sharply to above their maximum range limit as they were climbing the stairs to SL-1's floor level. They peered into the reactor room before withdrawing.*[14]

At 9:17 p.m., a health physicist arrived. He and a fireman, both wearing air tanks and masks with positive pressure in the mask to force out any potential contaminants, approached the reactor building stairs. Their detectors read 25 röntgens per hour (R/hr) as they started up the stairs, and they withdrew.

Some minutes later, a health physics response team arrived with radiation meters capable of measuring gamma radiation up to 500 R/hr—and full-body protective clothing. One health physicist and two firefighters ascended the stairs and, from the top, saw damage in the reactor room. With the meter showing maximum scale readings, they withdrew rather than approach the reactor more closely and risk further exposure.

Around 10:30 p.m. MST, the supervisor for the contractor running the site (Combustion Engineering) and the chief health physicist arrived. They entered the reactor building around 10:45 pm and found two mutilated men soaked with water: one clearly dead (Byrnes), the other moving slightly (McKinley) and moaning. With one entry per person and a 1-minute limit, a team of 5 men with stretchers recovered the operator who was still breathing around 10:50; he did not regain consciousness and died of his head injury at about 11 p.m. Even stripped, his body was so contaminated that it was emitting about 500 R/hr. Meanwhile, the third man was

The stretcher rig. Army volunteers from a special Chemical Radiological Unit at Dugway Proving Ground practiced before a crane inserted the rig into the SL-1 reactor building to collect the body of the man impaled to the ceiling directly above the reactor vessel.

discovered about 11 p.m., impaled to the ceiling. With all potential survivors now recovered, safety of rescuers took precedence and work was slowed to protect them.

On the night of January 4, a team of six volunteers used a plan involving teams of two to recover the body of Byrnes. Radioactive gold ^{198}Au from the man's gold watch buckle and copper ^{64}Cu from a screw in a cigarette lighter subsequently proved that the reactor had indeed gone prompt critical. Prior to the discovery of neutron-activated elements in the men's belongings, scientists had doubted that a nuclear excursion had occurred, believing the reactor was inherently safe. These findings ruled out early speculations that a chemical explosion caused the accident.[*][9]

The third man was discovered last because he was pinned to the ceiling above the reactor by a shield plug and not easily recognizable.[*][6] On January 9, in relays of two at a time, a team of ten men, allowed no more than 65 seconds exposure each, used sharp hooks on the end of long poles to pull Legg's body free of the shield plug, dropping it onto a 5-by-20-foot (1.5 by 6.1 m) stretcher attached to a crane.[*][6]

The bodies of all three were buried in lead-lined caskets sealed with concrete and placed in metal vaults with a concrete cover. Some highly radioactive body parts were buried in the Idaho desert as radioactive waste. Army Specialist Richard Leroy McKinley is buried in section 31 of Arlington National Cemetery.

Some sources and eyewitness accounts confuse the names and disposition of each victim.[*][6] In *Idaho Falls: The untold story of America's first nuclear accident*,[*][15] the author indicates that Byrnes was the man found alive initially, Legg's body was recovered the night after the accident, and that McKinley was impaled by the control rod.

22.3 Cause

One of the required maintenance procedures called for the central control rod to be manually withdrawn approximately 4 inches (10 cm) in order to attach it to the automated control mechanism from which it had been disconnected. Post-incident calculations estimate that the main control rod was actually withdrawn approximately 26 inches (66 cm), causing the reactor to go prompt critical, which resulted in the steam explosion. The fuel, portions of the fuel plates, and water surrounding the fuel plates vaporized in the extreme heat. The expansion caused by this heating process caused water hammer as water was accelerated upwards toward the reactor vessel head, producing approximately 10,000 pounds per square inch (69,000 kPa) of pressure on the head of the reactor vessel when water struck the head at 160 feet per second (50 m/s).[*][16]

The water hammer not only caused extreme physical damage and distortion of the reactor vessel, it also caused the shield plugs of the vessel to be ejected, one of which impaled Legg. The most surprising and unforeseen evidences of the steam explosion and water hammer were the impressions made on the ceiling above the reactor vessel when it jumped over 9 feet (2.7 m) in the air before settling back into its prior location. The post-incident analysis also concluded that the reactor vessel was dry, since most of the water and steam had been either ejected immediately or evaporated due to the heat inside the reactor.

It was water hammer that caused the physical damage to the reactor, the deaths of personnel who stood atop and nearby, and the release of radioactive isotopes to the environment. One of the lessons learned from SL-1 was that there is an extreme water hammer hazard whenever a shutdown reactor is cooled to room temperature and there is an air gap between the top of the water and the reactor vessel head. One of the recommendations in the analysis of the incident was that shutdown reactors be filled to the top with water so that a power excursion could not induce such a powerful water hammer. Air is not dense enough to appreciably slow water, while water (being nearly incompressible) is able to distribute explosive forces and limit peak pressure. The extra water is also a very effective radiation shield for those who are directly above the vessel. Written procedures at SL-1 had included a directive to pump down the level of water in the reactor prior to the maintenance procedure that destroyed it.

The most common theories proposed for the withdrawal of the rod so far are (1) sabotage or suicide by one of the operators, (2) a suicide-murder involving an affair with the wife of one of the other operators, (3) inadvertent withdrawal of the main control rod, or (4) an intentional attempt to "exercise" the rod (to make it travel more smoothly within its sheath).*[17]*[18] The maintenance logs do not address what the technicians were attempting to do, and thus the actual cause of the incident will never be known. The investigation took almost two years to complete.

Investigators analyzed the flux wires installed during the maintenance to determine the power output level. They also examined scratches on the central control rod. Using this data, they concluded that the central rod had been withdrawn 26.25 inches (66.7 cm).*[9] The reactor would have been critical at 23 inches (58.4 cm), and it took approximately 100 ms for the rod to travel the final 3.25 inches (8.3 cm). Once this was calculated, experiments were conducted with an identically weighted mock control rod to determine whether it was possible or feasible for one or two men to have performed this. Experiments included a simulation of the possibility that the 84-pound (38 kg) rod was stuck and one man freed it himself, reproducing the scenario that investigators considered the best explanation: Byrnes broke the control rod loose and withdrew it accidentally, killing all three men.*[6]

At SL-1, control rods would get stuck in the control rod channel sporadically. Numerous procedures were conducted to evaluate control rods to ensure they were operating properly. There were rod drop tests and scram tests of each rod, in addition to periodic rod exercising and rod withdrawals for normal operation. From February 1959 to November 18, 1960, there were 40 cases of a stuck control rod for scram and rod drop tests and about a 2.5% failure rate. From November 18, 1960 to December 23, 1960, there was a dramatic increase in stuck rods, with 23 in that time period and a 13.0% failure rate. Besides these test failures, there were an additional 21 rod sticking incidents from February 1959 to December 1960; 4 of these had occurred in the last month of operation during routine rod withdrawal. The central control rod, No. 9, had the best operational performance record even though it was operated more frequently than any of the other rods.

Rod sticking has been attributed to misalignment, corrosion product build-up, bearing wear, clutch wear, and drive mechanism seal wear. Many of the failure modes that caused a stuck rod during tests (like bearing and clutch wear) would only apply to a movement performed by the control rod drive mechanism. Since the No. 9 rod is centrally located, its alignment may have been better than Nos. 1, 3, 5, and 7 which were more prone to sticking. After the accident, logbooks and former plant operators were consulted to determine if there had been any rods stuck during the reassembly operation that Byrnes was performing. One person had performed this about 300 times, and another 250 times; neither had ever felt a control rod stick when being manually raised during this procedure. Furthermore, no one had ever reported a stuck rod during manual reconnection.

The mechanical and material evidence, combined with the nuclear and chemical evidence, forced them to believe that the central control rod had been withdrawn very rapidly. [···] The scientists questioned the [former operators of SL-1]: "Did you know that the reactor would go critical if the central control rod were removed?" Answer: "Of course! We often talked about what we would do if we were at a radar station and the Russians came. We'd yank it out."
—Susan M. Stacy, Proving the Principle, *[9]

22.4 Consequences

The remains of the SL-1 reactor are now buried near the original site at 43°31'02.9"N 112°49'22.2"W.*[19]

The incident caused this design to be abandoned and future reactors to be designed so that a single control rod removal would not have the ability to produce the very large excess reactivity which was possible with this design. Today this is known as the "one stuck rod" criterion and requires complete shutdown capability even with the most reactive rod stuck in the fully withdrawn position. The reduced excess reactivity limits the possible size and speed of the power surge. It should be pointed out that the "one stuck rod" criterion did not originate as a result of the SL-1 incident. It was, in fact, a hard and fast design criterion long before the SL-1, from the beginning of the Naval Reactors program, under the leadership of Admiral Hyman Rickover. This design criterion started with the USS Nautilus, and continued throughout subsequent submarine and surface ship designs, and with the Shippingport civilian nuclear plant. It continues to be a rigid requirement for all US reactor designs to this day.

The incident also showed that in a genuine, extreme accident, both the melting of the core and the water to steam conversion would shut down the nuclear reaction. This demonstrates in a real accident one aspect of inherent safety of the water-moderated design against the possibility of a nuclear explosion.

A nuclear explosion requires sufficient force to hold the reacting nuclear components together for a short but necessary time. This is achieved in a nuclear fission weapon by surrounding the core with a carefully engineered tamper (typically U-238), and a shaped explosive charge. This, along with other subsystems of the weapon keep the supercritical mass together long enough for sufficient generations of the fission reaction to produce the desired yield. Lacking these restraints to hold the vaporized core components together, the components of a reactor fly apart, as in this incident. The reaction ends, resulting in a steam explosion and a badly damaged reactor core, but not the type of explosion as would be achieved with a nuclear weapon.

Although portions of the center of the reactor core had been vaporized briefly, very little corium was recovered. The fuel plates showed signs of catastrophic destruction leaving voids, but "no appreciable amount of glazed molten material was recovered or observed." Additionally, "There is no evidence of molten material having flowed out between the plates." It is believed that rapid cooling of the core was responsible for the small amount of molten material. There was insufficient heat generated for any corium to reach or penetrate the bottom of the reactor vessel. The reactor vessel was removed on November 29, 1961 without incident. The only holes in the bottom of the vessel were the ones bored through with borescopes to determine the condition of the melted core.

Even without an engineered containment building like those used today, the SL-1 reactor building contained most of the radioactivity, though iodine-131 levels on plants during several days of monitoring reached fifty times background levels downwind. Radiation surveys of the Support Facilities Building, for example, indicated high contamination in halls, but light contamination in offices.

Radiation exposure limits prior to the incident were 100 röntgens to save a life and 25 to save valuable property. During the response to the incident, 22 people received doses of 3 to 27 Röntgens full-body exposure.[20] Removal of radioactive waste and disposal of the three bodies eventually exposed 790 people to harmful levels of radiation.[21] In March 1962, the Atomic Energy Commission awarded certificates of heroism to 32 participants in the response.

The documentation and procedures required for operating nuclear reactors expanded substantially, becoming far more formal as procedures which had previously taken two pages expanded to hundreds. Radiation meters were changed to allow higher ranges for emergency response activities.

After a pause for evaluation of procedures, the Army continued its use of reactors, operating the Mobile Low-Power Reactor (ML-1), which started full power operation on February 28, 1963, becoming the smallest nuclear power plant on record to do so. This design was eventually abandoned after corrosion problems. While the tests had shown that nuclear power was likely to have lower total costs, the financial pressures of the Vietnam War caused the Army to favor lower initial costs and it stopped the development of its reactor program in 1965, although the existing reactors continued operating (MH-1A until 1977).

22.5 Cleanup

The site was cleaned in 1961 to 1962, removing the bulk of the contaminated debris and burying it. The massive cleanup operation included the dismantling and disposal of the reactor and building. A burial ground was constructed approximately 1,600 feet (500 m) northeast of the original site of the reactor. This was done to minimize radiation exposure to the public and site workers that would have resulted from transport of contaminated debris from SL-1 to the Radioactive-Waste Management Complex over 16 miles (26 km) of public highway. Original cleanup of the site took about 18 months. The entire reactor building, contaminated materials from nearby buildings, and soil and gravel contaminated during cleanup operations were disposed of in the burial ground. The majority of buried materials consist of soils and gravel.[22][23]

SL-1 burial site in 2003, capped with rip rap

Recovered portions of the reactor core, including the fuel and all other parts of the reactor that were important to the incident investigation, were taken to the INEL's Test Area North for study. After the incident investigation was complete, the reactor fuel was sent to the Idaho Chemical Processing Plant for reprocessing. The reactor core minus the fuel, along with the other components sent to Test Area North for study, was eventually disposed of at the Radioactive Waste Management Complex.[22]

The SL-1 burial ground consists of three excavations, in which a total volume of 99,000 cubic feet (2800 m^3) of contaminated material was deposited. The excavations were dug as close to basalt as the equipment used would allow and ranges from 8 to 14 feet (2.4 to 4.3 m) in depth. At least 2 feet (0.6 m) of clean backfill was placed over each excavation. Shallow mounds of soil over the excavations were added at the completion of cleanup activities in September 1962. The site and burial mound are collectively known as United States Environmental Protection Agency Superfund Operable Unit 5-05.[22][24]

Numerous radiation surveys and cleanup of the surface

of the burial ground and surrounding area have been performed in the years since the SL-1 incident. Aerial surveys were performed by EG&G Las Vegas in 1974, 1982, 1990, and 1993. The Radiological and Environmental Sciences Laboratory conducted gamma radiation surveys every 3 to 4 years between 1973 and 1987 and every year between 1987 and 1994. Particle-picking at the site was performed in 1985 and 1993. Results from the surveys indicated that cesium-137 and its progeny (decay product) are the primary surface-soil contaminants. During a survey of surface soil in June 1994, "hot spots," areas of higher radioactivity, were found within the burial ground with activities ranging from 0.1 to 50 milliroentgen (mR)/hour. On November 17, 1994, the highest radiation reading measured at 2.5 feet (0.75 m) above the surface at the SL-1 burial ground was 0.5 mR/hour; local background radiation was 0.2 mR/hour. A 1995 assessment by the EPA recommended that a cap be placed over the burial mounds. The primary remedy for SL-1 was to be containment by capping with an engineered barrier constructed primarily of native materials.[22] This remedial action was completed in 2000 and first reviewed by the EPA in 2003.[24]

22.6   Movies and books

Animation of the film produced by the Atomic Energy Commission, available from The Internet Archive.

The U.S. Government produced a film about the incident for internal use in the 1960s. The video was subsequently released and can be viewed at The Internet Archive[25] and YouTube. *SL-1* is the title of a 1983 movie, written and directed by Diane Orr and C. Larry Roberts, about the nuclear reactor explosion.[21] Interviews with scientists, archival film, and contemporary footage, as well as slow-motion sequences, are used in the film.[26][27] The events of the incident are also the subject of one book: *Idaho Falls: The untold story of America's first nuclear accident* (2003)[15] and 2 chapters in *Proving the Principle -*

A History of The Idaho National Engineering and Environmental Laboratory, 1949-1999 (2000).[28]

In 1975, the anti-nuclear book *We Almost Lost Detroit*, by John G. Fuller was published, referring at one point to the Idaho Falls incident. *Prompt Critical* is the title of a 2012 short film, viewable on YouTube.com, written and directed by James Lawrence Sicard, dramatizing the events surrounding the SL-1 incident.[29] A documentary about the incident was shown on the History Channel.[30]

This image of the SL-1 core served as a reminder of the damage that a nuclear meltdown can cause.

Another author, Todd Tucker, studied the incident and published a book detailing the historical aspects of nuclear reactor programs of the U.S. military branches. Tucker used the Freedom of Information Act to obtain reports, including autopsies of the victims, writing in detail how each person died and how parts of their bodies were severed, analyzed, and buried as radioactive waste.[6] The autopsies were performed by the same pathologist known for his work following the Cecil Kelley criticality accident. Tucker explains the reasoning behind the autopsies and the severing of victims' body parts, one of which gave off 1,500 R/hour on contact. Because the SL-1 accident killed all three of the military operators on site, Tucker calls it "the deadliest nuclear reactor incident in U.S. history." [31]

22.7 See also

- BORAX Experiments, 1953-4, which proved that the transformation of water to steam would safely limit a boiling water reactor power excursion, similar to that in this incident.

- International Nuclear Event Scale

- List of civilian nuclear accidents

- List of civilian radiation accidents

- List of military nuclear accidents

- List of nuclear reactors

- Nuclear power debate

- Nuclear power

- Nuclear safety and security

- Radiation

- Radioactive contamination

22.8 References

[1] Stacy, Susan M. (2000). "Chapter 16: The Aftermath" (PDF). *Proving the Principle: A History of The Idaho National Engineering and Environmental Laboratory, 1949-1999* (PDF). U.S. Department of Energy, Idaho Operations Office. pp. 150–157. ISBN 0-16-059185-6.

[2] The Nuclear Power Deception Table 7: Some Reactor Accidents

[3] Horan, J. R., and J. B. Braun, 1993, *Occupational Radiation Exposure History of Idaho Field Office Operations at the INEL*, EGG-CS-11143, EG&G Idaho, Inc., October, Idaho Falls, Idaho.

[4] "Idaho: Runaway Reactor". *Time*. January 13, 1961. Retrieved July 30, 2010.

[5] Steve Wander (editor) (February 2007). "Supercritical" (PDF). *System Failure Case Studies*. NASA. **1** (4).

[6] Tucker, Todd (2009). *Atomic America: How a Deadly Explosion and a Feared Admiral Changed the Course of Nuclear History*. New York: Free Press. ISBN 978-1-4165-4433-3. See summary:

[7] LA-3611 *A Review of Criticality Accidents*, William R. Stratton, Los Alamos Scientific Laboratory, 1967

[8] LA-13638 *A Review of Criticality Accidents* (2000 Revision), Thomas P. McLaughlin, et al., Los Alamos National Laboratory, 2000.

[9] Stacy, Susan M. (2000). *Proving the Principle - A History of The Idaho National Engineering and Environmental Laboratory, 1949-1999* (PDF). U.S. Department of Energy, Idaho Operations Office. ISBN 0-16-059185-6. Chapter 15.

[10] "Nuclear Experts Probe Fatal Reactor Explosion". *Times Daily*. January 5, 1961. Retrieved July 30, 2010.

[11] "Richard Legg" (JPEG). *Find A Grave*. 14 May 2011. Retrieved 5 March 2013.

[12] Spokane Daily Chronicle - Jan 4, 1961. The article notes that Byrnes was a "Spec. 5" from Utica, New York, McKinley was a "Spec. 4" from Kenton, Ohio, Legg was a "Navy electrician L.C." from Roscommon, Michigan.

[13] Lamarsh, John R.; Baratta, Anthony J. (2001). *Introduction to Nuclear Engineering*. Upper Saddle River, New Jersey: Prentice Hall. p. 783. ISBN 0-201-82498-1.

[14] Berg, Sven (December 12, 2009). "Nuclear accident still mystery to rescue worker". *The Argus Observer*. Retrieved April 6, 2015.

[15] McKeown, William (2003). *Idaho Falls: The Untold Story of America's First Nuclear Accident*. Toronto: ECW Press. ISBN 978-1-55022-562-4.,

[16] *IDO-19313: ADDITIONAL ANALYSIS OF THE SL-1 EXCURSION; Final Report of Progress July through October 1962*, November 21, 1962, Flight Propulsion Laboratory Department, General Electric Company, Idaho Falls, Idaho, U.S. Atomic Energy Commission, Division of Technical Information.

[17] *ATOMIC CITY, by Justin Nobel* Tin House Magazine, Issue #51, Spring, 2012.

[18] *A Nuclear Family, By Maud Newton* The New York Times Magazine, April 1, 2012.

[19] Mahaffey, James (2012). *Nuclear Accidents and Disasters*. Facts on File. p. 40. ISBN 978-0-8160-7650-5.

[20] Johnston, Wm. Robert. "SL-1 reactor excursion, 1961". *Johnston's Archive*. Retrieved 30 July 2010.

[21] Maslin, Janet (March 21, 1984). "Sl-1 (1983): Looking at Perils of Toxicity". *The New York Times*. Retrieved July 30, 2010.

[22] EPA Superfund Record of Decision: Idaho National Engineering Laboratory (USDOE) EPA ID: ID4890008952, OU 24, Idaho Falls, ID, 12/01/1995

[23] Record of Decision, Stationary Low-Power Reactor-1 and Boiling Water Reactor Experiment-I Burial Grounds (Operable Units 5-05 and 6-01), and 10 No Action Sites (Operable Units 5-01, 5-03, 5-04, and 5-11), January 1996.

[24] 2003 Annual Inspection Summary for the Stationary Low-Power Reactor Burial Ground, Operable Unit 5-05

[25] "SL-1 The Accident: Phases I and II".

[26] *SL-1* at the Internet Movie Database

[27] SL-1 (1983)

[28] Stacy, Susan M. (2000). *Proving the Principle: A History of The Idaho National Engineering and Environmental Laboratory, 1949-1999*. U.S. Department of Energy, Idaho Operations Office. ISBN 0-16-059185-6.

[29] Prompt Critical on YouTube by James Lawrence Sicard.

[30] SL-1 Nuclear Accident on YouTube History Channel

[31] Secret Accidents and Lost Bombs

22.9   External links

- "SL-1 Reactor Accident on January 3, 1961, Interim Report" , May 1961. From the above page. 15.5 MB PDF.

- "IDO Report on the Nuclear Incident at the SL-1 Reactor on January 3, 1961, at the National Reactor Testing Station, January 1962. 16.5 MB PDF. From the above page. This report has more accurate times for the events.

- "The SL-1 Accident - AEC Educational Documentary" on YouTube

- "The SL-1 Accident: Briefing Film Report" 1961 AEC" on YouTube

- Department of Energy Document: Nuclear Reactor Testing

Chapter 23

Johnston Atoll

Johnston Atoll, also known as **Kalama Atoll** to Native Hawaiians, is an unincorporated territory of the United States currently administered by the United States Fish and Wildlife Service. Public entry is only by special-use permit from the U.S. Fish and Wildlife Service.

For nearly 70 years, the atoll was under the control of the American military. In that time it was used as a bird sanctuary, as a naval refueling depot, as an airbase, for nuclear and biological weapons testing, for space recovery, as a secret missile base, and as a chemical weapon and Agent Orange storage and disposal site. These activities left the area environmentally contaminated and remediation and monitoring continue.

23.1 Geography

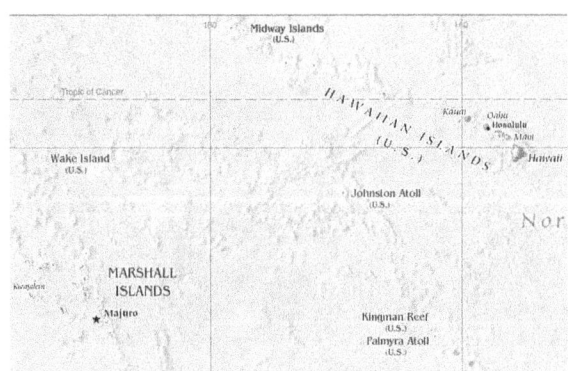

Johnston Atoll is located between the Marshall Islands and the Hawaiian Islands

The Johnston Atoll is a deserted 1,300 ha (3,200 acres) atoll in the North Pacific Ocean located about 750 nmi (860 mi; 1,390 km) southwest of the island of Hawai'i and is grouped as one of the United States Minor Outlying Islands.[1] The atoll, which is located on a coral reef platform, has four islands. Johnston (or Kalama) Island and Sand Island are both enlarged natural features, while *Akau* (North) and *Hikina* (East) are two artificial islands formed by coral dredging.[1] By 1964, dredge and fill operations had increased the size of Johnston Island to 596 acres (241 ha) from its original 46 acres (19 ha), also increased Sand Island from 10 to 22 acres (4.0 to 8.9 ha), and added two new islands, North and East, of 25 and 18 acres (10.1 and 7.3 ha).[2]

The four islands compose a total land area of 2.67 square kilometres (1.03 square miles).[1] Due to the atoll's tilt, much of the reef on the southeast portion has subsided. But even though it does not have an encircling reef crest, the reef crest on the northwest portion of the atoll does provide for a shallow lagoon, with depths ranging from 3–10 m (9.8–32.8 ft). The climate is tropical but generally dry. Northeast trade winds are consistent and there is little seasonal temperature variation.[1] With elevation ranging from sea level to 5 m (16 ft) at Summit Peak, the islands contain some low-growing vegetation and palm trees on mostly flat terrain and no natural fresh water resources.[1]

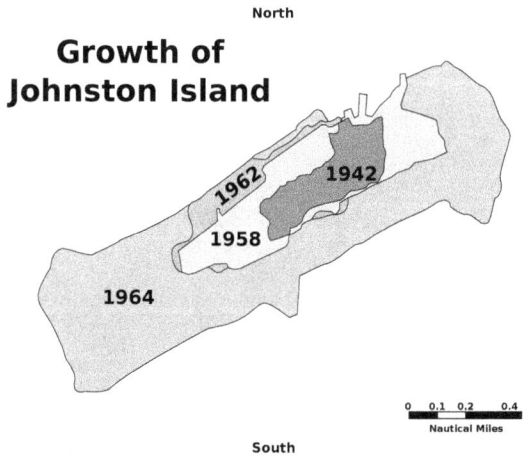

The size of Johnston Island was quadrupled by coral dredging.

The unofficial flag of Johnston Atoll was flown below the Stars and Stripes. The double bird holding 4 stars represents both the Air Force and the Fish and Wildlife Service, while the four stars denote the atoll's islands; the white is for coral and the aquamarine for the surrounding ocean.[3]

23.2 History

23.2.1 Early history

The first Western record of the atoll was on September 2, 1796 when the Boston-based American brig *Sally* accidentally grounded on a shoal near the islands. The ship's captain, Joseph Pierpont, published his experience in several American newspapers the following year giving an accurate position of Johnston and Sand Island along with part of the reef. However, he did not name or lay claim to the area.*[4] The islands were not officially named until Captain Charles J. Johnston of the Royal Naval ship HMS *Cornwallis* sighted them on December 14, 1807. In the following years, an occasional vessel stopped, but generally one look was enough.

The Guano Islands Act, enacted on August 18, 1856, was federal legislation passed by the U.S. Congress that enabled citizens of the U.S. to take possession of islands containing guano deposits. In 1858 William Parker and R. F. Ryan, chartered the schooner *Palestine* specifically to find Johnston Atoll. They located guano on the atoll in March 1858 and they proceeded to claim the island. By 1858, Johnston Atoll was claimed by both the United States and the Kingdom of Hawaii. In June 1858, Samuel Allen, sailing on the *Kalama*, tore down the U.S. flag and raised the Hawaiian flag renaming the atoll **Kalama**. The larger island was renamed Kalama Island, and the nearby smaller island was called Cornwallis.*[5]

Returning on July 27, 1858, the Captain of the *Palestine* again hoisted the American flag and reasserted the rights of the United States. The same day, the atoll was declared part of the domain of King Kamehameha IV.*[5] On this visit however the *Palestine* left two crew members on the island to gather phosphate. While *Palestine* was at the atoll

and these two men were still on the island, a July 27, 1858 proclamation of Kamehameha IV declared the annexation of this island to Hawaii stating that it was "derelict and abandoned." However, later that year King Kamehameha revoked the lease granted to Samuel Allen when the King learned that the atoll had been claimed previously by the United States.*[6] However, this did not prevent the Hawaiian Territory from making use of the Atoll or asserting ownership.

By 1890 the atoll's guano deposits had been almost entirely depleted (mined out) by U.S. interests operating under the Guano Islands Act. In 1892, HMS *Champion* made a survey and map of the island, hoping that it might be suitable as a telegraph cable station. On January 16, 1893, the Hawaiian Legation at London reported a diplomatic conference over this temporary occupation of the island. However, the Kingdom of Hawaii was overthrown on January 17, 1893. When in 1898 Hawaii became an integral part of the United States during the Spanish–American War, the name of Johnston Island was omitted from the list of Hawaiian Islands. On September 11, 1909, Johnston was leased by the Territory of Hawaii to a private citizen for fifteen years. A board shed was built on the southeast side of the larger island, and a small tramline run up onto the slope of the low hill, to facilitate the removal of guano. Apparently neither the quantity nor the quality of the guano was sufficient to pay for gathering it so that the project was soon abandoned.*[5]

23.2.2 National Wildlife Refuge since 1926

USS Tanager *with members of the 1923 Tanager Expedition*

The Tanager Expedition was a joint expedition sponsored by the U.S. Department of Agriculture and the Bishop Museum of Hawaii and visited the Atoll in 1923. The expedition to the atoll consisted of two teams accompanied by destroyer convoys, with the first departing Honolulu on July

7, 1923 aboard the USS *Whippoorwill*, which conducted the first survey of Johnston Island in the 20th century. Aerial survey and mapping flights over Johnston were conducted with a Douglas DT-2 floatplane carried on her fantail, which was hoisted into the water for take off. From July 10–22, 1923, the atoll was recorded in a pioneering aerial photography project. The USS *Tanager* left Honolulu on July 16 and joined up with the *Whippoorwill* to complete the survey and then traveled to Wake Island to complete surveys there.[7] Tents were pitched on the southwest beach of fine white sand, and a rather thorough biological survey was made of the island. Hundreds of sea birds, of a dozen kinds, were the principal inhabitants, together with lizards, insects, and hermit crabs. The reefs and shallow water abounded with fish and other marine life.[5]

USS Whippoorwill

On June 29, 1926 by Executive Order 4467, President Calvin Coolidge established **Johnston Island Reservation** as a federal bird refuge and placed it under the control of the U.S. Department of Agriculture as a "refuge and breeding ground for native birds."[8] Johnston Atoll was added to the United States National Wildlife Refuge system in 1926, and renamed the **Johnston Island National Wildlife Refuge** in 1940.[9] The Johnston Atoll National Wildlife Refuge was established to protect the tropical ecosystem and the wildlife that it harbors.[10] However, the Department of Agriculture had no ships, and the Navy was interested in the Atoll for strategic reasons, so with Executive Order 6935 on December 29, 1934, President Franklin D. Roosevelt placed the islands under the "control and jurisdiction of the Secretary of the Navy for administrative purposes," but subject to use as a refuge and breeding ground for native birds, under the Department of Interior.

On February 14, 1941, President Franklin Roosevelt issued Executive Order 8682 to create naval defenses areas in the central Pacific territories. The proclamation established "Johnston Island Naval Defensive Sea Area" which encompassed the territorial waters between the extreme high-water marks and the three-mile marine boundaries surrounding the atoll. "Johnston Island Naval Airspace Reservation" was also established to restrict access to the airspace over the naval defense sea area. Only U.S. government ships and aircraft were permitted to enter the naval defense areas at Johnston unless authorized by the Secretary of the Navy.

In 1990, two full-time U.S. Fish and Wildlife Service personnel, a Refuge Manager and a biologist, were stationed on Johnston Atoll to handle the increase in biological, contaminant, and resource conflict activities.[11]

After the military mission on the island ended in 2004, the Atoll was administered by the Pacific Remote Islands National Wildlife Refuge Complex. The outer islets and water rights were managed cooperatively by the Fish and Wildlife Service with some of the actual Johnston Island land mass remaining under control of the United States Air Force (USAF) for environmental remediation and the Defense Threat Reduction Agency (DTRA) for plutonium cleanup purposes. However, on January 6, 2009, under authority of section 2 of the Antiquities Act, The Pacific Remote Islands Marine National Monument was established by President George W. Bush to administer and protect Johnston Island along with six other Pacific islands.[12] The national monument includes Johnston Atoll National Wildlife Refuge within its boundaries and contains 696 acres (2.82 km^2) of land and over 800,000 acres (3,200 km^2) of water area.[13]

23.2.3 Military control 1934–2004

On December 29, 1934, President Franklin D. Roosevelt with Executive Order 6935 transferred control of Johnston Atoll to the United States Navy under the 14th Naval District, Pearl Harbor, in order to establish an air station, and also to the Department of the Interior to administer the bird refuge. In 1948 the USAF assumed control of the Atoll.[14] During the Operation Hardtack nuclear test series from April 22 – August 19, 1958 administration of Johnston Atoll was assigned to the Commander of Joint Task Force 7. After the tests were completed, the island reverted to the command of the US Air Force.[15]

From 1963 to 1970 the Navy's Joint Task force 8 and the Atomic Energy Commission (AEC) held joint operational control of the island during high altitude nuclear testing operations.[16]

In 1970 Operational control was handed back to the Air Force until July 1973 when Defense Special Weapons Agency was given host-management responsibility by the Secretary of Defense.[16] Over the years, sequential descendant organizations have been the Defense Atomic Support Agency (DASA) from 1959 to 1971, the Defense Nuclear Agency (DNA) from 1971 to 1996, and the Defense Special Weapons Agency (DSWA) from 1996 to

1998. In 1998, Defense Special Weapons Agency, and selected elements of the Office of Secretary of Defense were combined to form the Defense Threat Reduction Agency (DTRA).[17] In 1999 host-management responsibility transferred from the Defense Threat Reduction Agency once again to the Air Force until the Air Force mission ended in 2004 and the base was closed.[16]

23.2.4 Sand Island seaplane base

Aerial approach to the former base on Johnston Island (top). The ship channel is visible as a darker blue area starting at left and continuing up around the right side of Johnston Island, with Sand Island on the near side (bottom).

In 1935 personnel from the US Navy's Patrol Wing Two carried out some minor construction to develop the atoll for seaplane operation. In 1936, the Navy began the first of many changes to enlarge the atoll's land area. They erected some buildings and a boat landing on Sand Island and blasted coral to clear a 3,600 feet (1,100 m) seaplane landing.[18] Several seaplanes made flights from Hawaii to Johnston, such as that of a squadron of six aircraft in November, 1935. One of the most spectacular of these was on April 8, 1937, when two seaplanes from VP-6F made the round trip in ten and a half hours, to bring back a sick seaman.[5]

In November 1939 further work was commenced on Sand Island by civilian contractors to allow the operation of one squadron of patrol planes with tender support. Part of the lagoon was dredged and the excavated material was used to make a parking area connected by a 2,000-foot (610 m) causeway to Sand Island. Three seaplane landings were cleared, one 11,000 feet (3,400 m) by 1,000 feet (300 m) and two cross-landings each 7,000 feet (2,100 m) by 800 feet (240 m) and dredged to a depth of 8 feet (2.4 m). Sand Island had barracks built for 400 men, a mess hall, underground hospital, radio station, water tanks and a 100 feet (30 m) steel control tower.[18] In December 1943 an additional 10 acres (4.0 hectares) of parking was added to the

seaplane base.[18]

On May 26, 1942 a United States Navy Consolidated PBY-5 Catalina wrecked at Johnston Atoll. The Catalina pilot made a normal power landing and immediately applied throttle for take-off. At a speed of about fifty knots the plane swerved to the left and then continued into a violent waterloop. The hull of the plane was broken open and the Catalina sank immediately.[19] After the war on March 27, 1949 a PBY-6A Catalina had to make a forced landing during flight from Kwajalein to Johnston Island. The plane was damaged beyond repair and the crew of 11 was rescued nine hours later by a Navy ship which sank the plane by gunfire.[20] During 1958, a proposed support agreement for Navy Seaplane operations at Johnston Island was under discussion though it was never completed because a requirement for the operation failed to materialize.[11]

23.2.5 Airfield

Main article: Johnston Island Air Force Base

By September 1941 construction of an airfield on Johnston Island commenced. A 4,000 feet (1,200 m) by 500 feet (150 m) runway was built together with two 400-man barracks, two mess halls, a cold-storage building, an underground hospital, a fresh-water plant, shop buildings, and fuel storage. The runway was complete by December 7, 1941, though in December 1943 the 99th Naval Construction Battalion arrived at the atoll and proceeded to lengthen the runway to 6,000 feet (1,800 m).[18] The runway was subsequently lengthened and improved as the island was enlarged.

During WWII Johnston Atoll was used as a refueling base for submarines, and also as an aircraft refueling stop for American bombers transiting the Pacific Ocean, including the Boeing B-29 Enola Gay.[21] By 1944, the atoll was one of the busiest air transport terminals in the Pacific. Air Transport Command aeromedical evacuation planes stopped at Johnston en route to Hawaii. Following V-J Day on August 14, 1945, Johnston Atoll saw the flow of men and aircraft that had been coming from the mainland into the Pacific turn around. By 1947, over 1,300 B-29 and B-24 bombers had passed through the Marianas, Kwajalein, Johnston Island, and Oahu en route to Mather Field and civilian life.

Following WWII, Johnston Atoll Airport was used commercially by Continental Air Micronesia, touching down between Honolulu and Majuro. When an aircraft landed it was surrounded by armed soldiers and the passengers were not allowed to leave the aircraft. Aloha Airlines also made weekly scheduled flights to the island carrying civilian and military personnel, in the 1990s there were flights almost daily, and some days saw up to 3 arrivals.*[22] Just prior to movement of the chemical munitions to Johnston Atoll, the Surgeon General, Public Health Service, reviewed the shipment and the Johnston Atoll storage plans. His recommendations caused the Secretary of Defense in December 1970 to issue instructions suspending missile launches and all non-essential aircraft flights. As a result, Air Micronesia service was immediately discontinued, and missile firings were terminated with the exception of two 1975 satellite launches deemed critical to the islands mission.*[11]

There were many times when the runway was needed for emergency landings for both civil and military aircraft, including one landing by a Qantas Boeing 747. When the runway was decommissioned, it could no longer be used as a potential emergency landing place when planning flight routes across the Pacific Ocean. As of 2003, the airfield at Johnston Atoll consisted of an unmaintained closed single 9,000 feet (2,700 m) asphalt/concrete runway 5/23, a parallel taxiway, and a large paved ramp along the south-east side of the runway.*[22]

23.2.6   World War II 1941–1945

Main article: Shelling of Johnston and Palmyra

In February 1941 Johnston Atoll was designated as a Naval Defensive Sea Area and Airspace Reservation. On the day the Japanese struck Pearl Harbor, December 7, 1941, USS *Indianapolis* was out of her home port of Pearl Harbor, to make a simulated bombardment at Johnston Island. Japan's strike at Pearl Harbor occurred as the ship was unloading marines, civilians and stores on the atoll.*[23] On December 15, 1941 the atoll was shelled outside the reef by a Japanese submarine, which had been part of the attack on Pearl Harbor eight days earlier. Several buildings including the power station were hit, but no personnel were injured.*[24] Additional Japanese shelling occurred on December 22 and 23, 1941. On all occasions Johnston Atoll's coastal artillery guns returned fire, driving off the sub.

In July 1942 the civilian contractors at the atoll were replaced by 500 men from the 5th and 10th Naval Construction Battalions, who expanded the fuel storage and water production at the base and built additional facilities. The 5th Battalion departed in January 1943.*[24] In December 1943 the 99th Naval Construction Battalion arrived at the atoll and proceeded to lengthen the runway to 6,000 feet (1,800 m) and add an additional 10 acres (4.0 ha) of parking to the seaplane base.*[25]

Sand Island and former U.S. Coast Guard LORAN Station

23.2.7   Coast Guard mission 1957–1992

On January 25, 1957, the Department of Treasury was granted a 5-year permit for the United States Coast Guard (USCG) to operate and maintain a Long Range Aid to Navigation (LORAN) transmitting station on Johnston Atoll. Two years later in December 1959. The Secretary of Defense approved the Secretary of the Treasury's request to use Sand Island for U.S. Coast Guard LORAN A and C station sites. The USCG was granted permission to install a LORAN A and C station on Sand Island to be manned by U.S. Coast Guard personnel through June 30, 1992. The permit for a LORAN station to operate on Johnston Island was terminated in 1962. On November 1, 1957, a new United States Coast Guard LORAN-A station was commissioned. By 1958 the Coast Guard LORAN Station at Sand Island began transmitting on a 24-hour basis, thus establishing a new LORAN rate in the Central Pacific. The new rate between Johnston Island and French Frigate Shoals gave a higher order of accuracy for fixing positions in the steamship lanes from Oahu, Hawaii, to Midway Island. In the past, this was impossible in some areas along this important shipping route. The U.S. Coast Guard LORAN-A Station on Sand Island ceased transmitting on June 30, 1961 when the station began transmitting from the LORAN-C station and was using a larger 625-foot antenna by 1979.

The LORAN-C station was disestablished on July 1, 1992 and all Coast Guard personnel, electronic equipment, and property departed the atoll that month. Buildings on Sand Island were transferred to other activities. LORAN whip antennas on Johnston and Sand Islands were removed, and the 625-foot LORAN tower and antenna were demolished on December 3, 1992. The LORAN A and C station and buildings on Sand Island were then dismantled and removed.*[26]*[27]

23.2.8 National nuclear weapon test site 1958–1963

See also: Operation Fishbowl, Operation Dominic, and Operation Hardtack I

Successes

Between 1958 and 1975, Johnston Atoll area was used as an American national nuclear test site for atmospheric and high-altitude nuclear explosions. In 1958 Johnston Atoll participated in the "Hardtack I" nuclear test. The U.S. military used Johnston atoll for two nuclear tests during the series. The August 1, 1958 test codenamed "Teak" and the August 12, 1958 test codenamed "Orange" both involved 3.8-megaton explosions from rockets launched from Johnston Atoll. Johnston Island is also known to have been used to launch 124 sounding rockets reaching up to 1,158 kilometres (720 miles) altitude with scientific and data instrumentation either in support of nuclear tests or in experiments related to anti-satellite technology.[*][28][*][29]

Array of sounding rockets with instruments for making scientific measurements of high-altitude nuclear tests during liftoff preparations in the Scientific Row area on Johnston Island

Eight PGM-17 Thor missiles, the first operational ballistic missiles deployed by the U.S. Air Force (USAF), were to be launched from Johnston Island in 1962 as part of "Operation Fishbowl" under the "Operation Dominic" Pacific Ocean nuclear tests. The first "Operation Fishbowl" launch was a successful R&D flight with no warhead. In the end, "Operation Fishbowl" produced four successful high altitude explosions during the Dominic series: "Starfish Prime", "Checkmate", "Bluegill Triple Prime", and "Kingfish". In addition, it produced one near high alti-

tude nuclear explosion, "Tightrope."

One of these, "Starfish Prime" on July 9, 1962, was a 1.4 megaton explosion, created by a W49 warhead at an altitude of 400 kilometres (250 miles). It created a very brief fireball visible over a wide area and a bright artificial aurora that was visible for several minutes in Hawaii. "Starfish Prime" also produced an electromagnetic pulse that disrupted power and communications as far away as Hawaii. It also pumped enough radiation into the Van Allen belts to destroy or seriously degrade seven orbiting satellites.

The final Fishbowl launch that used a Thor missile carried the "Kingfish" 400 kiloton warhead up to its 98 kilometres (61 miles) detonation altitude. Although it was officially one of the Operation Fishbowl tests, it is sometimes not listed among high-altitude nuclear tests because of its lower detonation altitude. "Tightrope" was the final test of Operation Fishbowl and detonated on November 3, 1962. It launched on a Nike-Hercules missile, and detonated at a lower altitude than the other Fishbowl tests. "At Johnston Island, there was an intense white flash. Even with high-density goggles, the burst was too bright to view, even for a few seconds. A distinct thermal pulse was also felt on the bare skin. A yellow-orange disc was formed, which transformed itself into a purple doughnut. A glowing purple cloud was faintly visible for a few minutes." [*][30] The nuclear yield was reported in most official documents only as being less than 20 kilotons. One report by the U.S. federal government reported the "Tightrope" test yield as 10 kilotons.[*][31] Seven rockets carrying scientific instrumentation were launched from Johnston Island in support of the *Tightrope* test, which was the final atmospheric test conducted by the United States.

Failures

Nuclear-armed Thor missile explodes and burns on the launch pad at Johnston Island during the failed "Bluegill Prime" nuclear test, July 25, 1962

The "Fishbowl" series included four failures, all of which were deliberately disrupted by range safety officers when the missiles systems failed during launch and were aborted. The second launch of the Fishbowl series, "Bluegill", carried an active warhead. Bluegill was "lost" by a defective range safety tracking radar and had to be destroyed 10 minutes after liftoff even though it probably ascended successfully. The subsequent nuclear weapon launch failures from Johnston Atoll caused serious contamination to the island and surrounding areas with weapons-grade plutonium and Americium that remains an issue to this day.

The failure of the "Bluegill" launch created in effect a dirty bomb but did not release the nuclear warhead's plutonium debris onto Johnston Atoll as the missile fell into the ocean south of the island and was not recovered. However, the "Starfish", "Bluegill Prime", and "Bluegill Double Prime" test launch failures in 1962 scattered radioactive debris over Johnston Island contaminating it, the lagoon, and Sand Island with plutonium for decades.[15][32]

Johnston Island Launch Emplacement One (LE1) after a Thor missile launch failure and explosion contaminated the island with Plutonium during the Operation "Bluegill Prime" nuclear test, July, 1962

"Starfish", a high altitude Thor launched nuclear test scheduled for June 20, 1962, was the first to contaminate the atoll. The rocket with the 1.45 Mt Starfish device (W49 warhead and the MK-4 Re-entry vehicle) on its nose was launched that evening, but the Thor missile engine cut out only 59 seconds after launch. The range safety officer sent a destruct signal 65 seconds after launch, and the missile was destroyed at approximately 10.6 kilometres (6.6 miles) altitude. The warhead high explosive detonated in 1-point safe fashion, destroying the warhead without producing nuclear yield. Large pieces of the plutonium contaminated missile including pieces of the warhead, booster rocket, engine, re-entry vehicle and missile parts fell back on Johnston Island. More wreckage along with plutonium contamination was found on nearby Sand Island.

"Bluegill Prime," the second attempt to launch the payload

which failed last time was scheduled for 23:15 (local) on July 25, 1962. It too was a genuine disaster and caused the most serious plutonium contamination on the island. The Thor missile was carrying one pod, two re-entry vehicles and the W50 nuclear warhead. The missile engine malfunctioned immediately after ignition, and the range safety officer fired the destruct system while the missile was still on the launch pad. The Johnston Island launch complex was demolished in the subsequent explosions and fire which burned through the night. The launch emplacement and portions of the island were contaminated with radioactive plutonium spread by the explosion, fire and wind-blown smoke.

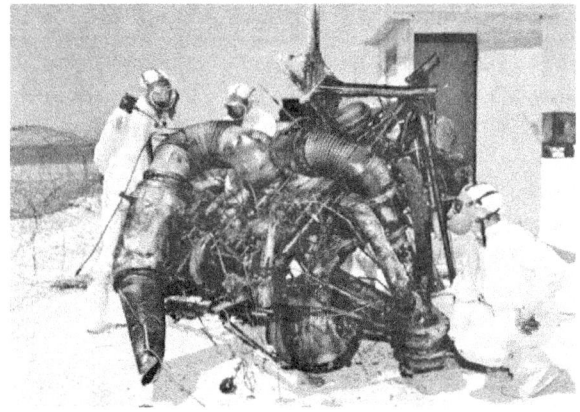

Inspection of Thor rocket engine remains on Johnston Island after failure of "Bluegill Prime" nuclear test attempt, July 1962

Afterward, the Johnston Island launch complex was heavily damaged and contaminated with plutonium. Missile launches and nuclear testing halted until the radioactive debris was dumped and soils were recovered and the launch emplacement rebuilt. Three months of repairs, decontamination, and rebuilding the LE1 as well as the backup pad LE2 were necessary before tests could resume. In an effort to continue with the testing program, U.S. troops were sent in to do a rapid cleanup. The troops scrubbed down the revetments and launch pad, carted away debris and removed the top layer of coral around the contaminated launch pad. The plutonium-contaminated rubbish was dumped in the lagoon, polluting the surrounding marine environment. More than 550 drums of contaminated material were dumped in the ocean off Johnston from 1964–1965. At the time of the Bluegill Prime disaster, the top fill around the launch pad was scraped by a bulldozer and grader. It was then dumped into the lagoon to make a ramp, so the rest of the debris could be loaded onto landing craft to be dumped out into the ocean. An estimated 10 percent of the plutonium from the test device was in the fill used to make the ramp. Then the ramp was covered and placed into a 25 acres (100,000 m²) landfill on the island during 1962 dredging to extend the island. The lagoon was again dredged in 1963–1964 and used to expand Johnston Island from 220 acres (89 ha) to 625 acres (253 ha) recontaminating additional portions of the island.

PGM-17 Thor missile at Johnston Island

On October 15, 1962 the "Bluegill Double Prime" test also misfired. During the test, the rocket was destroyed at a height of 109,000 feet after it malfunctioned 90 seconds into the flight. U.S. Defense Department officials confirm that when the rocket was destroyed, it contributed to the radioactive pollution on the island.

In 1963, the U.S. Senate ratified the Limited Test Ban Treaty, which contained a provision known as "Safeguard C". Safeguard C was the basis for maintaining Johnston Atoll as a "ready to test" above-ground nuclear testing site should atmospheric nuclear testing ever be deemed to be necessary again. In 1993, Congress appropriated no funds for the Johnston Atoll "Safeguard C" mission, bringing it to an end.

23.2.9   Anti-satellite mission 1962–1975

Main article: Program 437

Program 437 turned the PGM-17 Thor into an operational anti-satellite (ASAT) weapon system, a capability that was kept top secret even after it was deployed. The Program

437 mission was approved for development by U.S. Secretary of Defense Robert McNamara on November 20, 1962 and based at the Atoll. Program 437 used modified Thor missiles that had been returned from deployment in Great Britain and was the second deployed U.S. operational nuclear anti-satellite operation. Eighteen more suborbital Thor launches took place from Johnston Island during the 1964–1975 period in support of Program 437. In 1965–1966 four Program 437 Thors were launched with 'Alternate Payloads' for satellite inspection. This was evidently an elaboration of the system to allow visual verification of the target before destroying it. These flights may have been related to the late 1960s Program 922, a non-nuclear version of Thor with infrared homing and a high-explosive warhead. Thors were kept positioned and active near the two Johnston Island launch pads after 1964. However, partly because of the Vietnam War, in October 1970 the Department of Defense had transferred Program 437 to standby status as an economic measure. The Strategic Arms Limitation Talks led to Anti-Ballistic Missile Treaty that prohibited 'interference with national means of verification', which meant that ASAT's were not allowed, by treaty, to attack Russian spy satellites. Thors were removed from Johnston Atoll and were stored in mothballed war-reserve condition at Vandenberg Air Force Base from 1970 until the anti-satellite mission of Johnston Island facilities was ceased on August 10, 1974, and the program was officially discontinued on April 1, 1975, when any possibility of restoring the ASAT program was finally terminated. Eighteen Thor launches in support of the Program 437 Alternate Payload (AP) mission took place from Johnston Atoll's Launch emplacements.*[29]

23.2.10   Baker-Nunn satellite tracking camera station

See also: Project Space Track and United States Space Surveillance Network

The Space Detection and Tracking System or SPADATS was operated by North American Aerospace Defense Command (NORAD) along with the U.S. Air Force Spacetrack system, The Navy Space Surveillance System and Canadian Forces Air Defense Command Satellite Tracking Unit plus the Smithsonian Astrophysical Observatory operated a dozen 3.5 ton Baker-Nunn Camera systems for satellite cataloging of man-made satellites. The U.S. Air Force had ten Baker-Nunn camera stations around the world mostly from 1960 to 1977 with a phase-out beginning in 1964.*[33] The Baker-Nunn space camera station was constructed on Sand Island and was functioning by 1965.*[11] USAF 18th Surveillance Squadron operated the Baker-Nunn camera at a station built along the causeway on

Sand Island until 1975 when a contract to operate the four remaining Air Force stations was awarded to Bendix Field Engineering Corporation. In about 1977, the camera at Sand Island was moved to Daegu, South Korea.[*][5] Baker-Nunn were rendered obsolete with the IOC of 3 GEODSS optical tracking sites at Daegu, Korea; Mt Haleakala, Maui and White Sands Missile Range. A fourth site was operational in 1985 at Diego Garcia and a proposed 5th site in Portugal was cancelled. The Daegu, Korea site was closed due to encroaching city lights. GEODSS tracked satellites at night, though the MIT Lincoln Laboratory test site, co-located with Site 1 at White Sands did track asteroids in daytime as proof of concept in the early 1980s.[*][33]

23.2.11 Johnston Island Recovery Operations Center

See also: SAMOS (satellite) and Mid-air retrieval
 Satellite and Missile Observation System Project

A USAF JC-130 aircraft retrieving a SAMOS film capsule

(SAMOS-E) or "E-6" was a relatively short-lived series of United States visual reconnaissance satellites in the early 1960s. SAMOS was also known by the unclassified terms Program 101 and Program 201.[*][34] The Air Force program was used as a cover for the initial development of the Central Intelligence Agency's Key Hole (including Corona and Gambit) reconnaissance satellites systems.[*][35] Imaging was performed with film cameras and television surveillance from polar low Earth orbits with film canisters returning via capsule and parachute with mid-air retrieval. SAMOS was first launched in 1960, but not operational until 1963 with all of the missions being launched from Vandenberg AFB.[*][36]

Corona film capsule recovery sequence. Credit: CIA Directorate of Science and Technology

During the early months of the SAMOS program it was essential not only to hide the Corona and GAMBIT technical efforts under a screen of SAMOS activity, but also to make the orbital vehicle portions of the two systems resemble one another in outward appearance. Thus, some of the configuration details of SAMOS were decided less by engineering logic than by the need to camouflage GAMBIT and thus, in theory, a GAMBIT could be launched without alerting many people to its real nature. Problems relative to tracking networks, communications, and recovery were resolved with the decision in late February 1961 to use Johnston Island as the film capsule descent and recovery zone for the program.[*][37] On July 10, 1961 work was initiated on four buildings of the Johnston Island Recovery Operations Center for the National Reconnaissance Office. Men from the Johnston Atoll facility would recover the parachuting film canister capsules with a radar equipped JC-130 aircraft by capturing them in the air with a specialized recovery apparatus.[*][38] The recovery center was also responsible for collecting the radioactive scientific data pods dropped from missiles following launch and nuclear detonation.[*][39]

23.2.12 Biological warfare test site 1965

See also: Project SHAD, Project 112, and Deseret Test Center

In the lead up to biological warfare testing in the Pacific under Project 112 and Project SHAD, a new virus was discovered during the Pacific Ocean Biological Survey Program by teams from the Smithsonian's Division of Birds aboard a U.S Army tugboat involved in the program. Initially, the name of that effort was to be called the Pacific Ornithological Observation Project but this was changed for obvious reasons.*[40] First isolated in 1964 the tick-borne virus was discovered in *Ornithodores capensis* ticks, found in a nest of common noddy terns (*Anous stolidus*) at Sand Island, Johnston Atoll. It was designated *Johnston Atoll Virus* and is related to Influenza Viruses.*[41]

In February, March, and April 1965 Johnston Atoll was used to launch biological attacks against U.S. Army and Navy vessels 100 miles (160 kilometres) south-west of Johnston island in vulnerability, defense and decontamination tests conducted by the Deseret Test Center during Project SHAD under Project 112. Test DTC 64-4 (Deseret Test Center) was originally called "RED BEVA" (Biological EVAluation) though the name was later changed to "Shady Grove" likely for operational security reasons. The biological agents released during this test included *Francisella tularensis* (formerly called *Pasteurella tularensis*) (Agent UL), the causative agent of Tularemia; *Coxiella burnetii* (Agent OU), causative agent of Q fever; and *Bacillus globigii* (Agent BG).*[42] During Project SHAD, *Bacillus globigii* was used to simulate biological warfare agents (such as Anthrax), because it was then considered a contaminant with little health consequence to humans however, it is now considered a human pathogen.*[43] Ships equipped with the E-2 multi-head disseminator and A-4C aircraft equipped with Aero 14B spray tanks released live pathogenic agents in nine aerial and four surface trials in phase B of the test series on February 12 – March 15, 1965 and in four aerial trials in phase D of the test series on March 22 – April 3, 1965.*[42] According to Project SHAD veteran Jack Alderson who commanded the Army tugs, area three at Johnston Atoll was located at the most downwind part of the island and consisted of an collapsible Nissen hut to be used for weapons preparation and some communications.*[44]

23.2.13 Chemical weapon storage 1971–2001

In 1970, Congress redefined the island's military mission as the storage and destruction of chemical weapons. The United States Army leased 41 acres (17 ha) on the Atoll to store chemical weapons held in Okinawa, Japan. Johnston Atoll became a chemical weapons storage site in 1971 holding about 6.6 percent of the U.S. military chemical weapon arsenal.*[32] The chemical weapons were brought from Okinawa under Operation Red Hat with the re-deployment of the 267th Chemical Company and consisted of rockets, mines, artillery projectiles, and bulk 1-ton containers filled with Sarin, Agent VX, vomiting agent, and blister agent such as mustard gas. Chemical weapons from West Ger-

many and World War II era weapons from the Solomon Islands were also stored on the island after 1990.*[45] Chemical agents were stored in the high security Red Hat Storage Area (RHSA) which included hardened igloos in the weapon storage area, the Red Hat building (#850), two Red Hat hazardous waste warehouses (#851 and #852), an open storage area, and security entrances and guard towers.

Some of the other weapons stored at the site were shipped from U.S. stockpiles in West Germany in 1990. These shipments followed a 1986 agreement between the U.S. and West Germany to move the munitions.*[46] Merchant ships carrying the munitions left Germany under Operation Golden Python and Operation Steel Box in October 1990 and arrived at Johnston Island November 6, 1990. Although the ships were unloaded within nine days, the unpacking and storing of munitions continued into 1991.*[47] The remainder of the chemical weapons was a small number of World War II era weapons shipped from the Solomon Islands.*[48]

23.2.14 Agent Orange storage 1972–1977

Leaking Agent Orange Barrels in storage at Johnston Atoll circa 1973

See also: Operation Pacer IVY and Agent Orange § Johnston Atoll

Agent Orange was brought to Johnston Atoll from South Vietnam and Gulfport, Mississippi in 1972 under Operation Pacer IVY and stored on the northwest corner of the island known as the Herbicide Orange Storage site but dubbed the "Agent Orange Yard". The Agent Orange was eventually destroyed during Operation Pacer HO on the Dutch incineration ship *MT Vulcanus* in the Summer of 1977. The Environmental Protection Agency (EPA) reported that 1,800,000 gallons of Herbicide Orange were stored at Johnston Island in the Pacific and that an additional 480,000 gallons stored at Gulfport, Mississippi was brought to Johnston Atoll for destruction.[*][49] Leaking barrels during the storage and spills during re-drumming operations contaminated both the storage area and the lagoon with herbicide residue and its toxic contaminant 2,3,7,8-Tetrachlorodibenzodioxin.

23.2.15 Chemical weapon demilitarization mission 1990–2000

Johnston Atoll Chemical Agent Disposal System (JACADS) building

Main article: JACADS

The Army's Johnston Atoll Chemical Agent Disposal System (JACADS) was the first full-scale chemical weapons disposal facility. Built to incinerate chemical munitions on the island, planning started in 1981, construction began in 1985, and was completed five years later. Following completion of construction and facility characterization, JACADS began operational verification testing (OVT) in June 1990. From 1990 until 1993, the Army conducted four planned periods of Operational Verification Testing (OVT), required by Public Law 100-456. OVT was completed in March 1993, having demonstrated that the reverse assembly incineration technology was effective and that JACADS operations met all environmental parameters. The OVT process enabled the Army to gain critical insight into the factors that establish a safe and effective rate of destruction for all munitions and agent types. Only after this critical testing period did the Army proceed with full-scale disposal operations at JACADS. Transition to full-scale operations started in May 1993 but the facility did not begin full-scale operations until August 1993.

All of the chemical weapons once stored on Johnston Island were demilitarized and the agents incinerated at JACADS with the process completing in 2000 followed by the destruction of legacy hazardous waste material associ-

ated with chemical weapon storage and cleanup. JACADS was demolished by 2003 and the island was stripped of its remaining infrastructure and environmentally remediated.[*][45]

23.2.16 Closure and remaining structures

In 2003, structures and facilities, including those used in JACADS, were removed, and the runway was marked closed. The last flight out for official personnel was June 15, 2004. After this date, the base was completely deserted, with the only structures left standing being the Joint Operations Center (JOC) building at the east end of the runway, chemical bunkers in the weapon storage area and at least one Quonset hut.[*][50]

Built in 1964, the JOC is a 4-floor concrete and steel administration building for the island that has no windows and was built to withstand a category IV tropical cyclone as well as atmospheric nuclear tests. The building remains standing but was gutted entirely in 2004, during an asbestos abatement project. All doors of the JOC except one have been welded shut. The ground floor has a side building attached which served as a facility for decontamination that contained three long snaking corridors and 55 shower heads one could walk through during decontamination.[*][51]

Rows of bunkers in the Red Hat Storage Area remain intact however an agreement was established between the U.S. Army and EPA Region IX on August 21, 2003, that the Munitions Demilitarization Building (MDB) at JACADS would be demolished and the bunkers in the RHSA used for disposal of construction rubble and debris. After placement of the debris inside the bunkers, they were secured and the entries blocked with a concrete block barrier (a.k.a. King Tut Block) to prevent access to the bunker interior.[*][11]

23.2.17 Contamination and cleanup

Over the years, leaks of Agent Orange as well as chemical weapon leaks in the weapon storage area occurred where caustic chemicals such as sodium hydroxide were used to mitigate toxic agents during cleanup. Larger spills of nerve and mustard agent within the MCD at JACADS also took place. Small releases of chemical weapon components from JACADS were cited by the EPA. Multiple studies of the Johnston Atoll environment and ecology have been conducted and the Atoll is likely the most studied island in the Pacific.[*][11]

Dr. Lisa Lobel's work at the Atoll on the impact of PCB contamination in reef damselfish (*Abudefduf sordidus*) demonstrated that embryonic abnormalities could be utilized as a metric for comparing contaminated and uncontaminated areas.*[52] Some PCB contamination in the lagoon was traced to Coast Guard disposal practices of PCB laden electrical transformers.

In 1962 Plutonium pollution following three failed nuclear missile launches was heaviest near the destroyed launch emplacement, in the lagoon offshore of the launch pad and near Sand Island. The contaminated launch site was stripped, the debris gathered and buried in the island's 1962 expansion. A comprehensive radiological survey was completed in 1980 to record transuranic contamination remaining from the 1962 THOR missile aborts. The Air Force also initiated research on methods to remove dioxin contamination from soil resulting from leakage of the stored Herbicide Orange.*[11] Since then, U.S. defense authorities have surveyed the island in a series of studies.

Contaminated structures were dismantled and isolated within the former THOR Launch Emplacement No. 1 (LE-1) as a start for the cleanup program. About 45,000 tons of soil contaminated with radioactive isotopes was collected and placed into a fenced area covering 24 acres (10 ha) on the north of the island. The area was known as the Radiological Control Area, but dubbed "The Pluto' Yard" because its heavy contamination with highly radioactive Plutonium.*[15]*[53] The Pluto yard is on the site of the LE1 emplactment where the 1962 missile explosion occurred and also where a highly contaminated loading ramp was buried that was made for loading plutonium contaminated debris onto small boats that was dumped at sea. Remediation included a plutonium "mining" operation called the Johnston Atoll Plutonium Contaminated Soil Cleanup Project. The collected radioactive soil and other debris was buried in a landfill created within the former LE-1 area from June 2002 through November 11, 2002. Remediation at the Radiation Control Area included the construction of a 61 centimeters thick cap of coral sealing the landfill. Permanent markers were placed at each corner of the landfill to identify the landfill area.*[11]

23.2.18   After closing

The atoll was placed up for auction via the U.S. General Services Administration (GSA) in 2005 before it was withdrawn. The stripped Johnston Island was briefly offered for sale with several deed restrictions in 2005 as a "residence or vacation getaway," with potential usage for "eco-tourism" by the GSA's Office of Real Property Utilization and Disposal. The proposed sale included the unique postal zip code 96558, formerly assigned to the Armed Forces in the Pacific. The proposed sale did not include running water, electricity, or activation of the closed runway. The details of the offering were outlined on GSA's website and in a newsletter of the Center for Land Use Interpretation as un-

usual real estate listing # 6384, Johnston Island.*[54]*[55]

On August 22, 2006, Johnston Island was struck by Hurricane Ioke. The eastern eye-wall passed directly over the atoll, with winds exceeding 100 mph (160 km/h). Twelve people were on the island when the hurricane struck, part of a crew sent to the island to deliver a USAF contractor who sampled groundwater contamination levels. All 12 survived and one wrote a first hand account taking shelter from the storm in the JOC building.*[51]

On December 9, 2007, the United States Coast Guard swept the runway at Johnston Island of debris and used the runway in the removal and rescue of an ill Taiwanese fisherman to Oahu, Hawaii. The fisherman was transferred from the Taiwanese fishing vessel *Sheng Yi Tsai No. 166* to the Coast Guard buoy tender *Kukui* on December 6, 2007. The fisherman was transported to the island, and then picked up by a Coast Guard HC-130 Hercules rescue plane from Kodiak, Alaska.*[56]

Since the base was closed, the atoll has been visited by many vessels crossing the Pacific, as the deserted atoll has a strong lure due to the activities once performed there. Visitors have blogged about stopping there during a trip, or have posted photos of their visits.*[57]

In 2010, a Fish and Wildlife survey team identified a swarm of Anoplolepis ants that had invaded the island. The crazy ants are particularly destructive to the native wildlife, and needed to be eradicated. The "Crazy Ant Strike Team" project was led by the U.S. Fish and Wildlife Service who successfully reduced the ant numbers 99% by 2013 and continue to work towards a full eradication of the species. The team camped in a bunker that was previously used as a fallout shelter and office.*[58]*[59]

23.3   Demographics

Johnston Atoll has never had any indigenous inhabitants, although during the late part of the 20th century, there were averages of about 300 American military personnel and 1,000 civilian contractors present at any given time.*[1]

The primary means of transportation to this island was the airport which had a paved military runway or alternatively by ship via a pier and ship channel through the atoll's coral reef system. The islands were wired with 13 outgoing and 10 incoming commercial telephone lines, a 60-channel submarine cable, 22 DSN circuits by satellite, an Autodin with standard remote terminal, a digital telephone switch, the Military Affiliated Radio System (MARS station), a UHF/VHF air-ground radio, and a link to the Pacific Consolidated Telecommunications Network (PCTN) satellite. Amateur radio operators occasionally transmitted from the island, using the KH3 callsign prefix.*[5] The United States Undersea Cable Corporation were awarded contracts to lay underwater cable in the Pacific. A cable known as "Wet Wash C" was laid in 1966 between Makua, Oahu, Hawaii and the Johnston Island Air Force Base. USNS *Neptune* surveyed the route and laid 769 nm of cable and 45 repeaters. These cables were manufactured by the Simplex Wire and Cable Company with the repeaters being supplied by Felten and Guilleaume. In 1993 a satellite communication ground station was added to augment the atolls communications capability.

Johnston Atoll's economic activity was limited to providing services to American military and contractor personnel residing on the island. The Island was regularly resupplied by ship or barge and all foodstuffs and manufactured goods were imported. The base had six 2.5 megawatt (MW) electrical generators using diesel engines. The runway was also available to commercial airlines for emergency landings (a fairly common event), and for many years it was a regular stop on Continental Micronesia airline's "island hopper" service between Hawaii and the Marshall Islands.

There were no official license plates issued for use on Johnston Atoll. U.S. government vehicles were issued U.S. government license plates and private vehicles retained the plates from which they were registered. According to reputable license plate collectors, a number of "Johnston Atoll license plates" were created as souvenirs, and have even been sold on-line to collectors, but they were not officially issued.*[60]*[61]

23.4 Wildlife

About 300 species of fish have been recorded from the reefs and inshore waters of the atoll. It is also visited by green turtles and Hawaiian monk seals. Seabird species recorded as breeding on the atoll include Bulwer's petrel, wedge-tailed shearwater, Christmas shearwater, white-tailed tropicbird, red-tailed tropicbird, brown booby, red-footed booby, masked booby, great frigatebird, spectacled tern, sooty tern, brown noddy, black noddy and white tern. It is visited by migratory shorebirds, including the Pacific golden plover, wandering tattler, bristle-thighed curlew, ruddy turnstone and sanderling.*[62]

23.5 Areas

23.6 Launch facilities

23.7 See also

- 267th Chemical Company

- List of Guano Island claims

23.8 References

- This article incorporates public domain material from websites or documents of the United States Government.

- This article incorporates public domain material from the CIA World Factbook website https://www.cia.gov/library/publications/the-world-factbook/index.html.

[1] "United States Pacific Island Wildlife Refuges". Retrieved September 17, 2014.

[2] "Fish and Wildlife Service-Johnston Atoll, About the Refuge". Retrieved September 17, 2014.

[3] Maddish. "The Voice of Vexillology, Flags & Heraldry: Johnston Island Flag". Retrieved September 17, 2014.

[4] *American Polynesia and the Hawaiian Chain*, E. H. Bryan, Jr., 1941; Honolulu, Hawaii: Tongg Publishing Company p. 35

[5] "Johnston Island Memories Site". Retrieved September 17, 2014.

[6] "GAO/OGC-98-5 – U.S. Insular Areas: Application of the U.S. Constitution". U.S. Government Printing Office. November 7, 1997. Retrieved March 23, 2013.

[7] Thrum, Thos. G. (1923). "N. W. Pacific Exploration". *Hawaiian Almanac and Annual for 1924*. Honolulu, Hawaii. pp. 91–94.

[8] "JOHNSTON ISLAND". Office of Insular Affairs. January 11, 2007. Retrieved March 4, 2012.

[9] s:Proclamation 2416

[10] "JACADS Publications-U.S. Army's Chemical Materials Activity". Retrieved September 17, 2014.

[11] "Phase II Environmental Baseline Survey, Johnston Atoll, Appendix B" (PDF). Retrieved August 19, 2012.

[12] Bush, George W. (January 6, 2009). "Establishment of the Pacific Remote Islands Marine National Monument: A Proclamation by the President of the United States of America". White House. Retrieved March 4, 2012.

[13] "Johnston Atoll National Wildlife Refuge". U.S. Fish and Wildlife Service. Retrieved March 4, 2012.

[14] JACADS Timeline- U.S. Army's Chemical Materials Activity

[15] ""Cleaning up Johnston Atoll", APSNet Special Reports, November 25, 2005". *Nautilus Institute for Security and Sustainability*. Retrieved September 17, 2014.

[16] Global Security.org-Johnston Atoll

[17] "Defense's Nuclear Agency, 1947–1997" (PDF). DTRA History Series. 2002. Retrieved October 9, 2010.

[18] *Building the Navy's Bases in World War II History of the Bureau of Yards and Docks and the Civil Engineer Corps 1940–1946*. US Government Printing Office. 1947. pp. 158–159.

[19] "Aviation Safety Network Accident description 19420526". May 26, 1942. Retrieved September 17, 2014.

[20] "Aviation Safety Network Accident description 19490327". March 27, 1949. Retrieved September 17, 2014.

[21] Norman Polmar (2004). *The Enola Gay*. Potomac Books, Inc. pp. 20–. ISBN 978-1-59797-506-3.

[22] "Abandoned & Little-Known Airfields: Western Pacific Islands". Retrieved September 17, 2014.

[23] "Patrick J. Finneran,(Former) Executive Director USS INDIANAPOLIS CA-35 Survivors Memorial Organization, Inc.". Retrieved September 17, 2014.

[24] Bases, p.159

[25] Bases, p.160

[26] LORAN STATION JOHNSTON ISLAND

[27] "History of Johnston Atoll Timeline". Retrieved September 17, 2014.

[28] "Astronautix Web site, Johnston Island". Retrieved September 17, 2014.

[29] "Air Force Space and Missile Museum-Johnston Island". Retrieved September 17, 2014.

[30] Defense Nuclear Agency. Operation Dominic I 1962. Report DNA 6040F. (First published as an unclassified document on 1 February 1983.) Page 247.

[31] Allen, R.G., Jr., Project Officer. Report ADA995365. "Operation Dominic: Christmas and Fish Bowl Series. Project Officers Report. Project 4.1" 30 March 1965. Page 17.

[32] Kakesako, Gregg K. "The Army's disarming site Johnston Atoll once again soon will be strictly for the birds". *Honolulu Star-Bulletin*. Retrieved June 26, 2012.

[33] Curtis Peebles (1 June 1997). *High Frontier: The U.S. Air Force and the Military Space Program*. DIANE Publishing. pp. 39–. ISBN 978-0-7881-4800-2.

[34] Jonathan McDowell. "The history of spaceflight: SAMOS". Planet4589.org. Retrieved June 9, 2007.

[35] Gerald K. Haines (1997). "Development of the GAMBIT and HEXAGON Satellite Reconnaissance Systems" (PDF). National Reconnaissance Office.

[36] Yenne, Bill (1985). The Encyclopedia of US Spacecraft. Exeter Books (A Bison Book), New York. ISBN 0-671-07580-2.p.130 SAMOS

[37] HEXAGON (KH-9) Mapping Camera Program and Evolution, National Reconnaissance Office

[38] HISTORICAL CHRONOLOGY 1 July 1961 – 31 December 1961 Weapon System 117L, National Reconnaissance Office

[39] "Declassified U.S. Nuclear Test Film #65". *YouTube*. Retrieved September 17, 2014.

[40] Ed Regis (1 October 2000). *The Biology of Doom: America's Secret Germ Warfare Project*. Henry Holt and Company. pp. 41–. ISBN 978-0-8050-5765-2.

[41] "Johnston Atoll Virus".

[42] Deseret Test Center, Project SHAD, Shady Grove revised fact sheet

[43] The National Academies; The Center for Research Information, Inc. (2004). HEALTH EFFECTS OF PROJECT SHAD BIOLOGICAL AGENT: BACILLUS GLOBIGII, (Bacillus licheniformis), (Bacillus subtilis var. niger), (Bacillus atrophaeus) (PDF) (Report). Prepared for the National Academies. Contract No. IOM-2794-04-001. Retrieved January 14, 2014.

[44] Notes for Project SHAD presentation by Jack Alderson given to Institute of Medicine on April 19, 2012 for SHAD II study

[45] A Success Story, JACADS -U.S. Army's Chemical Materials Activity

[46] The Oceans and Environmental Security: Shared U.S. and Russian Perspectives. Retrieved September 17, 2014.

[47] "267th Unit History via Johnston Island Memories website". Retrieved September 9, 2012.{self}

[48] "A Success Story: Johnston Atoll Chemical Agent Disposal System". Retrieved 11 August 2012.

[49] Bourns, Charles T. (March 1, 1978). "Final report of the Federal Task Force for Hazardous Materials Management of the Western Federal Regional Council Region IX, August 1, 1973 to June 30, 1977" (PDF). US Environmental Protection Agency. Retrieved February 16, 2013.

[50] "Mark in the Pacific – The Last Day". Retrieved April 7, 2012.

[51] "Hurricane Island" .{self-published?}

[52] Lobel, Lisa K (2011). "Toxic Caviar: Using Fish Embryos to Monitor Contaminant Impacts". *In: Pollock NW, ed. Diving for Science 2011. Proceedings of the American Academy of Underwater Sciences 30th Symposium. Dauphin Island, AL: AAUS.* Retrieved March 8, 2013.

[53] TenBruggencate, Jan (March 3, 2002). "Feds want to bury Johnston Island's radioactive matter" . *Honolulu Advertiser.* Archived from the original on April 6, 2005. Retrieved January 9, 2014.

[54] "A Kaua'i Blog-Island for sale" . Retrieved September 17, 2014.

[55] "Unusual Real Estate Listing # 6384-Johnston Island" . Archived from the original on June 29, 2006. Retrieved September 17, 2014.

[56] "Coast Guard Successful on Risky Medevac from Johnston Island" . Retrieved September 17, 2014.

[57] "Blog Post from SV Sand Dollar" .

[58] "Johnston Atoll National Wildlife Refuge Pacific Remote Islands Marine National Monument Volunteer Powerpoint" (PDF). Fish And Wildlife Service.

[59] "Yellow Crazy Ant Eradication, Johnston Atoll Update, May 2011" (PDF). Project Fish and Wildlife Service.

[60] World License Plates: License Plates of Johnston Atoll (Accessed July 25, 2009)

[61] Plateshack.com: Johnston Atoll (Accessed July 25, 2009)

[62] U.S. Fish and Wildlife Service. 1995. *Bird list of Johnston Atoll National Wildlife Refuge.* Version 30DEC2002

23.9 External links

- Video from 1923 USS Tanager Expedition Northwestern Hawaiian Islands Multi-Agency Education Project of the University of Hawaii

- "Cleaning up Johnston Atoll" (2005), Plutonium contamination on the Island.

- Johnston Island Memories Site—the personal website of an AFRTS serviceman stationed there in 1975 to 1976

- Coast Guard Medevac from Johnston Island—photo from December 2007 medevac operation

- JACADS – Johnston Atoll Chemical Agent Disposal System —history of nuclear testing, and JACADS Sarin and VX nerve agent disposal

- CyberSarge—Pictorial evidence of chemical weapons disposal

- U.S. Fish & Wildlife Johnston Island National Wildlife Refuge—Contains additional information on wildlife and clean-up efforts

- Mark in the Pacific—website about the end of Johnston Atoll

- Flickr: Laysan at Johnston Island-Photographs of stop-over on abandoned Johnston Atoll in 2012

Chapter 24

Operation Fishbowl

Operation Fishbowl was a series of high-altitude nuclear tests in 1962 that were carried out by the United States as a part of the larger *Operation Dominic* nuclear test program. Flight-test vehicles were designed and manufactured by Avco Corporation.*[1]

Array of sounding rockets with instruments for making scientific measurements of high-altitude nuclear tests during liftoff preparations on Johnston Island

24.1   Introduction

The *Operation Fishbowl* nuclear tests were originally planned to be completed during the first half of 1962 with three tests named *Bluegill, Starfish* and *Urraca*.*[2]

The first test attempt was delayed until June. Planning for *Operation Fishbowl*, as well as many other nuclear tests in the region, was begun rapidly in response to the sudden Soviet announcement on 30 August 1961 that they were ending a three-year moratorium on nuclear testing.*[3] The rapid planning of very complex operations necessitated many changes as the project progressed.

All of the tests were to be launched on missiles from Johnston Island in the Pacific Ocean north of the equator. Johnston Island had already been established as a launch site for United States high-altitude nuclear tests, rather than the other locations in the Pacific Proving Grounds. In 1958, Lewis Strauss, then chairman of the United States Atomic Energy Commission, opposed doing any high-altitude tests at locations that had been used for earlier Pacific nuclear tests. His opposition was because of fears that the flash from the nighttime high-altitude detonations might blind civilians who were living on nearby islands. Johnston Island was a remote location, more distant from populated areas than other potential test locations.*[4] In order to protect residents of the Hawaiian Islands from flash blindness or permanent retinal injury from the bright nuclear flash, the nuclear missiles of *Operation Fishbowl* were launched generally toward the southwest of Johnston Island so that the detonations would be farther from Hawaii.

Urraca was to be a test of about 1 megaton yield at very high altitude (above 1000 km.).*[5] The proposed *Urraca* test was always controversial, especially after the damage caused to satellites by the *Starfish Prime* detonation, as described below. *Urraca* was finally canceled, and an extensive re-evaluation of the *Operation Fishbowl* plan was made during an 82-day operations pause after the *Bluegill Prime* disaster of 25 July 1962, as described below.

A test named *Kingfish* was added during the early stages of *Operation Fishbowl* planning. Two low-yield tests, *Checkmate* and *Tightrope*, were also added during the project, so the final number of tests in *Operation Fishbowl* were five.

See also: List of nuclear weapons tests of the United States

24.2   Research directions

The United States completed six high-altitude nuclear tests in 1958, but the high-altitude tests of that year raised a number of questions. According to U.S. Government Report

ADA955694 on the first successful test of the Fishbowl series, "Previous high-altitude nuclear tests: *Teak, Orange*, and *Yucca*, plus the three *ARGUS* shots were poorly instrumented and hastily executed. Despite thorough studies of the meager data, present models of these bursts are sketchy and tentative. These models are too uncertain to permit extrapolation to other altitudes and yields with any confidence. Thus there is a strong need, not only for better instrumentation, but for further tests covering a range of altitudes and yields." [6]

There were three phenomena in particular that required further investigation:

1. The electromagnetic pulse generated by a high-altitude nuclear explosion appeared to have very significant differences from the electromagnetic pulse generated by nuclear explosions closer to the surface.

2. The auroras associated with high-altitude nuclear explosions, especially the auroras that appeared almost instantaneously far away from the explosion in the opposite hemisphere, were not clearly understood. The nature of the possible radiation belts that were initially generated along the magnetic field lines connecting the areas of the auroral displays were also poorly understood.

3. Areas of blackout of radio communication needed to be understood in much more detail since that information would be critical for military operations during periods of possible nuclear explosions.

The *Fishbowl* tests were monitored by a large number of surface and aircraft-based stations in the wide area around the planned detonations and also in the region in the southern hemisphere in the Samoan Islands region, which was known in these tests as the **southern conjugate region**. Johnston Island is in the northern hemisphere, as were all of the planned *Operation Fishbowl* nuclear detonation locations. It was known from previous high altitude tests, as well as from theoretical work done in the late 1950s, that high-altitude nuclear tests produce a number of unique geophysical phenomena at the opposite end of the magnetic field line of the Earth's magnetic field.

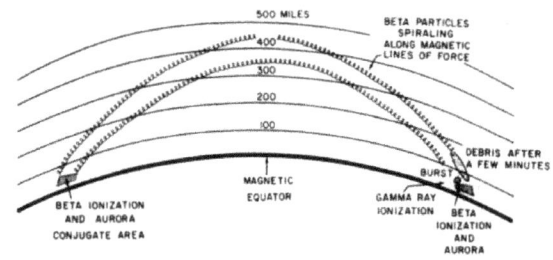

Explanation of magnetic conjugate regions

According to the standard reference book on nuclear

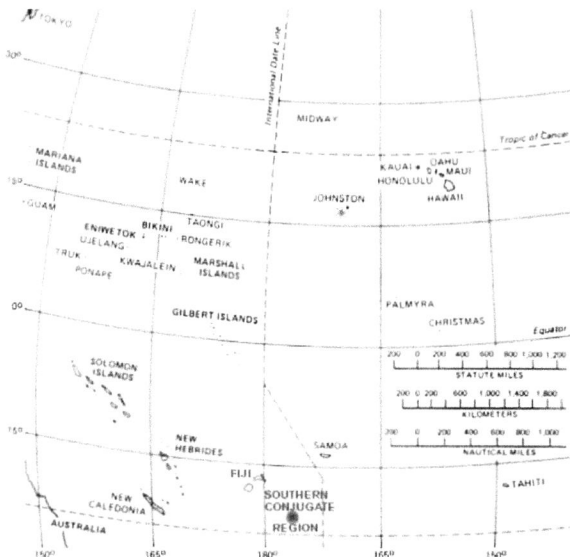

The red burst southwest of Johnston Island indicates the typical detonation point for the Operation Fishbowl *tests, and the blue round spot southeast of Fiji indicates the southern magnetic conjugate region where additional auroras were expected and were actually observed. The southern conjugate region is not directly south of the detonation point since the Earth's magnetic field lines do not run in a geographically north-south direction. Also, the magnetic equator of the Earth in this region is slightly to the south of the geographic equator. The detonation point and southern conjugate region were taken from Figure 3 of planning report ADA469481, which was originally a hand-drawn planning map for the* Bluegill *test.* [7] *The map shown here actually indicates the detonation point and southern conjugate point for* Starfish Prime. *The first* Bluegill *test failed, and the successful* Bluegill Triple Prime *test detonated at a point closer to Johnston Island.*

weapon effects by the United States Department of Defense, "For the high-altitude tests conducted in 1958 and 1962 in the vicinity of Johnston Island, the charged particles entered the atmosphere in the northern hemisphere between Johnston Island and the main Hawaiian Islands, whereas the **conjugate region** was in the vicinity of the Samoan, Fiji, and Tonga Islands. It is in these areas that auroras were actually observed, in addition to those in the areas of the nuclear explosions." [8]

Beta particles are charged particles (usually with a negative electrical charge) that are released from nuclear explosions. These particles travel in a spiral along the magnetic field lines in the Earth's magnetic field. The nuclear explosions also release heavier debris ions, which also carry an electrical charge, and which also travel in a spiral along the Earth's magnetic field lines.

The Earth's magnetic field lines arc high above the Earth until they reach the **magnetic conjugate area** in the opposite hemisphere.

According to the DOD nuclear weapon effects reference, "Because the beta particles have high velocities, the beta auroras in the remote (southern) hemisphere appeared within a fraction of a second of those in the hemisphere where the burst had occurred. The debris ions, however, travel more slowly and so the debris aurora in the remote hemisphere, if it is formed, appears at a somewhat later time. The beta auroras are generally most intense at an altitude of 30 to 60 miles, whereas the intensity of the debris auroras is greatest in the 60 to 125 miles range. Remote conjugate beta auroras can occur if the detonation is above 25 miles, whereas debris auroras appear only if the detonation altitude is in excess of some 200 miles." *[8]

Some of the charged particles traveling along the Earth's magnetic field lines cause auroras and other geophysical phenomena in the conjugate areas. Other charged particles are reflected back along the magnetic field lines, where they can persist for long periods of time (up to several months or longer), forming artificial radiation belts.*[9]

According to the *Operation Fishbowl* planning document of November 1961, "Since much valuable data can be obtained from time and spectrum resolved photography, this dictates that the test be performed at nighttime when auroral photographic conditions are best." *[7] As with all U.S. Pacific high-altitude nuclear tests, all of the *Operation Fishbowl* tests were completed at night. This is in contrast to the high-altitude nuclear tests of the Soviet Project K nuclear tests, which were done over the populated land region of central Kazakhstan, and therefore had to be done during the daytime to avoid eyeburn damage to the population from the very bright flash of high-altitude nuclear explosions (as discussed in the introduction to this article).

24.3   First attempts

According to the initial plan of *Operation Fishbowl*, the nuclear tests were to be *Bluegill*, *Starfish* and *Urraca*, in that order. If a test were to fail, the next attempt of the same test would be of the same name plus the word "prime." If *Bluegill* failed, the next attempt would be *Bluegill Prime*, and if *Bluegill Prime* failed, the next attempt would be *Bluegill Double Prime*, etc.

24.3.1   Bluegill

The first planned test of *Operation Fishbowl* was on 2 June 1962 when a nuclear warhead was launched from Johnston Island on a *Thor* missile just after midnight. Although the *Thor* missile appeared to be on a normal trajectory, the radar tracking system lost track of the missile. Because of the large number of ships and aircraft in the area, there was

The PGM-17 Thor missile shown here is very similar to the Thor missile used for the launch of the nuclear warhead in all attempts of the Bluegill, Starfish *and* Kingfish *nuclear tests of* Operation Fishbowl

no way to predict if the missile was on a safe trajectory, so the range safety officers ordered the missile with its warhead to be destroyed. No nuclear detonation occurred and no data were obtained, but subsequent investigation found that the Thor was actually following the proper flight trajectory.*[10]

24.3.2   Starfish

The second planned test of *Operation Fishbowl* was on 19 June 1962. The launch of a *Thor* missile with a nuclear warhead occurred just before midnight from Johnston Island.

The *Thor* missile flew a normal trajectory for 59 seconds; then the rocket engine suddenly stopped, and the missile began to break apart. The range safety officer ordered the destruction of the missile and the warhead. The missile was between 30,000 and 35,000 feet (between 9.1 and 10.7 km) in altitude when it was destroyed.

Some of the missile parts fell on Johnston Island, and a large amount of missile debris fell into the ocean in the vicinity of the island. Navy Explosive Ordnance Disposal and Underwater Demolition Team swimmers recovered approximately 250 pieces of the missile assembly during the next two weeks. Some of the debris was contaminated with plutonium. Nonessential personnel had been evacuated from Johnston Island during the test.

24.3.3 Starfish Prime

Main article: Starfish Prime

On 9 July 1962, at 09:00:09 Coordinated Universal Time, which was nine seconds after 10 p.m. on 8 July, Johnston Island local time, the *Starfish Prime* test was successfully detonated at an altitude of 400 kilometres (250 mi). The coordinates of the detonation were 16 degrees, 28 minutes North latitude, 169 degrees, 38 minutes West longitude (30 km, or about 18 mi, southwest of Johnston Island).*[11] The actual weapon yield was very close to the design yield, which has been described by various sources at different values in the very narrow range of 1.4 to 1.45 megatons (6.0 PJ).

The *Thor* missile carrying the Starfish Prime warhead actually reached an apogee (maximum height) of about 1100 km (just over 680 miles), and the warhead was detonated on its downward trajectory when it had fallen to the programmed altitude of 400 kilometres (250 mi). The nuclear warhead detonated at 13 minutes and 41 seconds after liftoff of the Thor missile.*[12]

Starfish Prime caused an electromagnetic pulse (EMP) which was far larger than expected, so much larger that it drove much of the instrumentation off scale, causing great difficulty in getting accurate measurements. The *Starfish Prime* electromagnetic pulse also made those effects known to the public by causing electrical damage in Hawaii, about 1,445 kilometres (900 mi) away from the detonation point, knocking out about 300 streetlights, setting off numerous burglar alarms and damaging a telephone company microwave link*[11] (the detonation time was nine seconds after 11 p.m. in Hawaii).

A total of 27 small rockets were launched from Johnston Island to obtain experimental data from the shot, with the first of the support rockets being launched 2 hours and 45 minutes before the launch of the Thor missile carrying the nuclear warhead. Most of these smaller instrumentation rockets were launched just after the time of the launch of the main Thor missile carrying the warhead. In addition, a large number of rocket-borne instruments were launched from a firing area at Barking Sands, Kauai in the Hawaiian Islands.*[13]

A very large number of United States military ships and aircraft were operating in support of *Starfish Prime* in the Johnston Island area and across the nearby North Pacific region. A few military ships and aircraft were also positioned in the southern conjugate region for the test, which was near the Samoan Islands. In addition, an uninvited scientific expeditionary ship from the Soviet Union was stationed near Johnston Island for the test and another Soviet scientific expeditionary ship was located in the southern conjugate region.*[14]

After the *Starfish Prime* detonation, bright auroras were observed in the detonation area as well as in the southern conjugate region on the other side of the equator from the detonation. According to one of the first technical reports, "The visible phenomena due to the burst were widespread and quite intense; a very large area of the Pacific was illuminated by the auroral phenomena, from far south of the south magnetic conjugate area (Tongatapu) through the burst area to far north of the north conjugate area (French Frigate Shoals). ... At twilight after the burst, resonant scattering of light from lithium and other debris was observed at Johnston and French Frigate Shoals for many days confirming the longtime presence of debris in the atmosphere. An interesting side effect was that the Royal New Zealand Air Force was aided in anti-submarine maneuvers by the light from the bomb." *[15]

The *Starfish Prime* radiation belt persisted at high altitude for many months and damaged the United States satellites *Traac*, *Transit 4B*, *Injun I* and *Telstar I*, as well as the United Kingdom satellite *Ariel*. It also damaged the Soviet satellite *Cosmos V*. All of these satellites failed completely within several months of the *Starfish* detonation.*[9] There is also evidence that the *Starfish Prime* radiation belt may have damaged the satellites *Explorer 14*, *Explorer 15* and *Relay 1*.*[16] *Telstar I* lasted the longest of the satellites that were clearly damaged by the *Starfish Prime* radiation, with its complete failure occurring on February 21, 1963.*[17]

In 2010, the United States Defense Threat Reduction Agency issued a report that had been written in support of the United States Commission to Assess the Threat to the United States from Electromagnetic Pulse Attack. The report, entitled "Collateral Damage to Satellites from an EMP Attack," discusses in great detail the satellite damage caused by the *Starfish Prime* artificial radiation belts as well as other historical nuclear events that caused artificial radiation belts and their effects on many satellites that were then in orbit. The same report also projects the effects of one or more present-day high altitude nuclear explosions upon the formation of artificial radiation belts and the probable resulting effects on satellites that were in orbit as of the year 2010.*[5]

24.3.4 Bluegill Prime

Inspection of Thor *engine parts after the radioactive contamination following the* Bluegill Prime *fire on Johnston Island.*

On 25 July 1962, a second attempt was made to launch the *Bluegill* device, but ended in disaster when the Thor suffered a stuck valve preventing the flow of LOX to the combustion chamber. The engine lost thrust and unburned RP-1 spilled down into the hot thrust chamber, igniting and starting a fire around the base of the missile. With the Thor engulfed in flames, the Range Safety Officer sent the destruct command, which split the rocket and ruptured both fuel tanks, completely destroying the missile and badly damaging the launch pad. The warhead charges also exploded asymmetrically and sprayed the area with the moderately radioactive core materials.

Although there was little danger of an accidental nuclear explosion, the destruction of the nuclear warhead on the launch pad caused contamination of the area by alpha-emitting core materials. Burning rocket fuel, flowing through the cable trenches, caused extensive chemical contamination of the trenches and the equipment associated with the cabling in the trenches.

The radioactive contamination on Johnston Island was determined to be a major problem, and it was necessary to decontaminate the entire area before the badly damaged launch pad could be rebuilt.*[18]

24.4 Operations pause

Operation Fishbowl test operations stopped after the disastrous failure of *Bluegill Prime*, and most of the personnel not directly involved in the radioactive cleanup and launch pad rebuild on Johnston Island returned to their home stations to await the resumption of tests.

According to the *Operation Dominic I* report, "The enforced pause allowed DOD to replan the remainder of the *Fishbowl* series. The *Urraca* event was canceled to avoid further damage to satellites and three new shots were added." *[19] A second launch pad was constructed during the operations pause so that *Operation Fishbowl* could continue in the event of another serious accident.

24.5 Continuation of the Fishbowl series

After a pause of nearly three months, *Operation Fishbowl* was ready to continue, beginning with another attempt at the *Bluegill* test.

24.5.1 Bluegill Double Prime

Eighty two days after the failure of *Bluegill Prime*, about 30 minutes before midnight on the night of 15 October 1962, local Johnston Island time (16 October UTC), another attempt was made at the *Bluegill* test. The *Thor* missile malfunctioned and began tumbling out of control about 85 seconds after launch, and the range safety officer ordered the destruction of the missile and its nuclear warhead about 95 seconds after launch.*[20]

24.5.2 Checkmate

On 19 October 1962, at about 90 minutes before midnight (local Johnston Island time), an XM-33 *Strypi* rocket launched a low-yield nuclear warhead which detonated successfully at an altitude of 147 kilometres (91 mi). It was reported that the yield and burst altitude were very close to those desired, but according to most official documents the exact nuclear yield remains classified. It is reported in the open literature as simply being less than 20 kilotons. One report by the U.S. federal government, however, reported the Checkmate test yield as 10 kilotons.*[21]

It was reported that, "Observers on Johnston Island saw a green and blue circular region surrounded by a blood-red ring formed overhead that faded in less than 1 minute. Blue-green streamers and numerous pink striations formed, the latter lasting for 30 minutes. Observers at Samoa saw a white flash, which faded to orange and disappeared in about 1 minute." *[20]

24.5.3 Bluegill Triple Prime

The fourth attempt at the *Bluegill* test was launched on a *Thor* missile on 25 October 1962 (Johnston Island time). It resulted in a successful detonation of a submegaton nuclear warhead at about one minute before midnight, local time (the official Coordinated Universal Time was 0959 on 26 October 1962). It was officially reported as being in the submegaton range (meaning more than 200 kilotons but less than one megaton), and most observers of the U.S. nuclear testing programs believe that the nuclear yield was about 400 kilotons.*[22] One report by the U.S. federal government reported the test yield as 200 kilotons.*[21]

Since all of the Operation Fishbowl tests were planned to occur during the night, the potential for eyeburn, especially for permanent retinal damage, was an important consideration at all levels of planning. Much research went into the potential eyeburn problem. One of the official reports for the project stated that, for the altitudes planned for the *Bluegill, Kingfish* and *Checkmate* tests, "the thermal-pulse durations are of the same order of magnitude or shorter than the natural blink period which, for the average person, is about 150 milliseconds. Furthermore, the atmospheric attenuation is normally much less for a given distance than in the case of sea-level or near-sea-level explosions. Consequently, the eye-damage hazard is more severe." *[9]

Two cases of retinal damage did occur with military personnel on Johnston Island during the **Bluegill Triple Prime** test. Neither individual had his protective goggles in place at the instant of the detonation. One official report stated, "In the first case, acuity for central vision was 20/400 initially, but returned to 20/25 by six months. The second victim was less fortunate, as central vision did not improve beyond 20/60. The lesion diameters were 0.35 and 0.50 mm respectively. Both individuals noted immediate visual disturbances, but neither was incapacitated." *[9]

There had been concern that eyeburn problems might occur during the earlier *Starfish Prime* test, since the countdown was rebroadcast by radio stations in Hawaii, and many civilians would be watching the thermonuclear detonation as it occurred,*[9] but no such problems in Hawaii were reported.

24.5.4 Kingfish

The *Kingfish* detonation occurred at 0210 (Johnston Island time) on 1 November 1962 and was the fourth successful detonation of the Fishbowl series. It was officially reported only as being a submegaton explosion (meaning in the range of more than 200 kilotons, but less than a megaton), but most independent observers believe that it used the same 400 kiloton warhead as the *Bluegill Triple Prime* test,*[22] although one report by the U.S. federal government reported the test yield as 200 kilotons.*[21]

As with the other Fishbowl tests, a number of small rockets with various scientific instrumentation were launched from Johnson Island to monitor the effects of the high-altitude explosion. In the case of the Kingfish test, 29 rockets were launched from Johnston Island in addition to the Thor rocket carrying the nuclear warhead.*[23]

According to the official report, at the time of the *Kingfish* detonation, "Johnston Island observers saw a yellow-white, luminous circle with intense purple streamers for the first minute. Some of the streamers displayed what appeared to be a rapid twisting motion at times. A large pale-green patch appeared somewhat south of the burst and grew, becoming the dominant visible feature after 5 minutes. By H+1 the green had become dull gray, but the feature persisted for 3 hours. At Oahu a bright flash was observed and after about 10 seconds a great white ball appeared to rise slowly out of the sea and was visible for about 9 minutes." *[23]

After most of the electromagnetic pulse measurements on Starfish Prime had failed because the EMP was so much larger than expected, extra care was taken to obtain accurate EMP measurements on the Bluegill Triple Prime and Kingfish tests. The EMP mechanism that had been hypothesized before *Operation Fishbowl* had been conclusively disproven by the *Starfish Prime* test. Prompt gamma ray output measurements on these later tests were also carefully obtained so that a new theory of the mechanism for high-altitude EMP could be developed and confirmed. That new theory about the generation of nuclear EMP was developed by Los Alamos physicist Conrad Longmire in 1963, and it is the high-altitude nuclear EMP theory that is still used today.*[24]

As of the beginning of 2011, the EMP waveforms and prompt gamma radiation outputs for *Bluegill Triple Prime* and *Kingfish* remain classified. An unclassified report, however, confirms that these measurements were successfully made and that a subsequent theory (which is the one now used) was developed which describes the mechanism by which the high-altitude EMP is generated. That new theory does give results which are consistent with both the *Bluegill Triple Prime* and *Kingfish* data.*[25] (The report actually using the *Bluegill Triple Prime* and *Kingfish* data to confirm the new EMP theory is the still-classified Part 2 of the unclassified report by Conrad Longmire.)*[25]

According to a Sandia National Laboratories report, EMP

generated during the *Operation Fishbowl* tests caused "... input circuit troubles in radio receivers during the *Starfish* and *Checkmate* bursts; the triggering of surge arresters on an airplane with a trailing-wire antenna during Starfish, Checkmate, and Bluegill; and the Oahu streetlight incident." [*][11] (The "Oahu streetlight incident" refers to the 300 streetlights in Honolulu extinguished by the Starfish Prime detonation.)

24.5.5   Tightrope

The final test of Operation Fishbowl was detonated at 2130 (9:30 p.m. local Johnston Island time) on 3 November 1962 (the time and date was officially recorded as 0730 UTC, 4 November 1962). It was launched on a Nike-Hercules missile, and detonated at a lower altitude than the other Fishbowl tests. Although it was officially one of the Operation Fishbowl tests, it is sometimes not listed among high-altitude nuclear tests because of its lower detonation altitude. The nuclear yield was reported in most official documents only as being less than 20 kilotons. One report by the U.S. federal government reported the *Tightrope* test yield as 10 kilotons.[*][21]

"At Johnston Island, there was an intense white flash. Even with high-density goggles, the burst was too bright to view, even for a few seconds. A distinct thermal pulse was also felt on the bare skin. A yellow-orange disc was formed, which transformed itself into a purple doughnut. A glowing purple cloud was faintly visible for a few minutes." [*][23]

Seven rockets carrying scientific instrumentation were launched from Johnston Island in support of the *Tightrope* test, which was the final atmospheric test conducted by the United States.

[1]   The US, France and Great Britain have code-named their test events, while the USSR and China did not, and therefore have only test numbers (with some exceptions – Soviet peaceful explosions were named). Word translations into English in parentheses unless the name is a proper noun. A dash followed by a number indicates a member of a salvo event. The US also sometimes named the individual explosions in such a salvo test, which results in "name1 – 1(with name2)". If test is canceled or aborted, then the row data like date and location discloses the intended plans, where known.

[2]   To convert the UT time into standard local, add the number of hours in parentheses to the UT time; for local daylight saving time, add one additional hour. If the result is earlier than 00:00, add 24 hours and subtract 1 from the day; if it is 24:00 or later, subtract 24 hours and add 1 to the day. All historical timezone data are derived from here:

[3]   Rough place name and a latitude/longitude reference; for rocket-carried tests, the launch location is specified before

the detonation location, if known. Some locations are extremely accurate; others (like airdrops and space blasts) may be quite inaccurate. "~" indicates a likely pro-forma rough location, shared with other tests in that same area.

[4]   Elevation is the ground level at the point directly below the explosion relative to sea level; height is the additional distance added or subtracted by tower, balloon, shaft, tunnel, air drop or other contrivance. For rocket bursts the ground level is "N/A". In some cases it is not clear if the height is absolute or relative to ground, for example, *Plumbbob/John*. No number or units indicates the value is unknown, while "0" means zero. Sorting on this column is by elevation and height added together.

[5]   Atmospheric, airdrop, balloon, gun, cruise missile, rocket, surface, tower, and barge are all disallowed by the Partial Nuclear Test Ban Treaty. Sealed shaft and tunnel are underground, and remained useful under the PTBT. Intentional cratering tests are borderline; they occurred under the treaty, were sometimes protested, and generally overlooked if the test was declared to be a peaceful use.

[6]   Include weapons development, weapon effects, safety test, transport safety test, war, science, joint verification and industrial/peaceful, which may be further broken down.

[7]   Designations for test items where known, "?" indicates some uncertainty about the preceding value, nicknames for particular devices in quotes. This category of information is often not officially disclosed.

[8]   Estimated energy yield in tons, kilotons, and megatons. A ton of TNT equivalent is defined as 4.184 gigajoules (1 gigacalorie).

[9]   Radioactive emission to the atmosphere aside from prompt neutrons, where known. The measured species is only iodine-131 if mentioned, otherwise it is all species. No entry means unknown, probably none if underground and "all" if not; otherwise notation for whether measured on the site only or off the site, where known, and the measured amount of radioactivity released.

24.6   See also

- Geomagnetic storm

- Starfish Prime

- Soviet Project K nuclear tests

- Nuclear electromagnetic pulse

- Electromagnetism

- Operation Argus

- Hardtack Teak

- Operation Dominic

- High-altitude nuclear explosion

- List of artificial radiation belts

24.7 References

[1] Project Fishbowl: 13 weeks from go-ahead to field delivery. // *Missiles and Rockets*, December 21, 1964, v. 15, no. 25, p. 48.

[2] Lewis, Jeffrey (2004). "The minimum means of reprisal: China's search for security in the nuclear age".

[3] "Operation Dominic". Nuclear Weapon Archive. Retrieved 2010-01-18.

[4] *Defense's Nuclear Agency 1947–1997*. Page 139. Defense Threat Reduction Agency, 2002

[5] Conrad, Edward E., et al. "Collateral Damage to Satellites from an EMP Attack" Report DTRA-IR-10-22, Defense Threat Reduction Agency. August 2010 (Retrieved 2014-01-20)

[6] Defense Atomic Support Agency. Project Officer's Interim Report: STARFISH Prime. Report ADA955694. August 1962

[7] Air Force Special Weapons Center. "Preliminary Plan for Operation Fishbowl." Report ADA469481, Kirtland Air Force Base, New Mexico. November 1961.

[8] Glasstone, Samuel and Dolan, Philip J., *The Effects of Nuclear Weapons*. Chapter 2, sections 2.144 and 2.145. United States Department of Defense. 1977.

[9] Hoerlin, Herman "United States High-Altitude Test Experiences: A Review Emphasizing the Impact on the Environment" Report LA-6405, Los Alamos Scientific Laboratory. October 1976 Retrieved 2010-01-12

[10] Defense Nuclear Agency. Operation Dominic I. 1962. Report DNA 6040F. (First published as an unclassified document on 1 February 1983.) Page 227.

[11] Vittitoe, Charles N., "Did High-Altitude EMP Cause the Hawaiian Streetlight Incident?" Sandia National Laboratories. June 1989.

[12] Dyal, P., Air Force Weapons Laboratory. Report ADA995428. "Operation Dominic. Fish Bowl Series. Debris Expansion Experiment." 10 December 1965. Page 15. Retrieved 2010-07-17

[13] United States Department of Defense. Report ADA955411. "A Quick Look at the Technical Results of Starfish Prime." August 1962.

[14] United States Central Intelligence Agency. National Intelligence Estimate. Number **11-2A-63**. "The Soviet Atomic Energy Program." page 44.

[15] United States Department of Defense. Report ADA955411. "A Quick Look at the Technical Results of Starfish Prime." August 1962.

[16] Defense Nuclear Agency. Wenaas, E.P., Jaycor Report RE-78-2044-057. DNA Report ADA191291. "Spacecraft Charging Effects on Satellites Following Starfish Event." Retrieved 2009-12-27

[17] National Space Science Data Center: Telstar 1 Retrieved 2009-12-28

[18] Defense Nuclear Agency. Operation Dominic I. 1962. Report DNA 6040F. (First published as an unclassified document on 1 February 1983.) Page 229-241.

[19] Defense Nuclear Agency. *Operation Dominic I*. 1962. Report DNA 6040F. (First published as an unclassified document on 1 February 1983.) Page 236.

[20] Defense Nuclear Agency. Operation Dominic I. 1962. Report DNA 6040F. (First published as an unclassified document on 1 February 1983.) Page 241.

[21] Allen, R.G., Jr., Project Officer. Report ADA995365. "Operation Dominic: Christmas and Fish Bowl Series. Project Officers Report. Project 4.1" 30 March 1965. Page 17.

[22] Johnston's Archive. High Altitude Nuclear Explosions

[23] Defense Nuclear Agency. Operation Dominic I 1962. Report DNA 6040F. (First published as an unclassified document on 1 February 1983.) Page 247.

[24] Longmire, Conrad L., "Fifty Odd Years of EMP", NBC Report, Fall/Winter, 2004. pp. 47-51. U.S. Army Nuclear and Chemical Agency

[25] Longmire, Conrad L., Theoretical Note 368. "Justification and Verification of High-Altitude EMP Theory, Part 1." Mission Research Corporation/Lawrence Livermore National Laboratory. June 1986. Page 3.

[26] "Timezone Historical Database". iana.com. Retrieved March 8, 2014.

[27] Hoerlin, Herman (October 1976), *United States High-Altitude Test Experiences: A Review Emphasizing the Impact on the Environment* (LA-6405), Los Alamos Scientific Laboratory, p. 4, retrieved 2014-02-26 Reference for timezone at Johnston Island 1958-1962.

[28] Conrad, Edward E.; et al. (August 2010), *Collateral Damage to Satellites from an EMP Attack* (PDF) (DTRA-IR-10-22), Defense Threat Reduction Agency, retrieved 2014-01-20

[29] *Operation Dominic I* (PDF) (DNA6040F), Washington, DC: Defense Nuclear Agency, 1983, retrieved 2014-01-12

[30] Hansen, Chuck (1995), *The Swords of Armageddon, Vol. 8*, Sunnyvale, CA: Chukelea Publications, ISBN 978-0-9791915-1-0

[31] *United States Nuclear Tests: July 1945 through September 1992* (PDF) (DOE/NV-209 REV15), Las Vegas, NV: Department of Energy, Nevada Operations Office, 2000-12-01, retrieved 2013-12-18

[32] Yang, Xiaoping; North, Robert; Romney, Carl (August 2000), *CMR Nuclear Explosion Database (Revision 3)*, SMDC Monitoring Research

[33] Norris, Robert Standish; Cochran, Thomas B. (1 February 1994), "United States nuclear tests, July 1945 to 31 December 1992 (NWD 94-1)" (PDF), *Nuclear Weapons Databook Working Paper*, Washington, DC: Natural Resources Defense Council, retrieved 2013-10-26

[34] Sublette, Carey, *Nuclear Weapons Archive*, retrieved 2014-01-06

- Lewis, Jeffrey (2004). "The minimum means of reprisal: China's search for security in the nuclear age"

.

24.8   External links

- Declassified U.S. Nuclear Film 65: Operation Dominic. Johnston Island. (Operation Fishbowl) YouTube video

Chapter 25

Lucens reactor

The **Lucens reactor** was a 6 MWe experimental nuclear power reactor built next to Lucens, Vaud, Switzerland. After its connection to the electrical grid on 29 January 1968, the reactor only operated for a few months before it suffered a loss-of-coolant accident on 21 January 1969, leading to a partial core meltdown and massive radioactive contamination of the cavern.*[1]

25.1 Description

In 1962 the construction of a Swiss-designed pilot nuclear power plant began.*[1]*[2]*[3]*[4] The heavy-water moderated, carbon dioxide gas-cooled reactor was built in an underground cavern*[5] and produced 30 megawatts of heat (which was used to generate 8.3 megawatts of electricity).*[6] It became critical 29 December 1966.*[1] It was fueled by 0.96% enriched uranium alloyed with chromium cased in magnesium alloy (magnesium with 0.6% zirconium) inserted into a graphite matrix. Carbon dioxide gas was pumped into the top of the channels at 6.28 MPa and 223 °C and exited the channels at a pressure of 5.79 MPa and at a temperature of 378 °C.*[7]

25.2 Nuclear accident

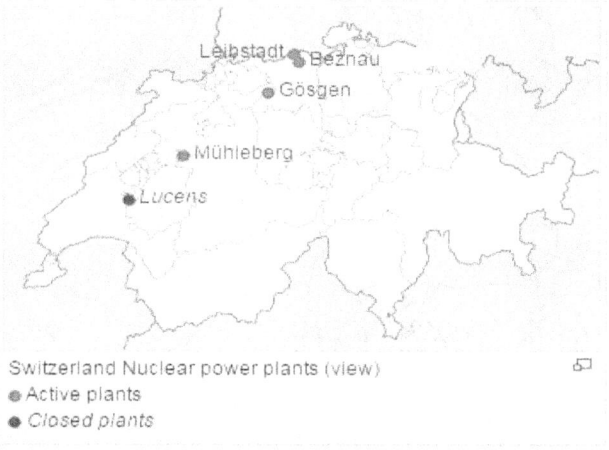

Switzerland Nuclear power plants (view)
● Active plants
● *Closed plants*

●Gösgen
●Leibstadt
●Mühleberg
●*Lucens*
Switzerland Nuclear power plants (view)
● Active plants
● *Closed plants*
  Beznau

It was intended to operate until the end of 1969, but during a startup on 21 January 1969, it suffered a loss-of-coolant accident, leading to a partial core meltdown and massive radioactive contamination of the cavern, which was then sealed. The accident was rated 4–5 on the International Nuclear Event Scale introduced in 1990 by the International Atomic Energy Agency.*[8]

The accident was caused by water condensation forming on some of the magnesium alloy fuel element components during shutdown and corroding them. The corrosion products from this accumulated in some of the fuel channels. One of the 73 vertical fuel channels was sufficiently blocked by it to impede the flow of carbon dioxide coolant so that the magnesium alloy cladding melted and further blocked the channel. The increase in temperature and exposure of the uranium metal fuel to the coolant eventually caused the

191

fuel to catch fire in the carbon dioxide coolant atmosphere. The pressure tube surrounding the fuel channel split because of overheating and bowing of the burning fuel assembly, and the carbon dioxide coolant leaked out of the reactor.*[9]*[10]

No irradiation of workers or the population occurred, though the cavern containing the reactor was seriously contaminated. The cavern was decontaminated and the reactor dismantled over the next few years. The plant was totally decommissioned in 1988 and the last radioactive waste was removed in 2003.*[11]*[12]

25.3   See also

- List of civilian nuclear accidents

25.4   References

[1] "LUCENS – Reactor Details". *IAEA Power Reactor Information System*. International Atomic Energy Agency. Retrieved 2016-10-08.

[2] Anthony, L. J. (1966). *Sources of Information on Atomic Energy – International Series of Monographs in Library and Information Science*. **2**. Elsevier. p. 85. ISBN 978-1-4831-5600-2.

[3] Wildi, Tobias (2003). "Der Traum vom eigenen Reaktor – die schweizerische Atomtechnologieentwicklung 1945–1969" (PDF). *PhD dissertation* (in German). Chronos. doi:10.3929/ethz-a-004459704. ISBN 3-0340-0594-6.

[4] Hug, Peter (2009). "Energie nucléaire". *Dictionnaire historique de la Suisse* (in French). Hauterive: Gilles Attinger. Retrieved 2016-10-09.

[5] Summary of Swiss nuclear reactors, SAPIERR Support Action: Pilot Initiative for European Regional Repositories Archived 19 July 2011 at the Wayback Machine.

[6] Swiss nuclear power, French Nuclear Energy Agency

[7] *Heavy water reactors : status and projected development* (PDF). Vienna: International Atomic Energy Agency. 2002. ISBN 9201115024.

[8] Ha-Duong, Minh; Journé, Venance (2014-05-14). "Calculating nuclear accident probabilities from empirical frequencies". *Environment Systems and Decisions*. **34** (2): 249–258. doi:10.1007/s10669-014-9499-0. ISSN 2194-5403.

[9] Description of events, Nuclear tourist

[10] Heavy water reactors: Status and projected development, IAEA, 2002

[11] "On-site disposal as a decommissioning strategy" (PDF). International Atomic Energy Agency. November 1999: 67. Retrieved 6 January 2013.

[12] "Switzerland's first nuclear plant decommissioned". *SWI swissinfo.ch*. Retrieved 2016-10-11.

25.5   External links

- Maps of Nuclear Power Reactors: Switzerland

- Major Nuclear Power Plant Accidents

Chapter 26

Three Mile Island accident

The **Three Mile Island accident** was a partial nuclear meltdown that occurred on March 28, 1979, in reactor number 2 of Three Mile Island Nuclear Generating Station (TMI-2) in Dauphin County, Pennsylvania, United States. It was the most significant accident in U.S. commercial nuclear power plant history.[2] The incident was rated a five on the seven-point International Nuclear Event Scale: Accident With Wider Consequences.[3][4]

The accident began with failures in the non-nuclear secondary system, followed by a stuck-open pilot-operated relief valve in the primary system, which allowed large amounts of nuclear reactor coolant to escape. The mechanical failures were compounded by the initial failure of plant operators to recognize the situation as a loss-of-coolant accident due to inadequate training and human factors, such as human-computer interaction design oversights relating to ambiguous control room indicators in the power plant's user interface. In particular, a hidden indicator light led to an operator manually overriding the automatic emergency cooling system of the reactor because the operator mistakenly believed that there was too much coolant water present in the reactor and causing the steam pressure release.[5]

The accident crystallized anti-nuclear safety concerns among activists and the general public, resulted in new regulations for the nuclear industry, and has been cited as a contributor to the decline of a new reactor construction program that was already underway in the 1970s.[6] The partial meltdown resulted in the release of radioactive gases and radioactive iodine into the environment. Worries were expressed by anti-nuclear movement activists;[7] however, epidemiological studies analyzing the rate of cancer in and around the area since the accident, determined there was a small statistically non-significant increase in the rate and thus no causal connection linking the accident with these cancers has been substantiated.[8][9][10][11][12][13] Cleanup started in August 1979, and officially ended in December 1993, with a total cleanup cost of about $1 billion.[14]

26.1 Accident

26.1.1 Stuck valve

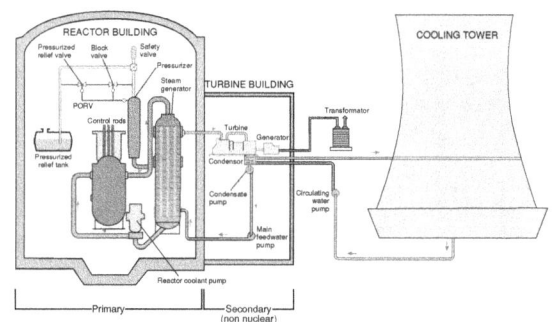

Simplified schematic diagram of the TMI-2 plant[15]

In the nighttime hours preceding the incident, the TMI-2 reactor was running at 97% of full power, while the companion TMI-1 reactor was shut down for refueling.[16] The main chain of events leading to the partial core meltdown began at 4:37 am EST on March 28, 1979, in TMI-2's secondary loop, one of the three main water/steam loops in a pressurized water reactor (PWR).

The initial cause of the accident happened eleven hours earlier, during an attempt by operators to fix a blockage in one of the eight condensate polishers, the sophisticated filters cleaning the secondary loop water. These filters are designed to stop minerals and impurities in the water from accumulating in the steam generators and increasing corrosion rates in the secondary side.

Blockages are common with these resin filters and are usually fixed easily, but in this case the usual method of forcing the stuck resin out with compressed air did not succeed. The operators decided to blow the compressed air into the water and let the force of the water clear the resin. When they forced the resin out, a small amount of water forced its way past a stuck-open check valve and found its way into an instrument air line. This would eventually cause the feedwater pumps, condensate booster pumps, and condensate pumps to turn off around 4:00 am, which would in turn cause a turbine trip.[17]

With the steam generators no longer receiving feedwater, heat and pressure increased in the reactor coolant system, causing the reactor to perform an emergency shutdown (SCRAM). Within eight seconds, control rods were inserted into the core to halt the nuclear chain reaction. The reactor continued to generate decay heat and, because steam was no longer being used by the turbine, heat was no longer being removed from the reactor's primary water loop.[18]

Once the secondary feedwater pumps stopped, three auxiliary pumps activated automatically. However, because the valves had been closed for routine maintenance, the system was unable to pump any water. The closure of these valves was a violation of a key Nuclear Regulatory Commission (NRC) rule, according to which the reactor must be shut down if all auxiliary feed pumps are closed for maintenance. This was later singled out by NRC officials as a key failure.[19]

The loss of heat removal from the primary loop and the failure of the auxiliary system to activate caused the primary loop pressure to increase, triggering the pilot-operated relief valve at the top of the pressurizer – a pressure active-regulator tank – to open automatically. The relief valve should have closed when the excess pressure had been released, and electric power to the solenoid of the pilot was automatically cut, but the relief valve stuck open because of a mechanical fault. The open valve permitted coolant water to escape from the primary system, and was the principal mechanical cause of the primary coolant system depressurization and partial core disintegration that followed.[20]

26.1.2   Human factors: confusion over valve status

Critical human factors and user interface engineering problems were revealed in the investigation of the reactor control system's user interface. Despite the valve being stuck open, a light on the control panel ostensibly indicated that the valve was *closed*. In fact the light did not indicate the position of the valve, only the status of the solenoid being powered or not, thus giving false evidence of a closed valve.[21] As a result, the operators did not correctly diagnose the problem for several hours.[22]

The design of the pilot-operated relief valve indicator light was fundamentally flawed. The bulb was simply connected in parallel with the valve solenoid, thus implying that the pilot-operated relief valve was shut when it went dark, without actually verifying the real position of the valve. When everything was operating correctly, the indication was true and the operators became habituated to rely on it. However, when things went wrong and the main relief valve stuck open, the unlighted lamp was actually misleading the operators by implying that the valve was shut. This caused the operators considerable confusion, because the pressure, temperature and coolant levels in the primary circuit, so far as they could observe them via their instruments, were not behaving as they would have if the pilot-operated relief valve were shut. This confusion contributed to the severity of the accident because the operators were unable to break out of a cycle of assumptions that conflicted with what their instruments were telling them. It was not until a fresh shift came in, who did not have the mind-set of the first shift of operators, that the problem was correctly diagnosed. By this time major damage had occurred.

The operators had not been trained to understand the ambiguous nature of the pilot-operated relief valve indicator and to look for alternative confirmation that the main relief valve was closed. There was a temperature indicator downstream of the pilot-operated relief valve in the tail pipe between the pilot-operated relief valve and the pressurizer that could have told them the valve was stuck open, by showing that the temperature in the tail pipe remained higher than it should have been had the pilot-operated relief valve been shut. This temperature indicator, however, was not part of the "safety grade" suite of indicators designed to be used after an incident, and the operators had not been trained to use it. Its location on the back of the seven-foot-high instrument panel also meant that it was effectively out of sight of the operators.[23]

26.1.3   Consequences of stuck valve

As the pressure in the primary system continued to decrease, reactor coolant continued to flow, but it was boiling inside the core. First, small bubbles of steam formed and immediately collapsed, known as nucleate boiling. As the system pressure decreased further, steam pockets began to form in the reactor coolant. This departure from nucleate boiling (DNB) into the regime of "film boiling" caused steam voids in coolant channels, blocking the flow of liquid coolant and greatly increasing the fuel cladding temperature. The overall water level inside the pressurizer was *rising* despite the loss of coolant through the open pilot-operated relief valve, as the volume of these steam voids increased much more quickly than coolant was lost. Because of the lack of a dedicated instrument to measure the level of water in the core, operators judged the level of water in the core solely by the level in the pressurizer. Since it was high, they assumed that the core was properly covered with coolant, unaware that because of steam forming in the reactor vessel, the indicator provided misleading readings.*[24] Indications of high water levels contributed to the confusion, as operators were concerned about the primary loop "going solid," (i.e. no steam pocket buffer existing in the pressurizer) which in training they had been instructed to never allow. This confusion was a key contributor to the initial failure to recognize the accident as a loss-of-coolant accident, and led operators to turn off the emergency core cooling pumps, which had automatically started after the pilot-operated relief valve stuck and core coolant loss began, due to fears the system was being overfilled.*[25]

With the pilot-operated relief valve still open, the pressurizer relief tank that collected the discharge from the pilot-operated relief valve overfilled, causing the containment building sump to fill and sound an alarm at 4:11 am. This alarm, along with higher than normal temperatures on the pilot-operated relief valve discharge line and unusually high containment building temperatures and pressures, were clear indications that there was an ongoing loss-of-coolant accident, but these indications were initially ignored by operators.*[26] At 4:15 am, the relief diaphragm of the pressurizer relief tank ruptured, and radioactive coolant began to leak out into the general containment building. This radioactive coolant was pumped from the containment building sump to an auxiliary building, outside the main containment, until the sump pumps were stopped at 4:39 am.*[27]

After almost 80 minutes of slow temperature rise, the primary loop's four main reactor coolant pumps began to cavitate as a steam bubble/water mixture, rather than water, passed through them. The pumps were shut down, and it was believed that natural circulation would continue the water movement. Steam in the system prevented flow through the core, and as the water stopped circulating it was converted to steam in increasing amounts. About 130 minutes after the first malfunction, the top of the reactor core was exposed and the intense heat caused a reaction to occur between the steam forming in the reactor core and the Zircaloy nuclear fuel rod cladding, yielding zirconium dioxide, hydrogen, and additional heat. This reaction melted the nuclear fuel rod cladding and damaged the fuel pellets, which released radioactive isotopes to the reactor coolant, and produced hydrogen gas that is believed to have caused a small explosion in the containment building later that afternoon.*[28]

At 6 am, there was a shift change in the control room. A new arrival noticed that the temperature in the pilot-operated relief valve tail pipe and the holding tanks was excessive and used a backup valve – called a "block valve" – to shut off the coolant venting via the pilot-operated relief valve, but around 32,000 US gal (120,000 l) of coolant had already leaked from the primary loop.*[29] It was not until 165 minutes after the start of the problem that radiation

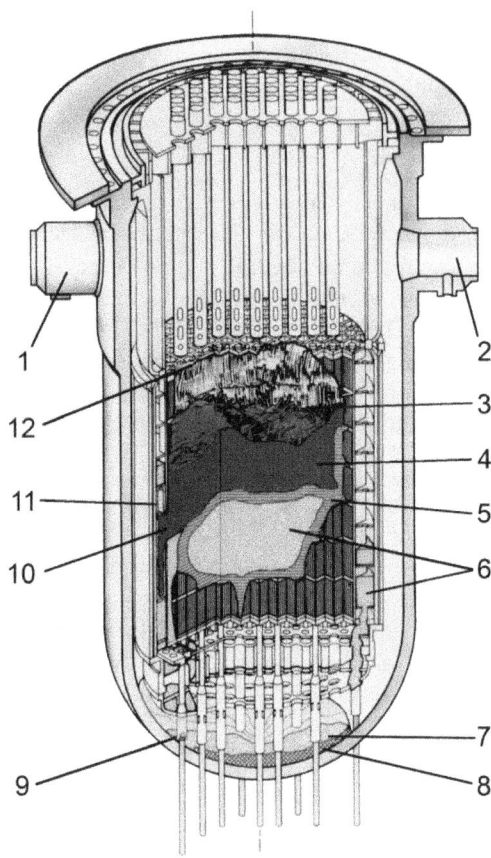

NRC graphic of TMI-2 core end-state configuration.

alarms activated as contaminated water reached detectors; by that time, the radiation levels in the primary coolant water were around 300 times expected levels, and the general containment building was seriously contaminated.

26.2 Emergency declared

At 6:57 am, a plant supervisor declared a site area emergency, and less than 30 minutes later station manager Gary Miller announced a general emergency, defined as having the "potential for serious radiological consequences" to the general public.*[30] Metropolitan Edison notified the Pennsylvania Emergency Management Agency (PEMA), which in turn contacted state and local agencies, Governor Richard L. Thornburgh and lieutenant governor William Scranton III, to whom Thornburgh assigned responsibility for collecting and reporting on information about the accident.*[31] The uncertainty of operators at the plant was reflected in fragmentary, ambiguous, or contradictory statements made by Met Ed to government agencies and to the press, particularly about the possibility and severity of off-site radioactivity releases. Scranton held a press conference in which he was reassuring, yet confusing, about this possibility, stating that though there had been a "small release of radiation...no increase in normal radiation levels" had been detected. These were contradicted by another official, and by statements from Met Ed, who both claimed that no radioactivity had been released.*[32] In fact, readings from instruments at the plant and off-site detectors had detected radioactivity releases, albeit at levels that were unlikely to threaten public health as long as they were temporary, and providing that containment of the then highly contaminated reactor was maintained.*[33]

Angry that Met Ed had not informed them before conducting a steam venting from the plant, and convinced that the company was downplaying the severity of the accident, state officials turned to the NRC.*[34] After receiving word of the accident from Met Ed, the NRC had activated its emergency response headquarters in Bethesda, Maryland and sent staff members to Three Mile Island. NRC chairman Joseph Hendrie and commissioner Victor Gilinsky*[35] initially viewed the accident, in the words of NRC historian Samuel Walker, as a "cause for concern but not alarm".*[36] Gilinsky briefed reporters and members of Congress on the situation and informed White House staff, and at 10:00 a.m. met with two other commissioners. However, the NRC faced the same problems in obtaining accurate information as the state, and was further hampered by being organizationally ill-prepared to deal with emergencies, as it lacked a clear command structure and the authority to tell the utility what to do, or to order an evacuation of the local area.*[37]

In a 2009 article, Gilinsky wrote that it took five weeks to learn that "the reactor operators had measured fuel temperatures near the melting point".*[38] He further wrote: "We didn't learn for years – until the reactor vessel was physically opened – that by the time the plant operator called the NRC at about 8:00 a.m., roughly half of the uranium fuel had already melted." *[38]

It was still not clear to the control room staff that the primary loop water levels were low and that over half of the core was exposed. A group of workers took manual readings from the thermocouples and obtained a sample of primary loop water. Seven hours into the emergency, new water was pumped into the primary loop and the backup relief valve was opened to reduce pressure so that the loop could be filled with water. After 16 hours, the primary loop pumps were turned on once again, and the core temperature began to fall. A large part of the core had melted, and the system was still dangerously radioactive.

On the third day following the accident, a hydrogen bubble was discovered in the dome of the pressure vessel, and became the focus of concern. A hydrogen explosion might not only breach the pressure vessel, but, depending on its

magnitude, might compromise the integrity of the containment vessel leading to large scale release of radioactive material. However, it was determined that there was no oxygen present in the pressure vessel, a prerequisite for hydrogen to burn or explode. Immediate steps were taken to reduce the hydrogen bubble, and by the following day it was significantly smaller. Over the next week, steam and hydrogen were removed from the reactor using a catalytic recombiner and, controversially, by venting straight to the atmosphere.

26.2.1   Release of radioactive material

This occurred when the cladding was damaged while the pilot-operated relief valve was still stuck open. Fission products were released into the reactor coolant. Since the pilot-operated relief valve was stuck open and the loss of coolant accident was still in progress, primary coolant with fission products and/or fuel was released, and ultimately ended up in the auxiliary building. This auxiliary building was outside the containment boundary.

This was evidenced by the radiation alarms that eventually sounded. However, since very little of the fission products released were solids at room temperature, very little radiological contamination was reported in the environment. No significant level of radiation was attributed to the TMI-2 accident outside of the TMI-2 facility. According to the Rogovin report, the vast majority of the radioisotopes released were the noble gases xenon and krypton. The report stated, "During the course of the accident, approximately 2.5 MCi (93 PBq) of radioactive noble gases and 15 Ci (560 GBq) of radioiodines were released." This resulted in an average dose of 1.4 mrem (14 μSv) to the two million people near the plant. The report compared this with the additional 80 mrem (800 μSv) per year received from living in a high altitude city such as Denver.*[39] As further comparison, a patient receives 3.2 mrem (32 μSv) from a chest X-ray – more than twice the average dose of those received near the plant.*[40] Measures of beta radiation were excluded from the report.

Within hours of the accident, the United States Environmental Protection Agency (EPA) began daily sampling of the environment at the three stations closest to the plant. Continuous monitoring at 11 stations was not established until April 1, and was expanded to 31 stations on April 3. An inter-agency analysis concluded that the accident did not raise radioactivity far enough above background levels to cause even one additional cancer death among the people in the area, but measures of beta radiation were not included. The EPA found no contamination in water, soil, sediment or plant samples.*[41]

Researchers at nearby Dickinson College – which had radiation monitoring equipment sensitive enough to detect Chinese atmospheric atomic weapons-testing – collected soil samples from the area for the ensuing two weeks and detected no elevated levels of radioactivity, except after rainfalls (likely due to natural radon plate-out, not the accident).*[42] Also, white-tailed deer tongues harvested over 50 mi (80 km) from the reactor subsequent to the accident were found to have significantly higher levels of cesium-137 than in deer in the counties immediately surrounding the power plant. Even then, the elevated levels were still below those seen in deer in other parts of the country during the height of atmospheric weapons testing.*[43] Had there been elevated releases of radioactivity, increased levels of iodine-131 and cesium-137 would have been expected to be detected in cattle and goat's milk samples. Yet elevated levels were not found.*[44] A later scientific study noted that the official emission figures were consistent with available dosimeter data,*[45] though others have noted the incompleteness of this data, particularly for releases early on.*[46]

According to the official figures, as compiled by the 1979 Kemeny Commission from Metropolitan Edison and NRC data, a maximum of 480 PBq (13 MCi) of radioactive noble gases (primarily xenon) were released by the event.*[47] However, these noble gases were considered relatively harmless,*[48] and only 481–629 GBq (13.0–17.0 Ci) of thyroid cancer-causing iodine-131 were released.*[47] Total releases according to these figures were a relatively small proportion of the estimated 370 EBq (10 GCi) in the reactor.*[48] It was later found that about half the core had melted, and the cladding around 90% of the fuel rods had failed,*[15]*[49] with 5 ft (1.5 m) of the core gone, and around 20 short tons (18 t) of uranium flowing to the bottom head of the pressure vessel, forming a mass of corium.*[50] The reactor vessel – the second level of containment after the cladding – maintained integrity and contained the damaged fuel with nearly all of the radioactive isotopes in the core.*[51]

Anti-nuclear political groups disputed the Kemeny Commission's findings, claiming that independent measurements provided evidence of radiation levels up to five times higher than normal in locations hundreds of miles downwind from TMI.*[52] Arnie Gundersen, a nuclear engineer and former nuclear industry executive, said "I think the numbers on the NRC's website are off by a factor of 100 to 1,000" .*[48]*[53]

Some other insiders, including Arnie Gundersen, a former nuclear industry executive who is now an expert witness in nuclear safety issues,*[54]*[55] make the same claim; Gundersen offers evidence, based on pressure monitoring data, for a hydrogen explosion shortly before 2:00 p.m. on March 28, 1979, which would have provided the means for a high dose of radiation to occur.*[48] Gundersen cites affidavits from four reactor operators according to which the plant manager was aware of a dramatic pressure spike, af-

ter which the internal pressure dropped to outside pressure. Gundersen also notes that the control room shook and doors were blown off hinges. However official NRC reports refer merely to a "hydrogen burn." *[48] The Kemeny Commission referred to "a burn or an explosion that caused pressure to increase by 28 pounds per square inch in the containment building".*[56] *The Washington Post* reported that "At about 2:00 pm, with pressure almost down to the point where the huge cooling pumps could be brought into play, a small hydrogen explosion jolted the reactor." *[57]

26.3 Aftermath

26.3.1 Voluntary evacuation

A sign dedicated in 1999 in Middletown, Pennsylvania near the plant describing the accident and the evacuation of the residents in the area.

Twenty-eight hours after the accident began, William Scranton III, the lieutenant governor, appeared at a news briefing to say that Metropolitan Edison, the plant's owner, had assured the state that "everything is under control" .*[58] Later that day, Scranton changed his statement, saying that the situation was "more complex than the company first led us to believe." *[58] There were conflicting statements about radioactivity releases.*[59] Schools were closed and residents were urged to stay indoors. Farmers were told to keep their animals under cover and on stored feed.*[58]*[59]

Governor Dick Thornburgh, on the advice of NRC chairman Joseph Hendrie, advised the evacuation "of pregnant women and pre-school age children...within a five-mile radius of the Three Mile Island facility." The evacuation zone was extended to a 20-mile radius on Friday, March 30.*[60] Within days, 140,000 people had left the area.[*][15][*][58][*][61] More than half of the 663,500 population[*][62] within the 20-mile radius remained in that area.[*][60] According to a survey conducted in April 1979, 98% of the evacuees had returned to their homes within three weeks.[*][60]

Post-TMI surveys have shown that less than 50% of the American public were satisfied with the way the accident was handled by Pennsylvania State officials and the NRC, and people surveyed were even less pleased with the utility (General Public Utilities) and the plant designer.[*][63]

26.3.2   Investigations

Several state and federal government agencies mounted investigations into the crisis, the most prominent of which was the **President's Commission on the Accident at Three Mile Island**, created by Jimmy Carter in April 1979.[*][64] The commission consisted of a panel of twelve people, specifically chosen for their lack of strong pro- or anti-nuclear views, and headed by chairman John G. Kemeny, president of Dartmouth College. It was instructed to produce a final report within six months, and after public hearings, depositions, and document collection, released a completed study on October 31, 1979.[*][65] The investigation strongly criticized Babcock & Wilcox, Met Ed, GPU, and the NRC for lapses in quality assurance and maintenance, inadequate operator training, lack of communication of important safety information, poor management, and complacency, but avoided drawing conclusions about the future of the nuclear industry.[*][66] The heaviest criticism from the Kemeny Commission concluded that "fundamental changes were necessary in the organization, procedures, practices 'and above all – in the attitudes' of the NRC [and the nuclear industry.]"[*][67] Kemeny said that the actions taken by the operators were "inappropriate" but that the workers "were operating under procedures that they were required to follow, and our review and study of those indicates that the procedures were inadequate" and that the control room "was greatly inadequate for managing an accident." [*][68]

The Kemeny Commission noted that Babcock & Wilcox's pilot-operated relief valve had previously failed on 11 occasions, nine of them in the open position, allowing coolant to escape. More disturbing, however, was the fact that the initial causal sequence of events at TMI had been duplicated 18 months earlier at another Babcock & Wilcox reactor, the Davis-Besse Nuclear Power Station owned at that time by Toledo Edison. The only difference was that the operators at Davis-Besse identified the valve failure after 20 minutes, where at TMI it took 80 minutes, and the Davis-Besse facility was operating at 9% power, against TMI's 97%. Although Babcock engineers recognized the prob-

lem, the company failed to clearly notify its customers of the valve issue.[*][69]

Upon his return to Dartmouth, Kemeny addressed Dartmouth college students. When asked what caused the meltdown, he replied that the proximate cause would probably never be known . The Government Affairs Vice President confirmed that the Metropolitan Edison Company, which operated the company, had shortly before received a warning from the Nuclear Regulatory Commission (NRC) that Babcock & Wilcox reactor valves were vulnerable to failure under certain conditions. He said he had sent it on to the Vice President of Engineering, who confirmed that he had read it . Shortly after that, the two men met at the water cooler where the Government Affairs VP asked the Engineering VP a question. The Government Affairs VP remembered the question as "Is there a problem here?" The Engineering VP thought the question was "Have you solved the problem?" Both VPs agreed that the answer was "no." One walked away believing that the problem was solved. The other believed that he had informed his bosses that there was a problem. The issue was never resolved . Kemeny told the students that he believed it never would be. The proximate cause of the meltdown remains unknown and no proof of negligence was ever uncovered.

The Pennsylvania House of Representatives conducted its own investigation, which focused on the need to improve evacuation procedures.

In 1985, a television camera was used to see the interior of the damaged reactor. In 1986, core samples and samples of debris were obtained from the corium layers on the bottom of the reactor vessel and analyzed.[*][70]

26.3.3   Effect on nuclear power industry

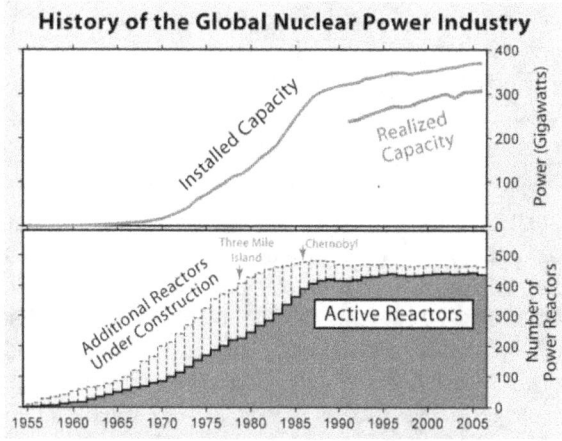

Global history of the use of nuclear power. The Three Mile Island accident is one of the factors cited for the decline of new reactor construction.

According to the IAEA, the Three Mile Island accident was a significant turning point in the global development of nuclear power.[*][71] From 1963–1979, the number of reactors under construction globally increased every year except 1971 and 1978. However, following the event, the number of reactors under construction in the U.S. declined every year from 1980–1998. Many similar Babcock & Wilcox reactors on order were canceled; in total, 51 U.S. nuclear reactors were canceled from 1980–1984.[*][72]

The 1979 TMI accident did not initiate the demise of the U.S. nuclear power industry, but it did halt its historic growth. Additionally, as a result of the earlier 1973 oil crisis and post-crisis analysis with conclusions of potential over-capacity in base load, forty planned nuclear power plants already had been canceled before the TMI accident. At the time of the TMI incident, 129 nuclear power plants had been approved, but of those, only 53 (which were not already operating) were completed. During the lengthy review process, complicated by the Chernobyl Disaster seven years later, Federal requirements to correct safety issues and design deficiencies became more stringent, local opposition became more strident, construction times were significantly lengthened and costs skyrocketed.[*][73] Until 2012,[*][74] no U.S. nuclear power plant had been authorized to begin construction since the year before TMI.

Globally, the end of the increase in nuclear power plant construction came with the more catastrophic Chernobyl disaster in 1986 (see graph).

26.3.4 Cleanup

Three Mile Island Unit 2 was too badly damaged and contaminated to resume operations; the reactor was gradually deactivated and permanently closed. TMI-2 had been online only 13 months but now had a ruined reactor vessel and a containment building that was unsafe to walk in. Cleanup started in August 1979 and officially ended in December 1993, with a total cleanup cost of about $1 billion.[*][14] Benjamin K. Sovacool, in his 2007 preliminary assessment of major energy accidents, estimated that the TMI accident caused a total of $2.4 billion in property damages.[*][75]

Initially, efforts focused on the cleanup and decontamination of the site, especially the defueling of the damaged reactor. Starting in 1985, almost 100 short tons (91 t) of radioactive fuel were removed from the site. The first major phase of the cleanup was completed in 1990, when workers finished shipping 150 short tons (140 t) of radioactive wreckage to Idaho for storage at the Department of Energy's National Engineering Laboratory. However, the contaminated cooling water that leaked into the containment building had seeped into the building's concrete, leaving the radioactive residue too impractical to remove. In 1988, the

A clean-up crew working to remove radioactive contamination at Three Mile Island

Nuclear Regulatory Commission announced that, although it was possible to further decontaminate the Unit 2 site, the remaining radioactivity had been sufficiently contained as to pose no threat to public health and safety. Accordingly, further cleanup efforts were deferred to allow for decay of the radiation levels and to take advantage of the potential economic benefits of retiring both Unit 1 and Unit 2 together.[*][14]

26.3.5 Health effects and epidemiology

Main article: Three Mile Island accident health effects

In the aftermath of the accident, investigations focused on the amount of radioactivity released by the accident. In total approximately 2.5 megacuries (93 PBq) of radioactive gases, and approximately 15 curies (560 GBq) of iodine-131 was released into the environment.[*][76] According to the American Nuclear Society, using the official radioactivity emission figures, "The average radiation dose to people living within ten miles of the plant was eight millirem, and no more than 100 millirem to any single individual. Eight millirem is about equal to a chest X-ray, and 100 millirem is about a third of the average background level of radiation received by US residents in a year." [*][51][*][77]

Based on these emission figures, early scientific publications, according to Mangano, on the health effects of the fallout estimated no additional cancer deaths in the 10 mi (16 km) area around TMI.[52] Disease rates in areas further than 10 miles from the plant were never examined.[52] Local activism in the 1980s, based on anecdotal reports of negative health effects, led to scientific studies being commissioned. A variety of epidemiology studies have concluded that the accident had no observable long term health effects.[8][12][78][79]

The Radiation and Public Health Project, an organization with little credibility amongst epidemiologists,[80] cited calculations by its member Joseph Mangano – who has authored 19 medical journal articles and a book on *Low Level Radiation and Immune Disease* – that reported a spike in infant mortality in the downwind communities two years after the accident.[52][81] Anecdotal evidence also records effects on the region's wildlife.[52] For example, according to one anti-nuclear activist, Harvey Wasserman, the fallout caused "a plague of death and disease among the area's wild animals and farm livestock" , including a sharp fall in the reproductive rate of the region's horses and cows, reflected in statistics from Pennsylvania's Department of Agriculture, though the Department denies a link with TMI.[82]

John Gofman used his own, non-peer reviewed low-level radiation health model to predict 333 excess cancer or leukemia deaths from the 1979 Three Mile Island accident.[7] A peer-reviewed research article by Dr. Steven Wing found a significant increase in cancers from 1979–1985 among people who lived within ten miles of TMI;[83] in 2009 Dr. Wing stated that radiation releases during the accident were probably "thousands of times greater" than the NRC's estimates. A retrospective study of Pennsylvania Cancer Registry found an increased incidence of thyroid cancer in counties south of TMI and in high-risk age groups but did not draw a causal link with these incidences and to the accident.[9][10] The Talbott lab at the University of Pittsburgh reported finding only a few, small, mostly statistically non-significant, increased cancer risks within the TMI population, such as a non-significant excess leukemia among males being observed.[11] The ongoing TMI epidemiological research has been accompanied by a discussion of problems in dose estimates due to a lack of accurate data, as well as illness classifications.[84]

26.3.6 Activism and legal action

See also: List of anti-nuclear groups in the United States § Three Mile Island Alert

The TMI accident enhanced the credibility of anti-nuclear groups, who had predicted an accident,[85] and triggered protests around the world.[86] (President Carter —who had specialized in nuclear power while in the United States Navy—told his cabinet after visiting the plant that the accident was minor, but reportedly declined to do so in public in

Anti-nuclear protest following the Three Mile Island accident, Harrisburg, 1979.

order to avoid offending left-wing Democrats who opposed nuclear power.[87])

Members of the American public, concerned about the release of radioactive gas from the accident, staged numerous anti-nuclear demonstrations across the country in the following months. The largest demonstration was held in New York City in September 1979 and involved 200,000 people, with speeches given by Jane Fonda and Ralph Nader.[88][89][90] The New York rally was held in conjunction with a series of nightly "No Nukes" concerts given at Madison Square Garden from September 19–23 by Musicians United for Safe Energy. In the previous May, an estimated 65,000 people – including California Governor Jerry Brown – attended a march and rally against nuclear power in Washington, D.C.[89]

In 1981, citizens' groups succeeded in a class action suit against TMI, winning $25 million in an out-of-court settlement. Part of this money was used to found the TMI Public Health Fund.[91] In 1983, a federal grand jury indicted Metropolitan Edison on criminal charges for the falsification of safety test results prior to the accident.[92]

Under a plea-bargaining agreement, Met Ed pleaded guilty to one count of falsifying records and no contest to six other charges, four of which were dropped, and agreed to pay a $45,000 fine and set up a $1 million account to help with emergency planning in the area surrounding the plant.[*][93]

According to Eric Epstein, chair of Three Mile Island Alert, the TMI plant operator and its insurers paid at least $82 million in publicly documented compensation to residents for "loss of business revenue, evacuation expenses and health claims".[*][94] Also according to Harvey Wasserman, hundreds of out-of-court settlements have been reached with alleged victims of the fallout, with a total of $15 million paid out to parents of children born with birth defects.[*][95] However, a class action lawsuit alleging that the accident caused detrimental health effects was rejected by Harrisburg U.S. District Court Judge Sylvia Rambo. The appeal of the decision to U.S. Third Circuit Court of Appeals also failed.[*][96]

26.3.7   Lessons learned

The Three Mile Island accident inspired Charles Perrow's Normal Accident Theory, in which an accident occurs, resulting from an unanticipated interaction of multiple failures in a complex system. TMI was an example of this type of accident because it was "unexpected, incomprehensible, uncontrollable and unavoidable." [*][97]

> Perrow concluded that the failure at Three Mile Island was a consequence of the system's immense complexity. Such modern high-risk systems, he realized, were prone to failures however well they were managed. It was inevitable that they would eventually suffer what he termed a 'normal accident'. Therefore, he suggested, we might do better to contemplate a radical redesign, or if that was not possible, to abandon such technology entirely.[*][98]

"Normal" accidents, or system accidents, are so-called by Perrow because such accidents are inevitable in extremely complex systems. Given the characteristic of the system involved, multiple failures which interact with each other will occur, despite efforts to avoid them.[*][99] Such events appear trivial to begin with before unpredictably cascading through the system to create a large event with severe consequences.[*][100]

> *Normal Accidents* contributed key concepts to a set of intellectual developments in the 1980s that revolutionized the conception of safety and risk. It made the case for examining technological failures as the product of highly interacting

systems, and highlighted organizational and management factors as the main causes of failures. Technological disasters could no longer be ascribed to isolated equipment malfunction, operator error or acts of God.[*][98]

26.3.8   Comparison to U.S. Navy operations

Following the Three Mile Island (TMI) power plant's partial core melt on March 28, 1979, President Jimmy Carter commissioned a study, *Report of the President's Commission on the Accident at Three Mile Island* (1979).[*][101] Subsequently, Admiral Hyman G. Rickover was asked to testify before Congress in the general context of answering the question as to why naval nuclear propulsion (as used in submarines) had succeeded in achieving a record of zero reactor-accidents (as defined by the uncontrolled release of fission products to the environment resulting from damage to a reactor core) as opposed to the dramatic one that had just taken place at Three Mile Island. In his testimony, he said:

> Over the years, many people have asked me how I run the Naval Reactors Program, so that they might find some benefit for their own work. I am always chagrined at the tendency of people to expect that I have a simple, easy gimmick that makes my program function. Any successful program functions as an integrated whole of many factors. Trying to select one aspect as the key one will not work. Each element depends on all the others.[*][102]

26.4   *The China Syndrome*

The accident at the plant occurred twelve days after the release of the movie *The China Syndrome*. In the film, television reporter Kimberly Wells (Jane Fonda) and her cameraman Richard Adams (Michael Douglas) secretly filmed a major accident at a nuclear power plant while taping a series on nuclear power. Plant supervisor Jack Godell (Jack Lemmon) discovers potentially catastrophic safety violations at the plant and with Wells' assistance attempts to raise public awareness of these violations.

After the release of the film, Fonda began lobbying against nuclear power. In an attempt to counter her efforts, the then elderly Edward Teller, a nuclear physicist and long-time government science adviser best known for contributing to the Teller-Ulam design breakthrough that made "hydrogen bombs" possible, personally lobbied in favor of nuclear power.[*][103] Teller, who suffered a heart attack from the stress of countering the increase in anti-nuclear momentum that followed the film, quipped that he was the only person whose health was affected by the TMI incident.[*][104]

26.5   Current status

Viewed from the east, Three Mile Island currently uses only one nuclear generating station, TMI-1, which is on the right. TMI-2, to the left, has not been used since the accident.

TMI-2 as of February 2014. The cooling towers are on the left. The spent fuel pool with containment building of the reactor are on the right.

Unit 1 had its license temporarily suspended following the incident at Unit 2. Although the citizens of the three counties surrounding the site voted by an overwhelming margin to retire Unit 1 permanently in an unbinding resolution in 1982, it was permitted to resume operations in 1985 following a 4-1 vote by the Nuclear Regulatory Commission.[*][105][*][106] General Public Utilities Corporation, the plant's owner, formed General Public Utilities Nuclear Corporation (GPUN) as a new subsidiary to own and operate the company's nuclear facilities, including Three Mile Island. The plant had previously been operated by Metropolitan Edison Company (Met-Ed), one of GPU's regional utility operating companies. In 1996, General Public Utilities shortened its name to GPU Inc. Three Mile Island Unit

1 was sold to AmerGen Energy Corporation, a joint venture between Philadelphia Electric Company (PECO), and British Energy, in 1998. In 2000, PECO merged with Unicom Corporation to form Exelon Corporation, which acquired British Energy's share of AmerGen in 2003. Today, AmerGen LLC is a fully owned subsidiary of Exelon Generation and owns TMI Unit 1, Oyster Creek Nuclear Generating Station, and Clinton Power Station. These three units, in addition to Exelon's other nuclear units, are operated by Exelon Nuclear Inc., an Exelon subsidiary.

General Public Utilities was legally obliged to continue to maintain and monitor the site, and therefore retained ownership of Unit 2 when Unit 1 was sold to AmerGen in 1998. GPU Inc. was acquired by FirstEnergy Corporation in 2001, and subsequently dissolved. FirstEnergy then contracted out the maintenance and administration of Unit 2 to AmerGen. Unit 2 has been administered by Exelon Nuclear since 2003, when Exelon Nuclear's parent company, Exelon, bought out the remaining shares of AmerGen, inheriting FirstEnergy's maintenance contract. Unit 2 continues to be licensed and regulated by the Nuclear Regulatory Commission in a condition known as Post Defueling Monitored Storage (PDMS).[*][107]

Today, the TMI-2 reactor is permanently shut down with the reactor coolant system drained, the radioactive water decontaminated and evaporated, radioactive waste shipped off-site, reactor fuel and core debris shipped off-site to a Department of Energy facility, and the remainder of the site is being monitored. The owner says it will keep the facility in long-term, monitored storage until the operating license for the TMI-1 plant expires, at which time both plants will be decommissioned.[*][15] In 2009, the NRC granted a license extension which allows the TMI-1 reactor to operate until April 19, 2034.[*][108][*][109]

26.6   Timeline

26.7   See also

- Forked River Nuclear Power Plant
- List of civilian nuclear accidents
- Lists of nuclear disasters and radioactive incidents
- Nuclear reactor accidents in the United States
- Nuclear and radiation accidents and incidents
- Nuclear energy policy of the United States
- Nuclear safety and security
- Nuclear safety in the United States

- Process control

- *Three Mile Island: A Nuclear Crisis in Historical Perspective*

- *Three Mile Island: Thirty Minutes to Meltdown*

26.8 References

[1] "PHMC Historical Markers Search" (Searchable database). *Pennsylvania Historical and Museum Commission*. Commonwealth of Pennsylvania. Retrieved January 25, 2014.

[2] Nuclear Regulatory Commission – Backgrounder on the Three Mile Island Accident

[3] Spiegelberg-Planer, Rejane. "A Matter of Degree: A revised International Nuclear and Radiological Event Scale (INES) extends its reach." . IAEA.org. Retrieved March 19, 2011.

[4] King, Laura; Kenji Hall and Mark Magnier (March 18, 2011). "In Japan, workers struggling to hook up power to Fukushima reactor". Los Angeles Times. Retrieved March 19, 2011.

[5] Minutes to Meltdown: Three Mile Island – National Geographic Archived April 29, 2011, at the Wayback Machine.

[6] Michael Levi on Nuclear Policy, in video "Tea with the Economist", 1:55–2:10, on http://audiovideo.economist.com/, retrieved April 6, 2011, 3.24pm.

[7] Gofman John W., Tamplin, Arthur R. (December 1, 1979). *Poisoned power: the case against nuclear power plants before and after Three Mile Island* (updated edition of Poisoned Power (1971) ed.). Emmaus, PA: Rodale Press. p. xvii. Retrieved October 1, 2013. (In 1979 Foreword:) "...we arrive at 333 fatal cancers or leukemias."

[8] Maureen C. Hatch; et al. (1990). "Cancer near the Three Mile Island Nuclear Plant: Radiation Emissions". *American Journal of Epidemiology*. Oxford Journals. **132** (3): 397–412. PMID 2389745.

[9] Levin, R. J. (2008). "Incidence of thyroid cancer in residents surrounding the three-mile island nuclear facility". *Laryngoscope*. **118** (4): 618–628. doi:10.1097/MLG.0b013e3181613ad2. PMID 18300710. Thyroid cancer incidence has not increased in Dauphin County, the county in which TMI is located. York County demonstrated a trend toward increasing thyroid cancer incidence beginning in 1995, approximately 15 years after the TMI accident. Lancaster County showed a significant increase in thyroid cancer incidence beginning in 1990. These findings, however, do not provide a causal link to the TMI accident.

[10] Levin R. J.; De Simone N. F.; Slotkin J. F.; Henson B. L. (August 2013). "Incidence of thyroid cancer surrounding Three Mile Island nuclear facility: the 30-year follow-up". *Laryngoscope*. **123** (8): 2064–71. doi:10.1002/lary.23953. PMID 23371046.

[11] Han YY; Youk AO; Sasser H; Talbott EO.; Youk, A. O.; Sasser, H; Talbott, E. O. (November 2011). "Cancer incidence among residents of the Three Mile Island accident area: 1982–1995". *Environ Res*. **111** (8): 1230–5. Bibcode:2011ER....111.1230H. doi:10.1016/j.envres.2011.08.005. PMID 21855866. Retrieved October 1, 2013.

[12] Hatch MC, Wallenstein S, Beyea J, Nieves JW, Susser M; Wallenstein; Beyea; Nieves; Susser (June 1991). "Cancer rates after the Three Mile Island nuclear accident and proximity of residence to the plant". *American Journal of Public Health*. **81** (6): 719–724. doi:10.2105/AJPH.81.6.719. PMC 1405170. PMID 2029040.

[13] http://www.uvm.edu/~{}vlrs/Energy/NuclearPower.pdf

[14] "14-Year Cleanup at Three Mile Island Concludes". New York Times. August 15, 1993. Retrieved March 28, 2011.

[15] "Fact Sheet on the Three Mile Island Accident". U.S. Nuclear Regulatory Commission. Retrieved November 25, 2008.

[16] Walker, p. 71.

[17] INPO ICES Report #4810 (Three Mile Island Unit 2) Small Break LOCA Results in Core Damage.

[18] Walker, pp. 72–73.

[19] "A Pump Failure and Claxon Alert". *www.washingtonpost.com*. The Washington Post Company. 1979. Retrieved 4 September 2016. Apparently the valves were closed for routine maintenance, in violation of one of the most stringent rules that the Nuclear Regulatory Commission has. The rule states simply that auxiliary feed pumps can never all be down for maintenance while the reactor is running.

[20] Walker, pp. 73–74.

[21] Norman, Donald (1988). *The Design of Everyday Things*. New York: Basic Books. pp. 43–44. ISBN 978-0-465-06710-7.

[22] Rogovin, pp. 14–15.

[23] Walker, J. Samuel (2004). *Three Mile Island : a nuclear crisis in historical perspective*. Berkeley, Calif. ; London: University of California Press. p. 74. ISBN 9780520239401.

[24] Kemeny, p. 94.

[25] Rogovin, p. 16, Walker, pp. 76–77.

[26] Kemeny, p. 96; Rogovin, pp. 17–18.

[27] Kemeny, p. 96.

[28] Kemeny, p. 99.

[29] Rogovin, p. 19; Walker, p. 78.

[30] Walker, p. 79.

[31] Walker, pp. 80–81.

[32] Walker, pp. 80–84.

[33] Walker, pp. 84–86.

[34] Walker, p. 87.

[35] NRC: Victor Gilinsky

[36] Walker, p. 89.

[37] Walker, pp. 90–91.

[38] Gilinsky, Victor (March 23, 2009). "Behind the scenes of Three Mile Island". *Bulletin of the Atomic Scientists*. Archived from the original on August 15, 2009. Retrieved March 31, 2009.

[39] Rogovin, pp. 25, 153.

[40] http://www.physics.isu.edu/radinf/risk.htm

[41] EPA's Role At Three Mile Island | EPA History | US EPA. Epa.gov. Retrieved on March 17, 2011.

[42] 3. (PDF). Retrieved on March 17, 2011.

[43] Field, RW (June 1993). "137Cs levels in deer following the Three Mile Island accident". *Health Phys.* **64** (6): 671–4. doi:10.1097/00004032-199306000-00015. PMID 8491625.

[44] http://www.threemileisland.org/downloads/210.pdf

[45] Hatch; et al. (1997). "Comments on "A Re-Evaluation of Cancer Incidence Near the Three Mile Island Nuclear Plant". *Environmental Health Perspectives.* **105** (1): 12. doi:10.1289/ehp.9710512. PMC 1469856. PMID 9074862.

[46] Wing, S; Richardson, D; Armstrong, D (March 1997). "Reply to comments on "A reevaluation of cancer incidence near the Three Mile Island". *Environ. Health Perspect.* **105**: 266–8. doi:10.2307/3433255. PMC 1469992. PMID 9171981.

[47] Walker, p. 231.

[48] Sturgis, Sue (2 April 2009). "Investigation: Revelations about Three Mile Island disaster raise doubts over nuclear plant safety". *www.facingsouth.org*. The Institute for Southern Studies. Archived from the original on 2016-08-14. Retrieved 4 September 2016. Arnie Gundersen —a nuclear engineer and former nuclear industry executive turned whistle-blower —has done his own analysis, which he shared for the first time at a symposium in Harrisburg last week. "I think the numbers on the NRC's website are off by a factor of 100 to 1,000," he said.

[49] Kemeny, p. 30.

[50] McEvily, Jr.; A. J.; Le May, I. (2002). "The accident at Three Mile Island". *Materials science research international.* **8** (1): 1–8.

[51] "What Happened and What Didn't in the TMI-2 Accident". American Nuclear Society. Archived from the original on April 2, 2011. Retrieved November 9, 2008.

[52] Mangano, Joseph (September–October 2004). "Three Mile Island: Health Study Meltdown". *Bulletin of the Atomic Scientists.* Metapress. **60** (5): 30–35. doi:10.2968/060005010. ISSN 0096-3402. Retrieved March 31, 2009. (subscription required (help)).

[53] Thompson; Bear (1995). "TMI Assessment (Part 2) - Releases of radiation to the environment" (PDF). *www.facingsouth.org*. The Institute for Southern Studies. Archived from the original (PDF) on 2016-04-16. Retrieved 4 September 2016.

[54] Video: TMI and Community Health – Gundersen 30th anniversary testimony for the Pennsylvania Legislature.

[55] Who We Are | Fairewinds Associates, Inc. Fairewinds.com. Retrieved on March 17, 2011. Archived May 17, 2010, at the Wayback Machine.

[56] Kemeny, John G., (Chairman, 1979), *President's Commission: The Need For Change: The Legacy Of TMI*

[57] *The Washington Post* The Tough Fight to Confine the Damage

[58] A Decade Later, TMI's Legacy Is Mistrust *The Washington Post*, March 28, 1989, p. A01.

[59] Stephanie Cooke (2009). *In Mortal Hands: A Cautionary History of the Nuclear Age*, Black Inc., p. 294.

[60] Susan Cutter and Barnes, Evacuation behavior and Three Mile Island, Disasters, vol. 6, 1982, pp. 116-124.

[61] People & Events: Dick Thornburgh

[62] 1975 estimate.

[63] Office of Technology Assessment. (1984). Public Attitudes Toward Nuclear Power p. 231.

[64] Walker, pp. 209–210

[65] Walker, p. 210.

[66] Walker, pp. 211–212.

[67] Kemeney Commission report to the President Overview, Overall Conclusion, 1st paragraph.

[68] "Three Mile Island, 1979 Year in Review."

[69] Hopkins, A. (2001), "Was Three Mile Island a 'normal accident'?", *Journal of contingencies and crisis management*, vol:9, iss. 2, pp. 65-72.

[70] "Examination of relocated fuel debris adjacent to the lower head of the TMI-2 reactor vessel"

[71] "50 Years of Nuclear Energy" (PDF). IAEA. Retrieved December 29, 2008.

[72] Cancelled Nuclear Units Ordered in the United States

[73] Jon Gertner, Atomic Balm?, *New York Times*, July 16, 2006.

[74] "NRC Approves Vogtle Reactor Construction – First New Nuclear Plant Approval in 34 Years".

[75] Sovacool, Benjamin K. (2008). "The costs of failure: A preliminary assessment of major energy accidents, 1907–2007". *Energy Policy*. **36**: 1807. doi:10.1016/j.enpol.2008.01.040.

[76] Rogovin, pp. 153.

[77] "Three-Mile Island cancer rates probed". *BBC News*. November 1, 2002. Retrieved November 25, 2008.

[78] R. J. Levin (2008), "Incidence of thyroid cancer in residents surrounding the three-mile island nuclear facility", Laryngoscope 118 (4), pp. 618–628 "These findings, however, do not provide a causal link to the TMI accident."

[79] http://www.scribd.com/doc/158526327/Settlement-of-Medical-Claims

[80] Newman, Andy (November 11, 2003). "In Baby Teeth, a Test of Fallout; A Long-Shot Search for Nuclear Peril in Molars and Cuspids". *The New York Times*.

[81] Teather, David (April 13, 2004). "US nuclear industry powers back into life". *The Guardian*. London. Retrieved December 29, 2008.

[82] Harvey Wasserman, *CounterPunch*, March 24, 2009, People Died at Three Mile Island. Retrieved September 2, 2015.

[83] Wing S, Richardson D, Armstrong D, Crawford-Brown D (January 1997). "A reevaluation of cancer incidence near the Three Mile Island nuclear plant: the collision of evidence and assumptions". *Environ Health Perspect*. 105(1). **105** (1): 52–7. doi:10.1289/ehp.9710552. PMC 1469835. PMID 9074881.

[84] Wing S, Richardson DB, Hoffmann W (April 2011). "Cancer risks near nuclear facilities: the importance of research design and explicit study hypotheses". *Environ Health Perspect*. 119(4). **119** (4): 417–21. doi:10.1289/ehp.1002853. PMID 21147606. Retrieved October 1, 2013.

[85] Luther J. Carter "Political Fallout from Three Mile Island", *Science*, 204, April 13, 1979, p. 154.

[86] Mark Hertsgaard (1983). *Nuclear Inc. The Men and Money Behind Nuclear Energy*, Pantheon Books, New York, pp. 95, 97.

[87] Evans, Rowland; Novak, Robert (April 6, 1979). "What Carter Found at Three Mile Island". *Pittsburgh Post-Gazette*. p. 9. Retrieved April 26, 2014.

[88] Interest Group Politics In America p. 149.

[89] Social Protest and Policy Change p. 45.

[90] Herman, Robin (September 24, 1979). "Nearly 200,000 Rally to Protest Nuclear Energy". *New York Times*. p. B1.

[91] Gayle Greene The Woman Who Knew Too Much: Alice Stewart

[92] "Three Mile Island operator falsified tests: jury". *Ottawa Citizen*. November 8, 1983.

[93] "Three Mile Island plant operator faked leak records but". *Ottawa Citizen*. February 29, 1984.

[94] Three Mile Island: 30 years of what if ... *Pittsburgh Tribune Review*, March 22, 2009.

[95] Harvey Wasserman, April 1, 2009, *CounterPunch*, Cracking the Media Silence on Three Mile Island

[96] "Three Mile Island: 1979". World Nuclear Association. Retrieved November 25, 2008.

[97] Perrow, C. (1982), 'The President's Commission and the Normal Accident', in Sils, D., Wolf, C. and Shelanski, V. (Eds), Accident at Three Mile Island: The Human Dimensions, Westview, Boulder, pp. 173–184.

[98] Nick Pidgeon (September 22, 2011). "In retrospect:Normal accidents". *Nature*.

[99] Perrow, Charles. *Normal Accidents: Living with High-Risk Technologies* New York: Basic Books, 1984. p. 5.

[100] Daniel E Whitney (2003). "Normal Accidents by Charles Perrow" (PDF). *Massachusetts Institute of Technology*.

[101] "The Accident at Three Mile Island" (PDF). Threemileisland.org. Retrieved 2014-12-12.

[102] http://www.navy.mil/navydata/testimony/safety/bowman031029.txt

[103] Melvin A Benarde (2007). *Our Precarious Habitat—It's in Your Hands*. Wiley InterScience. p. 256. ISBN 0-471-74065-9.

[104] Wikiquote:Edward Teller

[105] Walsh, Edward (1983). "Three Mile Island: meltdown of democracy?" (PDF). *Bulletin of the Atomic Scientists*.

[106] O'Toole, Thomas (May 30, 1985). "NRC Votes to Restart Three Mile Island". Retrieved 12/15/2016. Check date values in: |access-date= (help)

[107] "NRC: Three Mile Island – Unit 2". Retrieved January 29, 2009.

[108] "Three Mile Island 1 – Pressurized Water Reactor". Nuclear Regulatory Commission. Retrieved December 15, 2008.

[109] DiSavino, Scott (October 22, 2009). "NRC renews Exelon Pa. Three Mile Isl reactor license". Thomson Reuters. Retrieved October 23, 2009.

26.9   Bibliography

- Kemeny, John G. (October 1979). *Report of The President's Commission on the Accident at Three Mile Island: The Need for Change: The Legacy of TMI* (PDF). Washington, D.C.: The Commission. ISBN 0-935758-00-3.

- Rogovin, Mitchell (1980). *Three Mile Island: A report to the Commissioners and to the Public, Volume I* (PDF). Nuclear Regulatory Commission, Special Inquiry Group.

- Walker, J. Samuel (2004). *Three Mile Island: A Nuclear Crisis in Historical Perspective*. Berkeley: University of California Press. ISBN 0-520-23940-7. (Google Books)

- Ford, Daniel (1982). *Three Mile Island: Thirty Minutes to Meltdown*. Penguin. ISBN 978-0-14-006048-5.

- Osif, Bonnie A., Anthony Baratta, Thomas W. Conkling (2004). *TMI 25 Years Later: The Three Mile Island Nuclear Power Plant Accident and Its Impact*. Conkling. ISBN 0-271-02383-X.

- Dick Thornburgh (2010). *Where the Evidence Leads*. University of Pittsburgh Press. ISBN 978-0-8229-6112-3.

26.10   External links

- TMI web page from the US Department of Energy's Energy Information Administration

- "Three Mile Island 1979 Emergency" – website about the accident, with many reports and other documents relating to the accident, created by nearby Dickinson College

- Step-By-Step account of the accident with illustrations from pbs.org

- Three Mile Island Alert, the watchdog group that warned the public for nearly two years that reactor No. 2 was dangerously faulty. What's wrong with the "fact sheet" purports to correct errors in the NRC report.

- EFMR citizens radiation monitoring group for the Three Mile Island and Peach Bottom nuclear plants

- Annotated bibliography for Three Mile Island from the Alsos Digital Library for Nuclear Issues

- Video and audio relating to the Three Mile Island accident, from the Dick Thornburgh Papers at University of Pittsburgh.

- Killing Our Own a review of subsequent casualties by Harvey Wasserman and Norman Solomon with Robert Alveraez and Eleanore Walters

- Three Mile Island – Failure Of Science Or Spin?, Science Daily

- Crisis at Three Mile Island by The Washington Post

- Three Mile Island Research and Document Guide at Penn State University Libraries.

- "Three Mile Island: The Most Studied Nuclear Accident in History" (PDF). *www.gaonet.gov*. U.S. Government Accountability Office. September 9, 1980. OCLC 7975712.

Coordinates: 40°09′12″N 76°43′31″W / 40.153293°N 76.72534°W

Chapter 27

Three Mile Island accident health effects

The **health effects of the 1979 Three Mile Island nuclear accident** are widely, but not universally, agreed to be very low. The American Nuclear Society concluded that average local radiation exposure was equivalent to a chest X-ray, and maximum local exposure equivalent to less than a year's background radiation.[1] The U.S. BEIR report on the Biological Effects of Ionizing Radiation states that "[t]he collective dose equivalent resulting from the radioactivity released in the Three Mile Island accident was so low that the estimated number of excess cancer cases to be expected, if any were to occur, would be negligible and undetectable." [2] A variety of epidemiology studies have concluded that the accident has had no observable long term health effects.[3][4][5] One dissenting study is "A reevaluation of cancer incidence near the Three Mile Island nuclear plant" by Dr. Steven Wing of the University of North Carolina. In this study, Dr. Wing and his colleagues argue that earlier findings had "logical and methodological problems" and conclude that "cancer incidence, specifically lung cancer and leukemia, increased following the TMI accident in areas estimated to have been in the pathway of radioactive plumes than in other areas." [6] Other dissenting opinions can be found in the Radiation and Public Health Project, whose leader, Joseph Mangano, has questioned the safety of nuclear power since 1985.[7][8]

27.1 Initial investigations

In the aftermath of the accident, the investigations focused on the amounts of radioactivity released by the accident. According to the American Nuclear Society, using the official radioactivity emission figures, "The average radiation dose to people living within ten miles of the plant was eight millirem, and no more than 100 millirem to any single individual. Eight millirem is about equal to a chest X-ray, and 100 millirem is about a third of the average background level of radiation received by US residents in a year." .[1][9] To put this dose into context, while the average background radiation in the US is about 360 millirem per year, the Nuclear Regulatory Commission regulates all

workers' of any US nuclear power plant exposure to radiation to a total of 5000 millirem per year.[10] Based on these low emission figures, early scientific publications on the health effects of the fallout estimated one or two additional cancer deaths in the 10-mile area around TMI.[8]

27.1.1 Local resident reports

The official figures are too low to account for the acute health effects reported by some local residents and documented in two books;[11][12] such health effects require exposure to at least 100,000 millirems (100 rems) to the whole body - 1000 times more than the official estimates.[13] The reported health effects are consistent with high doses of radiation, and comparable to the experiences of cancer patients undergoing radio-therapy,.[14] but have many other potential causes.[13] The effects included "metallic taste, erythema, nausea, vomiting, diarrhea, hair loss, deaths of pets and farm and wild animals, and damage to plants." [15] Some local statistics showed dramatic one-year changes among the most vulnerable: "In Dauphin County, where the Three Mile Island plant is located, the 1979 death rate among infants under one year represented a 28 percent increase over that of 1978, and among infants under one month, the death rate increased by 54 percent." [8] Physicist Ernest Sternglass, a specialist in low-level radiation, noted these statistics in the 1981 edition of his book *Secret Fallout: low-level radiation from Hiroshima to Three-Mile Island*. In their final 1981 report, however, the Pennsylvania Department of Health, examining death rates within the 10-mile area around TMI for the 6 months after the accident, said that the TMI-2 accident did not cause local deaths of infants or fetuses.[16][17]

Scientific work continued in the 1980s, but focused heavily on the mental health effects due to stress,[8] as the Kemeny Commission had concluded that this was the sole public health effect.[18] A 1984 survey by a local psychologist of 450 local residents, documenting acute radiation health effects (as well as 19 cancers 1980-84 amongst the residents against an expected 2.6[15]), ultimately led the

TMI Public Health Fund reviewing the data*[19] and supporting a comprehensive epidemiological study by a team at Columbia University.*[14]

27.2 Columbia epidemiological study

In 1990-1 a Columbia University team, led by Maureen Hatch, carried out the first epidemiological study on local death rates before and after the accident, for the period 1975-1985, for the 10-mile area around TMI.*[3]*[18] Assigning fallout impact based on winds on the morning of March 28, 1979,*[3] the study found no link between fallout and cancer risk.*[8] The study found that cancer rates near the Three Mile Island plant peaked in 1982-3, but their mathematical model did not account for the observed increase in cancer rates, since they argued that latency periods for cancer are much longer than three years. From 1975 to 1979 there were 1,722 reported cases of cancer, and between 1981 and 1985 there were 2,831, signifying a 64 percent increase after the meltdown.*[20] The study concludes that stress may have been a factor (though no specific biological mechanism was identified), and speculated that changes in cancer screening were more important.*[18]

27.2.1 Wing review

Subsequently lawyers for 2000 residents asked epidemiologist Stephen Wing of the University of North Carolina at Chapel Hill, a specialist in nuclear radiation exposure, to re-examine the Columbia study. Wing was reluctant to get involved, later writing that "allegations of high radiation doses at TMI were considered by mainstream radiation scientists to be a product of radiation phobia or efforts to extort money from a blameless industry." *[15] Wing later noted that in order to obtain the relevant data, the Columbia study had to submit to what Wing called "a manipulation of research" in the form of a court order which prohibited "upper limit or worst case estimates of releases of radioactivity or population doses... [unless] such estimates would lead to a mathematical projection of less than 0.01 health effects." *[15] Wing found cancer rates raised within a 10-mile radius two years after the accident by 0.034% +/- 0.013%, 0.103% +/- 0.035%, and 0.139% +/- 0.073% for all cancer, lung cancer, and leukemia, respectively.*[6] An exchange of published responses between Wing and the Columbia team followed.*[8] Wing later noted a range of studies showing latency periods for cancer from radiation exposure between 1 and 5 years due to immune system suppression.*[15] Latencies between 1 and 9 years have been studied in a variety of contexts ranging from the Hiroshima

survivors and the fallout from Chernobyl to therapeutic radiation; a 5-10 year latency is most common.*[21]

27.3 Further studies

On the recommendation of the Columbia team, the TMI Public Health Fund followed up its work with a longitudinal study.*[22] The 2000-3 University of Pittsburgh study*[23] compared post-TMI death rates in different parts of the local area, again using the wind direction on the morning of 28 March to assign fallout impact, even though, according to Joseph Mangano in the *Bulletin of the Atomic Scientists*, the areas of lowest fallout by this criterion had the highest mortality rates.*[8] In contrast to the Columbia study, which estimated exposure in 69 areas, the Pittsburgh study drew on the TMI Population Registry, compiled by the Pennsylvania Department of Health. This was based on radiation exposure information on 93% of the population living within five miles of the nuclear plant - nearly 36,000 people, gathered in door-to-door surveys shortly after the accident.*[24] The study found slight increases in cancer and mortality rates but "no consistent evidence" of causation by TMI.*[23] Wing et al. criticized the Pittsburgh study for making the same assumption as Columbia: that the official statistics on low doses of radiation were correct - leading to a study "in which the null hypothesis cannot be rejected due to a priori assumptions." *[25] Hatch et al. noted that their assumption had been backed up by dosimeter data,*[22] though Wing et al. noted the incompleteness of this data, particularly for releases early on.*[26]

In 2005 R. William Field, an epidemiologist at the University of Iowa, who first described radioactive contamination of the wild food chain from the accident suggested that some of the increased cancer rates noted around TMI were related to the area's very high levels of natural radon, noting that according to a 1994 EPA study, the Pennsylvania counties around TMI have the highest regional screening radon concentrations in the 38 states surveyed.*[27] The factor had also been considered by the Pittsburgh study*[23] and by the Columbia team, which had noted that "rates of childhood leukemia in the Three Mile Island area are low compared with national and regional rates." *[3] A 2006 study on the standard mortality rate in children in 34 counties downwind of TMI found an increase in the rate (for cancers other than leukemia) from 0.83 (1979–83) to 1.17 (1984–88), meaning a rise from below the national average to above it.*[21]

A paper in 2008 studying thyroid cancer in the region found rates as expected in the county in which the reactor is located, and significantly higher than expected rates in two neighboring counties beginning in 1990 and 1995 respectively. The research notes that "These findings, however, do not provide a causal link to the TMI accident." *[28] According to Joseph Mangano (who is a member of The Radiation and Public Health Project, an organization with little credibility amongst epidemiologists,*[29]) three large gaps in the literature include: no study has focused on infant mortality data, or on data from outside the 10-mile zone, or on radioisotopes other than iodine, krypton, and xenon.*[8]

27.4 References

[1] "What Happened and What Didn't in the TMI-2 Accident" . American Nuclear Society. Retrieved 2008-11-09.

[2] Committee on the Biological Effects of Ionizing Radiation. "Health effects of exposure to low levels of ionizing radiation (BEIR V)". Washington DC: National Academy Press, 1990.

[3] Maureen C. Hatch; et al. (1990). "Cancer near the Three Mile Island Nuclear Plant: Radiation Emissions". *American Journal of Epidemiology*. Oxford Journals. **132** (3): 397–412. PMID 2389745.

[4] Hatch MC, Wallenstein S, Beyea J, Nieves JW, Susser M; Wallenstein; Beyea; Nieves; Susser (June 1991). "Cancer rates after the Three Mile Island nuclear accident and proximity of residence to the plant" . *American Journal of Public Health*. **81** (6): 719–724. doi:10.2105/AJPH.81.6.719. PMC 1405170⊙. PMID 2029040.

[5] R. J. Levin (2008), "Incidence of thyroid cancer in residents surrounding the three-mile island nuclear facility" , Laryngoscope 118 (4), pp. 618–628 "These findings, however, do not provide a causal link to the TMI accident."

[6] Wing, Steve; Richardson, David; Armstrong, Donna; Crawford, Douglas (January 1997). "A reevaluation of cancer incidence near the Three Mile Island nuclear plant: The collision of evidence and assumptions" (PDF). *Environmental Health Perspectives*. **105** (1): 52–57. doi:10.1289/ehp.9710552. JSTOR 3433062. PMC 1469835⊙. PMID 9074881. Retrieved 2 July 2015.

[7] Newman, Andy (November 11, 2003). "In Baby Teeth, a Test of Fallout; A Long-Shot Search for Nuclear Peril in Molars and Cuspids" . *The New York Times*.

[8] Mangano, Joseph (2004), "Three Mile Island: Health study meltdown" , Bulletin of the atomic scientists, 60(5), pp.31-35

[9] "Three-Mile Island cancer rates probed". *BBC News*. 2002-11-01. Retrieved 2008-11-25.

[10] "NRC: Fact Sheet on Biological Effects of Radiation" . Retrieved 2008-12-29.

[11] Katagiri Mitsuru and Aileen M. Smith (1989), *Three Mile Island: The People's Testament* - based on interviews with 250 residents between 1979 and 1988.

[12] Robert Del Tredici (1982) *The People of Three Mile Island*, Random House, Inc.

[13] GPU Utilities 91986), "Radiation And Health Effects: A Report on the TMI-2 Accident and Related Health Studies"

[14] Sue Sturgis (2009) "FOOLING WITH DISASTER? Startling revelations about Three Mile Island disaster raise doubts over nuclear plant safety", *Facing South: Online Magazine of the Institute for Southern Studies*, April 2009

[15] Wing, Steven (2003) "Objectivity and Ethics in Environmental Health Science", *Environmental Health Perspectives* Volume 111, Number 14, November 2003

[16] "Report doubts infant death rise from three mile island mishap" . *The New York Times*. 1981-03-21. Retrieved 2008-12-29.

[17] Walker, p234

[18] Hatch MC, Wallenstein S, Beyea J, Nieves JW, Susser M (June 1991). "Cancer rates after the Three Mile Island nuclear accident and proximity of residence to the plant" . *American Journal of Public Health*. **81** (6): 719–724. doi:10.2105/AJPH.81.6.719. PMC 1405170⊙. PMID 2029040.

[19] Moholdt B. Summary of acute symptoms by TMI area residents during accident. In:Proceedings of the workshop on Three Mile Island dosimetry. Philadelphia:Academy of Natural Sciences, 1985.

[20] Mangano, Joseph. "Three Mile Island: Health Study Meltdown" . *Bulletin of the Atomic Scientists*. Retrieved 1 May 2014.

[21] Mangano, Joseph (2006), "A short latency between radiation exposure from nuclear plants and cancer in young children" , *International journal of health services*, vol:36 iss:1 pg:113-135

[22] Hatch; et al. (1997). "Comments on "A Re-Evaluation of Cancer Incidence Near the Three Mile Island Nuclear Plant" . *Environmental Health Perspectives*. **105** (1): 1. doi:10.1289/ehp.9710512. PMC 1469856⊙. PMID 9074862.

[23] Talbott Evelyn O.; et al. (2000). "Mortality Among the Residents of the Three Mile Accident Area: 1979–1992" . *Environmental Health Perspectives*. **108** (6): 545–52. doi:10.1289/ehp.00108545.; Talbott Evelyn O.; et al. (2003). "Long-Term Follow-up of the Residents of the Three Mile Island Accident" . *Environmental Health Perspectives*. **111** (3): 341–48. doi:10.1289/ehp.5662.

[24] Holzman (2003), "Cancer and three mile island" , *Environmental health perspectives*, vol:111 iss:3 pg:A166 -A167

[25] Wing, S. and Richardson, D. (2000), "Collision of Evidence and Assumptions: TMI Déjà View" , *Environmental Health Perspectives*, 108(12), December 2000

[26] Wing, S.; Richardson, D.; Armstrong, D. (1997). "Reply to Comments on "A Reevaluation of Cancer Incidence near the Three Mile Island"". *Environmental Health Perspectives*. **105** (3): 266–268. doi:10.2307/3433255. JSTOR 3433255. PMC 1469992⊝. PMID 9171981.

[27] R. W. Field (2005), "Three Mile Island epidemiologic radiation dose assessment revisited: 25 years after the accident", Radiation Protection Dosimetry 113 (2), pp. 214-217

[28] RJ Levin (2008), "Incidence of thyroid cancer in residents surrounding the three mile island nuclear facility", Laryngoscope 118 (4), pp. 618-628

[29] Newman, Andy (2003-11-11). "In Baby Teeth, a Test of Fallout; A Long-Shot Search for Nuclear Peril in Molars and Cuspids". *The New York Times*.

Chapter 28

Church Rock uranium mill spill

The **Church Rock uranium mill spill** occurred in the US state of New Mexico on July 16, 1979, when United Nuclear Corporation's Church Rock uranium mill tailings disposal pond breached its dam.[1][2] Over 1,000 tons of solid radioactive mill waste and 93 million gallons of acidic, radioactive tailings solution flowed into the Puerco River, and contaminants traveled 80 miles (130 km) downstream to Navajo County, Arizona and onto the Navajo Nation.[2] The mill was located on privately owned land approximately 17 miles north of Gallup, New Mexico, and bordered to the north and southwest by Navajo Nation Tribal Trust lands.[3] Local residents, who were mostly Navajos, used the Puerco River for irrigation and livestock and were not immediately aware of the toxic danger.[2]

The accident is frequently described as having released more radioactivity than the Three Mile Island accident that occurred four months earlier and was the largest release of radioactive material in U.S. history.[2][4][5][6] The spill contaminated groundwater and rendered the Puerco unusable by local residents. The governor of New Mexico refused the Navajo Nation's request that the site be declared a federal disaster area, limiting aid to affected residents.[7] The event received less media coverage than that of Three Mile Island, likely because it occurred in a lightly populated, rural area.[8] Some scholars suggest there were elements of class and racism to the neglect as well, since it affected primarily poor Native Americans.[5]

In 2003 the Churchrock Chapter of the Navajo Nation began the Church Rock Uranium Monitoring Project to assess environmental impacts of abandoned uranium mines; it found significant radiation from both natural and mining sources in the area.[9] The EPA National Priorities List currently includes the Church Rock tailings storage site, where "groundwater migration is not under control." [10]

28.1 Dam failure

At around 5:30 am on July 16, 1979, a 20-foot breach formed in the south cell of United Nuclear Corporation's Church Rock uranium mill tailings disposal pond, and 1,100 tons of solid radioactive mill waste and approximately 93 million US gallons (350,000 m^3) of acidic, radioactive tailings solution flowed into Pipeline Arroyo, a tributary of the Puerco River.[2][5][11][12] Though the uranium mill only bordered the Navajo Nation, the spilled tailings spilled onto the Navajo Nation as they flowed down the Puerco River.[2]

The 20-foot breach in the tailings dam formed around 5:30 am on the morning of July 16, 1979.[2]

The tailings solution had a pH of 1.2[13] and a gross alpha particle activity of 128 nanocuries (4.7 kBq) per liter, and contained radioactive uranium, thorium, radium, polonium, metals including cadmium, aluminum, magnesium, manganese, molybdenum, nickel, selenium, sodium, vanadium, zinc, iron, lead, and high concentrations of sulfates.[14] The contaminated water from the Church Rock spill traveled 80 miles (130 km) downstream, through Gallup, New Mexico, and reached as far as Navajo County, Arizona. The flood backed up sewers, affected nearby aquifers, and left stagnating, contaminated pools on the riverside.[5][15][16]

28.1.1   Response

28.2   Causes

At 6:00 am, a United Nuclear Corporation employee noticed the breach and suspended further discharge of tailings solution to the holding pond.[5] By 8:00, a temporary dike had stopped the flow of residual tailings solution.[5]

The Indian Health Service and the Environmental Improvement Division of New Mexico warned local residents over the radio and with signs not to drink from, water livestock at, or enter the Puerco River, but many Navajos in the area didn't speak English or couldn't read and were unaware of the dangers of radiation.[17][18] United Nuclear Corporation employees were dispatched to warn Navajo-speaking residents downstream in accordance with a state contingency plan, but not until a few days after the spill.[4][19] The Navajo Nation asked the governor of New Mexico to request disaster assistance from the US government and have the site declared a disaster area, but he refused, limiting disaster relief assistance to the Navajo Nation.[5]

The New Mexico Environmental Improvement Division said the spill's "short-term and long-term impacts on people and the environment were quite limited."[20] United Nuclear denied claims that the spill caused livestock deaths. The company said in a statement issued by an attorney, "We just don't know of any substance to those claims. Some people aren't going to be satisfied no matter how thoroughly you show it."[20]

Under the "agreement state" legislative framework of the Uranium Mill Tailings Radiation Control Act, the Nuclear Regulatory Commission left New Mexico to handle the dam failure until October 12, 1979, when it was notified that the state would permit the uranium mill to resume operation that week. The NRC then suspended United Nuclear's operating license until it could be determined that the embankment was stable.[21] After fewer than four months of downtime following the dam failure, the mill resumed operations on November 2, 1979. This further contaminated the groundwater and resulted in the mill site's placement on the EPA's National Priorities List in 1983.[5][11] United Nuclear made a $525,000 out-of-court settlement with the Navajo Nation a year after the spill.[22]

In terms of the amount of radioactivity released, the accident was larger in magnitude than the Three Mile Island accident of the same year.[2][4][5] The spill has been called "the largest radioactive accident in U.S. history," but the Nuclear Regulatory Commission has said that this is "an overstatement," and that "there have been a number of other events that have been more significant in terms of radiological impact. The event was more significant from an environmental perspective than from a human one."[1]

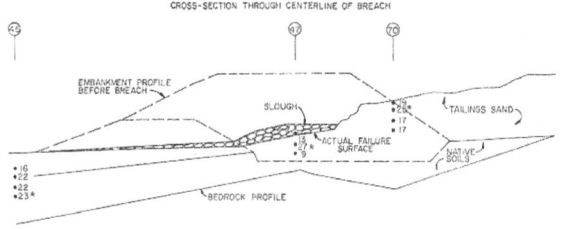

A diagrammed cross section of the breach from the report commissioned by the NRC. The "point" in the bedrock that UNC said acted as a fulcrum in the dam's breach is visible beneath the embankment.

The dam formed the southern wall of one of the mill's three holding ponds, which were used to evaporate tailings solution until the remaining solid waste could be buried.[23] During its operation from 1967 to 1982, the mill produced 3.5 million tons of tailings at the rate of 4,000 tons a day.[3][24][25] The 35-foot-high embankment was constructed on a deposit of collapsible clayey, silty sand a hundred feet deep.[13] United Nuclear used a new design, recommended by the Nuclear Regulatory Commission, that used earth rather than tailings themselves as building material.[17] The holding pond was not lined, in violation of standards in the Uranium Mill Tailings Radiation Control Act of 1978 that required that holding ponds be impermeable.[17] This lack of lining allowed tailings solution to seep into the ground, weakening the foundation of the dam and contaminating the groundwater.[13]

Horizontal and vertical cracks formed along the southern part of the embankment, allowing the acidic tailings solution to penetrate and weaken the embankment.[13] A sand beach was constructed to protect the face of the embankment from the tailings solution, but it was not properly maintained. Prior to the collapse, the level of the tailings solution in the holding pond had risen two feet higher than the dam's designed limit, and the sand beach no longer provided the dam protection.[13][26] The United States Army Corps of Engineers concluded in its report to Governor Bruce King of New Mexico that the principal cause of failure was differential settlement of the foundation beneath the dam wall,[27] and the report commissioned by the Nuclear Regulatory Commission corroborated this conclusion.[13] Critical variations in tailings pond operation practice from approved procedures contributed further to the dam failure.[17][27] United Nuclear's Chief Operating Officer, J. David Hann, blamed the failure of dam on the pointed shape of the bedrock beneath the embankment, which he said acted as a fulcrum and weakened the dam.[4]

Cracks were first noted in the dam wall by independent consultants in December 1977 and were sealed with bentonite and kerosene slurry in February 1978.[13] United Nuclear was notified of the cracking, but, aside from the initial seal, took little or no preventative action.[17] United Nuclear did not make regular inspections of the dam despite strong recommendations by independent consultants that it do so.[17] Further cracking was noted in October 1978. Neither the facility owner nor the State Engineer were formally notified of the episodes of cracking prior to the dam breach, though Arizona representative Morris K. Udall testified before Congress that at least three federal and state agencies had "ample opportunity" to predict that the dam's failure was likely.[27][28] At the same Congressional hearing, the United States Army Corps of Engineers testified that United Nuclear ignored warnings from the Corps that the dam was structurally unsound.[23]

Both New Mexico and Arizona were governed under the Uranium Mill Tailings Radiation Control Act as "agreement states," meaning that they, not the Nuclear Regulatory Commission, were responsible for ensuring compliance with NRC standards.[21] However, the states lacked the equipment and personnel to properly oversee tailings disposal sites statewide, and the ambiguity and drafting errors present in the act left them confused about how much regulatory power they had.[21][29]

28.3 Effects

A sign placed by the New Mexico Environmental Improvement Division discouraging use of the Puerco River.

Shortly after the breach, below the dam radioactivity levels of river water were 7000 times that of the allowable level of drinking water.[30] United Nuclear initially claimed that only one curie of radioactivity had been released in the spill, but that figure was later revised upward by the New Mexico Environmental Improvement Division.[31] In all, 46 curies (1.7 TBq) of radioactivity were released.[32]

The Puerco River in flood. After the tailings dam failure, 93 million gallons of acidic, radioactive tailings solution flowed into the Puerco River.[2]

Prior to the spill, local residents used the riverside for recreation and herb-gathering, and children often waded in the Puerco River.[5] Residents who waded in the river after the spill went to the hospital complaining of burning feet and were misdiagnosed with heat stroke.[15] Burns acquired by some of those who came into contact with the contaminated water developed serious infections and required amputations.[22] Herds of sheep and cattle died after drinking the contaminated water, and children played in pools of contaminated water.[23][33] The spill contaminated shallow aquifers near the river that residents drank and used to water livestock.[34] 1,700 people lost access to clean water after the spill.[23] United Nuclear Corporation distributed 600 gallon-jugs of clean water, but the affected area required more than 30,000 gallons of water daily.[2] The three community wells serving Church Rock had already been closed, one because of high radium levels and the other two for high levels of iron and bacteria.[35] The Indian Health Service advised the tribe to repair five shallow wells along the Puerco River and said that the wells "are not expected to show any contamination, if at all, for several years." [2] The Navajo Nation spent $100,000 on clean water,[36] and in 1981, the New Mexico and federal governments stopped providing water, which they had delivered by truck since the spill.[37]

An epidemiological study conducted by the NMEID in 1989 concluded that "the health risk to the public from eating exposed cattle is minimal, unless large amounts of this tissue, especially liver and kidney, are ingested." [38] An Indian Health Service study found significantly higher levels of radionuclides in Church Rock cattle compared to livestock from non-mining areas. The study's authors advised that contamination would not pose a risk as long as residents did not depend on livestock for food over long periods of time, but local Navajos did.[39] A few Navajo children were sent to Los Alamos to be checked for radiation exposure, but no long-term monitoring was undertaken, prompting a local writer to comment that the IHS spent more effort studying livestock than the people affected.[39] No ongoing epidemiological studies have been done at Church Rock.[5][40] Studies have shown since the 1950s that the Navajo have had significantly higher rates for some cancers than the national average, associated with contamination from the uranium mines and the exposure of workers to radiation.[41][42]

28.4   Cleanup

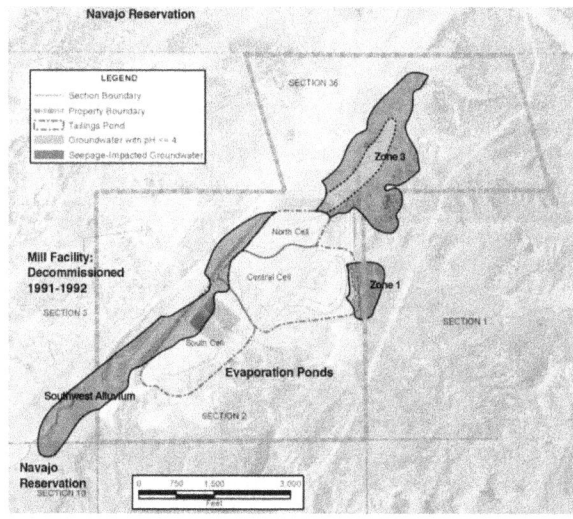

2009 map of remedial efforts at the United Nuclear Corporation site. The 1979 spill resulted from a breach of the South Cell dam.[3]

United Nuclear dispatched small crews with shovels and 55-gallon drums to begin cleanup, but expanded the workforce after complaints from local residents and pressure from the state.[26] The crews removed three inches of sediment from the river bed,[43] retrieving about 3,500 barrels (560 m^3) of waste materials over the course of three months, but this amount was estimated as only 1% of the solid waste spilled.[5] Groundwater remained contaminated by the spilled tailings solution, and rain transported leftover pollutants downstream into Arizona.[15][26] New Mexico ordered United Nuclear to monitor pools left behind by the spill along the Puerco River, but United Nuclear only measured uranium levels, ignoring the presence of ^{230}Th and ^{226}Ra.[26] The pools contained high levels of sulfuric acid and remained for more than a month after the spill, despite cleanup efforts by the New Mexico Environmental Improvement Division.[31] The NMEID ordered United Nuclear to control tailings seepage from the mill in 1979, and the company implemented a limited seepage collective program in 1981.[24]

The Navajo Nation appealed to the governor to request that the president declare the site a federal disaster area, but he refused, reducing the aid available to local residents.[5] United Nuclear continued operation of the uranium mill until 1982, when it closed because of the declining uranium market.[44]

United Nuclear neutralized the acidity of the tailings with ammonia and lime from 1979 to 1982.[45] In 1983, the site was entered on the National Priorities List of the Environmental Protection Agency's Superfund investigations and cleanup efforts, as radionuclides and chemical constituents were found to be contaminating local groundwater.[11] The EPA conducted a remedial investigation from 1984 to 1987, and in the NRC approved United Nuclear's closure and reclamation plan in 1988.[45]

Removal of radium-contaminated soil in Church Rock in 2007.

In 1994 the EPA extended its efforts with a study of all known uranium mines on the Navajo Nation.[46][47]

The EPA and United Nuclear removed 175,500 cubic feet (4,970 m^3) of radium-contaminated soil surrounding five buildings, some residential, in 2007.[25] The soil was moved to an off-site disposal facility.

In 2003 the Churchrock Chapter of the Navajo Nation began the Church Rock Uranium Monitoring Project to assess environmental impacts of abandoned uranium mines, and build capacity to conduct community-based research with policy implications.[9] Its May 2007 report found significant radiation remaining in the area, from both natural and mining sources.[9]

In 2008, the US Congress authorized a five-year plan for cleanup of contaminated uranium sites on the Navajo reservation.[48] The non-profit Groundswell Educational Films is hosting online videos of the cleanup effort.

28.5   See also

- Uranium mining and the Navajo people

- *The Return of Navajo Boy*

- *The Navajo People and Uranium Mining*

- Sequoyah Fuels Corporation

- Environmental racism

28.6 References

[1] "Navajos mark 20th anniversary of Church Rock spill", *The Daily Courier*, Prescott, Arizona, July 18, 1999

[2] Pasternak, Judy (2010). *Yellow Dirt: A Poisoned Land and a People Betrayed*. Free Press. p. 149. ISBN 1416594825.

[3] *United Nuclear Corporation (McKinley County)* (PDF), EPA, November 21, 2012

[4] US Congress, House Committee on Interior and Insular Affairs, Subcommittee on Energy and the Environment. *Mill Tailings Dam Break at Church Rock, New Mexico*, 96th Cong, 1st Sess (October 22, 1979):19–24.

[5] Brugge, D.; DeLemos, J.L.; Bui, C. (2007), "The Sequoyah Corporation Fuels Release and the Church Rock Spill: Unpublicized Nuclear Releases in American Indian Communities", *American Journal of Public Health*, **97** (9): 1595–600, doi:10.2105/ajph.2006.103044, PMC 1963288, PMID 17666688

[6] Quinones, Manuel (December 13, 2011), "As Cold War abuses linger, Navajo Nation faces new mining push", *E&E News*, retrieved December 28, 2012

[7] Pasternak 2010, p. 150.

[8] Dingmann, Tracy (July 16, 2009), "New attention to Church Rock uranium spill comes 30 years later", The New Mexico Independent

[9] Shuey, Chris; et al. (May 2007), *Report of the Church Rock Uranium Monitoring Project 2003-2007, Churchrock Chapter, Navajo Nation, Southwest Research and Information Center and Navajo Education and Scholarship Foundation* (PDF), Window Rock, Navajo Nation (Arizona), USA

[10] "EPA Superfund Program: UNITED NUCLEAR CORP., CHURCH ROCK, NM". EPA. Retrieved 2016-04-26.

[11] *Second Five-Year Review Report for the United Nuclear Corporation. Ground Water Operable Unit* (PDF), EPA, September 2003, archived from the original (PDF) on 2011-05-31

[12] *Church Rock tailings spill*, The Energy Library, retrieved December 9, 2012

[13] Nelson, John D.; Kane, Joseph D. (1980). *The Failure of the Church Rock Tailings Dam*. Nuclear Regulatory Commission. Wikisource.

[14] Rangel, Valerie (2010). "Church Rock Tailings Spill: July 16, 1979". New Mexico Office of the State Historian. Retrieved December 9, 2012.

[15] Giusti, Brendan (July 16, 2009), "Radiation Spill in Church Rock Still Haunts 30 Years Later", The Daily Times, Farmington, New Mexico

[16] Szasz, Ferenc Morton (2006). *'Larger Than Life: New Mexico in the Twentieth Century*. UNM Press. pp. 82–83. ISBN 0-8263-3883-6.

[17] Young, Lise (1981), "What Price Progress? Uranium Production on Indian Lands in the San Juan Basin", *American Indian Law Review*, **9** (1): 1–50, doi:10.2307/20068184, JSTOR 20068184

[18] Brugge, Doug; Benally, Timothy; Yazzie-Lewis, Esther, eds. (2006). *The Navajo People and Uranium Mining*. University of New Mexico. ISBN 978-0826337795.

[19] Kathie Saltzstein, "Navajos Ask $12.5 Million in UNC Suits," *Gallup Independent*, August 14, 1980

[20] "Uranium Spill Still Worries Navajos", *The New York Times*, July 21, 1983, retrieved December 9, 2012

[21] Grammer, Elisa J. (1981), "The Uranium Mill Tailings Radiation Control Act of 1978 and NRC's Agreement State Program", *Natural Resources Lawyer*, **13** (3): 469–522, JSTOR 40922651

[22] Kuletz, Valerie (1998). *The Tainted Desert*. New York: Routledge. p. 27.

[23] Mantonya, Kurt T. (January 1, 1999), *Contamination Nation*, University of Nebraska-Lincoln, p. 96

[24] *Administrative Order in the matter of United Nuclear Corporation* (PDF), EPA, June 29, 1989

[25] *Engineering Evaluation/Cost Analysis: Northeast Church Rock (NECR) Mine Site, Gallup, New Mexico*, San Francisco: EPA Region 9, May 30, 2009, p. 6

[26] Wasserman, Harry and Norman Solomon, *Killing Our Own: The Disaster of America's Experience with Atomic Radiation*, New York: Dell Publishing Co, 1980.

[27] Roth, Colonel Bernard J (9 October 1979). *Review Comments and Recommendations to Geotechnical Investigational Reports - Church Rock Tailings Dam*. Albuquerque, New Mexico: Army Corps of Engineers.

[28] Ward, Sinclair (October 23, 1979), "Uranium Spill Described as Preventable", *The Washington Post*, p. A5, retrieved December 9, 2012

[29] Robinson, Paul; Hector, Alice; Luis, Judy; Benavides, David; Hancock, Don (1979), "Uranium Mining and Milling: A Primer" (PDF), *The Workbook*, Albuquerque, New Mexico: Southwest Research & Information Center, **4** (6-7), retrieved December 9, 2012

[30] Bruce E. Johansen "The High Cost of Uranium in Nava-joland", *Akwesasne Notes New Series*, Spring - April May June - 1997, Volume 2 #2, pp. 10-12.

[31] Robertson, Bill (October 4, 1979), "Pediatrician says spill underplayed" (PDF), *New Mexico Daily Lobo*

[32] "Comparison of the Three Mile Island, Chernobyl, Sequoyah Fuels Corporation, and Church Rock Releases," *American Journal of Public Health*, 2007, 97(9) pp. 1595-1600.

[33] Brown, Jovana J.; Lambert, Lori (2010), *Blowing in the Wind: The Navajo Nation and Uranium*, The Evergreen State College

[34] *Technical Report on Technologically Enhanced Naturally Occurring Radioactive Materials from Uranium Mining Volume 2: Investigation of Potential Health, Geographic, And Environmental Issues of Abandoned Uranium Mines* (PDF), Washington, DC: EPA, Office of Radiation and Indoor Air Radiation Protection Division, April 2008, pp. Appendix IV, p. 7

[35] Pasternak 2010, p. 149-150.

[36] Nelkin, Dorothy (1981), "Native Americans and Nuclear Power", *Science, Technology & Human Values*, **6** (2): 2–13, doi:10.1177/016224398100600201, JSTOR 689554

[37] Chris Shuey, MPH "The Puerco River: Where Did the Water Go?", *Southwest Research and Information Center*, 1986

[38] Lapham, SC, JB Millard, and JM Samet. "Health implications of radionuclide levels in cattle raised near U mining and milling facilities in Ambrosia Lake, New Mexico. *Health Physics Journal*, 1989, 56(3) pp. 327-40.

[39] Pasternak 2010, p. 151.

[40] Gunter, Linda (September 2009), "Remembering the Forgotten Nuclear Accident", Z Magazine, pp. 6–7

[41] Chris Shuey, MPH Uranium Exposure and Public Health in New Mexico and the Navajo Nation: A Literature Summary *Southwest Research and Information Center*, 02.27.07, rev.10.14.08

[42] Pinderhughes, Raquel (1996), "The Impact of Race on Environmental Quality: An Empirical and Theoretical Discussion", *Sociological Perspectives*, **39** (2): 231–48, doi:10.2307/1389310, JSTOR 1389310

[43] Gault, Ramona (September 13, 1989), "Navajos inherity a legacy of radiation", *In These Times*, Chicago

[44] "UNC Resources Plans to Shut Last Uranium Mine Operation", The Washington Post, May 4, 1982

[45] *EPA Superfund Record of Decision: United Nuclear Corp.* (PDF), Church Rock, New Mexico: EPA, September 30, 1988

[46] "Abandoned Uranium Mines On The Navajo Nation", EPA Region 9: Superfund, retrieved July 30, 2010

[47] "Addressing Uranium Contamination in the Navajo Nation", Superfund - Region 9, EPA

[48] Felicia Fonseca, "Navajo woman helps prompt uranium mine cleanup", Associated Press, carried in *Houston Chronicle*, 5 September 2011, accessed 5 October 2011

28.7 Further reading

- Jamail, Dahr (20 June 2009). "Destroying Indigenous Populations". *truth-out.org*. Retrieved 27 June 2010.

- Christopher McLeod "Four Corners: A National Sacrifice Area?" *bullfrogfilms.com*

28.8 External links

- Media related to Church Rock Uranium Mill spill at Wikimedia Commons

- The Energy Library: "United Nuclear Corporation Superfund Site"

- *The Return of Navajo Boy* Webisodes: Clean-up of uranium contamination at Navajo Reservation

Chapter 29

Chernobyl disaster

This article is about the 1986 nuclear plant accident in Ukraine. For other uses, see Chernobyl (disambiguation).

Deceased liquidators' portraits used for an anti-nuclear power protest in Geneva.

The **Chernobyl disaster**, also referred to as the **Chernobyl accident**, was a catastrophic nuclear accident that occurred on 26 April 1986 in the No.4 light water graphite moderated reactor at the Chernobyl Nuclear Power Plant near Pripyat, in what was then part of the Ukrainian Soviet Socialist Republic of the Soviet Union (USSR).

During a hurried late night power-failure stress test, in which safety systems were deliberately turned off, a combination of inherent reactor design flaws, together with the reactor operators arranging the core in a manner contrary to the checklist for the stress test, eventually resulted in uncontrolled reaction conditions that flashed water into steam generating a destructive steam explosion and a subsequent open-air graphite "fire".*[note 1] This "fire" produced considerable updrafts for about 9 days, that lofted plumes of fission products into the atmosphere, with the estimated radioactive inventory that was released during this very hot "fire" phase, approximately equal in magnitude to the airborne fission products released in the initial destructive explosion.*[1] Practically all of this radioactive material would then go on to fall-out/precipitate onto much of the surface of the western USSR and Europe.

The Chernobyl accident dominates the Energy accidents sub-category, of most disastrous nuclear power plant accident in history, both in terms of cost and casualties. It is one of only two nuclear energy accidents classified as a level 7 event (the maximum classification) on the International Nuclear Event Scale, the other being the Fukushima Daiichi nuclear disaster in Japan in 2011.*[2] The struggle to safeguard against scenarios which were, at many times falsely,*[1] perceived as having the potential for greater catastrophe and the later decontamination efforts of the surroundings, ultimately involved over 500,000 workers and cost an estimated 18 billion rubles.*[3] During the accident, blast effects caused 2 deaths within the facility and later 29 firemen and employees died in the days-to-months afterward from acute radiation syndrome, with the potential for long-term cancers still being investigated.*[4]

The remains of the No.4 reactor building were enclosed in a large sarcophagus (radiation shield) by December 1986, at a time when what was left of the reactor was entering the cold shut-down phase; the enclosure was built quickly as occupational safety for the crews of the other undamaged reactors at the power station, with No.3 continuing to produce electricity into 2000.*[5]*[6]

The accident motivated safety upgrades on all remaining Soviet-designed reactors in the RBMK (Chernobyl No.4) family, of which eleven continued to power electric grids as of 2013.*[7]*[8]

29.1   Overview

The disaster began during a systems test on 26 April 1986 at reactor number four of the Chernobyl plant, which is near Pripyat and in proximity to the administrative border with Belarus and the Dnieper River. There was a sudden and unexpected power surge, and when an emergency shutdown was attempted, a much larger spike in power output occurred, which led to a reactor vessel rupture and a series of steam explosions. These events exposed the graphite moderator of the reactor to air, causing it to ignite.*[9] The

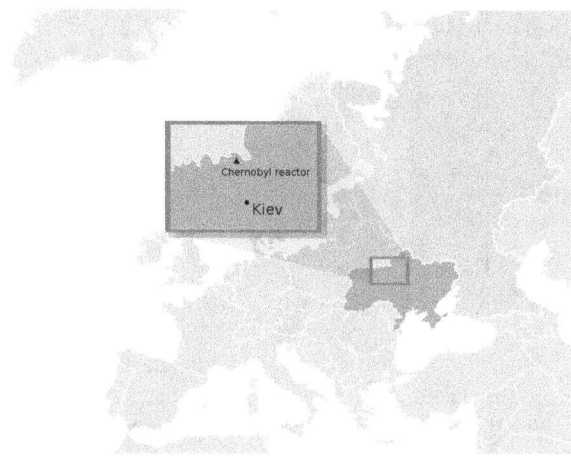

Location of Chernobyl nuclear power plant

The abandoned city of Pripyat with Chernobyl plant in the distance

resulting fire sent week long plumes of highly radioactive fallout into the atmosphere over an extensive geographical area, including Pripyat. The plumes drifted over large parts of the western Soviet Union and Europe. According to official post-Soviet data,[*][10][*][11] about 60% of the fallout landed in Belarus.

36 hours after the accident Soviet officials enacted a 10-kilometre exclusion zone which resulted in the rapid evacuation of 49,000 people and their animals, primarily from the largest population centre near the reactor, the town of Pripyat.[*][12] Although not communicated at the time, an immediate evacuation of the town following the accident was not advisable as the road leading out of the town had heavy nuclear fallout "hotspots" deposited on it, while up to around that point, the town itself was comparatively safe due to the more favourable wind direction, permitting shelter in place to be the best safety measure to take for the town, before the winds began to change direction.[*][12][*][13]

As the plumes and subsequent fallout continued to be generated, the evacuation zone was increased from 10 to 30 km about one week after the accident, resulting in a further 68,000 evacuated, including from the town of Chernobyl itself.[*][12] The surveying and detection of isolated fallout hotspots outside this zone over the following year eventually resulted in 135,000 "long-term evacuees" in total, accepting to be moved.[*][12] The near tripling in the total number of permanently resettled to some 350,000 over the decades, 1986 to 2000, from the most severely "contaminated" areas,[*][14][*][15] is regarded as largely political in nature, with the majority of the rest evacuated in an effort to redeem loss in trust in the government, which was most common around 1990.[*][16] Many thousands of these evacuees would have been "better off staying home" .[*][17] Risk analysis, supported by DNA biomarkers, has determined that the "people still living unofficially in the abandoned lands around Chernobyl" have a lower risk of dying as a result of the elevated doses of radiation in the rural areas than "if they were exposed to the air pollution health risk in a large city such as nearby Kiev" .[*][18][*][19]

Russia, Ukraine, and Belarus have been burdened with the continuing and substantial decontamination and monthly compensation costs,[*][16][*][17][*][20] of the Chernobyl accident. Although certain initiatives are legitimate, as the director of the UN Development Program Kalman Mizsei noted, "an industry has been built on this unfortunate event," with a "vast interest in creating a false picture." [*][17][*][21]

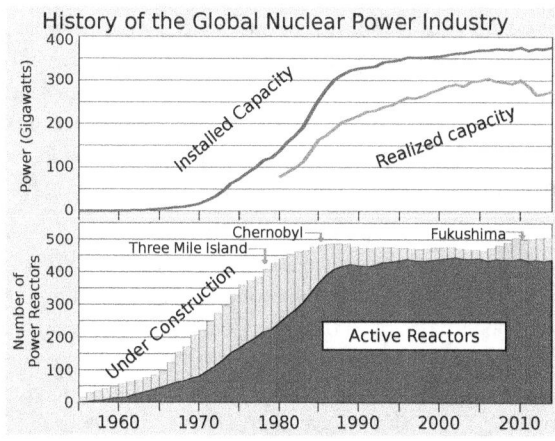

The rate of new construction builds for civilian fission-electric reactors dropped in the late 1980s, with the effects of accidents having a chilling effect. The World Association of Nuclear Operators was formed as a direct result of the accident with the aim of creating a greater exchange of information on safety and on techniques to increase the capacity of energy production.

The accident raised the already heightened concerns about fission reactors worldwide, and while most concern was focused on those of the same unusual design, hundreds of disparate electric-power reactor proposals, including those under construction at Chernobyl, reactor No.5 and 6, were eventually cancelled. With the worldwide issue generally being due to the ballooning in costs for new nuclear reactor safety system standards and the legal costs in dealing with the increasingly hostile/anxious public opinion. There was a precipitous drop in the prior rate of new "startups", after 1986.*[22]

The accident also raised concerns about the cavalier safety culture in the Soviet nuclear power industry, slowing industry growth and forcing the Soviet government to become less secretive about its procedures.*[23]*[notes 1] The government coverup of the Chernobyl disaster was a "catalyst" for glasnost, which "paved the way for reforms leading to the Soviet collapse".*[24]

29.1.1 Cancer projections

A report by the International Atomic Energy Agency examines the environmental consequences of the accident.*[11] Another UN agency, UNSCEAR, has estimated a global collective dose of radiation exposure from the accident "equivalent on average to 21 additional days of world exposure to natural background radiation"; individual doses were far higher than the global mean among those most exposed, including 530,000 primarily male recovery workers (the "Liquidators") who averaged an effective dose equivalent to an extra 50 years of typical natural background radiation exposure each.*[25]*[26]*[27] Estimates of the number of deaths that will eventually result from the accident vary enormously; disparities reflect both the lack of solid scientific data and the different methodologies used to quantify mortality—whether the discussion is confined to specific geographical areas or extends worldwide, and whether the deaths are immediate, short term, or long term.

Thirty-one deaths are directly attributed to the accident, all among the reactor staff and emergency workers.*[28] An UNSCEAR report places the total confirmed deaths from radiation at 64 as of 2008. The Chernobyl Forum predicts the eventual death toll could reach 4,000 among those exposed to the *highest levels of radiation* (200,000 emergency workers, 116,000 evacuees and 270,000 residents of the most contaminated areas); this figure is a total causal death toll prediction, combining the deaths of approximately 50 emergency workers who died soon after the accident from acute radiation syndrome, nine children who have died of thyroid cancer and a future predicted total of 3940 deaths from radiation-induced cancer and leukaemia.*[29]

In a peer-reviewed paper in the *International Journal of Cancer* in 2006, the authors expanded the discussion on those exposed to all of Europe (but following a different conclusion methodology to the Chernobyl Forum study,

which arrived at the total predicted death toll of 4,000 after cancer survival rates were factored in) they stated, without entering into a discussion on deaths, that in terms of total excess cancers attributed to the accident:*[30]

> The risk projections suggest that by now [2006] Chernobyl may have caused about 1000 cases of thyroid cancer and 4000 cases of other cancers in Europe, representing about 0.01% of all incident cancers since the accident. Models predict that by 2065 about 16,000 cases of thyroid cancer and 25,000 cases of other cancers may be expected due to radiation from the accident, whereas several hundred million cancer cases are expected from other causes.

In terms of non peer-reviewed estimates, based on a simple linear no-threshold model multiplication, that the International Commission on Radiological Protection (ICRP) states "should not be done",*[31] two anti-nuclear advocacy groups have publicised mortality estimates for those who were exposed to even smaller amounts of radiation. The Union of Concerned Scientists estimates that, among the hundreds of millions of people exposed worldwide, excluding thyroid cancer there will be an eventual 50,000 excess cancer cases resulting in 25,000 excess cancer deaths.*[32]

Along similar lines, the 2006 TORCH report, commissioned by the European Greens political party, predicts an eventual 30,000 to 60,000 excess cancer deaths in total, around the globe.*[33]

The Russian founder of that region's chapter of Greenpeace, authored a book titled *Chernobyl: Consequences of the Catastrophe for People and the Environment*, which suggests that among the billions of people worldwide who were exposed to radioactive contamination from the disaster, nearly a million premature cancer deaths occurred between 1986 and 2004.*[34] Greenpeace itself advocates a figure of at least 200,000 or more.*[35] The book was not peer reviewed prior to publication,*[36]*[37] and has been heavily criticized; of the five reviews published in the academic press, four considered the book severely flawed and contradictory, and one praised it while noting some shortcomings. The review by M. I. Balonov published by the New York Academy of Sciences concludes that the report is of negative value because it has very little scientific merit while being highly misleading to the lay reader. It characterized the estimate of nearly a million deaths as more in the realm of science fiction than science.*[38]

29.2   Accident

On 26 April 1986, at 01:23 (UTC+3), reactor four suffered a catastrophic power increase, leading to explosions in its core. This dispersed large quantities of radioactive isotopes into the atmosphere[*][39][*]:73 and caused an open-air fire. The fire increased the emission of radioactive particles, carried by the smoke, as the reactor had not been encased by any kind of hard containment vessel. The accident occurred during an experiment scheduled to test a potential safety emergency core cooling feature, which took place during a normal shutdown procedure.

29.2.1   Steam turbine tests

In steady state operation, a significant fraction (over 6%) of the power from a nuclear reactor is derived not from fission but from the decay heat of its accumulated fission products. This heat continues for some time after the chain reaction is stopped (e.g., following an emergency SCRAM) and active cooling may be required to prevent core damage.[*][40] RBMK reactors like those at Chernobyl use water as a coolant.[*][41][*][42] Reactor 4 at Chernobyl consisted of about 1,600 individual fuel channels; each required a coolant flow of 28 metric tons (28,000 litres or 7,400 US gallons) per hour.[*][39]

Since cooling pumps require electricity to cool a reactor after a SCRAM, in the event of a power grid failure, Chernobyl's reactors had three backup diesel generators; these could start up in 15 seconds, but took 60–75 seconds[*][39][*]:15 to attain full speed and reach the 5.5-megawatt (MW) output required to run one main pump.[*][39][*]:30

To solve this one-minute gap – considered an unacceptable safety risk – it had been theorized that rotational energy from the steam turbine (as it wound down under residual steam pressure) could be used to generate the required electrical power. Analysis indicated that this residual momentum and steam pressure might be sufficient to run the coolant pumps for 45 seconds,[*][39][*]:16 bridging the gap between an external power failure and the full availability of the emergency generators.[*][43]

This capability still needed to be confirmed experimentally, and previous tests had ended unsuccessfully. An initial test carried out in 1982 indicated that the excitation voltage of the turbine-generator was insufficient; it did not maintain the desired magnetic field after the turbine trip. The system was modified, and the test was repeated in 1984 but again proved unsuccessful. In 1985, the tests were attempted a third time but also yielded negative results. The test procedure would be repeated again in 1986, and it was scheduled to take place during the maintenance shutdown of Reactor Four.[*][43]

The test focused on the switching sequences of the electrical supplies for the reactor. The test procedure was expected to begin with an automatic emergency shutdown. No detrimental effect on the safety of the reactor was anticipated, so the test programme was not formally coordinated with either the chief designer of the reactor (NIKIET) or the scientific manager. Instead, it was approved only by the director of the plant (and even this approval was not consistent with established procedures).[*][44]

According to the test parameters, the thermal output of the reactor should have been *no lower* than 700 MW at the start of the experiment. If test conditions had been as planned, the procedure would almost certainly have been carried out safely; the eventual disaster resulted from attempts to boost the reactor output once the experiment had been started, which was inconsistent with approved procedure.[*][44]

The Chernobyl power plant had been in operation for two years without the capability to ride through the first 60–75 seconds of a total loss of electric power, and thus lacked an important safety feature. The station managers presumably wished to correct this at the first opportunity, which may explain why they continued the test even when serious problems arose, and why the requisite approval for the test had not been sought from the Soviet nuclear oversight regulator (even though there was a representative at the complex of 4 reactors).[*][notes 2][*]:18–20

The experimental procedure was intended to run as follows:

1. The reactor was to be running at a low power level, between 700 MW and 800 MW.

2. The steam-turbine generator was to be run up to full speed.

3. When these conditions were achieved, the steam supply for the turbine generator was to be closed off.

4. Turbine generator performance was to be recorded to determine whether it could provide the bridging power for coolant pumps until the emergency diesel generators were sequenced to start and provide power to the cooling pumps automatically.

5. After the emergency generators reached normal operating speed and voltage, the turbine generator would be allowed to continue to freewheel down.

29.2.2   Conditions before the accident

The conditions to run the test were established before the day shift of 25 April 1986. The day-shift workers had been instructed in advance and were familiar with the established procedures. A special team of electrical engineers was present to test the new voltage regulating system.[45] As planned, a gradual reduction in the output of the power unit was begun at 01:06 on 25 April, and the power level had reached 50% of its nominal 3200 MW thermal level by the beginning of the day shift.

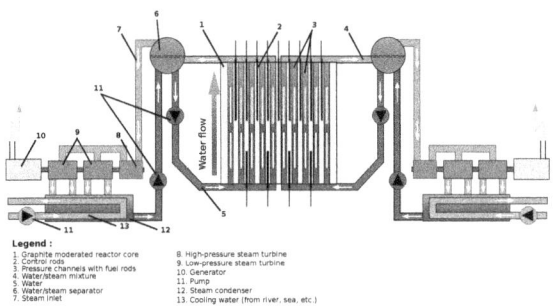

Legend :
1. Graphite moderated reactor core
2. Control rods
3. Pressure channels with fuel rods
4. Water/steam mixture
5. Water
6. Water/steam separator
7. Steam inlet
8. High-pressure steam turbine
9. Low-pressure steam turbine
10. Generator
11. Pump
12. Steam condenser
13. Cooling water (from river, sea, etc.)

A schematic diagram of the reactor

At this point, another regional power station unexpectedly went offline, and the Kiev electrical grid controller requested that the further reduction of Chernobyl's output be postponed, as power was needed to satisfy the peak evening demand. The Chernobyl plant director agreed, and postponed the test. Despite this delay, preparations for the test not affecting the reactor's power were carried out, including the disabling of the emergency core cooling system or ECCS, a passive/active system of core cooling intended to provide water to the core in a loss-of-coolant accident. Given the other events that unfolded, the system would have been of limited use, but its disabling as a "routine" step of the test is an illustration of the inherent lack of attention to safety for this test.[46] In addition, had the reactor been shut down for the day as planned, it is possible that more preparation would have been taken in advance of the test.

At 23:04, the Kiev grid controller allowed the reactor shutdown to resume. This delay had some serious consequences: the day shift had long since departed, the evening shift was also preparing to leave, and the night shift would not take over until midnight, well into the job. According to plan, the test should have been finished during the day shift, and the night shift would only have had to maintain decay heat cooling systems in an otherwise shut-down plant.[39]:36–38

The night shift had very limited time to prepare for and carry out the experiment. A further rapid decrease in the power level from 50% was executed during the shift change-over. Alexander Akimov was chief of the night shift, and Leonid Toptunov was the operator responsible for the reactor's operational regimen, including the movement of the control rods. Toptunov was a young engineer who had worked independently as a senior engineer for approxi-

mately three months.[39]:36–38

The test plan called for a gradual decrease in power output from reactor 4 to a thermal level of 700–1000 MW.[47] An output of 700 MW was reached at 00:05 on 26 April. Due to the reactor's production of a fission byproduct, xenon-135, which is a reaction-inhibiting neutron absorber, core power continued to decrease without further operator action—a process known as reactor poisoning. This continuing decrease in power occurred because in *steady state operation*, xenon-135 is "burned off" as quickly as it is created from decaying iodine-135 by absorbing neutrons from the ongoing chain reaction to become highly stable xenon-136. When the reactor power was lowered, previously produced high quantities of iodine-135 decayed into the neutron-absorbing xenon-135 faster than the reduced neutron flux could burn it off. As the reactor power output dropped further, to approximately 500 MW, Toptunov mistakenly inserted the control rods too far—the exact circumstances leading to this are unknown because Akimov and Toptunov both died in the hospital on 10 and 14 May respectively. This combination of factors put the reactor into an unintended near-shutdown state, with a power output of 30 MW thermal or less.

The reactor was now producing 5 percent of the minimum initial power level established as safe for the test.[44]:73 Control-room personnel decided to restore power by disabling the automatic system governing the control rods and manually extracting the majority of the reactor control rods to their upper limits.[48] Several minutes elapsed between their extraction and the point that the power output began to increase and subsequently stabilize at 160–200 MW (thermal), a much smaller value than the planned 700 MW. The rapid reduction in the power during the initial shutdown, and the subsequent operation at a level of less than 200 MW led to increased poisoning of the reactor core by the accumulation of xenon-135.[49][50] This restricted any further rise of reactor power, and made it necessary to extract additional control rods from the reactor core in order to counteract the poisoning.

The operation of the reactor at the low power level and high poisoning level was accompanied by unstable core temperature and coolant flow, and possibly by instability of neutron flux, which triggered alarms. The control room received repeated emergency signals regarding the levels in the steam/water separator drums, and large excursions or variations in the flow rate of feed water, as well as from relief valves opened to relieve excess steam into a turbine condenser, and from the neutron power controller. Between 00:35 and 00:45, emergency alarm signals concerning thermal-hydraulic parameters were ignored, apparently to preserve the reactor power level.[51]

When the power level of 200 MW was achieved, preparation for the experiment continued. As part of the test plan, extra water pumps were activated at 01:05 on 26 April, increasing the water flow. The increased coolant flow rate through the reactor produced an increase in the inlet coolant temperature of the reactor core (the coolant no longer having sufficient time to release its heat in the turbine and cooling towers), which now more closely approached the nucleate boiling temperature of water, reducing the safety margin.

The flow exceeded the allowed limit at 01:19, triggering an alarm of low steam pressure in the steam separators. At the same time, the extra water flow lowered the overall core temperature and reduced the existing steam voids in the core and the steam separators.[*][52] Since water weakly absorbs neutrons (and the higher density of liquid water makes it a better absorber than steam), turning on additional pumps decreased the reactor power further still. The crew responded by turning off two of the circulation pumps to reduce feedwater flow, in an effort to increase steam pressure, and also to remove more manual control rods to maintain power.[*][46][*][53]

All these actions led to an extremely unstable reactor configuration. Nearly all of the control rods were removed manually, including all but 18 of the "fail-safe" manually operated rods of the minimal 28 which were intended to remain fully inserted to control the reactor even in the event of a loss of coolant, out of a total 211 control rods.[*][54] While the emergency SCRAM system that would insert all control rods to shut down the reactor could still be activated manually (through the "AZ-5" button), the automated system that could do the same had been disabled to maintain the power level, and many other automated and even passive safety features of the reactor had been bypassed. Further, the reactor coolant pumping had been reduced, which had limited margin so any power excursion would produce boiling, thereby reducing neutron absorption by the water. The reactor was in an unstable configuration that was outside the safe operating envelope established by the designers. If anything pushed it into supercriticality, it was unable to recover automatically.

29.2.3 Experiment and explosion

At 1:23:04 a.m., the experiment began. Four of the main circulating pumps (MCP) were active; of the eight total, six are normally active during regular operation. The steam to the turbines was shut off, beginning a run-down of the turbine generator. The diesel generators started and sequentially picked up loads; the generators were to have completely picked up the MCPs' power needs by 01:23:43. In the interim, the power for the MCPs was to be supplied by the turbine generator as it coasted down. As the momentum

Radioactive steam plumes continued to be generated days after the initial explosion, as evidenced here on 3 May 1986 due to decay heat. The roof of the turbine hall is damaged (image centre). Roof of the adjacent reactor 3 (image lower left) shows minor fire damage.

of the turbine generator decreased, so did the power it produced for the pumps. The water flow rate decreased, leading to increased formation of steam voids (bubbles) in the core.

Because of the positive void coefficient of the RBMK reactor at low reactor power levels, it was now primed to embark on a positive feedback loop, in which the formation of steam voids reduced the ability of the liquid water coolant to absorb neutrons, which in turn increased the reactor's power output. This caused yet more water to flash into steam, giving a further power increase. During almost the entire period of the experiment the automatic control system successfully counteracted this positive feedback, inserting control rods into the reactor core to limit the power rise. This system had control of only 12 rods, and nearly all others had been manually retracted.

At 1:23:40, as recorded by the SKALA centralized control system, an emergency shutdown (SCRAM) of the reactor was initiated. The SCRAM was started when the EPS-5 button (also known as the AZ-5 button) of the reactor emergency protection system was pressed: this engaged the drive mechanism on all control rods to fully insert them, including the manual control rods that had been withdrawn earlier. The reason why the EPS-5 button was pressed is not known, whether it was done as an emergency measure in response to rising temperatures, or simply as a routine method of shutting down the reactor upon completion of the experiment.

There is a view that the SCRAM may have been ordered as a response to the unexpected rapid power increase, although there is no recorded data proving this. Some have suggested that the button was not pressed, and instead the signal was automatically produced by the emergency protection system; the SKALA registered a manual SCRAM signal. In spite of this, the question as to when or even whether the EPS-5 button was pressed has been the subject of debate. There are assertions that the pressure was caused by the rapid power acceleration at the start, and allegations that the button was not pressed until the reactor began to self-destruct but others assert that it happened earlier and in calm conditions.[*][55][*]:578[*][56]

After the EPS-5 button was pressed, the insertion of control rods into the reactor core began. The control rod insertion mechanism moved the rods at 0.4 m/s, so that the rods took 18 to 20 seconds to travel the full height of the core, about 7 metres. A bigger problem was the design of the RBMK control rods, which had graphite neutron moderators attached to boost reactor output when the rod was withdrawn. Those displacers had a 1.25 m column of water above and below them when the rods were at maximum extration, and lowering the rods displaced the neutron-absorbing water in the lower portion of the reactor with moderating graphite. As a result, the SCRAM increased the reaction rate in the lower part of the core as the graphite displaced the coolant. This behaviour was known after the initial insertion of control rods in another RBMK reactor at Ignalina Nuclear Power Plant in 1983 induced a power spike, but as the subsequent SCRAM of that reactor was successful, the information was disseminated but deemed of little importance.

A few seconds after the start of the SCRAM, a power spike occurred, and the core overheated, causing some of the fuel rods to fracture, blocking the control rod columns and jamming the control rods at one-third insertion, with the graphite displacers still in the lower part of the core. Within three seconds the reactor output rose above 530 MW.[*][39][*]:31

The subsequent course of events was not registered by instruments; it is known only as a result of mathematical simulation. Apparently, the power spike caused an increase in fuel temperature and steam buildup, leading to a rapid increase in steam pressure. This caused the fuel cladding to fail, releasing the fuel elements into the coolant, and rupturing the channels in which these elements were located.[*][57]

Then, according to some estimations, the reactor jumped to around 30,000 MW thermal, ten times the normal operational output. The last reading on the control panel was 33,000 MW. It was not possible to reconstruct the precise sequence of the processes that led to the destruction of the reactor and the power unit building, but a steam explosion, like the explosion of a steam boiler from excess vapour pressure, appears to have been the next event. There is a general understanding that it was explosive steam pressure from the damaged fuel channels escaping into the reactor's exterior cooling structure that caused the detonation that destroyed the reactor casing, tearing off and blasting the 2000-ton upper plate, to which the entire reactor assembly is fastened, through the roof of the reactor building. This is believed to be the first explosion that many heard.[*][58][*]:366 This explosion ruptured further fuel channels, as well as severing most of the coolant lines feeding the reactor chamber, and as a result the remaining coolant flashed to steam and escaped the reactor core. The total water loss in combination with a high positive void coefficient further increased the reactor's thermal power.

A second, more powerful explosion occurred about two or three seconds after the first; this explosion dispersed the damaged core and effectively terminated the nuclear chain reaction. This explosion also compromised more of the reactor containment vessel and ejected hot lumps of graphite moderator. The ejected graphite and the demolished channels still in the remains of the reactor vessel caught fire on exposure to air, greatly contributing to the spread of radioactive fallout and the contamination of outlying areas.[*][46]

According to observers outside Unit 4, burning lumps of material and sparks shot into the air above the reactor. Some of them fell onto the roof of the machine hall and started a fire. About 25 percent of the red-hot graphite blocks and overheated material from the fuel channels was ejected. Parts of the graphite blocks and fuel channels were out of the reactor building. As a result of the damage to the building an airflow through the core was established by the high temperature of the core. The air ignited the hot graphite and started a graphite fire.[*][39][*]:32

After the larger explosion a number of employees at the power station went outside to get a clearer view of the extent of the damage, one such survivor, Alexander Yuvchenko recounts that once he stopped outside and looked up towards the reactor hall he saw a "very beautiful" LASER-like beam of light bluish light, caused by the ionization of air, that appeared to "flood up into infinity".[*][59][*][60][*][61]

There were initially several hypotheses about the nature of the second explosion. One view was that the second explosion was caused by hydrogen, which had been produced either by the overheated steam-zirconium reaction or by the reaction of red-hot graphite with steam that produced hydrogen and carbon monoxide. Another hypothesis was that the second explosion was a thermal explosion of the reactor as a result of the uncontrollable escape of fast neutrons caused by the complete water loss in the reactor core.[*][62] A third hypothesis was that the explosion was a second steam explosion. According to this version, the first explosion was a more minor steam explosion in the circulating loop, causing a loss of coolant flow and pressure, that in turn caused the water still in the core to flash to steam. This second explosion then did the majority of the damage to the reactor and containment building.

The force of the second explosion, and the ratio of xenon radioisotopes released during the event, indicate that the second explosion could have been a nuclear power transient; the result of the melting core material, in the absence of its cladding, water coolant and moderator, undergoing runaway prompt criticality similar to the explosion of a fizzled nuclear weapon.[*][63] This nuclear excursion released 40 billion joules of energy, the equivalent of about ten tons of TNT. The analysis indicates that the nuclear excursion was limited to a small portion of the core.[*][63]

Contrary to safety regulations, bitumen, a combustible material, had been used in the construction of the roof of the reactor building and the turbine hall. Ejected material ignited at least five fires on the roof of the adjacent reactor 3, which was still operating. It was imperative to put those fires out and protect the cooling systems of reactor 3.[*][39][*]:42 Inside reactor 3, the chief of the night shift, Yuri Bagdasarov, wanted to shut down the reactor immediately, but chief engineer Nikolai Fomin would not allow this. The operators were given respirators and potassium iodide tablets and told to continue working. At 05:00, Bagdasarov made his own decision to shut down the reactor, leaving only those operators there who had to work the emergency cooling systems.[*][39][*]:44

Radiation levels

Approximate radiation intensity levels at different locations at Chernobyl reactor site shortly after the explosion are shown in the below table.[*][64] A dose of 500 roentgens (~5 Sv) delivered over five hours is usually lethal for human beings.

Plant layout

Based on the image of the plant[*][65]

Individual involvement

Main article: Individual involvement in the Chernobyl disaster

29.2.4   Immediate crisis management

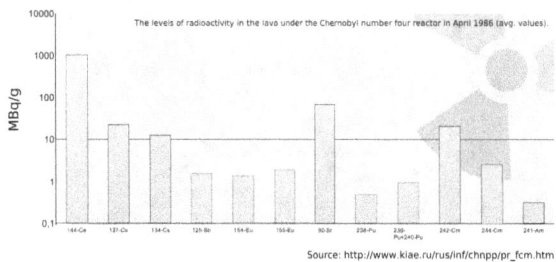

Extremely high levels of radioactivity in the lava under the Chernobyl number four reactor in 1986

Radiation levels

The radiation levels in the worst-hit areas of the reactor building have been estimated to be 5.6 roentgens per second (R/s), equivalent to more than 20,000 roentgens per hour. A lethal dose is around 500 roentgens (~5 Gy) over 5 hours, so in some areas, unprotected workers received fatal doses in less than a minute. However, a dosimeter capable of measuring up to 1000 R/s was buried in the rubble of a collapsed part of the building, and another one failed when turned on. All remaining dosimeters had limits of 0.001 R/s and therefore read "off scale". Thus, the reactor crew could ascertain only that the radiation levels were somewhere above 0.001 R/s (3.6 R/h), while the true levels were much higher in some areas.[*][39][*]:42–50

Because of the inaccurate low readings, the reactor crew chief Alexander Akimov assumed that the reactor was intact. The evidence of pieces of graphite and reactor fuel lying around the building was ignored, and the readings of another dosimeter brought in by 04:30 were dismissed under the assumption that the new dosimeter must have been defective.[*][39][*]:42–50 Akimov stayed with his crew in the reactor building until morning, sending members of his crew to try to pump water into the reactor. None of them wore any protective gear. Most, including Akimov, died from radiation exposure within three weeks.[*][54][*]:247–48

Fire containment

Shortly after the accident, firefighters arrived to try to extinguish the fires. First on the scene was a Chernobyl Power Station firefighter brigade under the command of Lieutenant Volodymyr Pravik, who died on 9 May 1986 of acute radiation sickness. They were not told how dangerously radioactive the smoke and the debris were, and may not even

Firefighter Leonid Telyatnikov, being decorated for bravery

have known that the accident was anything more than a regular electrical fire: "We didn't know it was the reactor. No one had told us." *[66]

Grigorii Khmel, the driver of one of the fire engines, later described what happened:

> We arrived there at 10 or 15 minutes to two in the morning.... We saw graphite scattered about. Misha asked: "Is that graphite?" I kicked it away. But one of the fighters on the other truck picked it up. "It's hot," he said. The pieces of graphite were of different sizes, some big, some small, enough to pick them up...
>
> We didn't know much about radiation. Even those who worked there had no idea. There was no water left in the trucks. Misha filled a cistern and we aimed the water at the top. Then those boys who died went up to the roof – Vashchik, Kolya and others, and Volodya Pravik.... They went up the ladder ... and I never saw them again.*[67]*:54

Anatoli Zakharov, a fireman stationed in Chernobyl since 1980, offers a different description in 2008:

> I remember joking to the others, "There must be an incredible amount of radiation here. We'll be lucky if we're all still alive in the morning." *[68]

He also said:

> Of course we knew! If we'd followed regulations, we would never have gone near the reactor.

But it was a moral obligation – our duty. We were like kamikaze.*[68]

The immediate priority was to extinguish fires on the roof of the station and the area around the building containing Reactor No. 4 to protect No. 3 and keep its core cooling systems intact. The fires were extinguished by 5:00, but many firefighters received high doses of radiation. The fire inside reactor 4 continued to burn until 10 May 1986; it is possible that well over half of the graphite burned out.*[39]*:73

The fire was extinguished by a combined effort of helicopters dropping over 5000 metric tons of sand, lead, clay, and neutron-absorbing boron onto the burning reactor and injection of liquid nitrogen. The Ukrainian filmmaker Vladimir Shevchenko captured film footage of an Mi-8 helicopter as its main rotor collided with a nearby construction crane cable, causing the helicopter to fall near the damaged reactor building and killing its four-man crew.*[69] It is now known that virtually none of the neutron absorbers reached the core.*[70]

From eyewitness accounts of the firefighters involved before they died (as reported on the CBC television series *Witness*), one described his experience of the radiation as "tasting like metal", and feeling a sensation similar to that of pins and needles all over his face. (This is similar to the description given by Louis Slotin, a Manhattan Project physicist who died days after a fatal radiation overdose from a criticality accident.)*[71]

The explosion and fire threw hot particles of the nuclear fuel and also far more dangerous fission products, radioactive isotopes such as caesium-137, iodine-131, strontium-90 and other radionuclides, into the air: the residents of the surrounding area observed the radioactive cloud on the night of the explosion.

Equipment assembled included remote-controlled bulldozers and robot-carts that could detect radioactivity and carry hot debris. Valery Legasov (first deputy director of the Kurchatov Institute of Atomic Energy in Moscow) said, in 1987: "But we learned that robots are not the great remedy for everything. Where there was very high radiation, the robot ceased to be a robot—the electronics quit working." *[72]

Timeline

- 1:26:03 – fire alarm activated

- 1:28 – arrival of local firefighters, Pravik's guard

- 1:35 – arrival of firefighters from Pripyat, Kibenok's guard

- 1:40 – arrival of Telyatnikov

- 2:10 – turbine hall roof fire extinguished

- 2:30 – main reactor hall roof fires suppressed

- 3:30 – arrival of Kiev firefighters[*][73]

- 4:50 – fires mostly localized

- 6:35 – all fires extinguished[*][‡][*][74]

[*]‡With the exception of the fire contained inside Reactor 4, which continued to burn for many days.[*][39][*]:73

Announcement and evacuation

The view of Chernobyl Nuclear Power Plant taken from the city of Pripyat

The nearby city of Pripyat was not immediately evacuated. The townspeople went about their usual business, completely oblivious to what had just happened. However, within a few hours of the explosion, dozens of people fell ill. Later, they reported severe headaches and metallic tastes in their mouths, along with uncontrollable fits of coughing and vomiting.[*][75]

As the plant was run by authorities in Moscow, the government of Ukraine did not receive prompt information on the accident.[*][76] Valentyna Shevchenko, then Chairman of the Presidium of Verkhovna Rada Supreme Soviet of the Ukrainian SSR, recalls that Ukraine's acting Minister of Internal Affairs Vasyl Durdynets phoned her at work at 9 am to report current affairs; only at the end of the conversation did he add that there had been a fire at the Chernobyl nuclear power plant, but it was extinguished and everything was fine. When Shevchenko asked "How are the people?", he replied that there was nothing to be concerned about: "Some are celebrating a wedding, others are gardening, and others are fishing in the Pripyat River".[*][76] Shevchenko then spoke over the phone to Volodymyr Shcherbytsky, Head of the Central Committee of the CPU and de facto head of state, who said he anticipated a delegation of the state commission headed by the deputy chairman of the Council of Ministers of USSR.[*][76]

A commission was set up the same day (26 April) to investigate the accident. It was headed by Valery Legasov, First Deputy Director of the Kurchatov Institute of Atomic Energy, and included leading nuclear specialist Evgeny Velikhov, hydro-meteorologist Yuri Izrael, radiologist Leonid Ilyin and others. They flew to Boryspil International Airport and arrived at the power plant in the evening of 26 April.[*][76] By that time two people had already died and 52 were hospitalized. The delegation soon had ample evidence that the reactor was destroyed and extremely high levels of radiation had caused a number of cases of radiation exposure. In the early hours of 27 April, over 24 hours after the initial blast, they ordered the evacuation of Pripyat. Initially it was decided to evacuate the population for three days; later this was made permanent.[*][76]

By 11:00 on 27 April, buses had arrived in Pripyat to start the evacuation.[*][76] The evacuation began at 14:00. A translated excerpt of the evacuation announcement follows:[*][77]

> For the attention of the residents of Pripyat! The City Council informs you that due to the accident at Chernobyl Power Station in the city of Pripyat the radioactive conditions in the vicinity are deteriorating. The Communist Party, its officials and the armed forces are taking necessary steps to combat this. Nevertheless, with the view to keep people as safe and healthy as possible, the children being top priority, we need to temporarily evacuate the citizens in the nearest towns of Kiev region. For these reasons, starting from April 27, 1986 2 pm each apartment block will be able to have a bus at its disposal, supervised by the police and the city officials. It is highly advisable to take your documents, some vital personal belongings and a certain amount of food, just in case, with you. The senior executives of public and industrial facilities of the city has decided on the list of employees needed to stay in Pripyat to maintain these facilities in a good working order. All the houses will be guarded by the police during the evacuation period. Comrades, leaving your residences temporarily please make sure you have turned off the lights, electrical equipment and water and shut the windows. Please keep calm and orderly in the process of this short-term evacuation.
>
> —Evacuation announcement in Pripyat, 27 April 1986 (14:00)

To expedite the evacuation, residents were told to bring only what was necessary, and that it would only last approximately three days. As a result, most personal belongings were left behind, and remain there today. By 15:00, 53,000 people were evacuated to various villages of the Kiev region.[*][76] The next day, talks began for evacuating people from the 10 km zone.[*][76] Ten days after the accident, the evacuation area was expanded to 30 km (19 mi).[*][78][*]:115,120–1 This "exclusion zone" has remained ever since, although its shape has changed and its size has been expanded.

These evacuations actually had some economic benefit, moving people to areas of labour shortage in Belarus and Ukraine.[*][78][*]:90

Evacuation began long before the accident was publicly known throughout the Union. Only on 28 April, after radiation levels set off alarms at the Forsmark Nuclear Power Plant in Sweden,[*][79] over 1,000 kilometres (620 mi) from the Chernobyl Plant, did the Soviet Union publicly admit that an accident had occurred. At 21:02 that evening a 20-second announcement was read in the TV news programme *Vremya*:[*][80][*][81]

> There has been an accident at the Chernobyl Nuclear Power Plant. One of the nuclear reactors was damaged. The effects of the accident are being remedied. Assistance has been provided for any affected people. An investigative commission has been set up.
> —*Vremya*, 28 April 1986 (21:00)[*][80]

This was the entirety of the announcement of the accident. The Telegraph Agency of the Soviet Union (TASS) then discussed Three Mile Island and other American nuclear accidents, an example of the common Soviet tactic of emphasizing foreign disasters when one occurred in the Soviet Union. The mention of a commission, however, indicated to observers the seriousness of the incident,[*][82] and subsequent state radio broadcasts were replaced with classical music, which was a common method of preparing the public for an announcement of a tragedy.[*][81]

Around the same time, ABC News released its report about the disaster.[*][83]

Shevchenko was the first of the Ukrainian state top officials to arrive at the disaster site early on 28 April. There she spoke with members of medical staff and people, who were calm and hopeful that they could soon return to their homes. Shevchenko returned home near midnight, stopping at a radiological checkpoint in Vilcha, one of the first that were set up soon after the accident.[*][76]

There was a notification from Moscow that there was no reason to postpone the 1 May International Workers' Day celebrations in Kiev (including the annual parade), but on 30 April a meeting of the Political bureau of the Central Committee of CPU took place to discuss the plan for the upcoming celebration. Scientists were reporting that the radiological background in Kiev city was normal. At the meeting, which was finished at 18:00, it was decided to shorten celebrations from the regular 3.5–4 to under 2 hours.[*][76]

Several buildings in Pripyat were officially kept open after the disaster to be used by workers still involved with the plant. These included the Jupiter Factory which closed in 1996 and the Azure Swimming Pool which closed in 1998.

Steam explosion risk

Two floors of bubbler pools beneath the reactor served as a large water reservoir for the emergency cooling pumps and as a pressure suppression system capable of condensing steam in case of a small broken steam pipe; the third floor above them, below the reactor, served as a steam tunnel. The steam released by a broken pipe was supposed to enter the steam tunnel and be led into the pools to bubble through a layer of water. After the disaster, the pools and the basement were flooded because of ruptured cooling water pipes and accumulated firefighting water, and constituted a serious steam explosion risk.

Chernobyl corium lava flows formed by fuel-containing mass in the basement of the plant[*][84]

The smoldering graphite, fuel and other material above, at more than 1200 °C,[*][85] started to burn through the reactor floor and mixed with molten concrete from the reactor lining, creating corium, a radioactive semi-liquid material comparable to lava.[*][84][*][86] If this mixture had melted through the floor into the pool of water, it was feared it could have created a serious steam explosion that would have ejected more radioactive material from the reactor. It became necessary to drain the pool.[*][87]

The bubbler pool could be drained by opening its sluice gates. However the valves controlling it were underwater, located in a flooded corridor in the basement. So volunteers in wetsuits and respirators (for protection against radioactive aerosols) and equipped with dosimeters, entered the knee-deep radioactive water and managed to open the valves.[*][88][*][89] These were the engineers Alexei Ananenko and Valeri Bezpalov (who knew where the valves were), accompanied by the shift supervisor Boris Baranov.[*][90] Upon succeeding and emerging from the water, according to many English language news articles, books and the prominent BBC docudrama *Surviving Disaster – Chernobyl Nuclear*, the three knew it was a suicide-mission and began suffering from radiation sickness and died soon after.[*][91] Some sources also incorrectly claimed that they died there in the plant.[*][90] However research by Andrew Leatherbarrow, author of the 2016 book *Chernobyl 01:23:40*, determined that the frequently recounted story is a gross exaggeration. Alexei Ananenko continues to work in the nuclear energy industry, and rebuffs the growth of the Chernobyl media sensationalism surrounding him.[*][92] While Valeri Bezpalov was found to still be alive by Leatherbarrow, the elderly 65 year old Baranov had lived until 2005 and died of heart failure.[*][93]

Once the bubbler pool gates were opened by the Ananenko team, fire brigade pumps were then used to drain the basement. The operation was not completed until 8 May, after 20,000 metric tons of highly radioactive water were pumped out.

With the bubbler pool gone, a meltdown was less likely to produce a powerful steam explosion. To do so, the molten core would now have to reach the water table below the reactor. To reduce the likelihood of this, it was decided to freeze the earth beneath the reactor, which would also stabilize the foundations. Using oil drilling equipment, the injection of liquid nitrogen began on 4 May. It was estimated that 25 metric tons of liquid nitrogen per day would be required to keep the soil frozen at -100 °C.[*][39][*]:59 This idea[*][94] was soon scrapped and the bottom room where the cooling system would have been installed was filled with concrete.

It is likely that intense alpha radiation hydrolysed the water, generating a low-pH hydrogen peroxide (H_2O_2) solution akin to an oxidizing acid.[*][95] Conversion of bubbler pool water to H_2O_2 is confirmed by the presence in the Chernobyl lavas of studtite and metastudtite,[*][96][*][97] the only minerals that contain peroxide.[*][98]

Debris removal

The worst of the radioactive debris was collected inside what was left of the reactor, much of it shoveled in by liquidators wearing heavy protective gear (dubbed "biorobots" by the military); these workers could only spend a

Chernobyl power plant in 2006 with the sarcophagus containment structure

maximum of 40 seconds at a time working on the rooftops of the surrounding buildings because of the extremely high doses of radiation given off by the blocks of graphite and other debris. The reactor itself was covered with bags of sand, lead and boric acid dropped from helicopters: some 5000 metric tons of material were dropped during the week that followed the accident.

At the time there was still fear that the reactor could re-enter a self-sustaining nuclear chain-reaction and explode again, and a new containment structure was planned to prevent rain entering and triggering such an explosion, and to prevent further release of radioactive material. This was the largest civil engineering task in history, involving a quarter of a million construction workers who all reached their official lifetime limits of radiation.[*][70] By December 1986, a large concrete sarcophagus had been erected to seal off the reactor and its contents.[*][99] A unique "clean up" medal was given to the workers.[*][100]

Many of the vehicles used by the liquidators remain parked in a field in the Chernobyl area.[*][101]

During the construction of the sarcophagus, a scientific team re-entered the reactor as part of an investigation dubbed "Complex Expedition", to locate and contain nuclear fuel in a way that could not lead to another explosion. These scientists manually collected cold fuel rods, but great heat was still emanating from the core. Rates of radiation in different parts of the building were monitored by drilling holes into the reactor and inserting long metal detector tubes. The scientists were exposed to high levels of radiation and radioactive dust.[*][70]

After six months of investigation, in December 1986, they discovered with the help of a remote camera an intensely radioactive mass in the basement of Unit Four, more than two metres wide and weighing hundreds of tons, which they called "the elephant's foot" for its wrinkled appearance.*[102] The mass was composed of sand, glass and a large amount of nuclear fuel that had escaped from the reactor. The concrete beneath the reactor was steaming hot, and was breached by solidified lava and spectacular unknown crystalline forms termed chernobylite. It was concluded that there was no further risk of explosion.*[70]

Liquidators worked under deplorable conditions, poorly informed and with poor protections. Many if not most of them exceeded radiation safety limits.*[78]*:177–183*[103]*:2 Some exceeded limits by over 100 times—leading to rapid death.*[78]*:187

The official contaminated zones became stage to a massive clean-up effort lasting seven months.*[78]*:177–183 The official reason for such early (and dangerous) decontamination efforts, rather than allowing time for natural decay, was that the land must be re-peopled and brought back into cultivation. Indeed, within fifteen months 75% of the land was under cultivation, even though only a third of the evacuated villages were resettled. Defence forces must have done much of the work. Yet this land was of marginal agricultural value. According to historian David Marples, the administration had a psychological purpose for the clean-up: they wished to forestall panic regarding nuclear energy, and even to restart the Chernobyl power station.*[78]*:78–9,87,192–3

29.3 Causes

There were two official explanations of the accident.

29.3.1 Operator error

The first official explanation of the accident, later acknowledged to be erroneous, was published in August 1986. It effectively placed the blame on the power plant operators. To investigate the causes of the accident the IAEA created a group known as the International Nuclear Safety Advisory Group (INSAG), which in its report of 1986, INSAG-1, on the whole also supported this view, based on the data provided by the Soviets and the oral statements of specialists.*[104] In this view, the catastrophic accident was caused by gross violations of operating rules and regulations. "During preparation and testing of the turbine generator under run-down conditions using the auxiliary load, personnel disconnected a series of technical protection systems and breached the most important operational safety provisions for conducting a technical exercise."*[105]*:311

The operator error was probably due to their lack of knowledge of nuclear reactor physics and engineering, as well as lack of experience and training. According to these allegations, at the time of the accident the reactor was being operated with many key safety systems turned off, most notably the Emergency Core Cooling System (ECCS), LAR (Local Automatic control system), and AZ (emergency power reduction system). Personnel had an insufficiently detailed understanding of technical procedures involved with the nuclear reactor, and knowingly ignored regulations to speed test completion.*[105]

The developers of the reactor plant considered this combination of events to be impossible and therefore did not allow for the creation of emergency protection systems capable of preventing the combination of events that led to the crisis, namely the intentional disabling of emergency protection equipment plus the violation of operating procedures. Thus the primary cause of the accident was the extremely improbable combination of rule infringement plus the operational routine allowed by the power station staff.*[105]*:312

In this analysis of the causes of the accident, deficiencies in the reactor design and in the operating regulations that made the accident possible were set aside and mentioned only casually. Serious critical observations covered only general questions and did not address the specific reasons for the accident. The following general picture arose from these observations. Several procedural irregularities also helped to make the accident possible. One was insufficient communication between the safety officers and the operators in charge of the experiment being run that night.

The reactor operators disabled safety systems down to the generators, which the test was really about. The main process computer, SKALA, was running in such a way that the main control computer could not shut down the reactor or even reduce power. Normally the reactor would have started to insert all of the control rods. The computer would have also started the "Emergency Core Protection System" that introduces 24 control rods into the active zone within 2.5 seconds, which is still slow by 1986 standards. All control was transferred from the process computer to the human operators.

On the subject of the disconnection of safety systems, Valery Legasov said, in 1987, that "[i]t was like airplane pilots experimenting with the engines in flight".*[72]

This view is reflected in numerous publications and also artistic works on the theme of the Chernobyl accident that appeared immediately after the accident,*[39] and for a long time remained dominant in the public consciousness and in popular publications.

29.3.2 Operating instructions and design deficiencies found

Reactor hall No. 1 of the Chernobyl Plant

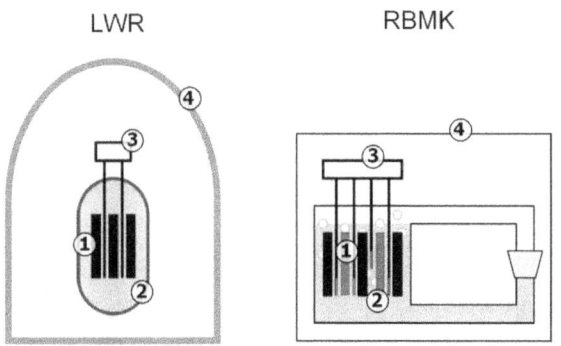

A simplified diagram of the major differences between the Chernobyl RBMK and the most common nuclear reactor design, the Light water reactor. 1. The use of a graphite moderator in a water cooled reactor. 2. A positive steam void coefficient that made the power excursion possible, which blew the reactor vessel. 3. The control rods were very slow, taking 18–20 seconds to be deployed. With the control rods having graphite tips that moderated, and therefore increased the fission rate in the beginning of the rod insertion 4. No reinforced containment building. [46] [106] [107]

Ukraine has declassified a number of KGB documents from the period between 1971 and 1988 related to the Chernobyl plant, mentioning for example previous reports of structural damages caused by negligence during construction of the plant (such as splitting of concrete layers) that were never acted upon. They document over 29 emergency situations in the plant during this period, 8 of which were caused by negligence or poor competence on the part of personnel. [108]

In 1991 a Commission of the USSR State Committee for the Supervision of Safety in Industry and Nuclear Power reassessed the causes and circumstances of the Chernobyl accident and came to new insights and conclusions. Based

Lumps of graphite moderator ejected from the core. The largest lump shows an intact control rod channel.

on it, in 1992 the IAEA Nuclear Safety Advisory Group (INSAG) published an additional report, INSAG-7, [44] which reviewed "that part of the INSAG-1 report in which primary attention is given to the reasons for the accident," and was included the USSR State Commission report as Appendix I. [44]

In this INSAG report, most of the earlier accusations against staff for breach of regulations were acknowledged to be either erroneous, based on incorrect information obtained in August 1986, or less relevant. This report reflected another view of the main reasons for the accident, presented in Appendix I. According to this account, the operators' actions in turning off the Emergency Core Cooling System, interfering with the settings on the protection equipment, and blocking the level and pressure in the separator drum did not contribute to the original cause of the accident and its magnitude, although they may have been a breach of regulations. Turning off the emergency system designed to prevent the two turbine generators from stopping was not a violation of regulations. [44]

Human factors contributed to the conditions that led to the disaster. These included operating the reactor at a low power level—less than 700 MW—a level documented in the run-down test programme, and operating with a small operational reactivity margin (ORM). The 1986 assertions of Soviet experts notwithstanding, regulations did not prohibit operating the reactor at this low power level. [44]:18

However, regulations did forbid operating the reactor with a small margin of reactivity. Yet "post-accident studies have shown that the way in which the real role of the ORM is reflected in the Operating Procedures and design documentation for the RBMK-1000 is extremely contradictory", and furthermore, "ORM was not treated as an operational safety limit, violation of which could lead to an accident"

.*[44]*:34–25

According to the INSAG-7 Report, the chief reasons for the accident lie in the peculiarities of physics and in the construction of the reactor. There are two such reasons:*[44]*:18

- The reactor had a dangerously large positive void coefficient of reactivity. The void coefficient is a measurement of how a reactor responds to increased steam formation in the water coolant. Most other reactor designs have a negative coefficient, i.e. the nuclear reaction rate slows when steam bubbles form in the coolant, since as the vapour phase in the reactor increases, fewer neutrons are slowed down. Faster neutrons are less likely to split uranium atoms, so the reactor produces less power (a negative feedback). Chernobyl's RBMK reactor, however, used solid graphite as a neutron moderator to slow down the neutrons, and the water in it, on the contrary, acts like a harmful neutron absorber. Thus neutrons are slowed down even if steam bubbles form in the water. Furthermore, because steam absorbs neutrons much less readily than water, increasing the intensity of vaporization means that more neutrons are able to split uranium atoms, increasing the reactor's power output. This makes the RBMK design very unstable at low power levels, and prone to suddenly increasing energy production to a dangerous level. This behaviour is counter-intuitive, and this property of the reactor was unknown to the crew.

- A more significant flaw was in the design of the control rods that are inserted into the reactor to slow down the reaction. In the RBMK reactor design, the lower part of each control rod was made of graphite and was 1.3 metres shorter than necessary, and in the space beneath the rods were hollow channels filled with water. The upper part of the rod, the truly functional part that absorbs the neutrons and thereby halts the reaction, was made of boron carbide. With this design, when the rods are inserted into the reactor from the uppermost position, the graphite parts initially displace some water (which absorbs neutrons, as mentioned above), effectively causing fewer neutrons to be absorbed initially. Thus for the first few seconds of control rod activation, reactor power output is increased, rather than reduced as desired. This behaviour is counter-intuitive and was not known to the reactor operators.

- Other deficiencies besides these were noted in the RBMK-1000 reactor design, as were its non-compliance with accepted standards and with the requirements of nuclear reactor safety.

29.3.3 Analysis of views

Both views were heavily lobbied by different groups, including the reactor's designers, power plant personnel, and the Soviet and Ukrainian governments. According to the IAEA's 1986 analysis, the main cause of the accident was the operators' actions. But according to the IAEA's 1993 revised analysis the main cause was the reactor's design.*[109] One reason there were such contradictory viewpoints and so much debate about the causes of the Chernobyl accident was that the primary data covering the disaster, as registered by the instruments and sensors, were not completely published in the official sources.

Once again, the human factor had to be considered as a major element in causing the accident. INSAG notes that both the operating regulations and staff handled the disabling of the reactor protection easily enough: witness the length of time for which the ECCS was out of service while the reactor was operated at half power. INSAG's view is that it was the operating crew's deviation from the test programme that was mostly to blame. "Most reprehensibly, unapproved changes in the test procedure were deliberately made on the spot, although the plant was known to be in a very different condition from that intended for the test." *[44]*:24

As in the previously released report INSAG-1, close attention is paid in report INSAG-7 to the inadequate (at the moment of the accident) "culture of safety" at all levels. Deficiency in the safety culture was inherent not only at the operational stage but also, and to no lesser extent, during activities at other stages in the lifetime of nuclear power plants (including design, engineering, construction, manufacture, and regulation). The poor quality of operating procedures and instructions, and their conflicting character, put a heavy burden on the operating crew, including the chief engineer. "The accident can be said to have flowed from a deficient safety culture, not only at the Chernobyl plant, but throughout the Soviet design, operating and regulatory organizations for nuclear power that existed at that time." *[44]*:24

29.4 Environmental effects

Main article: Chernobyl disaster effects

29.4.1 National and international spread of radioactive substances

Four hundred times more radioactive material was released from Chernobyl than by the atomic bombing of Hiroshima. The disaster released 1/100 to 1/1000 of the total amount of radioactivity released by nuclear weapons testing during the 1950s and 1960s.[110] Approximately 100,000 km² of land was significantly contaminated with fallout, with the worst hit regions being in Belarus, Ukraine and Russia.[111] Slighter levels of contamination were detected over all of Europe except for the Iberian Peninsula.[33][112][113]

The initial evidence that a major release of radioactive material was affecting other countries came not from Soviet sources, but from Sweden. On the morning of 28 April[114] workers at the Forsmark Nuclear Power Plant (approximately 1,100 km (680 mi) from the Chernobyl site) were found to have radioactive particles on their clothes.[115]

It was Sweden's search for the source of radioactivity, after they had determined there was no leak at the Swedish plant, that at noon on 28 April led to the first hint of a serious nuclear problem in the western Soviet Union. Hence the evacuation of Pripyat on 27 April, 36 hours after the initial explosions, was silently completed before the disaster became known outside the Soviet Union. The rise in radiation levels had at that time already been measured in Finland, but a civil service strike delayed the response and publication.[116]

Contamination from the Chernobyl accident was scattered irregularly depending on weather conditions, much of it deposited on mountainous regions such as the Alps, the Welsh mountains and the Scottish Highlands, where adiabatic cooling caused radioactive rainfall. The resulting patches of contamination were often highly localized, and waterflows across the ground contributed further to large variations in radioactivity over small areas. Sweden and Norway also received heavy fallout when the contaminated air collided with a cold front, bringing rain.[118]:43–44, 78

Rain was purposely seeded over 10,000 km² of the Belorussian SSR by the Soviet air force to remove radioactive particles from clouds heading toward highly populated areas. Heavy, black-coloured rain fell on the city of Gomel.[119] Reports from Soviet and Western scientists indicate that Belarus received about 60% of the contamination that fell on the former Soviet Union. However, the 2006 TORCH report stated that half of the volatile particles had landed outside Ukraine, Belarus, and Russia. A large area in Russia south of Bryansk was also contaminated, as were parts of northwestern Ukraine. Studies in surrounding countries indicate that over one million people could have been affected by radiation.[120]

Recently published data from a long-term monitoring program (The Korma Report II)[121] shows a decrease in internal radiation exposure of the inhabitants of a region in Belarus close to Gomel. Resettlement may even be possible in prohibited areas provided that people comply with appropriate dietary rules.

In Western Europe, precautionary measures taken in response to the radiation included seemingly arbitrary regulations banning the importation of certain foods but not others. In France some officials stated that the Chernobyl accident had no adverse effects.[122] Official figures in southern Bavaria in Germany indicated that some wild plant species contained substantial levels of caesium, which were believed to have been passed onto them during their consumption by wild boars, a significant number of which already contained radioactive particles above the allowed level.[123]

Piglet with dipygus on exhibit at the Ukrainian National Chornobyl Museum.

Mutations in both humans and other animals increased following the disaster. On farms in Narodychi Raion of Ukraine, for instance, in the first four years of the disaster nearly 350 animals were born with gross deformities such as missing or extra limbs, missing eyes, heads or ribs, or deformed skulls; in comparison, only three abnormal births had been registered in the five years prior.[124][125][126][127][128][129] Despite these claims, the World Health Organization states, "children conceived before or after their father's exposure showed no statistically significant differences in mutation frequencies".[130]

29.4.2 Radioactive release

Like many other releases of radioactivity into the environment, the Chernobyl release was controlled by the physical and chemical properties of the radioactive elements in the core. Particularly dangerous are the highly radioactive fission products, those with high nuclear decay rates that accumulate in the food chain, such as some of the isotopes of iodine, caesium and strontium. Iodine-131 and caesium-

137 are responsible for most of the radiation exposure received by the general population.[*][1]

Detailed reports on the release of radioisotopes from the site were published in 1989[*][131] and 1995,[*][132] with the latter report updated in 2002.[*][1]

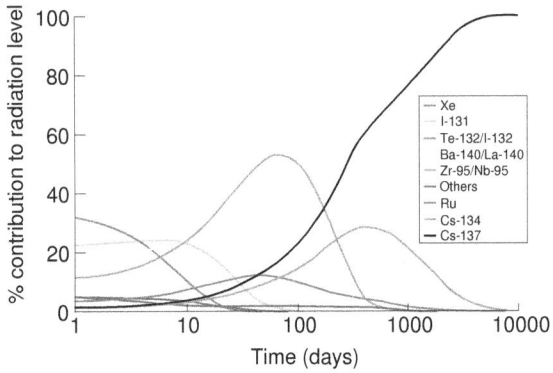

Contributions of the various isotopes to the external (atmospheric) absorbed dose in the contaminated area of Pripyat, from soon after the accident, to years after the accident.

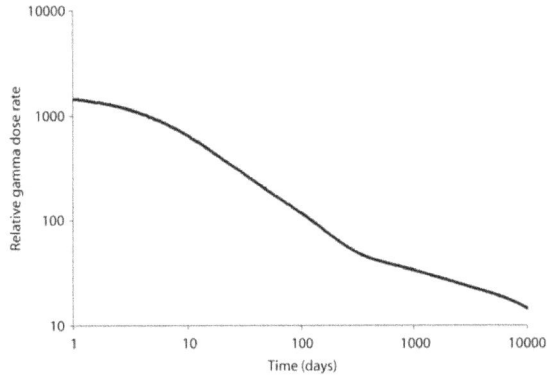

The external relative gamma dose for a person in the open near the Chernobyl disaster site.

At different times after the accident, different isotopes were responsible for the majority of the external dose. The remaining quantity of any radioisotope, and therefore the activity of that isotope, after 7 decay half-lives have passed, is less than 1% of its initial magnitude,[*][133] and it continues to reduce beyond 0.78% after 7 half-lives to 0.098% remaining after 10 half-lives have passed and so on.[*][134][*][135] (Some radionuclides have decay products that are likewise radioactive, which is not accounted for here.) The release of radioisotopes from the nuclear fuel was largely controlled by their boiling points, and the majority of the radioactivity present in the core was retained in the reactor.

- All of the noble gases, including krypton and xenon,

contained within the reactor were released immediately into the atmosphere by the first steam explosion.[*][1] The atmospheric release of xenon-133, with a half-life of 5 days, is estimated at 5200 PBq.[*][1]

- 50 to 60% of all core radioiodine in the reactor, about 1760 PBq (1760×10^{15} becquerels), or about 0.4 kg, was released, as a mixture of sublimed vapour, solid particles, and organic iodine compounds. Iodine-131 has a half-life of 8 days.[*][1]

- 20 to 40% of all core caesium-137 was released, 85 PBq in all.[*][1][*][136] Caesium was released in aerosol form; caesium-137, along with isotopes of strontium, are the two primary elements preventing the Chernobyl exclusion zone being re-inhabited.[*][137] 8.5×10^{16} Bq equals 24 kilograms of caesium-137.[*][137] Cs-137 has a half-life of 30 years.[*][1]

- Tellurium-132, half-life 78 hours, an estimated 1150 PBq was released.[*][1]

- An early estimate for total nuclear fuel material released to the environment was 3±1.5%; this was later revised to 3.5±0.5%. This corresponds to the atmospheric emission of 6 t of fragmented fuel.[*][132]

Two sizes of particles were released: small particles of 0.3 to 1.5 micrometres (aerodynamic diameter) and large particles of 10 micrometres. The large particles contained about 80% to 90% of the released nonvolatile radioisotopes zirconium-95, niobium-95, lanthanum-140, cerium-144 and the transuranic elements, including neptunium, plutonium and the minor actinides, embedded in a uranium oxide matrix.

The dose that was calculated is the relative external gamma dose rate for a person standing in the open. The exact dose to a person in the real world who would spend most of their time sleeping indoors in a shelter and then venturing out to consume an internal dose from the inhalation or ingestion of a radioisotope, requires a personnel specific radiation dose reconstruction analysis.

29.4.3 Residual radioactivity in the environment

Rivers, lakes and reservoirs

The Chernobyl nuclear power plant is located next to the Pripyat River, which feeds into the Dnieper reservoir system, one of the largest surface water systems in Europe, which at the time supplied water to Kiev's 2.4 million residents, and was still in spring flood when the accident occurred.[*][138][*]:60 The radioactive contamination of aquatic

Earth Observing-1 image of the reactor and surrounding area in April 2009

Map of radiation levels in 1996 around Chernobyl.

systems therefore became a major problem in the immediate aftermath of the accident.*[139] In the most affected areas of Ukraine, levels of radioactivity (particularly from radionuclides ^{131}I, ^{137}Cs and ^{90}Sr) in drinking water caused concern during the weeks and months after the accident,*[139] though officially it was stated that all contaminants had settled to the bottom "in an insoluble phase" and would not dissolve for 800–1000 years.*[138]*:64 Guidelines for levels of radioiodine in drinking water were temporarily raised to 3,700 Bq/L, allowing most water to be reported as safe,*[139] and a year after the accident it was announced that even the water of the Chernobyl plant's cooling pond was within acceptable norms. Despite this, two months after the disaster the Kiev water supply was abruptly switched from the Dnieper to the Desna River.*[138]*:64–5 Meanwhile, massive silt traps were constructed, along with an enormous 30m-deep underground barrier to prevent groundwater from the destroyed reactor entering the Pripyat River.*[138]*:65–67

Bio-accumulation of radioactivity in fish*[140] resulted in concentrations (both in western Europe and in the former Soviet Union) that in many cases were significantly above guideline maximum levels for consumption.*[139] Guideline maximum levels for radiocaesium in fish vary from country to country but are approximately 1000 Bq/kg in the European Union.*[141] In the Kiev Reservoir in Ukraine, concentrations in fish were several thousand Bq/kg during the years after the accident.*[140]

In small "closed" lakes in Belarus and the Bryansk region of Russia, concentrations in a number of fish species varied from 100 to 60,000 Bq/kg during the period 1990–92.*[142] The contamination of fish caused short-term concern in parts of the UK and Germany and in the long term (years rather than months) in the affected areas of Ukraine, Belarus, and Russia as well as in parts of Scandinavia.*[139]

Groundwater

Groundwater was not badly affected by the Chernobyl accident since radionuclides with short half-lives decayed away long before they could affect groundwater supplies, and longer-lived radionuclides such as radiocaesium and radiostrontium were adsorbed to surface soils before they could transfer to groundwater.*[143] However, significant transfers of radionuclides to groundwater have occurred from waste disposal sites in the 30 km (19 mi) exclusion zone around Chernobyl. Although there is a potential for transfer of radionuclides from these disposal sites off-site (i.e. out of the 30 km (19 mi) exclusion zone), the IAEA Chernobyl Report*[143] argues that this is not significant in comparison to current levels of washout of surface-deposited radioactivity.

Flora and fauna

After the disaster, four square kilometres of pine forest directly downwind of the reactor turned reddish-brown and died, earning the name of the "Red Forest".*[144] Some animals in the worst-hit areas also died or stopped reproducing. Most domestic animals were removed from the exclusion zone, but horses left on an island in the Pripyat River 6 km (4 mi) from the power plant died when their thyroid glands were destroyed by radiation doses of 150–200 Sv.*[145] Some cattle on the same island died and those that survived were stunted because of thyroid damage. The next generation appeared to be normal.*[145]

After the disaster, four square kilometres of pine forest directly downwind of the reactor turned reddish-brown and died, earning the name of the "Red Forest".[144]

A robot sent into the reactor itself has returned with samples of black, melanin-rich radiotrophic fungi that are growing on the reactor's walls.[146]

Of the 440,350 wild boar killed in the 2010 hunting season in Germany, over 1000 were found to be contaminated with levels of radiation above the permitted limit of 600 becquerels per kilogram, due to residual radioactivity from Chernobyl.[147]

The Norwegian Agricultural Authority reported that in 2009 a total of 18,000 livestock in Norway needed to be given uncontaminated feed for a period of time before slaughter in order to ensure that their meat was safe for human consumption. This was due to residual radioactivity from Chernobyl in the plants they graze on in the wild during the summer. 1,914 sheep needed to be given uncontaminated feed for a period of time before slaughter during 2012, and these sheep were located in just 18 of Norway's municipalities, a decrease of 17 from the 35 municipalities affected animals were located in during 2011 (117 municipalities were affected during 1986).[148]

The after-effects of Chernobyl were expected to be seen for a further 100 years, although the severity of the effects would decline over that period.[149] Scientists report this is due to radioactive caesium-137 isotopes being taken up

by fungi such as *Cortinarius caperatus* which is in turn eaten by sheep whilst grazing.[148]

The United Kingdom was forced to restrict the movement of sheep from upland areas when radioactive caesium-137 fell across parts of Northern Ireland, Wales, Scotland and northern England. In the immediate aftermath of the disaster in 1986, a total of 4,225,000 sheep had their movement restricted across a total of 9,700 farms, in order to prevent contaminated meat entering the human food chain.[150] The number of sheep and the number of farms affected has decreased since 1986, Northern Ireland was released from all restrictions in 2000 and by 2009 369 farms containing around 190,000 sheep remained under the restrictions in Wales, Cumbria and northern Scotland.[150] The restrictions applying in Scotland were lifted in 2010, whilst those applying to Wales and Cumbria were lifted during 2012, meaning no farms in the UK remain restricted because of Chernobyl fallout.[151][152]

The legislation used to control sheep movement and compensate farmers (farmers were latterly compensated per animal to cover additional costs in holding animals prior to radiation monitoring) was revoked during October and November 2012 by the relevant authorities in the UK.[153]

Demonstration on Chernobyl day near WHO in Geneva

29.5 Human impact

Main article: Chernobyl disaster effects
See also: Deaths due to the Chernobyl disaster

In the aftermath of the accident, 237 people suffered from acute radiation sickness (ARS), of whom 31 died within the first three months.[28][154] Most of the victims were fire and rescue workers trying to bring the accident under control, who were not fully aware of how dangerous the exposure to radiation in the smoke was.

In 2005 the Chernobyl Forum, composed of the IAEA, other UN organizations and the governments of Belarus, Russia and Ukraine, published a report on the radiological environmental and health consequences of the Chernobyl accident.

On the death toll of the accident, the report states that 28 emergency workers ("liquidators") died from acute radiation syndrome including beta burns and 15 patients died from thyroid cancer in the following years, and it roughly estimated that cancer deaths caused by Chernobyl may reach a total of about 4,000 among the 5 million persons residing in the contaminated areas, the report projected cancer mortality "increases of less than one per cent" (~0.3%) on a time span of 80 years, cautioning that this estimate was "speculative" since at this time only a few cancer deaths are linked to the Chernobyl disaster.[155] The report says it is impossible to reliably predict the number of fatal cancers arising from the incident as small differences in assumptions can result in large differences in the estimated health costs. The report says it represents the consensus view of the eight UN organizations.

Of all 66,000 Belarusian emergency workers, by the mid-1990s only 150 (roughly 0.2%) were reported by their government as having died. In contrast, 5,722 casualties were reported among Ukrainian clean-up workers up to the year 1995, by the National Committee for Radiation Protection of the Ukrainian Population.[111]

The four most harmful radionuclides spread from Chernobyl were iodine-131, caesium-134, caesium-137 and strontium-90, with half-lives of 8.02 days, 2.07 years, 30.2 years and 28.8 years respectively.[156]:8 The iodine was initially viewed with less alarm than the other isotopes, because of its short half-life, but it is highly volatile, and now appears to have travelled furthest and caused the most severe health problems in the short term.[111]:24 Strontium, on the other hand, is the least volatile of the four, and of main concern in the areas near Chernobyl itself.[156]:8 Iodine tends to become concentrated in thyroid and milk glands, leading, among other things, to increased incidence of thyroid cancers. Caesium tends to accumulate in vital organs such as the heart,[157]:133 while strontium accumulates in bones, and may thus be a risk to bone-marrow and lymphocytes.[156]:8 Radiation is most damaging to cells that are actively dividing. In adult mammals cell division is slow, except in hair follicles, skin, bone marrow and the gastrointestinal tract, which is why vomiting and hair loss are common symptoms of acute radiation sickness.[158]:42

29.5.1   Difficulties in assessment

Health in Belarus and Ukraine has shown disturbing trends following the Chernobyl disaster. In Belarus, incidence of congenital defects had risen by 40% within six years of the accident, to the point that it became the principal cause of infant mortality.[159]:52 There was a substantial increase in digestive, circulatory, nervous, respiratory and endocrine diseases and cancers, correlated with areas of high radioactive contamination, and in one especially contaminated district of Belarus, 95% of children were in 2005 reported to have at least one chronic illness.[157]:129, 199 The Ukrainian Ministry of Health estimated in 1993 that roughly 70% of its population were unwell, with large increases in respiratory, blood and nervous system diseases.[111]:27 By the year 2000, the number of Ukrainians claiming to be radiation 'sufferers' (*poterpili*) and receiving state benefits had jumped to 3.5 million, or 5% of the population. Many of these are populations resettled from contaminated zones, or former or current Chernobyl plant workers.[103]:4–5 According to IAEA-affiliated scientific bodies, these apparent increases of ill health result partly from economic strains on these countries and poor health-care and nutrition; also, they suggest that increased medical vigilance following the accident has meant that many cases that would previously have gone unnoticed (especially of cancer) are now being registered.[111]

Of the approximately 600,000 'liquidators' that were engaged in the Chernobyl clean-up, roughly 50,000 were required to work as 'bio-robots', in conditions of such extreme radiation that electronic robots ceased to operate. These bio-robots are well-known figures within every village, housing block and work-collective. Most are prematurely aged and many have died, and leukaemia rates among them are substantially higher than in the wider population.[103]:9–10,31 According to ethnographer Adriana Petryna, birth defects appear to have increased in Ukraine as well. She describes gross deformities in the Kyev city hospital's neonatal unit, including one infant born to a Chernobyl worker, who had an extra finger, a deformed ear, his trachea missing and his gut external to his body. Hospital staff were on the whole cooperative, but warned Petryna that she would be forbidden to access any statistics; she could therefore only treat these cases as anecdotal evidence.[103]:7–8 Poor or inaccessible statistics has meant that causal connections are very difficult to make in both Belarus and Ukraine. It has been observed that Belarus in particular actively suppresses or ignores health-related research,[103]:4–5 a false economy estimated to cost the country ten times more than it saves.[159]:51–2 One Belarusian villager describes: "We had a year once when almost every day there was a funeral. We must have buried about fifty people that year. Is it related to radiation? Who knows." [157]:259

Under Soviet rule, the extent of radiation injury was systematically covered up. Most cases of acute radiation sickness (ARS) were disguised as 'Vegetovascular dystonia' (VvD), a Soviet classification for a type of panic disorder with possible symptoms including heart palpitations, sweating, tremors, nausea, hypotension or hypertension, neurosis, spasms and seizures: symptoms which resemble the neurological effects of ARS. Declassified documents show that the Soviet Health Ministry ordered the systematic misdiagnosis of ARS as VvD, for all people who did not show gross signs of radiation sickness such as burns or hair loss, and for all 'liquidators' who had exceeded their maximum allowable dose. It appears that up to 17,500 people were intentionally misdiagnosed in this manner.[103]:43–4 Subsequent claims for health welfare were denied on the basis of this diagnosis or the application of other psychosocial medical categories (individual poor constitution; psychological self-induction).[103]:11 A key tool for Soviet denial was the '35 rem concept', whereby it was held that 35 rems was a safe radiation exposure for a lifetime, "based on international standards", and since most people near Chernobyl received less than that, their health complaints could be attributed to "radiophobia" .[157]:47

Both Belarus and Ukraine rely heavily on foreign aid and have been pressured to comply with international views of the disaster. For instance, in 2002 the World Bank advised Belarus to "shift its attention from calculating the impact of the accident to developing forward-looking activities directed at economic development and improvement in the quality of life of the affected people". Health-related government welfare was blamed for creating "the sense of victimization and dependency" and thus exacerbating psychosomatic disorders.[157]:98,101 Belarus in particular has complied by ignoring or suppressing scientific research.[103]:4–5 Historian David Marples attributes this more to that government's weakness and apathy than a simple desire to avoid health costs.[159]:51–2

29.5.2   Thyroid cancer

The 2005 Chernobyl Forum report revealed thyroid cancer among children to be one of the main health impacts from the Chernobyl accident. In that publication more than 4000 cases were reported, and that there was no evidence of an increase in solid cancers or leukemia. It said that there was an increase in psychological problems among the affected population.[155] Dr Michael Repacholi, manager of WHO's Radiation Program reported that the 4000 cases of thyroid cancer resulted in nine deaths.[160]

According to UNSCEAR, up to the year 2005, an excess of

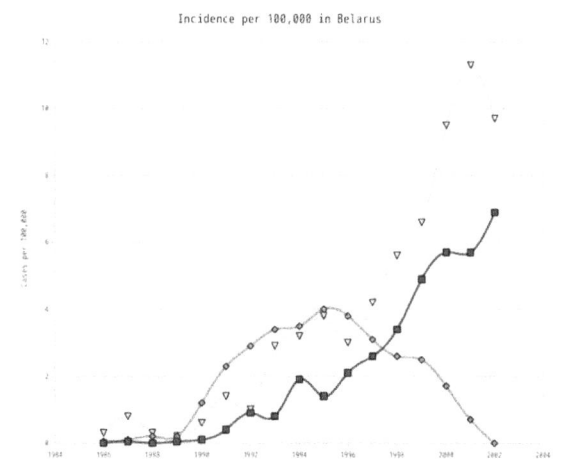

Thyroid cancer incidence in children and adolescents from Belarus after the Chernobyl accident
Yellow: Adults (19–34)
Blue: Adolescents (15–18)
Red: Children (0–14)

over 6000 cases of thyroid cancer have been reported. That is, over the estimated pre-accident baseline thyroid cancer rate, more than 6000 casual cases of thyroid cancer have been reported in children and adolescents exposed at the time of the accident, a number that is expected to increase. They concluded that there is no other evidence of major health impacts from the radiation exposure.[161]

Well-differentiated thyroid cancers are generally treatable,[162] and when treated the five-year survival rate of thyroid cancer is 96%, and 92% after 30 years.[163] UNSCEAR had reported 15 deaths from thyroid cancer in 2011.[164] The International Atomic Energy Agency (IAEA) also states that there has been no increase in the rate of birth defects or abnormalities, or solid cancers (such as lung cancer) corroborating UNSCEAR's assessments.[165] UNSCEAR does raise the possibility of long term genetic defects, pointing to a doubling of radiation-induced minisatellite mutations among children born in 1994.[166] However, the risk of thyroid cancer associated with the Chernobyl accident is still high according to published studies.[167][168]

The German affiliate of the International Physicians for the Prevention of Nuclear War (IPPNW) argued that more than 10,000 people are today affected by thyroid cancer and 50,000 cases are expected in the future.[169]

29.5.3   Other health disorders

Fred Mettler, a radiation expert at the University of New Mexico, puts the number of worldwide cancer deaths outside the highly contaminated zone at "perhaps" 5000, for a total of 9000 Chernobyl-associated fatal cancers, saying "the number is small (representing a few percent) relative to the normal spontaneous risk of cancer, but the numbers are large in absolute terms" .*[170] The same report outlined studies based in data found in the Russian Registry from 1991 to 1998 that suggested that "of 61,000 Russian workers exposed to an average dose of 107 mSv about 5% of all fatalities that occurred may have been due to radiation exposure." *[155]

The report went into depth about the risks to mental health of exaggerated fears about the effects of radiation.*[155] According to the IAEA the "designation of the affected population as "victims" rather than "survivors" has led them to perceive themselves as helpless, weak and lacking control over their future" . The IAEA says that this may have led to behaviour that has caused further health effects.*[171]

Fred Mettler commented that 20 years later: "The population remains largely unsure of what the effects of radiation actually are and retain a sense of foreboding. A number of adolescents and young adults who have been exposed to modest or small amounts of radiation feel that they are somehow fatally flawed and there is no downside to using illicit drugs or having unprotected sex. To reverse such attitudes and behaviours will likely take years although some youth groups have begun programs that have promise." *[172] In addition, disadvantaged children around Chernobyl suffer from health problems that are attributable not only to the Chernobyl accident, but also to the poor state of post-Soviet health systems.*[165]

The United Nations Scientific Committee on the Effects of Atomic Radiation (UNSCEAR), part of the Chernobyl Forum, have produced their own assessments of the radiation effects.*[173] UNSCEAR was set up as a collaboration between various United Nation bodies, including the World Health Organization, after the atomic bomb attacks on Hiroshima and Nagasaki, to assess the long-term effects of radiation on human health.*[174]

29.5.4   Deaths due to radiation exposure

The number of potential deaths arising from the Chernobyl disaster is heavily debated. The WHO's prediction of 4000 future cancer deaths in surrounding countries*[175] is based on the Linear no-threshold model (LNT), which assumes that the damage inflicted by radiation at low doses is directly proportional to the dose.*[176] Radiation epidemiologist Roy Shore contends that estimating health effects in a population from the LNT model "is not wise because of the uncertainties" .*[177]

According to the Union of Concerned Scientists the number

Radiation warning sign in Pripyat

of excess cancer deaths worldwide (including all contaminated areas) is approximately 27,000 based on the same LNT.*[178]

Another study critical of the Chernobyl Forum report was commissioned by Greenpeace, which asserted that the most recently published figures indicate that in Belarus, Russia and Ukraine the accident could have resulted in 10,000–200,000 additional deaths in the period between 1990 and 2004.*[35] The Scientific Secretary of the Chernobyl Forum criticized the report's reliance on non-peer-reviewed locally produced studies. Although most of the study's sources were from peer-reviewed journals, including many Western medical journals, the higher mortality estimates were from non-peer-reviewed sources,*[35] while Gregory Härtl (spokesman for the WHO) suggested that the conclusions were motivated by ideology.*[179]

Chernobyl: Consequences of the Catastrophe for People and the Environment is an English translation of the 2007 Russian publication *Chernobyl*. It was published in 2009 by the New York Academy of Sciences in their *Annals of the New York Academy of Sciences*. It presents an analysis of scientific literature and concludes that medical records between 1986, the year of the accident, and 2004 reflect 985,000 premature deaths as a result of the radioactivity released.*[180] Though, it was impossible to precisely determine what dose the affected people received, knowing the fact that the received doses varied strongly from one individual to the other in the population above which the radioactive cloud travelled, and also knowing the fact that one cannot tell for sure if a cancer in an individual from the former USSR is produced by radiation from Chernobyl accident or by other social or behavioural factors, such as smoking or alcohol drinking.*[181]

The authors suggest that most of the deaths were in Russia, Belarus and Ukraine, though others occurred worldwide throughout the many countries that were struck by radioactive fallout from Chernobyl. The literature analysis draws on over 1000 published titles and over 5000 internet and printed publications discussing the consequences of the Chernobyl disaster. The authors contend that those publications and papers were written by leading Eastern European authorities and have largely been downplayed or ignored by the IAEA and UNSCEAR.[*][180]

The estimated human health impacts were criticized by M. I. Balonov of the Institute of Radiation Hygiene, St. Petersburg, Russia who described them as biased, drawing from sources which were difficult to independently verify and lacking a proper scientific base. Balanov's expressed his opinion that "the authors unfortunately did not appropriately analyze the content of the Russian-language publications, for example, to separate them into those that contain scientific evidence and those based on hasty impressions and ignorant conclusions".[*][38]

29.5.5   Abortion requests

See also: Radiophobia

Following the accident, journalists mistrusted many medical professionals (such as the spokesman from the UK National Radiological Protection Board), and in turn encouraged the public to mistrust them.[*][182] Throughout the European continent, in nations where abortion is legal, many requests for induced abortions, of otherwise normal pregnancies, were obtained out of fears of radiation from Chernobyl, including an excess number of abortions in Denmark in the months following the accident.[*][183] In Greece, following the accident many obstetricians were unable to resist requests from worried pregnant mothers over fears of radiation. Although it was determined that the effective dose to Greeks would not exceed 1 mSv (100 mrem), a dose much lower than that which could induce embryonic abnormalities or other non-stochastic effects, there was an observed 2500 excess of otherwise wanted pregnancies being terminated, probably out of fear in the mother of radiation risk.[*][184] A "slightly" above the expected number of requested induced abortions occurred in Italy.[*][185][*][186]

Worldwide, an estimated excess of about 150,000 elective abortions may have been performed on otherwise healthy pregnancies out of unfounded fears of radiation from Chernobyl, according to Dr Robert Baker and ultimately a 1987 article published by Linda E. Ketchum in the *Journal of Nuclear Medicine* which mentions but does not reference an IAEA source on the matter.[*][187][*][188][*][189][*][190][*][184][*][191]

The available statistical data excludes the Soviet/Ukraine/Belarus abortion rates, as they are presently

unavailable. From the available data, an increase in the number of abortions in what were healthy developing human offspring in Denmark occurred in the months following the accident, at a rate of about 400 cases.[*][183] In Greece, there was an observed 2500 excess of otherwise wanted pregnancies being terminated.[*][184] In Italy, a "slightly" above the expected number of induced abortions occurred, approximately 100.[*][185][*][186]

> As the increase in radiation in Denmark was so low...the public debate and anxiety among the pregnant women and their husbands "caused" more fetal deaths in Denmark than the accident. This underlines the importance of public debate, the role of the mass media and of the way in which National Health authorities participate in this debate.

No evidence of changes in the prevalence of birth congenital anomalies(mutations) which might be associated with the accident are apparent in Belarus or the Ukraine, the two republics which had the highest exposure to fallout.[*][192] In Sweden,[*][193] and Finland where no increase in abortion rates occurred, it was likewise determined that "no association between the temporal and spatial variations in radioactivity and variable incidence of congenital malformations [was found]."[*][194] A similar null increase in the abortion rate and a, healthy baseline situation, of no increase in birth defects was determined from assessing the Hungarian Congenital Abnormality Registry,[*][195] Findings also mirrored in Austria.[*][196] Larger, "mainly western European" data sets approaching a million births in the EUROCAT database, divided into "exposed" and control groups were assessed in 1999. As no Chernobyl impacts were detected, the researchers conclude "in retrospect the widespread fear in the population about the possible effects of exposure on the unborn fetus was not justified".[*][197] Despite studies from Germany and Turkey, the only robust evidence of negative pregnancy outcomes that transpired after the accident were these elective abortion indirect effects, in Greece, Denmark, Italy etc., due to the anxieties created.[*][192]

In very high doses, it was known at the time that radiation can cause a physiological increase in the rate of pregnancy anomalies, but unlike the dominant linear-no threshold model of radiation and cancer rate increases, it was known, by researchers familiar with both the prior human exposure data and animal testing, that the "Malformation of organs appears to be a deterministic effect with a threshold dose" below which, no rate increase is observed.[*][198] This teratology (birth defects) issue was discussed by Frank Castronovo of the Harvard Medical School in 1999, publishing a detailed review of dose reconstructions and the available pregnancy data following the Chernobyl accident, inclusive of data from Kiev's two largest obstetrics hospitals.*[198] Castronovo concludes that "the lay press with newspaper reporters playing up anecdotal stories of children with birth defects" is, together with dubious studies that show selection bias, the two primary factors causing the persistent belief that Chernobyl increased the background rate of birth defects. When the vast amount of pregnancy data does not support this preception as no women took part in the most radioactive liquidator operations, no pregnant individuals were exposed to the threshold dose.*[198]

29.6 Speculation

According to Kenneth Mossman, a Professor of Health Physics and member of the U.S. Nuclear Regulatory Commission advisory committee,*[199] the "LNT philosophy is overly conservative, and low-level radiation may be less dangerous than commonly believed" .*[200] Yoshihisa Matsumoto, a radiation biologist at the Tokyo Institute of Technology, cites laboratory experiments on animals to suggest there must be a threshold dose below which DNA repair mechanisms can completely repair any radiation damage.*[177] Mossman suggests that the proponents of the current model believe that being conservative is justified due to the uncertainties surrounding low level doses and it is better to have a "prudent public health policy" .*[199]

Another significant issue is establishing consistent data on which to base the analysis of the impact of the Chernobyl accident. Since 1991 large social and political changes have occurred within the affected regions and these changes have had significant impact on the administration of health care, on socio-economic stability, and the manner in which statistical data is collected.*[201] Ronald Chesser, a radiation biologist at Texas Tech University, says that "the subsequent Soviet collapse, scarce funding, imprecise dosimetry, and difficulties tracking people over the years have limited the number of studies and their reliability" .*[177]

29.7 Economic and political consequences

It is difficult to establish the total economic cost of the disaster. According to Mikhail Gorbachev, the Soviet Union spent 18 billion rubles (the equivalent of US$18 billion at that time) on containment and decontamination, virtually bankrupting itself.*[3] In Belarus the total cost over 30 years is estimated at US$235 billion (in 2005 dollars).*[165] On-going costs are well known; in their 2003–2005 report, The Chernobyl Forum stated that between 5%

Abandoned buildings in Chernobyl

Russian President Medvedev and Ukrainian President Yanukovych lay wreathes at a memorial to victims of the Chernobyl disaster on 26 April 2011.

and 7% of government spending in Ukraine is still related to Chernobyl, while in Belarus over $13 billion is thought to have been spent between 1991 and 2003, with 22% of national budget having been Chernobyl-related in 1991, falling to 6% by 2002.*[165] Much of the current cost relates to the payment of Chernobyl-related social benefits to some 7 million people across the 3 countries.*[165]

A significant economic impact at the time was the removal of 784,320 ha (1,938,100 acres) of agricultural land and 694,200 ha (1,715,000 acres) of forest from production. While much of this has been returned to use, agricultural production costs have risen due to the need for special cultivation techniques, fertilizers and additives.*[165]

Politically, the accident gave great significance to the new Soviet policy of glasnost,*[202]*[203] and helped forge closer Soviet–US relations at the end of the Cold War, through bioscientific cooperation.*[204]*:44–48 The disaster also became a key factor in the Union's eventual 1991 dissolution, and a major influence in shaping the new Eastern Europe.*[204]*:20–21

Both Ukraine and Belarus, in their first months of indepen-dence, lowered legal radiation thresholds from the Soviet Union's previous, elevated thresholds (from 35 rems per lifetime under the USSR to 7 rems per lifetime in Ukraine and 0.1 rems per year in Belarus).[159]*:46–7,119–124 This required an expansion of territories that were consid-ered contaminated. In Ukraine, over 500,000 people have now been resettled, many of whom have become applicants for medical and other welfare. Ukraine also maintains the destroyed reactor, for which it employs a very large work-force in order to keep individual exposure times low. Many of these workers have since registered disabilities and en-rolled for welfare. In Ukraine, the Chernobyl disaster was an icon of the nationalist movement, symbolic of all that was wrong with the Soviet Union, and welfare became a key platform for winning independence. Ukraine has since developed a massive and burdensome welfare system that has become increasingly corrupt and ineffective.[204] It has presented its greatly increased welfare demands since 1991 as a demonstration of its own moral legitimacy, and as an argument for needing foreign aid.[204]*:24 Belarus, on the other hand, was politically weak when it gained in-dependence, and looked to Moscow for guidance; in many ways it has returned to the old Soviet policy of secrecy and denial.[159]*:46–7,119–124[204]*:22–24

29.8 Aftermath

Following the accident, questions arose about the future of the plant and its eventual fate. All work on the unfinished reactors 5 and 6 was halted three years later. However, the trouble at the Chernobyl plant did not end with the disas-ter in reactor 4. The damaged reactor was sealed off and 200 cubic meters (260 cu yd) of concrete was placed be-tween the disaster site and the operational buildings. The work was managed by Grigoriy Mihaylovich Naginskiy, the deputy chief engineer of Installation and Construction Di-rectorate – 90. The Ukrainian government continued to let the three remaining reactors operate because of an energy shortage in the country.

29.8.1 Decommissioning

Main article: Chernobyl Nuclear Power Plant § Decom-missioning

In October 1991, a fire broke out in the turbine building of reactor 2;[205] the authorities subsequently declared the reactor damaged beyond repair, and it was taken offline. Reactor 1 was decommissioned in November 1996 as part of a deal between the Ukrainian government and interna-tional organizations such as the IAEA to end operations at the plant. On 15 December 2000, then-President Leonid Kuchma personally turned off Reactor 3 in an official cer-emony, shutting down the entire site.[206]

29.8.2 Radioactive waste management

Containment of the reactor

The Chernobyl reactor is now enclosed in a large concrete sarcophagus, which was built quickly to allow continuing operation of the other reactors at the plant.[6]

New Safe Confinement in August 2016

A New Safe Confinement was to have been built by the end of 2005; however, it has suffered ongoing delays and as of 2010, when construction finally began, was expected to be completed in 2013. This was delayed again to 2016, the end of the 30-year lifespan of the sarcophagus. The structure is being built adjacent to the existing shelter and will be slid into place on rails. It is to be a metal arch 105 metres (344 ft) high and spanning 257 metres (843 ft), to cover both unit 4 and the hastily built 1986 structure. The Chernobyl Shelter Fund, set up in 1997, has received €810 million from international donors and projects to cover this project and previous work. It and the Nuclear Safety Account, also applied to Chernobyl decommissioning, are managed by the European Bank for Reconstruction and Development (EBRD).

As of 29 November 2016, Reactor No. 4 has been cov-ered by the New Safe Confinement that covers the reactor and the unstable "sarcophagus".[207] The huge steel arch was moved into place over several weeks, and the completion of this procedure was celebrated with a cere-mony at the site, attended by the Ukrainian president, Petro Poroshenko, diplomats and site workers.[208]

By 2002, roughly 15,000 Ukrainian workers were still working within the Zone of Exclusion, maintaining the plant and performing other containment- and research-related tasks, often in dangerous conditions.[204][*:2] A handful of Ukrainian scientists work inside the sarcophagus, but outsiders are rarely granted access. In 2006 an Australian *60 Minutes* team led by reporter Richard Carleton and producer Stephen Rice were allowed to enter the sarcophagus for 15 minutes and film inside the control room.[209]

On 12 February 2013, a 600 m^2 (6,500 sq ft) section of the roof of the turbine-building, adjacent to the sarcophagus, collapsed. At first it was assumed that the roof collapsed because of the weight of snow on it. However the amount of snow was not exceptional, and the report of a Ukrainian fact-finding panel concluded that the part collapse of the turbine-building was the result of sloppy repair work and aging of the structure. The report mentioned the possibility that the repaired part of the turbine-building added a larger strain on the total structure than expected, and the braces in the roof were damaged by corrosion and sloppy welding. Experts such as Valentin Kupny, former deputy director of the nuclear plant, did warn that the complex was on the verge of a collapse, leaving the building in an extremely dangerous condition. A proposed reinforcement in 2005 was cancelled by a superior official. After the 12 February incident, radioactivity levels were up to 19 becquerels per cubic meter of air: 12 times normal. The report assumed radioactive materials from inside the structure spread to the surroundings after the roof collapsed. All 225 workers employed by the Chernobyl complex and the French company that is building the new shelter were evacuated shortly after the collapse. According to the managers of the complex, radiation levels around the plant were at normal levels (between 5 and 6 μSv/h) and should not affect workers' health. According to Kupny the situation was underestimated by the Chernobyl nuclear complex managers, and information was kept secret.[210][211]

Radioactive materials and waste management

As of 2006, some fuel remained in the reactors at units 1 through 3, most of it in each unit's spent fuel pool, as well as some material in a small spent fuel interim storage facility pond (ISF-1).

In 1999 a contract was signed for construction of a radioactive waste management facility to store 25,000 used fuel assemblies from units 1–3 and other operational wastes, as well as material from decommissioning units 1–3 (which will be the first RBMK units decommissioned anywhere). The contract included a processing facility able to cut the RBMK fuel assemblies and to put the material in canisters, which were to be filled with inert gas and welded shut.

The canisters were to be transported to dry storage vaults, where the fuel containers would be enclosed for up to 100

years. This facility, treating 2500 fuel assemblies per year, would be the first of its kind for RBMK fuel. However, after a significant part of the storage structures had been built, technical deficiencies in the concept emerged, and the contract was terminated in 2007. The interim spent fuel storage facility (ISF-2) will now be completed by others by mid-2013.

Another contract has been let for a liquid radioactive waste treatment plant, to handle some 35,000 cubic meters of low- and intermediate-level liquid wastes at the site. This will need to be solidified and eventually buried along with solid wastes on site.

In January 2008, the Ukrainian government announced a 4-stage decommissioning plan that incorporates the above waste activities and progresses towards a cleared site.[120]

Lava-like fuel-containing materials (FCMs)

Main article: Corium (nuclear reactor)

According to official estimates, about 95% of the fuel in Reactor 4 at the time of the accident (about 180 metric tons) remains inside the shelter, with a total radioactivity of nearly 18 million curies (670 PBq). The radioactive material consists of core fragments, dust, and lava-like "fuel containing materials" (FCM)—also called "corium"—that flowed through the wrecked reactor building before hardening into a ceramic form.

Three different lavas are present in the basement of the reactor building: black, brown, and a porous ceramic. The lava materials are silicate glasses with inclusions of other materials within them. The porous lava is brown lava that dropped into water and thus cooled rapidly.

It is unclear how long the ceramic form will retard the release of radioactivity. From 1997 to 2002 a series of published papers suggested that the self-irradiation of the lava would convert all 1,200 metric tons into a submicrometre and mobile powder within a few weeks.[212] But it has been reported that the degradation of the lava is likely to be a slow and gradual process rather than sudden and rapid.[213] The same paper states that the loss of uranium from the wrecked reactor is only 10 kg (22 lb) per year; this low rate of uranium leaching suggests that the lava is resisting its environment.[213] The paper also states that when the shelter is improved, the leaching rate of the lava will decrease.[213]

Some of the surfaces of the lava flows have started to show new uranium minerals such as čejkaite (Na
4(UO
2)(CO
3)

3)[214] and uranyl carbonate. However, the level of radioactivity is such that during 100 years, the lava's self irradiation (2×10^{16} α decays per gram and 2 to 5×10^5 Gy of β or γ) will fall short of the level required to greatly change the properties of glass (10^{18} α decays per gram and 10^8 to 10^9 Gy of β or γ). Also the lava's rate of dissolution in water is very low (10^{-7} g·cm^{-2}·day^{-1}), suggesting that the lava is unlikely to dissolve in water.[213]

29.8.3 The Exclusion Zone

Main article: Chernobyl Exclusion Zone
An area originally extending 30 kilometres (19 mi) in all

Entrance to the zone of alienation around Chernobyl

directions from the plant is officially called the "zone of alienation". It is largely uninhabited, except for about 300 residents who have refused to leave. The area has largely reverted to forest, and has been overrun by wildlife because of a lack of competition with humans for space and resources. Even today, radiation levels are so high that the workers responsible for rebuilding the sarcophagus are only allowed to work five hours a day for one month before taking 15 days of rest. Ukrainian officials estimated the area would not be safe for human life again for another 20,000 years[75] (although by 2016, 187 local Ukrainians had returned and were living permanently in the zone[215]).

In 2011 Ukraine opened up the sealed zone around the Chernobyl reactor to tourists who wish to learn more about the tragedy that occurred in 1986.[216][217]

29.8.4 Forest fires

If the forests that have been contaminated by radioactive material catch on fire, they will spread the radioactive material further outwards in the smoke.[218][219]

29.9 Recovery projects

29.9.1 The Chernobyl Shelter Fund

Main articles: Chernobyl Shelter Fund and Chernobyl New Safe Confinement

The Chernobyl Shelter Fund was established in 1997 at the Denver 23rd G8 summit to finance the Shelter Implementation Plan (SIP). The plan calls for transforming the site into an ecologically safe condition by means of stabilization of the sarcophagus followed by construction of a New Safe Confinement (NSC). While the original cost estimate for the SIP was US$768 million, the 2006 estimate was $1.2 billion. The SIP is being managed by a consortium of Bechtel, Battelle, and Électricité de France, and conceptual design for the NSC consists of a movable arch, constructed away from the shelter to avoid high radiation, to be slid over the sarcophagus. The NSC is expected to be completed in 2015,[220] and will be the largest movable structure ever built.

Dimensions:

- Span: 270 m (886 ft)
- Height: 100 m (330 ft)
- Length: 150 m (492 ft)

29.9.2 The United Nations Development Programme

The United Nations Development Programme has launched in 2003 a specific project called the Chernobyl Recovery and Development Programme (CRDP) for the recovery of the affected areas.[221] The programme was initiated in February 2002 based on the recommendations in the report on Human Consequences of the Chernobyl Nuclear Accident. The main goal of the CRDP's activities is supporting the Government of Ukraine in mitigating long-term social, economic, and ecological consequences of the Chernobyl catastrophe. CRDP works in the four most Chernobyl-affected areas in Ukraine: Kyivska, Zhytomyrska, Chernihivska and Rivnenska.

29.9.3 The International Project on the Health Effects of the Chernobyl Accident

The International Project on the Health Effects of the Chernobyl Accident (IPEHCA) was created and received US

$20 million, mainly from Japan, in hopes of discovering the main cause of health problems due to ^{131}I radiation. These funds were divided between Ukraine, Belarus, and Russia, the three main affected countries, for further investigation of health effects. As there was significant corruption in former Soviet countries, most of the foreign aid was given to Russia, and no positive outcome from this money has been demonstrated.

29.9.4   Chernobyl Children International

Chernobyl Children International (CCI) is a United Nations-accredited, non-profit, international development, medical, and humanitarian organization that works with children, families and communities that continue to be affected by the Chernobyl nuclear disaster. The organization's founder and chief executive is Adi Roche, the Irish humanitarian and peace campaigner. The CCI was founded in 1991 in response to an appeal from Ukrainian and Belarusian doctors for aid. Roche then began organizing 'rest and recuperation' holidays for a few Chernobyl children. Recruiting Irish families who would welcome and care for them, CCI expanded into the United States in 2001.*[222]*[223]

Over its lifetime, the organization has grown in strength and numbers. It works closely with the Belarusian government, the United Nations, and many thousands of volunteers worldwide to deliver a broad range of supports to the children and the wider community. It also acts as an advocate for the rights of those affected by the Chernobyl explosion, and engages in research and outreach activities to encourage the rest of the world to remember the victims and understand the long-term impact on their lives.*[222]*[223]

29.10   Commemoration

The Front Veranda (1986), a lithograph by Susan Dorothea White in the National Gallery of Australia,*[224] exemplifies worldwide awareness of the event. *Heavy Water: A Film for Chernobyl* was released by Seventh Art in 2006 to commemorate the disaster through poetry and first-hand accounts.*[225] The film secured the Best Short Documentary at Cinequest Film Festival as well as the Rhode Island "best score" award*[226] along with a screening at Tate Modern.*[227]

Chernobyl Way is an annual rally run on 26 April by the opposition in Belarus as a remembrance of the Chernobyl disaster.

Soviet badge awarded to liquidators

200,000 karbovanets coin issued by the National Bank of Ukraine to commemorate the 10th anniversary of the Chernobyl disaster

29.11   Cultural impact

Main articles: Cultural impact of the Chernobyl disaster and Nuclear power debate

The Chernobyl accident attracted a great deal of interest. Because of the distrust that many people (both within and outside the USSR) had in the Soviet authorities, a great deal of debate about the situation at the site occurred in the First World during the early days of the event. Because of defective intelligence based on photographs taken from space, it was thought that unit number three had also suffered a dire accident.

Journalists mistrusted many professionals (such as the spokesman from the UK NRPB), and in turn encouraged the public to mistrust them.*[182]

In Italy, the Chernobyl accident was reflected in the outcome of the 1987 referendum. As a result of that referendum, Italy began phasing out its nuclear power plants in 1988, a decision that was effectively reversed in 2008. A referendum in 2011 reiterated Italians' strong objections to nuclear power, thus abrogating the government's decision of 2008.

In Germany, the Chernobyl accident led to the creation of a federal environment ministry, after several states had already created such a post. The minister was given the authority over reactor safety as well, which the current minister still holds as of 2015. The events are also credited with strengthening the anti-nuclear power movement, which culminated in the decision to end the use of nuclear power that was made by the 1998–2005 Schröder government.

29.12   See also

- Convention on Assistance in the Case of a Nuclear Accident or Radiological Emergency – An IAEA treaty created due to the accident.

- Chernobyl New Safe Confinement

- Chernobyl compared to other radioactivity releases

- Children of Chernobyl Benefit Concert

- Convention on Early Notification of a Nuclear Accident

- List of Chernobyl-related articles

- US public opinion on nuclear energy policy after Chernobyl

- National Geographic *Seconds From Disaster* episodes

- Threat of the Dnieper reservoirs

29.13   Notes

[1] Although most reports on the Chernobyl accident refer to a number of graphite fires, it is highly unlikely that the graphite itself burned. According to the General Atomics website <http://gt-mhr.ga.com/safety.php>: "It is often incorrectly assumed that the combustion behavior of graphite is similar to that of charcoal and coal. Numerous tests and calculations have shown that it is virtually impossible to burn high-purity, nuclear-grade graphites." On Chernobyl, the same source states: "Graphite played little or no role in the progression or consequences of the accident. The red glow observed during the Chernobyl accident was the expected color of luminescence for graphite at 700°C and not a large-scale graphite fire, as some have incorrectly assumed." Similarly, nuclear physicist Yevgeny Velikhov,<http://news.bbc.co.uk/2/hi/europe/4918742.stm> noted some 2 weeks after the accident that "Until now the possibility of a catastrophe really did exist: A great quantity of fuel and graphite of the reactor was in an incandescent state" That is, all the nuclear-decay heat that was generated inside the uranium fuel(heat which would normally be extracted by back-up coolant pumps, in an undamaged reactor) was instead responsible for making the fuel itself and any graphite in contact with it, to glow red-hot. This is contrary to the oft cited interpretation, which is that the graphite was red-hot chiefly because it was chemically oxidizing with the air.

29.14   References

Explanatory notes

[1] "No one believed the first newspaper reports, which patently understated the scale of the catastrophe and often contradicted one another. The confidence of readers was re-established only after the press was allowed to examine the events in detail without the original censorship restrictions. The policy of openness (glasnost) and 'uncompromising criticism' of outmoded arrangements had been proclaimed at the 27th Congress (of KPSS), but it was only in the tragic days following the Chernobyl disaster that glasnost began to change from an official slogan into an everyday practice. The truth about Chernobyl that eventually hit the newspapers opened the way to a more truthful examination of other social problems. More and more articles were written about drug abuse, crime, corruption and the mistakes of leaders of various ranks. A wave of 'bad news' swept over the readers in 1986–87, shaking the consciousness of society. Many were horrified to find out about the numerous calamities of which they had previously had no idea. It often seemed to people that there were many more outrages in the epoch of perestroika than before although, in fact, they had simply not been informed about them previously." -Kagarlitsky pp. 333–334

[2] "The mere fact that the operators were carrying out an experiment that had not been approved by higher officials indicates that something was wrong with the chain of command. The State Committee on Safety in the Atomic Power Industry is permanently represented at the Chernobyl station. Yet the engineers and experts in that office were not informed about the program. In part, the tragedy was the product of administrative anarchy or the attempt to keep everything secret." Medvedev, Z., pp. 18–20

Citations

[1] "Chernobyl: Assessment of Radiological and Health Impact, 2002 update; Chapter II – The release, dispersion and deposition of radionuclides" (PDF). OECD-NEA. 2002. Retrieved 2015-06-03.

[2] Black, Richard (12 April 2011). *"Fukushima: As Bad as Chernobyl?"*. BBC. Retrieved 20 August 2011.

[3] Gorbachev, Mikhail (1996), interview in Johnson, Thomas, *The Battle of Chernobyl on YouTube*, [film], Discovery Channel, retrieved 19 February 2014.

[4] "VIDEO: Ukraine remembers Chernobyl victims and heros". European Press Agancy. 30 April 2016. Retrieved 30 April 2016.

[5] Chernobyl Gallery timeline

[6] Чернобыль, Припять, Чернобыльская АЭС и зона отчуждения. ""Shelter" object description". Chornobyl.in.ua. Retrieved 8 May 2012.

[7] http://www.world-nuclear.org/information-library/ nuclear-fuel-cycle/nuclear-power-reactors/appendices/ rbmk-reactors.aspx RBMK Reactors Appendix to Nuclear Power Reactors. WNA.2016

[8] rbmk nuclear power plants: generic safety issues – IAEA 1996

[9] "Frequently Asked Chernobyl Questions". International Atomic Energy Agency – Division of Public Information. May 2005. Archived from the original on 23 February 2011. Retrieved 23 March 2011.

[10] ICRIN Project (2011). *International Chernobyl Portal cher-nobyl.info*. Retrieved 2011. Check date values in: |access-date= (help)

[11] *Environmental consequences of the Chernobyl accident and their remediation: Twenty years of experience. Report of the Chernobyl Forum Expert Group 'Environment'* (PDF). Vienna: International Atomic Energy Agency. 2006. p. 180. ISBN 92-0-114705-8. Retrieved 13 March 2011.

[12] [Nuclear Disasters & The Built Environment: A Report to the Royal Institute ...By Philip Steadman, Simon Hodgkinson pp 55]

[13] An introduction to serious nuclear accident chemistry. Mark Russell St. John Foreman.

[14] "Table 2.2 Number of people affected by the Chernobyl accident (to December 2000)" (PDF). *The Human Consequences of the Chernobyl Nuclear Accident.* UNDP and UNICEF. 22 January 2002. p. 32. Retrieved 17 September 2010.

[15] "Table 5.3: Evacuated and resettled people" (PDF). *The Human Consequences of the Chernobyl Nuclear Accident.* UNDP and UNICEF. 22 January 2002. p. 66. Retrieved 17 September 2010.

[16] IEEE. Chernobyl's Stressful After-Effects The first clear scientific findings are more surprising than is generally appreciated, and their meaning is more obscure By William Sweet Posted 1 Nov 1999

[17] Chernobyl in Perspective, James Peron, 2006

[18] Are passive smoking, air pollution and obesity a greater mortality risk than major radiation incidents? Jim T Smith, BMC Public Health. 20077:49 DOI: 10.1186/1471-2458-7-49

[19] An introduction to serious nuclear accident chemistry. Mark Russell St. John Foreman.*They also found that the level in urine of a marker for DNA damage (8-hydroxydeoxyguanosine) in radiation-exposed rural children was lower (8.5 nmol dm−3) than its level (18.8 nmol dm−3) in urban children whose annual radiation dose (0.25 mSv, 0.025 rem) is lower. They concluded that urban living had a greater effect on the oxidative stress level in children than the small radiation dose that rural children get as a result of the radioactivity released by Chernobyl years ago.*

[20] "Now, liquidators must go to court routinely to get their monthly payments adjusted so that they keep up with inflation. While there are laws dictating that liquidators are entitled to cost-of-living adjustments, the Federal Employment Service does not increase compensation payments until ordered to do so by a court, liquidators said." By Anastasiya Lebedev Apr. 25 2006

[21] Chernobyl's Myths and Misconceptions 2006, Mizsei

[22] Juhn, Poong-Eil; Kupitz, Juergen (1996). "Nuclear power beyond Chernobyl: A changing international perspective" (PDF). *IAEA Bulletin.* **38** (1): 2.

[23] Kagarlitsky, Boris (1989). "Perestroika: The Dialectic of Change". In Mary Kaldor; Gerald Holden; Richard A. Falk. *The New Detente: Rethinking East-West Relations.* United Nations University Press. ISBN 0-86091-962-5.

[24] "Chernobyl cover-up a catalyst for 'glasnost'". Associated Press. 24 April 2006. Retrieved 2015-06-21.

[25] "Assessing the Chernobyl Consequences". International Atomic Energy Agency.

[26] "UNSCEAR 2008 Report to the General Assembly, Annex D" (PDF). United Nations Scientific Committee on the Effects of Atomic Radiation. 2008.

[27] "UNSCEAR 2008 Report to the General Assembly" (PDF). United Nations Scientific Committee on the Effects of Atomic Radiation. 2008.

[28] Hallenbeck, William H (1994). *Radiation Protection.* CRC Press. p. 15. ISBN 0-87371-996-4. Reported thus far are 237 cases of acute radiation sickness and 31 deaths.

[29] "Chernobyl: the true scale of the accident". *Chernobyl's Legacy: Health, Environmental and Socio-Economic Impacts.* Retrieved 15 April 2011.

[30] Cardis, E.; Krewski, D.; Boniol, M.; Drozdovitch, V.; Darby, S. C.; Gilbert, E. S.; Akiba, S.; Benichou, J.; Ferlay, J.; Gandini, S.; Hill, C.; Howe, G.; Kesminiene, A.; Moser, M.; Sanchez, M.; Storm, H.; Voisin, L.; Boyle, P. (2006). "Estimates of the cancer burden in Europe from radioactive fallout from the Chernobyl accident". *International Journal of Cancer.* **119** (6): 1224. doi:10.1002/ijc.22037. PMID 16628547.

[31] Nuclear Law In Progress. INLA congress 2014. pg 4–5

[32] Chernobyl Cancer Death Toll Estimate More Than Six Times Higher Than the 4000 Frequently Cited, According to a New UCS Analysis Note: *"The UCS analysis is based on radiological data provided by UNSCEAR, and is consistent with the findings of the Chernobyl Forum and other researchers."*

[33] "Torch: The Other Report On Chernobyl—executive summary". European Greens and UK scientists Ian Fairlie PhD and David Sumner – Chernobylreport.org. April 2006. Retrieved 20 August 2011.

[34] Alexey V. Yablokov; Vassily B. Nesterenko; Alexey V. Nesterenko (2009). *Chernobyl: Consequences of the Catastrophe for People and the Environment (Annals of the New York Academy of Sciences)* (paperback ed.). Wiley-Blackwell. ISBN 978-1-57331-757-3.

[35] "The Chernobyl Catastrophe. Consequences on Human Health" (PDF). Greenpeace. 2006.

[36] Correspondence (see reference 17) to George Monbiot from Douglas Braaten, Director and Executive Editor, Annals of the New York Academy of Sciences, 2 April 2011: "In no sense did Annals of the New York Academy of Sciences or the New York Academy of Sciences commission this work; nor by its publication do we intend to independently validate the claims made in the translation or in the original publications cited in the work. The translated volume has not been peer-reviewed by the New York Academy of Sciences, or by anyone else."

[37] New York Academy of Sciences (2010-04-28). "Statement on Annals of the New York Academy of Sciences volume entitled "Chernobyl: Consequences of the Catastrophe for People and the Environment"". Retrieved 2011-09-15.

[38] M. I. Balonov (28 April 2010). "Review of Volume 1181". New York Academy of Sciences. Retrieved 15 September 2011. Full text PDF

[39] Medvedev, Zhores A. (1990). *The Legacy of Chernobyl* (Paperback. First American edition published in 1990 ed.). W. W. Norton & Company. ISBN 978-0-393-30814-3.

[40] M. Ragheb (22 March 2011). "Decay Heat Generation in Fission Reactors" (PDF). University of Illinois at Urbana-Champaign. Retrieved 26 January 2013.

[41] "DOE Fundamentals Handbook – Nuclear physics and reactor theory" (PDF). 1 of 2, module 1. United States Department of Energy. DOE–HDBK-1019/1-93 / Available to the public from the National Technical Information Services, U.S. Department of Commerce, 5285 Port Royal, Springfield, VA 22161. January 1996: 61. Retrieved 3 June 2010.

[42] "Standard Review Plan for the Review of Safety Analysis Reports for Nuclear Power Plants: LWR Edition (NUREG-0800)". *United States Nuclear Regulatory Commission.* May 2010. Retrieved 2 June 2010.

[43] Karpan 2006, pp. 312–13

[44] "IAEA Report INSAG-7 Chernobyl Accident: Updating of INSAG-1 Safety Series, No.75-INSAG-7" (PDF). Vienna: International Atomic Energy Agency. 1992.

[45] Dyatlov 2003, p. 30

[46] "Chernobyl: Assessment of Radiological and Health Impact, 2002 update; Chapter I – The site and accident sequence" (PDF). OECD-NEA. 2002. Retrieved 2015-06-03.

[47] "The official program of the test" (in Russian).

[48] Dyatlov 2003, p. 31

[49] "What Happened at Chernobyl?". Nuclear Fissionary. Retrieved 12 January 2011.

[50] The accumulation of Xe-135 in the core is burned out by neutrons. Higher power settings bring higher neutron flux and burn the xenon out more quickly. Conversely, low power settings result in the accumulation of xenon.

[51] The information on accident at the Chernobyl NPP and its consequences, prepared for IAEA, Atomic Energy, v. 61, 1986, pp. 308–320.

[52] The RBMK is a boiling water reactor, so in-core boiling is normal at higher power levels. The RBMK design has a negative void coefficient above 700 MW.

[53] "Physicians of Chernobyl Association" (in Russian). Association «Physicians of Chernobyl». Retrieved September 3, 2013.

[54] Medvedev, Grigori (1989). *The Truth About Chernobyl* (Hardcover. First American edition published by Basic Books in 1991 ed.). VAAP. ISBN 2-226-04031-5.

[55] E. O. Adamov; Yu. M. Cherkashov; et al. (2006). *Channel Nuclear Power Reactor RBMK* (in Russian) (Hardcover ed.). Moscow: GUP NIKIET. ISBN 5-98706-018-4.

[56] Dyatlov, Anatoly. "4". *Chernobyl. How did it happen?* (in Russian).

[57] "Chernobyl as it was – 2" (in Russian).

[58] Davletbaev, RI (1995). *Last shift Chernobyl. Ten years later. Inevitability or chance?* (in Russian). Moscow: Energoatomizdat. ISBN 5-283-03618-9.

[59] "Cheating Chernobyl This interview was first published in New Scientist print edition Source : New Scientist web site".

[60] "Chernobyl 20 years on".

[61] "Chernobyl: what happened and why? by CM Meyer, technical journalist" (PDF).

[62] Checherov, K.P. (25–27 November 1998). *Development of ideas about reasons and processes of emergency on the 4-th unit of Chernobyl NPP 26.04.1986* (in Russian). Slavutich, Ukraine: International conference "Shelter-98".

[63] Pakhomov, Sergey A.; Dubasov, Yuri V. (2009). "Estimation of Explosion Energy Yield at Chernobyl NPP Accident". *Pure and Applied Geophysics.* **167** (4–5): 575. doi:10.1007/s00024-009-0029-9.

[64] B. Medvedev (June 1989). "JPRS Report: Soviet Union Economic Affairs Chernobyl Notebook" (Republished by the Foreign Broadcast Information Service ed.). Novy Mir. Retrieved 27 March 2011.

[65] "Cross-sectional view of the RBMK-1000 main building". Archived from the original on 13 June 2011. Retrieved 11 September 2010.

[66] "Meltdown in Chernobyl". *National Geographic Channel* (Video). National Geographic. 2011-08-10. Retrieved 2015-06-21.

[67] Shcherbak, Y (1987). Medvedev, ed. "Chernobyl". **6**. Yunost: 44.

[68] Adam Higginbotham (26 March 2006). "Adam Higginbotham: Chernobyl 20 years on | World news | The Observer". *The Guardian.* London. Retrieved 22 March 2010.

[69] Mil Mi-8 crash near Chernobyl on YouTube 2006.

[70] "*Special Report: 1997: Chernobyl: Containing Chernobyl?*". BBC News. 21 November 1997. Retrieved 20 August 2011.

[71] Zeilig, Martin (August–September 1995). "Louis Slotin And 'The Invisible Killer'". *The Beaver.* **75** (4): 20–27. Retrieved 28 April 2008.

[72] National Geographic, VOL. 171, NO. 5, May 1987 (article "Chernobyl – One Year After")

[73] "Веб публикация статей газеты". Swrailway.gov.ua. Retrieved 22 March 2010.

[74] Методическая копилка (in Russian). Surkino.edurm.ru. Retrieved 22 March 2010.

[75] *Time: Disasters that Shook the World.* New York City: Time Home Entertainment. 2012. ISBN 1-60320-247-1.

[76] "Interview of Valentyna Shevchenko to "Young Ukraine" (Ukrainian Pravda)". Istpravda.com.ua. 25 April 2011. Archived from the original on 26 April 2016. Retrieved 20 August 2011.

[77] Director: Maninderpal Sahota; Narrator: Ashton Smith; Producer: Greg Lanning; Edited by: Chris Joyce (17 August 2004). "Meltdown in Chernobyl". *Seconds From Disaster.* Season 1. Episode 7. 30/40–50 minutes minutes in. National Geographic Channel.

[78] Marples, David R. (1988). *The Social Impact of the Chernobyl Disaster.* New York, NY: St Martin's Press.

[79] "Chernobyl haunts engineer who alerted world". *CNN Interactive World News.* Cable News Network, Inc. 26 April 1996. Retrieved 28 April 2008.

[80] (Russian) Video footage of Chernobyl disaster on 28 April on YouTube

[81] "Timeline". *The Chernobyl Gallery.*

[82] Schmemann, Serge (1986-04-29). "Soviet Announces Nuclear Accident at Electric Plant". *The New York Times.* pp. A1. Retrieved 26 April 2014.

[83] "1986: американський ТБ-сюжет про Чорнобиль. Порівняйте з радянським". *Історична правда.*

[84] Bogatov, S. A.; Borovoi, A. A.; Lagunenko, A. S.; Pazukhin, E. M.; Strizhov, V. F.; Khvoshchinskii, V. A. (2009). "Formation and spread of Chernobyl lavas". *Radiochemistry.* **50** (6): 650. doi:10.1134/S1066362208050131.

[85] Petrov, Yu. B.; Udalov, Yu. P.; Subrt, J.; Bakardjieva, S.; Sazavsky, P.; Kiselova, M.; Selucky, P.; Bezdicka, P.; Jorneau, C.; Piluso, P. (2009). "Behavior of melts in the UO2-SiO2 system in the liquid-liquid phase separation region". *Glass Physics and Chemistry.* **35** (2): 199. doi:10.1134/S1087659609020126.

[86] Journeau, C.; E. Boccaccio; C. Jégou; P. Piluso; G. Cognet (2001). "Flow and Solidification of Corium in the VULCANO facility" (PDF).

[87] Medvedev Z. (1990). *The Legacy of Chernobyl.* W W Norton & Co Inc. pp. 58–59. ISBN 0-393-30814-6.

[88] Kramer, Sarah (26 April 2016). "The amazing true story behind the Chernobyl 'suicide squad' that helped save Europe". *Business Insider.* Retrieved 7 October 2016.

[89] Samodelova, Svetlana (25 April 2011). "Белые пятна Чернобыля". *Московский комсомолец* (in Russian). Retrieved 7 October 2016.

[90] Chernobyl: The End of the Nuclear Dream, 1986, p.178, by Nigel Hawkes et al., ISBN 0-330-29743-0

[91] "Stephen McGinty: Lead coffins and a nation's thanks for the Chernobyl suicide squad". scotsman.com. 16 March 2011.

[92] Ukrainian translation of Alexei Ananenko memories, from original http://www.souzchernobyl.org/?id=2440

[93] "Человек широкой души: Вот уже девятнадцатая годовщина Чернобыльской катастрофы заставляет нас вернуться в своих воспоминаниях к апрельским дням 1986 года". Post Chernobyl. 2005-04-16. (autotranslation)

[94] Tom Burnett (28 March 2011). "When the Fukushima Meltdown Hits Groundwater". Hawai'i News Daily.

[95] Sattonnay, G.; Ardois, C.; Corbel, C.; Lucchini, J. F.; Barthe, M.-F.; Garrido, F.; Gosset, D. (2001). "Alpha-radiolysis effects on UO2 alteration in water". *Journal of Nuclear Materials.* **288**: 11–19. Bibcode:2001JNuM..288...11S. doi:10.1016/S0022-3115(00)00714-5.

[96] Clarens, F.; De Pablo, J.; Díez-Pérez, I.; Casas, I.; Giménez, J.; Rovira, M. (2004). "Formation of Studtite during the Oxidative Dissolution of UO_2 by Hydrogen Peroxide: A SFM Study". *Environmental Science & Technology*. **38** (24): 6656–6661. doi:10.1021/es0492891.

[97] Burakov, B. E.; Strykanova, E. E.; Anderson, E. B. (2012). "Secondary Uranium Minerals on the Surface of Chernobyl "Lava"". *MRS Proceedings*. **465**. doi:10.1557/PROC-465-1309.

[98] Burns, P. C; K. A. Hughes (2003). "Studtite, (UO2)(O2)(H2O)2(H2O)2: The first structure of a peroxide mineral". *American Mineralogist*. **88** (7): 1165–1168.

[99] The Social Impact of the Chernobyl Disaster, 1988, p. 166, by David R. Marples ISBN 0-333-48198-4

[100] Collecting History (1986-04-26). "Medal for Service at the Chernobyl Nuclear Disaster". Collectinghistory.net. Retrieved 2013-09-12.

[101] "Chernobyl's silent graveyards". *BBC News*. 20 April 2006.

[102] "Chernobyl's Hot Mess, "the Elephant's Foot," Is Still Lethal". *Nautilus magazine*. 4 December 2013.

[103] Petryna, Adriana (2002). *Life Exposed: Biological Citizens After Chernobyl*. Princeton, NJ: Princeton University Press.

[104] IAEA Report INSAG-1 (International Nuclear Safety Advisory Group). Summary Report on the Post-Accident Review on the Chernobyl Accident. Safety Series No. 75-INSAG-1.IAEA, Vienna, 1986.

[105] "Expert report to the IAEA on the Chernobyl accident" (in Russian). **61**. Atomic Energy. 1986.

[106] "INSAG-7 The Chernobyl Accident: Updating of INSAG-1" (PDF). Retrieved 2013-09-12.

[107] Masayuki Nakao. "Chernobyl Accident (Case details)".

[108] "Украина рассекретила документы, касающиеся аварии на Чернобыльской АЭС". Retrieved September 13, 2015.

[109] "NEI Source Book: Fourth Edition (NEISB_3.3.A1)". Insc.anl.gov. Archived from the original on 2 July 2016. Retrieved 31 July 2010.

[110] "Facts: The accident was by far the most devastating in the history of nuclear power". *Ten years after Chernobyl : What do we really know?*. International Atomic Energy Agency (IAEA). 21 September 1997. Retrieved 20 August 2011.

[111] Marples, David R. (May–June 1996). "The Decade of Despair". *The Bulletin of the Atomic Scientists*. **52** (3): 20–31.

[112] "Tchernobyl, 20 ans après" (in French). RFI. 24 April 2006. Retrieved 24 April 2006.

[113] "L'accident et ses conséquences: Le panache radioactif" [The accident and its consequences: The plume] (in French). Institut de Radioprotection et de Sûreté Nucléaire (IRSN). Retrieved 16 December 2006.

[114] Jensen, Mikael; Lindhé, John-Christer (Autumn 1986). "International Reports – Sweden: Monitoring the Fallout" (PDF). *IAEA Bulletin*. International Atomic Energy Agency (IAEA)

[115] Mould, Richard Francis (2000). *Chernobyl Record: The Definitive History of the Chernobyl Catastrophe*. CRC Press. p. 48. ISBN 0-7503-0670-X.

[116] Ikäheimonen, T.K. (ed.). *Ympäristön Radioaktiivisuus Suomessa – 20 Vuotta Tshernobylista* [*Environmental Radioactivity in Finland – 20 Years from Chernobyl*] (PDF). Säteilyturvakeskus Stralsäkerhetscentralen (STUK, Radiation and Nuclear Safety Authority)

[117] *3.1.5. Deposition of radionuclides on soil surfaces* (PDF). *Environmental Consequences of the Chernobyl Accident and their Remediation: Twenty Years of Experience, Report of the Chernobyl Forum Expert Group 'Environment'*. Vienna: International Atomic Energy Agency (IAEA). 2006. pp. 23–25. ISBN 92-0-114705-8. Retrieved 2013-09-12.

[118] Gould, Peter (1990). *Fire In the Rain: The Dramatic Consequences of Chernobyl*. Baltimore, MD: Johns Hopkins Press.

[119] Gray, Richard (22 April 2007). "How we made the Chernobyl rain". *Telegraph*. London. Retrieved 27 November 2009.

[120] "Chernobyl Accident 1986". World Nuclear Association. April 2015. Retrieved 21 April 2015.

[121] Zoriy, Pedro; Dederichs, Herbert; Pillath, Jürgen; Heuel-Fabianek, Burkhard; Hill, Peter; Lennartz, Reinhard (2016). "Long-term monitoring of radiation exposure of the population in radioactively contaminated areas of Belarus – Korma Study – The Korma Report II (1998–2015)". *Schriften des Forschungszentrums Jülich: Reihe Energie & Umwelt / Energy & Environment*. Forschungszentrum Jülich, Zentralbibliothek, Verlag. Retrieved 21 December 2016.

[122] fr:Conséquences de la catastrophe de Tchernobyl en France

[123] "'Radioactive boars' on loose in Germany". *Agence France Presse*. 10 August 2010. Retrieved 13 March 2015.

[124] Marples, David R. (1991). *Ukraine Under Perestroika: Ecology, Economics and the Workers' Revolt*. Basingstoke, Hampshire: MacMillan Press. pp. 50–51, 76.

[125] Wertelecki, W. (2010). "Malformations in a Chornobyl-Impacted Region". *Pediatrics*. **125** (4): e836–43. doi:10.1542/peds.2009-2219. PMID 20308207.

[126] Dancause, Kelsey Needham; Yevtushok, Lyubov; Lapchenko, Serhiy; Shumlyansky, Ihor; Shevchenko, Genadiy; Wertelecki, Wladimir; Garruto, Ralph M. (2010). "Chronic radiation exposure in the Rivne-Polissia

region of Ukraine: Implications for birth defects". *American Journal of Human Biology*. **22** (5): 667–74. doi:10.1002/ajhb.21063. PMID 20737614.

[127] Møller, Anders Pape; Pape, Anders (April 1998). "Developmental Instability of Plants and Radiation from Chernobyl". *Oikos*. Nordic Ecological Society. **81** (3): 444–48. doi:10.2307/3546765. JSTOR 3546765.

[128] Saino, N.; Mousseau, F.; De Lope, T. A.; Saino, A. P. (2007). "Elevated frequency of abnormalities in barn swallows from Chernobyl". *Biology Letters*. **3** (4): 414–17. doi:10.1098/rsbl.2007.0136. PMC 1994720⊙. PMID 17439847.

[129] Weigelt, E.; Scherb, H. (2004). "Spaltgeburtenrate in Bayern vor und nach dem Reaktorunfall in Tschernobyl". *Mund-, Kiefer- und Gesichtschirurgie*. **8** (2): 106. doi:10.1007/s10006-004-0524-1.

[130] Bennett, Burton; Repacholi, Michael; Carr, Zhanat, eds. (2006). *Health Effects of the Chernobyl Accident and Special Health Care Programmes: Report of the UN Chernobyl Forum, Expert Group "Health"* (PDF). Geneva: World Health Organization (WHO). p. 79. ISBN 978-92-4-159417-2. Retrieved 20 August 2011

[131] P. Gudiksen; et al. (1989). "Chernobyl Source Term, Atmospheric Dispersion, and Dose Estimation". *Health Physics*. **57** (5).

[132] "Chernobyl, Ten Years On: Assessment of Radiological and Health Impact" (PDF). OECD-NEA. 1995. Retrieved 2015-06-03.

[133] "The Society for Radiological Protection – SRP". Srp-uk.org. Retrieved 2013-09-12.

[134] "Applet for kids". Colorado.edu. 1999-09-20. Retrieved 2013-09-12.

[135] Ken Lyle. "Mathematical half life decay rate equations". Chem.purdue.edu. Retrieved 2013-09-12.

[136] "Unfall im japanischen Kernkraftwerk Fukushima". ZAMG. 24 March 2011. Retrieved 20 August 2011.

[137] Cesium-137: A Deadly Hazard. Large.stanford.edu (2012-03-20). Retrieved on 2013-02-13.

[138] Marples, David R. (1988). *The Social Impact of the Chernobyl Disaster*. New York, NY: St Martin's Press.

[139] Chernobyl: Catastrophe and Consequences, Springer, Berlin ISBN 3-540-23866-2

[140] Kryshev, I.I. (1995). "Radioactive contamination of aquatic ecosystems following the Chernobyl accident". *Journal of Environmental Radioactivity*. **27** (3): 207. doi:10.1016/0265-931X(94)00042-U.

[141] EURATOM Council Regulations No. 3958/87, No. 994/89, No. 2218/89, No. 770/90

[142] Fleishman, David G.; Nikiforov, Vladimir A.; Saulus, Agnes A.; Komov, Victor T. (1994). "137Cs in fish of some lakes and rivers of the Bryansk region and north-west Russia in 1990–1992". *Journal of Environmental Radioactivity*. **24** (2): 145. doi:10.1016/0265-931X(94)90050-7.

[143] ""Environmental consequences of the Chernobyl accident and their remediation"" (PDF). IAEA, Vienna

[144] *Wildlife defies Chernobyl radiation*, by Stefen Mulvey, BBC News

[145] The International Chernobyl Project Technical Report, IAEA, Vienna, 1991

[146] "'Radiation-Eating' Fungi Finding Could Trigger Recalculation Of Earth's Energy Balance And Help Feed Astronauts".

[147] "25 Jahre Tschernobyl: Deutsche Wildschweine immer noch verstrahlt – Nachrichten Wissenschaft – WELT ONLINE". *Die Welt* (in German). 18 March 2011. Retrieved 20 August 2011.

[148] "Record low number of radioactive sheep". *The Local*. The Local Europe AB. 23 September 2013. Retrieved 1 November 2013.

[149] "Fortsatt nedforing etter radioaktivitet i dyr som har vært på utmarksbeite – Statens landbruksforvaltning" (in Norwegian). SLF. 30 June 2010. Retrieved 21 June 2015.

[150] Macalister, Terry; Helen Carter (12 May 2009). "Britain's farmers still restricted by Chernobyl nuclear fallout". *The Guardian*. Retrieved 1 November 2013.

[151] Rawlinson, Kevin; Rachel Hovenden (7 July 2010). "Scottish sheep farms finally free of Chernobyl fallout". *The Independent*. Retrieved 1 November 2013.

[152] "Post-Chernobyl disaster sheep controls lifted on last UK farms". *BBC News*. BBC. 1 June 2012. Retrieved 1 November 2013.

[153] Food Standards Agency (29 November 2012). "Welsh sheep controls revoked". Retrieved 1 November 2013.

[154] Mould 2000, p. 29. "The number of deaths in the first three months were 31[.]"

[155] "Chernobyl's Legacy: Health, Environmental and Socio-Economic Impacts" (PDF). *Chernobyl Forum assessment report*. Chernobyl Forum. Retrieved 21 April 2012.

[156] Fairlie, Ian; Sumner, David (2006). *The Other Report on Chernobyl (TORCH)*. Berlin: The European Greens.

[157] Kuchinskaya, Olga (2007), *We Will Die and Become Science: The production of invisibility and public knowledge about Chernobyl radiation effects in Belarus (doctoral dissertation)*, San Diego: University of California

[158] Mycio, Mary (2005). *Wormwood Forest: A Natural History of Chernobyl*. Washington, D.C.: Joseph Henry Press.

[159] Marples, David R. (1996). *Belarus: From Soviet Rule to Nuclear Catastrophe*. Basingstoke, Hampshire: MacMillan Press.

[160] Chernobyl: the true scale of the accident, Joint News Release WHO/IAEA/UNDP, 5 SEPTEMBER 2005

[161] "UNSCEAR – Chernobyl health effects". Unscear.org. Retrieved 23 March 2011.

[162] Rosenthal, Elisabeth. (6 September 2005) Experts find reduced effects of Chernobyl. nytimes.com. Retrieved 14 February 2008.

[163] "Thyroid Cancer". Genzyme.ca. Retrieved 31 July 2010.

[164] "CHERNOBYL at 25th anniversary Frequently Asked Questions April 2011" (PDF). World Health Organisation. 23 April 2011. Retrieved 14 April 2012.

[165] "Chernobyl's Legacy: Health, Environmental and Socia-Economic Impacts and Recommendations to the Governments of Belarus, Russian Federation and Ukraine" (PDF). International Atomic Energy Agency – The Chernobyl Forum: 2003–2005. Retrieved 31 July 2010.

[166] "Excerpt from UNSCEAR 2001 REPORT ANNEX – Hereditary effects of radiation" (PDF). Retrieved 20 August 2011.

[167] Bogdanova, Tetyana I.; Zurnadzhy, Ludmyla Y.; Greenebaum, Ellen; McConnell, Robert J.; Robbins, Jacob; Epstein, Ovsiy V.; Olijnyk, Valery A.; Hatch, Maureen; Zablotska, Lydia B.; Tronko, Mykola D. (2006). "A cohort study of thyroid cancer and other thyroid diseases after the Chornobyl accident". *Cancer*. **107** (11): 2559–66. doi:10.1002/cncr.22321. PMC 2983485. PMID 17083123.

[168] Dinets, A.; Hulchiy, M.; Sofiadis, A.; Ghaderi, M.; Hoog, A.; Larsson, C.; Zedenius, J. (2012). "Clinical, genetic, and immunohistochemical characterization of 70 Ukrainian adult cases with post-Chornobyl papillary thyroid carcinoma". *European Journal of Endocrinology*. **166** (6): 1049–60. doi:10.1530/EJE-12-0144. PMC 3361791. PMID 22457234.

[169] "20 years after Chernobyl – The ongoing health effects". *IPPNW*. April 2006. Retrieved 24 April 2006.

[170] Mettler, Fred. "IAEA Bulletin Volume 47, No. 2 – Chernobyl's Legacy". Iaea.org. Retrieved 20 August 2011.

[171] "What's the situation at Chernobyl?". Iaea.org. Retrieved 20 August 2011.

[172] Mettler, Fred. "Chernobyl's living legacy". Iaea.org. Retrieved 20 August 2011.

[173] "UNSCEAR assessment of the Chernobyl accident". United Nations Scientific Committee of the Effects of Atomic Radiation. Retrieved 31 July 2010.

[174] "Historical milestones". United Nations Scientific Committee of the Effects of Atomic Radiation. Retrieved 14 April 2012.

[175] World Health Organisation "World Health Organization report explains the health impacts of the world's worst-ever civil nuclear accident", *WHO*, 26 April 2006. Retrieved 4 April 2011.

[176] Berrington De González, Amy; Mahesh, M; Kim, KP; Bhargavan, M; Lewis, R; Mettler, F; Land, C (2009). "Projected Cancer Risks from Computed Tomographic Scans Performed in the United States in 2007". *Archives of Internal Medicine*. **169** (22): 2071–77. doi:10.1001/archinternmed.2009.440. PMID 20008689.

[177] Normile, D. (2011). "Fukushima Revives the Low-Dose Debate". *Science*. **332** (6032): 908–10. doi:10.1126/science.332.6032.908. PMID 21596968.

[178] "How Many Cancers Did Chernobyl Really Cause?". UCSUSA.org. 17 April 2011.

[179] Hawley, Charles. "Greenpeace vs. the United Nations". *The Chernobyl Body Count Controversy*. SPIEGEL. Retrieved 15 March 2011.

[180] "Details". *Annals of the New York Academy of Sciences*. Annals of the New York Academy of Sciences. Retrieved 15 March 2011.

[181] - pp. 85–86, pp. 92–93 in "Radiation: What It Is, What You Need To Know" by Robert Peter Gale, M.D., Ph.D. and Eric Lax. Publisher: Alfred A. Knopf, New York, 2013. ("The correct number of Chernobyl-related cancers will never be known, in part because of the considerable uncertainties in estimating cancers and cancer deaths. Especially problematic is the controversy about whether very low doses of radiation, especially if given over a prolonged interval, increase cancer risk. There are other difficulties as well. For one, we do not know precisely what radiation dose most people received. People who were indoors when the radioactive plume passed received much less radiation than those who were outdoors. However, because most people did not know when the radioactive plume passed, they cannot accurately reconstruct their whereabouts at that time. Also, many people were evacuated from contaminated land at different times and thus received very different doses from ground and food contamination. Next we have the geopolitical reality that many of the exposed people no longer live in the Chernobyl area. Living elsewhere, even in other countries, they are lost to follow-up. The Chernobyl accident was relatively quickly followed by the dissolution of the Soviet Union, whereupon many people's lifestyles but perhaps not their lives changed, mostly for the worse. For example, cigarette smoking and alcohol consumption increased, resulting in a profound drop in life expectancy. Both activities are correlated with increased cancer risk independent of radiation exposure. Sorting out any changes in cancer incidence or prevalence will be difficult at best. [···] Fourth, high-quality cancer registries were absent before and even

after the accident, making it impossible to know with certainty the background rate of most cancers before the accident.")

[182] Kasperson, Roger E.; Stallen, Pieter Jan M. (1991). *Communicating Risks to the Public: International Perspectives.* Berlin: Springer Science and Media. pp. 160–162. ISBN 0-7923-0601-5.

[183] Knudsen, L. B. (1991). "Legally induced abortions in Denmark after Chernobyl". *Biomedicine & Pharmacotherapy.* **45** (6): 229–31. doi:10.1016/0753-3322(91)90022-L. PMID 1912378.

[184] Trichopoulos, D; Zavitsanos, X; Koutis, C.; Drogari, P; Proukakis, C.; Petridou, E. (1987). "The victims of chernobyl in Greece: Induced abortions after the accident". *British Medical Journal.* **295** (6606): 1100. doi:10.1136/bmj.295.6606.1100. PMC 1248180⊝. PMID 3120899.

[185] Parazzini, F.; Repetto, F.; Formigaro, M.; Fasoli, M.; La Vecchia, C . (1988). "Induced abortions after the Chernobyl accident". *British Medical Journal.* **296** (6615): 136. doi:10.1136/bmj.296.6615.136-a. PMC 2544742⊝. PMID 3122957.

[186] Perucchi, M; Domenighetti, G (1990). "The Chernobyl accident and induced abortions: Only one-way information". *Scandinavian Journal of Work, Environment & Health.* **16** (6): 443–44. doi:10.5271/sjweh.1761. PMID 2284594.

[187] Kasperson, Roger E.; Stallen, Pieter Jan M. (1991). *Communicating Risks to the Public: International Perspectives.* Berlin: Springer Science and Media. pp. 160–162. ISBN 0-7923-0601-5.

[188] Knudsen, LB (1991). "Legally induced abortions in Denmark after Chernobyl." . *Biomed Pharmacother.* **45**: 229–31. doi:10.1016/0753-3322(91)90022-l. PMID 1912378.

[189] Ketchum, Linda E. (1987). "Lessons of Chernobyl: SNM Members Try to Decontaminate World Threatened by Fallout". *Journal of Nuclear Medicine.* **28** (6): 933–42. PMID 3585500.

[190] Chernobyl's Hot Zone Holds Some Surprises

[191] [RadSafe   Chernobyl-related abortions Bjorn Cedervall, 2010]

[192] Little, J. (1993). "The Chernobyl accident, congenital anomalies and other reproductive outcomes". *Paediatric and Perinatal Epidemiology.* **7** (2): 121–51. doi:10.1111/j.1365-3016.1993.tb00388.x. PMID 8516187.

[193] Odlind, V; Ericson, A (1991). "Incidence of legal abortion in Sweden after the Chernobyl accident". *Biomedicine & Pharmacotherapy.* **45** (6): 225–8. doi:10.1016/0753-3322(91)90021-k. PMID 1912377.

[194] Harjulehto, T; Rahola, T; Suomela, M; Arvela, H; Saxén, L (1991). "Pregnancy outcome in Finland after the Chernobyl accident". *Biomedicine & Pharmacotherapy.* **45** (6): 263–6. doi:10.1016/0753-3322(91)90027-q. PMID 1912382.

[195] Czeizel, AE (1991). "Incidence of legal abortions and congenital abnormalities in Hungary". *Biomedicine & Pharmacotherapy.* **45** (6): 249–54. doi:10.1016/0753-3322(91)90025-o. PMID 1912381.

[196] Haeusler, MC; Berghold, A; Schoell, W; Hofer, P; Schaffer, M (1992). "The influence of the post-Chernobyl fallout on birth defects and abortion rates in Austria". *American Journal of Obstetrics and Gynecology.* **167** (4 Pt 1): 1025–31. PMID 1415387.

[197] Dolk, H.; Nichols, R. (1999). "Evaluation of the impact of Chernobyl on the prevalence of congenital anomalies in 16 regions of Europe. EUROCAT Working Group". *International Journal of Epidemiology.* **28** (5): 941–8. doi:10.1093/ije/28.5.941. PMID 10597995.

[198] Castronovo Jr, Frank P. (1999). "Teratogen Update: Radiation and Chernobyl" (PDF). *TERATOLOGY.* **60**: 100–106. doi:10.1002/(sici)1096-9926(199908)60:2<100::aid-tera14>3.3.co;2-8.

[199] ASU school of life scientist:Kenneth Mossman Archived 2 July 2012 at the Wayback Machine.

[200] Mossman, Kenneth L. (1998). "The linear no-threshold debate: Where do we go from here?". *Medical Physics.* **25** (3): 279–84; discussion 300. doi:10.1118/1.598208. PMID 9547494.

[201] Shkolnikov, V.; McKee, M.; Vallin, J.; Aksel, E.; Leon, D.; Chenet, L; Meslé, F (1999). "Cancer mortality in Russia and Ukraine: Validity, competing risks and cohort effects". *International Journal of Epidemiology.* **28** (1): 19–29. doi:10.1093/ije/28.1.19. PMID 10195659.

[202] Shlyakhter, Alexander; Wilson, Richard (1992). "Chernobyl andGlasnost: The Effects of Secrecy on Health and Safety". *Environment: Science and Policy for Sustainable Development.* **34** (5): 25. doi:10.1080/00139157.1992.9931445.

[203] Petryna, Adriana (1995). "Sarcophagus: Chernobyl in Historical Light". *Cultural Anthropology.* **10** (2): 196. doi:10.1525/can.1995.10.2.02a00030.

[204] Petryna, Adriana (2002). *Life Exposed: Biological Citizens after Chernobyl.* Princeton, NJ: Princeton University Press.

[205] "Information Notice No. 93–71". Nrc.gov. Retrieved 20 August 2011.

[206] IAEA's Power Reactor Information System polled in May 2008 reports shutdown for units 1, 2, 3 and 4 respectively at 30 November 1996, 11 October 1991, 15 December 2000 and 26 April 1986.

[207] Walker, Shaun (2016-11-29). "Chernobyl disaster site enclosed by shelter to prevent radiation leaks". *The Guardian.* ISSN 0261-3077. Retrieved 2016-12-23.

[208] Nechepurenko, Ivan; Fountain, Henry (2016-11-29). "Giant Arch, a Feat of Engineering, Now Covers Chernobyl Site in Ukraine". *The New York Times.* ISSN 0362-4331. Retrieved 2016-12-23.

[209] "Inside Chernobyl". 60 Minutes Australia, Nine Network Australia. 16 April 2006.

[210] "Collapse of Chernobyl nuke plant building attributed to sloppy repair work, aging". *The Mainichi Newspapers.* 25 April 2013. Archived from the original on 29 April 2013. Retrieved 26 April 2013.

[211] "Ukraine: Chernobyl nuclear roof collapse 'no danger'". *BBC News.* 2013-02-13. Retrieved 2016-12-23.

[212] Baryakhtar, V.; Gonchar, V.; Zhidkov, A.; Zhidkov, V. (2002). "Radiation damages and self-sputtering of high-radioactive dielectrics: spontaneous emission of submicronic dust particles" (PDF). *Condensed Matter Physics.* **5** (3{31}): 449–471. doi:10.5488/cmp.5.3.449.

[213] Borovoi, A. A. (2006). "Nuclear fuel in the shelter". *Atomic Energy.* **100** (4): 249. doi:10.1007/s10512-006-0079-3.

[214] Čejkaite

[215] Oliphant, Roland (24 April 2016). "30 years after Chernobyl disaster, wildlife is flourishing in radioactive wasteland". The Telegraph. Retrieved 27 April 2016.

[216] "News". Yahoo News. Associated Press. 13 December 2010. Retrieved 2 March 2012.

[217] "Tours of Chernobyl sealed zone officially begin". TravelSnitch. TravelSnitch. 18 March 2011.

[218] "Chernobyl's radioactive trees and the forest fire risk". *BBC News.*

[219] "History, Travel, Arts, Science, People, Places – Smithsonian".

[220] "NOVARKA and Chernobyl Project Management Unit confirm cost and time schedule for Chernobyl New Safe Confinement". 8 April 2011. Archived from the original on 18 September 2011. Retrieved 28 March 2012.

[221] "CRDP: Chernobyl Recovery and Development Programme (United Nations Development Programme)". Undp.org.ua. Retrieved 31 July 2010.

[222] Adi Roche – the early favourite, *BBC News,* October 29, 1997.

[223] The 2015 World of Children Award Honorees

[224] ""The Front Veranda" (1986)". Susandwhite.com.au. Retrieved 2013-09-12.

[225] "Processing the Dark: *Heavy Water – A Film for Chernobyl* | Movie Mail UK". Moviemail-online.co.uk. Retrieved 31 July 2010.

[226] "Heavy Water: A film for Chernobyl". *www.atomictv.com.* Retrieved 2015-06-21.

[227] "Heavy Water: a film for Chernobyl". Atomictv.com. 26 April 1986. Retrieved 6 August 2013.

Sources

The source documents relating to the emergency, published in unofficial sources:

- Technological Regulations on operation of 3 and 4 power units of Chernobyl NPP (in force at the moment of emergency)

- Tables and graphs of some parameters variation of the unit before the emergency

29.15 Further reading

- Abbott, Pamela (2006). *Chernobyl: Living With Risk and Uncertainty.* Health, Risk & Society 8.2. pp. 105–121.

- Cohen, Bernard Leonard (1990). "(7) The Chernobyl accident – can it happen here?". *The Nuclear Energy Option: An Alternative for the 90's.* Plenum Press. ISBN 978-0-306-43567-6.

- Dyatlov, Anatoly (2003). *Chernobyl. How did it happen.* (in Russian). Nauchtechlitizdat, Moscow. ISBN 5-93728-006-7.

- Hoffmann, Wolfgang (2001). *Fallout From the Chernobyl Nuclear Disaster and Congenital Malformations in Europe.* Archives of Environmental Health.

- Karpan, Nikolaj V. (2006). *Chernobyl. Vengeance of peaceful atom.* (in Russian). Dnepropetrovsk: IKK "Balance Club". ISBN 966-8135-21-0.

- Medvedev, Grigori (1989). *The Truth About Chernobyl.* VAAP. First American edition published by Basic Books in 1991. ISBN 2-226-04031-5.

- Medvedev, Zhores A. (1990). *The Legacy of Chernobyl* (Paperback. First American edition published in 1990 ed.). W. W. Norton & Company. ISBN 978-0-393-30814-3.

- Read, Piers Paul (1993). *Ablaze! The Story of the Heroes and Victims of Chernobyl.* Random House UK (paperback 1997). ISBN 978-0-7493-1633-4.

- Shcherbak, Yurii (1991). *Chernobyl* (in Russian and English). New York: Soviet Writers/St. Martin's Press. ISBN 0-312-03097-5.

- Yaroshinskaya, Alla A. *Chernobyl: Crime Without Punishment.* Piscataway, NJ: Transaction Publisher, 2015.

- Gerd Ludwig and Lois Lammerhuber: *Der lange Schatten von Tschernobyl – The Long Shadow of Chernobyl – L'ombre de Tchernobyl.* Edition Lammerhuber, 2014, ISBN 978-3901753664. Illustrated book containing photos taken in 2013 within the reactor hull. Interview with the photographer (in German)

- Vjačeslav Šestopalov, et al.: *Groundwater vulnerability: Chernobyl nuclear disaster.* American Geophysical Union, Washington 2015, ISBN 978-1-118-96219-0.

- Busby, Chris (January 2017). *The Scientific Hero of Chernobyl: Alexey V. Yablokov, the Man Who Dared to Speak the Truth*

29.16 External links

- Official UN Chernobyl site

- International Chernobyl Portal chernobyl.info, UN Inter-Agency Project ICRIN

- Frequently Asked Chernobyl Questions, by the IAEA

- Chernobyl Recovery and Development Programme (United Nations Development Programme)

- Photographs from inside the zone of alienation and City of Prypyat (2010)

- Photographs from inside the Chernobyl Reactor and City of Prypyat

- Photographs of those affected by the Chernobyl Disaster

- Photographs from the City of Pripyat, and of those affected by the disaster

- EnglishRussia Photos of a RBMK-based power plant, showing details of the reactor hall, pumps, and the control room

- 25 years of satellite imagery over Chernobyl

- Post-Soviet Pollution: Effects of Chernobyl from the Dean Peter Krogh Foreign Affairs Digital Archives

Coordinates: 51°23′23″N 30°05′57″E / 51.38972°N 30.09917°E

Chapter 30

List of Chernobyl-related articles

Abandoned living blocks of Pripyat, with a surviving tree

Entrance to the zone of alienation around Chernobyl

Chernobyl power plant in 2003 with the sarcophagus containment structure

This is a **list of Chernobyl-related articles**.

The radiation warning symbol (trefoil).

30.1 Disaster and effects

- Chernobyl compared to other radioactivity releases

- Chernobyl disaster

- Chernobyl disaster effects

- Chernobyl necklace
- Convention on Early Notification of a Nuclear Accident, adopted in direct response to Chernobyl
- Cultural impact of the Chernobyl disaster
- Deaths due to the Chernobyl disaster
- Individual involvement in the Chernobyl disaster
- Radiophobia

30.2   Places and geography

30.2.1   Power plant

- Chernobyl Nuclear Power Plant
- Chernobyl Nuclear Power Plant sarcophagus
- New Safe Confinement

30.2.2   Exclusion zone

- Chernobyl Nuclear Power Plant Exclusion Zone, also known as the Zone of Alienation
- Prypiat, abandoned city
- Chernobyl (city), semi-abandoned city
- Kopachi, abandoned village
- Poliske, abandoned town
- Red Forest

30.2.3   Other

- Slavutych, city established in 1986 after the disaster

30.3   Documents and media

See also: Cultural impact of the Chernobyl disaster

- *The Bell of Chernobyl*
- *Chernobyl: Consequences of the Catastrophe for People and the Environment*
- *Chernobyl Heart*
- *The Russian Woodpecker*

- TORCH report
- *The Truth About Chernobyl*
- *Voices from Chernobyl: The Oral History of a Nuclear Disaster*

30.3.1   Fiction

- Markiyan Kamysh novel about Chernobyl illegal trips "A Stroll to the Zone" . Is the confession of an illegal Chornobyl tourist and stalker.
- *White Horse*
- ru:Распад (фильм) (*Decay*) (Soviet fictional film)
- S.T.A.L.K.E.R: Shadow of Chernobyl (Video game)

30.4   Organizations

- Bellesrad
- Chernobyl Children's Project International
- Chernobyl Forum
- Chernobyl Recovery and Development Programme
- Chernobyl Shelter Fund
- Commission for Independent Research and Information on Radioactivity
- Friends of Chernobyl's Children
- List of Chernobyl-related charities
- Ukrainian National Chornobyl Museum

30.5   People

- Individual involvement in the Chernobyl disaster
- Alexander Akimov, block 4 shift leader
- Yury Bandazhevsky, Belarusian scientist who was jailed 4 years possibly because of his investigations on Chernobyl's consequences
- Anatoly Dyatlov, plant vice chief engineer, the experiment supervisor
- Elena Filatova, Ukrainian photographer known for her website, containing a photo-essay of purported solo motorcycle rides through Chernobyl's zone of alienation

- Valeri Legasov, chief of the investigation committee of the Chernobyl disaster

- Liquidator (Chernobyl), people who took part in the liquidation of the consequences of the disaster

- Vassili Nesterenko, physicist from Belarus involved as a liquidator, and working on the consequences of the Chernobyl disaster

- Wladimir Tchertkoff, Swiss journalist who made documentary films featuring the liquidators

- Leonid Telyatnikov, firefighter, head of the plant fire department

- Adi Roche, chief executive of the charity Chernobyl Children International

30.6   Other

- Chernobyl Way

- Chernobylite

30.7   See also

- Environmental effects of nuclear power

- Nuclear power debate

- List of civilian nuclear accidents

- List of books about nuclear issues

Chapter 31

Effects of the Chernobyl disaster

The 1986 Chernobyl disaster triggered the release of substantial amounts of radioactivity into the atmosphere in the form of both particulate and gaseous radioisotopes. It is one of the most significant unintentional releases of radioactivity into the environment to present.

The work of the Scientific Committee on Problems of the Environment (SCOPE), suggests that the Chernobyl incident cannot be directly compared to atmospheric tests of nuclear weapons through a single number, with one being simply *x* times larger than the other. This is partly due to the fact that the isotopes released at Chernobyl tended to be longer-lived than those released by the detonation of atomic bombs, thus producing radioactivity curves that vary in shape as well as size.

31.1 Radiation effects to humans

According to a 2009 United Nations Scientific Committee on the Effects of Atomic Radiation (UNSCEAR), the Chernobyl accident had by 2005 caused 61,200 man-Sv of radiation exposure to recovery workers and evacuees, 125,000 man-Sv to the populace of the Ukraine, Belarus, and Russia, and a dose to most of the more distant European countries amounting to 115,000 man-Sv. The same report estimated a further 25% more exposure would be received from residual radiosotopes after 2005.*[1] The total global collective dose from Chernobyl was earlier estimated by UNSCEAR in 1988 to be "600,000 man Sv, equivalent on average to 21 additional days of world exposure to natural background radiation." *[2]

31.1.1 Dose to the general public within 30 km of the plant

The inhalation dose (internal dose) for the public during the time of the accident and their evacuation from the area in what is now the 30 km evacuation zone around the plant

has been estimated (based on ground deposition of caesium-137) to be between 3 and 150 mSv.

Thyroid doses for adults around the Chernobyl area were estimated to be between 20 and 1000 mSv, while for one-year-old infants, these estimates were higher, at 20 to 6000 mSv. For those who left at an early stage after the accident, the internal dose due to inhalation was 8 to 13 times higher than the external dose due to gamma /beta emitters. For those who remained until later (day 10 or later), the inhalation dose was 50 to 70% higher than the dose due to external exposure. The majority of the dose was due to iodine-131 (about 40%) and tellurium and rubidium isotopes (about 20 to 30% for Rb and Te).*[3]

The ingestion doses in this same group of people have also been estimated using the cesium activity per unit of area, isotope ratios, an average day of evacuation, intake rate of milk and green vegetables, and what is known about the transfer of radioactivity via plants and animals to humans. For adults, the dose has been estimated to be between 3 and 180 mSv, while for one-year-old infants, a dose of between 20 and 1300 mSv has been estimated. Again, the majority of the dose was thought to be mostly due to iodine-131, and the external dose was much smaller than the internal dose due to the radioactivity in the diet.*[4]

31.2 Short-term health effects and immediate results

The explosion at the power station and subsequent fires inside the remains of the reactor resulted in the development and dispersal of a radioactive cloud which drifted not only over Russia, Belarus, and Ukraine, but also over most of Europe*[5] and as far as Canada.*[6]*[7]*[8] In fact, the initial evidence in other countries that a major release of radioactive material had occurred came not from Soviet sources, but from Sweden, where on 28 April*[9] workers at the Forsmark Nuclear Power Plant (approximately 1100 km from the Chernobyl site) were found to have radioactive particles on their clothing.

It was Sweden's search for the source of radioactivity (after they had determined there was no leak at the Swedish plant) that led to the first hint of a serious nuclear problem in the Western Soviet Union. In France, the government then claimed that the radioactive cloud had stopped at the Italian border. Therefore, while some kinds of food (mushrooms in particular) were prohibited in Italy because of radioactivity, the French authorities took no such measures, in an attempt to appease the population's fears (see below).

Contamination from the Chernobyl disaster was not evenly spread across the surrounding countryside, but scattered irregularly depending on weather conditions. Reports from Soviet and Western scientists indicate that Belarus received about 60% of the contamination that fell on the former Soviet Union. A large area in Russia south of Bryansk was also contaminated, as were parts of northwestern Ukraine.

203 people were hospitalized immediately, of whom 31 died (28 of them died from acute radiation exposure). Most of these were fire and rescue workers trying to bring the disaster under control, who were not fully aware of how dangerous the radiation exposure (from the smoke) was (for a discussion of the more important isotopes in fallout see fission products). 135,000 people were evacuated from the area, including 50,000 from the nearby town of Pripyat, Ukraine. Health officials have predicted that over the next 70 years there will be a 28% increase in cancer rates in much of the population which was exposed to the 5–12 EBq (depending on source) of radioactive contamination released from the reactor.

Soviet scientists reported that the Chernobyl Unit 4 reactor contained about 180–190 metric tons of uranium dioxide fuel and fission products. Estimates of the amount of this material that escaped range from 5 to 30%. Because of the intense heat of the fire, and with no containment building to stop it, part of the ejected fuel was vaporized or particulized and lofted high into the atmosphere, where it spread.

31.2.1 Workers and "Liquidators"

The workers involved in the recovery and cleanup after the disaster, called "liquidators", received high doses of radiation. In most cases, these workers were not equipped with individual dosimeters to measure the amount of radiation received, so experts could only estimate their doses. Even where dosimeters were used, dosimetric procedures varied - some workers are thought to have been given more accurate estimated doses than others. According to Soviet estimates, between 300,000 and 600,000 people were involved in the cleanup of the 30 km evacuation zone around the reactor, but many of them entered the zone two years after the dis-

Soviet medal awarded to 600,000+ liquidators.

aster.[*][10]

Estimates of the number of "liquidators" vary; the World Health Organization, for example, puts the figure at about 600,000; Russia lists as liquidators some people who did not work in contaminated areas. In the first year after the disaster, the number of cleanup workers in the zone was estimated to be 2,000. These workers received an estimated average dose of 165 millisieverts (16.5 REM).

A sevenfold increase in DNA mutations has been identified in children of liquidators conceived after the accident, when compared to their siblings that were conceived before. However, this effect has diminished sharply over time.[*][11]

31.2.2 Evacuation

Soviet authorities started evacuating people from the area around Chernobyl only on the second day after the disaster (after about 36 hours). By May 1986, about a month later, all those living within a 30 km (19 mi) radius of the plant (about 116,000 people) had been relocated. This area is often referred to as the zone of alienation. However, significant radiation affected the environment over a much wider scale than this 30 km radius encloses.

Map showing caesium-137 contamination in the Chernobyl area in 1996

According to reports from Soviet scientists, 28,000 square kilometers (km ^{2}, or 10,800 square miles, mi^2) were contaminated by caesium-137 to levels greater than 185 kBq per square meter. Roughly 830,000 people lived in this area. About 10,500 km ^{2} (4,000 mi^2) were contaminated by caesium-137 to levels greater than 555 kBq/m^2. Of this total, roughly 7,000 km^2 (2,700 mi^2) lie in Belarus, 2,000 km^2 (800 mi^2) in the Russian Federation and 1,500 km^2 (580 mi^2) in Ukraine. About 250,000 people lived in this area. These reported data were corroborated by the International Chernobyl Project.[12]

31.2.3   Civilians

Some children in the contaminated areas were exposed to high radiation doses of up to 50 gray (Gy), mostly due to an intake of radioactive iodine-131 (a relatively short-lived isotope with a half-life of 8 days) from contaminated milk produced locally. Several studies have found that the incidence of thyroid cancer among children in Belarus, Ukraine, and Russia has risen sharply since the Chernobyl disaster. The International Atomic Energy Agency (IAEA) notes "1800 documented cases of thyroid cancer in children who were between 0 and 14 years of age when the disaster occurred, which is far higher than normal", [13] although this source fails to note the expected rate. The childhood thyroid cancers that have appeared are of a large and aggressive type but, if detected early, can be treated. Treatment entails surgery followed by iodine-131 therapy for any metastases. To date, such treatment appears to have

been successful in the vast majority of cases.

Late in 1995, the World Health Organization (WHO) linked nearly 700 cases of thyroid cancer among children and adolescents to the Chernobyl disaster, and among these, some 10 deaths are attributed to radiation. However, the rapid increase in thyroid cancers detected suggests some of this increase may be an artifact of the screening process. Typical latency time of radiation-induced thyroid cancer is about 10 years, but the increase in childhood thyroid cancers in some regions was observed as early as 1987.

31.2.4   Plant and animal health

An exhibit of a piglet with dipygus at the Ukrainian National Chernobyl Museum. It is possible that birth defects are higher in this area.[14]

A large swath of pine forest killed by acute radiation was named the Red Forest. The dead pines were bulldozed and buried. Livestock were removed during the human evacuations.[15] Elsewhere in Europe, levels of radioactivity were examined in various natural foodstocks. In both Sweden and Finland, fish in deep freshwater lakes were banned for resale and landowners were advised not to consume certain types.[16] Information regarding physical deformities in the plant and animal populations in the areas affected by radioactive fallout require sampling and capture, along with DNA testing, of individuals to determine if abnormalities are the result of natural mutation, radiation poisoning, or exposure to other contaminants in the environment (i.e. pesticides, industrial waste, or agricultural run-off).

Animals living in contaminated areas in and around Chernobyl have suffered from a variety of side effects caused by radiation. Oxidative stress and low levels of antioxidants have had severe consequences on the development of the nervous system, including reduced brain size and impaired cognitive abilities. It has been found that birds living in areas with high levels of radiation have statistically significantly smaller brains, which has shown to be a deficit to viability in the wild.[*][17] Barn swallows (*Hirundo rustica*) that live in or around Chernobyl have displayed an increased rate of physical abnormalities compared to swallows from uncontaminated areas. Abnormalities included partially albinistic plumage, deformed toes, tumors, deformed tail feathers, deformed beaks, and deformed air sacs. Birds with these abnormalities have a reduced viability in the wild and a decrease in fitness. Moeller et al. claimed in 2007 that these effects were likely due to radiation exposure and elevated teratogenic effects of radioactive isotopes in the environment[*][18] although these conclusions have been challenged.[*][19] Various birds in the area appear to have adapted to lower levels of radiation by producing more antioxidants, such as glutathione, to help mitigate the oxidative stress.[*][20]

Invertebrate populations (including bumblebees, butterflies, grasshoppers, dragonflies, and spiders) significantly decreased. As of 2009, most radioactivity around Chernobyl was located in the top layer of soil, where many invertebrates live or lay their eggs. The reduced abundance of invertebrates could have negative implications for the entire ecosystem surrounding Chernobyl.[*][21]

Radionuclides migrate through either soil diffusion or transportation within the soil solution. The effects of ionizing radiation on plants and trees in particular depends on numerous factors, including climatic conditions, the mechanism of radiation deposition, and the soil type. In turn, radiated vegetation affects organisms further up the food chain. In general, the upper-level trophic organisms received less contamination, due to their ability to be more mobile and feed from multiple areas.[*][22]

The amount of radioactive nuclides found to have been deposited into surrounding lakes has increased the normal baseline radioactive amounts by 100 percent. Most of the radionuclides in surrounding water areas were found in the sediments at the bottom of the lakes. There has been a high incidence of chromosomal changes in plant and animal aquatic organisms, and this generally has correlated with the contamination and resulting genetic instability. Most of the lakes and rivers surrounding the Chernobyl exclusion zone are still highly contaminated with radionuclides (and will be for many years to come) as the natural decontamination processes of nucleotides with longer half-lives can take many years.[*][23]

One of the main mechanisms by which radionuclides were passed to humans was through the ingestion of milk from contaminated cows. Most of the rough grazing that the cows took part in contained plant species such as coarse grasses, sedges, rushes, and plants such as heather (also known as *calluna vulgaris*). These plant species grow in soils that are high in organic matter, low in pH, and are often very well

hydrated, thus making the storage and intake of these radionuclides much more feasible and efficient.[*][24] In the early stages following the Chernobyl accident, high levels of radionuclides were found in the milk and were a direct result of contaminated feeding. Within two months of banning most of the milk that was being produced in the affected areas, officials had phased out the majority of the contaminated feed that was available to the cows and much of the contamination was isolated. In humans, ingestion of milk containing abnormally high levels of iodine radionuclides was the precursor for thyroid disease, especially in children and in the immunocompromised.[*][24]

Some plants and animals were able to adapt to the increased radiation levels present in and around Chernobyl. Arabidopsis, a plant native to Chernobyl, was able to resist high concentrations of ionizing radiation and resist forming mutations. This species of plant has been able to develop mechanisms to tolerate chronic radiation that would otherwise be harmful or lethal to other species.[*][25]

Studies suggest the 19-mile (30 km) "exclusion zone" surrounding the Chernobyl disaster has become a wildlife sanctuary.[*][26][*][27] Animals have reclaimed the land including species such as the Przewalski's horse, Eurasian lynx, wild boar, grey wolf, elk, red deer, moose, brown bear, turtle,[*][28] voles, mice, shrews,[*][26] European badger, Eurasian beaver, raccoon dog, red fox, roe deer, European bison, black stork, golden eagle, white-tailed eagle[*][27] and eagle owl whose populations are all thriving. When the disaster first occurred, the health and reproductive ability of many animals and plants were negatively affected for the first six months.[*][29] However, 30 years later, animals and plants have reclaimed the abandoned zone to make it their habitat. Even the site of the explosion was flourishing with wildlife in 2012 as birds nested in the wrecked nuclear plant, and plants and mushrooms lived in and on the site.[*][30] A 2015 study found similar numbers of mammals in the zone compared to nearby similar nature reserves[*][29] and the wildlife population was probably higher than it had been before the accident.[*][31]

Due to the bioaccumulation of caesium-137, some mushrooms as well as wild animals which eat them, e.g. wild boars hunted in Germany and deer in Austria, may have levels which are not considered safe for human consumption.[*][32] Mandatory radioactivity testing of sheep in parts of the UK that graze on lands with contaminated peat was lifted in 2012.[*][33]

In 2016, 187 local Ukrainians had returned and were living permanently in the zone.[*][28]

31.2.5 Human pregnancy

Despite spurious studies from Germany and Turkey, the only robust evidence of negative pregnancy outcomes that transpired after the accident was the increase in elective abortions, these "indirect effects", in Greece, Denmark, Italy etc., have been attributed to "anxieties created" by the media.*[34]

In very high doses, it was known at the time that radiation can cause a physiological increase in the rate of pregnancy anomalies, but unlike the dominant linear-no threshold model of radiation and cancer rate increases, it was known, by select researchers who were familiar with both the prior human exposure data and animal testing, that the "Malformation of organs appears to be a deterministic effect with a threshold dose" below which, no rate increase is observed.*[35] This teratology(birth defects) issue was discussed by Frank Castronovo of the Harvard Medical School in 1999, publishing a detailed review of dose reconstructions and the available pregnancy data following the Chernobyl accident, inclusive of data from Kiev's two largest obstetrics hospitals.*[35] Castronovo concludes that "the lay press with newspaper reporters playing up anecdotal stories of children with birth defects" is, together with dubious studies that show "selection bias", the two primary factors causing the persistent belief that Chernobyl increased the background rate of birth defects. When the vast amount of pregnancy data simply does not support this preception as no pregnant individuals took part in the most radioactive liquidator operations, no pregnant individuals were exposed to the threshold dose.*[35]

- Down syndrome (trisomy 21). In West Berlin, Germany, prevalence of Down syndrome (trisomy 21) peaked 9 months following the main fallout.[11, 12] Between 1980 and 1986, the birth prevalence of Down syndrome was quite stable (i.e., 1.35–1.59 per 1,000 live births [27–31 cases]). In 1987, 46 cases were diagnosed (prevalence = 2.11 per 1,000 live births). Most of the excess resulted from a cluster of 12 cases among children born in January 1987. The prevalence of Down Syndrome in 1988 was 1.77, and in 1989, it reached pre-Chernobyl values. The authors noted that the isolated geographical position of West Berlin prior to reunification, the free genetic counseling, and complete coverage of the population through one central cytogenetic laboratory support completeness of case ascertainment; in addition, constant culture preparation and analysis protocols ensure a high quality of data.*[36]

31.3 Long-term health effects

31.3.1 Science and politics: the problem of epidemiological studies

An abandoned village near Pripyat, close to Chernobyl.

The issue of long-term effects of the Chernobyl disaster on civilians is very controversial. The number of people whose lives were affected by the disaster is enormous. Over 300,000 people were resettled because of the disaster; millions lived and continue to live in the contaminated area. On the other hand, most of those affected received relatively low doses of radiation; there is little evidence of increased mortality, cancers or birth defects among them; and when such evidence is present, existence of a causal link to radioactive contamination is uncertain.*[37]

An increased incidence of thyroid cancer among children in areas of Belarus, Ukraine and Russia affected by the Chernobyl disaster has been firmly established as a result of screening programs*[38] and, in the case of Belarus, an established cancer registry. The findings of most epidemiological studies must be considered interim, say experts, as analysis of the health effects of the disaster is an ongoing process.*[39]

Epidemiological studies have been hampered in the Ukraine, Russian Federation and Belarus by a lack of funds, an infrastructure with little or no experience in chronic disease epidemiology, poor communication facilities and an immediate public health problem with many dimensions. Emphasis has been placed on screening rather than on well-designed epidemiological studies. International efforts to organize epidemiological studies have been slowed by some of the same factors, especially the lack of a suitable scientific infrastructure.

Furthermore, the political nature of nuclear energy may have affected scientific studies. In Belarus, Yury Bandazhevsky, a scientist who questioned the official estimates of Chernobyl's consequences and the relevancy of the official maximum limit of 1,000 Bq/kg, was imprisoned from 2001 to 2005. Bandazhevsky and some human rights groups allege his imprisonment was a reprisal for his publication of reports critical of the official research being conducted into the Chernobyl incident.

The activities undertaken by Belarus and Ukraine in response to the disaster —remediation of the environment, evacuation and resettlement, development of uncontaminated food sources and food distribution channels, and public health measures —have overburdened the governments of those countries. International agencies and foreign governments have provided extensive logistic and humanitarian assistance. In addition, the work of the European Commission and World Health Organization in strengthening the epidemiological research infrastructure in Russia, Ukraine and Belarus is laying the basis for major advances in these countries' ability to carry out epidemiological studies of all kinds.

31.3.2 Caesium radioisotopes

Further information: Fission products

Immediately after the disaster, the main health concern involved radioactive iodine, with a half-life of eight days. Today, there is concern about contamination of the soil with strontium-90 and caesium-137, which have half-lives of about 30 years. The highest levels of caesium-137 are found in the surface layers of the soil where they are absorbed by plants, insects and mushrooms, entering the local food supply. Some scientists fear that radioactivity will affect the local population for the next several generations. Note that caesium is not mobile in most soils because it binds to the clay minerals.[*][40][*][41][*][42]

Tests (c. 1997) showed that caesium-137 levels in trees of the area were continuing to rise. It is unknown if this is still the case. There is some evidence that contamination is migrating into underground aquifers and closed bodies of water such as lakes and ponds (2001, Germenchuk). The main source of elimination is predicted to be natural decay of caesium-137 to stable barium−137, since runoff by rain and groundwater has been demonstrated to be negligible.

31.3.3 30 years after the incident

Twenty-five years after the incident, restriction orders had remained in place in the production, transportation and consumption of food contaminated by Chernobyl fallout. In the UK, only in 2012 the mandatory radioactivity testing of sheep in contaminated parts of the UK that graze on

lands was lifted. They covered 369 farms on 750 km^2 and 200,000 sheep. In parts of Sweden and Finland, restrictions are in place on stock animals, including reindeer, in natural and near-natural environments. "In certain regions of Germany, Austria, Italy, Sweden, Finland, Lithuania and Poland, wild game (including boar and deer), wild mushrooms, berries and carnivorous fish from lakes reach levels of several thousand Bq per kg of caesium-137", while "in Germany, caesium-137 levels in wild boar muscle reached 40,000 Bq/kg. The average level is 6,800 Bq/kg, more than ten times the EU limit of 600 Bq/kg", according to the TORCH 2006 report. The European Commission has stated that "The restrictions on certain foodstuffs from certain Member States must therefore continue to be maintained for many years to come".[*][43]

As of 2009, sheep farmed in some areas of the UK are still subject to inspection which may lead to them being prohibited from entering the human food chain because of contamination arising from the accident:

> Some of this radioactivity, predominantly radiocaesium-137, was deposited on certain upland areas of the UK, where sheep-farming is the primary land-use. Due to the particular chemical and physical properties of the peaty soil types present in these upland areas, the radiocaesium is still able to pass easily from soil to grass and hence accumulate in sheep. A maximum limit of 1,000 becquerels per kilogramme (Bq/kg) of radiocaesium is applied to sheep meat affected by the accident to protect consumers. This limit was introduced in the UK in 1986, based on advice from the European Commission's Article 31 group of experts. Under power provided under the Food and Environment Protection Act 1985 (FEPA), Emergency Orders have been used since 1986 to impose restrictions on the movement and sale of sheep exceeding the limit in certain parts of Cumbria, North Wales, Scotland and Northern Ireland... When the Emergency Orders were introduced in 1986, the Restricted Areas were large, covering almost 9,000 farms, and over 4 million sheep. Since 1986, the areas covered by restrictions have dramatically decreased and now cover 369 farms, or part farms, and around 200,000 sheep. This represents a reduction of over 95% since 1986, with only limited areas of Cumbria, South Western Scotland and North Wales, covered by restrictions.[*][44]

369 farms and 190,000 sheep are still affected, a reduction of 95% since 1986, when 9,700 farms and 4,225,000 sheep were under restriction across the United Kingdom.[*][45] Restrictions were finally lifted in 2012.[*][46]

In Norway, the Sami people were affected by contaminated food (the reindeer had been contaminated by eating lichen, which accumulates some types of radioactivity emitters).[47]

Data from a long-term monitoring program from 1998 to 2015 (The Korma Report II)[48] shows a significant decrease in internal radiation exposure of the inhabitants of small villages in Belarus 80 km north of Gomel. Resettlement may even be possible in parts of the prohibited areas provided that people comply with appropriate dietary rules.

31.3.4   Effect on the natural world

Earth Observing-1 image of the reactor and surrounding area in April 2009.

According to reports from Soviet scientists at the First International Conference on the Biological and Radiological Aspects of the Chernobyl Accident (September 1990), fallout levels in the 10 km zone around the plant were as high as 4.81 GBq/m^2. The so-called "Red Forest" of pine trees,[49][50] previously known as Wormwood Forest and located immediately behind the reactor complex, lay within the 10 km zone and was killed off by heavy radioactive fallout. The forest is so named because in the days following the disaster the trees appeared to have a deep red hue as they died because of extremely heavy radioactive fallout. In the post-disaster cleanup operations, a majority of the 4 km^2 forest was bulldozed and buried. The site of the Red Forest remains one of the most contaminated areas in the world.[51]

In recent years there have been many reports suggesting the zone may be a fertile habitat for wildlife.[52] For example, in the 1996 BBC Horizon documentary 'Inside Chernobyl's Sarcophagus', birds are seen flying in and out of large holes in the structure itself. Other casual observations suggest biodiversity around the massive radioactivity release has increased due to the removal of human influence (see the first hand account of the wildlife preserve). Storks, wolves,

beavers, and eagles have been reported in the area.[52]

Barn swallows sampled between 1991 and 2006 both in the Chernobyl exclusion zone had more physical abnormalities than control sparrows sampled elsewhere in Europe. Abnormal barn swallows mated with lower frequency, causing the percentage of abnormal swallows to decrease over time. This demonstrated the selective pressure against the abnormalities was faster than the effects of radiation that created the abnormalities.[53] "This was a big surprise to us," Dr. Mousseau said. "We had no idea of the impact." [52]

It is unknown whether fallout contamination will have any long-term adverse effect on the flora and fauna of the region, as plants and animals have significantly different and varying radiologic tolerance compared with humans. Some birds are reported with stunted tail feathers (which interferes with breeding). There are reports of mutations in plants in the area.[54] The Chernobyl area has not received very much biological study, although studies that have been done suggest that apparently healthy populations may be sink instead of source populations; in other words, that the apparently healthy populations are not contributing to the survival of species.[55]

Using robots, researchers have retrieved samples of highly melanized black fungus from the walls of the reactor core itself. It has been shown that certain species of fungus, such as *Cryptococcus neoformans* and *Cladosporium*, can actually thrive in a radioactive environment, growing better than non-melanized variants, implying that they use melanin to harness the energy of ionizing radiation from the reactor.[56][57][58]

31.3.5   Studies on wildlife in the Exclusion Zone

Main article: Chernobyl Exclusion Zone

The Exclusion Zone around the Chernobyl nuclear power station is reportedly a haven for wildlife.[59][60] As humans were evacuated from the area 25 years ago, existing animal populations multiplied and rare species not seen for centuries have returned or have been reintroduced, for example Eurasian lynx, wild boar, Eurasian wolf, Eurasian brown bear, European bison, Przewalski's horse, and Eurasian eagle owls.[59][60] Birds even nest inside the cracked concrete sarcophagus shielding the shattered remains of Reactor 4.[61] In 2007 the Ukrainian government designated the Exclusion Zone as a wildlife sanctuary,[62][63] and at 488.7 km^2 it is one of the largest wildlife sanctuaries in Europe.[60]

According to a 2005 U.N. report, wildlife has returned despite radiation levels that are presently 10 to 100 times higher than normal background radiation. Although radiation levels were significantly higher soon after the accident, they have fallen because of radioactive decay.[*][61]

Møller and Tim Mousseau have published the results of the largest census of animal life in the Chernobyl Exclusion Zone. It said, contrary to the Chernobyl Forum's 2005 report, that the biodiversity of insects, birds and mammals is declining. Møller and Mousseau have been criticized strongly by Sergey Gaschak, a Ukrainian biologist who did field work for the pair beginning in 2003. He regards their conclusions to be the result of a biased and unscientific anti-nuclear political agenda, unsupported by the data he collected for them. "I know Chernobyl Zone," he says. "I worked here many years. I can't believe their results."

Some researchers have said that by halting the destruction of habitat, the Chernobyl disaster helped wildlife flourish. Biologist Robert J. Baker of Texas Tech University was one of the first to report that Chernobyl had become a wildlife haven and that many rodents he has studied at Chernobyl since the early 1990s have shown remarkable tolerance for elevated radiation levels.[*][61][*][63]

Møller *et al.* (2005) suggested that the reproductive success and annual survival rates of barn swallows are much lower in the Exclusion Zone; 28% of barn swallows inhabiting Chernobyl return each year, while at a control area at Kanev, 250 km to the southeast, the return rate is around 40%.[*][64][*][65] A later study by Møller *et al.* (2007) furthermore claimed an elevated frequency of eleven categories of subtle physical abnormalities in barn swallows, such as bent tail feathers, deformed air sacs, deformed beaks, and isolated albinistic feathers.[*][66]

Smith et al. (2007) have disputed Møller's findings and instead proposed that a lack of human influence in the Exclusion Zone locally reduced the swallows' insect prey and that radiation levels across the vast majority of the exclusion zone are now too low to have an observable negative effect.[*][67] But the criticisms raised were responded to in Møller *et al.* (2008).[*][68] It is possible that barn swallows are particularly vulnerable to elevated levels of ionizing radiation because they are migratory; they arrive in the exclusion area exhausted and with depleted reserves of radioprotective antioxidants after their journey.[*][64]

Several research groups have suggested that plants in the area have adapted to cope with the high radiation levels, for example by increasing the activity of DNA cellular repair machinery and by hypermethylation.[*][25][*][69][*][70][*][71] Given the uncertainties, further research is needed to assess the long-term health effects of elevated ionizing radiation from Chernobyl on flora and fauna.[*][61]

In 2015, long-term empirical data showed no evidence of a negative influence of radiation on mammal abundance.[*][72]

31.3.6 Chernobyl Forum report and criticisms

In September 2005, a comprehensive report was published by the Chernobyl Forum, comprising a number of agencies including the International Atomic Energy Agency (IAEA), the World Health Organization (WHO), United Nations bodies and the Governments of Belarus, the Russian Federation and Ukraine. This report titled: "Chernobyl's legacy: Health, Environmental and Socio-Economic Impacts", authored by about 100 recognized experts from many countries, put the total predicted number of deaths due to the disaster around 4,000 (of which 2,200 deaths are expected to be in the ranks of 200,000 liquidators). This predicted death toll includes the 47 workers who died of acute radiation syndrome as a direct result of radiation from the disaster, nine children who died from thyroid cancer and an estimated 4000 people who could die from cancer as a result of exposure to radiation. This number was subsequently updated to 9000 excess cancer deaths.[*][73]

An IAEA press officer admitted that the 4000 figure was given prominence in the report "*...to counter the much higher estimates which had previously been seen. ... "It was a bold action to put out a new figure that was much less than conventional wisdom."* "[*][74]

The report also stated that, apart from a 30 kilometre area around the site and a few restricted lakes and forests, radiation levels had returned to acceptable levels.[*][75] For full coverage see the IAEA Focus Page.[*][76]

The methodology of the Chernobyl Forum report, supported by Elisabeth Cardis of the International Agency for Research on Cancer,[*][77] has been disputed by some advocacy organizations opposed to nuclear energy, such as Greenpeace and the International Physicians for Prevention of Nuclear Warfare (IPPNW), as well as some individuals such as Dr. Michel Fernex, retired medical doctor from the WHO and campaigner Dr. Christopher Busby (Green Audit, LLRC). The main criticism has been with regard to the restriction of the Forum's study to Belarus, Ukraine and Russia. Furthermore, it only studied the case of 200,000 people involved in the cleanup, and the 400,000 most directly affected by the released radioactivity. German Green Party Member of the European Parliament Rebecca Harms, commissioned a report on Chernobyl in 2006 (*TORCH, The Other Report on Chernobyl*). The 2006 TORCH report claimed that:

> In terms of their surface areas, Belarus (22% of its land area) and Austria (13%) were most affected by higher levels of contamination. Other countries were seriously affected; for example, more than 5% of Ukraine, Finland and Sweden were contaminated to high levels (>

40,000 Bq/m^2 caesium-137). More than 80% of Moldova, the European part of Turkey, Slovenia, Switzerland, Austria and the Slovak Republic were contaminated to lower levels (> 4,000 Bq/m^2 caesium-137). And 44% of Germany and 34% of the UK were similarly affected. (See map of radioactive distribution of caesium-137 in Europe)[43]

While the IAEA/WHO and UNSCEAR considered areas with exposure greater than 40,000 Bq/m^2, the TORCH report also included areas contaminated with more than 4,000 Bq/m^2 of Cs-137.

The TORCH 2006 report "estimated that more than half the iodine-131 from Chernobyl [which increases the risk of thyroid cancer] was deposited outside the former Soviet Union. Possible increases in thyroid cancer have been reported in the Czech Republic and the UK, but more research is needed to evaluate thyroid cancer incidences in Western Europe". It predicted about 30,000 to 60,000 excess cancer deaths, 7 to 15 Times greater than the figure of 4,000 in the IAEA press release; warned that predictions of excess cancer deaths strongly depend on the risk factor used; and predicted excess cases of thyroid cancer range between 18,000 and 66,000 in Belarus alone depending on the risk projection model.[78] However, elevated incidence thyroid cancer is still seen among Ukrainians who were exposed to radioactivity due to Chernobyl accident during their childhood, but who were diagnosed the malignancy as adults.[79]

Another study claims possible heightened mortality in Sweden.[80]

Greenpeace quoted a 1998 WHO study, which counted 212 dead from only 72,000 liquidators. The environmental NGO estimated a total death toll of 93,000 but cite in their report that "The most recently published figures indicate that in Belarus, Russia and the Ukraine alone the disaster could have resulted in an estimated 200,000 additional deaths in the period between 1990 and 2004." In its report, Greenpeace suggested there will be 270,000 cases of cancer alone attributable to Chernobyl fallout, and that 93,000 of these will probably be fatal compare with the IAEA 2005 report which claimed that "99% of thyroid cancers wouldn't be lethal" .[81]

According to the Union Chernobyl, the main organization of liquidators, 10% of the 600,000 liquidators are now dead, and 165,000 disabled.[82]

According to an April 2006 report by the International Physicians for Prevention of Nuclear Warfare (IPPNW), entitled "Health Effects of Chernobyl - 20 years after the reactor catastrophe" ,[83] more than 10,000 people are today affected by thyroid cancer and 50,000

cases are expected. In Europe, the IPPNW claims that 10,000 deformities have been observed in newborns because of Chernobyl's radioactive discharge, with 5,000 deaths among newborn children. They also state that several hundreds of thousands of the people who worked on the site after the disaster are now sick because of radiation, and tens of thousands are dead.[82]

Revisiting the issue for the 25th anniversary of the Chernobyl disaster, the Union of Concerned Scientists described the Forum's estimate of four thousand as pertaining only to "a much smaller subgroup of people who experienced the greatest exposure to released radiation" . Their estimates for the broader population are 50,000 excess cancer cases resulting in 25,000 excess cancer deaths.[84]

31.4 Controversy over human health effects

The majority of premature deaths caused by Chernobyl are expected to be the result of cancers and other diseases induced by radiation in the decades after the event. This will be the result of a large population (some studies have considered the entire population of Europe) exposed to relatively low doses of radiation increasing the risk of cancer across that population. Interpretations of the current health state of exposed populations vary. Therefore, estimates of the ultimate human impact of the disaster have relied on numerical models of the effects of radiation on health. Furthermore, the effects of low-level radiation on human health are not well understood, and so the models used, notably the linear no threshold model, are open to question.[85]

Given these factors, studies of Chernobyl's health effects have come up with different conclusions and are the subject of scientific and political controversy. The following section presents some of the major studies on this topic.

31.4.1 Chernobyl Forum report

In September 2005, a draft summary report by the Chernobyl Forum, comprising a number of UN agencies including the International Atomic Energy Agency (IAEA), the World Health Organization (WHO), the United Nations Development Programme (UNDP), other UN bodies and the Governments of Belarus, the Russian Federation and Ukraine, put the total predicted number of deaths due to the accident at 4000.[86] This death toll predicted by the WHO included the 47 workers who died of acute radiation syndrome as a direct result of radiation from the disaster and nine children who died from thyroid cancer, in the estimated 4000 excess cancer deaths expected among the

600,000 with the highest levels of exposure.[*][87]

The full version of the WHO health effects report adopted by the UN, published in April 2006, included the prediction of 5000 additional fatalities from significantly contaminated areas in Belarus, Russia and Ukraine and predicted that, in total, 9000 will die from cancer among the 6.9 million most-exposed Soviet citizens.[*][88] This report is not free of controversy, and has been accused of trying to minimize the consequences of the accident.[*][89]

31.4.2 TORCH report

Main article: TORCH report

In 2006 German Green Party Member of the European Parliament Rebecca Harms commissioned two UK scientists for an alternate report (**TORCH**, **T**he **O**ther **R**eport on **CH**ernobyl) in response to the UN report. The report included areas not covered by the Chernobyl forum report, and also lower radiation doses. It predicted about 30,000 to 60,000 excess cancer deaths and warned that predictions of excess cancer deaths strongly depend on the risk factor used, and urged more research stating that large uncertainties made it difficult to properly assess the full scale of the disaster.[*][43]

31.4.3 Greenpeace

Demonstration on Chernobyl day near WHO in Geneva

Greenpeace claimed contradictions in the Chernobyl Forum reports, quoting a 1998 WHO study referenced in the 2005 report, which projected 212 dead from 72,000 liquidators.[*][90] In its report, Greenpeace suggested there will be 270,000 cases of cancer attributable to Chernobyl fallout, and that 93,000 of these will probably be fatal, but

state in their report that "The most recently published figures indicate that in Belarus, Russia and Ukraine alone the accident could have resulted in an estimated 200,000 additional deaths in the period between 1990 and 2004." Blake Lee-Harwood, campaigns director at Greenpeace, believes that cancer was likely to be the cause of less than half of the final fatalities and that "intestinal problems, heart and circulation problems, respiratory problems, endocrine problems, and particularly effects on the immune system," will also cause fatalities. However, concern has been expressed about the methods used in compiling the Greenpeace report.[*][89][*][91] It is not peer reviewed nor does it rely on peer review science as the Chernobyl Forum report did.

31.4.4 April 2006 IPPNW report

According to an April 2006 report by the German affiliate of the International Physicians for Prevention of Nuclear Warfare (IPPNW), entitled "Health Effects of Chernobyl", more than 10,000 people are today affected by thyroid cancer and 50,000 cases are expected. The report projected tens of thousands dead among the liquidators. In Europe, it alleges that 10,000 deformities have been observed in newborns because of Chernobyl's radioactive discharge, with 5000 deaths among newborn children. They also claimed that several hundreds of thousands of the people who worked on the site after the accident are now sick because of radiation, and tens of thousands are dead.[*][92]

31.4.5 New York Academy of Sciences publication

Chernobyl: Consequences of the Catastrophe for People and the Environment is an English translation of the 2007 Russian publication *Chernobyl*. It was published online in 2009 by the New York Academy of Sciences in their *Annals of the New York Academy of Sciences*. It presents an analysis of scientific literature and concludes that medical records between 1986, the year of the accident, and 2004 reflect 985,000 deaths as a result of the radioactivity released. The authors suggest that most of the deaths were in Russia, Belarus and Ukraine, but others were spread through the many other countries the radiation from Chernobyl struck.[*][93]

The literature analysis draws on over 1,000 published titles and over 5,000 internet and printed publications discussing the consequences of the Chernobyl disaster. The authors contend that those publications and papers were written by leading Eastern European authorities and have largely been downplayed or ignored by the IAEA and UNSCEAR.[*][94] Author Alexy V. Yablokov was also one of the general editors on the Greenpeace commissioned report also criticizing the Chernobyl Forum finds published one year prior to the

Russian-language version of this report.

A critical review by Dr. Monty Charles in the journal *Radiation Protection Dosimetry* states that *Consequences* is a direct extension of the 2005 Greenpeace report, updated with data of unknown quality.[95] The New York Academy of Sciences also published a severely critical review by M. I. Balonov from the Institute of Radiation Hygiene (St. Petersburg, Russia) which stated that "The value of [*Consequences*] is not zero, but negative, as its bias is obvious only to specialists, while inexperienced readers may well be put into deep error." [96]

31.4.6   2008 UNSCEAR report

The United Nations Scientific Committee on the Effects of Atomic Radiation (UNSCEAR) produced a detailed report on the effects of Chernobyl for the General Assembly of the UN in 2011.[97] This report concluded that 134 staff and emergency workers suffered acute radiation syndrome and of those 28 died of radiation exposure within three months. Many of the survivors suffered skin conditions and radiation induced cataracts, and 19 had since died, but from conditions not necessarily associated with radiation exposure. Of the several hundred thousand liquidators, apart from some emerging indications of increased leukaemia, there was no other evidence of health effects.

In the general public in the affected areas, the only effect with 'persuasive evidence' was a substantial fraction of the 6,000 cases of thyroid cancer in adolescents of whom by 2005 15 cases had proved fatal. There was no evidence of increased rates of solid cancers or leukaemia among the general population. However, there was a widespread psychological worry about the effects of radiation.

The total deaths reliably attributable by UNSCEAR to the radiation produced by the accident therefore was 62.

The report concluded that 'the vast majority of the population need not live in fear of serious health consequences from the Chernobyl accident'.[98]

31.4.7   Other studies and claims

- The claim is made, by Collette Thomas, writing on 24 April 2006, that someone in the Ukrainian Health Ministry claimed in 2006 that more than 2.4 million Ukrainians, including 428,000 children, suffer from health problems related to the catastrophe.[5] The claim appears to have been invented by her through a very creative interpretation of a webpage of the Kyiv Regional Administration.[99] Psychological after-effects, as the 2006 UN report pointed out, have

also had adverse effects on internally displaced persons.

- In a recently published study scientists from Forschungszentrum Jülich, Germany, published the "Korma-Report" with data of radiological long-term measurements that were performed between 1998 and 2007 in a region in Belarus that was affected by the Chernobyl accident. The internal radiation exposure of the inhabitants in a village in Korma County/Belarus caused by the existing radioactive contamination has experienced a significant decrease from a very high level. The external exposure, however, reveals a different picture. Although an overall decrease was observed, the organic constituents of the soil show an increase in contamination. This increase was not observed in soils from cultivated land or gardens. According to the Korma Report the internal dose will decrease to less than 0.2 mSv/a in 2011 and to below 0.1 mSv/a in 2020. Despite this, the cumulative dose will remain significantly higher than "normal" values due to external exposure. Resettlement may even be possible in former prohibited areas provided that people comply with appropriate dietary rules.[100]

- Study of heightened mortality in Sweden.[80][101] But it must be pointed out that this study, and in particular the conclusions drawn has been very criticized.[102]

- One study reports increased levels of birth defects in Germany and Finland in the wake of the accident.[103]

- A change in the human sex ratio at birth from 1987 onward in several European countries has been linked to Chernobyl fallout.[104][105]

- In the Czech Republic, thyroid cancer has increased significantly after Chernobyl.[106]

- The Abstract of the April 2006 International Agency for Research on Cancer report *Estimates of the cancer burden in Europe from radioactive fallout from the Chernobyl accident* stated "It is unlikely that the cancer burden from the largest radiological accident to date could be detected by monitoring national cancer statistics. Indeed, results of analyses of time trends in cancer incidence and mortality in Europe do not, at present, indicate any increase in cancer rates – other than of thyroid cancer in the most contaminated regions – that can be clearly attributed to radiation from the Chernobyl accident." [107][108] They estimate, based on the linear no threshold model of cancer effects, that 16,000 excess cancer deaths could be expected from the effects of the Chernobyl accident up to 2065. Their estimates have very wide 95% confidence intervals from 6,700 deaths to 38,000.[*][109]

- The application of the linear no threshold model to predict deaths from low levels of exposure to radiation was disputed in a BBC (British Broadcasting Corporation) *Horizon* documentary, broadcast on 13 July 2006.[*][110] It offered statistical evidence to suggest that there is an exposure threshold of about 200 millisieverts, below which there is no increase in radiation-induced disease. Indeed, it went further, reporting research from Professor Ron Chesser of Texas Tech University, which suggests that low exposures to radiation can have a protective effect. The program interviewed scientists who believe that the increase in thyroid cancer in the immediate area of the explosion had been over-recorded, and predicted that the estimates for widespread deaths in the long term would be proved wrong. It noted the view of the World Health Organization scientist Dr Mike Rapacholi that, while most cancers can take decades to manifest, leukemia manifests within a decade or so: none of the previously expected peak of leukemia deaths has been found, and none is now expected. Identifying the need to balance the "fear response" in the public's reaction to radiation, the program quoted Dr Peter Boyle, director of the IARC: "Tobacco smoking will cause several thousand times more cancers in the [European] population." [*][111]

- An article in Der Spiegel in April 2016 also cast doubt on the use of the linear no threshold model to predict cancer rates from Chernobyl.[*][85] The article claimed that the threshold for radiation damage was over 100 millisieverts and reported initial results of large-scale trials in Germany by the GSI Helmholtz Centre for Heavy Ion Research and three other German institutes in 2016 showing beneficial results of decreasing inflammation and strengthening bones from lower radiation doses.

- Professor Wade Allison of Oxford University (a lecturer in medical physics and particle physics) gave a talk on ionising radiation 24 November 2006 in which he gave an approximate figure of 81 cancer deaths from Chernobyl (excluding 28 cases from acute radiation exposure and the thyroid cancer deaths which he regards as "avoidable"). In a closely reasoned argument using statistics from therapeutic radiation, exposure to elevated natural radiation (the presence of radon gas in homes) and the diseases of Hiroshima and Nagasaki survivors he demonstrated that the linear no-threshold model should not be applied to low-level exposure in humans, as it ignores the well-known natural repair mechanisms of the body.[*][112][*][113]

- A photographic essay by photojournalist Paul Fusco documents problems in the children in the Chernobyl region. No evidence is offered to suggest these problems are in any way related to the nuclear incident[*][114][*][115]

- The work of photojournalist Michael Forster Rothbart documents the human impact of the disaster on residents who stayed in the affected area.[*][116]

- Bandashevsky measured levels of radioisotopes in children who had died in the Minsk area that had received Chernobyl fallout, and the cardiac findings were the same as those seen in test animals that had been administered Cs-137.[*][117]

31.4.8 French legal action

Since March 2001, 400 lawsuits have been filed in France against "X" (the French equivalent of John Doe, an unknown person or company) by the French Association of Thyroid-affected People, including 200 in April 2006. These persons are affected by thyroid cancer or goitres, and have filed lawsuits alleging that the French government, at the time led by Prime Minister Jacques Chirac, had not adequately informed the population of the risks linked to the Chernobyl radioactive fallout. The complaint contrasts the health protection measures put in place in nearby countries (warning against consumption of green vegetables or milk by children and pregnant women) with the relatively high contamination suffered by the east of France and Corsica. Although the 2006 study by the French Institute of Radioprotection and Nuclear Safety said that no clear link could be found between Chernobyl and the increase of thyroid cancers in France, it also stated that papillary thyroid cancer had tripled in the following years.[*][118]

31.5 Comparisons to other radioactivity releases

Main article: Chernobyl compared to other radioactivity releases

31.6 See also

- Bellesrad

- Chernobyl Children's Project (UK)

- Chernobyl disaster

- Chernobyl compared to other radioactivity releases

- Chernobyl Heart

- Chernobyl in the popular consciousness

- Chernobyl necklace

- Chernobyl Shelter Fund

- Chernobyl Children's Project International

- Deaths due to the Chernobyl disaster

- Acute radiation syndrome

- Ionizing radiation

- Fission products, a more complete description of the radioactive byproducts of nuclear reactors

- Liquidator (Chernobyl)

- List of Chernobyl-related articles

- Nuclear and radiation accidents

- Nuclear power debate

- Radiophobia

- Red Forest

- Three Mile Island accident

- Three Mile Island accident health effects

- Yury Bandazhevsky, a Belarusian scientist imprisoned from 2001 to 2005 after his publication of a report critical of the official investigation on the consequences of the Chernobyl disaster

31.7 References

[1] "UNSCEAR 2008 Report to the General Assembly, Annex D" (PDF). United Nations Scientific Committee on the Effects of Atomic Radiation. 2008.

[2] "Assessing the Chernobyl Consequences". International Atomic Energy Agency.

[3] Mück, Konrad; Pröhl, Gerhard; Likhtarev, Ilya; Kovgan, Lina; Golikov, Vladislav; Zeger, Johann (2002). "Reconstruction of the Inhalation Dose in the 30-Km Zone After the Chernobyl Accident". *Health Physics.* **82** (2): 157–72. doi:10.1097/00004032-200202000-00003. PMID 11797891.

[4] Pröhl, Gerhard; Mück, Konrad; Likhtarev, Ilya; Kovgan, Lina; Golikov, Vladislav (2002). "Reconstruction of the Ingestion Doses Received by the Population Evacuated from the Settlements in the 30-Km Zone Around the Chernobyl Reactor". *Health Physics.* **82** (2): 173–81. doi:10.1097/00004032-200202000-00004. PMID 11797892.

[5] "Tchernobyl, 20 ans après" (in French). RFI. 24 April 2006. Retrieved 24 April 2006.

[6] Chernobyl: country by country A - H. Davistownmuseum.org. Retrieved 26 April 2012.

[7] "TORCH report executive summary" (PDF). European Greens and UK scientists Ian Fairlie PhD and David Sumner. April 2006. Retrieved 21 April 2006. (page 3)

[8] (French) Map of radioactive cloud with flash animation, French IRSN (Institut de Radioprotection et de Sûreté Nucléaire —Institute of Radioprotection and Nuclear Safety) "Accident de Tchernobyl : déplacement du nuage radioactif au dessus de l'Europe entre le 26 avril et le 10 mai 1986". IRSN. Retrieved 8 October 2015.

[9] Jensen, Mikael; Lindhé, John-Christer (Autumn 1986). "International Reports – Sweden: Monitoring the Fallout" (PDF). *IAEA Bulletin.* International Atomic Energy Agency (IAEA)

[10] Chapter IV: Dose estimates, *Nuclear Energy Agency*, 2002

[11] Weinberg, H. S.; Korol, A. B.; Kirzhner, V. M.; Avivi, A.; Fahima, T.; Nevo, E.; Shapiro, S.; Rennert, G.; Piatak, O.; Stepanova, E. I.; Skvarskaja, E. (2001). "Very high mutation rate in offspring of Chernobyl accident liquidators". *Proceedings of the Royal Society B: Biological Sciences.* **268** (1471): 1001–5. doi:10.1098/rspb.2001.1650. PMC 1088700. PMID 11375082.

[12] International Chernobyl Project. Ns.iaea.org. Retrieved 26 April 2012.

[13] Frequently Asked Chernobyl Questions Archived 23 February 2011 at the Wayback Machine.. Iaea.org. Retrieved 26 April 2012.

[14] Dancause, Kelsey Needham; Yevtushok, Lyubov; Lapchenko, Serhiy; Shumlyansky, Ihor; Shevchenko, Genadiy; Wertelecki, Wladimir; Garruto, Ralph M. (2010). "Chronic radiation exposure in the Rivne-Polissia region of Ukraine: Implications for birth defects". *American Journal of Human Biology.* **22** (5): 667–74. doi:10.1002/ajhb.21063. PMID 20737614.

[15] Mycio, Mary (2005). *Wormwood forest: A natural history of Chernobyl.* Washington, D.C.: Joseph Henry Press. p. 259. ISBN 0-309-09430-5.

[16] "Chernobyl - its impact on Sweden" (PDF). *SSI-rapport 86-12.* Staten Stralskydddinstitut. 1 August 1986. ISSN 0282-4434. Retrieved 3 June 2014.

[17] Møller, Anders Pape; Bonisoli-Alquati, Andea; Rudolfsen, Geir; Mousseau, Timothy A. (2011). Brembs, Björn, ed. "Chernobyl Birds Have Smaller Brains". *PLoS ONE*. **6** (2): e16862. doi:10.1371/journal.pone.0016862. PMC 3033907. PMID 21390202.

[18] Moeller, A.P; Mousseau, F.; De Lope, T.A.; Saino, N. (2007). "Elevated frequency of abnormalities in barn swallows from Chernobyl". *Biology Letters*. **3** (4): 414–7. doi:10.1098/rsbl.2007.0136. PMC 1994720. PMID 17439847.

[19] Smith, J.T. (23 February 2008). "Is Chernobyl radiation really causing negative individual and population-level effects on barn swallows?". *Biology Letters*. The Royal Society Publishing. **4** (1): 63–64. doi:10.1098/rsbl.2007.0430. PMC 2412919. PMID 18042513.

[20] Galván, Ismael; Bonisoli-Alquati, Andrea; Jenkinson, Shanna; Ghanem, Ghanem; Wakamatsu, Kazumasa; Mousseau, Timothy A.; Møller, Anders P. (2014-12-01). "Chronic exposure to low-dose radiation at Chernobyl favours adaptation to oxidative stress in birds". *Functional Ecology*. **28** (6): 1387–1403. doi:10.1111/1365-2435.12283. ISSN 1365-2435.

[21] Moeller, A. P.; Mousseau, T. A. (2009). "Reduced abundance of insects and spiders linked to radiation at Chernobyl 20 years after the accident". *Biology Letters*. **5** (3): 356–9. doi:10.1098/rsbl.2008.0778. PMC 2679916. PMID 19324644.

[22] Poiarkov, V.A.; Nazarov, A.N.; Kaletnik, N.N. (1995). "Post-Chernobyl radiomonitoring of Ukrainian forest ecosystems". *Journal of Environmental Radioactivity*. **26** (3): 259–271. doi:10.1016/0265-931X(94)00039-Y.

[23] Gudkov, DI; Kuz'Menko, MI; Kireev, SI; Nazarov, AB; Shevtsova, NL; Dziubenko, EV; Kaglian, AE (2009). "Radioecological problems of aquatic ecosystems of the Chernobyl exclusion zone". *Radiatsionnaia biologiia, radioecologiia*. **49** (2): 192–202. PMID 19507688.

[24] Voors, P.I.; Van Weers, A.W. (1991). "Transfer of Chernobyl radiocaesium (^{134}Cs and ^{137}Cs) from grass silage to milk in dairy cows". *Journal of Environmental Radioactivity*. **13** (2): 125–40. doi:10.1016/0265-931X(91)90055-K.

[25] Kovalchuk, I.; Abramov, V; Pogribny, I; Kovalchuk, O (2004). "Molecular Aspects of Plant Adaptation to Life in the Chernobyl Zone". *Plant Physiology*. **135** (1): 357–63. doi:10.1104/pp.104.040477. PMC 429389. PMID 15133154.

[26] Barras, Colin (22 April 2016). "The Chrenobyl exclusion zone is arguably a nature reserve". BBC Earth. Retrieved 27 April 2016.

[27] Wood, Mike; Beresford, Nick (2016). "The wildlife of Chernobyl: 30 years without man". *The Biologist*. London,UK: Royal Society of Biology. **63** (2): 16–19. Retrieved 27 April 2016.

[28] Oliphant, Roland (24 April 2016). "30 years after Chernobyl disaster, wildlife is flourishing in radioactive wasteland". The Telegraph. Retrieved 27 April 2016.

[29] Deryabina, T. G.; et al. (5 October 2015). "Long-term census data reveal abundant wildlife populations at Chernobyl". *Current Biology*. **25** (19): R824–R826. doi:10.1016/j.cub.2015.08.017. Retrieved 27 April 2016.

[30] Ravillious, Kate. "Despite Mutations, Chernobyl Wildlife is Thriving". National Geographic. Retrieved 16 April 2012.

[31] "What happened to wildlife when Chernobyl drove humans out? It thrived". The Guardian. 5 October 2015. Retrieved 28 April 2016.

[32] Juergen Baetz (1 April 2011). "Radioactive boars and mushrooms in Europe remain a grim reminder 25 years after Chernobyl". The Associated Press. Retrieved 7 June 2012.

[33] "Post-Chernobyl disaster sheep controls lifted on last UK farms". *BBC*. 1 June 2012. Retrieved 7 June 2012.

[34] Little, J. (1993). "The Chernobyl accident, congenital anomalies and other reproductive outcomes". *Paediatric and Perinatal Epidemiology*. **7** (2): 121–51. doi:10.1111/j.1365-3016.1993.tb00388.x. PMID 8516187.

[35] Teratogen Update: Radiation and Chernobyl, Frank P. Castronovo Jr.TERATOLOGY 60:100–106 (1999)

[36] Sperling, Karl; Neitzel, Heidemarie; Scherb, Hagen (2012). "Evidence for an increase in trisomy 21 (Down syndrome) in Europe after the Chernobyl reactor accident". *Genetic Epidemiology*. **36** (1): 48–55. doi:10.1002/gepi.20662. PMID 22162022.

[37] UNSCEAR (United Nations Scientific Committee on the Effects of Atomic Radiation). "Annex D: Health effects due to radiation from the Chernobyl accident" (PDF). *UNSCEAR 2008 Report to the General Assembly with Scientific Annexes*. UNSCEAR. Retrieved 5 April 2011.

[38] Brown, Valerie J. (2011). "Thyroid Cancer after Chornobyl: Increased Risk Persists Two Decades after Radioiodine Exposure". *Environmental Health Perspectives*. **119** (7): a306. doi:10.1289/ehp.119-a306a.

[39] Bogdanova, Tetyana I.; Zurnadzhy, Ludmyla Y.; Greenebaum, Ellen; McConnell, Robert J.; Robbins, Jacob; Epstein, Ovsiy V.; Olijnyk, Valery A.; Hatch, Maureen; Zablotska, Lydia B.; Tronko, Mykola D. (2006). "A cohort study of thyroid cancer and other thyroid diseases after the Chornobyl accident". *Cancer*. **107** (11): 2559–66. doi:10.1002/cncr.22321. PMC 2983485. PMID 17083123.

[40] Microsoft Word - !!MASTERDOC cesium dr3 mar2 ac.doc. (PDF). Retrieved 26 April 2012.

[41] http://ag.arizona.edu/swes/chorover_lab/pdf_papers/Bostick%20et%20al.,%202002.pdf

[42] Information Bridge: DOE Scientific and Technical Information - Sponsored by OSTI. Osti.gov. Retrieved 26 April 2012.

[43] "TORCH report executive summary" (PDF). European Greens and UK scientists Ian Fairlie PhD and David Sumner. April 2006. Retrieved 21 April 2006.

[44] "Post-Chernobyl Monitoring and Controls Survey Report" (PDF). UK Food Standards Agency. Retrieved 2006-04-19.

[45] MacAlister, Terry (12 May 2009). "Britain's farmers still restricted by Chernobyl nuclear fallout". *The Guardian*. London. Retrieved 28 April 2010.

[46] "Post-Chernobyl disaster sheep controls lifted on last UK farms". *BBC news Cumbria*. 1 June 2012. Retrieved 20 March 2015.

[47] Strand, P; Selnaes, TD; Bøe, E; Harbitz, O; Andersson-Sørlie, A (1992). "Chernobyl fallout: Internal doses to the Norwegian population and the effect of dietary advice". *Health physics*. **63** (4): 385–92. doi:10.1097/00004032-199210000-00001. PMID 1526778.

[48] Zoriy, Pedro; Dederichs, Herbert; Pillath, Jürgen; Heuel-Fabianek, Burkhard; Hill, Peter; Lennartz, Reinhard (2016). "Long-term monitoring of radiation exposure of the population in radioactively contaminated areas of Belarus - Korma Study - The Korma Report II (1998-2015)". *Schriften des Forschungszentrums Jülich: Reihe Energie & Umwelt / Energy & Environment*. Forschungszentrum Jülich, Zentralbibliothek, Verlag. Retrieved 21 December 2016.

[49] Energy Citations Database (ECD) - - Document #5012309. Osti.gov. Retrieved 26 April 2012.

[50] Archived 27 September 2006 at the Wayback Machine.

[51] "Chernobyl - Part One" publisher=BBC News | Last Updated: Tuesday, 4 April 2006

[52] "Did Chernobyl Leave an Eden for Wildlife?", by Henry Fountain, *New York Times*, 28 August 2007

[53] "Elevated frequency of abnormalities in barn swallows from Chernobyl", in *Biology Letters*, Volume 3, Number 4 / 22 August 2007

[54] "Wildlife defies Chernobyl radiation". BBC News. 20 April 2006.

[55] Moller, A; Mousseau, T (2006). "Biological consequences of Chernobyl: 20 years on". *Trends in Ecology & Evolution*. **21** (4): 200–7. doi:10.1016/j.tree.2006.01.008. PMID 16701086.

[56] Chernobyl Fungus Feeds On Radiation. Scienceagogo.com (23 May 2007). Retrieved 26 April 2012.

[57] Ionizing Radiation Changes the Electronic Properties of Melanin and Enhances the Growth of Melanized Fungi. Plos One. Retrieved 26 April 2012.

[58] Vember, VV; Zhdanova, NN (2001). "Peculiarities of linear growth of the melanin-containing fungi Cladosporium sphaerospermum Penz. And Alternaria alternata (Fr.) Keissler". *Mikrobiolohichnyi zhurnal (Kiev, Ukraine : 1993)*. **63** (3): 3–12. PMID 11785260.

[59] *BBC*, 20 April 2006, Wildlife defies Chernobyl radiation

[60] Mycio, Mary (9 September 2005). *Wormwood Forest: A Natural History of Chernobyl*. Joseph Henry Press. ISBN 0-309-09430-5. Retrieved 25 September 2009.

[61] *Washington Post*, 7 June 2007, Chernobyl Area Becomes Wildlife Haven

[62] Mother Nature Network, 7 May 2009, Scientists disagree over radiation effects

[63] Baker, Robert J.; Chesser, Roland K. "The Chernobyl Nuclear Disaster And Subsequent Creation of a Wildlife Preserve". *Environmental Toxicology and Chemistry*, Vol.19, No.5, pp.1231-1232, 2000. Archived from the original on 5 October 2003. Retrieved 14 August 2010.

[64] Ravilious, Kate (29 June 2009). "Despite Mutations, Chernobyl Wildlife Is Thriving". *National Geographic Magazine*. ISSN 0027-9358. Retrieved 23 September 2009.

[65] Moller, A. P.; Mousseau, T. A.; Milinevsky, G.; Peklo, A.; Pysanets, E.; Szep, T. (2005). "Condition, reproduction and survival of barn swallows from Chernobyl". *Journal of Animal Ecology*. **74** (6): 1102–1111. doi:10.1111/j.1365-2656.2005.01009.x.

[66] Saino, N.; Mousseau, F.; De Lope, T.A.; Saino, A.P. (2007). "Elevated frequency of abnormalities in barn swallows from Chernobyl". *Biology Letters*. **3** (4): 414–7. doi:10.1098/rsbl.2007.0136. PMC 1994720. PMID 17439847.

[67] Smith, J.T. (23 February 2008). "Is Chernobyl radiation really causing negative individual and population-level effects on barn swallows?" (PDF). *Biology Letters*. **4** (1): 63–64. doi:10.1098/rsbl.2007.0430. PMC 2412919. PMID 18042513. Retrieved 23 September 2009.

[68] Moller, A.P; Mousseau, T.A; De Lope, F; Saino, N (2008). "Anecdotes and empirical research in Chernobyl". *Biology Letters*. **4**: 65–66. doi:10.1098/rsbl.2007.0528.

[69] Danchenko, Maksym; Skultety, Ludovit; Rashydov, Namik M.; Berezhna, Valentyna V.; Mátel, L' Ubomír; Salaj, Terézia; Pret'Ová, Anna; Hajduch, Martin (2009). "Proteomic Analysis of Mature Soybean Seeds from the Chernobyl Area Suggests Plant Adaptation to the Contaminated Environment". *Journal of Proteome Research*. **8** (6): 2915–22. doi:10.1021/pr900034u. PMID 19320472.

[70] Kovalchuk, Olga; Burke, Paula; Arkhipov, Andrey; Kuchma, Nikolaj; James, S.Jill; Kovalchuk, Igor; Pogribny, Igor (2003). "Genome hypermethylation in Pinus silvestris of Chernobyl—a mechanism for radiation adaptation?".

Mutation Research/Fundamental and Molecular Mechanisms of Mutagenesis. **529**: 13–20. doi:10.1016/S0027-5107(03)00103-9.

[71] Boubriak, I. I.; Grodzinsky, D. M.; Polischuk, V. P.; Naumenko, V. D.; Gushcha, N. P.; Micheev, A. N.; McCready, S. J.; Osborne, D. J. (2007). "Adaptation and Impairment of DNA Repair Function in Pollen of Betula verrucosa and Seeds of Oenothera biennis from Differently Radionuclide-contaminated Sites of Chernobyl". *Annals of Botany.* **101** (2): 267–76. doi:10.1093/aob/mcm276. PMC 2711018⊙. PMID 17981881.

[72] "Long-term census data reveal abundant wildlife populations at Chernobyl". *Current Biology.* **25**: R824–R826. doi:10.1016/j.cub.2015.08.017.

[73] World Health Organisation "World Health Organization report explains the health impacts of the world's worst-ever civil nuclear accident", *WHO*, 26 April 2006. Retrieved 4 April 2011.

[74] BBC News "'Too little known on Chernobyl'", *BBC News*, 19 April 2006. Retrieved 4 April 2011.

[75] "IAEA Report". *In Focus: Chernobyl.* Retrieved 29 March 2006.

[76] and joint IAEA/WHO/UNDP press release Chernobyl: The True Scale of the Accident, *IAEA/WHO/UNDP*, 5 September 2005 (pdf file)

[77] "Special Report: Counting the dead". *Nature.* 19 April 2006. Retrieved 21 April 2006.

[78] TORCH report executive summary, op.cit., p.4

[79] Dinets, A.; Hulchiy, M.; Sofiadis, A.; Ghaderi, M.; Höög, A.; Larsson, C.; Zedenius, J. (2012). "Clinical, genetic, and immunohistochemical characterization of 70 Ukrainian adult cases with post-Chornobyl papillary thyroid carcinoma". *European Journal of Endocrinology.* **166** (6): 1049–60. doi:10.1530/EJE-12-0144. PMC 3361791⊙. PMID 22457234.

[80] Chernobyl 'caused Sweden cancers', *BBC News*, 20 November 2004

[81] "Greenpeace rejects Chernobyl toll". BBC News. 18 April 2006.

[82] "Selon un rapport indépendant, les chiffres de l'ONU sur les victimes de Tchernobyl ont été sous-estimés (According to an independent report, UN numbers on Chernobyl's victims has been underestimated)" (in French). Le Monde. 7 April 2006. and see also "'On n'a pas fini d'entendre parler de Tchernobyl', interview with Angelika Claussen, head of the German section of the IPPNW". Arte. 13 April 2006.

[83] http://www.ippnw-students.org/chernobyl/IPPNWStudy.pdf

[84] Chernobyl Cancer Death Toll Estimate More Than Six Times Higher Than the 4,000 Frequently Cited, According to a New UCS Analysis. Ucsusa.org. Retrieved 26 April 2012.

[85] Dworschak, Manfred (26 April 2016). "The Chernobyl Conundrum: Is Radiation As Bad As We Thought?". Spiegel Online International. Retrieved 27 April 2016.

[86] "IAEA Report". *In Focus: Chernobyl.* Archived from the original on 27 March 2006. Retrieved 29 March 2006.

[87] For full coverage see the IAEA Focus Page (*op.cit.*) and joint IAEA/WHO/UNDP 5 September 2005 press release Chernobyl: The True Scale of the Accident

[88] Peplow, M (2006). "Special Report: Counting the dead". *Nature.* **440** (7087): 982–3. doi:10.1038/440982a. PMID 16625167.

[89] "Spiegel, The Chernobyl body count controversy". *In Focus: Chernobyl.* Retrieved 25 August 2006.

[90] Burton Bennett; Michael Repacholi; Zhanat Carr, eds. (2006). *Health Effects of the Chernobyl Accident and Special Health Care Programmes: report of the UN Chernobyl Forum Expert Group "Health"* (PDF). Geneva: WHO. ISBN 92-4-159417-9. Retrieved May 2014. Check date values in: |access-date= (help)

[91] Bialik, Carl (27 April 2006). "Measuring Chernobyl's Fallout". *The Numbers Guy, The Wall Street Journal.* Retrieved 5 May 2014.

[92] "20 years after Chernobyl – The ongoing health effects". IPPNW. April 2006. Retrieved 24 April 2006.

[93] Alexey V. Yablokov; Vassily B. Nesterenko; Alexey V. Nesterenko (2009). *Chernobyl: Consequences of the Catastrophe for People and the Environment (Annals of the New York Academy of Sciences)* (paperback ed.). Wiley-Blackwell. ISBN 978-1-57331-757-3.

[94] "Details". *Annals of the New York Academy of Sciences.* Annals of the New York Academy of Sciences. Retrieved 15 March 2011.

[95] Charles, Monty (2010). "Chernobyl: consequences of the catastrophe for people and the environment (2010)" (PDF). *Radiation Protection Dosimetry.* **141** (1): 101–104. doi:10.1093/rpd/ncq185. *"During the production of the reports from the Chernobyl Forum and Greenpeace, a vast body of previously unknown data began to emerge in the form of publications, reports, theses, etc. from Belarus, Ukraine and Russia, much of it in Slavic languages. Little of these data appears to have been incorporated into the international literature. The quality of these publications and whether they would sustain critical peer-review in the western scientific literature is unknown. The book by Yablokov et al. is part of an attempt to summarise these new findings and include them to extend the findings of the Greenpeace report."*

[96] M. I. Balonov (28 April 2010). "Review of Volume 1181" . New York Academy of Sciences. Retrieved 15 September 2011.

[97] "Sources and Effects of Ionizing Radiation; 2008 Report to the General Assembly;" (PDF). **II** (Scientific Annexes C, D and E). New York, USA: United Nations Committee on the Effects of Atomic Radiation. 2011: 1–219. ISBN 978-92-1-142280-1. Retrieved 27 April 2016.

[98] "The Chernobyl accident: UNSCEAR's assessments of the radiation effects" . *United Nations Scientific Committee on the Effects of Atomic Radiation*. 16 July 2012. Retrieved 27 April 2016.

[99] "Chornobyl tragedy" .

[100] Dederichs, H.; Pillath, J.; Heuel-Fabianek, B.; Hill, P.; Lennartz, R. (2009): Langzeitbeobachtung der Dosisbelastung der Bevölkerung in radioaktiv kontaminierten Gebieten Weißrusslands - Korma-Studie. Vol. 31, series "Energy & Environment "by Forschungszentrum Jülich, ISBN 978-3-89336-562-3

[101] Tondel, M. (2004). "Increase of regional total cancer incidence in north Sweden due to the Chernobyl accident?". *Journal of Epidemiology & Community Health*. **58** (12): 1011–1016. doi:10.1136/jech.2003.017988.

[102] Inga hållpunkter för ökad cancerrisk i Sverige (article in Swedish from the Swedish doctors magazine)

[103] Scherb, Hagen; Weigelt, Eveline. "Congenital Malformation and Stillbirth in Germany and Europe Before and After the Chernobyl Nuclear Power Plant Accident" (PDF).

[104] Scherb, H; Voigt, K (2007). "Trends in the human sex odds at birth in Europe and the Chernobyl Nuclear Power Plant accident" . *Reproductive Toxicology*. **23** (4): 593–9. doi:10.1016/j.reprotox.2007.03.008. PMID 17482426.

[105] Scherb, Hagen; Voigt, Kristina (2011). "The human sex odds at birth after the atmospheric atomic bomb tests, after Chernobyl, and in the vicinity of nuclear facilities" . *Environmental Science and Pollution Research*. **18** (5): 697–707. doi:10.1007/s11356-011-0462-z. PMID 21336635.

[106] Mürbeth, S; Rousarova, M; Scherb, H; Lengfelder, E (2004). "Thyroid cancer has increased in the adult populations of countries moderately affected by Chernobyl fallout" . *Medical science monitor : international medical journal of experimental and clinical research*. **10** (7): CR300–6. PMID 15295858.

[107] Cardis, Elisabeth; Krewski, Daniel; Boniol, Mathieu; Drozdovitch, Vladimir; Darby, Sarah C.; Gilbert, Ethel S.; Akiba, Suminori; Benichou, Jacques; Ferlay, Jacques; Gandini, Sara; Hill, Catherine; Howe, Geoffrey; Kesminiene, Ausrele; Moser, Mirjana; Sanchez, Marie; Storm, Hans; Voisin, Laurent; Boyle, Peter (2006). "Estimates of the cancer burden in Europe from radioactive fallout from the Chernobyl accident" . *International Journal of Cancer*. **119** (6): 1224–1235. doi:10.1002/ijc.22037.

[108] IARC Press release on the report 'Estimates of the cancer burden in Europe from radioactive fallout from the Chernobyl accident' Archived 15 April 2007 at the Wayback Machine.

[109] Briefing document: Cancer burden in Europe following Chernobyl Archived 18 January 2007 at the Wayback Machine.

[110] Davidson, Nick (13 July 2006). "Chernobyl's 'nuclear nightmares'". *Horizon*. Retrieved 2 April 2008.

[111] "Inside Chernobyl's Sarcophagus" (13 July 1996), Horizon, *BBC*.

[112] Allison, Wade (24 November 2006). "How dangerous is ionising radiation?".

[113] Allison, Wade (2006). "The safety of nuclear radiation; a careful re-examination for a world facing climate change" (PDF). Physics Department of Oxford University. Retrieved 30 July 2007.

[114] A video of Fusco discussing his photo essay project on Chernobyl. Mediastorm.com. Retrieved 26 April 2012.

[115] information Paul Fusco's book on the Chernobyl legacy Archived 6 April 2008 at the Wayback Machine.. Magnumphotos.com (26 April 1986). Retrieved 26 April 2012.

[116] "Those who stayed in Chernobyl and Fukushima: An excerpt from the new TED Book brings you inside Control Room 4" . TED. October 31, 2013. Retrieved May 30, 2014.

[117] Bandashevsky, Y. I, "Pathology of Incorporated Ionizing Radiation" , Belarus Technical University, Minsk. 136 pp., 1999.

[118] "Nouvelles plaintes de malades français après Tchernobyl" (in French). RFI. 26 April 2006. Retrieved 26 April 2006. (includes Audio files, with an interview with Chantal Loire, president of the French Association of Thyroid-Affected People, as well as interviews with member of the CRIIRAD)

31.8   External links

- Animated map of radioactive cloud, French IRSN (official Institut de Radioprotection et de Sûreté Nucléaire —Institute of Radioprotection and Nuclear Safety) "Les leçons de Tchernobyl". IRSN. Retrieved 7 December 2009.

- Chernobyl animals worse affected than thought: study

- 25 years of satellite imagery over Chernobyl

Chapter 32

Goiânia accident

Coordinates: 16°40′29″S 49°15′51″W / 16.6746°S 49.2641°W

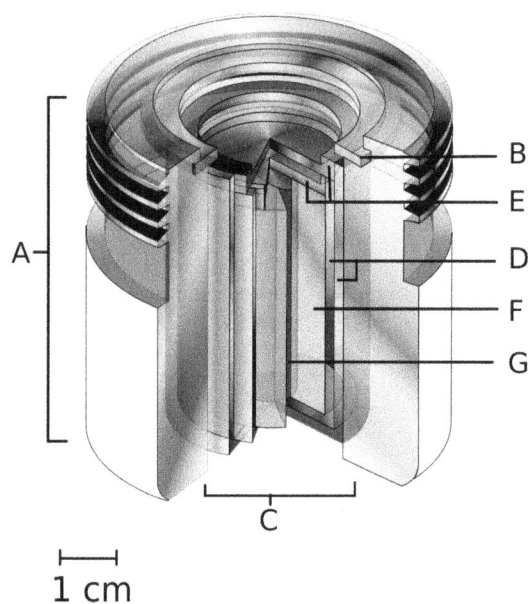

1 cm

A teletherapy radiation capsule composed of the following:
A.) an international standard source holder (usually lead),
B.) a retaining ring, and
C.) a teletherapy "source" composed of
D.) two nested stainless steel canisters welded to
E.) two stainless steel lids surrounding
F.) a protective internal shield (usually uranium metal or a tungsten alloy) and
G.) a cylinder of radioactive source material, often but not always cobalt-60. In the Goiânia incident it was caesium-137. The diameter of the "source" is 30 mm.

The **Goiânia accident** was a radioactive contamination accident that occurred on September 13, 1987, at Goiânia, in the Brazilian state of Goiás, after an old radiotherapy source was stolen from an abandoned hospital site in the city. It

was subsequently handled by many people, resulting in four deaths. About 112,000 people were examined for radioactive contamination and 249 were found to have significant levels of radioactive material in or on their bodies.[1][2]

In the cleanup operation, topsoil had to be removed from several sites, and several houses were demolished. All the objects from within those houses were removed and examined. *Time* magazine has identified the accident as one of the world's "worst nuclear disasters" and the International Atomic Energy Agency called it "one of the world's worst radiological incidents".[3][4]

32.1 Description of the source

The radiation source in the Goiânia accident was a small capsule containing about 93 grams (3.3 oz) of highly radioactive caesium chloride (a caesium salt made with a radioisotope, caesium-137) encased in a shielding canister made of lead and steel. The source was positioned in a container of the wheel type, where the wheel turns inside the casing to move the source between the storage and irradiation positions.[1]

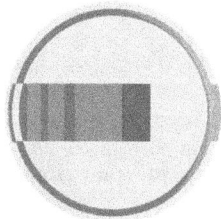

A wheel type radiotherapy device which has a long collimator to focus the radiation into a narrow beam. The caesium chloride radioactive source is the blue rectangle, and gamma rays are represented by the beam emerging from the aperture.

The activity of the source was 74 terabecquerels (TBq) in 1971. The International Atomic Energy Agency (IAEA) describes the container – 51 millimeters (2 inches) in diameter and 48 mm (1.8 inches) long – as an "international standard capsule". The specific activity of the active solid was about 814 TBq·kg[*]–1 of caesium-137 (half life of 30 years). The dose rate at one meter from the source was 4.56 grays per hour (456 rad·h[*]–1). While the serial number of the device was unknown, thus hindering definitive identification, the device was thought to have been made in the United States at Oak Ridge National Laboratory and was used as a radiation source for radiation therapy at the Goiânia hospital.[*][1]

The IAEA states that the source contained 50.9 TBq (1,380 Ci) when it was taken and that about 44 TBq (1200 Ci, 87%) of contamination had been recovered during the cleanup operation. This means that 7 TBq (190 Ci) remained in the environment; it will have decayed to about 3.5 TBq (95 Ci) by 2016.

32.2   Events

32.2.1   Hospital abandonment

Culture and Convention Center, built on the site where the IGR had been located

The Instituto Goiano de Radioterapia (IGR), a private radiotherapy institute in Goiânia,[*][1] was just 1 km (0.6 mi) northwest of Praça Cívica, the administrative center of the city. It moved to its new premises in 1985, leaving behind a caesium-137-based teletherapy unit that had been purchased in 1977.[*][5] The fate of the abandoned site was disputed in court between IGR and the Society of Saint Vincent de Paul, then owner of the premises.[*][6] On September 11, 1986, the Court of Goiás stated it had knowledge of the abandoned radioactive material in the building.[*][6]

Four months before the theft, on May 4, 1987, Saura Taniguti, then director of Ipasgo, the institute of insurance

for civil servants, used police force to prevent one of the owners of IGR, Carlos Figueiredo Bezerril, from removing the objects that were left behind.[*][6] Figueiredo then warned the president of Ipasgo, Lício Teixeira Borges, that he should take responsibility "for what would happen with the caesium bomb".[*][6]

The court posted a security guard to protect the hazardous abandoned equipment.[*][7] Meanwhile, the owners of IGR wrote several letters to the National Nuclear Energy Commission, warning them about the danger of keeping a teletherapy unit at an abandoned site, but they could not remove the equipment by themselves once a court order prevented them from doing so.[*][6][*][7]

32.2.2   Theft of the source

On September 13, 1987, taking advantage of the absence of the guard,[*][7] Roberto dos Santos Alves and Wagner Mota Pereira illegally entered the partially demolished facility. They partially disassembled the teletherapy unit, and placed the source assembly – which they thought might have some scrap value – in a wheelbarrow, taking it to Alves's home.[*][1] There, they began dismantling the equipment. That same evening, they both began to vomit. Nevertheless, they continued in their efforts. The following day, Pereira began to experience diarrhea and dizziness and his left hand began to swell. He soon developed a burn on this hand in the same size and shape as the aperture – he eventually underwent partial amputation of several fingers.[*][8]

On September 15, Pereira visited a local clinic where his symptoms were diagnosed as the result of something he had eaten, and he was told to return home and rest.[*][1] Alves, however, continued with his efforts to dismantle the equipment. In the course of this effort, he eventually freed the caesium capsule from its protective rotating head. His prolonged exposure to the radioactive material led to his right forearm becoming ulcerated, requiring amputation.[*][9]

32.2.3   Source is partially broken

On September 16, Alves succeeded in puncturing the capsule's aperture window with a screwdriver, allowing him to see a deep blue light coming from the tiny opening he had created.[*][1] He inserted the screwdriver and successfully scooped out some of the glowing substance. Thinking it was perhaps a type of gunpowder, he tried to light it, but the powder would not ignite. The exact mechanism by which the light was generated was not known at the time the IAEA report was written, though it was thought to be either ionized air glow, fluorescence or Cherenkov radiation associated with the absorption of moisture by the source; similar blue light was observed in 1988 at Oak Ridge National

Laboratory during the disencapsulation of a ^{137}Cs source.

32.2.4 Source is sold and dismantled

On September 18, Alves sold the items to a nearby scrapyard. That night, Devair Alves Ferreira (the owner of the scrapyard) noticed the blue glow from the punctured capsule. Thinking the capsule's contents were valuable or even supernatural, he immediately brought it into his house. Over the next three days, he invited friends and family to view the strange glowing substance.

On September 21 at the scrapyard, one of Ferreira's friends (given as EF1 in the IAEA report) succeeded in freeing several rice-sized grains of the glowing material from the capsule using a screwdriver. Alves Ferreira began to share some of them with various friends and family members. That same day, his wife, 37-year-old Gabriela Maria Ferreira, began to fall ill. On September 25, 1987, Devair Alves Ferreira sold the scrap metal to a second scrapyard.

32.2.5 Ivo and his daughter

The day before the sale to the second scrapyard, on September 24, Ivo, Devair's brother, successfully scraped some additional dust out of the source and took it to his house a short distance away. There he spread some of it on the concrete floor. His six-year-old daughter, Leide das Neves Ferreira, later ate a sandwich while sitting on this floor. She was also fascinated by the blue glow of the powder, applying it to her body and showing it off to her mother. Dust from the powder fell on the sandwich she was consuming; she eventually absorbed 1.0 GBq and received a total dose of 6.0 Gy, more than a fatal dose even with treatment.[10]

32.2.6 Gabriela Maria Ferreira notifies authorities

Gabriela Maria Ferreira had been the first to notice that many people around her had become severely ill at the same time.[11]

On September 28, 1987 – 15 days after the item was found —she reclaimed the materials from the rival scrapyard and transported them to a hospital. Because the remains of the source were kept in a plastic bag, the level of contamination at the hospital was low.

32.2.7 Source's radioactivity is detected

In the morning of September 29, 1987 a visiting medical physicist [12] used a scintillation counter to confirm the presence of radioactivity and persuaded the authorities to take immediate action. The city, state, and national governments were all aware of the incident by the end of the day.

32.3 Health outcomes

News of the radiation incident was broadcast on local, national, and international media. Within days, nearly 130,000 people swarmed local hospitals concerned that they might have been exposed.[2] Of those, 250 were indeed found to be contaminated—some with radioactive residue still on their skin—through the use of Geiger counters.[2] Eventually, 20 people showed signs of radiation sickness and required treatment.[2]

32.3.1 Fatalities

Ages in years are given, with dosages listed in grays (Gy).

- **Leide das Neves Ferreira**, age 6 (6.0 Gy), was the daughter of Ivo Ferreira. When an international team arrived to treat her, she was discovered confined to an isolated room in the hospital because the hospital staff were afraid to go near her. She gradually experienced swelling in the upper body, hair loss, kidney and lung damage, and internal bleeding. She died on October 23, 1987, of "septicemia and generalized infection" at the Marcilio Dias Navy Hospital, in Rio de Janeiro.[13] She was buried in a common cemetery in Goiânia, in a special fiberglass coffin lined with lead to prevent the spread of radiation. Despite these measures, news of her impending burial caused a riot of more than 2,000 people in the cemetery on the day of her burial, all fearing that her corpse would poison the surrounding land. Rioters tried to prevent her burial by using stones and bricks to block the cemetery roadway.[14] She was buried despite this interference.

- **Gabriela Maria Ferreira**, aged 37 (5.7 Gy), wife of scrapyard owner Devair Ferreira, became sick about three days after coming into contact with the substance. Her condition worsened, and she developed internal bleeding, especially in the limbs, eyes, and digestive tract, and hair loss. She suffered mental confusion, diarrhea, and acute renal insufficiency before also dying on October 23, 1987, of "septicemia and generalized infection",[13][15] about a month after exposure.

- **Israel Baptista dos Santos**, aged 22 (4.5 Gy), was an employee of Devair Ferreira who worked on the radioactive source primarily to extract the lead. He developed serious respiratory and lymphatic complications, was eventually admitted to hospital, and died six days later on October 27, 1987.

- **Admilson Alves de Souza**, aged 18 (5.3 Gy), was also an employee of Devair Ferreira who worked on the radioactive source. He developed lung damage, internal bleeding, and heart damage, and died October 18, 1987.

Devair Ferreira himself survived despite receiving 7 Gy of radiation. He died in 1994 of cirrhosis aggravated by depression and binge drinking.[16]

32.3.2   Other individuals

The outcomes for the 46 most contaminated people are shown in the bar chart below. Several people survived high doses of radiation. This is thought in some cases to be because the dose was fractionated. Given time, the body's repair mechanisms will reverse cell damage caused by radiation. If the dose is spread over a long time period, these mechanisms can mitigate the effects of radiation poisoning.

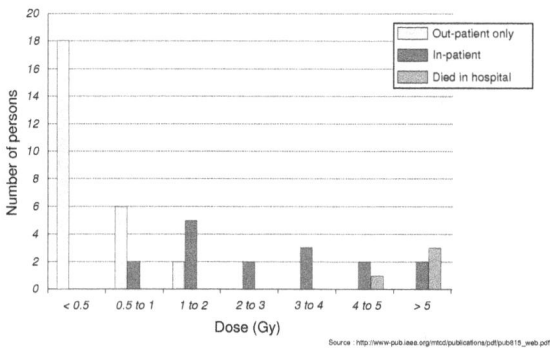

This is a barchart showing the outcome for the 46 most contaminated people for whom a dose estimate has been made. The people are divided into seven groups according to dose.

32.3.3   Other affected people

Afterwards, about 112,000 people were examined for radioactive contamination; 249 were found to have significant levels of radioactive material in or on their body.[1] Of this group, 129 people had internal contamination. The majority of the internally contaminated people only suffered small doses (< 50 mSv, less than a 1 in 400 risk of getting cancer as a result).

A thousand people were identified as having suffered a dose which was greater than one year of background radiation; it

is thought that 97% of these people had a dose of between 10 and 200 mSv (between 1 in 2,000 and 1 in 100 risk of developing cancer as a result).

In 2007, the Oswaldo Cruz Foundation determined that the rate of caesium-137 related diseases are the same in Goiânia accident survivors as they are in the population at large. Nevertheless, compensation is still distributed to survivors, who suffer radiation-related prejudices in everyday life.[17]

32.4   Legal matters

In light of the deaths caused, the three doctors who had owned and run IGR were charged with criminal negligence. Because the accidents occurred before the promulgation of the Federal Constitution of 1988 and because the substance was acquired by the clinic and not by the individual owners, the court could not declare the owners of IGR liable. One of the medical doctors owning IGR and the clinic's physicist were ordered to pay R$100,000 for the derelict condition of the building. The two thieves were not included as defendants in the public civil suit.

In 2000, CNEN, the National Nuclear Energy Commission, was ordered by the 8th Federal Court of Goiás to pay compensation of R$ 1.3 million (near US$750,000) and to guarantee medical and psychological treatment for the direct and indirect victims of the accident and their descendants down to the third generation.[18]

32.5   Cleanup

32.5.1   Objects and places

Topsoil had to be removed from several sites, and several houses were demolished. All the objects from within those houses were removed and examined. Those that were found to be free of radioactivity were wrapped in plastic bags, while those that were contaminated were either decontaminated or disposed of as waste. In industry, the choice between decontaminating or disposing objects is based on only the economic value of the object and the ease of decontamination. In this case, the IAEA recognized that to reduce the psychological impact of the event, greater effort should have been taken to clean up items of personal value, such as jewelry and photographs. It is not clear from the IAEA report to what degree this was practiced.

32.5.2 Means and methods

After the houses were emptied, vacuum cleaners were used to remove dust, and plumbing was examined for radioactivity. Painted surfaces could be scraped, while floors were treated with acid and Prussian blue mixtures. Roofs were vacuumed and hosed, but two houses had to have their roofs removed. The waste from the cleanup was moved out of the city to a remote place for storage.

Potassium alum dissolved in hydrochloric acid was used on clay, concrete, soil, and roofs. Caesium has a high affinity for many clays.

Organic solvents, followed by potassium alum dissolved in hydrochloric acid, were used to treat waxed/greased floors and tables. Sodium hydroxide solutions, also followed by dissolved potassium alum, were used to treat synthetic floors, machines and typewriters.

Prussian blue was used to internally decontaminate many people, although by the time it was applied, much of the radioactive material had already migrated from the bloodstream to the muscle tissue, greatly hampering its effectiveness. Urine from victims was treated with ion exchange resin to compact the waste for ease of storage.

32.5.3 Recovery considerations

The cleanup operation was much harder for this event than it could have been because the source was opened and the active material was water soluble. A sealed source need only be picked up, placed in a lead container, and transported to the radioactive waste storage. In the recovery of lost sources, the IAEA recommends careful planning and using a crane or other device to place shielding (such as a pallet of bricks or a concrete block) near the source to protect recovery workers.

32.5.4 Contamination locations

The Goiânia accident spread significant radioactive contamination throughout the Aeroporto, Central, and Ferroviários districts. Even after the cleanup, 7 TBq of radioactivity remained unaccounted for.

Some of the key contamination sites:

- **Goiânia's Instituto Goiano de Radioterapia (IGR)** (16°40′29″S 49°15′51″W / 16.6746°S 49.2641°W)[1] suffered no actual exposure or breach of radioactive contents, but the site is noteworthy as the source of deadly, unsecured material. The IGR clinic no longer exists, having been re-

placed around 2000 with the modernized Centro de Convenções de Goiânia (Goiânia Convention Center).

- **Roberto dos Santos' house** (16°40′07″S 49°15′48″W / 16.66848°S 49.26341°W)[1] on Rua 57. The radioactive source was here for about six days, and it was partially broken into.

- **Devair Ferreira's scrapyard** (16°40′02″S 49°15′59″W / 16.66713°S 49.26652°W),[1] on Rua 15A ("Junkyard I") in the Aeroporto section of the city, had possession of the items for 7 days. The caesium container was entirely dismantled, spreading significant contamination. Extreme radiation levels of up to 1.5 Sv·h⁻¹ were found by investigators in the middle of the scrapyard.

- **Ivo Ferreira's house** (16°39′50″S 49°16′09″W / 16.66401°S 49.26911°W)[1] ("Junkyard II"), at 1F Rua 6. Some of the contamination was spread about the house, causing the fatalities of Leide das Neves Ferreira and Gabriela Maria Ferreira. The adjacent junkyard scavenged the remainder of parts from the IGR facility. The premises were heavily contaminated, with radiation dose rates up to 2 Sv·h⁻¹.

- **Junkyard III** (16°40′09″S 49°16′48″W / 16.66915°S 49.28003°W).[1] This junkyard had possession of the items for 3 days until they were sent away.

- **Vigilância Sanitária** (16°40′30″S 49°16′23″W / 16.675°S 49.273°W).[1] Here, the substance was quarantined, and an official cleanup response began.

Other contamination was also found in or on:[19]

- 50,000 rolls of toilet paper
- Three buses
- 42 houses
- 14 cars
- five pigs

32.6 Legacy

32.6.1 Disposal of the capsule

The original teletherapy capsule was seized by the Brazilian military as soon as it was discovered, and since then the empty capsule has been on display at the *Escola de Instrução Especializada* ("School of Specialized Instruction") in Rio de Janeiro as a memento to those who participated in the cleanup of the contaminated area.

32.6.2 Research

In 1991, a group of researchers collected blood samples from highly exposed survivors of the incident. Subsequent analysis resulted in the publication of numerous scientific articles.[*][20][*][21][*][22][*][23]

32.6.3 Film

A 1990 film *Césio 137 – O Pesadelo de Goiânia* ("Caesium-137 – The Nightmare of Goiânia"), a dramatisation of the incident, was made by Roberto Pires.[*][24] It won several awards at the 1990 Festival de Brasília.[*][25]

32.6.4 Foundation

The state government of Goiás established the Fundação Leide das Neves Ferreira in February 1988, both to study the extent of contamination of the population as a result of the incident and to render aid to those affected.[*][26]

32.7 See also

- List of civilian radiation accidents

32.8 References

[1] *The Radiological accident in Goiânia* (PDF). Vienna: International Atomic Energy Agency. 1988. ISBN 92-0-129088-8.

[2] Foderaro, Lisa (July 8, 2010). "Columbia Scientists Prepare for a Threat: A Dirty Bomb" . *The New York Times*.

[3] The Worst Nuclear Disasters

[4] Yukiya Amano (March 26, 2012). "Time to better secure radioactive materials" . *Washington Post*.

[5] Puig, Diva E. "Apuntes sobre energia nuclear. Lo que los abogados deben saber sobre la tecnología nuclear" (in Spanish). Noticias Jurídicas. Retrieved May 2, 2014. |chapter= ignored (help)

[6] (Portuguese) Godinho, Iúri. "Os médicos e o acidente radioativo" . *Jornal Opção*. February 8, 2004. Archived September 18, 2009, at the Wayback Machine.

[7] (Portuguese) Borges, Weber. "O jornalista que foi vítima do césio" . *Jornal Opção*. May 27, 2007.

[8] Planeta Diário: July 2010

[9] Aint No Way to Go: All That Glitters

[10] "Brazil Deadly Glitter" . *Time*. October 19, 1987.

[11] "2 Die of Radiation Poisoning in Brazil" . *Los Angeles Times*. October 24, 1987.

[12] "País está preparado para atuar em acidente radioativo" [Country is prepared to act in radioactive incident] (in Portuguese). Ministry of Science, Technology and Innovation (MCTI). 13 September 2012. Retrieved 10 November 2013. Note: person named only as "WF" in the IAEA report.

[13] "Vida Verde" (in Portuguese). 1987. p. 15.

[14] "Memorial Césio 137" (in Portuguese). Greenpeace.

[15] Malheiros, Tania (1996). *Histórias secretas do Brasil nuclear* (in Portuguese). Rio de Janeiro: WVA. p. 122. ISBN 9788585644086.

[16] Irene, Mirelle (13 September 2012). "Goiânia, 25 anos depois: 'perguntam até se brilhamos', diz vítima" . *Terra*. Retrieved 5 December 2013.

[17] UOL. Vítimas do césio 137 voltam a receber remédios e pedem assistência médica para todos. September 25, 2012

[18] "Case Law and Administrative Decisions, Judgement of the Federal Court in the Public Civil Action concerning the Goiânia Accident" (PDF). OECD. 2000. Archived from the original on 2013-12-06. ()

[19] Steinhauser, Friedrich (November 2007). "Countering Radiological Terrorism: Consequences of the Radiation Exposure Incident in Goiania (Brazil)". "Volume 29 NATO Science for Peace and Security Series: Human and Societal Dynamics" : 7.

[20] Da Cruz, AD; Curry, J; Curado, MP; Glickman, BW (1996). "Monitoring hprt mutant frequency over time in T-lymphocytes of people accidentally exposed to high doses of ionizing radiation" . *Environmental and molecular mutagenesis*. **27** (3): 165–75. doi:10.1002/(SICI)1098-2280(1996)27:3<165::AID-EM1>3.0.CO;2-E. PMID 8625952.

[21] Saddi, V; Curry, J; Nohturfft, A; Kusser, W; Glickman, BW (1996). "Increased hprt mutant frequencies in Brazilian children accidentally exposed to ionizing radiation" . *Environmental and molecular mutagenesis*. **28** (3): 267–75. doi:10.1002/(SICI)1098-2280(1996)28:3<267::AID-EM11>3.0.CO;2-D. PMID 8908186.

[22] Da Cruz, AD; Volpe, JP; Saddi, V; Curry, J; Curadoc, MP; Glickman, BW (1997). "Radiation risk estimation in human populations: lessons from the radiological accident in Brazil" . *Mutation research*. **373** (2): 207–14. doi:10.1016/S0027-5107(96)00199-6. PMID 9042402.

[23] Skandalis, A; Da Cruz, AD; Curry, J; Nohturfft, A; Curado, MP; Glickman, BW (1997). "Molecular analysis of T-lymphocyte HPRT− mutations in individuals exposed to ionizing radiation in Goiânia, Brazil". *Environmental and molecular mutagenesis*. **29** (2): 107–16. doi:10.1002/(SICI)1098-2280(1997)29:2<107::AID-EM1>3.0.CO;2-B. PMID 9118962.

[24] *Césio 137 - O Pesadelo de Goiânia* at the Internet Movie Database

[25] UraniumFilmFestival.org: Roberto Pires

[26] Camargo Da Silva, T. (1997). Leibing, Annette, ed. *Biomedical Discourses and Health Care Experiences: The Goiâna Radiological Disaster*. *The Medical Anthropologies in Brazil. Curare* Sonderband. **12**. Berlin: Verlag für Wissenschaft Und Bildung. pp. 72–73. ISBN 9783861355687.

32.9 External links

- Detailed Report from the International Atomic Energy Agency, Vienna, 1988

- Similar accidents over the world (short overview)

- The Goiânia Radiation Incident

- Radioactive waste sold as scrap in India

- Q&A: Health effects of radiation exposure, *BBC News*, 21 July 2011.

Chapter 33

1990 Clinic of Zaragoza radiotherapy accident

The **1990 Clinic of Zaragoza radiotherapy accident** was a radiological accident that occurred from December 10–20, 1990, at the Clinic of Zaragoza, in Spain.

In the accident, at least 27 patients were injured, and 11 of them died, according to International Atomic Energy Agency (IAEA).[*][1] All of the injured were cancer patients receiving radiotherapy.[*][2]

On December 7, 1990, a technician performed maintenance on an electron accelerator at the Clinic of Zaragoza. On December 10, it returned to service after the repairs. On December 19,[*][3] the Spanish Nuclear Safety Board was scheduled to make its annual review to the device, but due to bureaucratic reasons this review was delayed. The Spanish Nuclear Safety Board found the electron accelerator power was too high. On December 20, 1990, the unit was stopped, and was restarted on March 8, 1991.

33.1 Chronology

Affected patients immediately suffered burns on the skin of the irradiated area, as well as inflammation of the internal organs and bone marrow. The first patient died on February 16, 1991, two months after irradiation. Fatalities increased until, on December 25, 1991, the last of a total of 25 patients died. However, the IAEA established that eleven of the deaths were due to the faulty maintenance.

The number affected might have been higher, because 31 other cancer patients were receiving treatment with the accelerator, but the other unit at the clinic was in perfect working condition.

33.2 The accident

The radiotherapy unit was repaired without following the correct instructions. The unit, in service 14 years at the time of the failure, had a breakdown in the electron beam accelerator control system ('deviator'). Repairs incorrectly increased output power, so patients that should have received therapy at 7 MeV were instead treated at 40 MeV.[*][4]

33.3 Responsibilities

Initially, the hospital was thought responsible for the accident, and specifically, the management of the radiological unit. The manager of the hospital said that the maintenance technician was responsible, and the Health Minister blamed General Electric (GE), the makers of the radiological unit, who had contracted out the maintenance.

Finally, on April 6, 1993, the hospital, its staff, and the Spanish National Institute of Health were acquitted. The court found the technician who performed the repair guilty, and secondarily, found General Electric guilty. GE had to compensate the affected families with 400 million pesetas (around 2.4 million euros).

33.4 Out of service

The device continued working until December 1996, when it was switched off and scrapped. This was done discreetly to avoid publicity.

33.5 See also

- List of civilian radiation accidents

- Ionizing radiation

33.6 References

[1] "IAEA Bulletin 413" (PDF). Archived from the original (PDF) on June 8, 2009.

[2] "El accidente del Clínico de Zaragoza, una cadena de fallos humanos única en el mundo, según los expertos". *El País* (in Spanish). October 12, 1991. Archived from the original on October 10, 2012.

[3] Garriga, Josep; Ortega, Javier (February 23, 1991). "El CSN no controló el aparato radiactivo del Clínico de Zaragoza en un año y medio". *El País* (in Spanish). Archived from the original on October 10, 2012.

[4] Serrano, Sebastian (February 26, 1991). "Un técnico revisó el acelerador del Clínico de Zaragoza tres días antes del accidente". *El País* (in Spanish). Archived from the original on October 10, 2012.

Coordinates: 41°38′N 0°54′W / 41.63°N 0.9°W

Chapter 34

1996 San Juan de Dios radiotherapy accident

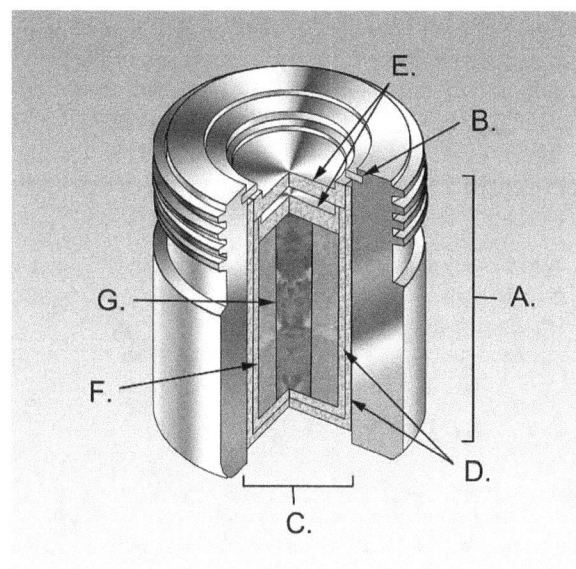

Cobalt-60 Teletherapy Capsule

- 1962 Mexico City radiation accident
- List of civilian radiation accidents
- Radiotherapy accident in Zaragoza
- X-ray
- Nuclear safety
- Nuclear whistleblowers

34.2 References

[1] Medical management of radiation accidents pp. 299 & 303.

[2] Strengthening the Safety of Radiation Sources p. 15.

Coordinates: 9°55′57″N 84°5′5″W / 9.93250°N 84.08472°W

The **radiotherapy accident in Costa Rica** occurred with the Alcyon II radiotherapy unit at San Juan de Dios Hospital in San José, Costa Rica. It was related to a cobalt-60 source that was being used for radiotherapy in 1996. An accidental overexposure of radiotherapy patients treated during August and September 1996 was detected. During the calibration process done after the change of ^{60}Co source on 22 August 1996, a mistake was made in calculating the dose rate, leading to severe overexposure of patients. The error of calibration was detected on 27 December 1997. In the course of the accident, 114 patients received an overdose of radiation and 13 died of radiation-related injuries.[*][1][*][2]

34.1 See also

- Goiânia accident

284

Chapter 35

Tokaimura nuclear accident

Coordinates: 36°28′47.00″N 140°33′13.24″E / 36.4797222°N 140.5536778°E

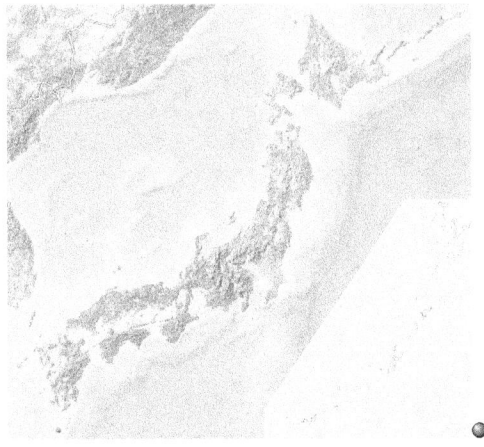
●Tōkai-mura

Location of the Tokaimura nuclear accident

There have been two **Tokaimura nuclear accidents** at the nuclear facility at Tōkai, Ibaraki:

1. On 11 March 1997, an explosion in a Dōnen plant.

2. On 30 September 1999, a serious criticality accident in a JCO plant.

When time is not specified, in most cases the 1999 incident is the one being referred to.[1]

35.1 In 1997

See also: Power Reactor and Nuclear Fuel Development Corporation

The first Tokaimura nuclear accident was the accident which occurred on 11 March 1997, in a nuclear reprocessing plant of the Dōnen (Power Reactor and Nuclear Fuel Development Corporation). Another name is the **Dōnen accident** (動燃事故 *Dōnen jiko*).

On the night of Tuesday 11 March 1997, a small explosion occurred in a nuclear reprocessing plant of the Dōnen. Windows were smashed and smoke escaped to the atmosphere.[2] On Thursday, workers repaired thirty broken windows and three doors with duct tape. They had been damaged during the blast. At least 37 workers were exposed to elevated levels of radiation during the incident.[3]

A week after the event, meteorological officials detected unusually high levels of caesium 40 kilometers (25 miles) south-west of the plant.[4]

35.2 In 1999

The second and more serious Tokaimura nuclear accident (Japanese: 東海村JCO臨界事故 *Tōkai-mura JCO-rinkai-jiko*) indicates the nuclear disaster which occurred on 30 September 1999,[5][6][7] resulting in two deaths.[8] It was the worst civilian nuclear radiation accident in Japan prior to the Fukushima Daiichi nuclear disaster of 2011.

The criticality accident occurred in a uranium reprocessing facility operated by JCO (formerly Japan Nuclear Fuel Conversion Co.), a subsidiary of Sumitomo Metal Mining Co. in the village of Tōkai, Naka District, Ibaraki Prefecture.[9]

The accident occurred as three workers, Hisashi Ouchi, Masato Shinohara, and Yutaka Yokokawa,[10] were preparing a small batch of fuel for the Jōyō experimental fast breeder reactor, using uranium enriched to 18.8% with the fissile radionuclide (radioisotope) known as U-235 (with the remainder being the fissionable-only U-238). It was JCO's first batch of fuel for that reactor in three years, and no proper qualification and training requirements appear to have been established to prepare those workers for the job. At around 10:35, a precipitation tank reached critical mass when its fill level, containing about 16 kilograms (35 pounds) of uranium, reached about 40 liters (11 U.S. gallons).[9]

35.2.1 Details

Criticality was reached upon the technicians adding a seventh bucket of an aqueous uranyl nitrate solution to the tank. The nuclear fission chain reaction became self-sustaining and began to emit intense gamma and neutron radiation. At the time of the criticality event, Ouchi had his body draped over the tank while Shinohara stood on a platform to pour the solution into it; Yokokawa was sitting at a desk four meters away. All three technicians observed a blue flash (possibly Cherenkov radiation) and gamma-radiation alarms sounded.[8][11]

Technicians Ouchi and Shinohara immediately experienced pain, nausea, difficulty breathing, and other symptoms. Ouchi then began to vomit in the decontamination room a few minutes later and lost consciousness shortly after.[12] There was no explosion, but fission products (fission fragments of U-235 with atomic masses typically around 95 and 137, such as yttrium-94 and barium-140) were progressively released inside the building.

Being a wet process with an intended liquid result, the water promoted the chain reaction by serving as a neutron moderator, whereby neutrons emitted from fissioned nuclei are slowed so they are more readily absorbed by neighboring nuclei, inducing them to fission in turn. The criticality continued intermittently for about 20 hours. As the solution boiled vigorously, steam bubbles attenuated the liquid water's action as a neutron moderator (see *Void coefficient*) and the solution lost criticality. However, the reaction resumed as the solution cooled and the voids disappeared.

The following morning, workers permanently stopped the reaction by draining water from a cooling jacket surrounding the precipitation tank since that water was serving as a neutron reflector. A boric acid solution (boron being a good neutron absorber) was then added to the tank to ensure that the contents remained subcritical. These operations exposed 27 workers to radioactivity.[9]

35.2.2 Cause

The cause of the accident was the workers adding a uranyl nitrate solution which contained about 16 kg of uranium into the precipitation tank. This greatly exceeded the tank's uranium limit of 2.4 kg and caused an instantaneous and uncontrolled nuclear fission. Under correct procedures, the uranyl nitrate would have been stored inside a buffer tank and then pumped from there into the precipitation tank at intervals of the correct volume level not exceeding 2.4 kg.

In this case, the workers bypassed using any buffer tanks en-

tirely and instead poured the uranyl nitrate directly into the precipitation tank with a stainless steel bucket rather than using a pump. The buffer tank would have actually held this solution safely, as it had a tall and narrow geometry and was designed to prevent criticality. The precipitation tank however had not been designed to hold this type of solution and was not configured to prevent criticality.[10]

35.2.3 Evacuation

Five hours after the start of the criticality, evacuation commenced of some 161 people from 39 households within a 350-meter radius from the conversion building. Residents were allowed home two days later with sandbags and other shielding to protect from residual gamma radiation. Twelve hours after the start of the incident residents within 10 km were asked to stay indoors as a precautionary measure, and this restriction was lifted the following afternoon.[9]

35.2.4 Aftermath

Dozens of emergency workers and nearby residents were hospitalized and hundreds of thousands of others were forced to remain indoors for 24 hours; 39 of the workers were exposed to the radiation.[13] At least 667 workers, emergency responders, and nearby residents were exposed to excess radiation as a result of the accident.[8]

By measuring the concentration of sodium-24, created by a neutron activation whereby sodium-23 nuclei were rendered radioactive by absorbing neutrons from the accident, it was possible to deduce the dose received by the technicians. According to the STA, Hisashi Ouchi was exposed to 17 sieverts (Sv) of radiation, Masato Shinohara received 10 Sv, and Yutaka Yokokawa 3 Sv.[8][10] By comparison, a dose of 50 mSv is the maximum allowable annual dose for Japanese nuclear workers.[9] A dose of 8 Sv (800 rem) is normally fatal and more than 10 Sv almost invariably so.[10] Normal background radiation amounts to an annual exposure of about 3 mSv.[8] There were 56 plant workers whose exposures ranged up to 23 mSv and a further 21 workers received elevated doses when draining the precipitation tank. Seven workers immediately outside the plant received doses estimated at 6-15 mSv (combined neutron and gamma effects).[9]

The two technicians who received the higher doses, Ouchi and Shinohara, died several months later. Ouchi suffered serious burns to most of his body, experienced severe damage to his internal organs, and had a near-zero white blood cell count.[8] Shinohara received numerous skin grafts, which were successful, but he ultimately succumbed to infection due to the damage his immune system sustained in the incident.

The cause of the accident was said to be "human error and serious breaches of safety principles", according to the International Atomic Energy Agency.*[9]

35.3 See also

- Nuclear power in Japan

- Tōkai Nuclear Power Plant

- Fukushima Daiichi nuclear disaster

- 5 yen coin#Use in nuclear accident investigation

- Rokkasho Reprocessing Plant, meant to be the successor to the Tokai reprocessing Plant (operating postponed to 2018)

35.4 References

[1] https://www.google.com.hk/search?q=Tokaimura+
    nuclear+accident&rls=com.microsoft:en-US&ie=
    UTF-8&oe=UTF-8&startIndex=&startPage=1&gfe_
    rd=cr&ei=8-NrVrioGsLU8AeE86PgDg&gws_rd=ssl

[2] "Fires damage Japanese nuclear facility, Tokaimura (1997) - on Newspapers.com". *Newspapers.com*. Retrieved 2015-08-10.

[3] "Japan acknowledges delays in dealing with accident at nuclear power plant, Tokaimura (1997) - on Newspapers.com". *Newspapers.com*. Retrieved 2015-08-10.

[4] "Greater radiation leak hinted - Tokaimura nuclear accident, Japan (1997) - on Newspapers.com". *Newspapers.com*. Retrieved 2015-08-09.

[5] "Timeline: Nuclear plant accidents". *BBC News*. 11 July 2006. Retrieved 17 March 2011.

[6] Charles Scanlon (30 September 2000). "Tokaimura: One year on". *BBC News*. Retrieved 17 March 2011.

[7] "Nuclear accident shakes Japan". *BBC News*. 30 September 1999. Retrieved 17 March 2011.

[8] Memorial University of Newfoundland: "The Tokaimura Accident (28 September 1999)"

[9] Tokaimura Criticality Accident

[10] Michael E. Ryan. "The Tokaimura Accident: Nuclear Energy and Reactor Safety". Department of Chemical Engineering, University at Buffalo, SUNY.

[11] Makoto Akashi, Director of Research Center for Radiation Emergency Medicine at Japan's National Institute of Radiological Sciences: "The Medical Basis for Radiation-Accident Preparedness", The Parthenon Publishing Group Inc., 2002, which states "All three workers saw a 'blue flash' and heard the gamma-radiation monitor alarm" (direct link to Google Book page). And "All three observed the Cherenkov light flash" (direct link to Google Book return).

[12] International Atomic Energy Agency: "Report on the preliminary fact finding mission following the accident at the nuclear fuel processing facility in Tokaimura, Japan", 1999 (See *External links*, below).

[13] In The Wake of Tokaimura, Japan Rethinks its Nuclear Picture

35.5 External links

- What Happened at Tokaimura?

- Tokaimura Criticality Accident – What happened in Japan

- International Atomic Energy Agency: "Report on the preliminary fact finding mission following the accident at the nuclear fuel processing facility in Tokaimura, Japan", 1999 (9.5 MB PDF, here)

- Criticality accident at Tokai nuclear fuel plant (Japan) Wise Uranium project

Chapter 36

Instituto Oncologico Nacional

The **National Oncologic Institute** or ION (Spanish: Instituto Oncologico Nacional) is a specialized hospital for cancer treatment, located in Panama City, Panama. Between August 2000 and March 2001, patients receiving radiation treatment for prostate cancer and cancer of the cervix received lethal doses of radiation, resulting in 17 fatalities.[1][2]

36.1 History

On 1936, President Juan Demóstenes Arosemena (a physician), conceived the creation of the National Radiologic Institute, an institution dedicated to treat cancer. The treatments were given on the Santo Tomas Hospital and on the former Panama Hospital.

On September 18, 1940, during the administration of President Augusto Boyd, the new facilities of the National Radiologic Institute were inaugurated, giving it its own building. The Institute was part of the Santo Tomas Hospital. This institute had 4 doctors, 3 nurses, and 40 beds. The treatments they had were radiotherapy, implantation of Radium needles, injections of hydrogenated mustard and surgery.

The institute had a passive handling of cancer, because the treatment of cancer given at the time was to ease pain in patients.

On 1965, a more active role in the battle against cancer is started, when the latest advancements of the time are applied in the detection and treatment of this illness.

Later that year the National Radiologic Institute was renamed Juan Demóstenes Arosemena Cancerologic Center, as a recognition of the work of this physician and creator of the institution. A cobalt-60 pump was acquired.

On 1980, the institution begins relations with the government of Japan, that was interested on the treatment of cancer in Panama, and by which a donation of medical and surgical equipment, including ultrasound, X-rays, and others are acquired.

On 1984, by law 11, the National Oncologic Institute (Instituto Oncologico Nacional) Juan Demóstenes Arosemena is created.

On June 3, 1999, the Panamanian Government, on President Ernesto Perez Balladares administration, gives buildings 242 and 254 of the former Gorgas Hospital to the Institute, and on July 23 the Institute moves to this location from the building on Justo Arosemena Avenue.

The Hospital has continued its growth and acquired new equipment, like a linear accelerator, a new CT and opened its ICU

36.2 Accident

As in most radiotherapy departments, the one at ION uses a treatment planning system (TPS) to calculate the resulting dose distributions and determine treatment times. The data for each shielding block should be entered into the TPS separately. The TPS allows a maximum of four shielding blocks per field to be taken into account when calculating treatment times and dose distributions. Shielding blocks are used to protect healthy tissue of patients undergoing radiotherapy at the Institute, as is the normal practice.

In order to satisfy the request of a radiation oncologist to include five blocks in the field, in August 2000 the method of digitizing shielding blocks was changed. It was found that it was possible to enter data into the TPS for multiple shielding blocks together as if they were a single block, thereby apparently overcoming the limitation of four blocks per field. As was found later, although the TPS accepted entry of the data for multiple shielding blocks as if they were a single block, at least one of the ways in which the data were entered the computer output indicated a treatment time substantially longer than it should have been. The result was that patients received a proportionately higher dose than that prescribed. The modified treatment protocol was used for 28 patients, who were treated between August 2000 and March 2001 for prostate cancer and cancer of the cervix.

There were 17 deaths and 11 injuries.*[2]

The modified protocol was used without a verification test, i.e. a manual calculation of the treatment time for comparison with the computer calculated treatment time, or a simulation of treatment by irradiating a water phantom and measuring the dose delivered. In spite of the treatment times being about twice those required for correct treatment, the error went unnoticed. Some early symptoms of excessive exposure were noted in some of the irradiated patients. The seriousness, however, was not realized, with the consequence that the accidental exposure went unnoticed for a number of months. The continued emergence of these symptoms, however, eventually led to the accidental exposure being detected in March 2001.

In May 2001, the Government of Panama requested assistance under the terms of the Convention on Assistance in the Case of a Nuclear Accident or Radiological Emergency. In its response, the International Atomic Energy Agency sent a team of five medical doctors and two physicists to Panama to perform a dosimetric and medical assessment of the accidental exposure and a medical evaluation of the affected patients' prognosis and treatment. The team was complemented by a physicist from the Pan American Health Organization (PAHO), also at the request of the Government of Panama.

The accidental exposures at the ION in Panama were very serious. Many patients have suffered severe radiation effects due to excessive dose. Both morbidity and mortality have increased significantly.

The IAEA report was consistent with the report made by local investigators. It was found that the radiotherapy equipment was properly calibrated and worked properly. The error was on the data entry, using a protocol not validated to enter more shielding blocks, that resulted in increased dose in the treatment. Most of the exposed patients have died, some radiation related, others by means of their advanced cancer. The Government of Panama agreed to share urgently the conclusions of the report to help prevent similar accidents. The physicists of ION involved were taken to trial by the patients' families.

36.3 References

[1] Investigation of an accidental Exposure of radiotherapy patients in Panama - International Atomic Energy Agency

[2] Johnston, Robert (September 23, 2007). "Deadliest radiation accidents and other events causing radiation casualties". Database of Radiological Incidents and Related Events.

36.4 External links

- Official Website

- Investigation of an accidental Exposure of radiotherapy patients in Panama – International Atomic Energy Agency

Chapter 37

Fukushima Daiichi nuclear disaster

"Fukushima nuclear disaster" redirects here. For the incidents at Fukushima Daini (Fukushima II), see Fukushima Daini Nuclear Power Plant.

"2011 Japanese nuclear accidents" redirects here. For other 2011 Japanese nuclear accidents/incidents, see Fukushima Daini Nuclear Power Plant, Onagawa Nuclear Power Plant, Tōkai Nuclear Power Plant, and Rokkasho Reprocessing Plant.

IAEA Experts at Fukushima Daiichi Nuclear Power Plant Unit 4, 2013

The **Fukushima Daiichi nuclear disaster** (福島第一原子力発電所事故 *Fukushima Dai-ichi* (🔊 pronunciation) *genshiryoku hatsudensho jiko*) was an energy accident at the Fukushima I Nuclear Power Plant in Fukushima, initiated primarily by the tsunami following the Tōhoku earthquake on 11 March 2011.[6] Immediately after the earthquake, the active reactors automatically shut down their sustained fission reactions. However, the tsunami destroyed the emergency generators that would have provided power to cool the reactors. The insufficient cooling led to three nuclear meltdowns, hydrogen-air chemical explosions, and the release of radioactive material in Units 1, 2 and 3 from 12 March to 15 March. Loss of cooling also caused the pool for storing spent fuel from Reactor 4 to overheat on 15 March due to the decay heat from the fuel rods.

On 5 July 2012, the Fukushima Nuclear Accident Independent Investigation Commission (NAIIC) found that the causes of the accident had been foreseeable, and that the plant operator, Tokyo Electric Power Company (TEPCO), had failed to meet basic safety requirements such as risk assessment, preparing for containing collateral damage, and developing evacuation plans. On 12 October 2012, TEPCO admitted for the first time that it had failed to take necessary measures for fear of inviting lawsuits or protests against its nuclear plants.[7][8][9][10]

The Fukushima disaster is the largest nuclear disaster since the 1986 Chernobyl disaster and the second disaster to be given the Level 7 event classification of the International Nuclear Event Scale.[11] Though there have been no fatalities linked to radiation due to the accident, the eventual number of cancer deaths, according to the linear no-threshold theory of radiation safety, that will be caused by the accident is expected to be around 130–640 people in the years and decades ahead.[12][13][14] The United Nations Scientific Committee on the Effects of Atomic Radiation[15] and World Health Organization report that there will be no increase in miscarriages, stillbirths or physical and mental disorders in babies born after the accident.[16] However, an estimated 1,600 deaths are believed to have occurred due to the resultant evacuation conditions.[17][18] There are no clear plans for decommissioning the plant, but the plant management estimate is 30 or 40 years.[19] A frozen soil barrier has been constructed in an attempt to prevent further contamination of seeping groundwater,[20] but in July 2016 TEPCO revealed that the ice wall had failed to stop groundwater from flowing in and mixing with highly radioactive water inside the wrecked reactor buildings, adding that they are "technically incapable of blocking off groundwater with the frozen wall".[21]

In February 2017, TEPCO released images taken inside reactor 2 by a remote-controlled camera, that show there is a 2-meter hole in the metal grating under the pressure vessel in the reactor's primary containment vessel[22], which could have been caused by fuel escaping the pressure vessel. Radiation levels of about 650 sieverts per hour have been detected in the containment vessel of reactor No. 2.

37.1 Overview

The Fukushima I Nuclear Power Plant comprised six separate boiling water reactors originally designed by General Electric (GE) and maintained by the Tokyo Electric Power Company (TEPCO). At the time of the Tōhoku earthquake on 11 March 2011, reactors 4, 5 and 6 were shut down in preparation for re-fueling.[23] However, their spent fuel pools still required cooling.[24]

Immediately after the earthquake, the electricity-producing reactors 1, 2 and 3 automatically shut down their sustained fission reactions by inserting control rods in a legally-mandated safety procedure referred to as SCRAM, which ceases the reactors' normal running conditions. As the reactors were unable to generate power to run their own coolant pumps, emergency diesel generators came online, as designed, to power electronics and coolant systems. These operated nominally until the tsunami destroyed the generators for reactors 1–5. The two generators cooling reactor 6 were undamaged and were sufficient to be pressed into service to cool the neighboring reactor 5 along with their own reactor, averting the overheating issues that reactor 4 suffered.[24]

The largest tsunami wave was 13 meters high and hit 50 minutes after the initial earthquake, overwhelming the plant's seawall, which was 10 m high.[6] The moment of impact was recorded by a camera.[25] Water quickly flooded the low-lying rooms in which the emergency generators were housed.[26] The flooded diesel generators failed soon afterwards, resulting in a loss of power to the critical coolant water pumps. These pumps needed to continuously circulate coolant water through a Generation II reactor for several days to keep the fuel rods from melting, as the fuel rods continued to generate decay heat after the SCRAM event. The fuel rods would become hot enough to melt during the fuel decay time period if an adequate heat sink was not available. After the secondary emergency pumps (run by back-up electrical batteries) ran out, one day after the tsunami, 12 March,[27] the water pumps stopped and the reactors began to overheat. The insufficient cooling eventually led to meltdowns in reactors 1, 2, and 3, where the resulting corium is believed to have melted through the bottom of each reactor pressure vessel.

Meanwhile, as workers struggled to supply power to the reactors' coolant systems and restore power to their control rooms, a number of hydrogen-air chemical explosions occurred, the first in Unit 1, on 12 March and the last in Unit 4, on 15 March.[27][28][29] It is estimated that the hot zirconium fuel cladding-water reaction in reactors 1–3 produced 800 to 1000 kilograms of hydrogen gas each. The pressurized gas was vented out of the reactor pressure vessel where it mixed with the ambient air, and eventually reached explosive concentration limits in units 1 and 3. Due to piping connections between units 3 and 4, or alternatively from the same reaction occurring in the spent fuel pool in unit 4 itself,[30] unit 4 also filled with hydrogen, resulting in an explosion. In each case, the hydrogen-air explosions occurred at the top of each unit, that was in their upper secondary containment buildings.[31][32] Drone overflights on 20 March and afterwards captured clear images of the effects of each explosion on the outside structures, while the view inside was largely obscured by shadows and debris.[1]

There have been no fatalities linked to short term overexposure to radiation reported due to the Fukushima accident, while approximately 18,500 people died due to the earthquake and tsunami. The maximum cancer mortality and morbidity estimate according to the Linear no-threshold theory is 1,500 and 1,800 but with most estimates considerably lower, in the range of a few hundred.[33] In addition, the rates of psychological distress among evacuated people rose fivefold compared to the Japanese average due to the experience of the disaster and evacuation.[34]

In 2013, the World Health Organization (WHO) indicated that the residents of the area who were evacuated were exposed to low amounts of radiation and that radiation-induced health impacts are likely to be low.[35][36] In particular, the 2013 WHO report predicts that for evacuated infant girls, their 0.75% pre-accident lifetime risk of developing thyroid cancer is calculated to be increased to 1.25% by being exposed to radioiodine, with the increase being slightly less for males. The risks from a number of additional radiation-induced cancers are also expected to be elevated due to exposure caused by the other low boiling point fission products that were released by the safety failures. The single greatest increase is for thyroid cancer, but in total, an overall 1% higher lifetime risk of developing cancers of all types, is predicted for infant females, with the risk slightly lower for males, making both some of the most radiation-sensitive groups.[36] Along with those within the womb, which the WHO predicted, depending on their gender, to have the same elevations in risk as the infant groups.[37]

A screening program a year later in 2012 found that more than a third (36%) of children in Fukushima Prefecture have abnormal growths in their thyroid glands.[38] As of August 2013, there have been more than 40 children newly diagnosed with thyroid cancer and other cancers in Fukushima prefecture as a whole. In 2015, the number of thyroid cancers or detections of developing thyroid cancers numbered 137.[39] However whether these incidences of cancer are elevated above the rate in un-contaminated areas and therefore were due to exposure to nuclear radiation is unknown at this stage. Data from the Chernobyl accident showed that an unmistakable rise in thyroid cancer rates following the disaster in 1986 only began after a cancer incubation period of 3–5 years,[40] however whether this data can be directly compared to the Fukushima nuclear disaster is still yet to be determined.[41]

A survey by the newspaper Mainichi Shimbun computed that of some 300,000 people who evacuated the area, approximately 1,600 deaths related to the evacuation conditions, such as living in temporary housing and hospital closures have occurred as of August 2013, a number comparable to the 1,599 deaths directly caused by the earthquake and tsunami in the Fukushima Prefecture in 2011. With the exact cause of the majority of these evacuation related deaths not being specified, as according to the municipalities, that would hinder application for condolence money compensation[17][18] by the relatives of the deceased.

On 5 July 2012, the Japanese National Diet-appointed Fukushima Nuclear Accident Independent Investigation Commission (NAIIC) submitted its inquiry report to the Japanese Diet.[42] The Commission found the nuclear disaster was "manmade", that the direct causes of the accident were all foreseeable prior to 11 March 2011. The report also found that the Fukushima Daiichi Nuclear Power Plant was incapable of withstanding the earthquake and tsunami. TEPCO, the regulatory bodies (NISA and NSC) and the government body promoting the nuclear power industry (METI), all failed to correctly develop the most basic safety requirements—such as assessing the probability of damage, preparing for containing collateral damage from such a disaster, and developing evacuation plans for the public in the case of a serious radiation release. Meanwhile, the government-appointed Investigation Committee on the Accident at the Fukushima Nuclear Power Stations of Tokyo Electric Power Company submitted its final report to the Japanese government on 23 July 2012.[43] A separate study by Stanford researchers found that Japanese plants operated by the largest utility companies were particularly unprotected against potential tsunami.[6]

TEPCO admitted for the first time on 12 October 2012 that it had failed to take stronger measures to prevent disasters for fear of inviting lawsuits or protests against its nuclear plants.[7][8][9][10] There are no clear plans for decommissioning the plant, but the plant management estimate is thirty or forty years.[19]

A frozen soil barrier has been constructed in an attempt to prevent further contamination of seeping groundwater by melted-down nuclear fuel [20], but in July 2016 TEPCO revealed that the ice wall had failed to stop groundwater from flowing in and mixing with highly radioactive water inside the wrecked reactor buildings, adding that they are "technically incapable of blocking off groundwater with the frozen wall" [21]

37.2   Plant description

Main article: Fukushima Daiichi Nuclear Power Plant

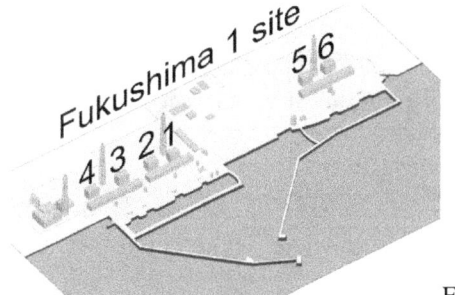

- Fukushima I (Daiichi) nuclear powerplant site close-up.

- Map of Japan's electricity distribution network, showing incompatible systems between regions. Fukushima is in the 50 hertz Tohoku region.

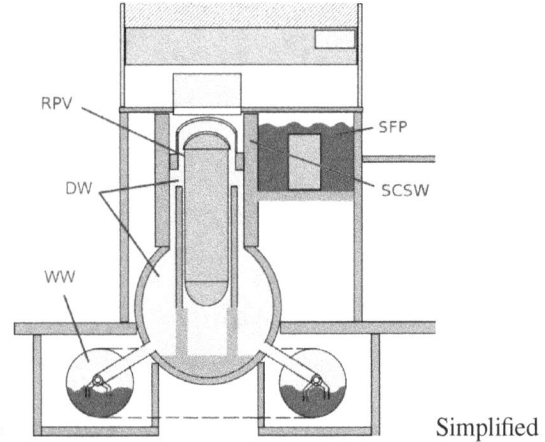

• Simplified cross-section sketch of a typical BWR Mark I containment as used in units 1 to 5.

Key:

RPV: reactor pressure vessel.

DW: dry well enclosing reactor pressure vessel.

WW: wet well - torus-shaped all around the base enclosing steam suppression pool. Excess steam from the dry well enters the wet well water pool via downcomer pipes.

SFP: spent fuel pool area.

SCSW: secondary concrete shield wall.

The Fukushima I (*Daiichi*) Nuclear Power Plant consisted of six GE light water boiling water reactors (BWRs) with a combined power of 4.7 gigawatts, making it one of the world's 25 largest nuclear power stations. It was the first GE-designed nuclear plant to be constructed and run entirely by the Tokyo Electric Power Company (TEPCO). Reactor 1 was a 439 MWe type (BWR-3) reactor constructed in July 1967, and commenced operation on 26 March 1971.[44] It was designed to withstand an earthquake with a peak ground acceleration of 0.18 g (1.74 m/s^2) and a response spectrum based on the 1952 Kern County earthquake.[45] Reactors 2 and 3 were both 784 MWe type BWR-4s. Reactor 2 commenced operation in July 1974, and Reactor 3 in March 1976. The earthquake design basis for all units ranged from 0.42 g (4.12 m/s^2) to 0.46 g (4.52 m/s^2).[46][47] After the 1978 Miyagi earthquake, when the ground acceleration reached 0.125 g (1.22 m/s^2) for 30 seconds, no damage to the critical parts of the reactor was found.[45] Units 1–5 have a Mark-1 type (light bulb torus) containment structure; unit 6 has Mark 2-type (over/under) containment structure.[45] In September 2010, Reactor 3 was partially fueled by mixed-oxides (MOX).[48]

At the time of the accident, the units and central storage facility contained the following numbers of fuel assemblies:[49]

There is no MOX fuel in any of the cooling ponds. The only MOX fuel is loaded in the Unit 3 reactor.

37.2.1 Cooling

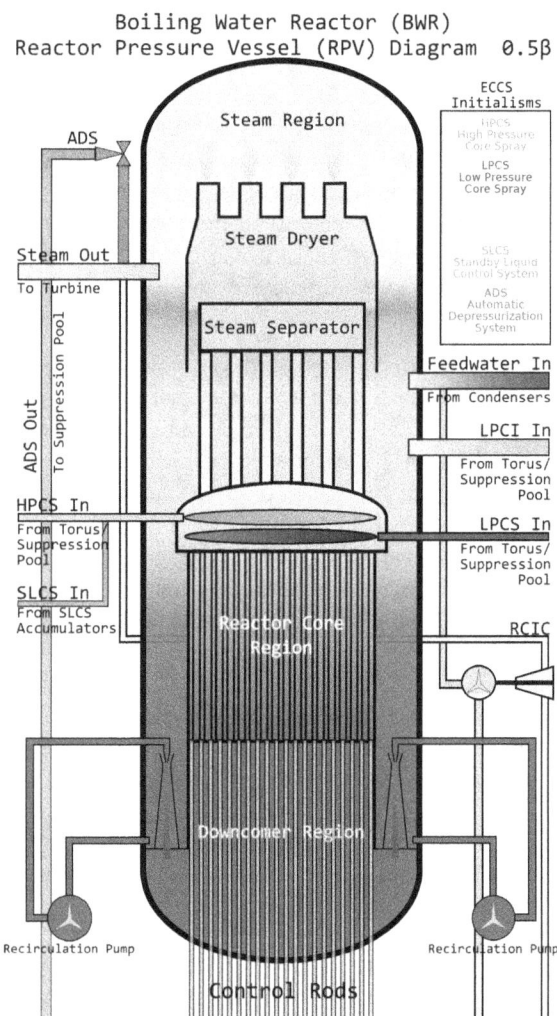

Diagrammatic representation of the cooling systems of a BWR.

See also: Decay heat § Power reactors in shutdown, and Nuclear reactor safety systems

Nuclear reactors generate electricity by using the heat of the fission reaction to create steam. When the reactor stops operating, the radioactive decay of unstable isotopes in the fuel continues to generate heat (decay heat) for a time, and so require continued cooling.[53][54] Initially this decay heat amounts to approximately 6.5% of the amount produced by fission,[53] decreasing over several days before reaching shutdown levels.[55] Afterwards, spent fuel rods typically require several years in a spent fuel pool before they can be safely transferred to dry cask storage vessels.*[56] The decay heat in the Unit 4 spent fuel pool had the capacity to boil about 70 metric tons of water per day (12 gallons per minute).*[57]

In the reactor core, high-pressure systems cycle water between the reactor pressure vessel and heat exchangers. These systems transfer heat to a secondary heat exchanger via the essential service water system, using water pumped out to sea or an onsite cooling tower.*[58] Units 2 and 3 were equipped with steam turbine-driven emergency core cooling systems that could be directly operated by steam produced by decay heat, and which could inject water directly into the reactor.*[59] Some electrical power was needed to operate valves and monitoring systems.

Unit 1 was equipped with a different, entirely passive cooling system, the "Isolation Condenser" ("IC"). It consisted of a series of pipes run from the reactor core to the inside of a large tank of water. When the valves are opened, steam flows upward to the IC where the cool water in the tank condenses the steam back to water, and it runs under gravity back to the reactor core. For unknown reasons, Unit 1's IC was operated only intermittently during the emergency. However, during a 25 March 2014 presentation to the TVA, Dr Takeyuki Inagaki explained that the IC was being operated intermittently to maintain reactor vessel level and to prevent the core from cooling too quickly which can increase reactor power. Unfortunately, as the tsunami engulfed the station, the IC valves were closed and could not be reopened automatically due to the loss of electrical power, but could have been opened manually.*[60] On 16 April 2011, TEPCO declared that cooling systems for Units 1–4 were beyond repair.*[61]

37.2.2 Backup generators

When the reactor is not producing electricity, cooling pumps can be powered by other reactor units, the grid or by diesel generators or batteries.*[62]*[63]

Two emergency diesel generators were available for each of units 1–5 and three for unit 6.*[64]

In the late 1990s, three additional backup generators for Units 2 and 4 were placed in new buildings located higher on the hillside, to comply with new regulatory requirements. All six units were given access to these generators, but the switching stations that sent power from these backup generators to the reactors' cooling systems for Units 1 through 5 were still in the poorly protected turbine buildings. The switching station for Unit 6 was protected inside the only GE Mark II reactor building and continued to function.*[65] All three of the generators added in the late 1990s

were operational after the tsunami. If the switching stations had been moved to inside the reactor buildings or to other flood-proof locations, power would have been provided by these generators to the reactors' cooling systems.*[66]

The reactor's emergency diesel generators and DC batteries, crucial components in powering cooling systems after a power loss, were located in the basements of the reactor turbine buildings, in accordance with GE's specifications. Mid-level engineers expressed concerns that this left them vulnerable to flooding.*[67]

Fukushima I was not designed for such a large tsunami,*[68]*[69] nor had the reactors been modified when concerns were raised in Japan and by the IAEA.*[70]

Fukushima II was also struck by the tsunami. However, it had incorporated design changes that improved its resistance to flooding, reducing flood damage. Generators and related electrical distribution equipment were located in the watertight reactor building, so that power from the electricity grid was being used by midnight.*[71] Seawater pumps for cooling were protected from flooding, and although 3 of 4 initially failed, they were restored to operation.*[72]

37.2.3 Central fuel storage areas

Used fuel assemblies taken from reactors are initially stored for at least 18 months in the pools adjacent to their reactors. They can then be transferred to the central fuel storage pond.*[4] Fukushima I's storage area contains 6375 fuel assemblies. After further cooling, fuel can be transferred to dry cask storage, which has shown no signs of abnormalities.*[73]

37.2.4 Zircaloy

Many of the internal components and fuel assembly cladding are made from zircaloy because it is relatively transparent to neutrons. At normal operating temperatures of approximately 300 °C (572 °F), zircaloy is inert. However, above 1200 degrees Celsius, zirconium metal can react exothermically with water to form free hydrogen gas.*[74] The reaction between zirconium and the coolant produces more heat, accelerating the reaction.*[75] In addition zircaloy can react with uranium dioxide to form zirconium dioxide and uranium metal. This exothermic reaction together with the reaction of boron carbide with stainless steel can release additional heat energy thus contributing to the overheating of a reactor.*[76]

37.3 Prior safety concerns

37.3.1 1967: Layout of the emergency-cooling system

The Fukushima reactor control room in 1999

On 27 February 2012, the Nuclear and Industrial Safety Agency ordered TEPCO to report its reasoning for changing the piping layout for the emergency cooling system.

The original plans separated the piping systems for two reactors in the isolation condenser from each other. However, the application for approval of the construction plan showed the two piping systems connected outside the reactor. The changes were not noted, in violation of regulations.[*][77]

After the tsunami, the isolation condenser should have taken over the function of the cooling pumps, by condensing the steam from the pressure vessel into water to be used for cooling the reactor. But the condenser did not function properly and TEPCO could not confirm whether a valve was opened.

37.3.2 1991: Backup generator of reactor 1 flooded

On 30 October 1991, one of two backup generators of Reactor 1 failed, after flooding in the reactor's basement. Seawater used for cooling leaked into the turbine building from a corroded pipe at 20 cubic meters per hour, as reported by former employees in December 2011. An engineer was quoted as saying that he informed his superiors of the possibility that a tsunami could damage the generators. TEPCO installed doors to prevent water from leaking into the generator rooms.

The Japanese Nuclear Safety Commission stated that it would revise its safety guidelines and would require the in-

stallation of additional power sources. On 29 December 2011, TEPCO admitted all these facts: its report mentioned that the room was flooded through a door and some holes for cables, but the power supply was not cut off by the flooding, and the reactor was stopped for one day. One of the two power sources was completely submerged, but its drive mechanism had remained unaffected.[*][78]

37.3.3 2008: Tsunami study ignored

In 2007, TEPCO set up a department to supervise its nuclear facilities. Until June 2011 its chairman was Masao Yoshida, the Fukushima Daiichi chief. A 2008 in-house study identified an immediate need to better protect the facility from flooding by seawater. This study mentioned the possibility of tsunami-waves up to 10.2 metres (33 ft). Headquarters officials insisted that such a risk was unrealistic and did not take the prediction seriously.[*][79]

A Mr. Okamura of the Active Fault and Earthquake Research Center urged TEPCO and NISA to review their assumption of possible tsunami heights based on a tenth century earthquake, but it was not seriously considered at that time.[*][80] The U.S. Nuclear Regulatory Commission warned of a risk of losing emergency power in 1991 (NUREG-1150) and NISA referred to the report in 2004. No action to mitigate the risk was taken.[*][81]

37.3.4 Vulnerability to earthquakes

Japan, like the rest of the Pacific Rim, is in an active seismic zone, prone to earthquakes. The International Atomic Energy Agency (IAEA) had expressed concern about the ability of Japan's nuclear plants to withstand earthquakes. At a 2008 meeting of the G8's Nuclear Safety and Security Group in Tokyo, an IAEA expert warned that a strong earthquake with a magnitude above 7.0 could pose a "serious problem" for Japan's nuclear power stations.[*][82] The region had experienced three earthquakes of magnitude greater than 8, including the 869 Jogan Sanriku earthquake, the 1896 Meiji-Sanriku earthquake, and the 1933 Sanriku earthquake.

37.4 Events

Further information: Timeline of the Fukushima I nuclear accidents and 2011 Tōhoku earthquake and tsunami

37.4.1   Tōhoku earthquake

The 9.0 M_W Tōhoku earthquake occurred at 14:46 on Friday, 11 March 2011, with the epicenter near Honshu, the largest island of Japan.[83] It produced maximum ground g-forces of 0.56, 0.52, 0.56 (5.50, 5.07 and 5.48 m/s^2) at units 2, 3 and 5 respectively. This exceeded the earthquake tolerances of 0.45, 0.45 and 0.46 g (4.38, 4.41 and 4.52 m/s^2). The shock values were within the design tolerances at units 1, 4 and 6.[47]

When the earthquake struck, units 1, 2 and 3 were operating, but units 4, 5 and 6 had been shut down for a scheduled inspection.[46][84] Reactors 1, 2 and 3 immediately shut down automatically;[85][86] this meant the plant stopped generating electricity and could no longer use its own power.[87] One of the two connections to off-site power for units 1–3 also failed,[87] so 13 on-site emergency diesel generators began providing power.[88]

37.4.2   Tsunami and flooding

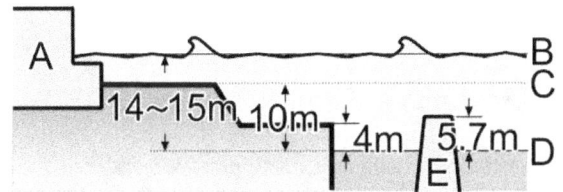

The height of the tsunami that struck the station approximately 50 minutes after the earthquake. A:Power station buildings B:peak height of tsunami C:Ground level of site D:average sea level E: Sea Wall to block waves.

The earthquake triggered a 13-to-15-metre (43 to 49 ft)-high tsunami that arrived approximately 50 minutes later. The waves overtopped the plant's 5.7 metres (19 ft) seawall,[89][90][91] flooding the basements of the power plant's turbine buildings and disabling the emergency diesel generators[64][92][93] at approximately 15:41.[87][94] TEPCO then notified authorities of a "first-level emergency" .[85] The switching stations that provided power from the three backup generators located higher on the hillside failed when the building that housed them flooded.[66] Power for the plant's control systems switched to batteries designed to provide power for about eight hours.[95] Further batteries and mobile generators were dispatched to the site, but were delayed by poor road conditions; the first arrived at 21:00 11 March,[88][96] almost six hours after the tsunami struck.

Multiple unsuccessful attempts were made to connect portable generating equipment to power water pumps. The failure was attributed to flooding at the connection point

in the Turbine Hall basement and the absence of suitable cables.[92] TEPCO switched its efforts to installing new lines from the grid.[97] One generator at unit 6 resumed operation on 17 March, while external power returned to units 5 and 6 only on 20 March.[98]

37.4.3   Evacuation

The government initially set in place a four-stage evacuation process: a prohibited access area out to 3 km, an on-alert area 3–20 km and an evacuation prepared area 20–30 km. On day one, an estimated 170,000 people[99] were evacuated from the prohibited access and on-alert areas. Prime Minister Kan instructed people within the on-alert area to leave and urged those in the prepared area to stay indoors.[100][101] The latter groups were urged to evacuate on 25 March.[102] The 20 kilometer exclusion zone was guarded by roadblocks to ensure that fewer people would be affected by the radiation.[103]

The earthquake and tsunami damaged or destroyed more than one million buildings leading to a total of 470,000 people needing evacuation. Of the 470,000, the nuclear accident was responsible for 154,000 being evacuated.[104]

As of March 2016, of the original 470,000 evacuees, 174,000 evacuees remain.[105]

37.4.4   Units 1, 2 and 3

See also: Fukushima Daiichi nuclear disaster (Unit 1 Reactor), Fukushima Daiichi nuclear disaster (Unit 2 Reactor), and Fukushima Daiichi nuclear disaster (Unit 3 Reactor)

In Reactors 1, 2 and 3, overheating caused a reaction between the water and the zircaloy, creating hydrogen gas.[106][107][108] On 12 March, an explosion in Unit 1 was caused by the ignition of the hydrogen, destroying the upper part of the building. On 14 March, a similar explosion occurred in the Reactor 3 building, blowing off the roof and injuring eleven people. On the 15th, there was an explosion in the Reactor 2 building due to a shared vent pipe with Reactor 3.

Core meltdowns

The amount of damage sustained by the reactor cores during the accident, and the location of molten nuclear fuel ("corium") within the containment buildings, is unknown; TEPCO has revised its estimates several times.[109] On 16 March 2011, TEPCO estimated that 70% of the fuel in Unit 1 had melted and 33% in Unit 2, and that Unit

3's core might also be damaged.*[110] As of 2015 it can be assumed that most fuel melted through the reactor pressure vessel (RPV, commonly known as the "reactor core") and is resting on the bottom of the primary containment vessel (PCV), having been stopped by the PCV concrete.*[111]*[112]*[113]*[114]

In the November 2011 TEPCO report of the Modular Accident Analysis Program (MAAP), further estimates are made to the state and location of the fuel.*[115] The report concluded that the Unit 1 RPV was damaged during the disaster and that "significant amounts" of molten fuel had fallen into the bottom of the PCV. The erosion of the concrete of the PCV by the molten fuel after the core meltdown was estimated to stop in approx. 0.7 metres (2 ft 4 in) in depth, while the thickness of the containment is 7.6 metres (25 ft) thick. Gas sampling carried out before the report detected no signs of an ongoing reaction of the fuel with the concrete of the PCV and all the fuel in Unit 1 was estimated to be "well cooled down, including the fuel dropped on the bottom of the reactor". Fuel in Units 2 and 3 had melted, however less than Unit 1, and fuel was presumed to be still in the RPV, with no significant amounts of fuel fallen to the bottom of the PCV. The report further suggested that "there is a range in the evaluation results" from "all fuel in the RPV (none fuel fallen to the PCV)" in Unit 2 and Unit 3, to "most fuel in the RPV (some fuel in PCV)". For Unit 2 and Unit 3 it was estimated that the "fuel is cooled sufficiently". The larger damage in Unit 1 in comparison with the other two units was according to the report due to longer time that no cooling water was injected in Unit 1, which resulted in much more decay heat to accumulate – for about 1 day there was no water injection for Unit 1, while Unit 2 and Unit 3 had only a quarter of a day without water injection.*[115]

In November 2013, Mari Yamaguchi reported for Associated Press that there are computer simulations which suggest that "the melted fuel in Unit 1, whose core damage was the most extensive, has breached the bottom of the primary containment vessel and even partially eaten into its concrete foundation, coming within about 30 centimeters (one foot) of leaking into the ground" – a Kyoto University nuclear engineer said with regards to these estimates: "We just can't be sure until we actually see the inside of the reactors." *[109]

According to a December 2013 report, TEPCO estimated for Unit 1 that "the decay heat must have decreased enough, the molten fuel can be assumed to remain in PCV (Primary container vessel)".*[111]

In August 2014, TEPCO released a new revised estimate that reactor 3 had a complete melt through in the initial phase of the accident. According to this new estimate within the first three days of the accident the entire core

content of reactor 3 had melted through the RPV and fallen to the bottom of the PCV.*[113]*[114]*[116] These estimates were based on a simulation, which indicated that reactor 3's melted core penetrated through 1.2 metres (3 ft 11 in) of the PCV's concrete base, and came close to 26–68 centimetres (10–27 in) of the PCV's steel wall.*[112]

In February 2015, TEPCO started the "Muon scanning" process for Units 1, 2 and 3.*[117]*[118] With this scanning setup it will be possible to determine the approximate amount and location of the remaining nuclear fuel within the reactor pressure vessel (RPV), but not the amount and resting place of the Corium in the PCV. In March 2015 TEPCO released the result of the Muon scan for Unit 1 which showed that no fuel was visible in the RPV, which would suggest that most if not all of the molten fuel had dropped onto the bottom of the PCV – this will change the plan for the removal of the fuel from Unit 1.*[119]*[120]

In February 2017 six years after the disaster radiation levels inside the Unit 2 containment building were estimated to be about 650 sieverts per hour.*[121] Images showed a hole in metal grating beneath the reactor pressure vessel (RPV), suggesting that melted nuclear fuel had escaped the RPV in that area.*[122]

37.4.5 Units 4, 5 and 6

Main article: Fukushima Daiichi units 4, 5 and 6

Unit 4

Reactor 4 was not operating when the earthquake struck. All fuel rods from Unit 4 had been transferred to the spent fuel pool on an upper floor of the reactor building prior to the tsunami. On 15 March, an explosion damaged the fourth floor rooftop area of Unit 4, creating two large holes in a wall of the outer building. It was reported that water in the spent fuel pool might be boiling. Radiation inside the Unit 4 control room prevented workers from staying there for long periods. Visual inspection of the spent fuel pool on 30 April revealed no significant damage to the rods. A radiochemical examination of the pond water confirmed that little of the fuel had been damaged.*[123]

In October 2012, the former Japanese Ambassador to both Switzerland and Senegal, Mitsuhei Murata, said that the ground under Fukushima unit 4 was sinking, and the structure may collapse.*[124]*[125]

In November 2013, TEPCO started the process of moving the 1533 fuel rods in the Unit 4 cooling pool to the central pool. This process was completed on 22 December 2014.*[126]

Aerial view of the station in 1975, showing separation between units 5 and 6, and 1–4.
· *Unit 6, not completed until 1979, is seen under construction.*

37.4.6   Central fuel storage areas

On 21 March, temperatures in the fuel pond had risen slightly, to 61 °C and water was sprayed over the pool.[*][4] Power was restored to cooling systems on 24 March and by 28 March, temperatures were reported down to 35 °C.[*][130]

37.4.7   Contamination

Main article: Radiation effects from Fukushima Daiichi nuclear disaster

Sub article: Comparison of Fukushima and Chernobyl nuclear accident with detailed tables inside

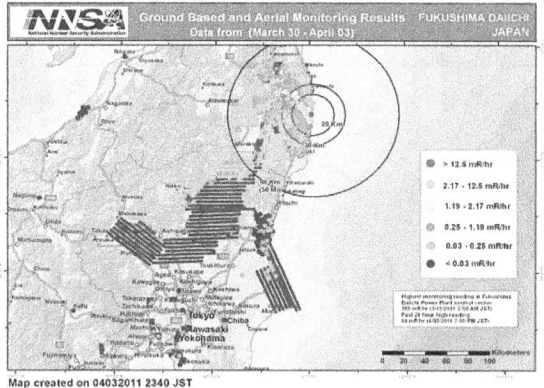

Map of contaminated areas around the plant (22 March – 3 April 2011).

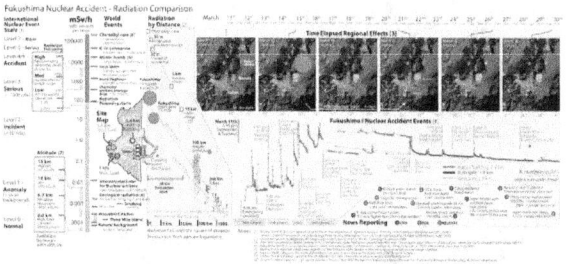

Fukushima dose rate comparison to other incidents and standards, with graph of recorded radiation levels and specific accident events from 11 to 30 March.

Units 5 and 6

Reactors 5 and 6 were also not operating when the earthquake struck. Unlike Reactor 4, their fuel rods remained in the reactor.   The reactors had been closely monitored, as cooling processes were not functioning well.[*][127] Both Unit 5 and Unit 6 shared a working generator and switchgear during the emergency and achieved a successful cold shutdown nine days later on 20 March.[*][128][*][129]

Radioactive material was released from the containment vessels for several reasons: deliberate venting to reduce gas pressure, deliberate discharge of coolant water into the sea, and uncontrolled events. Concerns about the possibility of a large scale release led to a 20-kilometre (12 mi) exclusion zone around the power plant and recommendations that

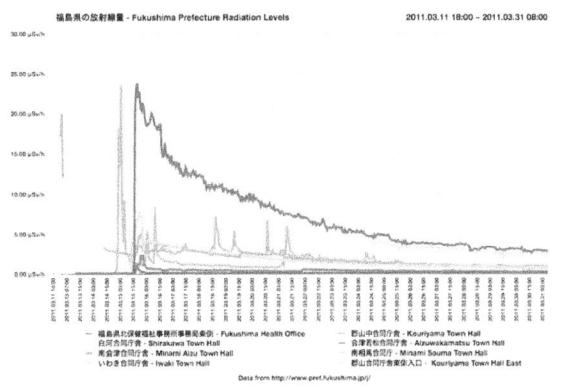

Radiation measurements from Fukushima Prefecture, March 2011

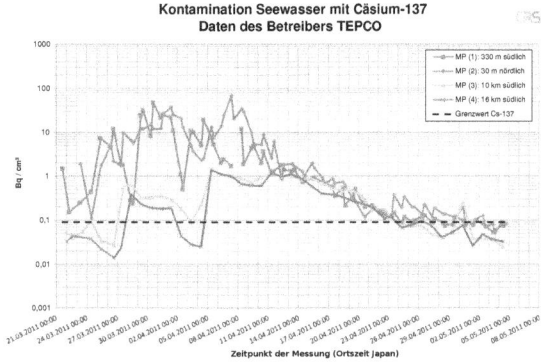

Seawater-contamination along coast with Caesium-137, from 21 March until 5 May 2011 (Source: GRS)

Radiation hotspot in Kashiwa, February 2012.

people within the surrounding 20–30 km zone stay indoors. Later, the UK, France and some other countries told their nationals to consider leaving Tokyo, in response to fears of spreading contamination.[131] In 2015, the tap water contamination was still higher in Tokyo compared to other cities in Japan.[132] Trace amounts of radioactivity, in-

cluding iodine-131, caesium-134 and caesium-137, were widely observed.[133][134][135]

Between 21 March and mid-July, around 2.7×10^{16} Bq of caesium-137 (about 8.4 kg) entered the ocean, with about 82 percent having flowed into the sea before 8 April.[136] However, the Fukushima coast has some of the world's strongest currents and these transported the contaminated waters far into the Pacific Ocean, thus causing great dispersion of the radioactive elements. The results of measurements of both the seawater and the coastal sediments led to the supposition that the consequences of the accident, in terms of radioactivity, would be minor for marine life as of autumn 2011 (weak concentration of radioactivity in the water and limited accumulation in sediments). On the other hand, significant pollution of sea water along the coast near the nuclear plant might persist, due to the continuing arrival of radioactive material transported towards the sea by surface water running over contaminated soil. Organisms that filter water and fish at the top of the food chain are, over time, the most sensitive to caesium pollution. It is thus justified to maintain surveillance of marine life that is fished in the coastal waters off Fukushima. Despite caesium isotopic concentration in the waters off of Japan being 10 to 1000 times above concentration prior to the accident, radiation risks are below what is generally considered harmful to marine animals and human consumers.[137]

A monitoring system operated by the Preparatory Commission for the Comprehensive Nuclear-Test-Ban Treaty Organization (CTBTO) tracked the spread of radioactivity on a global scale. Radioactive isotopes were picked up by over 40 monitoring stations.[138]

On 12 March, radioactive releases first reached a CTBTO monitoring station in Takasaki, Japan, around 200 km away. The radioactive isotopes appeared in eastern Russia on 14 March and the west coast of the United States two days later. By day 15, traces of radioactivity were detectable all across the northern hemisphere. Within one month, radioactive particles were noted by CTBTO stations in the southern hemisphere.[139][140]

Estimates of radioactivity released ranged from 10–40%[141][142][143][144] of that of Chernobyl's. The significantly contaminated area was 10[141]−12%[142] that of Chernobyl.[141][145][146]

In March 2011, Japanese officials announced that "radioactive iodine-131 exceeding safety limits for infants had been detected at 18 water-purification plants in Tokyo and five other prefectures".[147] On 21 March, the first restrictions were placed on the distribution and consumption of contaminated items.[148] As of July 2011, the Japanese government was unable to control the spread of radioactive material into the nation's food supply. Radioactive material was detected in food produced in 2011, including spinach, tea leaves, milk, fish and beef, up to 320 kilometres from the plant. 2012 crops did not show signs of radioactivity contamination. Cabbage, rice*[149] and beef showed insignificant levels of radioactivity. A Fukushima-produced rice market in Tokyo was accepted by consumers as safe.*[149]

On 24 August 2011, the Nuclear Safety Commission (NSC) of Japan published the results of the recalculation of the total amount of radioactive materials released into the air during the accident at the Fukushima Daiichi Nuclear Power Station. The total amounts released between 11 March and 5 April were revised downwards to 130 PBq (petabecquerels, 3.5 megacuries) for iodine-131 and 11 PBq for caesium-137, which is about 11% of Chernobyl emissions. Earlier estimations were 150 PBq and 12 PBq.*[150]*[151]

In 2011, scientists working for the Japan Atomic Energy Agency, Kyoto University and other institutes, recalculated the amount of radioactive material released into the ocean: between late March through April they found a total of 15 PBq for the combined amount of iodine-131 and caesium-137, more than triple the 4.72 PBq estimated by TEPCO. The company had calculated only the direct releases into the sea. The new calculations incorporated the portion of airborne radioactive substances that entered the ocean as rain.*[152]

In the first half of September 2011, TEPCO estimated the radioactivity release at some 200 MBq (megabecquerels, 5.4 millicuries) per hour. This was approximately one four-millionth that of March.*[153] Traces of iodine-131 were detected in several Japanese prefectures in November*[154] and December 2011.*[155]

According to the French Institute for Radiological Protection and Nuclear Safety, between 21 March and mid-July around 27 PBq of caesium-137 entered the ocean, about 82 percent before 8 April. This emission represents the most important individual oceanic emissions of artificial radioactivity ever observed. The Fukushima coast has one of the world's strongest currents (Kuroshio Current). It transported the contaminated waters far into the Pacific Ocean, dispersing the radioactivity. As of late 2011 measurements of both the seawater and the coastal sediments suggested that the consequences for marine life would be minor. Significant pollution along the coast near the plant might persist, because of the continuing arrival of radioactive material transported to the sea by surface water crossing contaminated soil. The possible presence of other radioactive substances, such as strontium-90 or plutonium, has not been sufficiently studied. Recent measurements show persistent contamination of some marine species (mostly fish) caught along the Fukushima coast.*[156] Migratory pelagic species are highly effective and rapid transporters of radioactivity throughout the ocean. Elevated levels of caesium-134 appeared in migratory species off the coast of California that were not seen pre-Fukushima.*[157]

As of March 2012, no cases of radiation-related ailments had been reported. Experts cautioned that data was insufficient to allow conclusions on health impacts. Michiaki Kai, professor of radiation protection at Oita University of Nursing and Health Sciences, stated, "If the current radiation dose estimates are correct, (cancer-related deaths) likely won't increase." *[158]

In May 2012, TEPCO released their estimate of cumulative radioactivity releases. An estimated 538.1 PBq of iodine-131, caesium-134 and caesium-137 was released. 520 PBq was released into the atmosphere between 12–31 March 2011 and 18.1 PBq into the ocean from 26 March – 30 September 2011. A total of 511 PBq of iodine-131 was released into both the atmosphere and the ocean, 13.5 PBq of caesium-134 and 13.6 PBq of caesium-137.*[159] TEPCO reported that at least 900 PBq had been released "into the atmosphere in March last year [2011] alone".*[160]*[161]

In 2012 researchers from the Institute of Problems in the Safe Development of Nuclear Energy, Russian Academy of Sciences, and the Hydrometeorological Center of Russia concluded that "on March 15, 2011, ~400PBq iodine, ~100PBq cesium, and ~400PBq inert gases entered the atmosphere" on that day alone.*[162]

In August 2012, researchers found that 10,000 nearby residents had been exposed to less than 1 millisievert of radiation, significantly less than Chernobyl residents.*[163]

As of October 2012, radioactivity was still leaking into the ocean. Fishing in the waters around the site was still prohibited, and the levels of radioactive ^{134}Cs and ^{137}Cs in the fish caught were not lower than immediately after the disaster.*[164]

On 26 October 2012, TEPCO admitted that it could not stop radioactive material entering the ocean, although emission rates had stabilized. Undetected leaks could not be ruled out, because the reactor basements remained flooded. The company was building a 2,400-foot-long steel and concrete wall between the site and the ocean, reaching 100 feet below ground, but it would not be finished before mid-2014. Around August 2012 two greenling were caught close to shore. They contained more than 25,000 becquerels (0.67 millicuries) of caesium-137 per kilogram, the highest measured since the disaster and 250 times the government's safety limit.*[165]*[166]

On 22 July 2013, it was revealed by TEPCO that the plant continued to leak radioactive water into the Pacific Ocean, something long suspected by local fishermen and independent investigators.*[167] TEPCO had previously denied that this was happening. Japanese Prime Minister Shinzō Abe ordered the government to step in.*[168]

On 20 August, in a further incident, it was announced that 300 metric tons of heavily contaminated water had leaked from a storage tank,*[169] approximately the same amount of water as one eighth (1/8) of that found in an Olympic-size swimming pool.*[170] The 300 metric tons of water was radioactive enough to be hazardous to nearby staff, and the leak was assessed as Level 3 on the International Nuclear Event Scale.*[171]

On 26 August, the government took charge of emergency measures to prevent further radioactive water leaks, reflecting their lack of confidence in TEPCO.*[172]

As of 2013, about 400 metric tons per day of cooling water was being pumped into the reactors. Another 400 metric tons of groundwater was seeping into the structure. Some 800 metric tons of water per day was removed for treatment, half of which was reused for cooling and half diverted to storage tanks.*[173] Ultimately the contaminated water, after treatment to remove radionuclides other than tritium, may have to be dumped into the Pacific.*[19] TEPCO intend to create an underground ice wall to reduce the rate contaminated groundwater reaches the sea.*[174]

In February 2014, NHK reported that TEPCO was reviewing its radioactivity data, after finding much higher levels of radioactivity than was reported earlier. TEPCO now says that levels of 5 million becquerels (0.12 millicuries) of strontium per liter were detected in groundwater collected in July 2013 and not 900,000 becquerels (0.02 millicuries), as initially reported.*[175]*[176]*[177]

On 10 September 2015, floodwaters driven by Typhoon Etau prompted mass evacuations in Japan and overwhelmed the drainage pumps at the stricken Fukushima nuclear plant. A TEPCO spokesperson said that hundreds of metric tons of radioactive water had entered the ocean as a result.*[178] Plastic bags filled with contaminated soil and grass were also swept away by the flood waters.*[179]

Contamination in the eastern Pacific

In March 2014, numerous news sources, including NBC,*[180] began predicting that the radioactive underwater plume traveling through the Pacific Ocean would reach the western seaboard of the continental United States. The common story was that the amount of radioactivity would be harmless and temporary once it arrived. The National Oceanic and Atmospheric Administration measured cesium-134 at points in the Pacific Ocean and models were cited in predictions by several government agencies to announce that the radiation would not be a health hazard for North American residents. Groups, including Beyond Nuclear and the Tillamook Estuaries Partnership, challenged these predictions on the basis of continued isotope releases after 2011, leading to a demand for more re-

cent and comprehensive measurements as the radioactivity made its way east. These measurements were taken by a cooperative group of organizations under the guidance of a marine chemist with the Woods Hole Oceanographic Institution, and it was revealed that total radiation levels, of which only a fraction bore the fingerprint of Fukushima, were not high enough to pose any direct risk to human life and in fact were far less than Environmental Protection Agency guidelines or several other sources of radiation exposure deemed safe.*[181] Integrated Fukushima Ocean Radionuclide Monitoring project (InFORM) also failed to show any significant amount of radiation*[182] and as a result authors received death threats from supporters of a Fukushima-induced "wave of cancer deaths across North America" theory.*[183]

37.5 Response

See also: Investigations into the Fukushima Daiichi nuclear disaster

Government agencies and TEPCO were unprepared for the "cascading nuclear disaster".*[184] The tsunami that "began the nuclear disaster could and should have been anticipated and that ambiguity about the roles of public and private institutions in such a crisis was a factor in the poor response at Fukushima".*[184] In March 2012, Prime Minister Yoshihiko Noda said that the government shared the blame for the Fukushima disaster, saying that officials had been blinded by a false belief in the country's "technological infallibility", and were taken in by a "safety myth". Noda said "Everybody must share the pain of responsibility." *[185]

According to Naoto Kan, Japan's prime minister during the tsunami, the country was unprepared for the disaster, and nuclear power plants should not have been built so close to the ocean.*[186] Kan acknowledged flaws in authorities' handling of the crisis, including poor communication and coordination between nuclear regulators, utility officials and the government. He said the disaster "laid bare a host of an even bigger man-made vulnerabilities in Japan's nuclear industry and regulation, from inadequate safety guidelines to crisis management, all of which he said need to be overhauled." *[186]

Physicist and environmentalist Amory Lovins said that Japan's "rigid bureaucratic structures, reluctance to send bad news upwards, need to save face, weak development of policy alternatives, eagerness to preserve nuclear power's public acceptance, and politically fragile government, along with TEPCO's very hierarchical management culture, also contributed to the way the accident unfolded. Moreover, the information Japanese people receive about nuclear energy and its alternatives has long been tightly controlled by both TEPCO and the government." *[187]

37.5.1 Poor communication and delays

The Japanese government did not keep records of key meetings during the crisis.*[188] Data from the SPEEDI network were emailed to the prefectural government, but not shared with others. Emails from NISA to Fukushima, covering 12 March 11:54 PM to 16 March 9 AM and holding vital information for evacuation and health advisories, went unread and were deleted. The data was not used because the disaster countermeasure office regarded the data as "useless because the predicted amount of released radiation is unrealistic." *[189] On 14 March 2011 TEPCO officials were instructed not to use the phrase "core meltdown" at press conferences.*[190]

On the evening of March 15, Prime Minister Kan called Seiki Soramoto, who used to design nuclear plants for Toshiba, to ask for his help in managing the escalating crisis. Soramoto formed an impromptu advisory group, which included his former professor at the University of Tokyo, Toshiso Kosako, a top Japanese expert on radiation measurement. Mr. Kosako, who studied the Soviet response to the Chernobyl crisis, said he was stunned at how little the leaders in the prime minister's office knew about the resources available to them. He quickly advised the chief cabinet secretary, Yukio Edano, to use SPEEDI, which used measurements of radioactive releases, as well as weather and topographical data, to predict where radioactive materials could travel after being released into the atmosphere.*[191]

The Investigation Committee on the Accident at the Fukushima Nuclear Power Stations of Tokyo Electric Power Company's interim report stated that Japan's response was flawed by "poor communication and delays in releasing data on dangerous radiation leaks at the facility". The report blamed Japan's central government as well as TEPCO, "depicting a scene of harried officials incapable of making decisions to stem radiation leaks as the situation at the coastal plant worsened in the days and weeks following the disaster".*[192] The report said poor planning worsened the disaster response, noting that authorities had "grossly underestimated tsunami risks" that followed the magnitude 9.0 earthquake. The 12.1 metre (40 ft) high tsunami that struck the plant was double the height of the highest wave predicted by officials. The erroneous assumption that the plant's cooling system would function after the tsunami worsened the disaster. "Plant workers had no clear instructions on how to respond to such a disaster, causing miscommunication, especially when the disaster destroyed

backup generators." *[192]

In February 2012, the Rebuild Japan Initiative Foundation described how Japan's response was hindered by a loss of trust between the major actors: Prime Minister Kan, TEPCO's Tokyo headquarters and the plant manager. The report said that these conflicts "produced confused flows of sometimes contradictory information".*[193]*[194] According to the report, Kan delayed the cooling of the reactors by questioning the choice of seawater instead of fresh water, accusing him of micromanaging response efforts and appointing a small, closed, decision-making staff. The report stated that the Japanese government was slow to accept assistance from U.S. nuclear experts.*[195]

A 2012 report in *The Economist* said: "The operating company was poorly regulated and did not know what was going on. The operators made mistakes. The representatives of the safety inspectorate fled. Some of the equipment failed. The establishment repeatedly played down the risks and suppressed information about the movement of the radioactive plume, so some people were evacuated from more lightly to more heavily contaminated places".*[196]

From 17 to 19 March 2011, US military aircraft measured radiation within a 45-km radius of the site. The data recorded 125 microsieverts per hour of radiation as far as 25 km (15.5 mi) northwest of the plant. The US provided detailed maps to the Japanese Ministry of Economy, Trade, and Industry (METI) on 18 March and to the Ministry of Education, Culture, Sports, Science and Technology (MEXT) two days later, but officials did not act on the information.*[197]

The data were not forwarded to the prime minister's office or the Nuclear Safety Commission (NSC), nor were they used to direct the evacuation. Because a substantial portion of radioactive materials reached ground to the northwest, residents evacuated in this direction were unnecessarily exposed to radiation. According to NSC chief Tetsuya Yamamoto, "It was very regrettable that we didn't share and utilize the information." Itaru Watanabe, from the Science and Technology Policy Bureau, blamed the US for not releasing the data.*[198]

Data on the dispersal of radioactive materials were provided to the U.S. forces by the Japanese Ministry for Science a few days after March 11; however, the data was not shared publicly until the Americans published their map on March 23, at which point Japan published fallout maps compiled from ground measurements and SPEEDI the same day.*[199] According to Watanabe's testimony before the Diet, the US military was given access to the data "to seek support from them" on how to deal with the nuclear disaster. Although SPEEDI's effectiveness was limited by not knowing the amounts released in the disaster, and thus was considered "unreliable", it was still able to forecast dispersal routes and could have been used to help local governments designate more appropriate evacuation routes.[*][200]

On 19 June 2012, science minister Hirofumi Hirano stated that his "job was only to measure radiation levels on land" and that the government would study whether disclosure could have helped in the evacuation efforts.[*][199]

On 28 June 2012 Nuclear and Industrial Safety Agency officials apologized to mayor Yuko Endo of Kawauchi Village for NISA having failed to release the American-produced radiation maps in the first days after the meltdowns. All residents of this village were evacuated after the government designated it a no-entry zone. According to a Japanese government panel, authorities had shown no respect for the lives and dignity of village people. One NISA official apologized for the failure and added that the panel had stressed the importance of disclosure; however, the mayor said that the information would have prevented the evacuation into highly polluted areas, and that apologies a year too late had no meaning.[*][201]

In June 2016, it was revealed that TEPCO officials had been instructed on 14 March 2011 not to describe the reactor damage using the word "meltdown". Officials at that time were aware that 25–55% of the fuel had been damaged, and the threshold for which the term "meltdown" became appropriate (5%) had been greatly exceeded. TEPCO President Naomi Hirose told the media: "I would say it was a cover-up... It's extremely regrettable." [*][202]

37.6 Event rating

Main article: Accident rating of the Fukushima Daiichi nuclear disaster

The incident was rated 7 on the International Nuclear

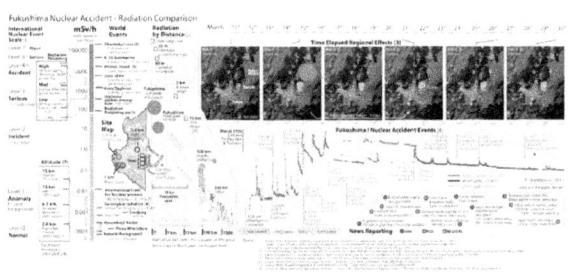

Comparison of radiation levels for different nuclear events.

Event Scale (INES).[*][203] This scale runs from 0, indicating an abnormal situation with no safety consequences, to 7, indicating an accident causing widespread contamination with serious health and environmental effects. Prior to Fukushima, the Chernobyl disaster was the only level 7 event on record, while the Three Mile Island accident was

rated as level 5.

A 2012 analysis of the intermediate and long-lived radioactivity released found about 10–20% of that released from the Chernobyl disaster.[*][204][*][205] Approximately 15 PBq of caesium-137 was released,[*][206] compared with approximately 85 PBq of caesium-137 at Chernobyl,[*][207] indicating the release of 26.5 kilograms (58 lb) of caesium-137.

Unlike Chernobyl, all Japanese reactors were in concrete containment vessels, which limited the release of strontium-90, americium-241 and plutonium, which were among the radioisotopes released by the earlier incident.[*][204][*][207]

Some 500 PBq of iodine-131 were released,[*][206] compared to approximately 1,760 PBq at Chernobyl.[*][207] Iodine-131 has a half life of 8.02 days, decaying into a stable nuclide. After ten half lives (80.2 days), 99.9% has decayed to xenon-131, a stable isotope.[*][208]

37.7 Aftermath

Main article: Fukushima Daiichi nuclear disaster casualties

No deaths followed short term radiation exposure,[*][209] though there were a number of deaths in the evacuation of the nearby population,[*][210] while 15,884 died (as of 10 February 2014[*][211]) due to the earthquake and tsunami.

37.7.1 Risks from radiation

Very few cancers would be expected as a result of accumulated radiation exposures,[*][212][*][213][*][214] even though people in the area worst affected by Japan's Fukushima nuclear accident have a slightly higher risk of developing certain cancers such as leukemia, solid cancers, thyroid cancer and breast cancer.[*][215][*][216]

Estimated effective doses from the accident outside Japan are considered to be below (or far below) the dose levels regarded as very small by the international radiological protection community.[*][217][*][182]

In 2013, WHO reported that area residents who were evacuated were exposed to so little radiation that radiation induced health impacts were likely to be below detectable levels.[*][218][*][219] The health risks were calculated by applying conservative assumptions, including the conservative linear no-threshold model of radiation exposure, a model that assumes even the smallest amount of radiation exposure will cause a negative health effect.[*][220][*][221] The report indicated that for those infants in the most affected areas, lifetime cancer risk would increase by about

1%.*[219]*[222] It predicted that populations in the most contaminated areas faced a 70% higher relative risk of developing thyroid cancer for females exposed as infants, and a 7% higher relative risk of leukemia in males exposed as infants and a 6% higher relative risk of breast cancer in females exposed as infants.*[36] One-third of involved emergency workers would have increased cancer risks.*[36]*[223] Cancer risks for fetuses were similar to those in 1 year old infants.*[37] The estimated cancer risk to children and adults was lower than infants.*[224]

> These percentages represent estimated relative increases over the baseline rates and are not absolute risks for developing such cancers. Due to the low baseline rates of thyroid cancer, even a large relative increase represents a small absolute increase in risks. For example, the baseline lifetime risk of thyroid cancer for females is just three-quarters of one percent and the additional lifetime risk estimated in this assessment for a female infant exposed in the most affected location is one-half of one percent.
> —World Health Organization. "Health Risk Assessment from the nuclear accident after the 2011 Great East Japan Earthquake and Tsunami based on a preliminary dose estimation" (PDF). Archived from the original (PDF) on 2013-10-22.

According to a linear no-threshold model (LNT model), the accident would most likely cause 130 cancer deaths.*[12]*[13] However, radiation epidemiologist Roy Shore countered that estimating health effects from the LNT model "is not wise because of the uncertainties" *[225] Darshak Sanghavi noted that to obtain reliable evidence of the effect of low-level radiation would require an impractically large number of patients, Luckey reported that the body's own repair mechanisms can cope with small doses of radiation*[226] and Aurengo stated that "The LNT model cannot be used to estimate the effect of very low doses···" *[227]

In April 2014, studies confirmed the presence of radioactive tuna off the coasts of the pacific U.S.*[228] Researchers carried out tests on 26 albacore tuna caught prior to the 2011 power plant disaster and those caught after. However, the amount of radioactivity is less than that found naturally in a single banana.*[229]*[230]

As of June 2016, dispersed nuclear fallout and associated radiation contamination continue to pollute the environment. Tilman Ruff, a professor at the University of Melbourne, stated that every day 300 tons of contaminated water leak from the crippled nuclear plant.*[231] The Recon-struction Agency states that 174,000 people have been unable to return to their homes. Ecological diversity has decreased and malformations have been found in trees, birds, and mammals.*[231]

37.7.2   Thyroid screening program

The World Health Organization stated that a 2013 thyroid ultrasound screening program was, due to the screening effect, likely to lead to an increase in recorded thyroid cases due to early detection of non-symptomatic disease cases.*[232] The overwhelming majority of thyroid growths are benign growths that will never cause symptoms, illness or death, even if nothing is ever done about the growth. Autopsy studies on people who died from other causes show that more than one third of adults technically have a thyroid growth/cancer.*[233] As a precedent, in 1999 in South Korea, the introduction of advanced ultrasound thyroid examinations resulted in an explosion in the rate of benign thyroid cancers being detected and needless surgeries occurring.*[234] Despite this, the death rate from thyroid cancer has remained the same.*[234]

According to the Tenth Report of the Fukushima Prefecture Health Management Survey released in February 2013, more than 40% of children screened around Fukushima prefecture were diagnosed with thyroid nodules or cysts. Ultrasonographic detectable thyroid nodules and cysts are extremely common and can be found at a frequency of up to 67% in various studies.*[235] 186 (0.5%) of these had nodules larger than 5.1 mm and/or cysts larger than 20.1 mm and underwent further investigation, while none had thyroid cancer. A Russia Today report into the matter was highly misleading.*[236] Fukushima Medical University give the number of children diagnosed with thyroid cancer, as of December 2013, as 33 and concluded "it is unlikely that these cancers were caused by the exposure from I-131 from the nuclear power plant accident in March 2011" .*[237]

In October 2015, 137 children from the Fukushima Prefecture were described as either being diagnosed with or showing signs of developing thyroid cancer. The study's lead author Toshihide Tsuda from Okayama University has stated that the increased detection could not be accounted for by attributing it to the screening effect. He described the screening results to be "20 times to 50 times what would be normally expected." *[39] By the end of 2015, the number had increased to 166 children.*[238]

However, despite his paper being widely reported by the media,*[234] an undermining error, according to teams of other epidemiologists who point out Tsuda's remarks are fatally wrong, is that Tsuda did an apples and oranges comparison by comparing the Fukushima surveys, which uses advanced ultrasound devices that detect otherwise unnoticeable thyroid growths, with data from traditional non-advanced clinical examinations, to arrive at his "20 to 50 times what would be expected" conclusion. In the critical words of epidemiologist Richard Wakeford, "It is inappropriate to compare the data from the Fukushima screening program with cancer registry data from the rest of Japan where there is, in general, no such large-scale screening,". Wakeford's criticism was one of seven other author's letters that were published criticizing Tsuda's paper.*[234] According to Takamura, another epidemiologist, who examined the results of small scale advanced ultrasound tests on Japanese children not near Fukushima, "The prevalence of thyroid cancer [using the same detection technology] does not differ meaningfully from that in Fukushima Prefecture,".*[234]

Thyroid cancer is one of the most survivable cancers, with an approximate 94% survival rate after first diagnosis. That rate increases to a nearly 100% survival rate if caught early.*[239]

Chernobyl comparison

Radiation deaths at Chernobyl were also statistically undetectable. Only 0.1% of the 110,645 Ukraninian cleanup workers, included in a 20-year study out of over 500,000 former Soviet clean up workers, had as of 2012 developed leukemia, although not all cases resulted from the accident.*[240]*[241]

Data from Chernobyl showed that there was a steady then sharp increase in thyroid cancer rates following the disaster in 1986, but whether this data can be directly compared to Fukushima is yet to be determined.*[40]*[41]

Chernobyl thyroid cancer incidence rates did not begin to increase above the prior baseline value of about 0.7 cases per 100,000 people per year until 1989 to 1991, 3–5 years after the incident in both adolescent and child age groups.*[40]*[41] The rate reached its highest point so far, of about 11 cases per 100,000 in the decade of the 2000s, approximately 14 years after the accident.*[40] From 1989 to 2005, an excess of 4,000 children and adolescent cases of thyroid cancer were observed. Nine of these had died as of 2005, a 99% survival rate.*[242]

37.7.3 Effects on evacuees

In the former Soviet Union, many patients with negligible radioactive exposure after the Chernobyl disaster displayed extreme anxiety about radiation exposure. They developed many psychosomatic problems, including radiophobia along with an increase in fatalistic alcoholism. As Japanese health and radiation specialist Shunichi Ya-

mashita noted:*[243]

> We know from Chernobyl that the psychological consequences are enormous. Life expectancy of the evacuees dropped from 65 to 58 years – not because of cancer, but because of depression, alcoholism and suicide. Relocation is not easy, the stress is very big. We must not only track those problems, but also treat them. Otherwise people will feel they are just guinea pigs in our research.*[243]

A survey by the Iitate local government obtained responses from approximately 1,743 evacuees within the evacuation zone. The survey showed that many residents are experiencing growing frustration, instability and an inability to return to their earlier lives. Sixty percent of respondents stated that their health and the health of their families had deteriorated after evacuating, while 39.9% reported feeling more irritated compared to before the disaster.*[244]

> Summarizing all responses to questions related to evacuees' current family status, one-third of all surveyed families live apart from their children, while 50.1% live away from other family members (including elderly parents) with whom they lived before the disaster. The survey also showed that 34.7% of the evacuees have suffered salary cuts of 50% or more since the outbreak of the nuclear disaster. A total of 36.8% reported a lack of sleep, while 17.9% reported smoking or drinking more than before they evacuated.*[244]

Stress often manifests in physical ailments, including behavioral changes such as poor dietary choices, lack of exercise and sleep deprivation. Survivors, including some who lost homes, villages and family members, were found likely to face mental health and physical challenges. Much of the stress came from lack of information and from relocation.*[245]

A survey computed that of some 300,000 evacuees, approximately 1,600 deaths related to the evacuation conditions, such as living in temporary housing and hospital closures that had occurred as of August 2013, a number comparable to the 1,599 deaths directly caused by the earthquake and tsunami in the Prefecture. The exact causes of these evacuation related deaths were not specified, because according to the municipalities, that would hinder relatives applying for compensation.*[17]*[246]

37.7.4 Radioactivity releases

In June 2011, TEPCO stated the amount of contaminated water in the complex had increased due to substantial rainfall.[247] On 13 February 2014, TEPCO reported 37,000 becquerels (1.0 microcurie) of cesium-134 and 93,000 becquerels (2.5 microcuries) of cesium-137 were detected per liter of groundwater sampled from a monitoring well.[248]

37.7.5 Insurance

According to reinsurer Munich Re, the private insurance industry will not be significantly affected by the disaster.[249] Swiss Re similarly stated, "Coverage for nuclear facilities in Japan excludes earthquake shock, fire following earthquake and tsunami, for both physical damage and liability. Swiss Re believes that the incident at the Fukushima nuclear power plant is unlikely to result in a significant direct loss for the property & casualty insurance industry." [250]

37.7.6 Compensation

The amount of compensation to be paid by TEPCO is expected to reach 7 trillion yen.[251]

Costs to Japanese taxpayers are likely to exceed 12 trillion yen ($100 billion).[252] In December 2016 the government estimated decontamination, compensation, decommissioning and radioactive waste storage costs at 21.5 trillion yen ($187 billion), nearly double the 2013 estimate.[253]

37.7.7 Energy policy implications

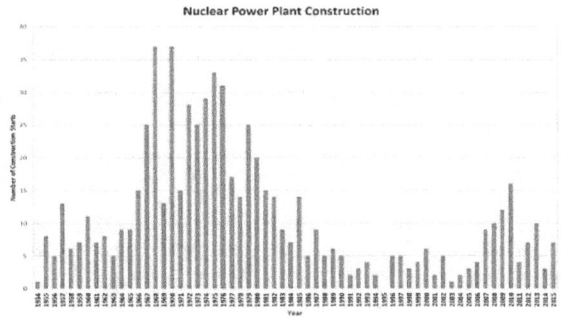

The number of nuclear power plant constructions started each year worldwide, from 1954 to 2013. Following an increase in new constructions from 2007 to 2010, there was a decline after the Fukushima nuclear disaster.

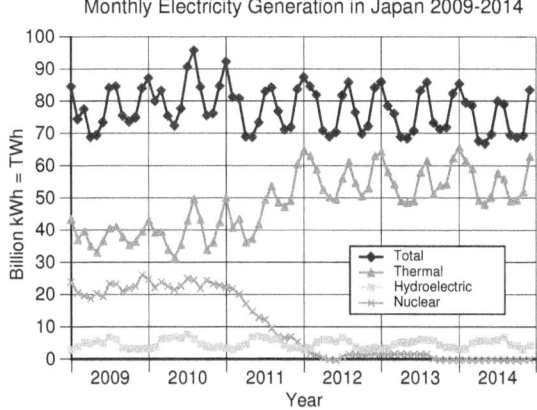

Electricity generation by source in Japan (month-level data). Nuclear energy's contribution declined steadily throughout 2011 due to shutdowns and has been mainly replaced with thermal power stations such as fossil gas and coal power plants.

Komekurayama Solar Power Plant owned and operated by TEPCO in Kofu, Yamanashi Prefecture

By March 2012, one year after the disaster, all but two of Japan's nuclear reactors had been shut down; some had been damaged by the quake and tsunami. Authority to restart the others after scheduled maintenance throughout the year was given to local governments, who in all cases decided against. According to The Japan Times, the disaster changed the national debate over energy policy almost overnight. "By shattering the government's long-pitched safety myth about nuclear power, the crisis dramatically raised public awareness about energy use and sparked strong anti-nuclear sentiment". An energy white paper, approved by the Japanese Cabinet in October 2011, says "public confidence in safety of nuclear power was greatly damaged" by the disaster and called for a reduction in the nation's reliance on nuclear power. It also omitted a section on nuclear power expansion that was in the previous year's policy review.[254]

Part of the Seto Hill Windfarm in Japan, one of several windfarms that continued generating without interruption after the 2011 earthquake and tsunami and the Fukushima nuclear disaster.

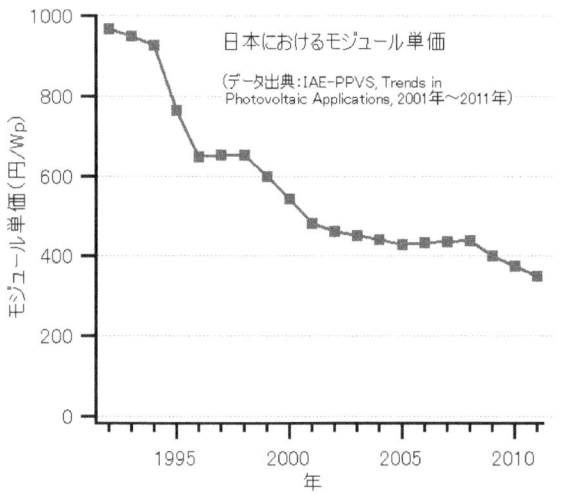

日本におけるモジュール単価

(データ出典：IAE-PPVS, Trends in Photovoltaic Applications, 2001年～2011年)

Price of PV modules (yen/Wp) in Japan

Anti-nuclear power plant rally on 19 September 2011 at the Meiji Shrine complex in Tokyo.

Michael Banach, the current Vatican representative to the IAEA, told a conference in Vienna in September 2011 that the disaster created new concerns about the safety of nuclear plants globally. Auxiliary Bishop of Osaka Michael Goro Matsuura said this incident should cause Japan and other countries to abandon nuclear projects. He called on the worldwide Christian community to support this anti-nuclear campaign. Statements from Bishops' conferences in Korea and the Philippines called on their governments to abandon atomic power. Author Kenzaburō Ōe, who received a Nobel prize in literature, urged Japan to abandon its reactors.[255]

The nuclear plant closest to the epicenter of the earthquake, the Onagawa Nuclear Power Plant, successfully withstood

the cataclysm. According to Reuters it may serve as a "trump card" for the nuclear lobby, providing evidence that it is possible for a correctly designed and operated nuclear facility to withstand such a cataclysm.[256]

The loss of 30% of the country's generating capacity led to much greater reliance on liquified natural gas and coal.[257] Unusual conservation measures were undertaken. In the immediate aftermath, nine prefectures served by TEPCO experienced power rationing.[258] The government asked major companies to reduce power consumption by 15%, and some shifted their weekends to weekdays to smooth power demand.[259] Converting to a nuclear-free gas and oil energy economy would cost tens of billions of dollars in annual fees. One estimate is that even including the disaster, more lives would have been lost if Japan had used coal or gas plants instead of nuclear.[12]

Many political activists have begun calling for a phase-out of nuclear power in Japan, including Amory Lovins, who claimed, "Japan is poor in *fuels*, but is the richest of all major industrial countries in renewable *energy* that can meet the entire long-term energy needs of an energy-efficient Japan, at lower cost and risk than current plans. Japanese industry can do it faster than anyone —*if* Japanese policymakers acknowledge and allow it" .[187] Benjamin K. Sovacool asserted that Japan could have exploited instead its renewable energy base. Japan has a total of "324 GW of achievable potential in the form of onshore and off-shore wind turbines (222 GW), geothermal power plants (70 GW), additional hydroelectric capacity (26.5 GW), solar energy (4.8 GW) and agricultural residue (1.1 GW)."[260]

In contrast, others have said that the zero mortality rate from the Fukushima incident confirms their opinion that nuclear fission is the only viable option available to replace fossil fuels. Journalist George Monbiot wrote "Why Fukushima made me stop worrying and love nuclear power." In it he said "As a result of the disaster at Fukushima, I am no longer nuclear-neutral. I now support the technology." *[261]*[262]

He continues "A crappy old plant with inadequate safety features was hit by a monster earthquake and a vast tsunami. The electricity supply failed, knocking out the cooling system. The reactors began to explode and melt down. The disaster exposed a familiar legacy of poor design and corner-cutting. Yet, as far as we know, no one has yet received a lethal dose of radiation." *[263]*[264]

In September 2011, Mycle Schneider said that the disaster can be understood as a unique chance "to get it right" on energy policy. "Germany – with its nuclear phase-out decision based on a renewable energy program – and Japan – having suffered a painful shock but possessing unique technical capacities and societal discipline – can be at the forefront of an authentic paradigm shift toward a truly sustainable, low-carbon and nuclear-free energy policy" .*[265]

On the other hand, climate and energy scientists James Hansen, Ken Caldeira, Kerry Emanuel and Tom Wigley released an open letter calling on world leaders to support development of safer nuclear power systems, stating "There is no credible path to climate stabilization that does not include a substantial role for nuclear power." *[266] In December 2014, an open letter from 75 climate and energy scientists concluding "nuclear power has lowest impact on wildlife and ecosystems —which is what we need given the dire state of the world's biodiversity." *[267]

As of September 2011, Japan planned to build a pilot offshore floating wind farm, with six 2 MW turbines, off the Fukushima coast.*[268] The first became operational in November 2013.*[269] After the evaluation phase is complete in 2016, "Japan plans to build as many as 80 floating wind turbines off Fukushima by 2020." *[268] In 2012, Prime Minister Kan said the disaster made it clear to him that "Japan needs to dramatically reduce its dependence on nuclear power, which supplied 30% of its electricity before the crisis, and has turned him into a believer of renewable energy". Sales of solar panels in Japan rose 30.7% to 1,296 MW in 2011, helped by a government scheme to promote renewable energy. Canadian Solar received financing for its plans to build a factory in Japan with capacity of 150 MW, scheduled to begin production in 2014.*[270]

As of September 2012, the Los Angeles Times reported that "Prime Minister Yoshihiko Noda acknowledged that the vast majority of Japanese support the zero option on nuclear power" ,*[271] and Prime Minister Noda and the Japanese government announced plans to make the country nuclear-free by the 2030s. They announced the end to construction of nuclear power plants and a 40-year limit on existing nuclear plants. Nuclear plant restarts must meet safety standards of the new independent regulatory authority. The plan requires investing $500 billion over 20 years.*[272]

On 16 December 2012, Japan held its general election. The Liberal Democratic Party (LDP) had a clear victory, with Shinzō Abe as the new Prime Minister. Abe supported nuclear power, saying that leaving the plants closed was costing the country 4 trillion yen per year in higher costs.*[273] The comment came after Junichiro Koizumi, who chose Abe to succeed him as premier, made a recent statement to urge the government to take a stance against using nuclear power.*[274] A survey on local mayors by the Yomiuri Shimbun newspaper in January 2013 found that most of them from cities hosting nuclear plants would agree to restarting the reactors, provided the government could guarantee their safety.*[275] More than 30,000 people marched on 2 June 2013, in Tokyo against restarting nuclear power plants. Marchers had gathered more than 8 million petition signatures opposing nuclear power.*[276]

In October 2013, it was reported that TEPCO and eight other Japanese power companies were paying approximately 3.6 trillion yen (37 billion dollars) more in combined imported fossil fuel costs compared to 2010, before the accident, to make up for the missing power.*[277]

37.7.8   Equipment, facility and operational changes

As the crisis unfolded, the Japanese government sent a request for robots developed by the U.S. military. The robots went into the plants and took pictures to help assess the situation, but they couldn't perform the full range of tasks usually carried out by human workers.*[278] The Fukushima disaster illustrated that robots lacked sufficient dexterity and robustness to perform critical tasks. In response to this shortcoming, a series of competitions were hosted by DARPA to accelerate the development of humanoid robots that could supplement relief efforts.*[279]*[280]

A number of nuclear reactor safety system lessons emerged from the incident. The most obvious was that in tsunami-prone areas, a power station's sea wall must be adequately tall and robust.*[6] At the Onagawa Nuclear Power Plant, closer to the epicenter of 11 March earthquake and tsunami,*[281] the sea wall was 14 meters tall and successfully withstood the tsunami, preventing serious damage and radioactivity releases.*[282]*[283]

Nuclear power station operators around the world began to install Passive Auto-catalytic hydrogen Recombiners ("PARs"), which do not require electricity to operate.*[284]*[285]*[286] PARs work much like the catalytic converter on the exhaust of a car to turn potentially explosive gases such as hydrogen into water. Had such devices been positioned at the top of Fukushima I's reactor and containment buildings, where hydrogen gas collected, the explosions would not have occurred and the releases of radioactive isotopes would arguably have been much less.[*][26]

Unpowered filtering systems on containment building vent lines, known as Filtered Containment Venting Systems (FCVS), can safely catch radioactive materials and thereby allow reactor core de-pressurization, with steam and hydrogen venting with minimal radioactivity emissions.[*][26][*][287] Filtration using an external water tank system is the most common established system in European countries, with the water tank positioned outside the containment building.[*][288] In October 2013, the owners of Kashiwazaki-Kariwa nuclear power station began installing wet filters and other safety systems, with completion anticipated in 2014.[*][289][*][290]

For generation II reactors located in flood or tsunami prone areas, a 3+ day supply of back-up batteries has become an informal industry standard.[*][291][*][292] Another change is to harden the location of back-up diesel generator rooms with water-tight, blast-resistant doors and heat sinks, similar to those used by nuclear submarines.[*][26] The oldest operating nuclear power station in the world, Beznau, which has been operating since 1969, has a 'Notstand' hardened building designed to support all of its systems independently for 72 hours in the event of an earthquake or severe flooding. This system was built prior to Fukushima Daiichi.[*][293][*][294]

Upon a station blackout, similar to the one that occurred after Fukushima's back-up battery supply was exhausted,[*][295] many that had constructed Generation III reactors adopt the principle of passive nuclear safety. They take advantage of convection (hot water tends to rise) and gravity (water tends to fall) to ensure an adequate supply of cooling water to handle the decay heat, without the use of pumps.[*][296][*][297]

37.8 Reactions

37.8.1 Japan

Main article: Japanese reaction to Fukushima Daiichi nuclear disaster

Japanese authorities later admitted to lax standards and poor oversight.[*][300] They took fire for their handling of the emergency and engaged in a pattern of withholding and denying damaging information.[*][300][*][301][*][302][*][303] Authorities allegedly wanted to "limit the size of costly

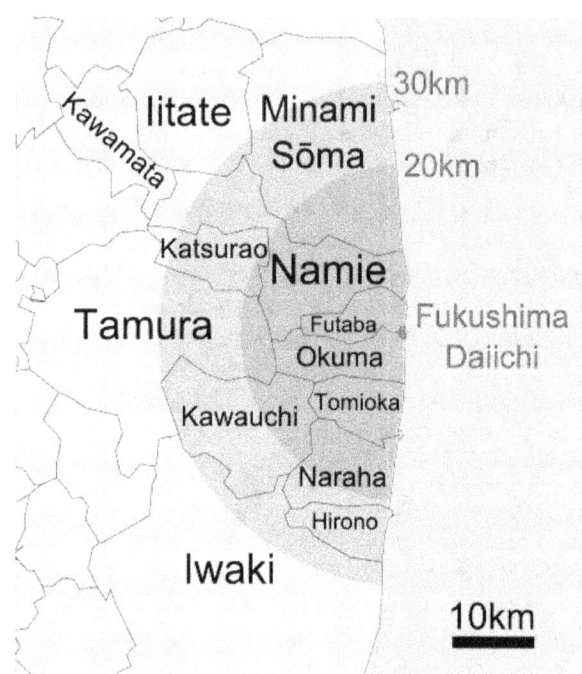

Japan towns, villages, and cities in and around the Daiichi nuclear plant exclusion zone. The 20 km and 30 km areas had evacuation and shelter in place orders, and additional administrative districts that had an evacuation order are highlighted. However the above map's factual accuracy is called into question as only the southern portion of Kawamata district had evacuation orders. More accurate maps are available.[*][298][*][299]

and disruptive evacuations in land-scarce Japan and to avoid public questioning of the politically powerful nuclear industry". Public anger emerged over an "official campaign to play down the scope of the accident and the potential health risks".[*][302][*][303][*][304]

In many cases, the Japanese government's reaction was judged to be less than adequate by many in Japan, especially those who were living in the region. Decontamination equipment was slow to be made available and then slow to be utilized. As late as June 2011, even rainfall continued to cause fear and uncertainty in eastern Japan because of its possibility of washing radioactivity from the sky back to earth.

To assuage fears, the government enacted an order to decontaminate over a hundred areas with a level contamination greater than or equivalent to one millisievert of radiation. This is a much lower threshold than is necessary for protecting health. The government also sought to address the lack of education on the effects of radiation and the extent to which the average person was exposed.[*][305]

Previously a proponent of building more reactors, Kan took an increasingly anti-nuclear stance following the disaster. In May 2011, he ordered the aging Hamaoka Nuclear Power

Plant closed over earthquake and tsunami concerns, and said he would freeze building plans. In July 2011, Kan said, "Japan should reduce and eventually eliminate its dependence on nuclear energy" .*[306] In October 2013, he said that if the worst-case scenario had been realized, 50 million people within a 250-kilometer radius would have had to evacuate.*[307]

On 22 August 2011, a government spokesman mentioned the possibility that some areas around the plant "could stay for some decades a forbidden zone" . According to Yomiuri Shimbun the Japanese government was planning to buy some properties from civilians to store waste and materials that had become radioactive after the accidents.*[308]*[309] Chiaki Takahashi, Japan's foreign minister, criticized foreign media reports as excessive. He added that he could "understand the concerns of foreign countries over recent developments at the nuclear plant, including the radioactive contamination of seawater" .*[310]

Due to frustration with TEPCO and the Japanese government "providing differing, confusing, and at times contradictory, information on critical health issues" *[311] a citizen's group called "Safecast" recorded detailed radiation level data in Japan.*[312]*[313] The Japanese government "does not consider nongovernment readings to be authentic" . The group uses off-the-shelf Geiger counter equipment. A simple Geiger counter is a contamination meter and not a dose rate meter. The response differs too much between different radioisotopes to permit a simple GM tube for dose rate measurements when more than one radioisotope is present. A thin metal shield is needed around a GM tube to provide energy compensation to enable it to be used for dose rate measurements. For gamma emitters either an ionization chamber, a gamma spectrometer or an energy compensated GM tube are required. Members of the Air Monitoring station facility at the Department of Nuclear Engineering at the University of Berkeley, California have tested many environmental samples in Northern California.*[314]

37.8.2   International

Main article: International reaction to the Fukushima Daiichi nuclear disaster
 The international reaction to the disaster was diverse and widespread. Many inter-governmental agencies immediately offered help, often on an ad hoc basis. Responders included IAEA, World Meteorological Organization and the Preparatory Commission for the Comprehensive Nuclear Test Ban Treaty Organization.*[315]

In May 2011, UK chief inspector of nuclear installations Mike Weightman traveled to Japan as the lead of an International Atomic Energy Agency (IAEA) expert mission.

Evacuation flight departs Misawa.

U.S. Navy humanitarian flight undergoes radioactive decontamination

The main finding of this mission, as reported to the IAEA ministerial conference that month, was that risks associated with tsunamis in several sites in Japan had been underestimated.*[316]

In September 2011, IAEA Director General Yukiya Amano said the Japanese nuclear disaster "caused deep public anxiety throughout the world and damaged confidence in nuclear power" .*[317]*[318] Following the disaster, it was reported in the *The Economist* that the IAEA halved its estimate of additional nuclear generating capacity to be built by 2035.*[319]

In the aftermath, Germany accelerated plans to close its nuclear power reactors and decided to phase the rest out by 2022.*[320] Italy held a national referendum, in which 94 percent voted against the government's plan to build new nuclear power plants.*[321] In France, President Hollande announced the intention of the government to reduce nuclear usage by one third. So far, however, the government has only earmarked one power station for closure – the aging plant at Fessenheim on the German border – which prompted some to question the government's commitment to Hollande's promise. Industry Minister Arnaud Montebourg is on record as saying that Fessenheim will be the only nuclear power station to close.

On a visit to China in December 2014 he reassured his audience that nuclear energy was a "sector of the future" and would continue to contribute "at least 50%" of France's electricity output.*[322]

Another member of Hollande's Socialist Party, the MP Christian Bataille, says the plan to curb nuclear was hatched as a way of securing the backing of his Green coalition partners in parliament.*[323]

Nuclear power plans were not abandoned in Malaysia, the Philippines, Kuwait and Bahrain, or radically changed, as in Taiwan. China suspended its nuclear development program briefly, but restarted it shortly afterwards. The initial plan had been to increase the nuclear contribution from 2 to 4 percent of electricity by 2020, with an escalating program after that. Renewable energy supplies 17 percent of China's electricity, 16% of which is hydroelectricity. China plans to triple its nuclear energy output to 2020, and triple it again between 2020 and 2030.*[324]

New nuclear projects were proceeding in some countries. KPMG reports 653 new nuclear facilities planned or proposed for completion by 2030.*[325] By 2050, China hopes to have 400–500 gigawatts of nuclear capacity – 100 times more than it has now.*[326] The Conservative Government of the United Kingdom is planning a major nuclear expansion despite widespread public objection. So is Russia. India are also pressing ahead with a large nuclear program, as is South Korea.*[327] Indian Vice President M Hamid Ansari said in 2012 that "nuclear energy is the only option" for expanding India's energy supplies,*[328] and Prime Minister Modi announced in 2014 that India intended to build 10 more nuclear reactors in a collaboration with Russia.*[329]

37.8.3 Investigations

Three investigations into the Fukushima disaster showed the man-made nature of the catastrophe and its roots in regulatory capture associated with a "network of corruption, collusion, and nepotism." *[330]*[331] Regulatory capture refers to the "situation where regulators charged with promoting the public interest defer to the wishes and advance the agenda of the industry or sector they ostensibly regulate." Those with a vested interest in specific policy or regulatory outcomes lobby regulators and influence their choices and actions. Regulatory capture explains why some of the risks of operating nuclear power reactors in Japan were systematically downplayed and mismanaged so as to compromise operational safety.*[331]

Many reports say that the government shares blame with

the regulatory agency for not heeding warnings and for not ensuring the independence of the oversight function.*[332] The New York Times said that the Japanese nuclear regulatory system sided with and promoted the nuclear industry because of amakudari ('descent from heaven') in which senior regulators accepted high paying jobs at companies they once oversaw. To protect their potential future position in the industry, regulators sought to avoid taking positions that upset or embarrass the companies. TEPCO's position as the largest electrical utility in Japan made it the most desirable position for retiring regulators. Typically the "most senior officials went to work at TEPCO, while those of lower ranks ended up at smaller utilities." *[333]

In August 2011, several top energy officials were fired by the Japanese government; affected positions included the Vice-minister for Economy, Trade and Industry; the head of the Nuclear and Industrial Safety Agency, and the head of the Agency for Natural Resources and Energy.*[334]

In 2016 three former TEPCO executives, chairman Tsunehisa Katsumata and two vice presidents, were indicted for negligence resulting in death and injury.*[210]*[335]

NAIIC

Main article: National Diet of Japan Fukushima Nuclear Accident Independent Investigation Commission

The Fukushima Nuclear Accident Independent Investigation Commission (NAIIC) was the first independent investigation commission by the National Diet in the 66-year history of Japan's constitutional government.

Fukushima "cannot be regarded as a natural disaster," the NAIIC panel's chairman, Tokyo University professor emeritus Kiyoshi Kurokawa, wrote in the inquiry report. "It was a profoundly man-made disaster – that could and should have been foreseen and prevented. And its effects could have been mitigated by a more effective human response." *[336] "Governments, regulatory authorities and Tokyo Electric Power [TEPCO] lacked a sense of responsibility to protect people's lives and society," the Commission said. "They effectively betrayed the nation's right to be safe from nuclear accidents.*[337]

The Commission recognized that the affected residents were still struggling and facing grave concerns, including the "health effects of radiation exposure, displacement, the dissolution of families, disruption of their lives and lifestyles and the contamination of vast areas of the environment".

Investigation Committee

Main article: Investigation Committee on the Accident at the Fukushima Nuclear Power Stations of Tokyo Electric Power Company

The purpose of the Investigation Committee on the Accident at the Fukushima Nuclear Power Stations (ICANPS) was to identify the disaster's causes and propose policies designed to minimize the damage and prevent the recurrence of similar incidents.*[338] The 10 member, government-appointed panel included scholars, journalists, lawyers and engineers.*[339]*[340] It was supported by public prosecutors and government experts*[341] and released its final, 448-page*[342] investigation report on 23 July 2012.*[43]*[343]

The panel's report faulted an inadequate legal system for nuclear crisis management, a crisis-command disarray caused by the government and TEPCO, and possible excess meddling on the part of the Prime Minister's office in the crisis' early stage.*[344] The panel concluded that a culture of complacency about nuclear safety and poor crisis management led to the nuclear disaster.*[339]

37.9   See also

- Comparison of Fukushima and Chernobyl nuclear accidents

- Environmental issues in Japan

- Fukushima disaster cleanup

- Japanese nuclear incidents

- Japanese Nuclear Safety Commission

- List of civilian nuclear accidents

- Lists of nuclear disasters and radioactive incidents

- Nuclear power in Japan

- Radiation effects from the Fukushima Daiichi nuclear disaster

- Timeline of the Fukushima Daiichi nuclear disaster

37.10   References

37.10.1   Notes

[1] "High-resolution photos of Fukushima Daiichi" (Press release). Air Photo Service Co. Ltd., Japan. 24 March 2011. Retrieved 14 January 2014.

[2] Negishi, Mayumi (12 April 2011). "Japan raises nuclear crisis severity to highest level". *Reuters.*

[3] "Fukushima accident upgraded to severity level 7". *IEEE Spectrum*. 12 April 2011.

[4] "IAEA Update on Japan Earthquake". Archived from the original on 12 March 2011. Retrieved 16 March 2011. As reported earlier, a 400 millisieverts (mSv) per hour radiation dose observed at Fukushima Daiichi occurred between 1s 3 and 4. This is a high dose-level value, but it is a local value at a single location and at a certain point in time. The IAEA continues to confirm the evolution and value of this dose rate. It should be noted that because of this detected value, non-indispensable staff was evacuated from the plant, in line with the Emergency Response Plan, and that the population around the plant is already evacuated.

[5] "Radiation-exposed workers to be treated at Chiba hospital". *Kyodo News*. 17 April 2011. Retrieved February 12, 2016.

[6] Phillip Lipscy, Kenji Kushida, and Trevor Incerti. 2013. "The Fukushima Disaster and Japan's Nuclear Plant Vulnerability in Comparative Perspective." *Environmental Science and Technology* 47 (May), 6082–6088.

[7] Fackler, Martin (12 October 2012). "Japan Power Company Admits Failings on Plant Precautions". *The New York Times*. Archived from the original on 6 October 2014. Retrieved 13 October 2012.

[8] Sheldrick, Aaron (12 October 2012). "Fukushima operator must learn from mistakes, new adviser says". Reuters. Archived from the original on 2014-03-09. Retrieved 13 October 2012.

[9] Yamaguchi, Mari (12 October 2012). "Japan utility agrees nuclear crisis was avoidable". Boston.com. Associated Press. Archived from the original on 2013-10-05. Retrieved 13 October 2012.

[10] "Japanese nuclear plant operator admits playing down risk". *CNN Wire Staff*. CNN. 12 October 2012. Archived from the original on 2014-03-09. Retrieved 13 October 2012.

[11] "Analysis: A month on, Japan nuclear crisis still scarring" Archived 15 August 2012 at the Wayback Machine. *International Business Times* (Australia). 9 April 2011, retrieved 12 April 2011

[12] Dennis Normile (27 July 2012). "Is Nuclear Power Good for You?". *Science*. **337** (6093): 395–396. doi:10.1126/science.337.6093.395-b. Archived from the original on 13 February 2013.

[13] John E. Ten Hoeve; Mark Z. Jacobson (2012). "Worldwide health effects of the Fukushima Daiichi nuclear accident" (PDF). *Energy & Environmental Science*. **5** (9): 8743. doi:10.1039/c2ee22019a. Retrieved 18 July 2012.

[14] Predictions on the contamination levels from various fission products released from the accident and updates on the risk

assessment for solid and thyroid cancers. Science of The Total EnvironmentVolumes 500–501, 1 December 2014, Pages 155–172 doi:10.1016/j.scitotenv.2014.08.102

[15] UNIS/OUS/237 2 April 2014 Increase in Cancer Unlikely following Fukushima Exposure – says UN Report Low Risk of Thyroid Cancer Among Children Most Exposed

[16] Fukushima disaster predicted to raise cancer rates slightly, 5 March 2013, Ned Stafford

[17] Smith, Alexander (10 September 2013). "Fukushima evacuation has killed more than earthquake and tsunami, survey says". Archived from the original on 2013-10-27. Retrieved 11 September 2013.

[18] "Stress-induced deaths in Fukushima top those from 2011 natural disasters.". Archived from the original on 27 September 2013.

[19] Justin Mccurry (10 March 2014). "Fukushima operator may have to dump contaminated water into Pacific". *The Guardian*. Archived from the original on 2014-03-18. Retrieved 10 March 2014.

[20] Peter Fairley (20 Oct 2015). "Startup Time for Fukushima's Frozen Wall. Here's Why it Should Work". *IEEE*. Retrieved 13 Nov 2015.

[21] Otake, Tomoko (2016-07-20). "In first, Tepco admits ice wall can't stop Fukushima No. 1 groundwater". *The Japan Times Online*. ISSN 0447-5763. Retrieved 2017-02-12.

[22] "Highest radiation reading since 3/11 detected at Fukushima No. 1 reactor". *The Japan Times Online*. 2017-02-03. ISSN 0447-5763. Retrieved 2017-02-12.

[23] Black, Richard (15 March 2011). "Reactor breach worsens prospects". *BBC Online*. Retrieved 23 March 2011.

[24] IAEA press release Japanese Earthquake Update (19 March 2011, 4:30 UTC) 19 March 2011. Archive.org

[25] W. Maschek; A. Rineiski; M. Flad; V. Kriventsev; F. Gabrielli; K. Morita. "Recriticality, a Key Phenomenon to Investigate in Core Disruptive Accident Scenarios of Current and Future Fast Reactor Designs" (PDF). IAEA & Institute for Nuclear and Energy Technologies (IKET). Note: See picture in the upper left corner of page 2.

[26] 24 Hours at Fukushima A blow-by-blow account of the worst nuclear accident since Chernobyl By Eliza Strickland Posted 31 Oct 2011 Archived 14 November 2013 at the Wayback Machine.

[27] "OECD Timeline for the Fukushima Daiichi nuclear power plant accident". Archived from the original on 2013-10-29.

[28] "Fukushima nuclear accident update log, updates of 15 March 2011". *IAEA*. 15 March 2011. Archived from the original on 9 April 2011. Retrieved 8 May 2011.

[29] Hydrogen explosions Fukushima nuclear plant: what happened?

[30] "MELCOR Model of the Spent Fuel Pool of Fukushima Dai-ichi Unit 4" (PDF). Oak Ridge National Laboratory.

[31] page 6

[32] http://eetd-seminars.lbl.gov/sites/eetd-seminars.lbl.gov/files/Fukushima1_Technical_Perspective_LBL_EEDT_04052011-1.pdf *What happened at Fukushima a Technical Perspective*. Nuclear Regulatory Commission page 11, 26, 29.

[33] Sadiq Aliyu, Abubakar; et al. (2015). "An overview of current knowledge concerning the health and environmental consequences of the Fukushima Daiichi Nuclear Power Plant (FDNPP) accident". *Environment International*. **85**: 213–228. doi:10.1016/j.envint.2015.09.020.

[34] Hasegawa; et al. (2015). "From Hiroshima and Nagasaki to Fukushima 2. Health effects of radiation and other health problems in the aftermath of nuclear accidents, with an emphasis on Fukushima". *The Lancet*. **386** (9992): 479–488. doi:10.1016/S0140-6736(15)61106-0.

[35] WHO report, page 92.

[36] Walsh, Bryan. (2013-03-01) WHO Report Says That Fukushima Nuclear Accident Posed Minimal Risk to Health | TIME.com. Science.time.com. Retrieved on 2013-09-06. Archived 4 November 2013 at the Wayback Machine.

[37] WHO 2013, pp. 70, 79–80.

[38] Ryall, Julian (19 July 2012). "Nearly 36pc of Fukushima children diagnosed with thyroid growths". *The Daily Telegraph*.

[39] "Experts link higher incidence of children's cancer to Fukushima radiation". *ScienceAlert*. Retrieved 2016-01-15.

[40] "Radioactivity and thyroid cancer* Christopher Reiners Clinic and Polyclinic of Nuclear Medicine University of Würzburg. See Figure 1. Thyroid cancer Incidence in children and adolescents from Belarus after the Chernobyl accident". Archived from the original on 2013-10-15.

[41] "Disturbing thyroid cancer rise in Fukushima minors". *RT*. 21 August 2013. Archived from the original on 2014-03-27.

[42] National Diet of Japan Fukushima Nuclear Accident Independent Investigation Commission. "国会事故調 | 東京電力福島原子力発電所事故調査委員会のホームページ". National Diet of Japan Fukushima Nuclear Accident Independent Investigation Commission. Archived from the original on 2013-01-19. Retrieved 9 July 2012.

[43] "UPDATE: Government panel blasts lack of 'safety culture' in nuclear accident". *The Asahi Shimbun*. 23 July 2012. Archived from the original on 2014-04-13. Retrieved 29 July 2012.

[44] "Fukushima Daiichi Information Screen". *Icjt.org*. Archived from the original on 2013-07-12. Retrieved 15 March 2011.

[45] Brady, A. Gerald (1980). Ellingwood, Bruce, ed. *An Investigation of the Miyagi-ken-oki, Japan, earthquake of June 12, 1978. United States Department of Commerce, National Bureau of Standards.* NBS special publication. **592**. p. 123.

[46] "The record of the earthquake intensity observed at Fukushima Daiichi Nuclear Power Station and Fukushima Daini Nuclear Power Station (Interim Report)". *TEPCO* (Press release). 1 April 2011. Archived from the original on 2014-05-06.

[47] "Fukushima faced 14-metre tsunami". *World Nuclear News.* 24 March 2011. Archived from the original on 2011-06-16. Retrieved 24 March 2011.

[48] "Fukushima to Restart Using MOX Fuel for First Time". *Nuclear Street.* 17 September 2010. Archived from the original on 2014-04-29. Retrieved 12 March 2011.

[49] Martin, Alex, "Lowdown on nuclear crisis and potential scenarios", *Japan Times*, 20 March 2011, p. 3.

[50] "Fukushima: Background on Fuel Ponds" (PDF). Archived from the original on 2013-10-16. Retrieved 23 November 2013.

[51] "No. 1 fuel pool power to be restored: Tepco". Archived from the original on 2014-01-07. Retrieved 20 March 2013.

[52] "NISA – The 2011 off the Pacific coast of Tohoku Pacific Earthquake and the seismic damage to the NPPs, pg 35" (PDF). Archived from the original (PDF) on 1 May 2011. Retrieved 24 April 2011.

[53] Grier, Peter (16 March 2011). "Meltdown 101: Why is Fukushima crisis still out of control?". *Christian Science Monitor.* Archived from the original on 2014-05-06. Retrieved 27 March 2011.

[54] Helman, Christopher (15 March 2011). "Explainer: What caused the incident at Fukushima-Daiichi". *Forbes.* Archived from the original on 16 March 2011. Retrieved 7 April 2011.

[55] DOE fundamentals handbook – Decay heat, Nuclear physics and reactor theory at the Wayback Machine (archived 16 March 2011), Vol. 2, module 4, p. 61.

[56] "What if it happened here?". *Somdnews.com.* Retrieved 7 April 2011.

[57] "More on spent fuel pools at Fukushima". *Allthingsnuclear.org.* 21 March 2011. Archived from the original on 9 April 2011. Retrieved 7 April 2011.

[58] Pre-construction safety report – Sub-chapter 9.2 – Water Systems. AREVA NP / EDF, published 2009-06-29, Retrieved 23 March 2011.

[59] "Why has it become impossible for Fukushima Daiichi Nuclear Power Station to cool reactor core?". *Shimbun.denki.or.jp.* Archived from the original on 27 April 2011. Retrieved 7 April 2011.

[60] https://www.nfb.ca/film/meltdown_doc

[61] Higgins, Andrew, "disorder intensified Japan's crisis", *The Washington Post*, 19 April 2011, Retrieved 21 April 2011.

[62] Mike Soraghan (24 March 2011). "Japan disaster raises questions about backup power at US nuclear plants". *The New York Times.* Greenwire. Retrieved 7 April 2011.

[63] "Regulatory effectiveness of the station blackout rule" (PDF). Retrieved 7 April 2011.

[64] "The 2011 off the Pacific coast of Tohoku Pacific Earthquake and the seismic damage to the NPPs" (PDF). Archived from the original (PDF) on 22 May 2011. Retrieved 13 July 2011.

[65] http://www.wsj.com/articles/ SB10001424052702304887904576395580035481822 Design Flaw Fueled Nuclear Disaster accessed may 2016

[66] Shirouzu, Norihiko (1 July 2011). "Wall Street Journal: Design Flaw Fueled Nuclear Disaster". *Online.wsj.com.* Archived from the original on 1 July 2011. Retrieved 13 July 2011.

[67] Yoshida, Reiji, "GE plan followed with inflexibility", *Japan Times*, 14 July 2011, p. 1. Archived 13 July 2011 at WebCite

[68] Arita, Eriko, "Disaster analysis you may not hear elsewhere Archived 29 August 2011 at the Wayback Machine.", *Japan Times*, 20 March 2011, p. 12.

[69] Agence France-Presse/Jiji Press, "Tsunami that knocked out nuke plant cooling systems topped 14 meters", *Japan Times*, 23 March 2011, p. 2.

[70] "IAEA warned Japan over nuclear quake risk: WikiLeaks". *physorg.com.* Archived from the original on 2012-01-17. Retrieved 26 March 2011.

[71] "Plant Status of Fukushima Daini Nuclear Power Station (as of 0 AM 12 March)", *TEPCO*, end of day 11 April. Archived 15 February 2014 at the Wayback Machine.

[72] Fukushima No. 1 plant designed on 'trial-and-error' basis, *Asahi Shimbun*, 7 April 2011. Archived 13 April 2011 at WebCite

[73] "Spraying continues at Fukushima Daiichi". 18 March 2011. Archived from the original on 18 March 2011. Retrieved 19 March 2011.

[74] "The Japanese Nuclear Emergency – Sydney Technical Presentation". *Engineers Australia.* 6 June 2011. Archived from the original on 2011-09-30. Retrieved 22 August 2011.

[75] B. Cox, JOURNAL OF NUCLEAR MATERIALS, PELLET CLAD INTERACTION (PCI) FAILURES OF ZIRCONIUM ALLOY FUEL CLADDING – A REVIEW, 1990, volume 172, pages 249–292

[76] "An introduction to serious nuclear accident chemistry", M.R.StJ. Foreman, *Cogent Chemistry* 2015, **1**(1), retrieved 13 Feb 2016

[77] The Mainichi Shimbun (28 28 February 2012)TEPCO ordered to report on change in piping layout at Fukushima plant

[78] NHK-world (29 December 2011) Fukushima plant's backup generator failed in 1991.
JAIF (30 December 2011)Earthquake report 304:Fukushima plant's backup generator failed in 1991.
The Mainichi Daily News (30 December 2011) TEPCO neglected anti-flood measures at Fukushima plant despite knowing risk.

[79] "TEPCO did not act on tsunami risk projected for nuclear plant |". Jagadees.wordpress.com. 2012-02-13. Archived from the original on 2014-04-12. Retrieved 2013-12-30.

[80] "AFERC urged to review assumption on Tsunami in 2009". *Yomiuri News Paper*. 11 March 2011. Archived from the original on 2014-02-16. Retrieved 14 September 2013.

[81] "Fukushima Nuclear Accident – U.S. NRC warned a risk on emergency power 20 years ago". *Bloomberg L.P.* 16 March 2011. Archived from the original on 2014-02-16. Retrieved 14 September 2013.

[82] "IAEA warned Japan over nuclear quake risk: WikiLeaks". *physorg.com*. Daily Telegraph. 17 March 2011. Archived from the original on 2012-01-17.

[83] "Magnitude 9.0 – near the East coast of Honshu, Japan". *Earthquake.usgs.gov*. Archived from the original on 12 March 2011. Retrieved 17 March 2011.

[84] "Plant Status of Fukushima Daiichi Nuclear Power Station (as of 0AM March 12th)". *TEPCO* (Press release). 12 March 2011. Archived from the original on 10 May 2011. Retrieved 13 March 2011.

[85] "Occurrence of a specific incident stipulated in Article 10, Clause 1 of the Act on "Special measures concerning nuclear emergency preparedness (Fukushima Daiichi)"". *TEPCO* (Press release). 11 March 2011. Archived from the original on 3 April 2011. Retrieved 13 March 2011.

[86] Associated Press, "How the first 24 hours shaped Fukushima nuclear crisis", *Japan Times*, 7 July 2011, p. 3. Archived 7 July 2011 at WebCite

[87] "TEPCO press release 3". *Tepco* (Press release). 11 March 2011. Archived from the original on 25 April 2011.

[88] TEPCO tardy on N-plant emergency: National: Daily Yomiuri Online (The Daily Yomiuri). Yomiuri.co.jp (12 April 2011). Retrieved 30 April 2011. Archived 13 April 2011 at WebCite

[89] Eric Talmadge (1 July 2011). "AP: First 24 hours shaped Japan nuke crisis". *Google.com*. Retrieved 13 July 2011.

[90] Japan Meteorological Agency|Tsunami Warnings/Advisories, Tsunami Information. Jma.go.jp. Retrieved 30 April 2011. Archived 18 April 2011 at WebCite

[91] Bloomberg, "Tepco revises tsunami's height to 15 meters", 10 April 2011, Archived 29 December 2013 at the Wayback Machine.

[92] David Sanger and Matthew Wald, Radioactive releases in Japan could last months, experts say. *The New York Times* 13 March 2011 Archived 25 September 2012 at the Wayback Machine.

[93] "Massive earthquake hits Japan". *World Nuclear News*. 11 March 2011. Archived from the original on 11 April 2011. Retrieved 13 March 2011.;

[94] Bloomberg L.P., "Time not on workers' side as crisis raced on", *Japan Times*, 5 May 2011, p. 3. Archived 6 May 2011 at WebCite

[95] Inajima, Tsuyoshi; Okada, Yuji (11 March 2011). "Japan Orders Evacuation From Near Nuclear Plant After Quake". *Bloomberg BusinessWeek*. Archived from the original on 2 February 2012. Retrieved 11 March 2011.

[96] "Japan Earthquake Update (2210 CET)". *International Atomic Energy Agency* (Press release). 11 March 2011. Archived from the original on 14 March 2011. Retrieved 12 March 2011.

[97] Magnier, Mark; et al. (16 March 2011). "New power line could restore cooling systems at Fukushima Daiichi plant". *Los Angeles Times*. Archived from the original on 17 March 2011. Retrieved 19 March 2011.

[98] "Stabilisation at Fukushima Daiichi". *World-nuclear-news.org*. 20 March 2011. Archived from the original on 11 April 2011. Retrieved 24 April 2011.

[99] AP, "IAEA: 170,000 Evacuated near Japan nuclear plant," March 12, 2011. https://news.yahoo.com/iaea-170-000-evacuated-near-japan-nuclear-plant-20110312-133929-90lhtml accessed Feb. 23, 2016.

[100] Richard Black (15 March 2011). "Japan quake: Radiation rises at Fukushima nuclear plant". *BBC Online*. Archived from the original on 15 March 2011. Retrieved 15 March 2011.

[101] "Japan's PM urges people to clear 20-km zone around Fukushima NPP (Update-1)". *RIA Novosti*. Archived from the original on 2013-05-11. Retrieved 15 March 2011.

[102] Makinen, Julie (25 March 2011). "Japan steps up nuclear plant precautions; Kan apologizes". *Los Angeles Times*.

[103] Herman, Steve (12 April 2011). "VOA Correspondent Reaches Crippled Fukushima Daiichi Nuclear Plant". *VOA*. Archived from the original on 2013-06-01. Retrieved 5 March 2014.

[104] "Reconstruction Agency". www.reconstruction.go.jp. Retrieved 2016-06-02.

[105] "Reconstruction Agency". www.reconstruction.go.jp. Retrieved 2016-06-02.

[106] Takahashi, Hideki, and Shinya Kokubun, "Workers grappled with darkness at start of Fukushima nuclear crisis", *Japan Times*, 3 September 2014, p. 3

[107] Takahashi, Hideki, Shinya Kokubun, and Yukiko Maeda, "Response stymied by loss of electricity", *Japan Times*, 3 September 2014, p. 3

[108] Takahashi, Hideki, and Hisashi Ota, "Fukushima workers tried to save reactor 1 through venting", *Japan Times*, 3 September 2014, p. 3

[109] Uncertainties abound in Fukushima decommissioning. Phys.org. 19 Nov 2013. Archived 14 March 2014 at the Wayback Machine.

[110] Fukushima Timeline scientificamerican.com Archived 6 March 2014 at the Wayback Machine.

[111] Most of fuel NOT remaining in reactor1 core / Tepco "but molten fuel is stopped in the concrete base" Fukushima-Diary.com Archived 25 March 2014 at the Wayback Machine.

[112] "Reactor 3 fuel is assumed to have melted concrete base up to 26cm to the wall of primary vessel.". *Fukushima Diary*. Retrieved 12 June 2015.

[113] "TEPCO Admits Unit 3 Had Total Melt Through". *SimplyInfo*. Retrieved 12 June 2015.

[114] "Fukushima Unit 3 Reactor Vessel Failure Preceded Explosion". *SimplyInfo*. Retrieved 12 June 2015.

[115] The Evaluation Status of Reactor Core Damage at Fukushima Daiichi Nuclear Power Station Units 1 to 3 30 November 2011 Tokyo Electric Power Company

[116] Report on the Investigation and Study of Unconfirmed/Unclear Matters in the Fukushima Nuclear Accident – Progress Report No.2 – 6 August 2014 Tokyo Electric Power Company, Inc.

[117] TEPCO to start "scanning" inside of Reactor 1 in early February by using muon – Fukushima Diary

[118] Muon Scans Begin At Fukushima Daiichi – SimplyInfo

[119] Muon Scan Finds No Fuel In Fukushima Unit 1 Reactor Vessel – SimplyInfo

[120] IRID saw no fuel or water remaining in reactor core of Reactor 1 – Fukushima Diary

[121] "High radiation readings at Fukushima's No. 2 reactor complicate robot-based probe". *The Japan Times*. 10 February 2017. Retrieved 11 February 2017.

[122] Justin McCurry (3 February 2017). "Fukushima nuclear reactor radiation at highest level since 2011 meltdown". *The Guardian*. Retrieved 3 February 2017.

[123] "Most fuel in Fukushima 4 pool undamaged". *world nuclear news*. 14 April 2011. Archived from the original on 15 April 2011. Retrieved 27 January 2012.

[124] "Japan Diplomat: Ground underneath Fukushima Unit 4 is sinking —More than 30 inches in some areas —Now in danger of collapse". ENENews. Archived from the original on 2014-03-26. Retrieved 24 October 2012. Due to its ground has been sinking, reactor 4 is now endangered in collapse. ···According to secretary of former Prime Minister Kan, the ground level of the building has been sinking 80 cm ···unevenly. Because the ground itself has the problem, whether the building can resist a quake bigger than M6 still remains a question.

[125] "Gundersen: Japan ambassador confirms Fukushima Unit 4 is sinking unevenly —Building "may begin to be tilting"". ENENews. Archived from the original on 2014-03-11. Retrieved 24 October 2012. So I have been able to confirm that there is unequal sinking at Unit 4, not just the fact the site sunk by 36 inches immediately after the accident, but also that Unit 4 continues to sink something on the order of 0.8 meters, or around 30 inches.

[126] "FUEL REMOVAL FROM UNIT 4 REACTOR BUILDING COMPLETED AT FUKUSHIMA DAIICHI". TEPCO. 22 December 2014. Retrieved 24 December 2014.

[127] http://edition.cnn.com/2011/WORLD/asiapcf/03/15/japan.nuclear.reactors/?hpt=T1

[128] https://web.archive.org/web/20110523050825/http://www.nisa.meti.go.jp/english/files/en20110322-1-1.pdf NISA news release

[129] http://www.wsj.com/articles/SB10001424052702304887904576395580035481822

[130] "Seismic Damage Information (the 61st Release)" (PDF). *Nuclear and Industrial Safety Agency*. 29 March 2011. Archived from the original (PDF) on 11 April 2011. Retrieved 12 April 2011.

[131] Cresswell, Adam (16 March 2011). "Stealthy, silent destroyer of DNA". *The Australian*

[132] "Fukushima Radiation Found In Tap Water Around Japan".

[133] Fukushima radioactive fallout nears Chernobyl levels – 24 March 2011. New Scientist. Retrieved 30 April 2011. Archived 25 March 2011 at WebCite

[134] Report: Emissions from Japan plant approach Chernobyl levels, *USA Today*, 24 March 2011 Archived 18 August 2013 at the Wayback Machine.

[135] Doughton, Sandi. (5 April 2011) Local News|Universities come through in monitoring for radiation|Seattle Times Newspaper. Seattletimes.nwsource.com. Retrieved 30 April 2011. Archived 21 September 2011 at the Wayback Machine.

[136] IRSN (26 October 2011). "Synthèse actualisée des connaissances relatives à l'impact sur le milieu marin des rejets radioactifs du site nucléaire accidenté de Fukushima Dai-ichi" (PDF). Retrieved 3 January 2012.

[137] Buesseler, Ken O.; Jayne, Steven R.; Fisher, Nicholas S.; Rypina, Irina I.; Baumann, Hannes; Baumann, Zofia; Breier, Crystaline F.; Douglass, Elizabeth M.; George, Jennifer; MacDonald, Alison M.; Miyamoto, Hiroomi; Nishikawa, Jun; Pike, Steven M.; Yoshida, Sashiko (2012). "Fukushima-derived radionuclides in the ocean and biota off Japan". *Proceedings of the National Academy of Sciences*. **109** (16): 5984–8. doi:10.1073/pnas.1120794109. PMC 3341070◌. PMID 22474387.

[138] "CTBTO to Share Data with IAEA and WHO". *CTBTO Press Release 18 March 2011*. Archived from the original on 2013-12-24. Retrieved 17 May 2012.

[139] "Fukushima-Related Measurements by the CTBTO". *CTBTO Press Release 13 April 2011*. Archived from the original on 6 May 2011. Retrieved 17 May 2012.

[140] "CTBTO Trakcs Fukushima's Radioactive Release". *Animation CTBTO YouTube Channel*. Archived from the original on 2012-05-23. Retrieved 17 May 2012.

[141] Frank N. von Hippel (September/October 2011 vol. 67 no. 5). "The radiological and psychological consequences of the Fukushima Daiichi accident". *Bulletin of the Atomic Scientists*. pp. 27–36. Archived from the original on 20 December 2011. Check date values in: |date= (help)

[142] *No-Man's Land Attests to Japan's Nuclear Nightmare*. ABC News, 27. December 2011. Archived 28 December 2011 at the Wayback Machine.

[143] "Reactor accident Fukushima – New international study". *Norwegian Institute for Air Research*. 21 October 2011. Archived from the original on 2014-01-06. Retrieved 20 January 2012.

[144] David Guttenfelder (27 December 2011). "No-man's land attests to Japan's nuclear nightmare". *theStar.com*. Toronto. Archived from the original on 2012-01-10. Retrieved 20 January 2012.

[145] Kyodo News, "Radioactivity Dispersal Distance From Fukushima 1/10th Of Chernobyl's", 13 March 2012, (wire service report), "The data showed, for example, more than 1.48 million becquerels (40 microcuries) of radioactive caesium per square meter was detected in soil at a location some 250 kilometers away from the Chernobyl plant. In the case of the Fukushima Daiichi plant, the distance was much smaller at about 33 km, the officials said."

[146] Hongo, Jun, "Fukushima soil fallout far short of Chernobyl", *Japan Times*, 15 March 2012, p. 1. Archived 16 March 2012 at the Wayback Machine.

[147] Michael Winter (24 March 2011). "Report: Emissions from Japan plant approach Chernobyl levels". *USA Today*. Archived from the original on 2013-08-18.

[148] Hamada, Nobuyuki. "Safety regulations of food and water implemented in the first year following the Fukushima nuclear accident". Oxford Journals. Retrieved 30 November 2013.

[149] "福島産の新米、東京で販売開始全袋検査に合格". 共同 *Nikkei Kyodo news*. 2012-09-01. Archived from the original on 2013-12-03. Retrieved 18 April 2013.

[150] JAIF (5 September 2011) NSC Recalculates Total Amount of Radioactive Materials Released Archived 20 December 2011 at WebCite

[151] *INES (the International Nuclear and Radiological Event Scale) Rating on the Events in Fukushima Dai-ichi Nuclear Power Station by the Tohoku District – off the Pacific Ocean Earthquake*. NISA/METI, 12 April 2011, archived from Original.

[152] JAIF (9 September 2011) Radioactive release into sea estimated triple Archived 11 December 2011 at WebCite

[153] JAIF 20 September 2011 Earthquake-report 211: A new plan set to reduce radiation emissions

[154] Possibility of recriticality again, *Fukushima Diary* Archived 24 December 2013 at the Wayback Machine.

[155] Increasing leakage of Iodine-131, *Fukushima Diary* Archived 24 December 2013 at the Wayback Machine.

[156] IRSN (26 October 2011). "Synthèse actualisée des connaissances relatives à l'impact sur le milieu marin des rejets radioactifs du site nucléaire accidenté de Fukushima Dai-ichi" (PDF). Retrieved 3 January 2012

[157] Daniel J. Madigan; Zofia Baumann; Nicholas S. Fisher (29 May 2012). "Pacific bluefin tuna transport Fukushima-derived radionuclides from Japan to California". *Proceedings of the National Academy of Sciences of the United States of America*. **109** (24): 9483–9486. doi:10.1073/pnas.1204859109. PMC 3386103◌. PMID 22645346

[158] Aoki, Mizuho, "Tohoku fears nuke crisis evacuees gone for good", *Japan Times*, 8 March 2012, p. 1. Archived 7 March 2012 at the Wayback Machine.

[159] TEPCO Press Release. "The Estimated Amount of Radioactive Materials Released into the Air and the Ocean Caused by Fukushima Daiichi Nuclear Power Station Accident Due to the Tohoku-Chihou-Taiheiyou-Oki Earthquake (As of May 2012)". *TEPCO*. Archived from the original on 2014-02-15. Retrieved 24 May 2012.

[160] Kevin Krolicki (24 May 2012). "Fukushima radiation higher than first estimated". *Reuters*. Archived from the original on 2013-10-15. Retrieved 24 May 2012.

[161] "TEPCO puts radiation release early in Fukushima crisis at 900 PBq". *Kyodo News*. 24 May 2012. Archived from the original on 2012-05-24. Retrieved 24 May 2012.

[162] Estimation of radionuclide emission during the 15 March 2011 accident at the fukushima-1 npp (japan)", R. V. Arutyunyan, L. A. Bolshov, D. A. Pripachkin, V. N. Semyonov, O. S. Sorokovikova, A. L. Fokin, K. G. Rubinstein, R. Yu. Ignatov, M. M. Smirnova, Atomnaya Énergiya, Vol. 112, No. 3, pp. 159–163, March, 2012, as reported in Atomic Energy, July 2012, Volume 112, Issue 3, pp 188-193s

[163] Boytchev, Hristio, "First study reports very low internal radioactivity after Fukushima disaster", *Washington Post*, 15 August 2012

[164] Ken O. Buesseler (26 October 2012). "Fishing for Answers off Fukushima". *Science*. **338** (6106): 480–482. doi:10.1126/science.1228250. PMID 23112321. Archived from the original on 2013-08-17.

[165] Tabuchi, Hiroko (25 October 2012). "Fish Off Japan's Coast Said to Contain Elevated Levels of Cesium". New York Times Asia Pacific. Retrieved 28 October 2012.

[166] (Dutch) Nu.nl (26 oktober 2012) Tepco sluit niet uit dat centrale Fukushima nog lekt Archived 8 January 2014 at the Wayback Machine.

[167] Fukushima Plant Admits Radioactive Water Leaked To Sea. Huffingtonpost.com. Retrieved on 2013-09-06. Archived 17 April 2014 at the Wayback Machine.

[168] Adelman, Jacob. (2013-08-07) Abe Pledges Government Help to Stem Fukushima Water Leaks. Bloomberg. Retrieved on 2013-09-06. Archived 2 December 2013 at the Wayback Machine.

[169] "Wrecked Fukushima storage tank leaking highly radioactive water". *Reuters*. 20 August 2013. Archived from the original on 2014-04-29. Retrieved 21 August 2013.

[170] SI Units: Volume

[171] "Japan nuclear agency upgrades Fukushima alert level". BBC. 21 August 2013. Archived from the original on 2014-04-29. Retrieved 21 August 2013.

[172] Takashi Hirokawa; Jacob Adelman; Peter Langan; Yuji Okada (26 August 2013). "Fukushima Leaks Prompt Government to 'Emergency Measures' (1)". *Businessweek*. Bloomberg. Archived from the original on 2013-09-30. Retrieved 27 August 2013.

[173] "Japan seeks outside help for contaminated water". World Nuclear News. 26 September 2013. Archived from the original on 2014-04-02. Retrieved 1 October 2013.

[174] "How TEPCO plans to build an ice wall at Fukushima". Nuclear Engineering International. 18 February 2014. Archived from the original on 2014-02-24. Retrieved 19 February 2014.

[175] Varma, Subodh (10 February 2014). "Fukushima radiation data is wildly wrong, management apologizes". TNN. Archived from the original on 11 February 2014. Retrieved 11 March 2016.

[176] "TEPCO to review erroneous radiation data". *NKH World*. NHK. 9 February 2014. Retrieved 2014-02-09. Tokyo Electric Power Company, or TEPCO, says it has detected a record high 5 million becquerels (0.13 millicuries)per liter of radioactive strontium in groundwater collected last July from one of the wells close to the ocean. ... Based on the result, levels of radioactive substances that emit beta particles are estimated to be 10 million becquerels (0.26 millicuries) per liter, which is more than 10 times the initial reading.

[177] "TEPCO to Review Erroneous Radiation Data". *Yomiuri Online*. Yomiuri Shimbun. 9 February 2014. Retrieved 2014-02-09. On February 6, TEPCO announced that 5 million Bq/Liter of radioactive strontium was detected from the groundwater sample taken on June 5 last year from one of the observation wells on the embankment of Fukushima I Nuclear Power Plant. The density is 160,000 times that of the legal limit for release into the ocean, and it is about 1,000 times that of the highest density in the groundwater that had been measured so far (5,100 Bq/L). TEPCO didn't disclose the result of measurement of strontium alone, as the company believed there was a possibility that the result of measurement was wrong. As to this particular sample, TEPCO had announced on July last year that the sample had contained 900,000 Bq/L of all-beta including strontium. On February 6, TEPCO explained that they had "underestimated all of the results of high-density all-beta, which exceeded the upper limit of measurement." This particular sample may contain about 10 million Bq/L of all-beta, according to TEPCO. The company recently switched to a different method of analysis that uses diluted samples when the density of radioactive materials is high.

[178] Fernquest, John. "Japan floods: After typhoon, rivers overflow, nuclear water". Retrieved 2015-09-10.

[179] "Flooding swept away radiation cleanup bags in Fukushima". *The Japan Times Online*. 2015-09-12. ISSN 0447-5763. Retrieved 2015-09-13.

[180] Fukushima's radioactive ocean plume due to reach US waters in 2014 – NBC News Archived 26 March 2014 at the Wayback Machine.

[181] Sherwood, Courtney (11 November 2014). "Fukushima radiation nears California coast, judged harmless". Science.

[182] "British Columbia I Home". *fukushimainform.ca*. Retrieved 2015-11-02.

[183] "Canadian researcher targeted by hate campaign over Fukushima findings". *The Globe and Mail*. Retrieved 2015-11-02.

[184] Yoichi Funabashi; Kay Kitazawa (1 March 2012). "Fukushima in review: A complex disaster, a disastrous response" (PDF). *Bulletin of the Atomic Scientists.*

[185] Hiroko Tabuchi (3 March 2012). "Japanese Prime Minister Says Government Shares Blame for Nuclear Disaster". *The New York Times.*

[186] "AP Interview: Japan woefully unprepared for nuclear disaster, ex-prime minister says". *ctv.ca.* 17 February 2012.

[187] Amory Lovins (2011). "Soft Energy Paths for the 21st Century". Archived from the original on 2013-12-24.

[188] "Japan did not keep records of nuclear disaster meetings". *BBC Online.* 27 January 2012. Archived from the original on 2014-02-20.

[189] "Fukushima Pref. deleted 5 days of radiation dispersion data just after meltdowns". *The Mainichi Shimbun.* 22 March 2012. Archived from the original on 2012-03-25.

[190] "Tepco concealed core meltdowns during Fukushima accident". Nuclear Engineering International. 24 June 2016. Retrieved 25 June 2016.

[191] "Japan Held Nuclear Data, Leaving Evacuees in Peril". *Herald-Tribune.* 8 August 2011. Retrieved 8 August 2011.

[192] "Report: Japan, utility at fault for response to nuclear disaster". *LA Times.* 26 December 2011. Archived from the original on 2014-01-23.

[193] Martin Fackler (27 February 2012). "Japan Weighed Evacuating Tokyo in Nuclear Crisis". *The New York Times.* Archived from the original on 26 June 2012.

[194] Yoshida, Reiji (17 March 2012). "Kan hero, or irate meddler?". *Japan Times.* p. 2. Archived from the original on 2012-11-01.

[195] Hongo, Jun (29 February 2012). "Panel lays bare Fukushima recipe for disaster". *Japan Times.* p. 1. Archived from the original on 2012-02-29.

[196] "Blow-ups happen: Nuclear plants can be kept safe only by constantly worrying about their dangers". *The Economist.* 10 March 2012. Archived from the original on 2014-04-12.

[197] Kyodo News (20 June 2012). "Japan sat on U.S. radiation maps showing immediate fallout from nuke crisis". *Japan Times.* p. 1. Archived from the original on 2012-11-01.

[198] "Japan failed to use U.S. radiation data gathered after nuke crisis". *The Mainichi Shimbun.* 18 June 2012.

[199] Japan Atomic Industrial Forum, Inc. (JAIF) (19 June 2012). "Earthquake report 447" (PDF)

[200] The Japan Times (17 January 2012) U.S. forces given SPEEDI data early

[201] JIAF (29 June 2012)Earthequake-report 455: NISA "sorry" for withholding US radiation maps

[202] "Japanese utility admits to 'coverup' during Fukushima nuclear meltdown | Toronto Star". *thestar.com.* Retrieved 2016-07-01.

[203] "NISA News Release April 12, 2011" (PDF). Archived from the original (PDF) on 23 July 2012. Retrieved 24 April 2011.

[204] Directly comparing Fukushima to Chernobyl : Nature News Blog. Blogs.nature.com (2013-01-31). Retrieved on 2013-02-13. Archived 28 October 2013 at the Wayback Machine.

[205] Austria (12 April 2011). "IAEA Fukushima Nuclear Accident Update Log – Updates of 12 April 2011". *Iaea.org.* Archived from the original on 15 April 2011. Retrieved 24 April 2011.

[206] Press Release | The Estimated Amount of Radioactive Materials Released into the Air and the Ocean Caused by Fukushima Daiichi Nuclear Power Station Accident Due to the Tohoku-Chihou-Taiheiyou-Oki Earthquake (As of May 2012). TEPCO. Retrieved on 2013-02-13. Archived 15 February 2014 at the Wayback Machine.

[207] Chapter II The release, dispersion and deposition of radionuclides – Chernobyl: Assessment of Radiological and Health Impact. Oecd-nea.org. Retrieved on 2013-02-13. Archived 20 April 2011 at WebCite

[208] Isotopic ratio of radioactive iodine (129I/131I) released from Fukushima Daiichi NPP accident

[209] Johnson, George (21 September 2015). "When Radiation Isn't the Real Risk". The New York Times. Retrieved 30 November 2015.

[210] "Fukushima disaster: Ex-Tepco executives charged with negligence". *BBC News.* 29 February 2016. Retrieved 13 March 2016.

[211] *Japan Earthquake – Tsunami Fast Facts*, CNN, 2014-02-20, retrieved 2014-04-06 Archived 31 October 2013 at the Wayback Machine.

[212] Brumfiel, Geoffrey (23 May 2012). "World Health Organization weighs in on Fukushima". *Nature (journal).* Archived from the original on 2013-10-06. Retrieved 20 March 2013.

[213] Brumfiel, Geoff (Jan 2013). "Fukushima: Fallout of fear". *Nature.* **493** (7432): 290–293. doi:10.1038/493290a. PMID 23325191.

[214] Brumfiel, Geoff (May 2012). "PRINT – FUKUSHIMA". *Nature.* **485** (7399): 423–424. doi:10.1038/485423a. PMID 22622542.

[215] Nebehay, Stephanie (28 February 2013). "Higher cancer risk after Fukushima nuclear disaster: WHO". *Reuters.* Archived from the original on 2013-10-15.

[216] Rojavin, Y; Seamon, MJ; Tripathi, RS; Papadimos, TJ; Galwankar, S; Kman, N; Cipolla, J; Grossman, MD; Marchigiani, R; Stawicki, SP (Apr 2011). "Civilian nuclear incidents: An overview of historical, medical, and scientific aspects". *J Emerg Trauma Shock*. **4** (2): 260–72. doi:10.4103/0974-2700.82219.

[217] WHO 2013, p. 42.

[218] WHO 2013, p. 92.

[219] "Global report on Fukushima nuclear accident details health risks". Archived from the original on 2014-04-12. Retrieved 28 April 2014.

[220] Frequently asked questions on the Fukushima health risk assessment, questions 3 & 4 Archived 13 February 2014 at the Wayback Machine.

[221] WHO 2013, p. 83.

[222] "WHO: Slight cancer risk after Japan nuke accident". Archived from the original on 2013-03-03.

[223] "WHO report: cancer risk from Fukushima is low". Nuclear Engineering International. 1 March 2013. Archived from the original on 2013-10-14. Retrieved 6 March 2013.

[224] WHO 2013, p. 13.

[225] Normile, D. (2011). "Fukushima Revives the Low-Dose Debate". *Science*. **332** (6032): 908–910. doi:10.1126/science.332.6032.908. PMID 21596968.

[226] Luckey, T. J. (27 September 2006). "Radiation Hormesis: The Good, the Bad, and the Ugly". *Dose Response*. **4** (3): 189–190. doi:10.2203/dose-response.06-102.Luckey. PMC 2477686. PMID 18648595.

[227] Aurengo, A.; et al. (2005). "Dose-Effect Relationships and Estimation of the Carcinogenic Effects of Low Doses of Ionizing Radiation". *Academies of Sciences and Medicine, Paris*. **2**: 135. doi:10.1504/IJLR.2006.009510.

[228] "Radioactive Tuna Fish From Fukushima Reactor Spotted Off U.S. Shores". Fox Weekly. 2014-04-30. Archived from the original on 2014-05-02.

[229] Sebens, Shelby (2014-04-29). "Study finds Fukushima radioactivity in tuna off Oregon, Washington". Yahoo. Archived from the original on 2014-05-03.

[230] Worstall, Tim (2013-11-16). "Fukushima Radiation In Pacific Tuna Is Equal To One Twentieth Of A Banana". Forbes. Archived from the original on 2014-05-02.

[231] Tilman Ruff. Fukushima: The Misery Piles up, *Pursuit magazine*, University of Melbourne, 2016.

[232] WHO 2013, p. 87-88.

[233] Welch, H. Gilbert; Woloshin, Steve; Schwartz, Lisa A. (2011). *Overdiagnosed: Making People Sick in the Pursuit of Health*. Beacon Press. pp. 61–34. ISBN 978-0-8070-2200-9.

[234] Screening effect? Examining thyroid cancers found in Fukushima children. Jonathan Kellogg PhD, 2016

[235] Guth, S; Theune, U; Aberle, J; Galach, A; Bamberger, CM (2009). "Very high prevalence of thyroid nodules detected by high frequency (13 MHz) ultrasound examination". *Eur. J. Clin. Invest*. **39**: 699–706. doi:10.1111/j.1365-2362.2009.02162.x. PMID 19601965.

[236] "Fukushima kids have skyrocketing number of thyroid abnormalities – report". *Russia Times*. 18 February 2013. Archived from the original on 2014-04-15.

[237] http://www.fmu.ac.jp/radiationhealth/workshop201402/presentation/Co-Chairs_Summary_E.pdf

[238] "Experts divided on causes of high thyroid cancer rates among Fukushima children – The Mainichi". Retrieved 2016-07-06.

[239] cancer.org Thyroid Cancer By the American Cancer Society. In turn citing: AJCC Cancer Staging Manual (7th ed). Archived 18 October 2013 at the Wayback Machine.

[240] Brumfiel, Geoff (10 September 2012). "Fukushima's doses tallied". Archived from the original on 2014-02-14. Retrieved 23 May 2013.

[241] Zablotska, Lydia (8 November 2012). "Chernobyl Cleanup Workers Had Significantly Increased Risk of Leukemia". *UCSF*. Archived from the original on 2014-01-04.

[242] "Chernobyl: the true scale of the accident. 20 Years Later a UN Report Provides Definitive Answers and Ways to Repair Lives". Archived from the original on 2007-10-03.

[243] Studying the Fukushima Aftermath: 'People Are Suffering from Radiophobia' – SPIEGEL ONLINE. Spiegel.de (2011-08-19). Retrieved on 2013-09-06. Archived 16 January 2014 at the Wayback Machine.

[244] "Evacuees of Fukushima village report split families, growing frustration" (PDF). *Mainichi Daily News*. 30 January 2012.

[245] Katherine Harmon (2 March 2012). "Japan's Post-Fukushima Earthquake Health Woes Go Beyond Radiation Effects". *Nature*. Archived from the original on 2013-10-13.

[246] "Stress-induced deaths in Fukushima top those from 2011 natural disasters". Archived from the original on 2013-09-27.

[247] "Rain raises fear of more contamination at Fukushima". *CNN*. 4 Jun 2011. Archived from the original on 2013-12-24.

[248] "about the situation at the Fukushima Daiichi nuclear power plant". 3 Feb 2014. Archived from the original on 2014-02-22.

[249] "estimates claims burden from earthquake in Japan at around €1.5bn" . *Munich Re.* 22 March 2011. Archived from the original on 15 May 2011. Retrieved 24 April 2011.

[250] Swiss Re provides estimate of its claims costs from Japan earthquake and tsunami, Swiss Re, news release, 21 March 2011

[251] "UPDATE 1-Fukushima operator's mounting legal woes to fuel nuclear opposition" . *Reuters.* 2015-08-17. Retrieved 2016-02-02.

[252] Robin Harding (6 March 2016). "Japan taxpayers foot $100bn bill for Fukushima disaster" . *Financial Times.* Retrieved 20 March 2016.

[253] Justin McCurry (30 January 2017). "Possible nuclear fuel find raises hopes of Fukushima plant breakthrough" . *The Guardian.* Retrieved 3 February 2017.

[254] Tsuyoshi Inajima; Yuji Okada (28 Oct 2011). "Nuclear Promotion Dropped in Japan Energy Policy After Fukushima" . *Bloomberg.* Archived from the original on 2013-12-28.

[255] Mari Yamaguchi (September 2011). "Kenzaburo Oe, Nobel Winner Urges Japan To Abandon Nuclear Power" . *Huffington Post.* Archived from the original on 2013-12-20.

[256] Japanese nuclear plant survived tsunami, offers clues. Reuters. Retrieved on 2013-09-06. Archived 25 October 2011 at the Wayback Machine.

[257] "Fukushima Starts Long Road To Recovery" . NPR. 2012-03-10. Archived from the original on 2013-12-19. Retrieved 2012-04-16.

[258] "Neon city goes dim as power shortage threatens traffic lights and telephones in Tokyo" . *news.com.au.* 15 March 2011. Archived from the original on 15 March 2011.

[259] Yuri Kageyama, dealing with power shortage. Associated Press, 22 May 2011

[260] Benjamin K. Sovacool (2011). *Contesting the Future of Nuclear Power: A Critical Global Assessment of Atomic Energy,* World Scientific, p. 287.

[261] George Monbiot. "Why Fukushima made me stop worrying and love nuclear power" . *the Guardian.* Retrieved 12 June 2015.

[262] "Why This Matters" . Retrieved 12 June 2015.

[263] "The Moral Case for Nuclear Power" . Retrieved 12 June 2015.

[264] "How the Greens Were Misled" . Retrieved 12 June 2015.

[265] Mycle Schneider (9 September 2011). "Fukushima crisis: Can Japan be at the forefront of an authentic paradigm shift?". *Bulletin of the Atomic Scientists.* Archived from the original on 2013-01-06.

[266] Dr. Ken Caldeira, Senior Scientist, Department of Global Ecology, Carnegie Institution, Dr. Kerry Emanuel, Atmospheric Scientist, Massachusetts Institute of Technology, Dr. James Hansen, Climate Scientist, Columbia University Earth Institute, Dr. Tom Wigley, Climate Scientist, University of Adelaide and the National Center for Atmospheric Research. "There is no credible path to climate stabilization that does not include a substantial role for nuclear power" . cnn.com.

[267] Barry W. Brook – Professor of Environmental Sustainability at University of Tasmania, Corey Bradshaw Professor and Director of Ecological Modelling at University of Adelaide. "It's time for environmentalists to give nuclear a fair go" . theconversation.com.

[268] "Japan Plans Floating Wind Power Plant" . *Breakbulk.* 16 September 2011. Archived from the original on 2012-05-21. Retrieved 12 October 2011.

[269] Elaine Kurtenbach. "Japan starts up offshore wind farm near Fukushima" The Sydney Morning Herald, 12 November 2013. Accessed: 11 November 2013. Archived 30 December 2013 at the Wayback Machine.

[270] Joshua S Hill (2013-12-11). "Canadian Solar Signs Loan Agreement For Japan Development" . CleanTechnica. Retrieved 2013-12-30.

[271] Carol J. Williams (14 September 2012). "In wake of Fukushima disaster, Japan to end nuclear power by 2030s" . *LA Times.* Archived from the original on 2014-01-23.

[272] Gerhardt, Tina (22 July 2012). "After Fukushima, Nuclear Power on Collision Course With Japanese Public" . *Alternet.* Archived from the original on 2013-10-14. Retrieved 8 August 2013.

[273] "Abe dismisses Koizumi's call for zero nuclear power plants" . *Asahi Shimbun.* 2013-10-25. Archived from the original on 2014-04-13. Retrieved 2013-12-30.

[274] "Supporters of zero nuclear power "irresponsible": Abe" . Archived from the original on 2013-10-29.

[275] "Most Japan cities hosting nuclear plants OK restart: survey" . *Bangkok Post.* Retrieved 2013-12-30.

[276] United Press International (2 June 2013). "60,000 protest Japan's plan to restart nuclear power plants" . *UPI Asia.* Archived from the original on 2013-10-29.

[277] "Japan's Fuel Costs May Rise to 7.5 Trillion Yen, Meti Estimates" . Archived from the original on 2013-10-09.

[278] "Disaster response robots" , Open Minds blog, featuring BBC Documentary Archived 22 February 2014 at the Wayback Machine.

[279] Seiji Iwata; Ryuichi Kanari (26 May 2011). "Japanese robots long gone before Fukushima accident" . *Asahi Shimbun.* Retrieved 27 August 2014.

[280] "DARPA Robotics Challenge". *DRC*. DARPA. Archived from the original on 28 April 2016. Retrieved 27 April 2016.

[281] Maeda, Risa (20 October 2011). "Japanese nuclear plant survived tsunami, offers clues". Reuters. Archived from the original on 2011-10-25. Retrieved 2013-10-27.

[282] IAEA Expert Team Concludes Mission to Onagawa NPP Archived 29 October 2013 at the Wayback Machine.

[283] Japanese nuclear plant 'remarkably undamaged' in earthquake – UN atomic agency. Archived 29 October 2013 at the Wayback Machine.

[284] Hydrogen fix for Japanese reactors Archived 14 February 2014 at the Wayback Machine.

[285] Hydrogen recombiners at all 20 NPC plants to avoid Fukushima. Sanjay Jog | Mumbai 7 April 2011 Last Updated at 00:29 IST Archived 29 October 2013 at the Wayback Machine.

[286] CFD analysis of passive autocatalytic recombiner interaction with atmosphere. Archive Kerntechnik – Issue 2011/02. Archived 29 October 2013 at the Wayback Machine.

[287] Daly, Matthew (10 March 2013). "Nuclear chief: U.S. plants safer after Japan crisis. March 10, 2013". *USA Today*.

[288] "Vents and Filtering Strategies Come to Forefront in Fukushima Response Nuclear Energy Insight. Fall 2012".

[289] "TEPCO implements new safety measures in bid to restart Niigata reactors". Archived from the original on 2014-04-13.

[290] "Kashiwazaki-Kariwa plant shown to reporters". Archived from the original on 2013-10-29.

[291] Nuclear power plant operator in China orders backup batteries for installation at plants 7 September 2012 Archived 29 October 2013 at the Wayback Machine.

[292] China's Guangdong Nuclear Power Corp Announces Orders for BYD Battery Back-up for Nuclear Plants Archived 29 October 2013 at the Wayback Machine.

[293] Epstein, Woody (7 May 2012). "Not losing to the rain". *Quantitative Risk Assessment*. woody.com. Retrieved 26 February 2016. The Notstand building, a bunkered facility which could support all of the plant systems for at least 72 hours given a severe flood or earthquake which could take out the normal power and cooling facilities. I asked Martin Richner, the head of risk assessment, why Beznau spent so much money on the Notstand building when there was no regulation or government directive to do so. Martin answered me, "Woody, we live here." Archived 14 October 2013 at the Wayback Machine.

[294] "A PRA Practitioner Looks at the Fukushima Daiichi Accident" (PDF).

[295] "2012 20th International Conference on Nuclear Engineering and the ASME 2012 Power Conference". Archived from the original on 2013-12-15.

[296] "Gen III reactor design 04/06/2011 By Brian Wheeler Associate Editor". Archived from the original on 2013-12-14.

[297] "Nuclear Science and Techniques 24 (2013) 040601 Study on the long-term passive cooling extension of AP1000 reactor". Archived from the original on 2013-12-14.

[298] "Areas to which evacuation orders have been issued" (PDF). 7 August 2013.

[299] "Designating and Rearranging the Areas of Evacuation (pg 7)" (PDF).

[300] Dahl, Fredrik (15 August 2011). "U.N. atom body wants wider nuclear safety checks". *Reuters*. Archived from the original on 2014-05-06.

[301] Brasor, Philip, "Public wary of official optimism", *Japan Times*, 11 March 2012, p. 11. Archived 3 January 2013 at the Wayback Machine.

[302] Norimitsu Onishi (8 August 2011). "Japan Held Nuclear Data, Leaving Evacuees in Peril". *The New York Times*. Archived from the original on 21 August 2011.

[303] Charles Digges (10 August 2011). "Japan ignored its own radiation forecasts in days following disaster, imperiling thousands". *Bellona*. Archived from the original on 18 March 2012.

[304] "Analysis: A month on, Japan nuclear crisis still scarring," Archived 18 April 2011 at WebCite *International Business Times* (Australia). 9 April 2011, retrieved 12 April 2011; excerpt, According to James Acton, Associate of the Nuclear Policy Program at the Carnegie Endowment for International Peace, "Fukushima is not the worst nuclear accident ever but it is the most complicated and the most dramatic ... This was a crisis that played out in real time on TV. Chernobyl did not."

[305] Hasegawa, Koichi (2012). "Facing Nuclear Risks: Lessons from the Fukushima Nuclear Disaster". *International Journal of Japanese Sociology*. **21** (1): 84–91. doi:10.1111/j.1475-6781.2012.01164.x.

[306] Hiroko Tabuchi (13 July 2011). "Japan Premier Wants Shift Away From Nuclear Power". *The New York Times*.

[307] Naoto Kan (2013-10-28). "Encountering the Fukushima Daiichi Accident". *The Huffington Post*. Archived from the original on 2014-01-25. Retrieved 2013-11-09.

[308] (dutch)*Nu.nl* (22 August 2011)Area around Fukushima maybe a forbidden zone for decades to come Archived 23 October 2013 at the Wayback Machine.

[309] *The Guardian* (22 August 2011)residents may never return to radiation-hit homes Archived 23 August 2011 at the Wayback Machine.

[310] *Earthquake Report – JAIF, No. 45: 20:00, 7 April.* JAIF / NHK, 7 April 2011, archived from original on 9 April 2011, Retrieved 9 April 2011.

[311] Al-Jazeera English: *Citizen group tracks down Japan's radiation* (10 August 2011) Archived 31 August 2011 at the Wayback Machine.

[312] Safecast Organization Official Blog Archived 15 April 2014 at the Wayback Machine.

[313] Franken, Pieter (17 January 2014). "Volunteers Crowdsource Radiation Monitoring to Map Potential Risk on Every Street in Japan". *Democracy Now!* (Interview). Interview with Amy Goodman. Tokyo, Japan. Archived from the original on 2014-04-25. Retrieved 17 January 2014.

[314] UC Berkeley Nuclear Engineering Air Monitoring Station | The Nuclear Engineering Department at UC Berkely web site Archived 1 April 2011 at WebCite

[315] "USS Ronald Reagan Exposed to Radiation". Navy Handbook. 14 March 2011. Archived from the original on 2013-11-10. Retrieved 18 March 2011.

[316] Grimes, Robin (2014-06-16). "The UK Response to Fukushima and Anglo-Japanese Relations". *Science & Diplomacy.* **3** (2).

[317] "IAEA sees slow nuclear growth post Japan". *UPI.* 23 September 2011. Archived from the original on 2014-03-09.

[318] Nucléaire : une trentaine de réacteurs dans le monde risquent d'être fermés Les Échos, published 12 April 2011, accessed 15 April 2011

[319] "Gauging the pressure". *The Economist.* 28 April 2011. Archived from the original on 31 August 2012.

[320] RAFAEL POCH (2011-05-31). "Merkel se despide de lo nuclear y anuncia una revolución en renovables" (in Spanish). LAVANGUARDIA.com. Retrieved 26 January 2014.

[321] "Italy nuclear: Berlusconi accepts referendum blow". *BBC News.* 14 June 2011. Archived from the original on 12 June 2011. Retrieved 26 January 2014.

[322] "France struggles to cut down on nuclear power". *BBC News.* Retrieved 12 June 2015.

[323] Rob Broomby (11 January 2014). "France struggles to cut down on nuclear power". *BBC News Magazine.* Archived from the original on 2014-02-07. Retrieved 26 January 2014.

[324] "China Nuclear Power – Chinese Nuclear Energy". Retrieved 12 June 2015.

[325] http://www.kpmg.com/Global/en/ IssuesAndInsights/ArticlesPublications/Documents/ nuclear-power-role-in-shaping-energy-policies-v3.pdf

[326] Shannon Tiezzi; The Diplomat. "Why China Will Go All-In on Nuclear Power". *The Diplomat.* Retrieved 12 June 2015.

[327] "Nuclear Power in South Korea". Retrieved 12 June 2015.

[328] "Nuclear energy only option before country: Ansari". *The Indian Express.* 20 October 2012. Retrieved 12 June 2015.

[329] "Modi: India to build 10 more nuclear reactors with Russia". *International Business Times.* 12 December 2014.

[330] Richard Tanter (October–December 2013). *After Fukushima: A Survey of Corruption in the Global Nuclear Power Industry. Asian Perspective.* Vol. 37, No. 4.

[331] Jeff Kingston (10 September 2012). "Japan's Nuclear Village". *Japan Focus.* Archived from the original on 2014-03-29.

[332] Kaufmann, Daniel; Veronika Penciakova (17 March 2011). "Japan's triple disaster: Governance and the earthquake, tsunami and nuclear crises". *Brookings Institution.* Archived from the original on 2012-05-03.

[333] Culture of complicity tied to stricken nuclear plant, NYTimes, 27 April 2011

[334] "Japan to fire 3 top nuclear officials – CNN". *Articles.cnn.com.* 4 August 2011. Archived from the original on 2011-08-19. Retrieved 11 August 2011.

[335] "3 former TEPCO executives face criminal trial over Fukushima crisis". *The Asahi Shimbun.* 31 July 2015. Archived from the original on 14 March 2016. Retrieved 13 March 2016.

[336] "Fukushima nuclear accident 'man-made', not natural disaster". *Bloomberg L.P.* The Sydney Morning Herald. Archived from the original on 2013-11-03. Retrieved 9 July 2012.

[337] "Japan says Fukushima disaster was 'man-made'". *Al Jazeera and agencies.* AL Jazeera English. 5 July 2012. Archived from the original on 2014-01-30. Retrieved 9 July 2012.

[338] "Official website of the Investigation Committee on the Accident at the Fukushima Nuclear Power Stations of Tokyo Electric Power Company". Archived from the original on 31 July 2011. Retrieved 29 July 2012. This committee was established with the aim of conducting an investigation to determine the causes of the accident that occurred at Fukushima Daiichi and Daini Nuclear Power Stations of Tokyo Electric Power Company, and those of the damages generated by the accident, and thereby making policy proposals designed to prevent the expansion of the damages and the recurrence of similar accidents in the future.

[339] "Japan nuclear plants 'still not safe'". Al Jazeera Online. 23 July 2012. Archived from the original on 2014-04-16. Retrieved 29 July 2012.

[340] "Japan, TEPCO ignored atomic accident risks due to 'myth of nuclear safety': Report". *Asian News International (ANI)*. News Track India. 23 July 2012. Archived from the original on 2013-12-25. Retrieved 29 July 2012.

[341] Mitsuru Obe; Eleanor Warnock (23 July 2012). "Japan Panel Says Plant Operator Falls Short on Nuclear Safety". *The Wall Street Journal*. Archived from the original on 2013-09-27. Retrieved 30 July 2012.

[342] Tsuyoshi Inajima; Yuji Okada (23 July 2012). "Fukushima Investigators Say More Study Needed on What Went Wrong". *Bloomsberg Businessweek*. Archived from the original on 2013-09-28. Retrieved 29 July 2012.

[343] Hancocks, Paula (23 July 2012). "New report criticizes TEPCO over Fukushima nuclear crisis". CNN. Archived from the original on 2013-12-26. Retrieved 29 July 2012.

[344] Kazuaki Nagata (24 July 2012). "Government, Tepco again hit for nuke crisis". *The Japan Times*. Archived from the original on 2012-11-01. Retrieved 29 July 2012.

37.10.2 Sources

Cited

- WHO (2013). *Health risk assessment from the nuclear accident after the 2011 Great East Japan Earthquake and Tsunami* (PDF). ISBN 978 92 4 150513 0. Retrieved 2016-09-07.

Other

- Caldicott, Helen [ed.]: *Crisis Without End: The Medical and Ecological Consequences of the Fukushima Nuclear Catastrophe.* [From the "Symposium at the New York Academy of Medicine, March 11–12, 2013"]. The New Press, 2014. ISBN 978-1-59558-970-5 (eBook)

37.11 External links

37.11.1 Investigation

- The Fukushima Nuclear Accident Independent Investigation Commission Report website in English
- Investigation Committee on the accidents at the Fukushima Nuclear Power Station of Tokyo Electric Power Company
- The Radioactive Waters of Fukushima
- Lessons Learned From Fukushima Dai-ichi – Report & Movie

37.11.2 Video, drawings and images

- Webcam Fukushima nuclear power plant I, Unit 1 through Unit 4
- Inside the slow and dangerous clean up of the Fukushima nuclear crisis
- TerraFly Timeline Aerial Imagery of Fukushima Nuclear Reactor after 2011 Tsunami and Earthquake
- In graphics: Fukushima nuclear alert, as provided by the BBC, 9 July 2012
- Analysis by IRSN of the Fukushima Daiichi accident

37.11.3 Artwork

- Ah humanity!- a film essay by Lucien Castaign-Taylor, Ernst Karel and Véréna Paravel

37.11.4 Other

- Fukushima Revitalization Station (Fukushima Prefectural Government) in English
- TEPCO News Releases, Tokyo Electric Power Company
- "Reassessment of Fukushima Nuclear Accident and Outline of Nuclear Safety Reform Plan(Interim Report)" by TEPCO Nuclear Reform Special Task Force.14 December 2012

Chapter 38

Russian submarine Ekaterinburg (K-84)

K-84 *Ekaterinburg* (Russian: К−84 *Екатеринбург*) is a Project 667BDRM *Delfin* class (NATO reporting name: Delta IV) nuclear-powered ballistic missile submarine. The submarine was laid down on 17 February 1982 at the Russian Northern Machine-Building Enterprise (Sevmash).[*][2] It was commissioned into the Soviet Navy on 30 December 1985.[*][2] After the collapse of the Soviet Union, the submarine continued to serve in the Russian Navy. Initially known only by her hull number, in February 1999 she was renamed after the city of Yekaterinburg.[*][2][*][3]

38.1 Construction

Construction of the nuclear submarine *Ekaterinburg* (K-84) began at the Northern Machinebuilding Enterprise (Sevmash) in Severodvinsk on 17 February 1982, before being commissioned into the Soviet Navy on 30 December 1985.[*][2] She was the second of the seven-boat Project 667BDRM *Delfin* class, which was developed at the Rubin Design Bureau in September 1975.[*][4] A ballistic missile submarine, she was designed primarily to carry up to 16 R-29RM *Shtil* (NATO designation: SS-N-23 Skiff) SLBM for use against military and industrial facilities in the case of a nuclear war.[*][2][*][4] Each *Shtil* missile carries ten 100 kt multiple independently targeted reentry vehicles, and has a circular error probable of 500 metres (1,600 ft).[*][5] She is also equipped with RPK-7 *Veter* (NATO designation: SS-N-16 Stallion) anti-ship missiles for use against large surface vessels, and self-defense torpedoes.[*][5]

38.2 Operational history

After commissioning, *Ekaterinburg* was deployed to the base at Olenya Bay, and during the second half of 1986 underwent acoustic trials.[*][2] In August 1989, *Ekaterinburg* conducted a failed launch of all its missile under Operation Behemoth. Four months later, in December 1989, she was the first submarine to attempt to launch all her missiles while underwater; the first launch was successful, though the second was not.[*][2] In 1993, she was transferred to the base at Sayda-Guba.[*][2] On 3 December 1996, *Ekaterinburg* entered the Zvezdochka shipyard in Severodvinsk for an overhaul, though work did not begin until March 1998.[*][2] She re-entered service in 2003, based in Yagelnaya Bay.[*][2] She test-fired R-29RMU Sineva missiles in December 2003 and June 2004, and during Northern Fleet exercises in August 2005 fired missiles at the Kamchatka range.[*][2] Also in 2005, *Ekaterinburg* was awarded the Navy Commander's Prize for her missile launches.[*][2] In 2006, she successfully fired missiles at the Chizha Range from the North Pole.[*][2] On 20 May 2011, the boat fired the first R-29RMU2 Liner SLBM, aimed at the Kura Test Range.[*][6]

38.3 Fire incident and near nuclear catastrophe

On 29 December 2011 around 1220 UTC, *Ekaterinburg* caught fire while in drydock in Murmansk, and after several hours of firefighting efforts, she was partially sunk in an effort to control the fire.[*][7] Initial statements from Russian authorities indicate there were no injuries or radiation leakage, and that the vessel was not carrying any weapons as she was drydocked for repairs. The fire apparently began when sparks from welding being done on the boat's hull ignited wooden scaffolding around the ship, then spread to the flammable rubber coating covering the hull.[*][8] Russian President Dmitry Medvedev ordered the repair of the submarine and a thorough investigation of the incident on 30 December 2011.[*][9] The boat's hydro acoustic system was disabled in the fire.[*][9] Some sources speculate that the submarine's pressure hull suffered possible structural damage due to the intense heat; the temperature inside the torpedo room allegedly rose to 60-70 °C.[*][10]

A commission was to study the damage to the submarine and determine whether it was economical to repair it.[*][3] A Zvezdochka shipyard spokesperson said that the repairs would take more than a year.[*][11]

On 12 January 2012 ITAR-TASS reported that the repair of the submarine would take three to four years.[*][12] The repair would be combined with a scheduled refit that was to start in 2013. As it would be some months before the submarine can be transferred to the shipyard due to winter sea ice, the repairs would begin in May–June, 2012, so the submarine would not be expected to return to service before 2015.

On 14 February 2012, *Vlast* reported that the submarine had been carrying 16 R-29RM Shtil (NATO designation SS-N-23 Skiff) SLBMs, armed with four nuclear warheads in each missile, at the time of the fire, though officials had said at the time of the fire that no nuclear weapons were on board, as they had been unloaded before the fire broke out.[*][13] According to *Vlast*, the presence of nuclear weapons on the burning vessel would have meant that "Russia, for a day, was on the brink of the biggest catastrophe since the time of Chernobyl." [*][13]

Viktor Litovkin, editor-in-chief of Nezavisimoye Voyennoye Obozreniye, seriously doubted that a nuclear submarine could undergo maintenance with missiles and torpedoes on board because an off-duty submarine cannot have any weapons on board due to the Russian-American Prague agreement on nuclear arms. Litovkin explains that "when such a submarine is in dock, all missile pits are opened so that the Americans could check from surveillance satellites that there are no missiles in them... Moscow and Washington notify each other when nuclear missile carriers are going off duty for maintenance and Russia does the same surveillance of American subs." [*][14]

38.4   References

[1] http://flot.com/2014/181145/

[2] "K-84 Yekaterinburg" . Rusnavy.com. Archived from the original on 1 January 2012. Retrieved 1 January 2012.

[3] "После пожара АПЛ "Екатеринбург", возможно, утилизируют, агрегаты пострадали от высоких температур". Gazeta.ru. 30 December 2011. Retrieved 30 December 2011.

[4] "67BDRM Dolphin Delta IV" . Federation of American Scientists. 13 July 2000. Retrieved 2 January 2012.

[5] "Delta IV class" . Military-today.com. Retrieved 2 January 2012.

[6] "Внезапный "Лайнер"" [Sudden "Liner"]. *Lenta.Ru*. Lenta.ru. 10 August 2011. Retrieved 10 February 2012.

[7] "Russia submerges nuclear submarine to douse blaze" . Reuters. 29 December 2011. Retrieved 30 December 2011.

[8] "UPDATE: Fire aboard Russian nuclear submarine said to be extinguished" . Bellona Foundation. 30 December 2011. Retrieved 30 December 2011.

[9] "Medvedev orders repair of fire-damaged sub" . RIA Novosti. 30 December 2011. Retrieved 30 December 2011.

[10] http://www.novayagazeta.ru/inquests/50842.html

[11] "Repairs of fire-damaged nuclear sub to take at least one year" . Ria Novosti. 31 December 2011. Retrieved 31 December 2011.

[12] "Подлодка "Екатеринбург" вернется в боевой состав через 3 - 4 года". ITAR_TASS. 12 January 2012.

[13] "Russia faced major nuclear disaster in 2011-report" . Reuters. 14 February 2012. Archived from the original on 14 February 2012. Retrieved 15 February 2012.

[14] "Armageddon averted? Nukes 'on board' blazing sub (VIDEO)". Russia Today. 14 February 2012.

38.5   External links

- "667BDRM Dolphin Delta IV"

Chapter 39

International Nuclear Event Scale

The **International Nuclear and Radiological Event Scale** (**INES**) was introduced in 1990[*][1] by the International Atomic Energy Agency (IAEA) in order to enable prompt communication of safety-significant information in case of nuclear accidents.

The scale is intended to be logarithmic, similar to the moment magnitude scale that is used to describe the comparative magnitude of earthquakes. Each increasing level represents an accident approximately ten times more severe than the previous level. Compared to earthquakes, where the event intensity can be quantitatively evaluated, the level of severity of a man-made disaster, such as a nuclear accident, is more subject to interpretation. Because of the difficulty of interpreting, the INES level of an incident is assigned well after the incident occurs. Therefore, the scale has a very limited ability to assist in disaster-aid deployment.

As INES ratings are not assigned by a central body, high-profile nuclear incidents are sometimes assigned INES ratings by the operator, by the formal body of the country, but also by scientific institutes, international authorities or other experts which may lead to confusion as to the actual severity.

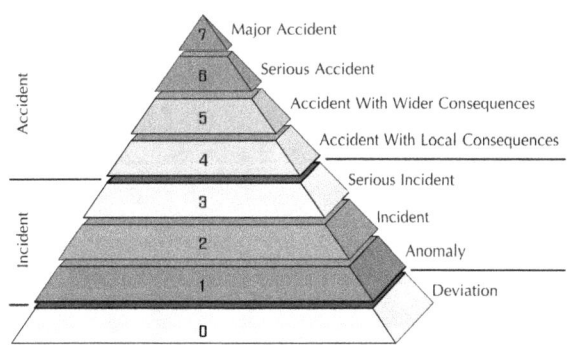

39.1 Details

A number of criteria and indicators are defined to assure coherent reporting of nuclear events by different official authorities. There are seven nonzero levels on the INES scale: three *incident*-levels and four *accident*-levels. There is also a level 0.

The level on the scale is determined by the highest of three scores: off-site effects, on-site effects, and defence in depth degradation.

39.1.1 Level 7: Major accident

See also: Comparison of Fukushima and Chernobyl nuclear accidents

Impact on people and environment Major release of radioactive material with widespread health and environmental effects requiring implementation of planned and extended countermeasures

There have been two such events to date:

- Chernobyl disaster, 26 April 1986. A power surge during a test procedure resulted in a criticality accident, leading to a powerful steam explosion and fire that released a significant fraction of core material into the environment, resulting in an eventual death toll of 56 at the time of the UNSCEAR report, as well as a projected 4,000 additional cancer fatalities (official WHO estimate) that will eventually occur among the people exposed to the highest doses of radiation. As a result of the plumes of radio-isotopes, the city of Chernobyl (pop. 14,000) was largely abandoned, the larger city of Pripyat (pop. 49,400) was completely abandoned, and a 30 kilometres

(19 mi) exclusion zone around the reactor was established.

- Fukushima Daiichi nuclear disaster, a series of events beginning on 11 March 2011. A month later the Japanese government's nuclear safety agency rated it level 7.[2][3] Major damage to the backup power and containment systems caused by the 2011 Tōhoku earthquake and tsunami resulted in overheating and leaking from some of the Fukushima I nuclear plant's reactors. Each reactor accident was rated separately; out of the six reactors, three were rated level 5, one was rated at a level 3, and the situation as a whole was rated level 7.[4] A temporary exclusion zone of 20 kilometres (12 mi) was established around the plant as well as a 30 kilometres (19 mi) voluntary evacuation zone;[5] In addition, the evacuation of Tokyo – Japan's capital and the world's most populous metropolitan area, 225 kilometres (140 mi) away – was at one point considered.[6]

39.1.2 Level 6: Serious accident

Impact on people and environment Significant release of radioactive material likely to require implementation of planned countermeasures.

There has been only one such event to date:

- Kyshtym disaster at Mayak Chemical Combine (MCC) Soviet Union, 29 September 1957. A failed cooling system at a military nuclear waste reprocessing facility caused an explosion with a force equivalent to 70-100 tons of TNT. About 70 to 80 metric tons of highly radioactive material were carried into the surrounding environment. The impact on the local population is not fully known, however reports of a unique condition known as Chronic radiation syndrome is reported due to the moderately high dose rates that 66 locals were continually exposed to. At least 22 villages were evacuated.[7]

39.1.3 Level 5: Accident with wider consequences

Impact on people and environment Limited release of radioactive material likely to require implementation of some planned countermeasures.

Several deaths from radiation.

Impact on radiological barriers and control Severe damage to reactor core.

Release of large quantities of radioactive material within an installation with a high probability of significant public exposure. This could arise from a major criticality accident or fire.

Examples:[7]

- Windscale fire (United Kingdom), 10 October 1957.[8] Annealing of graphite moderator at a military air-cooled reactor caused the graphite and the metallic uranium fuel to catch fire, releasing radioactive pile material as dust into the environment.

- Three Mile Island accident near Harrisburg, Pennsylvania (United States), 28 March 1979.[9] A combination of design and operator errors caused a gradual loss of coolant, leading to a partial meltdown. An unknown amount of radioactive gases were released into the atmosphere, so injuries and illnesses that have been attributed to this accident can be deduced from epidemiological studies but can never be proven.

- First Chalk River accident,[10][11] Chalk River, Ontario (Canada), 12 December 1952. Reactor core damaged.

- Lucens partial core meltdown (Switzerland), 21 January 1969. A test reactor built in an underground cavern suffered a loss-of-coolant accident during a startup, leading to a partial core meltdown and massive radioactive contamination of the cavern, which was then sealed.[12]

- Goiânia accident (Brazil), 13 September 1987. An unsecured caesium chloride radiation source left in an abandoned hospital was recovered by scavenger thieves unaware of its nature and sold at a scrapyard. 249 people were contaminated and 4 died.

39.1.4 Level 4: Accident with local consequences

Impact on people and environment Minor release of radioactive material unlikely to result in implementation of planned countermeasures other than local food controls.

At least one death from radiation.

Impact on radiological barriers and control Fuel melt or damage to fuel resulting in more than 0.1% release of core inventory.

Release of significant quantities of radioactive material within an installation with a high probability of significant public exposure.

Examples:*[7]

- Sellafield (United Kingdom) – five incidents from 1955 to 1979.*[13]
- SL-1 Experimental Power Station (United States) – 1961, reactor reached prompt criticality, killing three operators.
- Saint-Laurent Nuclear Power Plant (France) – 1969, partial core meltdown; 1980, graphite overheating.
- Buenos Aires (Argentina) – 1983, criticality accident on research reactor RA-2 during fuel rod rearrangement killed one operator and injured two others.
- Jaslovské Bohunice (Czechoslovakia) – 1977, contamination of reactor building.
- Tokaimura nuclear accident (Japan) – 1999, three inexperienced operators at a reprocessing facility caused a criticality accident; two of them died.

39.1.5 Level 3: Serious incident

Impact on people and environment Exposure in excess of ten times the statutory annual limit for workers.

Non-lethal deterministic health effect (e.g., burns) from radiation.

Impact on radiological barriers and control Exposure rates of more than 1 Sv/h in an operating area.

Severe contamination in an area not expected by design, with a low probability of significant public exposure.

Impact on defence-in-depth Near-accident at a nuclear power plant with no safety provisions remaining.

Lost or stolen highly radioactive sealed source.

Misdelivered highly radioactive sealed source without adequate procedures in place to handle it.

Examples:

- THORP plant, Sellafield (United Kingdom) – 2005.
- Paks Nuclear Power Plant (Hungary), 2003; fuel rod damage in cleaning tank.
- Vandellos Nuclear Power Plant (Spain), 1989; fire destroyed many control systems; the reactor was shut down.
- Davis-Besse Nuclear Power Station (United States), 2002; negligent inspections resulted in corrosion through 6 inches (15.24 cm) of the carbon steel reactor head leaving only 3/8 inch (9.5 mm) of stainless steel cladding holding back the high-pressure (~2500 psi, 17 MPa) reactor coolant.

39.1.6 Level 2: Incident

Impact on people and environment. Exposure of a member of the public in excess of 10 mSv.

Exposure of a worker in excess of the statutory annual limits.

Impact on radiological barriers and control Radiation levels in an operating area of more than 50 mSv/h.

Significant contamination within the facility into an area not expected by design.

Impact on defence-in-depth Significant failures in safety provisions but with no actual consequences.

Found highly radioactive sealed orphan source, device or transport package with safety provisions intact.

Inadequate packaging of a highly radioactive sealed source.

Examples:

- Blayais Nuclear Power Plant flood (France) December 1999
- Ascó Nuclear Power Plant (Spain) April 2008; radioactive contamination.

- Forsmark Nuclear Power Plant (Sweden) July 2006; backup generator failure; two were online but fault could have caused all four to fail.

- Gundremmingen Nuclear Power Plant (Germany) 1977; weather caused short-circuit of high-tension power lines and rapid shutdown of reactor

- Shika Nuclear Power Plant (Japan) 1999; criticality incident caused by dropped control rods, covered up until 2007.*[14]

39.1.7   Level 1: Anomaly

Impact on defence-in-depth   Overexposure of a member of the public in excess of statutory annual limits.

Minor problems with safety components with significant defence-in-depth remaining.

Low activity lost or stolen radioactive source, device or transport package.

(Arrangements for reporting minor events to the public differ from country to country. It is difficult to ensure precise consistency in rating events between INES Level-1 and Below scale/Level-0)

Examples:

- Penly (Seine-Maritime, France) 5 April 2012; an abnormal leak on the primary circuit of the reactor n°2 was found in the evening of 5 April 2012 after a fire in reactor n°2 around noon was extinguished.*[15]

- Gravelines (Nord, France), 8 August 2009; during the annual fuel bundle exchange in reactor #1, a fuel bundle snagged on to the internal structure. Operations were stopped, the reactor building was evacuated and isolated in accordance with operating procedures.*[16]

- TNPC (Drôme, France), July 2008; leak of 18,000 litres (4,000 imp gal; 4,800 US gal) of water containing 75 kilograms (165 lb) of unenriched uranium into the environment.*[17]

39.1.8   Level 0: Deviation

No safety significance.

Examples:

- 4 June 2008: Krško, Slovenia: Leakage from the primary cooling circuit.*[18]

- 17 December 2006, Atucha, Argentina: Reactor shutdown due to tritium increase in reactor compartment.*[19]

- 13 February 2006:   Fire in Nuclear Waste Volume Reduction Facilities of the Japanese Atomic Energy Agency (JAEA) in Tokaimura.*[20]

Out of scale

There are also events of no safety relevance, characterized as "out of scale" .*[21]

Examples:

- 17 November 2002, Natural Uranium Oxide Fuel Plant at the Nuclear Fuel Complex in Hyderabad, India: A chemical explosion at a fuel fabrication facility.*[22]

- 29 September 1999:   H.B. Robinson, United States: A tornado sighting within the protected area of the nuclear power plant.*[23]*[24]*[25]

- 5 March 1999: San Onofre, United States: Discovery of suspicious item, originally thought to be a bomb, in nuclear power plant.*[26]

39.2   Criticism

Deficiencies in the existing INES have emerged through comparisons between the 1986 Chernobyl disaster and 2011 Fukushima nuclear disaster. Firstly, the scale is essentially a discrete qualitative ranking, not defined beyond event level 7. Secondly, it was designed as a public relations tool, not an objective scientific scale. Thirdly, its most serious shortcoming is that it conflates magnitude with intensity. David Smythe has proposed a new quantitative nuclear accident magnitude scale (NAMS) to address these issues.*[27] One example given, is that it would deal with what some regard as having the potential to cause confusion, such as the underground and contained Lucens reactor meltdown presently residing in same event category as the Three Mile Island accident, which in contrast, did cause releases to the public.

One study found that the INES scale of the IAEA is highly inconsistent, and the scores provided by the IAEA incomplete, with many events not having an INES rating. Further, the actual accident damage values do not reflect the INES scores. For example, the Fukushima disaster should have an INES level of 10 or 11, rather than the top level of 7. A quantifiable, continuous scale might be preferable to the INES, in the same way that the antiquated Mercalli scale for earthquake magnitudes was superseded by the continuous physically-based Richter scale. When such a framework is established, data on nuclear incidents and accidents can made more rigorous, and transparent, accident risks can be better understood, and perhaps even minimised.*[28]

Nuclear experts say that the "INES emergency scale is very likely to be revisited" given the confusing way in which it was used in the 2011 Japanese nuclear accidents.*[29]

39.3 See also

39.4 Notes and references

[1] "Event scale revised for further clarity". World-nuclear-news.org. 6 October 2008. Retrieved 13 September 2010.

[2] "Japan to raise Fukushima crisis level to worst". Retrieved 12 April 2011.

[3] "Japan raises nuclear crisis to same level as Chernobyl". *Reuters*. 12 April 2011.

[4] "Japan: Nuclear crisis raised to Chernobyl level". *BBC News*. 12 April 2011. Retrieved 12 April 2011.

[5] "Japan's government downgrades its outlook for growth". *BBC News*. 13 April 2011. Retrieved 13 April 2011. The death toll rose to over 15,000 with 8,206 missing and 5,363 injured the numbers are still rising.

[6] Krista Mahr (29 February 2012). "Fukushima Report: Japan Urged Calm While It Mulled Tokyo Evacuation". *Time*.

[7] "The world's worst nuclear power disasters". *Power Technology*. 7 October 2013.

[8] Richard Black (18 March 2011). "Fukushima - disaster or distraction?". BBC. Retrieved 7 April 2011.

[9] Spiegelberg-Planer, Rejane. "A Matter of Degree" (PDF). *IAEA Bulletin*. IAEA. Retrieved 24 May 2016.

[10] *Canadian Nuclear Society* (1989) The NRX Incident by Peter Jedicke

[11] *The Canadian Nuclear FAQ* What are the details of the accident at Chalk River's NRX reactor in 1952?

[12] FlohEinstein. "Versuchsatomkraftwerk Lucens". *ENSI Bericht*. ENSI. Retrieved 12 May 2014.

[13] Webb, G A M; Anderson, R W; Gaffney, M J S (2006). "Classification of events with an off-site radiological impact at the Sellafield site between 1950 and 2000, using the International Nuclear Event Scale". *Journal of Radiological Protection*. IOP. **26** (1): 33–49. doi:10.1088/0952-4746/26/1/002. PMID 16522943.

[14] Information on Japanese criticality accidents,

[15] (ASN) - 5 April 2012. "ASN:ASN has decided to lift its emergency crisis organisation and has temporarily classified the event at the level 1". ASN. Retrieved 6 April 2012.

[16] (AFP) – 10 août 2009. "AFP: Incident "significatif" à la centrale nucléaire de Gravelines, dans le Nord". Google.com. Retrieved 13 September 2010.

[17] River use banned after French uranium leak | Environment. The Guardian (2008-07-10). Retrieved on 2013-08-22.

[18] News | Slovenian Nuclear Safety Administration

[19] http://200.0.198.11/comunicados/18_12_2006.pdf (Spanish)

[20] http://www.jaea.go.jp/02/press2005/p06021301/index.html (Japanese)

[21] IAEA: "This event is rated as out of scale in accordance with Part I-1.3 of the 1998 Draft INES Users Manual, as it did not involve any possible radiological hazard and did not affect the safety layers."

[22] Archived 21 July 2011 at the Wayback Machine.

[23] "NRC: SECY-01-0071 – Expanded NRC Participation in the Use of the International Nuclear Event Scale". US Nuclear Regulatory Commission. 25 April 2001. p. 8. Archived from the original (PDF) on 27 October 2010. Retrieved 13 March 2011.

[24] "SECY-01-0071-Attachment 5 - INES Reports, 1995-2000". US Nuclear Regulatory Commission. 25 April 2001. p. 1. Archived from the original (PDF) on 27 October 2010. Retrieved 13 March 2011.

[25] Tornado sighting within protected area | Nuclear power in Europe. Climatesceptics.org. Retrieved on 2013-08-22.

[26] Discovery of suspicious item in plant | Nuclear power in Europe. Climatesceptics.org. Retrieved on 2013-08-22.

[27] David Smythe (12 December 2011). "An objective nuclear accident magnitude scale for quantification of severe and catastrophic events". *Physics Today*.

[28] Spencer Wheatley, Benjamin Sovacool, and Didier Sornette Of Disasters and Dragon Kings: A Statistical Analysis of Nuclear Power Incidents & Accidents, *Physics Society*, 7 April 2015.

[29] Geoff Brumfiel (26 April 2011). "Nuclear agency faces reform calls". *Nature*.

39.5   External links

- Nuclear Events Web-based System (NEWS), IAEA

- International Nuclear Event Scale factsheet, IAEA

- "International Nuclear Event Scale, User's manual". Archived from the original (PDF) on 19 March 2011. Retrieved 19 March 2011. International Nuclear Event Scale, User's manual, IAEA, 2008

Chapter 40

Nuclear safety and security

A clean-up crew working to remove radioactive contamination after the Three Mile Island accident.

Nuclear safety is defined by the International Atomic Energy Agency (IAEA) as "The achievement of proper operating conditions, prevention of accidents or mitigation of accident consequences, resulting in protection of workers, the public and the environment from undue radiation hazards". The IAEA defines **nuclear security** as "The prevention and detection of and response to, theft, sabotage, unauthorized access, illegal transfer or other malicious acts involving nuclear material, other radioactive substances or their associated facilities".*[1]

This covers nuclear power plants and all other nuclear facilities, the transportation of nuclear materials, and the use and storage of nuclear materials for medical, power, industry, and military uses.

The nuclear power industry has improved the safety and performance of reactors, and has proposed new and safer reactor designs. However, a perfect safety cannot be guaranteed. Potential sources of problems include human errors and external events that have a greater impact than anticipated: The designers of reactors at Fukushima in Japan did not anticipate that a tsunami generated by an earthquake would disable the backup systems that were supposed to stabilize the reactor after the earthquake.*[2]*[3]*[4]*[5] According to UBS AG, the Fukushima I nuclear accidents have cast doubt on whether even an advanced economy like Japan can master nuclear safety.*[6] Catastrophic scenarios involving terrorist attacks, insider sabotage, and cyberattacks are also conceivable.*[7]

In his book, *Normal accidents*, Charles Perrow says that multiple and unexpected failures are built into society's complex and tightly-coupled nuclear reactor systems. Such accidents are unavoidable and cannot be designed around.*[8] To date, there have been three serious accidents (core damage) in the world since 1970, involving five reactors (one at Three Mile Island in 1979; one at Chernobyl in 1986; and three at Fukushima-Daiichi in 2011), corresponding to the beginning of the operation of generation II reactors.

Nuclear weapon safety, as well as the safety of military research involving nuclear materials, is generally handled by agencies different from those that oversee civilian safety, for various reasons, including secrecy. There are ongoing concerns about terrorist groups acquiring nuclear bomb-making material.*[9]

40.1 Overview of nuclear processes and safety issues

As of 2011, nuclear safety considerations occur in a number of situations, including:

- Nuclear fission power used in nuclear power stations, and nuclear submarines and ships
- Nuclear weapons

- Fissionable fuels such as uranium and plutonium and their extraction, storage and use

- Radioactive materials used for medical, diagnostic, batteries for some space projects, and research purposes

- Nuclear waste, the radioactive waste residue of nuclear materials

- Nuclear fusion power, a technology under long-term development

- Unplanned entry of nuclear materials into the biosphere and food chain (living plants, animals and humans) if breathed or ingested.

With the exception of thermonuclear weapons and experimental fusion research, all safety issues specific to nuclear power stems from the need to limit the biological uptake of committed dose (ingestion or inhalation of radioactive materials), and external radiation dose due to radioactive contamination.

Nuclear safety therefore covers at minimum: -

- Extraction, transportation, storage, processing, and disposal of fissionable materials

- Safety of nuclear power generators

- Control and safe management of nuclear weapons, nuclear material capable of use as a weapon, and other radioactive materials

- Safe handling, accountability and use in industrial, medical and research contexts

- Disposal of nuclear waste

- Limitations on exposure to radiation

40.2   Responsible agencies

40.2.1   International

Internationally the International Atomic Energy Agency "works with its Member States and multiple partners worldwide to promote safe, secure and peaceful nuclear technologies." *[10] Some scientists say that the 2011 Japanese nuclear accidents have revealed that the nuclear industry lacks sufficient oversight, leading to renewed calls to redefine the mandate of the IAEA so that it can better police nuclear power plants worldwide.*[11]

The IAEA Convention on Nuclear Safety was adopted in Vienna on 17 June 1994 and entered into force on 24 October 1996. The objectives of the Convention are to achieve

IAEA headquarters in Vienna, Austria

The International Atomic Energy Agency was created in 1957 to encourage peaceful development of nuclear technology while providing international safeguards against nuclear proliferation.

and maintain a high level of nuclear safety worldwide, to establish and maintain effective defences in nuclear installations against potential radiological hazards, and to prevent accidents having radiological consequences.*[12]

The Convention was drawn up in the aftermath of the Three Mile Island and Chernobyl accidents at a series of expert level meetings from 1992 to 1994, and was the result of considerable work by States, including their national regulatory and nuclear safety authorities, and the International Atomic Energy Agency, which serves as the Secretariat for the Convention.

The obligations of the Contracting Parties are based to a large extent on the application of the safety principles for nuclear installations contained in the IAEA document Safety Fundamentals 'The Safety of Nuclear Installations' (IAEA Safety Series No. 110 published 1993). These obligations cover the legislative and regulatory framework, the regulatory body, and technical safety obligations related to, for instance, siting, design, construction, operation, the availability of adequate financial and human resources, the assessment and verification of safety, quality assurance and emergency preparedness.

The convention was amended in 2015 by the Vienna Declaration on Nuclear Safety *[13] This resulted in the following principles:

1. New nuclear power plants are to be designed, sited, and constructed, consistent with the objective of preventing accidents in the commissioning and operation and, should an accident occur, mitigating possible releases of radionuclides causing long-term off site contamination and avoiding early radioactive releases or radioactive releases large enough to require long-term protective measures and actions.

2. Comprehensive and systematic safety assessments are to be carried out periodically and regularly for existing installations throughout their lifetime in order to identify safety improvements that are oriented to meet the above objective. Reasonably practicable or achievable safety improvements are to be implemented in a timely manner.

3. National requirements and regulations for addressing this objective throughout the lifetime of nuclear power plants are to take into account the relevant IAEA Safety Standards and, as appropriate, other good practices as identified inter alia in the Review Meetings of the CNS.

There are several problems with the IAEA, says Najmedin Meshkati of University of Southern California, writing in 2011:

> "It recommends safety standards, but member states are not required to comply; it promotes nuclear energy, but it also monitors nuclear use; it is the sole global organization overseeing the nuclear energy industry, yet it is also weighed down by checking compliance with the Nuclear Non-Proliferation Treaty (NPT)".*[11]

40.2.2 National

Many nations utilizing nuclear power have specialist institutions overseeing and regulating nuclear safety. Civilian nuclear safety in the U.S. is regulated by the Nuclear Regulatory Commission (NRC). However, critics of the nuclear industry complain that the regulatory bodies are too intertwined with the inustries themselves to be effective. The book *The Doomsday Machine* for example, offers a series of examples of national regulators, as they put it 'not regulating, just waving' (a pun on *waiving*) to argue that, in Japan, for example, "regulators and the regulated have long been friends, working together to offset the doubts of a public brought up on the horror of the nuclear bombs" .*[14]

Other examples offered *[15] include:

- in the United States, a dangerous custom whereby only supporters of the nuclear industry are allowed to supervise it and lobbyists have been allowed to have an effective veto over regulators.

- in China, where Kang Rixin, former general manager of the state-owned China National Nuclear Corporation, was sentenced to life in jail in 2010 for accepting bribes (and other abuses), a verdict raising questions about the quality of his work on the safety and trustworthiness of China' s nuclear reactors.

- in India, where the nuclear regulator reports to the national Atomic Energy Commission, which champions the building of nuclear power plants there and the chairman of the Atomic Energy Regulatory Board, S. S. Bajaj, was previously a senior executive at the Nuclear Power Corporation of India, the company he is now helping to regulate.

- in Japan, where the regulator reports to the Ministry of Economy, Trade and Industry, which overtly seeks to promote the nuclear industry and ministry posts and top jobs in the nuclear business are passed among the same small circle of experts.

The book argues that nuclear safety is compromised by the suspicion that, as Eisaku Sato, formerly a governor of Fukushima province (with its infamous nuclear reactor complex), has put it of the regulators: "They' re all birds of a feather" .*[15]

The safety of nuclear plants and materials controlled by the U.S. government for research, weapons production, and those powering naval vessels is not governed by the NRC.*[16]*[17] In the UK nuclear safety is regulated by the Office for Nuclear Regulation (ONR) and the Defence Nuclear Safety Regulator (DNSR). The Australian Radiation Protection and Nuclear Safety Agency (ARPANSA) is the Federal Government body that monitors and identifies solar radiation and nuclear radiation risks in Australia. It is the main body dealing with ionizing and non-ionizing radiation*[18] and publishes material regarding radiation protection.*[19]

Other agencies include:

- Autorité de sûreté nucléaire

- Canadian Nuclear Safety Commission

- Radiological Protection Institute of Ireland

- Federal Atomic Energy Agency in Russia

- Kernfysische dienst, (NL)

- Pakistan Nuclear Regulatory Authority

- Bundesamt für Strahlenschutz, (DE)

- Atomic Energy Regulatory Board (India)

40.3 Nuclear power plant safety and security

See also: Nuclear power plant

40.3.1 Complexity

Nuclear power plants are some of the most sophisticated and complex energy systems ever designed.[*][20] Any complex system, no matter how well it is designed and engineered, cannot be deemed failure-proof.[*][4] Veteran journalist and author Stephanie Cooke has argued:

> The reactors themselves were enormously complex machines with an incalculable number of things that could go wrong. When that happened at Three Mile Island in 1979, another fault line in the nuclear world was exposed. One malfunction led to another, and then to a series of others, until the core of the reactor itself began to melt, and even the world's most highly trained nuclear engineers did not know how to respond. The accident revealed serious deficiencies in a system that was meant to protect public health and safety.[*][21]

The 1979 Three Mile Island accident inspired Perrow's book *Normal Accidents*, where a nuclear accident occurs, resulting from an unanticipated interaction of multiple failures in a complex system. TMI was an example of a normal accident because it was "unexpected, incomprehensible, uncontrollable and unavoidable".[*][22]

> Perrow concluded that the failure at Three Mile Island was a consequence of the system's immense complexity. Such modern high-risk systems, he realized, were prone to failures however well they were managed. It was inevitable that they would eventually suffer what he termed a 'normal accident'. Therefore, he suggested, we might do better to contemplate a radical redesign, or if that was not possible, to abandon such technology entirely.[*][23]

A fundamental issue contributing to a nuclear power system's complexity is its extremely long lifetime. The time-frame from the start of construction of a commercial nuclear power station through the safe disposal of its last radioactive waste, may be 100 to 150 years.[*][20]

40.3.2 Failure modes of nuclear power plants

There are concerns that a combination of human and mechanical error at a nuclear facility could result in significant harm to people and the environment:[*][24]

> Operating nuclear reactors contain large amounts of radioactive fission products which, if dispersed, can pose a direct radiation hazard, contaminate soil and vegetation, and be ingested by humans and animals. Human exposure at high enough levels can cause both short-term illness and death and longer-term death by cancer and other diseases.[*][25]

It is impossible for a commercial nuclear reactor to explode like a nuclear bomb since the fuel is never sufficiently enriched for this to occur.[*][26]

Nuclear reactors can fail in a variety of ways. Should the instability of the nuclear material generate unexpected behavior, it may result in an uncontrolled power excursion. Normally, the cooling system in a reactor is designed to be able to handle the excess heat this causes; however, should the reactor also experience a loss-of-coolant accident, then the fuel may melt or cause the vessel in which it is contained to overheat and melt. This event is called a nuclear meltdown.

After shutting down, for some time the reactor still needs external energy to power its cooling systems. Normally this energy is provided by the power grid to which that plant is connected, or by emergency diesel generators. Failure to provide power for the cooling systems, as happened in Fukushima I, can cause serious accidents.

Nuclear safety rules in the United States "do not adequately weigh the risk of a single event that would knock out electricity from the grid and from emergency generators, as a quake and tsunami recently did in Japan", Nuclear Regulatory Commission officials said in June 2011.[*][27]

As a safeguard against mechanical failure, many nuclear plants are designed to shut down automatically after two days of continuous and unattended operation.

40.3.3 Vulnerability of nuclear plants to attack

Nuclear reactors become preferred targets during military conflict and, over the past three decades, have been repeatedly attacked during military air strikes, occupations, invasions and campaigns:*[28]

- In September 1980, Iran bombed the Al Tuwaitha nuclear complex in Iraq in Operation Scorch Sword.

- In June 1981, an Israeli air strike completely destroyed Iraq's Osirak nuclear research facility in Operation Opera.

- Between 1984 and 1987, Iraq bombed Iran's Bushehr nuclear plant six times.

- On 8 January 1982, Umkhonto we Sizwe, the armed wing of the ANC, attacked South Africa's Koeberg nuclear power plant while it was still under construction.

- In 1991, the U.S. bombed three nuclear reactors and an enrichment pilot facility in Iraq.

- In 1991, Iraq launched Scud missiles at Israel's Dimona nuclear power plant

- In September 2007, Israel bombed a Syrian reactor under construction.*[28]

In the U.S., plants are surrounded by a double row of tall fences which are electronically monitored. The plant grounds are patrolled by a sizeable force of armed guards.*[29] The NRC's "Design Basis Threat" criterion for plants is a secret, and so what size of attacking force the plants are able to protect against is unknown. However, to scram (make an emergency shutdown) a plant takes fewer than 5 seconds while unimpeded restart takes hours, severely hampering a terrorist force in a goal to release radioactivity.

Attack from the air is an issue that has been highlighted since the September 11 attacks in the U.S. However, it was in 1972 when three hijackers took control of a domestic passenger flight along the east coast of the U.S. and threatened to crash the plane into a U.S. nuclear weapons plant in Oak Ridge, Tennessee. The plane got as close as 8,000 feet above the site before the hijackers' demands were met.*[30]*[31]

The most important barrier against the release of radioactivity in the event of an aircraft strike on a nuclear power plant is the containment building and its missile shield. Current NRC Chairman Dale Klein has said "Nuclear power plants are inherently robust structures that our studies show provide adequate protection in a hypothetical attack by an airplane. The NRC has also taken actions that require nuclear power plant operators to be able to manage large fires or explosions—no matter what has caused them." *[32]

In addition, supporters point to large studies carried out by the U.S. Electric Power Research Institute that tested the robustness of both reactor and waste fuel storage and found that they should be able to sustain a terrorist attack comparable to the September 11 terrorist attacks in the U.S. Spent fuel is usually housed inside the plant's "protected zone" *[33] or a spent nuclear fuel shipping cask; stealing it for use in a "dirty bomb" would be extremely difficult. Exposure to the intense radiation would almost certainly quickly incapacitate or kill anyone who attempts to do so.*[34]

40.3.4 Threat of terrorist attacks

Nuclear power plants are considered to be targets for terrorist attacks.*[35] Even during the construction of the first nuclear power plants, this issue has been advised by security bodies. Concrete threats of attack against nuclear power plants by terrorists or criminals are documented from several states.*[36] While older nuclear power plants were built without special protection against air accidents in Germany, the later nuclear power plants built with a massive concrete buildings are partially protected against air accidents. They are designed against the impact of combat aircraft at a speed of about 800 km / h.*[37] It was assumed as a basis of assessment of the impact of an aircraft of type Phantom II with a mass of 20 tonnes and speed of 215 m / s.*[38]

The dangers arising from a terrorist caused large aircraft crash on a nuclear power plant *[37] is currently being discussed. Such a terrorist attack could have catastrophic consequences.*[39] For example, the German government has confirmed that the nuclear power plant Biblis A not against the crash had secured a military aircraft.*[40] Following the terrorist attacks in Brussels in 2016 several nuclear power plants have been partially evacuated. At the same time it became known that the terrorists had spied on the nuclear power plants. Several employees access privileges has been withdrawn.*[41]

Moreover, even "nuclear terrorism", for instance with a so-called "Dirty bomb" pose a considerable potantial hazard.*[42] For their production would come any radioactive waste or enriched for nuclear power plants uranium in question.*[43]

40.3.5 Plant location

In many countries, plants are often located on the coast, in order to provide a ready source of cooling water for the essential service water system. As a consequence the de-

earthquake map

Fort Calhoun Nuclear Generating Station surrounded by the 2011 Missouri River Floods on June 16, 2011

Angra Nuclear Power Plant in Rio de Janeiro state, Brazil

sign needs to take the risk of flooding and tsunamis into account. The World Energy Council (WEC) argues disaster risks are changing and increasing the likelihood of disasters such as earthquakes, cyclones, hurricanes, typhoons, flooding.[44] High temperatures, low precipitation levels and severe droughts may lead to fresh water shortages.[44] Failure to calculate the risk of flooding correctly lead to

a Level 2 event on the International Nuclear Event Scale during the 1999 Blayais Nuclear Power Plant flood,[45] while flooding caused by the 2011 Tōhoku earthquake and tsunami lead to the Fukushima I nuclear accidents.[46]

The design of plants located in seismically active zones also requires the risk of earthquakes and tsunamis to be taken into account. Japan, India, China and the USA are among the countries to have plants in earthquake-prone regions. Damage caused to Japan's Kashiwazaki-Kariwa Nuclear Power Plant during the 2007 Chūetsu off-shore earthquake[47][48] underlined concerns expressed by experts in Japan prior to the Fukushima accidents, who have warned of a *genpatsu-shinsai* (domino-effect nuclear power plant earthquake disaster).[49]

40.3.6 Multiple reactors

The Fukushima nuclear disaster illustrated the dangers of building multiple nuclear reactor units close to one another. Because of the closeness of the reactors, Plant Director Masao Yoshida "was put in the position of trying to cope simultaneously with core meltdowns at three reactors and exposed fuel pools at three units" .[50]

40.3.7 Nuclear safety systems

Main article: Nuclear safety systems

The three primary objectives of nuclear safety systems as defined by the Nuclear Regulatory Commission are to shut down the reactor, maintain it in a shutdown condition, and prevent the release of radioactive material during events and accidents.[51] These objectives are accomplished using a variety of equipment, which is part of different systems, of which each performs specific functions.

40.3.8 Routine emissions of radioactive materials

For the controversial debate on the health effects by the routine emissions, see Nuclear power debate § Health effects on population near nuclear power plants and workers, and Environmental impact of nuclear power § Risk of cancer.

During everyday routine operations, emissions of radioactive materials from nuclear plants are released to the outside of the plants although they are quite slight amounts.[52][53][54][55] The daily emissions go into the air, water and soil.[53][54]

NRC says, "nuclear power plants sometimes release radioactive gases and liquids into the environment under controlled, monitored conditions to ensure that they pose no danger to the public or the environment" ,[56] and "routine emissions during normal operation of a nuclear power plant are never lethal" .[57]

According to the United Nations (UNSCEAR), regular nuclear power plant operation including the nuclear fuel cycle amounts to 0.0002 millisieverts (mSv) annually in average public radiation exposure; the legacy of the Chernobyl disaster is 0.002 mSv/a as a global average as of a 2008 report; and natural radiation exposure averages 2.4 mSv annually although frequently varying depending on an individual's location from 1 to 13 mSv.[58]

40.3.9 Japanese public perception of nuclear power safety

In March 2012, Prime Minister Yoshihiko Noda said that the Japanese government shared the blame for the Fukushima disaster, saying that officials had been blinded by an image of the country's technological infallibility and were "all too steeped in a safety myth." [59]

Japan has been accused by authors such as journalist Yoichi Funabashi of having an "aversion to facing the potential threat of nuclear emergencies." According to him, a national program to develop robots for use in nuclear emergencies was terminated in midstream because it "smacked too much of underlying danger." Though Japan is a major power in robotics, it had none to send in to Fukushima during the disaster. He mentions that Japan's Nuclear Safety Commission stipulated in its safety guidelines for light-water nuclear facilities that "the potential for extended loss of power need not be considered." However, this kind of extended loss of power to the cooling pumps caused the Fukushima meltdown.[60] In other countries such as the UK, nuclear plants have not been claimed to be absolutely safe. It is instead claimed that a major accident has a likelihood of occurrence lower than (for example) 0.0001/year.

Incidents such as the Fukushima Daiichi nuclear disaster could have been avoided with stricter regulations over nuclear power. In 2002, TEPCO, the company that operated the Fukushima plant, admitted to falsifying reports on over 200 occasions between 1997 and 2002. TEPCO faced no fines for this. Instead, they fired four of their top executives. Three of these four later went on to take jobs at companies that do business with TEPCO.[61]

40.4 Hazards of nuclear material

Main article: High-level radioactive waste management
There is currently a total of 47,000 tonnes of high-level

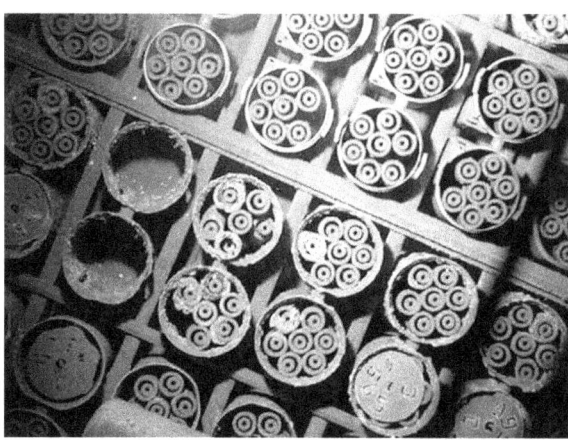

Spent nuclear fuel stored underwater and uncapped at the Hanford site in Washington, USA.

nuclear waste stored in the USA. Nuclear waste is approximately 94% Uranium, 1.3% Plutonium, 0.14% other Actinides, and 5.2% fission products.[62] About 1.0% of this waste consists of long-lived isotopes ^{79}Se, ^{93}Zr, ^{99}Te, ^{107}Pd, ^{126}Sn, ^{129}I and ^{135}Cs. Shorter lived isotopes including ^{89}Sr, ^{90}Sr, ^{106}Ru, ^{125}Sn, ^{134}Cs, ^{137}Cs, and ^{147}Pm constitute 0.9% at one year, decreasing to 0.1% at 100 years. The remaining 3.3-4.1% consists of non-radioactive isotopes.[63][64][65] There are technical challenges, as it is preferable to lock away the long-lived fission products, but the challenge should not be exaggerated. One tonne of waste, as described above, has measurable radioactivity of approximately 600 TBq equal to the natural radioactivity in one km^3 of the Earth's crust, which if buried, would add only 25 parts per trillion to the total radioactivity.

The difference between short-lived high-level nuclear waste and long-lived low-level waste can be illustrated by the following example. As stated above, one mole of both ^{131}I and ^{129}I release $3x10^{23}$ decays in a period equal to one half-life. ^{131}I decays with the release of 970 keV whilst ^{129}I decays with the release of 194 keV of energy. 131gm of ^{131}I would therefore release 45 Gigajoules over eight days beginning at an initial rate of 600 EBq releasing 90 Kilowatts with the last radioactive decay occurring inside two years.[66] In contrast, 129gm of ^{129}I would therefore release 9 Gigajoules over 15.7 million years beginning at an initial rate of 850 MBq releasing 25 microwatts with the radioactivity decreasing by less than 1% in 100,000 years.[67]

One tonne of nuclear waste also reduces CO_2 emission by 25 million tonnes.[62]

[68] Radionuclides such as ^{129}I or ^{131}I, may be highly radioactive, or very long-lived, but they cannot be both. One mole of ^{129}I (129 grams) undergoes the same number of decays (3×10^{23}) in 15.7 million years, as does one mole of ^{131}I (131 grams) in 8 days. ^{131}I is therefore highly radioactive, but disappears very quickly, whilst ^{129}I releases a very low level of radiation for a very long time. Two long-lived fission products, Technetium-99 (half-life 220,000 years) and Iodine-129 (half-life 15.7 million years), are of somewhat greater concern because of a greater chance of entering the biosphere.[69] The transuranic elements in spent fuel are Neptunium-237 (half-life two million years) and Plutonium-239 (half-life 24,000 years).*[70] will also remain in the environment for long periods of time. A more complete solution to both the problem of both Actinides and to the need for low-carbon energy may be the integral fast reactor. One tonne of nuclear waste after a complete burn in an IFR reactor will have prevented 500 million tonnes of CO_2 from entering the atmosphere.*[62] Otherwise, waste storage usually necessitates treatment, followed by a long-term management strategy involving permanent storage, disposal or transformation of the waste into a nontoxic form.*[71]

Governments around the world are considering a range of waste management and disposal options, usually involving deep-geologic placement, although there has been limited progress toward implementing long-term waste management solutions.*[72] This is partly because the timeframes in question when dealing with radioactive waste range from 10,000 to millions of years,*[73]*[74] according to studies based on the effect of estimated radiation doses.*[75]

Since the fraction of a radioisotope's atoms decaying per unit of time is inversely proportional to its half-life, the relative radioactivity of a quantity of buried human radioactive waste would diminish over time compared to natural radioisotopes (such as the decay chain of 120 trillion tons of thorium and 40 trillion tons of uranium which are at relatively trace concentrations of parts per million each over the crust's $3 * 10^{19}$ ton mass).*[76]*[77]*[78] For instance, over a timeframe of thousands of years, after the most active short half-life radioisotopes decayed, burying U.S. nuclear waste would increase the radioactivity in the top 2000 feet of rock and soil in the United States (10 million km^2) by $\approx$ 1 part in 10 million over the cumulative amount of natural radioisotopes in such a volume, although the vicinity of the site would have a far higher concentration of artificial radioisotopes underground than such an average.*[79]

40.5 Safety culture and human errors

One relatively prevalent notion in discussions of nuclear safety is that of safety culture. The International Nuclear Safety Advisory Group, defines the term as "the personal dedication and accountability of all individuals engaged in any activity which has a bearing on the safety of nuclear power plants" .*[80] The goal is "to design systems that use human capabilities in appropriate ways, that protect systems from human frailties, and that protect humans from hazards associated with the system" .*[80]

At the same time, there is some evidence that operational practices are not easy to change. Operators almost never follow instructions and written procedures exactly, and "the violation of rules appears to be quite rational, given the actual workload and timing constraints under which the operators must do their job". Many attempts to improve nuclear safety culture "were compensated by people adapting to the change in an unpredicted way" .*[80]

According to Areva's Southeast Asia and Oceania director, Selena Ng, Japan's Fukushima nuclear disaster is "a huge wake-up call for a nuclear industry that hasn't always been sufficiently transparent about safety issues" . She said "There was a sort of complacency before Fukushima and I don't think we can afford to have that complacency now" .*[81]

An assessment conducted by the *Commissariat à l' Énergie Atomique* (CEA) in France concluded that no amount of technical innovation can eliminate the risk of human-induced errors associated with the operation of nuclear power plants. Two types of mistakes were deemed most serious: errors committed during field operations, such as maintenance and testing, that can cause an accident; and human errors made during small accidents that cascade to complete failure.*[82]

According to Mycle Schneider, reactor safety depends above all on a 'culture of security', including the quality of maintenance and training, the competence of the operator and the workforce, and the rigour of regulatory oversight. So a better-designed, newer reactor is not always a safer one, and older reactors are not necessarily more dangerous than newer ones. The 1979 Three Mile Island accident in the United States occurred in a reactor that had started operation only three months earlier, and the Chernobyl disaster occurred after only two years of operation. A serious loss of coolant occurred at the French Civaux-1 reactor in 1998, less than five months after start-up.*[83]

However safe a plant is designed to be, it is operated by humans who are prone to errors. Laurent Stricker, a nuclear engineer and chairman of the World Association of

Nuclear Operators says that operators must guard against complacency and avoid overconfidence. Experts say that the "largest single internal factor determining the safety of a plant is the culture of security among regulators, operators and the workforce —and creating such a culture is not easy" .*[83]

40.6 Risks

The routine health risks and greenhouse gas emissions from nuclear fission power are small relative to those associated with coal, but there are several "catastrophic risks":*[84]

> The extreme danger of the radioactive material in power plants and of nuclear technology in and of itself is so well known that the US government was prompted (at the industry's urging) to enact provisions that protect the nuclear industry from bearing the full burden of such inherently risky nuclear operations. The Price-Anderson Act limits industry's liability in the case of accidents, and the 1982 Nuclear Waste Policy Act charges the federal government with responsibility for permanently storing nuclear waste.*[85]

Population density is one critical lens through which other risks have to be assessed, says Laurent Stricker, a nuclear engineer and chairman of the World Association of Nuclear Operators:*[83]

> The KANUPP plant in Karachi, Pakistan, has the most people —8.2 million —living within 30 kilometres of a nuclear plant, although it has just one relatively small reactor with an output of 125 megawatts. Next in the league, however, are much larger plants —Taiwan's 1,933-megawatt Kuosheng plant with 5.5 million people within a 30-kilometre radius and the 1,208-megawatt Chin Shan plant with 4.7 million; both zones include the capital city of Taipei.*[83]

172,000 people living within a 30 kilometre radius of the Fukushima Daiichi nuclear power plant, have been forced or advised to evacuate the area. More generally, a 2011 analysis by *Nature* and Columbia University, New York, shows that some 21 nuclear plants have populations larger than 1 million within a 30-km radius, and six plants have populations larger than 3 million within that radius.*[83]

Black Swan events are highly unlikely occurrences that have big repercussions. Despite planning, nuclear power will always be vulnerable to black swan events:*[5]

A rare event – especially one that has never occurred – is difficult to foresee, expensive to plan for and easy to discount with statistics. Just because something is only supposed to happen every 10,000 years does not mean that it will not happen tomorrow.*[5] Over the typical 40-year life of a plant, assumptions can also change, as they did on September 11, 2001, in August 2005 when Hurricane Katrina struck, and in March, 2011, after Fukushima.*[5]

The list of potential black swan events is "damningly diverse":*[5]

> Nuclear reactors and their spent-fuel pools could be targets for terrorists piloting hijacked planes. Reactors may be situated downstream from dams that, should they ever burst, could unleash massive floods. Some reactors are located close to earthquake faults or shorelines, a dangerous scenario like that which emerged at Three Mile Island and Fukushima – a catastrophic coolant failure, the overheating and melting of the radioactive fuel rods, and a release of radioactive material.*[5]

- International Nuclear Events Scale
- Comparative Risk Assessment*[86]
- Statistical Risk Assessment*[87]
- Probabilistic risk assessment
  - *Severe Accident Risks: An Assessment for Five U.S. Nuclear Power Plants* NUREG-1150 1991
  - *Calculation of Reactor Accident Consequences* CRAC-II 1982
  - Rasmussen Report: *Reactor Safety Study* WASH-1400 1975
  - The Brookhaven Report: *Theoretical Possibilities and Consequences of Major Accidents in Large Nuclear Power Plants* WASH-740 1957

The AP1000 has a maximum core damage frequency of $5.09 \times 10^{*}-7$ per plant per year. The Evolutionary Power Reactor (EPR) has a maximum core damage frequency of $4 \times 10^{*}-7$ per plant per year. General Electric has recalculated maximum core damage frequencies per year per plant for its nuclear power plant designs:*[88]

BWR/4 -- $1 \times 10^{*}-5$

BWR/6 -- $1 \times 10^{*}-6$

ABWR -- $2 \times 10^{*}-7$

ESBWR -- $3 \times 10^{*}-8$

40.7 Beyond design basis events

The Fukushima I nuclear accident was caused by a "beyond design basis event," the tsunami and associated earthquakes were more powerful than the plant was designed to accommodate, and the accident is directly due to the tsunami overflowing the too-low seawall.*[2]*[89] Since then, the possibility of unforeseen beyond design basis events has been a major concern for plant operators.*[83]

40.8 Transparency and ethics

According to journalist Stephanie Cooke, it is difficult to know what really goes on inside nuclear power plants because the industry is shrouded in secrecy. Corporations and governments control what information is made available to the public. Cooke says "when information is made available, it is often couched in jargon and incomprehensible prose".*[90]

Kennette Benedict has said that nuclear technology and plant operations continue to lack transparency and to be relatively closed to public view:*[91]

> Despite victories like the creation of the Atomic Energy Commission, and later the Nuclear Regular Commission, the secrecy that began with the Manhattan Project has tended to permeate the civilian nuclear program, as well as the military and defense programs.*[91]

In 1986, Soviet officials held off reporting the Chernobyl disaster for several days. The operators of the Fukushima plant, Tokyo Electric Power Co, were also criticised for not quickly disclosing information on releases of radioactivity from the plant. Russian President Dmitry Medvedev said there must be greater transparency in nuclear emergencies.*[92]

Historically many scientists and engineers have made decisions on behalf of potentially affected populations about whether a particular level of risk and uncertainty is acceptable for them. Many nuclear engineers and scientists that have made such decisions, even for good reasons relating to long term energy availability, now consider that doing so without informed consent is wrong, and that nuclear power safety and nuclear technologies should be based fundamentally on morality, rather than purely on technical, economic and business considerations.*[93]

Non-Nuclear Futures: The Case for an Ethical Energy Strategy is a 1975 book by Amory B. Lovins and John H. Price.*[94]*[95] The main theme of the book is that the most important parts of the nuclear power debate are not

technical disputes but relate to personal values, and are the legitimate province of every citizen, whether technically trained or not.*[96]

40.9 Nuclear and radiation accidents

The nuclear industry has an excellent safety record and the deaths per megawatt hour are the lowest of all the major energy sources.*[97] According to Zia Mian and Alexander Glaser, the "past six decades have shown that nuclear technology does not tolerate error". Nuclear power is perhaps the primary example of what are called 'high-risk technologies' with 'catastrophic potential', because "no matter how effective conventional safety devices are, there is a form of accident that is inevitable, and such accidents are a 'normal' consequence of the system." In short, there is no escape from system failures.*[98]

Whatever position one takes in the nuclear power debate, the possibility of catastrophic accidents and consequent economic costs must be considered when nuclear policy and regulations are being framed.*[99]

40.9.1 Accident liability protection

See also: Price-Anderson Nuclear Industries Indemnity Act

Kristin Shrader-Frechette has said "if reactors were safe, nuclear industries would not demand government-guaranteed, accident-liability protection, as a condition for their generating electricity".*[100] No private insurance company or even consortium of insurance companies "would shoulder the fearsome liabilities arising from severe nuclear accidents".*[101]

40.9.2 Hanford Site

The Hanford Site is a mostly decommissioned nuclear production complex on the Columbia River in the U.S. state of Washington, operated by the United States federal government. Plutonium manufactured at the site was used in the first nuclear bomb, tested at the Trinity site, and in Fat Man, the bomb detonated over Nagasaki, Japan. During the Cold War, the project was expanded to include nine nuclear reactors and five large plutonium processing complexes, which produced plutonium for most of the 60,000 weapons in the U.S. nuclear arsenal.*[102]*[103] Many of the early safety procedures and waste disposal practices were inadequate, and government documents have

The Hanford site represents two-thirds of America's high-level radioactive waste by volume. Nuclear reactors line the riverbank at the Hanford Site along the Columbia River in January 1960.

Map showing Caesium-137 contamination in Belarus, Russia, and Ukraine as of 1996.

since confirmed that Hanford's operations released significant amounts of radioactive materials into the air and the Columbia River, which still threatens the health of residents and ecosystems.[104] The weapons production reactors were decommissioned at the end of the Cold War, but the decades of manufacturing left behind 53 million US gallons ($200{,}000$ m^3) of high-level radioactive waste,[105] an additional 25 million cubic feet ($710{,}000$ m^3) of solid radioactive waste, 200 square miles (520 km^2) of contaminated groundwater beneath the site[106] and occasional discoveries of undocumented contaminations that slow the pace and raise the cost of cleanup.[107] The Hanford site represents two-thirds of the nation's high-level radioactive waste by volume.[108] Today, Hanford is the most contaminated nuclear site in the United States[109][110] and is the focus of the nation's largest environmental cleanup.[102]

40.9.3 1986 Chernobyl disaster

Main articles: Chernobyl disaster and Effects of the Chernobyl disaster

The Chernobyl disaster was a nuclear accident that occurred on 26 April 1986 at the Chernobyl Nuclear Power Plant in Ukraine. An explosion and fire released large quantities of radioactive contamination into the atmosphere, which spread over much of Western USSR and Europe. It is considered the worst nuclear power plant accident in history, and is one of only two classified as a level 7 event on the International Nuclear Event Scale (the other being the Fukushima Daiichi nuclear disaster).[111] The battle to contain the contamination and avert a greater catastrophe ultimately involved over 500,000 workers and cost an estimated 18 billion rubles, crippling the Soviet economy.[112] The accident raised concerns about the safety of the nuclear power industry, slowing its expansion for a number of years.[113]

UNSCEAR has conducted 20 years of detailed scientific and epidemiological research on the effects of the Chernobyl accident. Apart from the 57 direct deaths in the accident itself, UNSCEAR predicted in 2005 that up to 4,000 additional cancer deaths related to the accident would appear "among the 600 000 persons receiving more significant exposures (liquidators working in 1986–87, evacuees, and residents of the most contaminated areas)".[114] Russia, Ukraine, and Belarus have been burdened with the continuing and substantial decontamination and health care costs of the Chernobyl disaster.[115]

Eleven of Russia's reactors are of the RBMK 1000 type, similar to the one at Chernobyl Nuclear Power Plant. Some of these RBMK reactors were originally to be shut down but have instead been given life extensions and uprated in output by about 5%. Critics say that these reactors are of an "inherently unsafe design", which cannot be improved through upgrades and modernization, and some reactor parts are impossible to replace. Russian environmental groups say that the lifetime extensions "violate Russian law, because the projects have not undergone environmental assessments".[116]

40.9.4 2011 Fukushima I accidents

See also: Fukushima I nuclear accidents and Timeline of the Fukushima nuclear accidents

Fukushima reactor control room.

Following the 2011 Japanese Fukushima nuclear disaster, authorities shut down the nation's 54 nuclear power plants. As of 2013, the Fukushima site remains highly radioactive, with some 160,000 evacuees still living in temporary housing, and some land will be unfarmable for centuries. The difficult cleanup job will take 40 or more years, and cost tens of billions of dollars.[117][118]

Despite all assurances, a major nuclear accident on the scale of the 1986 Chernobyl disaster happened again in 2011 in Japan, one of the world's most industrially advanced countries. Nuclear Safety Commission Chairman Haruki Madarame told a parliamentary inquiry in February 2012 that "Japan's atomic safety rules are inferior to global standards and left the country unprepared for the Fukushima nuclear disaster last March". There were flaws in, and lax enforcement of, the safety rules governing Japanese nuclear power companies, and this included insufficient protection against tsunamis.[119]

A 2012 report in *The Economist* said: "The reactors at Fukushima were of an old design. The risks they faced had not been well analysed. The operating company was poorly regulated and did not know what was going on. The operators made mistakes. The representatives of the safety inspectorate fled. Some of the equipment failed. The estab-

lishment repeatedly played down the risks and suppressed information about the movement of the radioactive plume, so some people were evacuated from more lightly to more heavily contaminated places".[120]

The designers of the Fukushima I Nuclear Power Plant reactors did not anticipate that a tsunami generated by an earthquake would disable the backup systems that were supposed to stabilize the reactor after the earthquake.[2] Nuclear reactors are such "inherently complex, tightly coupled systems that, in rare, emergency situations, cascading interactions will unfold very rapidly in such a way that human operators will be unable to predict and master them".[3]

Lacking electricity to pump water needed to cool the atomic core, engineers vented radioactive steam into the atmosphere to release pressure, leading to a series of explosions that blew out concrete walls around the reactors. Radiation readings spiked around Fukushima as the disaster widened, forcing the evacuation of 200,000 people. There was a rise in radiation levels on the outskirts of Tokyo, with a population of 30 million, 135 miles (210 kilometers) to the south.[46]

Back-up diesel generators that might have averted the disaster were positioned in a basement, where they were quickly overwhelmed by waves. The cascade of events at Fukushima had been predicted in a report published in the U.S. several decades ago:[46]

The 1990 report by the U.S. Nuclear Regulatory Commission, an independent agency responsible for safety at the country's power plants, identified earthquake-induced diesel generator failure and power outage leading to failure of cooling systems as one of the "most likely causes" of nuclear accidents from an external event.[46]

The report was cited in a 2004 statement by Japan's Nuclear and Industrial Safety Agency, but it seems adequate measures to address the risk were not taken by TEPCO. Katsuhiko Ishibashi, a seismology professor at Kobe University, has said that Japan's history of nuclear accidents stems from an overconfidence in plant engineering. In 2006, he resigned from a government panel on nuclear reactor safety, because the review process was rigged and "unscientific".[46]

According to the International Atomic Energy Agency, Japan "underestimated the danger of tsunamis and failed to prepare adequate backup systems at the Fukushima Daiichi nuclear plant". This repeated a widely held criticism in

Japan that "collusive ties between regulators and industry led to weak oversight and a failure to ensure adequate safety levels at the plant" .*[118] The IAEA also said that the Fukushima disaster exposed the lack of adequate backup systems at the plant. Once power was completely lost, critical functions like the cooling system shut down. Three of the reactors "quickly overheated, causing meltdowns that eventually led to explosions, which hurled large amounts of radioactive material into the air" .*[118]

Louise Fréchette and Trevor Findlay have said that more effort is needed to ensure nuclear safety and improve responses to accidents:

> The multiple reactor crises at Japan's Fukushima nuclear power plant reinforce the need for strengthening global instruments to ensure nuclear safety worldwide. The fact that a country that has been operating nuclear power reactors for decades should prove so alarmingly improvisational in its response and so unwilling to reveal the facts even to its own people, much less the International Atomic Energy Agency, is a reminder that nuclear safety is a constant work-in-progress. *[121]

David Lochbaum, chief nuclear safety officer with the Union of Concerned Scientists, has repeatedly questioned the safety of the Fukushima I Plant's General Electric Mark 1 reactor design, which is used in almost a quarter of the United States' nuclear fleet.*[122]

A report from the Japanese Government to the IAEA says the "nuclear fuel in three reactors probably melted through the inner containment vessels, not just the core". The report says the "inadequate" basic reactor design —the Mark-1 model developed by General Electric —included "the venting system for the containment vessels and the location of spent fuel cooling pools high in the buildings, which resulted in leaks of radioactive water that hampered repair work" .*[123]

Following the Fukushima emergency, the European Union decided that reactors across all 27 member nations should undergo safety tests.*[124]

According to UBS AG, the Fukushima I nuclear accidents are likely to hurt the nuclear power industry' s credibility more than the Chernobyl disaster in 1986:

> The accident in the former Soviet Union 25 years ago 'affected one reactor in a totalitarian state with no safety culture,' UBS analysts including Per Lekander and Stephen Oldfield wrote in a report today. 'At Fukushima, four reactors have been out of control for weeks -- casting doubt on

whether even an advanced economy can master nuclear safety.'*[125]

The Fukushima accident exposed some troubling nuclear safety issues:*[126]

> Despite the resources poured into analyzing crustal movements and having expert committees determine earthquake risk, for instance, researchers never considered the possibility of a magnitude-9 earthquake followed by a massive tsunami. The failure of multiple safety features on nuclear power plants has raised questions about the nation's engineering prowess. Government flip-flopping on acceptable levels of radiation exposure confused the public, and health professionals provided little guidance. Facing a dearth of reliable information on radiation levels, citizens armed themselves with dosimeters, pooled data, and together produced radiological contamination maps far more detailed than anything the government or official scientific sources ever provided.*[126]

As of January 2012, questions also linger as to the extent of damage to the Fukushima plant caused by the earthquake even before the tsunami hit. Any evidence of serious quake damage at the plant would "cast new doubt on the safety of other reactors in quake-prone Japan" .*[127]

Two government advisers have said that "Japan's safety review of nuclear reactors after the Fukushima disaster is based on faulty criteria and many people involved have conflicts of interest" . Hiromitsu Ino, Professor Emeritus at the University of Tokyo, says "The whole process being undertaken is exactly the same as that used previous to the Fukushima Dai-Ichi accident, even though the accident showed all these guidelines and categories to be insufficient" .*[128]

In March 2012, Prime Minister Yoshihiko Noda acknowledged that the Japanese government shared the blame for the Fukushima disaster, saying that officials had been blinded by a false belief in the country's "technological infallibility" , and were all too steeped in a "safety myth" .*[129]

40.9.5 Other accidents

See also: List of civilian nuclear accidents, List of civilian radiation accidents, and List of military nuclear accidents

Serious nuclear and radiation accidents include the Chalk River accidents (1952, 1958 & 2008), Mayak disaster

(1957), Windscale fire (1957), SL-1 accident (1961), Soviet submarine K-19 accident (1961), Three Mile Island accident (1979), Church Rock uranium mill spill (1979), Soviet submarine K-431 accident (1985), Goiânia accident (1987), Zaragoza radiotherapy accident (1990), Costa Rica radiotherapy accident (1996), Tokaimura nuclear accident (1999), Sellafield THORP leak (2005), and the Flerus IRE cobalt-60 spill (2006).*[130]*[131]

40.10   Health impacts

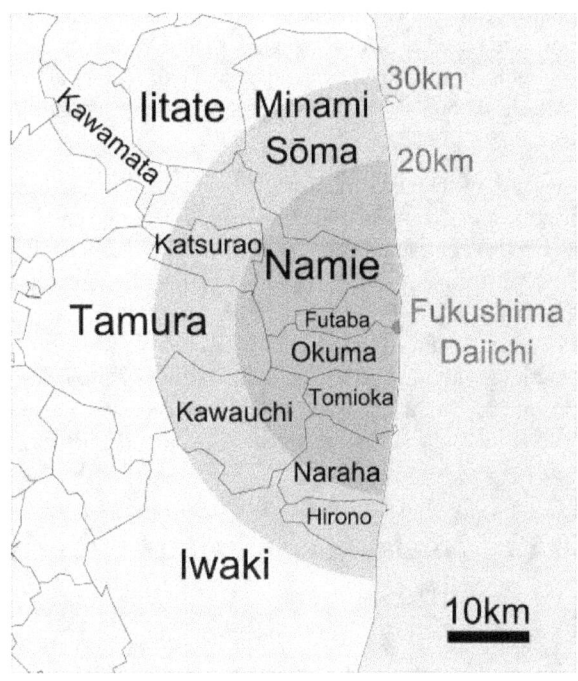

Japan towns, villages, and cities around the Fukushima Daiichi nuclear plant. The 20km and 30km areas had evacuation and sheltering orders, and additional administrative districts that had an evacuation order are highlighted.

See also: Effects of the Chernobyl disaster and Nuclear power debate

Four hundred and thirty-seven nuclear power stations are presently in operation but, unfortunately, five major nuclear accidents have occurred in the past. These accidents occurred at Kyshtym (1957), Windscale (1957), Three Mile Island (1979), Chernobyl (1986), and Fukushima (2011). A report in *Lancet* says that the effects of these accidents on individuals and societies are diverse and enduring:*[132]

"Accumulated evidence about radiation health effects on atomic bomb survivors and other radiation-exposed people has formed the basis

for national and international regulations about radiation protection. However, past experiences suggest that common issues were not necessarily physical health problems directly attributable to radiation exposure, but rather psychological and social effects. Additionally, evacuation and long-term displacement created severe healthcare problems for the most vulnerable people, such as hospital inpatients and elderly people."*[132]

In spite of accidents like these, studies have shown that nuclear deaths are mostly in uranium mining and that nuclear energy has generated far fewer deaths than the high pollution levels that result from the use of conventional fossil fuels.*[133] However, the nuclear power industry relies on uranium mining, which itself is a hazardous industry, with many accidents and fatalities.*[134]

Journalist Stephanie Cooke says that it is not useful to make comparisons just in terms of number of deaths, as the way people live afterwards is also relevant, as in the case of the 2011 Japanese nuclear accidents:*[135]

"You have people in Japan right now that are facing either not returning to their homes forever, or if they do return to their homes, living in a contaminated area for basically ever... It affects millions of people, it affects our land, it affects our atmosphere ... it's affecting future generations ... I don't think any of these great big massive plants that spew pollution into the air are good. But I don't think it's really helpful to make these comparisons just in terms of number of deaths".*[135]

The Fukushima accident forced more than 80,000 residents to evacuate from neighborhoods around the plant.*[123]

A survey by the Iitate, Fukushima local government obtained responses from some 1,743 people who have evacuated from the village, which lies within the emergency evacuation zone around the crippled Fukushima Daiichi Plant. It shows that many residents are experiencing growing frustration and instability due to the nuclear crisis and an inability to return to the lives they were living before the disaster. Sixty percent of respondents stated that their health and the health of their families had deteriorated after evacuating, while 39.9 percent reported feeling more irritated compared to before the disaster.*[136]

"Summarizing all responses to questions related to evacuees' current family status, one-third of all surveyed families live apart from their children, while 50.1 percent live away from other family members (including elderly parents) with whom they lived before the disaster. The survey also showed that 34.7 percent of the evacuees have suffered salary cuts of 50 percent or more since the outbreak of the nuclear disaster. A total of 36.8 percent reported a lack of sleep, while 17.9 percent reported smoking or drinking more than before they evacuated." *[136]

Chemical components of the radioactive waste may lead to cancer. For example, Iodine 131 was released along with the radioactive waste when Chernobyl disaster and Fukushima disasters occurred. It was concentrated in leafy vegetation after absorption in the soil. It also stays in animals' milk if the animals eat the vegetation. When Iodine 131 enters the human body, it migrates to the thyroid gland in the neck and can cause thyroid cancer.*[137]

Other elements from nuclear waste can lead to cancer as well. For example, Strontium 90 causes breast cancer and leukemia, Plutonium 239 causes liver cancer.*[138]

40.11 Improvements to nuclear fission technologies

Newer reactor designs intended to provide increased safety have been developed over time. These designs include those that incorporate passive safety and Small Modular Reactors. While these reactor designs "are intended to inspire trust, they may have an unintended effect: creating distrust of older reactors that lack the touted safety features".*[139]

The next nuclear plants to be built will likely be Generation III or III+ designs, and a few such are already in operation in Japan. Generation IV reactors would have even greater improvements in safety. These new designs are expected to be passively safe or nearly so, and perhaps even inherently safe (as in the PBMR designs).

Some improvements made (not all in all designs) are having three sets of emergency diesel generators and associated emergency core cooling systems rather than just one pair, having quench tanks (large coolant-filled tanks) above the core that open into it automatically, having a double containment (one containment building inside another), etc.

However, safety risks may be the greatest when nuclear systems are the newest, and operators have less experience with them. Nuclear engineer David Lochbaum explained that almost all serious nuclear accidents occurred with what was at the time the most recent technology. He argues that "the problem with new reactors and accidents is twofold: scenarios arise that are impossible to plan for in simulations; and humans make mistakes" .*[82] As one director of a U.S.

research laboratory put it, "fabrication, construction, operation, and maintenance of new reactors will face a steep learning curve: advanced technologies will have a heightened risk of accidents and mistakes. The technology may be proven, but people are not" .*[82]

40.12 Developing countries

There are concerns about developing countries "rushing to join the so-called nuclear renaissance without the necessary infrastructure, personnel, regulatory frameworks and safety culture" .*[121] Some countries with nuclear aspirations, like Nigeria, Kenya, Bangladesh and Venezuela, have no significant industrial experience and will require at least a decade of preparation even before breaking ground at a reactor site.*[121]

The speed of the nuclear construction program in China has raised safety concerns. The challenge for the government and nuclear companies is to "keep an eye on a growing army of contractors and subcontractors who may be tempted to cut corners" .*[140] China is advised to maintain nuclear safeguards in a business culture where quality and safety are sometimes sacrificed in favor of cost-cutting, profits, and corruption. China has asked for international assistance in training more nuclear power plant inspectors.*[140]

40.13 Nuclear security and terrorist attacks

Main article: Vulnerability of nuclear plants to attack

Nuclear power plants, civilian research reactors, certain naval fuel facilities, uranium enrichment plants, and fuel fabrication plants, are vulnerable to attacks which could lead to widespread radioactive contamination. The attack threat is of several general types: commando-like ground-based attacks on equipment which if disabled could lead to a reactor core meltdown or widespread dispersal of radioactivity; and external attacks such as an aircraft crash into a reactor complex, or cyber attacks.*[141]

The United States 9/11 Commission has said that nuclear power plants were potential targets originally considered for the September 11, 2001 attacks. If terrorist groups could sufficiently damage safety systems to cause a core meltdown at a nuclear power plant, and/or sufficiently damage spent fuel pools, such an attack could lead to widespread radioactive contamination. The Federation of American Scientists have said that if nuclear power use is to expand significantly, nuclear facilities will have to be made extremely safe from attacks that could release massive quantities of radioactivity into the community. New reactor designs have features of passive safety, which may help. In the United States, the NRC carries out "Force on Force" (FOF) exercises at all Nuclear Power Plant (NPP) sites at least once every three years.[*][141]

Nuclear reactors become preferred targets during military conflict and, over the past three decades, have been repeatedly attacked during military air strikes, occupations, invasions and campaigns.[*][28] Various acts of civil disobedience since 1980 by the peace group Plowshares have shown how nuclear weapons facilities can be penetrated, and the groups actions represent extraordinary breaches of security at nuclear weapons plants in the United States. The National Nuclear Security Administration has acknowledged the seriousness of the 2012 Plowshares action. Non-proliferation policy experts have questioned "the use of private contractors to provide security at facilities that manufacture and store the government's most dangerous military material".[*][142] Nuclear weapons materials on the black market are a global concern,[*][143][*][144] and there is concern about the possible detonation of a small, crude nuclear weapon by a militant group in a major city, with significant loss of life and property.[*][145][*][146] *Stuxnet* is a computer worm discovered in June 2010 that is believed to have been created by the United States and Israel to attack Iran's nuclear facilities.[*][147]

40.14   Nuclear fusion research

Main articles: Fusion power and Fusion power § Safety and the environment

Nuclear fusion power is a developing technology still under research. It relies on fusing rather than fissioning (splitting) atomic nuclei, using very different processes compared to current nuclear power plants. Nuclear fusion reactions have the potential to be safer and generate less radioactive waste than fission.[*][148][*][149] These reactions appear potentially viable, though technically quite difficult and have yet to be created on a scale that could be used in a functional power plant. Fusion power has been under theoretical and experimental investigation since the 1950s.

Construction of the International Thermonuclear Experimental Reactor facility began in 2007, but the project has run into many delays and budget overruns. The facility is now not expected to begin operations until the year 2027 – 11 years after initially anticipated.[*][150] A follow on commercial nuclear fusion power station, DEMO, has been proposed.[*][151][*][152] There is also suggestions for a power plant based upon a different fusion approach, that of a Inertial fusion power plant.

Fusion powered electricity generation was initially believed to be readily achievable, as fission power had been. However, the extreme requirements for continuous reactions and plasma containment led to projections being extended by several decades. In 2010, more than 60 years after the first attempts, commercial power production was still believed to be unlikely before 2050.[*][151]

40.15   More stringent safety standards

Matthew Bunn, the former US Office of Science and Technology Policy adviser, and Heinonen, the former Deputy Director General of the IAEA, have said that there is a need for more stringent nuclear safety standards, and propose six major areas for improvement:[*][99]

- operators must plan for events beyond design bases;
- more stringent standards for protecting nuclear facilities against terrorist sabotage;
- a stronger international emergency response;
- international reviews of security and safety;
- binding international standards on safety and security; and
- international co-operation to ensure regulatory effectiveness.

Coastal nuclear sites must also be further protected against rising sea levels, storm surges, flooding, and possible eventual "nuclear site islanding".[*][99]

40.16   See also

- Lists of nuclear disasters and radioactive incidents
- Deep geological repository
- Design basis accident
- Environmental impact of nuclear power
- International Nuclear Events Scale
- *Journey to the Safest Place on Earth*
- Nuclear 9/11
- Nuclear accidents in the United States

- Nuclear criticality safety

- RELAP5-3D A reactor design and simulation tool to prevent accidents.

- Nuclear fuel response to reactor accidents

- Nuclear holocaust

- Nuclear power debate

- Nuclear power plant emergency response team

- Nuclear whistleblowers

- Nuclear weapon

- Micro nuclear reactor

- Passive nuclear safety

- Yucca Mountain nuclear waste repository

- Safety code (nuclear reactor)

- World Association of Nuclear Operators

40.17 References

[1] IAEA safety Glossary - Version 2.0 September 2006

[2] Phillip Lipscy, Kenji Kushida, and Trevor Incerti. 2013. "The Fukushima Disaster and Japan's Nuclear Plant Vulnerability in Comparative Perspective." *Environmental Science and Technology* 47 (May), 6082–6088.

[3] Hugh Gusterson (16 March 2011). "The lessons of Fukushima". *Bulletin of the Atomic Scientists.*

[4] Diaz Maurin, François (26 March 2011). "Fukushima: Consequences of Systemic Problems in Nuclear Plant Design". *Economic & Political Weekly.* **46** (13): 10–12.

[5] Adam Piore (June 2011). "Nuclear energy: Planning for the Black Swan p.32". *Scientific American.* Retrieved 2014-05-15.

[6] James Paton (April 4, 2011). "Fukushima Crisis Worse for Atomic Power Than Chernobyl, UBS Says". *Bloomberg Businessweek.*

[7] Jacobson, Mark Z. & Delucchi, Mark A. (2010). "Providing all Global Energy with Wind, Water, and Solar Power, Part I: Technologies, Energy Resources, Quantities and Areas of Infrastructure, and Materials" (PDF). *Energy Policy.* p. 6.

[8] Daniel E Whitney (2003). "Normal Accidents by Charles Perrow" (PDF). *Massachusetts Institute of Technology.*

[9] "Nuclear Terrorism: Frequently Asked Questions". Belfer Center for Science and International Affairs. September 26, 2007.

[10] Vienna International Centre (March 30, 2011). "About IAEA: The "Atoms for Peace" Agency". *iaea.org.*

[11] Stephen Kurczy (March 17, 2011). "Japan nuclear crisis sparks calls for IAEA reform". *CSMonitor.com.*

[12] IAEA Convention on Nuclear Safety

[13] Vienna Declaration on Nuclear Safety

[14] *The Doomsday Machine*, by Martin Cohen and Andrew Mckillop, Palgrave 2012, page 74

[15] *The Doomsday Machine*, by Martin Cohen and Andrew Mckillop, Palgrave 2012, page 72

[16] About NRC, U.S. Nuclear Regulatory Commission, Retrieved 2007-06-01.

[17] Our Governing Legislation, U.S. Nuclear Regulatory Commission, Retrieved 2007-06-01.

[18] Health and Safety http://www.australia.gov.au

[19] Radiation Protection http://www.arpansa.gov.au

[20] Jan Willem Storm van Leeuwen (2008). Nuclear power – the energy balance

[21] Stephanie Cooke (2009). *In Mortal Hands: A Cautionary History of the Nuclear Age*, Black Inc., p. 280.

[22] Perrow, C. (1982), 'The President's Commission and the Normal Accident', in Sils, D., Wolf, C. and Shelanski, V. (Eds), *Accident at Three Mile Island: The Human Dimensions*, Westview, Boulder, pp.173–184.

[23] Pidgeon, N. (2011). "In retrospect: Normal Accidents". *Nature.* **477** (7365): 404–405. doi:10.1038/477404a.

[24] Union of Concerned Scientists: Nuclear safety

[25] Globalsecurity.org: *Nuclear Power Plants: Vulnerability to Terrorist Attack* p. 3.

[26] Safety of Nuclear Power Reactors, World Nuclear Association, http://www.world-nuclear.org/info/inf06.html

[27] Matthew Wald (June 15, 2011). "U.S. Reactors Unprepared for Total Power Loss, Report Suggests". *New York Times.*

[28] Benjamin K. Sovacool (2011). *Contesting the Future of Nuclear Power: A Critical Global Assessment of Atomic Energy*, World Scientific, p. 192.

[29] U.S. NRC: "Nuclear Security – Five Years After 9/11". Accessed 23 July 2007

[30] Threat Assessment: U.S. Nuclear Plants Near Airports May Be at Risk of Airplane Attack, *Global Security Newswire*, June 11, 2003.

[31] Newtan, Samuel Upton (2007). *Nuclear War 1 and Other Major Nuclear Disasters of the 20th Century*, AuthorHouse, p.146.

[32] "STATEMENT FROM CHAIRMAN DALE KLEIN ON COMMISSION'S AFFIRMATION OF THE FINAL DBT RULE". Nuclear Regulatory Commission. Retrieved 2007-04-07.

[33] "The Nuclear Fuel Cycle". *Information and Issue Briefs.* World Nuclear Association. 2005. Retrieved 2006-11-10.

[34] Lewis Z Koch (2004). "Dirty Bomber? Dirty Justice". Bulletin of the Atomic Scientists. Retrieved 2006-11-10.

[35] Julia Mareike Neles, Christoph Pistner (Hrsg.), *Kernenergie. Eine Technik für die Zukunft?*, Berlin – Heidelberg 2012, S. 114 f.

[36] Julia Mareike Neles, Christoph Pistner (Hrsg.), *Kernenergie. Eine Technik für die Zukunft?*, Berlin – Heidelberg 2012, S. 114 f.

[37] Julia Mareike Neles, Christoph Pistner (Hrsg.), *Kernenergie. Eine Technikkk für die Zukunft?*, Berlin – Heidelberg 2012, S. 115.

[38] Manfred Grathwohl, *Energieversorgung*, Berlin – New York 1983, S. 429.

[39] Terroranschlag auf Atomkraftwerk Biblis würde Berlin bedrohen. In: Der Spiegel

[40] In: Der Spiegel: Biblis nicht gegen Flugzeugabsturz geschützt

[41] Tihange-Mitarbeiter gesperrt, Terroristen spähen Wissenschaftler aus, Aachener Zeitung, 24.3.2016

[42] Wolf-Georg Schärf, *Europäisches Atomrecht. Recht der Nuklearenergie* Berlin – Boston 2012, S. 1.

[43] spiegel.de: Experten warnen vor neuen Terrorgefahren durch Atom-Comeback

[44] Dr. Frauke Urban and Dr. Tom Mitchell 2011. Climate change, disasters and electricity generation Archived September 20, 2012, at the Wayback Machine.. London: Overseas Development Institute and Institute of Development Studies

[45] COMMUNIQUE N°7 - INCIDENT SUR LE SITE DU BLAYAIS Archived May 27, 2013, at the Wayback Machine. ASN, published 1999-12-30, accessed 2011-03-22

[46] Jason Clenfield (March 17, 2011). "Japan Nuclear Disaster Caps Decades of Faked Reports, Accidents". *Bloomberg Businessweek.*

[47] ABC News. Strong Quake Rocks Northwestern Japan. July 16, 2007.

[48] Xinhua News. Two die, over 200 injured in strong quake in Japan. July 16, 2007.

[49] Genpatsu-Shinsai: Catastrophic Multiple Disaster of Earthquake and Quake-induced Nuclear Accident Anticipated in the Japanese Islands (Abstract), Katsuhiko Ishibashi, 23rd. General Assembly of IUGG, 2003, Sapporo, Japan, accessed 2011-03-28

[50] Yoichi Funabashi and Kay Kitazawa (March 1, 2012). "Fukushima in review: A complex disaster, a disastrous response" (PDF). *Bulletin of the Atomic Scientists.*

[51] "Glossary: Safety-related". Retrieved 2011-03-20.

[52] "What you can do to protect yourself: Be Informed". *Nuclear Power Plants | RadTown USA | US EPA.* United States Environmental Protection Agency. Retrieved March 12, 2012.

[53] Nuclear Information and Resource Service (NIRS): ROUTINE RADIOACTIVE RELEASES FROM NUCLEAR REACTORS - IT DOESN'T TAKE AN ACCIDENT at the Wayback Machine (archived May 14, 2011)

[54] "Nuclear Power: During normal operations, do commercial nuclear power plants release radioactive material?". *Radiation and Nuclear Power | Radiation Information and Answers.* Radiation Answers. Retrieved March 12, 2012.

[55] "Radiation Dose". *Factsheets & FAQs: Radiation in Everyday Life.* International Atomic Energy Agency (IAEA). Retrieved March 12, 2012.

[56] "What happens to radiation produced by a plant?". *NRC: Frequently Asked Questions (FAQ) About Radiation Protection.* Nuclear Regulatory Commission. Retrieved March 12, 2012.

[57] "Is radiation exposure from a nuclear power plant always fatal?". *NRC: Frequently Asked Questions (FAQ) About Radiation Protection.* Nuclear Regulatory Commission. Retrieved March 12, 2012.

[58] "UNSCEAR 2008 Report to the General Assembly" (PDF). United Nations Scientific Committee on the Effects of Atomic Radiation. 2008.

[59] Hiroko Tabuchi (March 3, 2012). "Japanese Prime Minister Says Government Shares Blame for Nuclear Disaster". *The New York Times.* Retrieved 2012-04-13.

[60] Yoichi Funabashi (March 11, 2012). "The End of Japanese Illusions". *New York Times.* Retrieved 2012-04-13.

[61] Wang, Qiang, Xi Chen, and Xu Yi-Chong. "Accident like the Fukushima Unlikely in a Country with Effective Nuclear Regulation: Literature Review and Proposed Guidelines." Renewable and Sustainable Energy Reviews 16.1 (2012): 126-46. Web. 3 July 2016. <http://www.egi.ac.cn/xwzx/kydt/201211/W020121101676826557345.pdf>.

[62] "What is Nuclear Waste?". What is Nuclear?.

[63] "Fission 235U". US Nuclear Data Program.

[64] "Fission 233U" . US Nuclear Data Program.

[65] "Fission 239Pu" . US Nuclear Data Program.

[66] "131I" . US Nuclear Data Program.

[67] "129I" . US Nuclear Data Program.

[68] "Natural Radioactivity" . Idaho State University.

[69] "Environmental Surveillance, Education and Research Program" . Idaho National Laboratory. Archived from the original on 2008-11-21. Retrieved 2009-01-05.

[70] Vandenbosch 2007, p. 21.

[71] Ojovan, M. I.; Lee, W.E. (2005). *An Introduction to Nuclear Waste Immobilisation.* Amsterdam: Elsevier Science Publishers. p. 315. ISBN 0-08-044462-8.

[72] Brown, Paul (2004-04-14). "Shoot it at the sun. Send it to Earth's core. What to do with nuclear waste?". *The Guardian.* London.

[73] National Research Council (1995). *Technical Bases for Yucca Mountain Standards.* Washington, D.C.: National Academy Press. p. 91. ISBN 0-309-05289-0.

[74] "The Status of Nuclear Waste Disposal" . The American Physical Society. January 2006. Retrieved 2008-06-06.

[75] "Public Health and Environmental Radiation Protection Standards for Yucca Mountain, Nevada; Proposed Rule" (PDF). United States Environmental Protection Agency. 2005-08-22. Retrieved 2008-06-06.

[76] Sevior M. (2006). "Considerations for nuclear power in Australia" (PDF). *International Journal of Environmental Studies.* **63** (6): 859–872. doi:10.1080/00207230601047255.

[77] Thorium Resources In Rare Earth Elements

[78] American Geophysical Union, Fall Meeting 2007, abstract #V33A-1161. Mass and Composition of the Continental Crust

[79] Interdisciplinary Science Reviews 23:193-203;1998. Dr. Bernard L. Cohen, University of Pittsburgh. Perspectives on the High Level Waste Disposal Problem

[80] M.V. Ramana. Nuclear Power: Economic, Safety, Health, and Environmental Issues of Near-Term Technologies, *Annual Review of Environment and Resources*, 2009. 34, pp.139-140.

[81] David Fickling (April 20, 2011). "Areva Says Fukushima A Huge Wake-Up Call For Nuclear Industry" . *Fox Business.*

[82] Benjamin K. Sovacool. A Critical Evaluation of Nuclear Power and Renewable Electricity in Asia, *Journal of Contemporary Asia*, Vol. 40, No. 3, August 2010, p. 381.

[83] Declan Butler (21 April 2011). "Reactors, residents and risk" . *Nature.*

[84] International Panel on Fissile Materials (September 2010). "The Uncertain Future of Nuclear Energy" (PDF). *Research Report 9.* p. 1.

[85] Kennette Benedict (13 October 2011). "The banality of death by nuclear power" . *Bulletin of the Atomic Scientists.*

[86] "Paul Scherrer Institut (PSI) :: Severe Accidents in the Energy Sector (see pages 287,310,317)" (PDF). gabe.web.psi.ch. Retrieved 2015-02-07.

[87] Hofert, Wüthrich (2011) Statistical Review of Nuclear Power Accidents

[88] "Next-generation nuclear energy: The ESBWR" (PDF). ans.org. Retrieved 2015-02-07.

[89] "Genesis of a disaster: Moment tsunami swamps Japan's doomed Fukushima nuclear plant" . *Daily Mail.* London. 2011-05-19.

[90] Stephanie Cooke (March 19, 2011). "Nuclear power is on trial" . *CNN.com.*

[91] Kennette Benedict (26 March 2011). "The road not taken: Can Fukushima put us on a path toward nuclear transparency?". *Bulletin of the Atomic Scientists.*

[92] "Anti-nuclear protests in Germany and France". *BBC News.* 25 April 2011.

[93] Pandora's box, A is for Atom- Adam Curtis

[94] Lovins, Amory B. and Price, John H. (1975). *Non-nuclear Futures: The Case for an Ethical Energy Strategy* (Cambridge, Mass.: Ballinger Publishing Company, 1975. xxxii + 223pp. ISBN 0-88410-602-0, ISBN 0-88410-603-9).

[95] Weinberg, Alvin M. (December 1976). "Book review. Non-nuclear futures: the case for an ethical energy strategy" . *Energy Policy.* Elsevier Science Ltd. **4** (4): 363–366. doi:10.1016/0301-4215(76)90031-8. ISSN 0301-4215.

[96] *Non-Nuclear Futures*, pp. xix-xxi.

[97] Brian Wang (16 March 2011). "Deaths from electricity generation" .

[98] Zia Mian & Alexander Glaser (June 2006). "Life in a Nuclear Powered Crowd" (PDF). *INESAP Information Bulletin No.26.*

[99] European Environment Agency (Jan 23, 2013). "Late lessons from early warnings: science, precaution, innovation: Full report" . p. 28.

[100] Kristin Shrader-Frechette (19 August 2011). "Cheaper, safer alternatives than nuclear fission". *Bulletin of the Atomic Scientists.* Archived from the original on 2012-01-21.

[101] Arjun Makhijani (21 July 2011). "The Fukushima tragedy demonstrates that nuclear energy doesn't make sense" . *Bulletin of the Atomic Scientists.* Archived from the original on 2012-01-21.

[102] "Hanford Site: Hanford Overview". United States Department of Energy. Archived from the original on 2012-06-05. Retrieved 2012-02-13.

[103] "Science Watch: Growing Nuclear Arsenal". *The New York Times*. April 28, 1987. Retrieved 2007-01-29.

[104] "An Overview of Hanford and Radiation Health Effects". Hanford Health Information Network. Archived from the original on 2010-01-06. Retrieved 2007-01-29.

[105] "Hanford Quick Facts". Washington Department of Ecology. Archived from the original on 2008-06-24. Retrieved 2010-01-19.

[106] "Hanford Facts". psr.org. Retrieved 2015-02-07.

[107] Stang, John (December 21, 2010). "Spike in radioactivity a setback for Hanford cleanup". *Seattle Post-Intelligencer*.

[108] Harden, Blaine; Dan Morgan (June 2, 2007). "Debate Intensifies on Nuclear Waste". *Washington Post*. p. A02. Retrieved 2007-01-29.

[109] Dininny, Shannon (April 3, 2007). "U.S. to Assess the Harm from Hanford". *Seattle Post-Intelligencer*. Associated Press. Retrieved 2007-01-29.

[110] Schneider, Keith (February 28, 1989). "Agreement for a Cleanup at Nuclear Site". *The New York Times*. Retrieved 2008-01-30.

[111] Black, Richard (2011-04-12). ""Fukushima: As Bad as Chernobyl?"". Bbc.co.uk. Retrieved 2011-08-20.

[112] From interviews with Mikhail Gorbachev, Hans Blix and Vassili Nesterenko. *The Battle of Chernobyl*. Discovery Channel. Relevant video locations: 31:00, 1:10:00.

[113] Kagarlitsky, Boris (1989). "Perestroika: The Dialectic of Change". In Mary Kaldor; Gerald Holden; Richard A. Falk. *The New Detente: Rethinking East-West Relations*. United Nations University Press. ISBN 0-86091-962-5.

[114] "IAEA Report". *In Focus: Chernobyl*. International Atomic Energy Agency. Archived from the original on 2007-12-17. Retrieved 2006-03-29.

[115] Hallenbeck, William H (1994). *Radiation Protection*. CRC Press. p. 15. ISBN 0-87371-996-4. Reported thus far are 237 cases of acute radiation sickness and 31 deaths.

[116] Igor Koudrik & Alexander Nikitin (13 December 2011). "Second life: The questionable safety of life extensions for Russian nuclear power plants". *Bulletin of the Atomic Scientists*.

[117] Richard Schiffman (12 March 2013). "Two years on, America hasn't learned lessons of Fukushima nuclear disaster". *The Guardian*. London.

[118] Martin Fackler (June 1, 2011). "Report Finds Japan Underestimated Tsunami Danger". *New York Times*.

[119] "Nuclear Safety Chief Says Lax Rules Led to Fukushima Crisis". *Bloomberg*. 16 February 2012.

[120] "Blow-ups happen: Nuclear plants can be kept safe only by constantly worrying about their dangers". *The Economist*. 10 March 2012.

[121] Louise Fréchette & Trevor Findlay (March 28, 2011). "Nuclear safety is the world's problem". *Ottawa Citizen*.

[122] Hannah Northey (March 28, 2011). "Japanese Nuclear Reactors, U.S. Safety to Take Center Stage on Capitol Hill This Week". *New York Times*.

[123] "Japan says it was unprepared for post-quake nuclear disaster". *Los Angeles Times*. June 8, 2011. Archived from the original on June 8, 2011.

[124] James Kanter (March 25, 2011). "Europe to Test Safety of Nuclear Reactors". *New York Times*.

[125] James Paton (April 4, 2011). "Fukushima Crisis Worse for Atomic Power Than Chernobyl, UBS Says". *Bloomberg Businessweek*. Archived from the original on 2011-05-15.

[126] Dennis Normile (28 November 2011). "In Wake of Fukushima Disaster, Japan's Scientists Ponder How to Regain Public Trust". *Science*.

[127] Hiroko Tabuchi (January 15, 2012). "Panel Challenges Japan's Account of Nuclear Disaster". *New York Times*.

[128] "Japan Post-Fukushima Reactor Checks 'Insufficient,' Advisers Say". *Businessweek*. January 27, 2012.

[129] Hiroko Tabuchi (March 3, 2012). "Japanese Prime Minister Says Government Shares Blame for Nuclear Disaster". *The New York Times*.

[130] Newtan, Samuel Upton (2007). *Nuclear War 1 and Other Major Nuclear Disasters of the 20th Century*, AuthorHouse.

[131] "The Worst Nuclear Disasters - Photo Essays - TIME". time.com. 2009-03-25. Retrieved 2015-02-07.

[132] Arifumi Hasegawa, Koichi Tanigawa, Akira Ohtsuru, Hirooki Yabe, Masaharu Maeda, et. al. "Health effects of radiation and other health problems in the aftermath of nuclear accidents, with an emphasis on Fukushima", *Lancet*, Volume 386, No. 9992, pp. 479–488, 1 August 2015.

[133] "Fossil fuels are far deadlier than nuclear power - tech - 23 March 2011 - New Scientist". Archived from the original on 2011-03-25. Retrieved 2015-02-07.

[134] Doug Brugge; Jamie L. deLemos & Cat Bui (September 2007). "The Sequoyah Corporation Fuels Release and the Church Rock Spill: Unpublicized Nuclear Releases in American Indian Communities". *American Journal of Public Health*. **97**: 1595–600. doi:10.2105/AJPH.2006.103044. PMC 1963288. PMID 17666688.

[135] Annabelle Quince (30 March 2011). "The history of nuclear power". *ABC Radio National*.

[136] "Evacuees of Fukushima village report split families, growing frustration". *Mainichi Daily News*. January 30, 2012.

[137] http://science.time.com/2013/03/01/ meltdown-despite-the-fear-the-health-risks-from-the-fukushima-accident-are-minimal/ #ixzz2MnbjhPmv

[138] "Medical Hazards of Radioactive Waste" (PDF). *PNFA*.

[139] M. V. Ramana (July 2011). "Nuclear power and the public". *Bulletin of the Atomic Scientists*. p. 48.

[140] Keith Bradsher (December 15, 2009). "Nuclear Power Expansion in China Stirs Concerns". New York Times. Retrieved 2010-01-21.

[141] Charles D. Ferguson & Frank A. Settle (2012). "The Future of Nuclear Power in the United States" (PDF). *Federation of American Scientists*.

[142] Kennette Benedict (9 August 2012). "Civil disobedience". *Bulletin of the Atomic Scientists*.

[143] Jay Davis. After A Nuclear 9/11 *The Washington Post*, March 25, 2008.

[144] Brian Michael Jenkins. A Nuclear 9/11? *CNN.com*, September 11, 2008.

[145] Orde Kittrie. Averting Catastrophe: Why the Nuclear Nonproliferation Treaty is Losing its Deterrence Capacity and How to Restore It May 22, 2007, p. 338.

[146] Nicholas D. Kristof. A Nuclear 9/11 *The New York Times*, March 10, 2004.

[147] Zetter, Kim (25 March 2013). "Legal Experts: Stuxnet Attack on Iran Was Illegal 'Act of Force'". Wired.

[148] *Introduction to Fusion Energy*, J. Reece Roth, 1986.

[149] T. Hamacher & A.M. Bradshaw (October 2001). "Fusion as a Future Power Source: Recent Achievements and Prospects" (PDF). World Energy Council. Archived from the original (PDF) on 2004-05-06.

[150] W Wayt Gibbs (30 December 2013). "Triple-threat method sparks hope for fusion". *Nature*.

[151] "Beyond ITER". *The ITER Project*. Information Services, Princeton Plasma Physics Laboratory. Archived from the original on 2006-11-07. Retrieved 2011-02-05. - Projected fusion power timeline

[152] "Overview of EFDA Activities". *EFDA*. European Fusion Development Agreement. Archived from the original on 2006-10-01. Retrieved 2006-11-11.

40.18 External links

- International Atomic Energy Agency website
- Nuclear Safety Info Resources
- Nuclear Safety Discussion Forums
- The Nuclear Energy Option, online book by Bernard L. Cohen. Emphasis on risk estimates of nuclear.

Chapter 41

Atomic spies

Klaus Fuchs, arguably the most important of the identified "atomic spies" for his extensive access to high-level scientific data and his ability to make sense of it through his technical training.

"**Atomic spies**" or "**atom spies**" were people in the United States, Great Britain, and Canada who are known to have illicitly given information about nuclear weapons production or design to the Soviet Union during World War II and the early Cold War. Exactly what was given, and whether everyone on the list gave it are still matters of some scholarly dispute, and in some cases, what were originally seen as strong testimonies or confessions were admitted as fabricated in later years. Their work constitutes the most publicly well-known and well-documented case of nuclear espionage in the history of nuclear weapons. There was a move-

ment among nuclear scientists to share the information with the world scientific community, but it was firmly quashed by the U.S. government.

Confirmation about espionage work came from the Venona project, which intercepted and decrypted Soviet intelligence reports sent during and after World War II. They provided clues to the identity of several spies at Los Alamos and elsewhere, some of whom have never been identified. Some of this information was available but not usable in court for secrecy reasons during the 1950s trials. Additionally, records from Soviet archives, which were briefly opened to researchers after the fall of the Soviet Union, included more information about some spies.

41.1   Importance

Before World War II, the theoretical possibility of nuclear fission occasioned intense discussion among leading physicists world-wide. Scientists from the Soviet Union were later recognized for their contributions to the understanding of a nuclear reality, and won several Nobel Prizes. Soviet scientists such as Igor Kurchatov, L. D. Landau, and Kirill Sinelnikov helped establish the idea of, and prove the existence of, a splittable atom. Dwarfed by and lost in the scale of the Manhattan Project, the significance of the Soviet contributions is rarely understood or credited outside of the field of physics. According to several sources, it was understood on a theoretical level that the atom provided for extremely powerful and novel releases of energy, and could possibly be utilized in the future for military purposes.[*][1] In recorded comments, the physicists themselves lamented over their inability to achieve any kind of practical application from the discoveries. This would show that the scientists thought the creation of an atomic weapon was a pipedream and untenable. According to a US Congressional joint committee, although the scientists could conceivably have been first to generate a man-made fission reaction, in reality they lacked the ambition, funding, engineering capability, leadership, and ultimately, the capability to do so.

The undertaking would be of an unimaginable scale, and the resources required to engineer for such use as a nuclear bomb, and nuclear power were deemed too great to pursue.*[2]

At the urging of Albert Einstein and Leo Szilard through their Einstein–Szilárd letter of August 2, 1939, the United States - in collaboration with Britain and Canada - recognized the potential significance of an atomic bomb and embarked in 1942 upon work to achieve a usable device. Estimates suggest that during the quest to create the atomic bomb $2 billion, eighty-six thousand tons of silver, and twenty-four thousand skilled workers drove the research-and-development phase of the project.*[3] Those skilled workers included the people to maintain and operate the machinery necessary for research. The largest Western facility had five hundred scientists working on the project, as well as a team of fifty to derive the equations for the cascade of neutrons required to drive the reaction. The fledgling equivalent Soviet program was quite different: the program consisted of fifty scientists, and a mere two mathematicians trying to work out the equations for the particle cascade.*[4] The research and development of techniques to produce sufficiently enriched uranium and plutonium were beyond the scope and efforts of the Soviet group. The knowledge of techniques and strategies which the Allied programs used - and of which Soviet espionage obtained - might have played a role in the rapid development of the Soviet bomb after the war.

The research and development of methods suitable for doping and separating the highly reactive isotopes needed to create the payload for a nuclear warhead took years, and consumed a vast amount of resources. The United States and Great Britain dedicated their best scientists to this cause and constructed three plants, each with a different isotope-extraction method.*[5] The Allied program decided to use gas-phase extraction to obtain the pure uranium necessary for an atomic detonation.*[2] Using this method, it took large quantities of uranium ore and other rare materials such as graphite to successfully purify the U-235 isotope. The quantities required for the development were also beyond the scope and purview of the Soviet program.

The Soviet Union did not have natural uranium-ore mines at the start of the nuclear arms race. A lack of materials made it very difficult to conduct novel research or to map out a clear pathway to achieving the fuel they needed. The Soviet scientists became frustrated with the difficulties of producing uranium fuel cheaply, and they found their industrial techniques for refinement lacking. The use of information stolen from the Manhattan Project eventually rectified the problem.*[6] Without such information, the problems the Soviet atomic team experienced would have taken many years to correct, affecting the production of a Soviet atomic weapon significantly.

The missing link that explains the great leaps in the Soviets Union's atomic program is the espionage information and technical data which Moscow succeeded in obtaining from the Manhattan Project. Once the Soviets had learned of the American plans to develop an atomic bomb during the 1940s, Moscow began actively seeking agents to get information.*[7] Moscow sought very specific information from its intelligence cells in America, and demanded updates on the progress of the Allied project. Moscow was also greatly concerned with the procedures being used for U-235 separation, what method of detonation was being used, and what industrial equipment was being used for these techniques.*[8]

To obtain this information from the Manhattan Project, the Soviet Union needed spies who had security clearance high enough to have access to classified information and who could understand and interpret what they were stealing. Moscow also needed reliable spies who believed in the communist cause and who would provide accurate information. One such Soviet spy, Theodore Hall, had worked on the development of the bombs dropped in Japan.*[9] Hall provided the specifications of the bomb dropped on Nagasaki. This information allowed the Soviet scientists a first-hand look at the successful set up of an atomic weapon built by the Manhattan Project.

Although Hall's information proved helpful to the Soviet cause, the most influential of the atomic spies was Klaus Fuchs. Fuchs, a German-born British physicist, went to America to work on the atomic project and became one of its lead scientists. Fuchs had become a member of the Communist Party in 1932 while still a student in Germany. At the onset the Third Reich Fuchs fled to Great Britain (1933), where he eventually became one of the lead nuclear physicists in the British program. In 1943 he moved to the United States to collaborate on the Manhattan Project.*[10] Due to Fuchs's position in the atomic program, he had access to most, if not all, of the material Moscow desired. Fuchs was also able to interpret and understand the information he was stealing, which made him an invaluable resource. Fuchs provided the Soviets with detailed information on the gas-phase separation process. He also provided specifications for the payload, calculations and relationships for setting of the fission reaction, and schematics for labs producing weapons-grade isotopes.*[11] This information helped the smaller under-manned and under-supplied Soviet group with a hard push in the direction of the successful detonation of a nuclear weapon.

The Soviet nuclear program would have eventually been able to develop a nuclear weapon without the aid of espionage. But a basic understanding of the usefulness of an atomic weapon, the sheer resources, and the talent did not develop until much later. Espionage helped the Soviet scientists identify which methods worked and prevented wasting valuable resources on techniques which the development of the American bomb had proven ineffective. The speed at which the Soviet nuclear program achieved a working bomb with so few resources depended on the amount of information acquired through espionage. During the Cold War trials that espionage was touted as one of the most significant intelligence coups in human history.*[12]

41.2 Notable examples

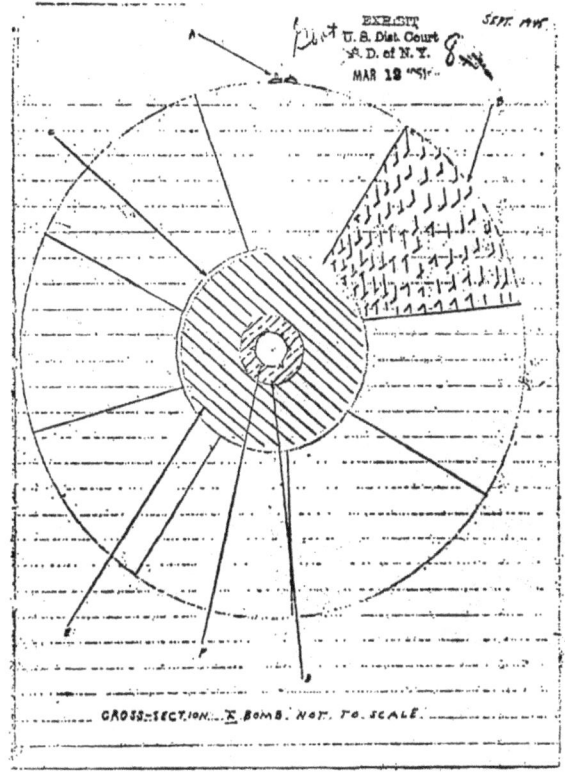

Sketch of an implosion-type nuclear weapon design made by David Greenglass as state's evidence, illustrating what he gave the Rosenbergs to pass on to the Soviet Union.

- Morris Cohen – an American, "Thanks to Cohen, designers of the Soviet atomic bomb got piles of technical documentation straight from the secret laboratory in Los Alamos," the newspaper Komsomolskaya Pravda said. Morris and his wife, Lona, served eight years in prison, less than half of their sentences before being released in a prisoner swap with the Soviet Union. He died without revealing the name of the American scientist who helped pass vital information about the United States atomic bomb project.*[13]

- Klaus Fuchs – the German-born British theoretical physicist who worked with the British delegation at

Los Alamos during the Manhattan Project. After Fuchs' confession, there was a trial that lasted less than 90 minutes, Lord Goddard sentenced him to fourteen years' imprisonment, the maximum for violating the Official Secrets Act. He escaped the charge of espionage because of the lack of independent evidence and because, at the time of his activities, the Soviet Union was not an enemy of Great Britain.*[14] In December 1950 he was stripped of his British citizenship. He was released on June 23, 1959, after serving nine years and four months of his sentence at Wakefield prison. He was allowed to emigrate to Dresden, then in the German Democratic Republic.*[15]*[16]

- Harry Gold – an American, confessed to acting as a courier for Greenglass and Fuchs. He was sentenced in 1951 to thirty years imprisonment. He was paroled in May 1966, after serving just over half of his sentence.*[17]

- David Greenglass – an American machinist at Los Alamos during the Manhattan Project. Greenglass confessed that he gave crude schematics of lab experiments to the Russians during World War II. Some aspects of his testimony against his sister and brother-in-law (the Rosenbergs, see below) are now thought to have been fabricated in an effort to keep his own wife, Ruth, from prosecution. Greenglass was sentenced to 15 years in prison, served 10 years, and later reunited with his wife.*[18]

- Theodore Hall – an American, the youngest physicist at Los Alamos who gave a detailed description of the *Fat Man* plutonium bomb, and of several processes for purifying plutonium, to Soviet intelligence. His identity as a spy was not revealed until very late in the 20th century. He was never tried for his espionage work, though he admitted to it in later years to reporters and to his family.*[19]

- George Koval – the American-born son of a Belorussian emigrant family that returned to the Soviet Union where he was inducted into the Red Army and recruited into the GRU intelligence service. He infiltrated the US Army and became a radiation health officer in the Special Engineering Detachment. Acting under the code name *Delmar* he obtained information from Oak Ridge National Laboratory and the Dayton Project about the *Urchin* detonator used on the *Fat Man* plutonium bomb. His work was not known to the west until he was posthumously recognized as a hero of the Russian Federation by Vladimir Putin in 2007.

- Irving Lerner – an American film director, he was caught photographing the cyclotron at the University of California, Berkeley in 1944.*[20] After the war he was blacklisted.

- Alan Nunn May – a British citizen, he was one of the first Soviet spies uncovered during the Cold War. He worked on the Manhattan Project and was betrayed by a Soviet defector in Canada. His was uncovered in 1946 and it led the United States to restrict the sharing of atomic secrets with Britain. On May 1, 1946, he was sentenced to ten years hard labour. He was released in 1952, after serving six and a half years.*[21]

- Ethel and Julius Rosenberg – Americans who were involved in coordinating and recruiting an espionage network that included Ethel's brother, David Greenglass. Julius and Ethel Rosenberg were tried for conspiracy to commit espionage, since the prosecution seemed to feel that there was not enough evidence to convict on espionage. Treason charges were not applicable, since the United States and the Soviet Union were allies at the time. The Rosenbergs denied all the charges but were convicted in a trial in which the prosecutor Roy Cohn said he was in daily secret contact with the judge, Irving Kaufman. Despite an international movement demanding clemency, and appeals to President Dwight D. Eisenhower by leading European intellectuals and the Pope, the Rosenbergs were executed at the height of the Korean War. President Eisenhower wrote to his son, serving in Korea, that if he spared Ethel (presumably for the sake of her children), then the Soviets would simply recruit their spies from among women.*[22]*[23]*[24]

- Saville Sax – an American, acted as the courier for Klaus Fuchs and Theodore Hall.*[19]

- Morton Sobell – the American engineer tried and convicted along with the Rosenbergs who was sentenced to 30 years imprisonment, but released in 1969, after serving 17 years and 9 months.*[25] After proclaiming his innocence for over half a century, Sobell admitted spying for the Soviets, and implicated Julius Rosenberg, in an interview with the New York Times published on September 11, 2008.*[26]

41.3 Gallery

- K. E. J. Fuchs Klaus Fuchs ID badge photo from

Los Alamos National Laboratory

- Police photograph of Julius Rosenberg after his arrest

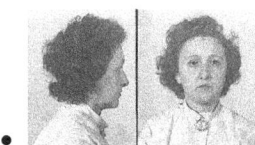
- Mugshot of Ethel Rosenberg

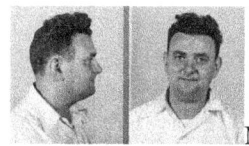
- Mugshot of David Greenglass

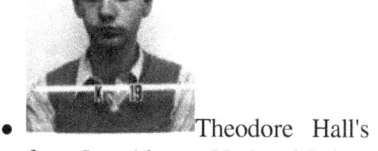
- Theodore Hall's ID badge photo from Los Alamos National Laboratory

41.4 See also

- Nuclear espionage
- Venona project

41.5 References

[1] Schwartz, Michael. The Russian-A(merican) Bomb: The Role of Espionage in the Soviet Atomic Bomb Project. J. Undergrad. Sci. 3: 103–108 (Summer 1996) http://www.hcs.harvard.edu/~{}jus/0302/schwartz.pdf

[2] Joint Committee on Atomic Energy. Soviet Atomic Espionage. Chapters 2–3 United States Government Printing Office, Washington 1951. https://archive.org/stream/sovietatomicespi1951unit#page/n3/mode/2up

[3] Schwartz, Michael. Russian Bomb 103-08

[4] Schwartz, Michael. Russian Bomb 103-08f

[5] Allen Weinstein and Alexander Vassiliev, "Atomic Espionage: from Fuchs to the Rosenburgs" in The Haunted Wood, (New York: Random House Inc, 1999), 172–222.

[6] Weinstein and Vassiliev,Atomic Espionage, 180–85

[7] Weinstein and Vassiliev,Atomic Espionage, 190–200

[8] Weinstein and Vassiliev,Atomic Espionage,180

[9] Holmes,Marian.   "Spies Who Spilled Atomic Bomb Secrets".   Smithsonian.com,April 20, 2009. http://www.smithsonianmag.com/history-archaeology/Spies-Who-Spilled-Atomic-Bomb-Secrets.html

[10] Holmes, Spies Who Spilled, 1–2

[11] Weinstein and Vassiliev, Atomic Espionage, 200–10

[12] Joint Committee Chapter 2

[13] "Morris Cohen, 84, Soviet Spy Who Passed Atom Plans in 40's". New York Times. 5 July 1995. Retrieved 2008-07-07. Morris Cohen, an American who spied for the Soviet Union and was instrumental in relaying atomic bomb secrets to the Kremlin in the 1940s, has died, Russian newspapers reported today. Mr. Cohen, best known in the West as Peter Kroger, died of heart failure in a Moscow hospital on June 23 at age 84, according to news reports.

[14] A.M. Hornblum, 'The Invisible Harry Gold' (Yale University Press, 2010) kindle edition. locations 4030-37

[15] Pace, Eric (January 29, 1988). "Klaus Fuchs, Physicist Who Gave Atom Secrets to Soviet, Dies at 76". New York Times. Retrieved 2008-07-07. Klaus Fuchs, the German-born physicist who was imprisoned in the 1950s in Britain after being convicted of passing nuclear secrets to the Soviet Union, died yesterday, the East German press agency A.D.N. reported. He was 76 years old.

[16] "Klaus Fuchs". TruTV. Retrieved 2008-07-07. His name was Klaus Emil Fuchs, and he was, as it has been shown by history, the most important atom spy in history. Not any of the notorious names in the saga of the theft of the atom bomb secrets Alan Nunn May, Julius and Ethel Rosenberg, and David Greenglass had been as important to the Russian effort as Klaus Fuchs.

[17] "1972 Death of Harry Gold Revealed". New York Times. February 14, 1974. Retrieved 2008-07-07. Harry Gold, who served 15 years in Federal prison as a confessed atomic spy courier, for Klaus Fuchs, a Soviet agent, and who was a key Government witness in the Julius and Ethel Rosenberg espionage case in 1951, died 18 months ago in Philadelphia.

[18] "Greenglass, in Prison, Vows to Kin He Told Truth About Rosenbergs". New York Times. March 19, 1953. Retrieved 2008-07-07. David Greenglass, serving fifteen years as a confessed atom spy, denied to members of his family recently that he had been coached by the Federal Bureau of Investigation in the drawing of segments of the atom bomb, or that he had given perjured testimony against his sister, Mrs. Ethel Rosenberg, and her husband, Julius.

[19] Cowell, Alan (November 10, 1999). "Theodore Hall, Prodigy and Atomic Spy, Dies at 74". New York Times. Retrieved 2008-06-26. Theodore Alvin Hall, who was the youngest physicist to work on the atomic bomb project at Los Alamos during World War II and was later identified as a Soviet spy, died on Nov. 1 in Cambridge, England, where he had become a leading, if diffident, pioneer in biological research. He was 74. ... Mr. Albright and Ms. Kunstel say Mr. Hall and a former Harvard roommate, Saville Sax, approached a Soviet trade company in New York in late 1944 and began supplying critical information about the atomic project.

[20] https://books.google.com/books?id=nIYC5pd1XQoC&pg=PA325&lpg=PA325&dq=irving+lerner+cyclotron&source=bl&ots=Jz1jHDDGws&sig=8d5l3qC1QdpiDv9PQjY4b5vAjU8&hl=en&ei=cvUrTcg_xN-WB_qL0coK&sa=X&oi=book_result&ct=

[21] "Alan Nunn May, 91, Pioneer In Atomic Spying for Soviets". New York Times. 25 January 2003. Retrieved 2008-07-07. Alan Nunn May, a British atomic scientist who spied for the Soviet Union, died on Jan. 12 in Cambridge. He was 91. ... One of the first Soviet spies uncovered during the cold war, Dr. Nunn May worked on the Manhattan Project and was betrayed by a Soviet defector in Canada. His unmasking in 1946 led the United States to restrict the sharing of atomic secrets with Britain.

[22] "Execution of the Rosenbergs". London: The Guardian. June 20, 1953. Retrieved 2008-06-24. Julius and Ethel Rosenberg were executed early this morning at Sing Sing Prison for conspiring to pass atomic secrets to Russia in World War II.

[23] "The Rosenbergs: A Case of Love, Espionage, Deceit and Betray". TruTV. Retrieved 2008-07-07. Julius and Ethel Rosenberg were charged with the crime of conspiracy to commit espionage, and tried under the Espionage Act of 1917.

[24] "Execution of the Rosenbergs". The Guardian. London. June 20, 1953. Retrieved 2008-06-24. Julius and Ethel Rosenberg were executed early this morning at Sing Sing Prison for conspiring to pass atomic secrets to Russia in World War II.

[25] "Morton Sobell Free As Spy Term Ends". New York Times. January 15, 1969. Retrieved 2008-07-07. Morton Sobell, sentenced to 30 years for a wartime espionage conspiracy to deliver vital national secrets to the Soviet Union, was released from prison yesterday after serving 17 years and 9 months.

[26] Roberts, Sam (September 11, 2008). "For First Time, Figure in Rosenberg Case Admits Spying for Soviets". *New York Times.* Retrieved 2008-09-11. In an interview on Thursday, Mr. Sobell, who served nearly 19 years in Alcatraz and other federal prisons, admitted for the first time that he had been a Soviet spy.

41.6   Further reading

- Alexei Kojevnikov, *Stalin's Great Science: The Times and Adventures of Soviet Physicists* (Imperial College Press, 2004). ISBN 1-86094-420-5 (use of espionage data by Soviets)

- Gregg Herken, *Brotherhood of the Bomb: The Tangled Lives and Loyalties of Robert Oppenheimer, Ernest Lawrence, and Edward Teller* (New York: Henry Holt and Co., 2002). ISBN 0-8050-6588-1 (details on Fuchs)

- Richard Rhodes, *Dark Sun: The Making of the Hydrogen Bomb* (New York: Simon and Schuster, 1995). ISBN 0-684-80400-X (general overview of Fuchs and Rosenberg cases)

Chapter 42

Nuclear terrorism

Nuclear terrorism refers to an act of terrorism in which a person or persons belonging to a terrorist organization detonates a nuclear device.*[1] Some definitions of nuclear terrorism include the sabotage of a nuclear facility and/or the detonation of a radiological device, colloquially termed a dirty bomb, but consensus is lacking. In legal terms, nuclear terrorism is an offense committed if a person unlawfully and intentionally "uses in any way radioactive material ···with the intent to cause death or serious bodily injury; or with the intent to cause substantial damage to property or to the environment; or with the intent to compel a natural or legal person, an international organization or a State to do or refrain from doing an act", according to the 2005 United Nations International Convention for the Suppression of Acts of Nuclear Terrorism.*[2]

The possibility of terrorist organizations using nuclear weapons (including those of a relatively smaller size, such as those contained within suitcases (suitcase nuclear device), is something which is known of within U.S. culture, and at times previously discussed within the political settings of the U.S. It is considered plausible that terrorists could acquire a nuclear weapon.*[3] However, despite thefts and trafficking of small amounts of fissile material, all low-concern and less than Category III Special nuclear material (SNM), there is no credible evidence that any terrorist group has succeeded in obtaining Category I SNM, the necessary multi-kilogram critical mass amounts of weapons grade plutonium required to make a nuclear weapon.*[4]*[5]

42.1 Scope

Main article: Vulnerability of nuclear plants to attack

Nuclear terrorism could include:

- Acquiring or fabricating a nuclear weapon

- Fabricating a dirty bomb

- Attacking a nuclear reactor, e.g., by disrupting critical inputs (e.g. water supply)

- Attacking or taking over a nuclear-armed submarine, plane or base.*[6]

Nuclear terrorism, according to a 2011 report published by the Belfer Center for Science and International Affairs at Harvard University, can be executed and distinguished via four pathways:*[7]

- The use of a nuclear weapon that has been stolen or purchased on the black market

- The use of a crude explosive device built by terrorists or by nuclear scientists who the terrorist organization has furtively recruited

- The use of an explosive device constructed by terrorists and their accomplices using their own fissile material

- The acquisition of fissile material from a nation-state.

Former U.S. President Barack Obama called nuclear terrorism "the single most important national security threat that we face". In his first speech to the U.N. Security Council, President Obama said that "Just one nuclear weapon exploded in a city -- be it New York or Moscow, Tokyo or Beijing, London or Paris -- could kill hundreds of thousands of people". It would "destabilize our security, our economies, and our very way of life".*[8]

42.2 History

As early as December 1945, politicians worried about the possibility of smuggling nuclear weapons into the United States, though this was still in the context of a battle between the superpowers of the Cold War. Congressmen quizzed the "father of the atomic bomb," J. Robert Oppenheimer, about the possibility of detecting a smuggled atomic bomb:

Sen. Millikin: We... have mine-detecting devices, which are rather effective... I was wondering if anything of that kind might be available to use as a defense against that particular type of use of atomic bombs.

Dr. Oppenheimer: If you hired me to walk through the cellars of Washington to see whether there were atomic bombs, I think my most important tool would be a screwdriver to open the crates and look. I think that just walking by, swinging a little gadget would not give me the information.*[9]

This sparked further work on the question of smuggled atomic devices during the 1950s.

Discussions of non-state nuclear terrorism among experts go back at least to the 1970s. In 1975 The Economist warned that "You can make a bomb with a few pounds of plutonium. By the mid-1980s the power stations may easily be turning out 200,000 lb of the stuff each year. And each year, unless present methods are drastically changed, many thousands of pounds of it will be transferred from one plant to another as it proceeds through the fuel cycle. The dangers of robbery in transit are evident.... Vigorous co-operation between governments and the International Atomic Energy Agency could, even at this late stage, make the looming perils loom a good deal smaller." *[10] And the New York Times commented in 1981 that The Nuclear Emergency Search Team's "origins go back to the aftershocks of the Munich Olympic massacre in mid-1972. Until that time, no one in the United States Government had thought seriously about the menace of organized, international terrorism, much less nuclear terrorism. There was a perception in Washington that the value of what is called 'special nuclear material' - plutonium or highly enriched uranium (HEU) - was so enormous that the strict financial accountability of the private contractors who dealt with it would be enough to protect it from falling into the wrong hands. But it has since been revealed that the physical safeguarding of bomb-grade material against theft was almost scandalously neglected." *[11]

This discussion took on a larger public character in the 1980s after NBC aired Special Bulletin, a television dramatization of a nuclear terrorist attack on the United States.*[12] In 1986 a private panel of experts known as the International Task Force on the Prevention of Terrorism released a report urging all nuclear-armed states to beware the dangers of terrorism and work on equipping their nuclear arsenals with permissive action links. "The probability of nuclear terrorism," the experts warned, "is increasing and the consequences for urban and industrial societies could be catastrophic." *[13]

The World Institute for Nuclear Security is an organization which seeks to prevent nuclear terrorism and improve world nuclear security. It works alongside the International Atomic Energy Agency. WINS was formed in 2008, less than a year after a break-in at the Pelindaba nuclear facility in South Africa, which contained enough enriched uranium to make several nuclear bombs.

The Global Initiative to Combat Nuclear Terrorism (GICNT) is an international partnership of 86 nations and 4 official observers working to improve capacity on a national and international level for prevention, detection, and response to a nuclear terrorist event. Partners join the GICNT by endorsing the Statement of Principles, a set of broad nuclear security objectives. GICNT partner nations organize and host workshops, conferences, and exercises to share best practices for implementing the Statement of Principles. The GICNT also holds Plenary meetings to discuss improvements and changes to the partnership.

42.3 Militant groups

Nuclear weapons materials on the black market are a global concern,*[14]*[15] and there is concern about the possible detonation of a small, crude nuclear weapon by a militant group in a major city, with significant loss of life and property.*[16]*[17]

It is feared that a terrorist group could detonate a dirty bomb, a type of radiological weapon. A dirty bomb is made of any radioactive source and a conventional explosive. There would be no nuclear blast and likely no fatalities, but the radioactive material is dispersed and can cause extensive fallout depending on the material used. A foot-long stick of radioactive cobalt could be taken from a food irradiation plant and combined with ten pounds of explosives to contaminate 1,000 square kilometers and make some areas uninhabitable for decades.*[17] There are other radiological weapons called radiological exposure devices where an explosive is not necessary. A radiological weapon may be very appealing to terrorist groups as it is highly successful in instilling fear and panic among a population (particularly because of the threat of radiation poisoning) and would contaminate the immediate area for some period of time, disrupting attempts to repair the damage and subsequently inflicting significant economic losses.

42.3.1 al-Qaeda

According to *Bunn & Wier*, Osama bin Laden requested a ruling (a *fatwa*), and was subsequently informed via a cleric of Saudi Arabia during 2003, of it being in accordance with Islamic law for him to use a nuclear device against civilians if it were the only course of action available to him in a situation of defending Muslims against the actions of the U.S. military.*[18]

According to leaked diplomatic documents, al-Qaeda can produce radiological weapons, after sourcing nuclear material and recruiting rogue scientists to build "dirty bombs" .*[19] Al-Qaeda, along with some North Caucasus terrorist groups that seek to establish an Islamic Caliphate in Russia, have consistently stated they seek nuclear weapons and have tried to acquire them.*[7] Al-Qaeda has sought nuclear weapons for almost two decades by attempting to purchase stolen nuclear material and weapons and has sought nuclear expertise on numerous occasions. Osama bin Laden stated that the acquisition of nuclear weapons or other weapons of mass destruction is a "religious duty." *[20] While pressure from a wide range of counter-terrorist activity has hampered Al-Qaeda's ability to manage such a complex project, there is no sign that it has jettisoned its goals of acquiring fissile material. Statements made as recently as 2008 indicate that Al-Qaeda's nuclear ambitions are still very strong.*[7]

42.3.2   ISIS

ISIS has demonstrated ambition to use weapons of mass destruction,*[21] although the chances of them obtaining a nuclear bomb are small, the group have been trying/suspected to be trying to obtain a nuclear dirty bomb.*[22] In July 2014, ISIS militants captured nuclear materials from Mosul University, In a letter to UN Secretary-General Ban Ki-moon, Iraq's UN Ambassador Mohamed Ali Alhakim said that the materials had been kept at the university and "can be used in manufacturing weapons of mass destruction" , however Nuclear experts regarded the threat as insignificant. International Atomic Energy Agency spokeswoman Gill Tudor said that the seized materials were "low grade and would not present a significant safety, security or nuclear proliferation risk" .*[23]*[24]

In October 2015 it was reported that Moldovan authorities working with the FBI have stopped four attempts from 2010 to 2015 by gangs with suspected connections to Russia's intelligence services that sought to sell radioactive material to ISIS and other Middle Eastern extremists. The last reported case came in February 2015-a smuggler with a large amount of radioactive caesium specifically sought a buyer from ISIS. The Criminal organizations are thriving on black market nuclear materials in Moldova, since relations between Russia and the West deteriorated, it is difficult to know whether smugglers are succeeding in selling radioactive material originating from Russia to Islamist terrorists and elsewhere.*[21]*[25]*[26]

In March 2016, it was reported that a senior Belgian nuclear

official was being monitored by ISIS suspects linked to the November 2015 Paris attacks leading Belgium authorities to suspect that ISIS was planning on abducting the official to obtain nuclear materials for a dirty bomb.*[27]

In April 2016, EU and NATO security chiefs warned that ISIS are plotting to carry out nuclear attacks on the UK and Europe.*[28]

42.3.3   North Caucasus terrorists

North Caucasus terrorists have attempted to seize a nuclear submarine armed with nuclear weapons. They have also engaged in reconnaissance activities on nuclear storage facilities and have repeatedly threatened to sabotage nuclear facilities. Similar to Al-Qaeda, these groups' activities have been hampered by counter-terrorism activity; nevertheless they remain committed to launching such a devastating attack within Russia.*[7]

42.3.4   Aum Shinrikyo

The Japanese terror cult Aum Shinrikyo, which used nerve gas to attack a Tokyo subway in 1995, has also tried to acquire nuclear weapons. However, according to nuclear terrorism researchers at Harvard University's Belfer Center for Science and International Affairs, there is no evidence that they continue to do so.*[7]

42.4   Incidents involving nuclear material

Information reported to the International Atomic Energy Agency (IAEA) shows "a persistent problem with the illicit trafficking in nuclear and other radioactive materials, thefts, losses and other unauthorized activities" .*[29] The IAEA Illicit Nuclear Trafficking Database notes 1,266 incidents reported by 99 countries over the last 12 years, including 18 incidents involving HEU or plutonium trafficking:*[30]

- There have been 18 incidents of theft or loss of highly enriched uranium (HEU) and plutonium confirmed by the IAEA.*[20]

- Security specialist Shaun Gregory argued in an article that terrorists have attacked Pakistani nuclear facilities three times in the recent past; twice in 2007 and once in 2008.*[31]

- In November 2007, burglars with unknown intentions infiltrated the Pelindaba nuclear research facility near

Pretoria, South Africa. The burglars escaped without acquiring any of the uranium held at the facility.*[32]*[33]

- In June 2007, the Federal Bureau of Investigation released to the press the name of Adnan Gulshair el Shukrijumah, allegedly the operations leader for developing tactical plans for detonating nuclear bombs in several American cities simultaneously.*[34]

- In November 2006, MI5 warned that al-Qaida were planning on using nuclear weapons against cities in the United Kingdom by obtaining the bombs via clandestine means.*[35]

- In February 2006, Oleg Khinsagov of Russia was arrested in Georgia, along with three Georgian accomplices, with 79.5 grams of 89 percent HEU.*[20]

- The Alexander Litvinenko poisoning with radioactive polonium "represents an ominous landmark: the beginning of an era of nuclear terrorism," according to Andrew J. Patterson.*[36]

- In June 2002, U.S. citizen José Padilla was arrested for allegedly planning a radiological attack on the city of Chicago; however, he was never charged with such conduct. He was instead convicted of charges that he conspired to "murder, kidnap and maim" people overseas.

42.5 Pakistan

After several incidents in Pakistan in which terrorists attacked three of its military nuclear facilities, it became clear that there emerged a serious danger that they would gain access to the country's nuclear arsenal, according to a journal published by the US Military Academy at West Point.*[37] In January 2010, it was revealed that the US army was training a specialised unit "to seal off and snatch back" Pakistani nuclear weapons in the event that militants would obtain a nuclear device or materials that could make one. Pakistan supposedly possesses about 80 nuclear warheads. US officials refused to speak on the record about the American safety plans.*[38]

A study by the Belfer Center for Science and International Affairs at Harvard University titled "Securing the Bomb 2010," found that Pakistan's stockpile "faces a greater threat from Islamic extremists seeking nuclear weapons than any other nuclear stockpile on earth." *[39]

According to Rolf Mowatt-Larssen, a former investigator with the CIA and the US Department of Energy, there is "a greater possibility of a nuclear meltdown in Pakistan than

anywhere else in the world. The region has more violent extremists than any other, the country is unstable, and its arsenal of nuclear weapons is expanding." *[40]

Nuclear weapons expert David Albright and author of "Peddling Peril" has also expressed concerns that Pakistan's stockpile may not be secure despite assurances by both Pakistan, U.S. and Southeast Asia government. He stated that Pakistan "has had many leaks from its program of classified information and sensitive nuclear equipment, and so you have to worry that it could be acquired in Pakistan," *[41]

A 2010 study by the Congressional Research Service titled 'Pakistan's Nuclear Weapons: Proliferation and Security Issues' noted that even though Pakistan had taken several steps to enhance nuclear security in recent years, "instability in Pakistan has called the extent and durability of these reforms into question." *[42]

42.6 United States

President Barack Obama has reviewed Homeland Security policy and concluded that "attacks using improvised nuclear devices ... pose a serious and increasing national security risk" .*[43] In their presidential contest, President George W. Bush and Senator John Kerry both agreed that the most serious danger facing the United States is the possibility that terrorists could obtain a nuclear bomb.*[4] Most nuclear-weapon analysts agree that "building such a device would pose few technological challenges to reasonably competent terrorists" . The main barrier is acquiring highly enriched uranium.*[44]

In 2004, Graham Allison, U.S. Assistant Secretary of Defense during the Clinton administration, wrote that "on the current path, a nuclear terrorist attack on America in the decade ahead is more likely than not" .*[45] In 2004, Bruce Blair, president of the Center for Defense Information stated: "I wouldn't be at all surprised if nuclear weapons are used over the next 15 or 20 years, first and foremost by a terrorist group that gets its hands on a Russian nuclear weapon or a Pakistani nuclear weapon" .*[17] In 2006, Robert Galluccii, Dean of the Georgetown University School of Foreign Service, estimated that, "it is more likely than not that al-Qaeda or one of its affiliates will detonate a nuclear weapon in a U.S. city within the next five to ten years." *[45] Despite a number of claims,*[46]*[47] there is no credible evidence that any terrorist group has yet succeeded in obtaining a nuclear bomb or the materials needed to make one.*[4]*[5]

Detonation of a nuclear weapon in a major U.S. city could kill more than 500,000 people and cause more than a trillion dollars in damage.*[16]*[17] Hundreds of thousands could die from fallout, the resulting fires and collapsing buildings. In this scenario, uncontrolled fires would burn for days and emergency services and hospitals would be completely overwhelmed.[*][4][*][48][*][49] The likely socio-economic consequences in the United States outside the immediate vicinity of an attack, and possibly in other countries, would also likely be far-reaching. A Rand Corporation report speculates that there may be an exodus from other urban centers by populations fearful of another nuclear attack.[*][50]

The Obama administration will focus on reducing the risk of high-consequence, non-traditional nuclear threats. Nuclear security is to be strengthened by enhancing "nuclear detection architecture and ensuring that our own nuclear materials are secure," and by "establishing well-planned, well-rehearsed, plans for co-ordinated response." [*][43] According to senior Pentagon officials, the United States will make "thwarting nuclear-armed terrorists a central aim of American strategic nuclear planning." [*][51] Nuclear attribution is another strategy being pursued to counter terrorism. Led by the National Technical Nuclear Forensics Center, attribution would allow the government to determine the likely source of nuclear material used in the event of a nuclear attack. This would prevent terrorist groups, and any states willing to help them, from being able to pull off a covert attack without assurance of retaliation.[*][52]

In July 2010 medical personnel from the U.S. Army practiced the techniques they would use to treat people injured by an atomic blast. The exercises were carried out at a training center in Indiana, and were set up to "simulate the aftermath of a small nuclear bomb blast, set off in a U.S. city by terrorists." [*][53]

Stuxnet is a computer worm discovered in June 2010 that is believed to have been created by the United States and Israel to attack Iran's nuclear facilities.[*][54]

42.6.1 Nuclear power plants

After 9/11, nuclear power plants were to be prepared for an attack by a large, well-armed terrorist group. But the Nuclear Regulatory Commission, in revising its security rules, decided not to require that plants be able to defend themselves against groups carrying sophisticated weapons. According to a study by the Government Accountability Office, the N.R.C. appeared to have based its revised rules "on what the industry considered reasonable and feasible to defend against rather than on an assessment of the terrorist threat itself" .[*][55][*][56] If terrorist groups could sufficiently damage safety systems to cause a core meltdown at a nuclear power plant, and/or sufficiently damage spent fuel pools, such an attack could lead to widespread radioactive contamination. The Federation of American Scientists have said that if nuclear power use is to expand sig-

nificantly, nuclear facilities will have to be made extremely safe from attacks that could release massive quantities of radioactivity into the community. New reactor designs have features of passive safety, which may help. In the United States, the NRC carries out "Force on Force" (FOF) exercises at all Nuclear Power Plant (NPP) sites at least once every three years.[*][57]

The peace group Plowshares have shown how nuclear weapons facilities can be penetrated, and the groups actions represent extraordinary breaches of security at nuclear weapons plants in the United States. The National Nuclear Security Administration has acknowledged the seriousness of the 2012 Plowshares action. Non-proliferation policy experts have questioned "the use of private contractors to provide security at facilities that manufacture and store the government's most dangerous military material" .[*][58]

42.6.2 Hoaxes

In late 1974, President Gerald Ford was warned that the FBI received a communication from an extortionist wanting $200,000 ($1,000,000 today) after claiming that a nuclear weapon had been placed somewhere in Boston. A team of experts rushed in from the United States Atomic Energy Commission but their radiation detection gear arrived at a different airport. Federal officials then rented a fleet of vans to carry concealed radiation detectors around the city but forgot to bring the tools they needed to install the equipment. The incident was later found to be a hoax. However, the government's response made clear the need for an agency capable of effectively responding to such threats in the future. Later that year, President Ford created the Nuclear Emergency Search Team (NEST), which under the Atomic Energy Act is tasked with investigating the "illegal use of nuclear materials within the United States, including terrorist threats involving the use of special nuclear materials" .[*][59]

One of its first responses by the Nuclear Emergency Search/Support Team was in Spokane, Washington on November 23, 1976. An unknown group called the "Days of Omega" had mailed an extortion threat claiming it would explode radioactive containers of water all over the city unless paid $500,000 ($2,100,000 today). Presumably, the radioactive containers had been stolen from the Hanford Site, less than 150 miles to the southwest. Immediately, NEST flew in a support aircraft from Las Vegas and began searching for non-natural radiation, but found nothing. No one ever responded despite the elaborate instructions given, or made any attempt to claim the (fake) money which was kept under surveillance. Within days, the incident was deemed a hoax, though the case was never solved. To avoid panic, the public was not notified until a few years later.[*][60][*][61]

42.7 Policy landscape

42.7.1 Recovery

The Cooperative Threat Reduction Program (CTR), which is also known as the Nunn–Lugar Cooperative Threat Reduction, is a 1992 law sponsored by Senators Sam Nunn and Richard Lugar. The CTR established a program that gave the U.S. Department of Defense a direct stake in securing loose fissile material inside the since-dissolved USSR. According to Graham Allison, director of Harvard University's Belfer Center for Science and International Affairs, this law is a major reason why not a single nuclear weapon has been discovered outside the control of Russia' s nuclear custodians.*[62] The Belfer Center is itself running the *Project on Managing the Atom,* Matthew Bunn is a co-principal investigator of the project, Martin B. Malin is its executive director (circa. 2014).*[63]

In August 2002, the United States launched a program to track and secure enriched uranium from 24 Soviet-style reactors in 16 countries, in order to reduce the risk of the materials falling into the hands of terrorists or "rogue states". The first such operation was *Project Vinca,* "a multinational, public-private effort to remove nuclear material from a poorly-secured Yugoslav research institute." The project has been hailed as "a nonproliferation success story" with the "potential to inform broader 'global cleanout' efforts to address one of the weakest links in the nuclear nonproliferation chain: insufficiently secured civilian nuclear research facilities." *[64]

In 2004, the U.S. Global Threat Reduction Initiative (GTRI) was established in order to consolidate nuclear stockpiles of highly enriched uranium (HEU), plutonium, and assemble nuclear weapons at fewer locations.*[65] Additionally, the GTRI converted HEU fuels to low-enriched uranium (LEU) fuels, which has prevented their use in making a nuclear bomb within a short amount of time. HEU that has not been converted to LEU has been shipped back to secure sites, while amplified security measures have taken hold around vulnerable nuclear facilities.*[66]

42.7.2 Options

Robert Gallucci, President of the John D. and Catherine T. MacArthur Foundation, argues that traditional deterrence is not an effective approach toward terrorist groups bent on causing a nuclear catastrophe.*[67] Henry Kissinger, stating the wide availability of nuclear weapons makes deterrence "decreasingly effective and increasingly hazardous." *[68] Preventive strategies, which advocate the elimination of an enemy before it is able to mount an attack, are risky and controversial, therefore difficult to implement. Gal-

lucci believes that "the United States should instead consider a policy of expanded deterrence, which focuses not on the would-be nuclear terrorists but on those states that may deliberately transfer or inadvertently lead nuclear weapons and materials to them. By threatening retaliation against those states, the United States may be able to deter that which it cannot physically prevent." .*[67]

Graham Allison makes a similar case, arguing that the key to expanded deterrence is coming up with ways of tracing nuclear material to the country that forged the fissile material. "After a nuclear bomb detonates, nuclear forensic cops would collect debris samples and send them to a laboratory for radiological analysis. By identifying unique attributes of the fissile material, including its impurities and contaminants, one could trace the path back to its origin." *[69] The process is analogous to identifying a criminal by fingerprints. "The goal would be twofold: first, to deter leaders of nuclear states from selling weapons to terrorists by holding them accountable for any use of their own weapons; second, to give every leader the incentive to tightly secure their nuclear weapons and materials." *[69]

42.7.3 Nuclear skeptics

John Mueller, a scholar of international relations at the Ohio State University, is a prominent nuclear skeptic. He makes three claims: (1) the nuclear intent and capability of terrorist groups such as Al Qaeda has been "fundamentally exaggerated;" (2) "the likelihood a terrorist group will come up with an atomic bomb seems to be vanishingly small;"and (3) policymakers are guilty of an "atomic obsession" that has led to "substantively counterproductive" policies premised on "worst case fantasies." *[70] In his book *Atomic Obsession*: *Nuclear Alarmism from Hiroshima to Al-Qaeda* he argues that: "anxieties about terrorists obtaining nuclear weapons are essentially baseless: a host of practical and organizational difficulties make their likelihood of success almost vanishingly small" .*[71]

Intelligence officials have pushed back, testifying before Congress that the inability to recognize the shifting modus oparandi of terrorist groups was part of the reason why members of Aum Shinrikyo, for example, were "not on anybody' s radar screen." *[72] Matthew Bunn, associate professor at Harvard University's John F. Kennedy School of Government, argues that "Theft of HEU and plutonium is not a hypothetical worry, it is an ongoing reality." *[30] Almost all of the stolen HEU and plutonium that has been seized over the years had never been missed before it was seized. The IAEA Illicit Nuclear Trafficking Database notes 1,266 incidents reported by 99 countries over the last 12 years, including 18 incidents involving HEU or plutonium trafficking.*[30]

Keir Lieber and Daryl Press argue that despite the prominent U.S. focus on nuclear terrorism, "the fear of terrorist transfer [of nuclear weapons] seems greatly exaggerated... [and] the dangers of a state giving nuclear weapons to terrorists have been overstated." A decade of terrorism statistics show a strong correlation between attack fatalities and the attribution of the attack, and Lieber and Press assert that "neither a terror group nor a state sponsor would remain anonymous after a nuclear terror attack." About 75 percent of attacks with 100 or more fatalities were traced to the culprits; also, 97 percent of attacks on U.S. soil or that of a major ally (resulting in 10 or more deaths) were attributed to the guilty party. Lieber and Press conclude that the lack of anonymity would deter a state from providing terrorist groups with nuclear weapons.*[73]

The use of HEU and plutonium in satellites has raised the concern that a sufficiently motivated rogue state could retrieve materials from a satellite crash (notably on land as occurred with Kosmos-954, Mars-96 and Fobos-Grunt) and then use these to supplement the yield of an already working nuclear device. This has been discussed recently in the UN and the Nuclear Emergency Search Team regularly consults with Roscosmos and NASA about satellite re-entries that may have contained such materials. As yet no parts were verifiably recovered from Mars 96 but recent Wikileaks releases suggest that one of the "cells" may have been recovered by mountain climbers in Chile.

42.7.4 Security summits

On April 12–13, 2010, President of the United States Barack Obama initiated and hosted the first-ever nuclear security summit in Washington D.C., commonly known as the Washington Nuclear Security Summit. The goal was to strengthen international cooperation to prevent nuclear terrorism. President Obama, along with nearly fifty world leaders, discussed the threat of nuclear terrorism, what steps needed to be taken to mitigate illicit nuclear trafficking, and how to secure nuclear material. The Summit was successful in that it produced a consensus delineating nuclear terrorism as a serious threat to all nations. Finally, the Summit produced over four-dozen specific actions embodied in commitments by individual countries and the Joint Work Plan.*[74] However, world leaders at the Summit failed to agree on baseline protections for weapons-usable material, and no agreement was reached on ending the use of highly enriched uranium (HEU) in civil nuclear functions. Many of the shortcomings of the Washington Nuclear Security Summit were addressed at the Seoul Nuclear Security Summit in March 2012.

According to Graham Allison, director of Harvard University's Belfer Center for Science and International Affairs, the objectives of the Nuclear Security Summit in Seoul are to continue to, "assess the progress made since the Washington Summit and propose additional cooperation measures to (1) Combat the threat of nuclear terrorism, (2) protect nuclear materials and related facilities, and (3) prevent illicit trafficking in nuclear materials." *[75]

42.8 Media coverage

In 2011, the British news agency, the *Telegraph*, received leaked documents regarding the Guantanamo Bay interrogations of Khalid Sheikh Mohammed. The documents cited Khalid saying that, if Osama bin Laden is captured or killed by the Coalition of the Willing, an al-Qaeda sleeper cell will detonate a "weapon of mass destruction" in a "secret location" in Europe, and promised it would be "a nuclear hellstorm".*[76]*[77] *[78]*[79]*[80] No such attack occurred.

42.9 See also

- The Apollo Affair - allegations of theft of HEU from the US' NUMEC facility by Israel, losses later recovered from pipes in facility and additional large amounts were lost after alleged theft was discovered and security enhanced.
- Atomic spies
- Crimes involving radioactive substances
- Guantanamo Bay files leak
- International Project - Forum «Nuclear Security – Counteraction Measures to Acts of Nuclear Terrorism»
- Lists of nuclear disasters and radioactive incidents
- Nuclear espionage
- Nuclear warfare
- Pelindaba
- Superphénix
- Terrorism
- 2014 Nuclear Security Summit
- Vulnerability of nuclear plants to attack
- War on Terror
- Weapons of mass destruction
- World Institute for Nuclear Security

42.10 References

[1] "Nuclear Security Dossier: Nuclear Terrorism Fact Sheet" . *Harvard Kennedy School, Belfer Center for Science and International Affairs*. Retrieved 28 January 2013.

[2] "International Convention for the Suppression of Acts of Nuclear Terrorism - Article 1" (PDF). United Nations. 2005. Retrieved 13 April 2012.

[3] *Nuclear Terrorism: Frequently Asked Questions*, Belfer Center for Science and International Affairs, September 26, 2007

[4] Matthew Bunn. Preventing a Nuclear 9/11 *Issues in Science and Technology*, Winter 2005, p. v.

[5] Ajay Singh. Nuclear terrorism —Is it real or the stuff of 9/11 nightmares? *UCLA Today*, February 11, 2009.

[6] Ruff, Tilman (November 2006), *Nuclear terrorism* (PDF), energyscience.org.au

[7] Bunn, Matthew, Colonel Yuri Morozov, Rolf Mowatt-Larssen, Simon Saradzhyan, William Tobey, Colonel General (ret.) Viktor I. Yesin, and Major General (ret.) Pavel S. Zolotarev (2011). "The U.S.-Russia Joint Threat Assessment on Nuclear Terrorism" (PDF). Belfer Center for Science and International Affairs, Harvard University. Retrieved July 26, 2012.

[8] Graham Allison (January 26, 2010). "A Failure to Imagine the Worst" . *Foreign Policy*.

[9] Alex Kingsbury, "History's Troubling Lessons", *U.S. News and World Report* (February 18, 2007).

[10] "Nuclear Terrorism," *The Economist* (January 25, 1975) p. 38.

[11] Larry Collins, "Combating Nuclear Terrorism," *New York Times* (December 14, 1980) Sec. 6 pg. 37.

[12] Sally Bedell, "A Realistic Film Stirs NBC Debate," *New York Times* (March 17, 1983) B13; Sally Bedell, "NBC Nuclear Terror Show Criticized," *New York Times (March 22, 1983) C15; Aljean Harmetz, "NBC Film on Terror Wins Prize,"* New York Times *(July 8, 1983) C19.*

[13] D. Costello, "Experts Warn on Nuclear Terror," *Courier-Mail* (June 26, 1986).

[14] Jay Davis. After A Nuclear 9/11 *The Washington Post*, March 25, 2008.

[15] Brian Michael Jenkins. A Nuclear 9/11? *CNN.com*, September 11, 2008.

[16] Orde Kittrie. Averting Catastrophe: Why the Nuclear Nonproliferation Treaty is Losing its Deterrence Capacity and How to Restore It May 22, 2007, p. 338.

[17] Nicholas D. Kristof. A Nuclear 9/11 *The New York Times*, March 10, 2004.

[18] M. BUNN & A. WIER. *The Seven Myths of Nuclear Terrorism* (PDF). Belfer Centre for Science and International Affairs. Retrieved 2015-08-08. External link in |publisher= (help)

[19] "al-Qaeda moving world toward 'nuclear 9/11'". *The Age*. Melbourne. February 3, 2011.

[20] Bunn, Matthew & Col-Gen. E.P. Maslin (2010). "All Stocks of Weapons-Usable Nuclear Materials Worldwide Must be Protected Against Global Terrorist Threats" (PDF). Belfer Center for Science and International Affairs, Harvard University. Retrieved July 26, 2012.

[21] "Smugglers Tried to Sell Nuclear Material to ISIS" . NBC News. 7 October 2015.

[22] "The Risk of a Nuclear ISIS Grows" . the Huffington post. 8 October 2015.

[23] Cowell, Alan (10 July 2014). "Low-Grade Nuclear Material Is Seized by Rebels in Iraq, U.N. Says" . *The New York Times*. Retrieved 15 July 2014.

[24] Sherlock, Ruth (10 July 2014). "Iraq jihadists seize 'nuclear material', says ambassador to UN". *The Telegraph*. London. Retrieved 15 July 2014.

[25] "Nuclear smuggling deals 'thwarted' in Moldova" . BBC News. 7 October 2015.

[26] "FBI foils smugglers' plot to sell nuclear material to Isis" . the independent. 7 October 2015.

[27] "Brussels attacks: Belgium fears Isis seeking to make 'dirty' nuclear bomb" . the independent. 25 March 2016.

[28] "Nato raises 'justified concern' that Isil is plotting nuclear attack on Britain" . MSN. 19 April 2016.

[29] IAEA Illicit Trafficking Database (ITDB) p. 3.

[30] Bunn, Matthew. "Securing the Bomb 2010: Securing All Nuclear Materials in Four Years" (PDF). President and Fellows of Harvard College. Retrieved 28 January 2013.

[31] Rhys Blakeley, "Terrorists 'have attacked Pakistan nuclear sites three times'," *Times Online* (August 11, 2009).

[32] "IOL - Pretoria News" . *IOL*.

[33] Washington Post, December 20, 2007, Op-Ed by Micah Zenko

[34] "Feds Hoped to Snag Bin Laden Nuke Expert in JFK Bomb Plot" . *Fox News*. June 4, 2007.

[35] Teather, David; Younge, Gary (January 5, 2005). "Briton accused of trying to sell missiles" . *The Guardian*. London.

[36] "Ushering in the era of nuclear terrorism," by Patterson, Andrew J. MD, PhD, *Critical Care Medicine*, v. 35, p.953-954, 2007.

[37] Blakely, Rhys (August 11, 2009), "Terrorists 'have attacked Pakistan nuclear sites three times'", *Times Online*, London

[38] "Login".

[39] Pakistan nuclear weapons at risk of theft by terrorists, US study warns, The Guardian, 2010-04-12

[40] Could terrorists get hold of a nuclear bomb?, BBC, 2010-04-12

[41] Official: Terrorists seek nuclear material, but lack ability to use it, CNN, 2010-04-13

[42] Pakistan's Nuclear Weapons: Proliferation and Security Issues, Congressional Research Service, 2010-02-23

[43] The White House. Homeland Security

[44] Charles D. Ferguson. Preventing a nuclear 9/11 : First, secure the highly enriched uranium *The New York Times*, September 24, 2004.

[45] Orde Kittrie. Averting Catastrophe: Why the Nuclear Nonproliferation Treaty is Losing its Deterrence Capacity and How to Restore It May 22, 2007, p. 342.

[46] Paul Williams (2005). *The Al Qaeda Connection : International Terrorism, Organized Crime, and the Coming Apocalypse*, Prometheus Books, pp. 192–194.

[47] Nuclear 9/11: Interview with Dr. Paul L. Williams *Global Politician*, September 11, 2007.

[48] Controlling Nuclear Warheads and Materials p. 16.

[49] Bleek, Philipp, Anders Corr, and Micah Zenko. Nuclear 9/11: What if Port is Ground Zero? *The Houston Chronicle*, May 1, 2005.

[50] *Considering the Effects of a Catastrophic Terrorist Attack* by Charles Meade & Roger C. Molander p 9, Retrieved March 11, 2013 - this report uses smuggled nuclear weapons in container ships at a US port as an example, so speculates an exodus from coastal cities

[51] Thom Shanker and Eric Scmitt. U.S. to Make Stopping Nuclear Terror Key Aim *The New York Times*, December 18, 2009.

[52] Richelson, Jeffrey. "U.S. Nuclear Detection and Counterterrorism, 1998-2009". George Washington University.

[53] Deborah Block. US Military Practices Medical Response to Nuclear Attack *Voice of America*, 26 July 2010.

[54] Zetter, Kim (25 March 2013). "Legal Experts: Stuxnet Attack on Iran Was Illegal 'Act of Force'". Wired.

[55] Elizabeth Kolbert (28 March 2011). "The Nuclear Risk". *The New Yorker*.

[56] Daniel Hirsch et al. The NRC's Dirty Little Secret, *Bulletin of the Atomic Scientists*, May 1, 2003, vol. 59 no. 3, pp. 44-51.

[57] Charles D. Ferguson & Frank A. Settle (2012). "The Future of Nuclear Power in the United States" (PDF). *Federation of American Scientists*.

[58] Kennette Benedict (9 August 2012). "Civil disobedience". *Bulletin of the Atomic Scientists*.

[59] "Nuclear Emergency Support Team (NEST)" (PDF). U.S. Department of Energy. Retrieved 2012-10-21.

[60] Peck, Chris (1981-02-08). "The day they said they'd nuke Spokane-Part 1" (scan). *The Spokesman-Review*. p. 17. Retrieved 2012-10-21.

[61] Peck, Chris (1981-02-08). "The day they said they'd nuke Spokane-Part 2" (scan). *The Spokesman-Review*. p. 24. Retrieved 2012-10-21.

[62] Allison, Graham (December 29, 2011). "Washington Can Work: Celebrating Twenty Years With Zero Nuclear Terrorism". *The Huffington Post*. Retrieved July 26, 2012.

[63] "Managing the Atom - Harvard - Belfer Center for Science and International Affairs".

[64] Philipp C. Bleek, "Project Vinca: Lessons for Securing Civil Nuclear Material Stockpiles," *The Nonproliferation Review* (Fall-Winter 2003) p. 1.

[65] Bunn, Matthew & Eben Harrell (2012). "Consolidation: Thwarting Nuclear Theft" (PDF). Belfer Center for Science and International Affairs, Harvard University. Retrieved July 26, 2012.

[66] Wier, Anthony and Matthew Bunn (November 19, 2006). "Bombs That Won't Go Off". *The Washington Post*. Retrieved July 26, 2012.

[67] Gallucci, Robert (September 2006). "Averting Nuclear Catastrophe: Contemplating Extreme Responses to U.S. Vulnerability". *Annals of the American Academy of Political and Social Science*. **607**: 51–58. doi:10.1177/0002716206290457. Retrieved 28 January 2013.

[68] Kissinger, Henry (15 January 2008). "Toward a Nuclear-Free World". *NTI*. Retrieved 28 January 2013.

[69] Allison, Graham (13 March 2009). "How to Keep the Bomb From Terrorists". *Newsweek*. Retrieved 28 January 2013.

[70] Mueller, John (15 January 2008). *The Atomic Terrorist: Assessing the Likelihood, prepared for presentation at the Program on International Security Policy* (PDF). University of Chicago.

[71] http://www.oup.com/us/catalog/general/subject/HistoryWorld/?view=usa&ci=9780195381368 Atomic Obsession: Nuclear Alarmism from Hiroshima to Al-Qaeda: Oxford University Press

[72] Allison, Graham (2004). *Nuclear Terrorism: The Ultimate Preventable Catastrophe*. New York: Macmillan. p. 15. ISBN 9781429945516.

[73] Lieber, Keir; Press, Daryl (Summer 2013). "Why States Won't Give Nuclear Weapons to Terrorists". *International Security*. **38** (1): 80–84, 104. doi:10.1162/isec_a_00127.

[74] Tobey, William (2011). "Planning for Success at the 2012 Seoul Nuclear Security Summit" (PDF). The Stanley Foundation. Retrieved July 26, 2012.

[75] "2012 Seoul Nuclear Security Summit Q&A with Professor Graham Allison" (PDF). Belfer Center for Science and International Affairs, Harvard University. 2012. Retrieved July 26, 2012.

[76] Hope, Christopher (April 25, 2011). "WikiLeaks: Guantanamo Bay terrorist secrets revealed". London: Telegraph.co.uk. Retrieved April 27, 2011.

[77] Gould, Martin. "WikiLeaks: Al-Qaida Already Has Nuclear Capacity". NewsMax. Retrieved April 27, 2011.

[78] "'Nuclear hellstorm' if bin Laden caught - 9/11 mastermind". News.com.au. April 25, 2011. Retrieved April 27, 2011.

[79] "'Nuclear hellstorm' if bin Laden caught: 9/11 mastermind". News.Yahoo.com. 2011-04-25. Retrieved April 27, 2011.

[80] http://newstabulous.com/al-qaeda-hid-bomb-in-europe-wikileaks-releases-secret-files/9722/

42.11 Further reading

- Allison, Graham (9 August 2004). *Nuclear Terrorism: The Ultimate Preventable Catastrophe*. New York, New York: Times Books. ISBN 978-0-8050-7651-6.

- Byrne, John and Steven M. Hoffman (1996). *Governing the Atom: The Politics of Risk*, Transaction Publishers.

- Cooke, Stephanie (2009). *In Mortal Hands: A Cautionary History of the Nuclear Age*, Black Inc.

- Ferguson, Charles D., and William C. Potter, with Amy Sands, Leonard S. Spector and Fred L. Wehling (2004). *The Four Faces of Nuclear Terrorism*. Monterey, California: Center for Nonproliferation Studies. ISBN 1-885350-09-0.

- Jones, Ishmael (2010) [2008]. *The Human Factor: Inside the CIA's Dysfunctional Intelligence Culture*. Encounter Books. ISBN 978-1-59403-382-7.

- Levi, Michael (2007). *On Nuclear Terrorism*. Cambridge, Massachusetts: Harvard University Press. ISBN 978-0-674-02649-0.

- Lovins, Amory B. and John H. Price (1975). *Non-Nuclear Futures: The Case for an Ethical Energy Strategy*, Ballinger Publishing Company, 1975, ISBN 0-88410-602-0

- Schell, Jonathan (2007). *The Seventh Decade: The New Shape of Nuclear Danger*. New York, New York: Metropolitan Books.

42.12 External links

- Nuclear Terrorism publications from Harvard Kennedy School faculty and fellows

- What if the terrorists go nuclear?, Center for Defense Information

- Preventing Catastrophic Nuclear Terrorism, Council on Foreign Relations

- Use of nuclear and radiological weapons by terrorists?, International Review of the Red Cross

- "Can Terrorists Build Nuclear Weapons?", Nuclear Control Institute

- Annotated bibliography, Alsos Digital Library for Nuclear Issues

- Fallout: After a Nuclear Attack - slideshow by *Life magazine*

- Nuclear Emergency and Radiation Resources

Chapter 43

United States military nuclear incident terminology

The United States Armed Forces uses a number of terms to define the magnitude and extent of nuclear incidents.

43.1 Origin

United States Department of Defense directive 5230.16, *Nuclear Accident and Incident Public Affairs (PA) Guidance*,[*][1] Chairman Joint Chiefs of Staff Manual 3150.03B *Joint Reporting Structure Event and Incident Reporting*, and the United States Air Force Operation Reporting System, as set out in Air Force Instruction 10-206[*][2] detail a number of terms for internally and externally (including press releases) reporting nuclear incidents. They are used by the United States of America, and are neither NATO nor global standards.

43.2 Terminology

43.2.1 Pinnacle

Pinnacle is a Chairman of the Joint Chiefs of Staff OPREP-3 (Operational Event/Incident Report) reporting flagword used in the United States National Command Authority structure. The term "Pinnacle" denotes an incident of interest to the Major Commands, Department of Defense and National Command Authority, in that it:

- Generates a higher level of military action

- Causes a national reaction

- Affects international relationships

- Causes immediate widespread coverage in news media

- Is clearly against the national interest

- Affects current national policy

All of the following reporting terms are classified *Pinnacle*, with the exception of *Bent Spear*, *Faded Giant* and *Dull Sword*. AFI 10-206 notes that the flagword *Pinnacle* may be added to *Bent Spear* or *Faded Giant* to expedite reporting to the National Military Command Center (NMCC).[*][3]

43.2.2 Bent Spear

Bent Spear refers to incidents involving nuclear weapons, warheads, components or vehicles transporting nuclear material that are of significant interest but are not categorized as *Pinnacle - Nucflash* or *Pinnacle - Broken Arrow*. *Bent Spear* incidents include violations or breaches of handling and security regulations.

A recent *Bent Spear* example is the August 2007 flight of a B-52 bomber from Minot AFB to Barksdale AFB which mistakenly carried six cruise missiles with live nuclear warheads.[*][4]

43.2.3 Broken Arrow

Broken Arrow refers to an accidental event that involves nuclear weapons, warheads or components which does not create the risk of nuclear war. These include:

- Accidental or unexplained nuclear detonation.

- Non-nuclear detonation or burning of a nuclear weapon.

- Radioactive contamination.

- Loss in transit of nuclear asset with or without its carrying vehicle.

- Jettisoning of a nuclear weapon or nuclear component.

- Public hazard, actual or implied.

Broken Arrow incidents

Main article: List of military nuclear accidents

As of September 2013, the US Department of Defense has officially recognized 32 "Broken Arrow" incidents.[5] Examples of these events include:

- 1950 British Columbia B-36 crash
- 1956 B-47 disappearance
- 1958 Mars Bluff B-47 nuclear weapon loss incident
- 1958 Tybee Island mid-air collision
- 1961 Yuba City B-52 crash
- 1961 Goldsboro B-52 crash
- 1964 Savage Mountain B-52 crash
- 1964 Bunker Hill AFB runway accident
- 1965 Philippine Sea A-4 incident
- 1966 Palomares B-52 crash[6]
- 1968 Thule Air Base B-52 crash
- 1980 Damascus, Arkansas incident

43.2.4 Nucflash

Nucflash refers to detonation or possible detonation of a nuclear weapon which creates a risk of an outbreak of nuclear war. Events which may be classified *Pinnacle - Nucflash* include:

- Accidental, unauthorized, or unexplained nuclear detonation or possible detonation.
- Accidental or unauthorized launch of a nuclear-armed or nuclear-capable missile in the direction of, or having the capability to reach, another nuclear-capable country.
- Unauthorized flight of, or deviation from an approved flight plan by, a nuclear-armed or nuclear-capable aircraft with the capability to penetrate the airspace of another nuclear-capable country.
- Detection of unidentified objects by a missile warning system or interference (experienced by such a system or related communications) that appears threatening and could create a risk of nuclear war.

This term is a report that has the highest precedence in the OPREP-3 reporting structure. All other reporting terms such as *Broken Arrow*, *Empty Quiver*, etc., while very important, are secondary to this report.[3]

43.2.5 Emergency Disablement

Emergency Disablement refers to operations involving the emergency destruction of nuclear weapons.

43.2.6 Emergency Evacuation

Emergency Evacuation refers to operations involving the emergency evacuation of nuclear weapons.

43.2.7 Empty Quiver

Empty Quiver refers to the seizure, theft, or loss of a functioning nuclear weapon.

43.2.8 Faded Giant

Faded Giant refers to an event involving a military nuclear reactor or other radiological accident not involving nuclear weapons.

43.2.9 Dull Sword

Dull Sword is the term that describes reports of minor incidents involving nuclear weapons, components or systems, or which could impair their deployments. This could include actions involving vehicles capable of carrying nuclear weapons but with no nuclear weapons on board at the time of the accident. This also is used to report damage or deficiencies with equipment, tools, or diagnostic testers that are designed for use on nuclear weapons or the nuclear weapon release systems of nuclear-capable aircraft.

43.3 See also

- Lists of nuclear disasters and radioactive incidents
- List of military nuclear accidents
- Nuclear and radiation accidents
- United States and weapons of mass destruction

43.4   Notes and references

[1] "DoD Directive 5230.16, "Nuclear Accident and Incident Public Affairs (PA) Guidance", 12/20/1993". www.dtic. mil. Archived from the original on 7 Nov 2009. Retrieved 2010-02-05.

[2] "Air Force E-Publishing - Home". www.e-publishing.af. mil. Retrieved 2010-02-05.

[3] "Air Force Instruction 10-206" (PDF). United States Air Force. 11 June 2014.

[4] Warrick, Joby; Pincus, Walter.   "Missteps in the Bunker - washingtonpost.com". washingtonpost.com. Retrieved 2010-02-05.

[5] Nuclear weapons: an accident waiting to happen 2013Sep14

[6] Palomares Nuclear Weapons Accident: Revised Dose Evaluation Report (PDF) (Report). Bolling Air Force Base, Washington, D.C.: Office of the Surgeon General, United States Air Force. April 2001. Retrieved 2011-06-15.

43.5   External links

- Annotated bibliography from the Alsos Digital Library for Nuclear Issues

- Department of Defense directive 5230.16

- AFMAN 10-206

- Taylor's Nuke Site - Broken Arrow Investigations

43.6 Text and image sources, contributors, and licenses

43.6.1 Text

- **Nuclear and radiation accidents and incidents** *Source:* https://en.wikipedia.org/wiki/Nuclear_and_radiation_accidents_and_incidents? oldid=766077114 *Contributors:* Bryan Derksen, Alex.tan, Rmhermen, Phil Bordelon, SimonP, Maury Markowitz, Montrealais, Ewen, N8chz, Rickyrab, Yves Junqueira, Patrick, RTC, Infrogmation, JakeVortex, Bobby D. Bryant, Bcrowell, Sannse, Yann, Eric119, Minesweeper, Aho-erstemeier, Mac, Snoyes, Aarchiba, Iammaxus, Jll, Punkche, Cimon Avaro, Sandstorm~enwiki, Jengod, Mulad, Adam Bishop, Vanished user 5zariu3jisj0j4irj, Reddi, Viajero, Zoicon5, Katana0182, DJ Clayworth, Robertb-dc, Maximus Rex, Alleycat, Tempshill, Omegatron, Thue, Mor-ven, Topbanana, JonathanDP81, Pstudier, Ortonmc, Qertis, Pollinator, Owen, Jni, Robbot, Dale Arnett, Korath, RedWolf, Altenmann, Naddy, Modulatum, Enceladus, Texture, Argonaut, Mervyn, Catbar, Robinh, Stay cool~enwiki, Rasputin1072, Robartin, Cyrius, Elde, Carnildo, Alan Liefting, Zzyzx, Reubenbarton, Beefman, Fastfission, Zigger, Karn, Wwoods, Anville, Alison, Jonathan O'Donnell, Epokha, Bird, Brendan-Ryan, DO'Neil, Mboverload, Brockert, Matthäus Wander, Bobblewik, Edcolins, Golbez, ALargeElk, Wmahan, Gadfium, Utcursch, Andycjp, Farside~enwiki, Breez, ConradPino, Quadell, WhiteDragon, Icairns, Sam Hocevar, Gscshoyru, Neutrality, Fg2, Klemen Kocjancic, Karl Dick-man, Eisnel, Corti, CALR, AlexPU, Guanabot, Zombiejesus, Ahkond, Bender235, Calair, Neko-chan, Ignignot, Kaszeta, Pt, J-Star, JRM, Net-Bot, Foobaz, Arcadian, ParticleMan, SpeedyGonsales, Cavrdg, Kjkolb, Xoddam, Bobbis, PiccoloNamek, Pearle, Mpulier, Orzetto, Alansohn, Uncle.bungle, QVanillaQ, Wjbean, Jeffhos, Riana, Ashley Pomeroy, Fritzpoll, Spangineer, SeanDuggan, Rwendland, Radical Mallard, Klaser, BRW, Wtshymanski, Evil Monkey, Jtrainor, DV8 2XL, Gene Nygaard, Ultramarine, Pmberry, Camw, TomTheHand, JFG, ^demon, MONGO, Triddle, Scm83x, Wayward, NightOnEarth, Christopher Thomas, Rocafella, Mandarax, Rast, Deltabeignet, BD2412, Sjö, Rjwilmsi, Meepzorp, Koavf, Avochelm, Miros~enwiki, Pleiotrop3, Jivecat, Urbane Legend, Alban, Dougluce, Farcical, Sgkay, Bobstay, Naraht, Survivor, Nitrogen-sixteen, Joseph398, Kolbasz, Vonspringer, Intgr, Lithiyum, Zotel, Spencerk, Mstroeck, Knife Knut, Simesa, YurikBot, Wavelength, Sceptre, RussBot, Rxnd, Petiatil, Schol-R-LEA, WritersCramp, Allister MacLeod, Xihr, Lexi Marie, Limulus, Epolk, Ytrottier, Shawn81, Shaddack, Shanel, NawlinWiki, Bachrach44, Welsh, Esthurin, PhilipO, Desk Jockey, Neil.steiner, Htonl, Bota47, Phgao, Petri Krohn, Vicarious, That Guy, From That Show!, MacsBug, SmackBot, DocS, Hydrogen Iodide, WikiuserNI, Kilo-Lima, Eskimbot, Timeshifter, Elk Salmon, Gilliam, Keegan, Ottawakismet, Cadmium, Digitalhen, Mjl0509, Danielnez1, Ascentury, Superdix, MyNameIsVlad, HoodedMan, V1adis1av, Scarlet-smith, Dripp, Bereza, Theanphibian, Enr-v, Paul 012, John, Mike1901, Shrew, Gang65, Hvn0413, Dugbrown, Ubaierbhat, Jc37, Alessandro57, Joseph Solis in Australia, Hurricanefloyd, JoeBot, AStudent, Marysunshine, Plexus2, Xcentaur, CmdrObot, Ethnopunk, Krobison13, Myasuda, Mikehead, Cydebot, Caparn, Daniel J. Leivick, BetacommandBot, Headbomb, PaperTruths, Nick Number, Uruiamme, Natalie Erin, Amlz, CPWinter, Fayenatic london, JAnDbot, Artificiel, Nicransby, Magioladitis, Bongwarrior, Allstarecho, Aldenrw, Vedder110, Ccmonty, Mschel, MapleTree, J.delanoy, Maurice Carbonaro, Wikip rhyre, Milo03, Icseaturtles, Jaysarvage, Rwessel, Hanacy, KudzuVine, ACSE, VolkovBot, DMcMPO11AAUK, Johnfos, RingtailedFox, Martinevans123, Mark v1.0, Nsougia2, Masaqui, Natg 19, Steve3849, Goodmapd, Kaiketsu, ANogin, Thisismyrofl, Waseh123321, BodegasAmbite, Slick023, Happysailor, Janopus, Thebuff91, NihilNominis, RMB1987, RW Marloe, Ufinne, Sunrise, Afernand74, Randy Kryn, ClueBot, Binksternet, Fasettle, Nailedtooth, The Thing That Should Not Be, Gamer Eek, Syn-taxBlitz, WDavis1911, Mild Bill Hiccup, Eiland, Polyamorph, Grey noise, Ybtcphk, Pittsburgh Poet, Excirial, Redthoreau, Mlaffs, Layla27, Gxs3, Delta422869, Aitias, W345thn, SoxBot III, Apparition11, DumZiBoT, Knowngangster, Spitfire, FellGleaming, Addbot, Roentgenium111, Tryanmax, Olli Niemitalo, CarsracBot, FiriBot, Torla42, Doniago, Dayewalker, Isaac-Gaffer, Tide rolls, Jan eissfeldt, Luckas-bot, ZX81, Yobot, Canuslatrans, Ayrton Prost, Mindbuilder, Juliancolton Alternative, AnomieBOT, Kingpin13, Materialscientist, Citation bot, Cardio-ceras, Thing94, Grim23, Tyrol5, Nasa-verve, Tbliss558, Elemarth, Green Cardamom, GliderMaven, Abcdefgy2, FrescoBot, LucienBOT, Reing, DivineAlpha, Pinethicket, Abductive, Jonesey95, English06, Watchpup, Jauhienij, Robvanvee, Mercy11, Trappist the monk, Lotje, MarkFore-man, Baslerchr, RjwilmsiBot, Trofobi, Boundarylayer, Dewritech, Ianbrettcooper, GoingBatty, Moswento, Nissenbaum, Wikipelli, Dcirovic, Trinidade, Redhanker, Dagashin, SCStrikwerda, Raaskal, Noggo, HandsomeFella, Papg2010, Whoop whoop pull up, Emc2fred83, Isocliff, Ivolocy, Miradre, ClueBot NG, Coachcapt, Chester Markel, Frietjes, Philadelphia 2009, Widr, Dougmcdonell, Jk2q3jrklse, MerlIwBot, Dewey-oxberg, Harsimaja, Jooles6, Helpful Pixie Bot, Vagobot, OliverHargreaves, Subanshu, Asauers, Bikkader, Energy4All, BattyBot, Netherzone, David.moreno72, StarryGrandma, Cyberbot II, Khazar2, EuroCarGT, IjonTichyIjonTichy, Dexbot, 30 SW, TwoTwoHello, Pinfix, Vanishe-dUser 2313214sad1, Knobeeoldben, Jamesmcmahon0, Infamous Castle, Comp.arch, Samhpes, Finnusertop, D2211basu, Arujunan, Limnalid, Fixture, Brasnow, Monkbot, Allerenenfer, Filedelinkerbot, Fer48, Trackteur, Mahboubx, Iicl2.0001, NFSPRO, ProprioMe OW, James Hare (NIOSH), InternetArchiveBot, VladyslavGubin, GreenC bot, B2ubried, Bender the Bot, Scottythom, Hustrab and Anonymous: 434

- **Crimes involving radioactive substances** *Source:* https://en.wikipedia.org/wiki/Crimes_involving_radioactive_substances?oldid=753100910 *Contributors:* Wereon, Alan Liefting, Ich, Nlaporte, Rich Farmbrough, Bender235, Ryanmcdaniel, Velella, Saga City, Vuo, Gene Nygaard, Kay Dekker, Miss Madeline, Rjwilmsi, Itinerant1, Kolbasz, Quuxplusone, Spencerk, Random user 39849958, Phantomsteve, RussBot, Conscious, Matt Fitzpatrick, Gaius Cornelius, Trovatore, Haikz, Chris52131, Abune, Petri Krohn, Crystallina, SmackBot, Kjaergaard, Edgar181, Cadmium, Colonies Chris, Shalom Yechiel, Robofish, Shattered, Beetstra, Eastlaw, CmdrObot, A876, Grahamec, Hughdbrown, SvetBeard, Jessicapierce, Cgingold, D.h, LinuxPickle, Hodja Nasreddin, Nikki311, G force11, Athomic69, Johnfos, Jous~enwiki, RW Marloe, Fratrep, Heracletus, Rock-fang, Addbot, Yobot, ThinkingTwice, AnomieBOT, Evan.oltmanns, FrescoBot, Citation bot 1, RjwilmsiBot, Boundarylayer, H3llBot, ClueBot NG, Northamerica1000, Allecher, MrBill3, Cyberbot II, Catavar, Kamikazee69, Frosty, Conundrum1947, Haminoon, Kalloman, Jerodlycett, GreenC bot and Anonymous: 25

- **Vulnerability of nuclear plants to attack** *Source:* https://en.wikipedia.org/wiki/Vulnerability_of_nuclear_plants_to_attack?oldid=765690921 *Contributors:* Bender235, Art LaPella, Jprg1966, Cydebot, Johnfos, Asa Zernik, Niceguyedc, Callinus, AnomieBOT, Trappist the monk, Dcirovic, BattyBot, Cyberbot II, ChrisGualtieri, Yilku1, Bksovacool, Filedelinkerbot, Fer48, Graemem56, Elmeter, CaptainCarlosdeCorona, *Treker and Anonymous: 6

- **Criticality accident** *Source:* https://en.wikipedia.org/wiki/Criticality_accident?oldid=764878358 *Contributors:* Roadrunner, Tedernst, Nealmcb, Patrick, Karada, Andrewa, Aarchiba, Julesd, Charles Matthews, David Newton, Quoth-22, JonathanDP81, Raul654, Pstudier, Pakaran, Jamesday, Sanders muc, Fastfission, Karn, Jabowery, Toytoy, Geni, Chirlu, ConradPino, Zantolak, Xtreambar, Icairns, Deglr6328, Rich Farm-brough, R6144, Bender235, Cmdrjameson, Adrian~enwiki, Nesnad, Ikester8, Disneyfreak96, Atlant, Rwendland, Uucp, Stephan Leeds, Jheald, TenOfAllTrades, DV8 2XL, SteinbDJ, Gene Nygaard, Axeman89, Dan100, Jävligsvengelska, Bobrayner, Richard Arthur Norton (1958-), Anilocra, Graham87, Nanite, Anty, Meepzorp, Dougluce, Bobstay, Old Moonraker, Kolbasz, Limulus, Ytrottier, Shaddack, NawlinWiki, Rat-tleMan, Trovatore, Bayle Shanks, Adamrush, Mortein, Anetode, Light current, Deville, Omtay38, Ageekgal, Petri Krohn, Swuboo, Fram,

Smurrayinchester, MacsBug, SmackBot, Pwt898, Elminster Aumar, Hugorudd, Cla68, Brianski, Gregjgrose, Chris the speller, Ottawakismet, Cadmium, Hibernian, Sbharris, Extremesanity, Bodysurf, John, Loadmaster, BillFlis, Makyen, Kyoko, Beefyt, Gegnome, Ewulp, DKqwerty, MightyWarrior, HDCase, Cydebot, A876, Boardhead, The real dan, Gproud, Jmg38, Keraunos, Brokencog, Davidhorman, Miller17CU94, Greg L, WinBot, T-r-davies, Ignacio Egea, Phil153, Sharon.Silver, Dreadengineer, AniRaptor2001, CosineKitty, Coolhandscot, Dricherby, Wasell, Magioladitis, Nyttend, LorenzoB, Matt B., Yggdrasil42, Wikianon, Read-write-services, Johnny.cache, BJ Axel, SaxicolousOne, Tarotcards, DadaNeem, FJPB, PBIPhotobug, Cs302b, ACSE, AlwinFW, Johnfos, Liko81, Rolandschulz, Chuck Sirloin, Wushi-En, Xenophon777, ViennaUK, Bgordski, Vanished user qkqknjitkcse45u3, SummerWithMorons, Trivialist, A.Lingo, Joe N, Humanengr, DumZiBoT, Little Mountain 5, Addbot, Kurkudjikul, Ocdnctx, Vyom25, Guffydrawers, Tide rolls, Yobot, Evilschmoo, AnomieBOT, KDS4444, LilHelpa, 4twenty42o, Patrick4130, Nij90, Rainald62, LOLthulu, Íazak, Nagualdesign, Dmm8409, FrescoBot, Surv1v4l1st, LucienBOT, Mpsunxhckr, Bravo Foxtrot, ZéroBot, Mdeen, Medeis, Wingman4l7, MinistryOfLostCauses, HandsomeFella, Whoop whoop pull up, Cbbkr, Catlemur, Snotbot, CopperSquare, Nick O'Sea, BG19bot, Kendall-K1, DesertRat262, Ejbrown1949, Asauers, Qasaur, Bachware, Andyhowlett, Mark viking, Jwclough, Limnalid, Wassermaus, Babybluesplash, Monkbot, InternetArchiveBot and Anonymous: 143

- **Nuclear meltdown** *Source:* https://en.wikipedia.org/wiki/Nuclear_meltdown?oldid=761909406 *Contributors:* Bryan Derksen, Roadrunner, Europrobe, Heron, Tedernst, Patrick, Michael Hardy, Kwertii, Iluvcapra, Mbessey, Mkweise, Aarchiba, Julesd, Rob Hooft, RandySpears, Hike395, Katana0182, IceKarma, E23~enwiki, Furrykef, Pstudier, Pollinator, Dale Arnett, Argonaut, Alan Liefting, Javidjamae, Sj, Fastfission, BalthCat, Utcursch, Andycjp, Shibboleth, Jodamiller, Beland, WhiteDragon, Oscar, H Padleckas, Sam Hocevar, Gscshoyru, Ulflarsen, Rich Farmbrough, Hydrox, FT2, Cfailde, User2004, Berkut, Bender235, Kbh3rd, J-Star, Cap'n Refsmmat, Viriditas, Brutulf, StYxXx, Linuxlad, Burzum, Freekie, Radical Mallard, Wtmitchell, Benson85, Wtshymanski, DV8 2XL, MIT Trekkie, Axeman89, Martian, Dan100, UFu, MatthewJ, Robert K S, Commander Keane, Tabletop, TotoBaggins, SDC, Icydid, Mandarax, Sparkit, Buxtehude, BD2412, Rjwilmsi, Urbane Legend, Strait, Bhadani, Whitlock, Ffaarr, Ysangkok, Nihiltres, Kolbasz, Simishag, Chobot, Simesa, WriterHound, YurikBot, RobotE, Ankerl~enwiki, Sillybilly, Hede2000, Limulus, Gaius Cornelius, Shaddack, NawlinWiki, Ospalh, MrIntegrity, Bdell555, StuRat, Sefarkas, Moogsi, Josh3580, CapitalLetterBeginning, Bondegezou, Wsiegmund, Petri Krohn, Tobixen, Vicarious, Fram, Eaefremov, That Guy, From That Show!, SmackBot, FocalPoint, Mikesheffler, Gilliam, Nzd, Chris the speller, LEC20, Cadmium, Ynotswim, Kostmo, Danielnez1, Audriusa, Modest Genius, WikiPedant, Nahum Reduta, Theanphibian, Giancarlo Rossi, Richard0612, Rodeosmurf, Fxg97873, Will Beback, Moeburn, John, Gobonobo, Jaganath, Scetoaux, Mr. Lefty, Dio1982~enwiki, Nicetomeetyou, Dcflyer, Ahering@cogeco.ca, TPIRFanSteve, Dl2000, Beefyt, JoeBot, Chamberlian, Tawkerbot2, SeanMD80, Petr Matas, Kendroche, JForget, Van helsing, Dgw, MarsRover, WeggeBot, Logical2u, T23c, RagingR2, Brendanbailey, Cydebot, KPbIC, Reywas92, UncleBubba, Islander, Tawkerbot4, Garik, Cielovista, Aldis90, Thijs!bot, Epbr123, Qwyrxian, Trappleton, Pjvpjv, CharlotteWebb, Uruiamme, Seaphoto, Prolog, SkoreKeep, Kcowolf, Dreadengineer, Photobiker, AlmostReadytoFly, Savant13, Rothorpe, Repku, Bongwarrior, Faizhaider, CTF83!, Cgingold, Frigax, Timbert, Philg88, MartinBot, Andre.holzner, V-Man737, J.delanoy, Pharaoh of the Wizards, Maurice Carbonaro, Tdadamemd, Belovedfreak, DadaNeem, Spellcast, ACSE, Hammersoft, Johnfos, ABF, Martinevans123, Anynobody, Khutuck, DrewP, LenTheWhiteCat, Lexington50, JhsBot, PDFbot, Kilmer-san, Joseph A. Spadaro, Mspritch, Fanatix, SieBot, Scarian, M.thoriyan, Chromaticity, Infestor, Lightmouse, Hobartimus, Dravecky, Chrisrus, Martarius, ClueBot, The Thing That Should Not Be, Alksentrs, Brookfield53045, Eiland, Shinpah1, Farras Octara, Timberframe, Niceguyedc, Alexbot, Anon lynx, Layla27, Containment Dome, Mustufailed, Dome Shield, Togokill, DumZiBoT, InternetMeme, Power2084, Graham999999999, MystBot, JCDenton2052, Addbot, Queenmomcat, Walkness, Leszek Jańczuk, Mac Dreamstate, Chzz, Debresser, Rent A Troop, Sheffy94, Numbo3-bot, LongLiveRomania, Luckas-bot, Yobot, Bunnyhop11, Dmarquard, AnomieBOT, DemocraticLuntz, Astolmar, ScienceOfThePS3, BluegillTriplePrime, Materialscientist, Citation bot, ArthurBot, LilHelpa, Quazgaa, Drilnoth, Nasnema, BritishWatcher, Tyrol5, Chasethesky, Coenen, Mark Schierbecker, Kiwistag, Middle 8, GliderMaven, FrescoBot, Anna Roy, Scoutstr295, Clevercheetah123, Pinethicket, Arctic Night, RedBot, FHMRUSSIA, White Shadows, Trappist the monk, Intentium, Gejyspa, Jeffrd10, Andrea105, RjwilmsiBot, Jamie314, EmausBot, George H. Harvey, Boundarylayer, RenamedUser01302013, Mtzsch, Dcirovic, Bravo Foxtrot, Stympkin, Drishtant12, L Kensington, Donner60, ChuispastonBot, Teapeat, Bowen10000, Mjbmrbot, Sonicyouth86, ClueBot NG, Prioryman, Satellizer, Widr, Zeus011, ROTORHEAD77, Alikash, Ferdinandrock, Heimdall2011, Helpful Pixie Bot, Onejina2011, Curb Chain, 355D54, BG19bot, PhnomPencil, MusikAnimal, Sonasonic, Yowanvista, FormerNukeSubmariner, Snow Blizzard, HMman, Asauers, NoelyNoel, Achowat, BattyBot, Cyberbot II, ChrisGualtieri, GoShow, Isaidnoway, Soni, Dissident93, FoCuSandLeArN, Lugia2453, MichelleKay, Vintovka Dragunova, Partycat11, KingSupernova, Gcampane, Finnusertop, SnoozeKing, Monkbot, DaveyHume, Trackteur, Kiwuser, Jacob Gotts, Allenh205, Cactuslynx, Ethanbas, SamTDM, Blitzice and Anonymous: 364

- **Three Mile Island (disambiguation)** *Source:* https://en.wikipedia.org/wiki/Three_Mile_Island_(disambiguation)?oldid=712991764 *Contributors:* Andrewman327, Jerzy, Koavf, Ken Gallager, Ahecht, Johnfos, Addbot, I dream of horses, HiW-Bot, BG19bot and Anonymous: 1

- **Nagasaki** *Source:* https://en.wikipedia.org/wiki/Nagasaki?oldid=765874708 *Contributors:* AxelBoldt, Trelvis, Brion VIBBER, Bryan Derksen, Guppie, Ed Poor, Scipius, Branden, William Avery, Ktsquare, Perique des Palottes, Hephaestos, Kerberos, Olivier, Ericd, Robertolyra, Stevertigo, Cointyro, Infrogmation, MartinHarper, Fruge~enwiki, TakuyaMurata, Iluvcapra, Ellywa, Ahoerstemeier, Synthetik, ToastyKen, Andrewa, Whkoh, Netsnipe, Palmpilot900, Wfeidt, Iseeaboar, Emperorbma, Dino, WhisperToMe, Jimbreed, Topbanana, Raul654, Dpbsmith, Nnh, Jason M, Astronautics~enwiki, PBS, Altenmann, Severdrup, Sekicho, Bkell, Ghaz~enwiki, Wikibot, Kzhr, Profoss, Netjeff, Cyrius, DocWatson42, Isam, Sj, Abigail-II, Lethe, Tom harrison, Fastfission, Everyking, Rick Block, Per Honor et Gloria, Jason Quinn, Get-back-world-respect, KirbyMeister, Stevietheman, Andycjp, R. fiend, Antandrus, Robert Brockway, Kusunose, Karl-Henner, Lumidek, Neutrality, Acad Ronin, Fg2, Klemen Kocjancic, CES~enwiki, Spiffy sperry, Rich Farmbrough, Guanabot, Supercoop, Ranma9617, Dbachmann, Paul August, SpookyMulder, Bender235, Violetriga, Brian0918, Crunchy Frog, Karmafist, El C, Kross, Tverbeek, Aude, Tom, Just zis Guy, you know?, Bendono, Tachitsuteto, Oarih, Pharos, Nsaa, Ranveig, Jonojet, Alansohn, Anthony Appleyard, Arthena, Photojpn.org, Kurt Shaped Box, Historian, BanyanTree, Cmapm, LordAmeth, HenryLi, Deror avi, Kelly Martin, Billpike, Woohookitty, Mr Tan, EnSamulili, Nameneko, Astrowob, Eras-mus, Karmosin, Kralizec!, Ggonnell, Lusitana, Wrh2, FreplySpang, RxS, Sjakkalle, Eyu100, BlueMoonlet, Voretus, Notorious4life, MChew, DirkvdM, FlaBot, Daderot, Hottentot, Ewlyahoocom, Gurch, Wars, Bmicomp, Chobot, DVdm, YurikBot, Wavelength, Kafziel, RussBot, Longbow4u, Fabartus, Bachrach44, The Ogre, Badagnani, Mmccalpin, Dake~enwiki, Thiseye, Nick, TimDuncan, DAJF, Bobak, Dyre, Mkill, Cheeser1, Bota47, Mddake, Empty2005, CLW, Intershark, Paul Magnussen, Nikkimaria, Chase me ladies, I'm the Cavalry, Jwissick, Bamse, DGaw, Rande M Sefowt, Kaicarver, Eaefremov, Attilios, Neier, SmackBot, YellowMonkey, Nihonjoe, Hydrogen Iodide, Kintetsubuffalo, Hmains, Betacommand, Skizzik, SLU, Saros136, Endroit, Ottawakismet, MalafayaBot, Akanemoto, Keith Paynter, Xchbla423, MJCdetroit, John C PI, Rrburke, Soosed, Nakon, Jellyfisho, The PIPE, Jóna Þórunn, Curly Turkey, Geofrog~enwiki, CorvetteZ51, J 1982, Shlomke, Padrhig, JHunterJ, Luokehao, MTSbot~enwiki, Bwpach, Cerealkiller13, Violncello, HisSpaceResearch, Iridescent, Laurens-af, Jason7825, Fusebokme, Igoldste, Courcelles, Tawkerbot2, Fow, ChrisCork, JForget, RSido, JFMATLOCK, Friendlystar, BigBang19, Cydebot, 663highland, Yawja, Njames-

Otashiro, Rowellcf, Guitarmankev1, Liu Bei, Karenjc, Cydebot, MC10, Steel, Gogo Dodo, Bellerophon5685, CurtisJohnson, Travelbird, Soup Blazer, Tawkerbot4, HitroMilanese, Omicronpersei8, PsychoSmith, Casliber, BetacommandBot, Sabbre, Thijs!bot, Epbr123, Biruitorul, Erich Schmidt, Bobabobabo, Mojo Hand, Marek69, RickinBaltimore, Ram4eva, JCam, TangentCube, DonaldoKun, Natalie Erin, Grandin, Mentifisto, AntiVandalBot, Yonatan, Whats up skip, Luna Santin, CodeWeasel, Seaphoto, Opelio, QuiteUnusual, Quintote, FFCecil, Kevinkeegan, RapidR, Gh5046, Kzaral, John Boxer, Myanw, Canadian-Bacon, JAnDbot, Pettiebone, Dan D. Ric, Wicket1, Jezza125, MER-C, CosineKitty, The Transhumanist, Instinct, Charles01, Hut 8.5, Flying tiger, Acroterion, Magioladitis, Karlhahn, Bongwarrior, VoABot II, Dekimasu, Wikidudeman, Alexultima, JNW, Mbarbier, Cmacris, CTF83!, Bluespaceoddity2, Tsyoshi, Foochar, MetsBot, Allstarecho, Gomm, Glen, DerHexer, Pan Dan, Charitwo, AliaGemma, Jackson Peebles, MartinBot, Arjun01, Paracel63, Rettetast, Liam Patrick, Kiyokun, Kostisl, CommonsDelinker, ASD-FGH, Johnluisocasio, J.delanoy, DrKay, Bogey97, Uncle Dick, Dbiel, Textangel, Katalaveno, Almudena.bonaplata, DarkFalls, Cfcdance, Omega Archdoom, Jayden54, Gurchzilla, NewEnglandYankee, Carewser, Spruceforest, Shoessss, Kingbobs, Bob, Juliancolton, Cometstyles, Jamesofur, Bonadea, Vinsfan368, Idioma-bot, ArchetypeRyan, ACSE, Diluvial, Hugo999, VolkovBot, CWii, Kvasi, Johnfos, Pleasantville, Meaningful Username, Lt. Col. Cole, Enderminh, Jeff G., Barneca, Philip Trueman, TXiKiBoT, Maximillion Pegasus, Technopat, Afluent Rider, 4444sd-jhf4444, Ohyrsmsnbsnmr, Qxz, Bsharvy, Piperh, Lexington50, Melsaran, Leafyplant, Bentley4, Sushiya, Mannafredo, Sentineneve, Wiae, Pishogue, Maxim, Madhero88, Coching, Lerdthenerd, BilabialBoxing, SQL, Redacteur, Lova Falk, Falcon8765, Enviroboy, Sylent, Jo0r0wn3d, Kajiakira7777, SRHighSchool, Why Not A Duck, Gillesp 7, Kehrbykid, Battlecity0, PWInsider, Are2dee2, GreaterWikiholic, EJF, Ghedrick, SieBot, Nubiatech, K. Annoyomous, Tresiden, Spartan, Work permit, BotMultichill, Quasirandom, Triwbe, Smsarmad, Ketone16, Nummer29, Jgoldmanhall, Keilana, Happysailor, Flyer22 Reborn, Oda Mari, Sohelpme, Wilson44691, Iuoydfy, Teketime, Joeyc2156, Oxymoron83, Harry-, Targeman, Steven Crossin, Boldlyman, Stephenreebs, Poindexter Propellerhead, Techman224, Icedevil14, Macy, Evgervtervgdfg, Maelgwnbot, Dear Reader, Gtadoc, ShadyS14, Pink Surfer, StaticGull, Mygerardromance, Bpeps, Regushee, Escape Orbit, Nygorilla, Gr8opinionater, ImageRemovalBot, Ocdcntx, SallyForth123, WikipedianMarlith, Sfan00 IMG, ClueBot, LAX, Shafeghati, Sonictrey, Binksternet, Kyoww, The Thing That Should Not Be, Meisterkoch, Googners, Allgoodnamesalreadytaken, Jan1nad, Mijalo, Ndenison, Gaia Octavia Agrippa, Snuggleguns, Uncle Milty, SuperHamster, Boing! said Zebedee, Jarmanwalsh, Ohsoh, Sethl25, Rockfang, 718 Bot, Iamstephhxx, Anakin230, Imboredincisco, Excirial, Jusdafax, Maple Tsumori, Crywalt, UrsoBR, SpudHawg948, Putrid76, Razorflame, A sandwich, IamNotU, Ottawa4ever, ChrisHodges-sUK, Kakofonous, Stepheng3, Sitrep, Jfioeawfjdls453, Yoda227, Thingg, Aitias, 7, Yoda789, Versus22, Teleomatic, SoxBot III, Loosmark, Myspace69, DumZiBoT, LordJesseD, Fastily, Spitfire, Wertuose, Tbsdy lives, 21stCenturyGreenstuff, Sergay, NellieBly, PL290, Badgernet, WikiDao, Johnkatz1972, Kbdankbot, HexaChord, Wyatt915, Jhendin, Addbot, Proofreader77, Idobtcare, Willking1979, Glaphaductile, Guoguo12, Betterusername, Non-dropframe, Captain-tucker, Xcoolpersonx, Gregz08, Tdubb7, Rofl1234, Fieldday-sunday, Humma77, CanadianLinuxUser, Yume-doll, Frozenpineapple, CactusWriter, Avengerlachy, NjardarBot, Cst17, MrOllie, LaaknorBot, Sparten556, Auzwitch Panda, Ld100, Debresser, FCSundae, Cress Arvein, Favonian, Kyle1278, Doniago, West.andrew.g, Scholar1234590, Omgsh1234, Goodwill455, Nickcape, Nickcape2, Numbo3-bot, Tide rolls, Lightbot, MuZemike, David0811, Xenobot, Takkyon, Ben Ben, Legobot, Luckas-bot, Yobot, NeoQuintessence, Kinouya, Washburnmav, Reenem, QueenCake, Maxí, KamikazeBot, Eric-Wester, Fjkelfeimvvn, Backslash Forwardslash, AnomieBOT, Numujoe, Paralympic, Gnomeselby, Jim1138, Hadrian89, Piano non troppo, AdjustShift, JaredInsanity, RandomAct, Bluerasberry, Cheez123, Materialscientist, Jstnhughes93, RobertEves92, Citation bot, Bob Burkhardt, ArthurBot, Quebec99, Andrewmc123, MauritsBot, Xqbot, Jayarathina, Patsboy12, Sionus, Patsboy1212, Capricorn42, Jayhawk of Justice, Bihco, Jsharpminor, Grim23, Xiagon, ミスターカーブ, GrouchoBot, Ipromise, Frosted14, Tomballguy, Frankie0607, SassoBot, Amaury, Shadowjams, Chaheel Riens, Elemesh, Erik9, A.amitkumar, Dougofborg, Loganp26, Dave3457, Scarborg94, Dalton3101, Jenmegheg, JohnPeterAltgeld, Tobby72, WPANI, KazukiMiyajima, The midnight, M2545, Purpleturple, PigFlu Oink, T man 2012, Trueshow111, Pinethicket, I dream of horses, Oxana879, Trijnstel, Faaat, Xfansd, Serols, Full-date unlinking bot, Choccybic, December21st2012Freak, ActivExpression, Wayne Riddock, Pigeoninhirosimasky, FoxBot, TobeBot, Ffbear, Lotje, PorkHeart, SeoMac, January, Reaper Eternal, Surf5270, Suffusion of Yellow, PleaseStand, Stroppolo, Randomman919, TheGreenMartian, Spursnik, Mean as custard, Dolpheen1, Ripchip Bot, Smvbmf, CalicoCatLover, DASHBot, EmausBot, Takid123, Adherent of the Enlightenment 10.0, Acather96, Zima56, Shotgun5559, Sophie, Filipdr, Boundarylayer, Spindrift23, RA0808, XinaNicole, Thizzakazam, Bryce1313, Will9609, Asshoka, Wikipelli, Dcirovic, Bob 3195, BenZellmer21, Gordos S.A, Likkert, JDDJS, ZéroBot, John Cline, Traxs7, Baseballclarity, Rockman331, 07joe2k7, Terramorphous, Stemoc, Lolitsaidan, Butface101, Figureofnine, Thomas5433, Isarra, Brandmeister, L Kensington, Donner60, R.K. Keswani, VandTrack, Kmoney72, Chandlerbusing23, Sazmayo, Ebehn, ClueBot NG, Funny4fingers, Adgraff, Bchorrible, Corusant, Parcly Taxel, 123Hedgehog456, O.Koslowski, Widr, Liam mr7, Jhoeflich14, Fltyingpig, Loneaircadet, Helpful Pixie Bot, Theill6, HMSSolent, Phill McCracken, Hiroshima32123, KLBot2, Yoruboi808, Hiroshima123, DBigXray, BG19bot, Fluctuator, Pravin20111, TCN7JM, MasashiInoue, Bbhhgg, Sleeping is fun, Mifter Public, Kagundu, Mark Arsten, AdventurousSquirrel, Tazerdadog, Aranea Mortem, Min.neel, GenesisxVII, Lzy881114, Challemoni, Chris.w.braun, David.moreno72, Slowski, Pratyya Ghosh, Earthly123, Eraserman123, Greenvalleycatalls, Ali1485, Anderson, XXzoonamiXX, Yogwi21, Morfusmax, Lugia2453, Graphium, Maxliam111, Radar Holds, Cadillac000, Fidelledo, Ejfwfjw, Timothy.hoag, Epicgenius, Coppercross, Ghettoangelz, Red-eyed demon, Daniel992011, TechnicianGB, Melonkelon, Eyesnore, Madtrolls, Master of Time, Kiruning, 1STAPRIL2013, Metadox, TDOGGOMAD, LieutenantLatvia, Chessdude101, Zenibus, NottNott, Theface97, George8211, Halocon720, Grfr12345, Jianhui67, Ropstar, Kenny904, Depthdiver, Goldstein1946, JaconaFrere, Grantavery, Lovkal, Patient Zero, C1776M, Nigerica, HMSLavender, Libertarian12111971, Wzozm, Narenko, KH-1, Mateusz Malinowski, 0xF8E8, Liance, Julietdeltalima, Kukookskan, NekoKatsun, Fibrahim88888888, Nanosec2, Wheredidninjago, GeneralizationsAreBad, ShadowX1397, XXXPersonXXX, Huntrer, KasparBot, Lteroy, Huejanis, Cinder7, Adam9007, KSFT, Poop255, MicahHerr, WikiPancake, Buttermcbuttermcbuttermc, Thacatinthehat, Circusfreakkk, HarryKernow, CLCStudent, ELkacimi ELidrissi, Japanese sincerity, Jigglemaster10, GSS-1987, Baking Soda, Arvind Jiji Antony, JJMC89 bot, GreenC bot, TheBiggestLoserPartTwo, Gymdonut, Motivação, UserNumber, T-Tommy90, Wakemeup6969696, The good dank edited man, Arbor Fici, 6God JohnHalltheGOAT, Trickortreatment, Claycini, XXxpussyslayer42069xXX, Paxton silbs, What cat?, Off the Record, LeftenantAl, Joemcguire227, Hollis Munson, Zoms, Jeffyjeffyjeffy123 and Anonymous: 1399

- **Pollution of Lake Karachay** *Source:* https://en.wikipedia.org/wiki/Pollution_of_Lake_Karachay?oldid=735042032 *Contributors:* Alex Bakharev, Ohnoitsjamie, Tec15, The Anomebot2, Hugo999, Johnfos, Arjayay, Jarble, Fmrauch, AnomieBOT, Ὁ οἶστρος, Widr, Northamerica1000, Hergilei, Tom.zeimet, Bladesmulti, CarnivorousBunny and Anonymous: 15

- **Techa River** *Source:* https://en.wikipedia.org/wiki/Techa_River?oldid=749521697 *Contributors:* AaronSw, Twang, Alan Liefting, Tweenk, Rjwilmsi, Kolbasz, Wavelength, Alex Bakharev, Anomalocaris, Kkmurray, Arthur Rubin, SmackBot, Roger.lee, Frokor, Joostvandeputte~enwiki, Nyttend, The Anomebot2, Eldumpo, Maurice Carbonaro, Hugo999, VolkovBot, Johnfos, Finetooth, Stepheng3, DumZiBoT, Addbot, Benjamin Trovato, Legobot, Yobot, AnomieBOT, LilHelpa, Xqbot, GrouchoBot, RibotBOT, LucienBOT, Dinamik-bot, EmausBot, Bravo Foxtrot, Trymybestwikipedia, Helpful Pixie Bot, Cyberbot II, So categorical, GreenC bot and Anonymous: 9

- **Rocky Flats Plant** *Source:* https://en.wikipedia.org/wiki/Rocky_Flats_Plant?oldid=765030736 *Contributors:* Maury Markowitz, Fred Bauder, Tannin, Prz, Docu, Mulad, DJ Clayworth, Nv8200pa, Topbanana, Aetheling, DocWatson42, Fastfission, Rick Block, Deglr6328, Abdull, Jayjg, Kjkolb, Danthemankhan, Klestrob44, Woohookitty, Descendall, Kbdank71, Rjwilmsi, Vegaswikian, Daderot, 121a0012, Wavelength, MattWright, Skirandonee, Epolk, Gaius Cornelius, Shaddack, SEWilcoBot, Rronalds, CecilWard, Everyguy, IceCreamAntisocial, Groyolo, That Guy, From That Show!, SmackBot, Federalist51, AndySayler, Hmains, CSWarren, Dual Freq, Backspace, Soarhead77, Glacier109, Ser Amantio di Nicolao, John, Radiant chains, Eastlaw, Islandius, Nabokov, DonFB, Ebyabe, WVhybrid, OrenBochman, Nick Number, Tillman, MECU, McGhiever, Dricherby, VoABot II, Jllm06, Nyttend, The Anomebot2, Cgingold, Fulvius~enwiki, Jonwiener, Xenonice, Hugo999, Johnfos, Ian Struan, Plazak, LanceBarber, Eskovan, Marine-Blue, Lightmouse, Beerkeg420, TrufflesTheLamb, Senor Cuete, Kumioko (renamed), Sjl822, Hamiltondaniel, Mr. Stradivarius, Certayne, Xnatedawgx, Martarius, Pakaraki, VQuakr, Mild Bill Hiccup, Polyamorph, Isthisthingworking, MelonBot, Kbdankbot, Addbot, Download, Lightbot, Yobot, AnomieBOT, RevelationDirect, Mfbear, Jack B108, FrescoBot, Watchpup, Serols, Footwarrior, Wwjandrodo, Full-date unlinking bot, Julien1978, RjwilmsiBot, Oconnor181, Red Raevyn, Bomazi, NM Firefighter, Lukewarm revenge, Electriccatfish2, BG19bot, Reynold Gorden, Cyberpower678, Gorthian, FormerNukeSubmariner, Netherzone, Cablv1968, Mariahmevissen, Cyberbot II, Mers91, ColoradoArt, JJMC89 bot, GreenC bot, Bender the Bot and Anonymous: 49

- **Radioactive contamination from the Rocky Flats Plant** *Source:* https://en.wikipedia.org/wiki/Radioactive_contamination_from_the_Rocky_Flats_Plant?oldid=765010958 *Contributors:* Andrewman327, Topbanana, Bender235, Laurascudder, Eric Kvaalen, Bobrayner, SKopp, BD2412, Rjwilmsi, Ground Zero, Bgwhite, Wavelength, Messier110, RL0919, Federalist51, Ottawakismet, Colonies Chris, Nbound, DonFB, JustAGal, The Anomebot2, R'n'B, Hugo999, Johnfos, EricSerge, Mark v1.0, Senor Cuete, Mr. Stradivarius, VQuakr, XLinkBot, The Sage of Stamford, Yobot, AnomieBOT, Leavit2stever, PeaceLoveHarmony, Green Cardamom, FrescoBot, Redrose64, Tom.Reding, Trappist the monk, RjwilmsiBot, John of Reading, GA bot, Frietjes, Helpful Pixie Bot, Curb Chain, BG19bot, FormerNukeSubmariner, CitationCleanerBot, BattyBot, Khazar2, Kumioko, Cerabot~enwiki, Jodosma, Tom Prangnell, ZoomJag, Monkbot, Bender the Bot and Anonymous: 19

- **Hanford Site** *Source:* https://en.wikipedia.org/wiki/Hanford_Site?oldid=765749363 *Contributors:* Trelvis, BenBaker, Danny, Maury Markowitz, Heron, Mintguy, Fred Bauder, Isomorphic, Kosebamse, Andrewa, Lukobe, Zoicon5, Tempshill, RanchoRosco, David.Monniaux, Twang, Fastfission, Anville, Rick Block, Niteowlneils, BigBen212, Jason Quinn, Bobblewik, Woofles, Bumm13, TobinFricke, Sword~enwiki, Deglr6328, Abdull, Discospinster, Brianhe, Rich Farmbrough, Vsmith, Bender235, Ylee, Kwamikagami, Diomidis Spinellis, Cacophony, Bobo192, Scgallafent, Smalljim, Davidruben, Duk, Kjkolb, NickSchweitzer, Merenta, Alansohn, Neuhaus~enwiki, Amadeust, Axl, Suruena, Gene Nygaard, Blaxthos, Saxifrage, Daranz, Bushytails, Woohookitty, Camw, GregorB, John Hill, Atomicarchive, Jdill, Waninoco, Kbdank71, Canderson7, Rjwilmsi, Knave, Lockley, Graibeard, DoubleBlue, Williamborg, Nihiltres, Kolbasz, Sbove, 121a0012, Bgwhite, Simesa, Whosasking, The Rambling Man, RussBot, Red Slash, Cliffb, Limulus, Epolk, ScottMainwaring, Kirill Lokshin, Vanished user kjdioejh329io3rksdkj, Hawkeye7, Trovatore, Rjensen, Dogcow, Ragesoss, Dhollm, Tony1, Doncram, Elkman, FoxholeAtheist, BrianWhite, Cartwarmark, Fsiler, Curpsbot-unicodify, Andrew73, Tom Morris, SmackBot, Cla68, Yamaguchi 先生, Hmains, Ottawakismet, EncMstr, CSWarren, William Allen Simpson, Can't sleep, clown will eat me, Quartermaster, MJCdetroit, Prmacn, Peteforsyth, Valfontis, John, JohnI, Senrable, Vamoose, Mr Stephen, Meco, SandyGeorgia, Iridescent, Eastlaw, JForget, CmdrObot, Deathbob, Agathman, D3j409, Xila~enwiki, Cydebot, Hebrides, Kozuch, Brad101, Ebyabe, Thijs!bot, Epbr123, PaulLambert, Sturm55, Uruiamme, Dawnseeker2000, Cyclonenim, AntiVandalBot, Gerberb, Yellowdesk, BeefRendang, MER-C, Ph.eyes, Greg Comlish, Acroterion, Dragonnas, Exerda, Bongwarrior, VoABot II, Jllm06, Felix Stember, Steven Walling, Ciaccona, Schumi555, Chuckwatson, R'n'B, Dr Almost, Brothejr, J.delanoy, DrKay, Tlim7882, Nothingofwater, Brenman3, Plasticup, Funandtrvl, Hugo999, VolkovBot, Johnfos, Murderbike, Philip Trueman, GimmeBot, SoulOfChAoS, Oh Snap, Raymondwinn, Bbadjosh75, Milkbreath, Cmcnicoll, HiDrNick, Michael Frind, Logan, 1013-josh, JulieFlute, Dawn Bard, Caltas, Green-eyed girl, Lightmouse, RayAdomaitis, Afernand74, LonelyMarble, Dabomb87, Finetooth, Toliar, ClueBot, Kennvido, Yoshi Canopus, Niceguyedc, Piledhigheranddeeper, Paulcmnt, Ktr101, Excirial, Nymf, Socrates2008, Vivio Testarossa, Shinkolobwe, Sun Creator, Kaecyy, SounderBruce, Dana boomer, SoxBot III, DumZiBoT, Jamesflint, Dthomsen8, Northwesterner1, MarmadukePercy, RP459, Kbdankbot, HexaChord, Cabayi, Addbot, Rolland Goossens, Non-dropframe, Blechnic, LaaknorBot, Karl gregory jones, Aunva6, AtheWeatherman, Erutuon, Lightbot, Legobot, Yobot, EchetusXe, Fraggle81, TaBOT-zerem, Aboalbiss, AnomieBOT, VanishedUser sdu9aya9fasdsopa, Galoubet, Kimstring63, Chuckiesdad, Materialscientist, Citation bot, LilHelpa, Xqbot, Jolly Janner, RadiX, Irsobeastin, Mister shithead, Tacotastic13, Frijoledor, Msstreets, ConfusedOneOhOne, RightCowLeftCoast, Riventree, Sławomir Biały, Berriochoa, Zunerune, Citation bot 1, PeterCowan, Tannem7, Julien1978, Ruzihm, Trappist the monk, UltimatesocCer, Allen4names, Wdcraven, RjwilmsiBot, 1947enkidu, Salvio giuliano, DASHBot, WikitanvirBot, Ghostofnemo, Dewritech, Dcirovic, K6ka, Alfredo ougaowen, August571, Blackcatmeow, Arbnos, H3llBot, SporkBot, Jsayre64, Orange Suede Sofa, ClueBot NG, BrekekekexKoaxKoax, Loginnigol, CaroleHenson, Dougmcdonell, Tholme, Kndimov, Frze, Hacker254411, BattyBot, Guanaco55, Netherzone, Cyberbot II, FoCuSandLeArN, 237lawrence, Mogism, 331dot, Makecat-bot, Howpper, Nimetapoeg, Epicgenius, Wuerzele, Sagehugger, Acalycine, Stevelijek, Monkbot, Makkadkda, Kuc147, Erwigg, Gfgbb, Brecapla000, Idasod, Historygirl1414, Jimbo8391, DABurbank, Shorty smalld, GreenC bot, Abenbelkacem, Bender the Bot and Anonymous: 184

- **Bikini Atoll** *Source:* https://en.wikipedia.org/wiki/Bikini_Atoll?oldid=766079875 *Contributors:* Mav, Zundark, The Anome, Michael Hardy, Ixfd64, Skysmith, Julesd, Andres, Kaihsu, Agtx, Doradus, DJ Clayworth, Head, Robbot, Tomchiukc, Postdlf, Texture, Hadal, David Edgar, Isopropyl, Davidcannon, Jyril, YanA, Folks at 137, Fastfission, Karn, Gilgamesh~enwiki, Peter Ellis, Btphelps, Gadfium, J. 'mach' wust, Beland, PFHLai, Cornischong, B.d.mills, D6, Discospinster, Rich Farmbrough, Vsmith, Bender235, Goplat, Sfahey, Gertjan R., Summer Song, Art LaPella, Mdhowe, Robotje, Bollar, .:Ajvol:., Apoltix, Darwinek, Polylerus, CKlunck, Diego Moya, Stillnotelf, MattWade, Wtmitchell, RPH, TenOfAllTrades, RandomWalk, Kitch, Urbster1, Before My Ken, Ratzer, Former user 2, Duncan.france, Zilog Jones, Firien, Zzyzx11, Emerson7, Rtcpenguin, GrundyCamellia, Rjwilmsi, Fernando Reis, Bill37212, Xanderall, BrickParade, MChew, Hottentot, Kolbasz, Drumguy8800, Travis.Thurston, Hibana, Chobot, Bgwhite, Crosstimer, YurikBot, Huw Powell, JarrahTree, Test-tools~enwiki, Czyrko, Vivaldi, BOT-Superzerocool, Wknight94, Caroline Sanford, AjaxSmack, Takethemud, FF2010, Nikkimaria, Closedmouth, Chanheigeorge, Th1rt3en, Little Savage, Petri Krohn, SmackBot, Brick Thrower, Delldot, Hmains, ERcheck, Icemuon, Squiddy, Chris the speller, MalafayaBot, Droll, CrimsonFlash, CSWarren, JRPG, MJCdetroit, Exec8, Xiner, Konczewski, Niranjan108, WestA, Memming, Khoikhoi, Theo10, Nakon, Lambiam, Nareek, John, J 1982, SilkTork, XinJeisan, Gnevin, RomanSpa, Loadmaster, Lady of the dead, Stwalkerster, Andyroo316, Neddyseagoon, Marcusclancypearl, Udibi, Dl2000, Hu12, Pqrstuv, Pimlottc, Bruinfan12, Coffee Atoms, SeanMD80, George100, Sakurambo, CRGreathouse, David@sickmiller.com, Wafulz, JohnCD, Kieranmullen, Themightyquill, Cydebot, Corpx, Editor at Large, Stephhunt, Thijs!bot, Marek69, Horologium, James086, Anggerik, JustAGal, Uruiamme, Elert, Mentifisto, Salgueiro~enwiki, Mutt Lunker, SkoreKeep, Jaredroberts, Ironiridis, ClassicSC, Tigga, Scottcmu, X4096, Wasell, Magioladitis, Connormah, Vaylen, Hullaballoo Wolfowitz, Nyttend, The Anomebot2, Gabriel Kielland, Lloyd borrett, Edward321, Pikazilla, Ga1lyons, Jdauie, J.delanoy, OhNoPeedyPeebles, Coexerj145, Victuallers, Collegebookworm,

Naniwako, Ipigott, Emanuel Harper, Chikinsawsage, Tiyoringo, Evb-wiki, Chpfeiffer, Mjzwick, Funandtrvl, ACSE, Hugo999, VolkovBot, Johnfos, TXiKiBoT, Tavix, GDonato, Oh Snap, Jonpetty, GlobeGores, Xsarahrahx, Masaqui, Bikinijack, Shama595, Efilipek, Obaidz96, SieBot, MuzikJunky, Trackinfo, BotMultichill, Undead Herle King, Christian.cantrell, Oxymoron83, SilverbackNet, Lightmouse, Senor Cuete, Benon-iBot~enwiki, Eugen Simion 14, Jaan, Canglesea, Bekuletz, Sfan00 IMG, MBK004, ClueBot, PipepBot, Darthveda, Neverquick, Deselliers, Nymf, Bruceanthro, Germanicus Alan, Millionsandbillions, DMG-42, Proxy User, Thehelpfulone, Stepheng3, Jellyfish dave, AP500, Wnt, Miami33139, Finalnight, Mjbauer95, RexxS, Runnynose47, Facts707, SilvonenBot, Cmr08, MystBot, Dubmill, Addbot, Xp54321, Dianearmitage, Sir Foley, Numbo3-bot, Darkness3123, Lightbot, مانی, Bebestbe, Luckas-bot, Yobot, AnomieBOT, Rubinbot, Jim1138, Xufanc, Purcellj, Citation bot, ArthurBot, Quebec99, Xqbot, Night w, I am Me true, Anonymous from the 21st century, GrouchoBot, Armbrust, Abce2, GhalyBot, Krj373, BenzolBot, Freebirds, Pinethicket, Jonesey95, RedBot, MastiBot, Encyclopedist1, Kgrad, Yunshui, Lotje, Bushzm, Dinamik-bot, Jeffrd10, Ripchip Bot, Beyond My Ken, DASHBot, EmausBot, Racerx11, GoingBatty, RA0808, The Mysterious El Willstro, Dcirovic, AvicBot, Brandmeister, Mjbmrbot, Christian.nuccio, ClueBot NG, CactusBot, MIKHEIL, Mateo Flecha, The Goiter*Guru, Widr, Newyorkadam, Helpful Pixie Bot, BG19bot, Metricopolus, Ollieinc, LoneWolf1992, Tsar Bomba, ZappaOMati, Mediran, Runlevel1, Scotboy321, Earth100, Mogism, Cam9477, LemonDevo, Разрывные, Capture-n-Kill Kommando, Jehornalis, Jamesmcmahon0, Jakec, Court Appointed Shrub, Asmetr, Avi8tor, Volker Alexander, Ugog Nizdast, SarahRMadden, Nationalfannz, CatcherStorm, Goldwright, Concord hioz, Monkbot, Qwertyxp2000, ChamithN, DangerousJXD, Matt ford18, CBHusky, Wzrogers, KasparBot, Randonman, Ola Språkgubbe, Dilidor, LumberJack.357, Arthurteb303, InternetArchiveBot, JJMC89 bot, KGirlTrucker81, GreenC bot, MordeKyle, ColouredFrames, Jhumphries9, Bender the Bot and Anonymous: 268

- **Desert Rock exercises** *Source:* https://en.wikipedia.org/wiki/Desert_Rock_exercises?oldid=750558745 *Contributors:* Tomekpe, Cydebot, Crowish, SkoreKeep, XavierGreen, Addbot, Archon 2488, Tucoxn, Boundarylayer, ZéroBot, Bomazi, ClueBot NG, Helpful Pixie Bot, Dainomite, 220 of Borg, Curtgreeneyes, Atomicvetkid and Anonymous: 8

- **Totskoye nuclear exercise** *Source:* https://en.wikipedia.org/wiki/Totskoye_nuclear_exercise?oldid=751282293 *Contributors:* Maury Markowitz, Cherkash, Altenmann, Wwoods, Cloud200, Mzajac, Maximaximax, Rich Farmbrough, Igny, Tckma, Tabletop, Jno, Quuxplusone, Bgwhite, Jimp, RussBot, Bleakcomb, Hydrargyrum, Alex Bakharev, Czyrko, Welsh, Chesnok, SmackBot, PeterReid, Hmains, EncMstr, AndySimpson, Qqs83, RookZERO, Hi There, Cydebot, Tec15, Cancun771, Emandel, Thijs!bot, Kubanczyk, Superzohar, TAnthony, Catslash, Buckshot06, Manuelch10, Aleksandr Grigoryev, Maurice Carbonaro, Hodja Nasreddin, Asharidu, Hugo999, Johnfos, Constantine Gorov, Janggeom, ImageRemovalBot, NickCT, MYury, John Nevard, Coinmanj, DumZiBoT, Luke Ilott, Aleclitvinov, Addbot, Mckinley99, Alandeus, Fireaxe888, Yobot, AnomieBOT, Xqbot, GenQuest, Metricmike, MastiBot, Widar23, My very best wishes, Aporio, EmausBot, The Hollow Man2010, Uploadvirus, The Madras, Eriba-Marduk, Aulacopalpus256, ChuispastonBot, ClueBot NG, BG19bot, Netherzone, Monkbot, Mhhossein and Anonymous: 36

- **Operation Plumbbob** *Source:* https://en.wikipedia.org/wiki/Operation_Plumbbob?oldid=759071834 *Contributors:* Wesley, Bryan Derksen, The Anome, Chuckhoffmann, Michael Hardy, Crenner, CORNELIUSSEON, Wolfkeeper, Fastfission, Rich Farmbrough, Pavel Vozenilek, Night Gyr, Bender235, Femto, Hektor, AjAldous, Ashley Pomeroy, Fourthords, Middenface, Gene Nygaard, Sandius, Robertwharvey, GregorB, Offtherails, Tim!, Capnez, Vegaswikian, Kolbasz, Jrtayloriv, Bgwhite, Pigman, Bullzeye, Howcheng, Ospalh, I'm me 101, NorsemanII, Jmackaerospace, Groyolo, SmackBot, 1dragon, Sam8, Auton1, Hmains, Modest Genius, Thomas Connor, Filpaul, NeilFraser, Will Beback, JunCTionS, John, Jidanni, Mgiganteus1, Kvng, Boardhead, JamesLucas, Cancun771, Rspeed, PamD, Kushami, Woody, Nick Number, Alexduric, Dawnseeker2000, SkoreKeep, Emax0, Magioladitis, The Anomebot2, Cgingold, CommonsDelinker, Dinkytown, Ops101ex, Cs302b, Johnfos, TXiKiBoT, Petebutt, Rdfox 76, Thunderbird2, Fratrep, Hyperionsteel, Haydn likes carpet, Binksternet, Niceguyedc, Ktr101, Coinmanj, Razorflame, Johnuniq, InternetMeme, Rreagan007, Jim Sweeney, Kristianrj, Addbot, Pkkphysicist, Lightbot, OlEnglish, Luckas-bot, Yobot, Uniwersalista, WikUWerHere, Benzen, FrescoBot, D'ohBot, Sso1976, RjwilmsiBot, Boundarylayer, Dewritech, ERRORHUNT, A2soup, Caspertheghost, Peterh5322, Bomazi, -xwingsx-, One.Ouch.Zero, Ebehn, Whoop whoop pull up, Mikhail Ryazanov, ClueBot NG, CherryX, Frietjes, Helpful Pixie Bot, Whytk, BG19bot, AvocatoBot, Trevayne08, FormerNukeSubmariner, CitationCleanerBot, Siderz, BattyBot, Mattholomew, Guywholikesca2+, Cerabot~enwiki, Joshtaco, Vladkrapp, Retu K, Monkbot, Airkingaid, StjJackson and Anonymous: 67

- **Windscale fire** *Source:* https://en.wikipedia.org/wiki/Windscale_fire?oldid=765042059 *Contributors:* Bryan Derksen, PierreAbbat, Maury Markowitz, Edward, Aarchiba, Jll, Jengod, Pstudier, Wereon, Danceswithzerglings, Alan Liefting, DocWatson42, Beefman, Dirtside, Mzajac, Georgesch4, Corti, Mike Rosoft, Rupertslander, Bender235, Neko-chan, Ylee, Orlady, R. S. Shaw, Kjkolb, Mh26, Linuxlad, TheParanoidOne, Rwendland, Rapscallion, Gene Nygaard, Dan100, BlueCanoe, Pol098, Ardfern, Kmg90, Mgolden, Jacj, Brownsteve, BD2412, Rjwilmsi, Tim!, Urbane Legend, Strait, Jdhowens90, DVdm, Bgwhite, Ahpook, YurikBot, Wavelength, Rapido, RussBot, Limulus, Gaius Cornelius, Lavenderbunny, Bullzeye, Dysmorodrepanis~enwiki, Ino5hiro, Afiler, Andreaskem, Schnob Reider, Pash, Closedmouth, Vicarious, HereToHelp, Whobot, Mais oui!, Nekura, PKnight, Jsnx, SmackBot, Fireworks, Esradekan, Gigs, Mauls, Hmains, Grassynoel, Squiddy, Bluebot, Cadmium, ProtocolOH, John, Mathel, Rubisco~enwiki, IanOfNorwich, Vanisaac, Pmyteh, Van helsing, Joelholdsworth, Cydebot, Trident13, Nabokov, Rads190, Sukisuki, Sobreira, Roggg, Stevvvv4444, Kbthompson, Maj0r Tom, AlmostReadytoFly, Acroterion, VoABot II, Roches, The Anomebot2, DogNewTricks, Cabmille, AntiSpamBot, Felix Han, Serge925, ArmadilloProcess, Kd4dcy, Krispopsa, Hugo999, Johnfos, PeterGHughes, Martinevans123, R898713, Piperh, PDFbot, Jamesfett, Andy Dingley, Go-in, Tomaxer, The Equalizer, TinribsAndy, Cj1340, Overlord11001001, Digwuren, Messagetolove, Colfer2, Scorpion451, ViennaUK, Steven Crossin, Moletrouser, SoxSweepAgain, Ei2g, Nailedtooth, Billwilson5060, Eddiehig16, Daniel D.L., KWSMaster, Northernhenge, PixelBot, Darkicebot, Tarheel95, FellGleaming, Lstanley1979, Roentgenium111, DOI bot, ExecPE, Leszek Jańczuk, JLWinkler, Yobot, AnomieBOT, Captain Quirk, Citation bot, J G Campbell, Gigemag76, Seligne, Elemarth, FrescoBot, LucienBOT, Bob mullins, Aliotra, Jjfreem, Louperibot, SHtev, Citation bot 1, Jonesey95, Beck daross, Sykotnik, Aanabar, Gnomus, Fastilysock, RjwilmsiBot, Ripchip Bot, Phlegat, EmausBot, Boundarylayer, Primefac, RenamedUser01302013, Wikipelli, Dcirovic, Bravo Foxtrot, ZéroBot, A2soup, HeyStopThat, Noggo, ChuispastonBot, Whoop whoop pull up, Autodidact1, Mikhail Ryazanov, ClueBot NG, Benjis1999, Ahnrenene, Helpful Pixie Bot, Bigfun137, Bibcode Bot, BG19bot, Wdzinc, 123Beaverboy, Siphon06, Dexbot, Joeinwiki, Tom Prangnell, CHADTHOMASSIE, ZX John, Monkbot, Tigercompanion25, Von Callay, My Chemistry romantic and Anonymous: 179

- **Kyshtym disaster** *Source:* https://en.wikipedia.org/wiki/Kyshtym_disaster?oldid=764726036 *Contributors:* Gabbe, Robertb-dc, Wetman, Dimadick, Twang, Cdang, Alan Liefting, Rafaelgr, RScheiber, Mzajac, Abdull, Bender235, Orlady, Axe-Lander, Darwinek, Obradovic Goran, Bobrayner, Daniel Case, Rjwilmsi, Lockley, Guinness2702, Mindme, Alex Sims, Kolbasz, Ytrottier, Eleassar, Anomalocaris, Arjunasbow, Arthur Rubin, Bondegezou, Petri Krohn, SmackBot, Accurimbono, Gilliam, Hmains, Ottawakismet, Jwillbur, Squigish, Cybercobra, ArgentAngel, Ser Amantio di Nicolao, John, Joeylawn, MikeWazowski, RekishiEJ, Lavateraguy, Cydebot, Tec15, Doug Weller, Jmg38, Bethpage89,

Vhorvat, Sluzzelin, Cgingold, LorenzoB, DandyDan2007, Maurice Carbonaro, Nothingofwater, Serge925, ACSE, Hugo999, VolkovBot, John-fos, Wrev, Oshwah, Andy Dingley, DavidAndersen, ClueBot, Nyctea, Piledhigheranddeeper, Avoran, Nymf, Alexbot, PixelBot, Trulystand700, Addbot, AVand, The Bushranger, Ben Ben, Luckas-bot, Amirobot, AnomieBOT, Wikieditoroftoday, Materialscientist, Yonseca, Citation bot, Xqbot, Пипумбрик, SassoBot, Kyng, Uusijani, Qurozon, YOKOTA Kuniteru, Prisonermonkeys, Dankarl, Cordovao, Citation bot 1, Calistemon, Jaykaybay, Pompous Trihedron, Andrewmyles, 564dude, Lhollo, RjwilmsiBot, Undescribed, WikitanvirBot, The Mysterious El Willstro, Bravo Foxtrot, AvicBot, Erlenmayr, Whoop whoop pull up, Mikhail Ryazanov, ClueBot NG, Dakaminski, Helpful Pixie Bot, Leopd, Aretheysafe, Wzrd1, Neøn, WebTV3, Bartonsenior, Dexbot, HeavyBinary, Frosty, Dristarg, Court Appointed Shrub, Kennethaw88, YiFeiBot, Tharvik, Os-bournehutch, Monkbot, Nonstopmaximum, Tigercompanion25, Trackteur, JudeccaXIII, Sam Brown, Costock, Andy Evans of Michi-gama, Troy Oakes, JokerDaTroll and Anonymous: 60

- **Santa Susana Field Laboratory** *Source:* https://en.wikipedia.org/wiki/Santa_Susana_Field_Laboratory?oldid=763968141 *Contributors:* Ax-elBoldt, Mac, Docu, PaulinSaudi, WhisperToMe, Robertb-dc, Nv8200pa, David.Monniaux, Madelinefelkins, Rfc1394, Timrollpickering, DocWatson42, Rpyle731, Andycjp, Abdull, User2004, Szyslak, Unquietwiki, Krellis, Wjbean, Ricky81682, Apoc2400, Woohookitty, Kb-dank71, Rjwilmsi, The wub, NekoDaemon, Paul foord, Gurch, Kolbasz, Tedder, Bgwhite, Wavelength, TexasAndroid, Epolk, Hydrargyrum, C777, Gaius Cornelius, Shaddack, SEWilcoBot, RFBailey, Smvans7, Mike Dillon, Fang Aili, JLaTondre, Revengeofthynerd, That Guy, From That Show!, SmackBot, Yamaguchi 先生, Hmains, TimBentley, Droll, OrphanBot, Backspace, Zvar, Will Beback, CmdrObot, Cy-debot, Fnlayson, BillySharps, Brian1078, Nick Number, K7aay, Alphachimpbot, Jllm06, Animum, Cgingold, LorenzoB, Jeffconn, Jacobst, BigrTex, Ohms law, Happytrombonist16, Johnfos, Niceley, Sdsds, Freebiegrabber, Rocketdynewatch, Michael Frind, Lmharnisch, Jj ee bb, WereSpielChequers, Cbl62, Lightmouse, Afernand74, Dravecky, ImageRemovalBot, SalineBrain, Robin731, Mild Bill Hiccup, Jhapeman, Jeremiestrother, DumZiBoT, WikHead, Addbot, Debresser, Tassedethe, Lightbot, The Bushranger, Ben Ben, Yobot, AnomieBOT, Bluegill-TriplePrime, Citation bot, MauritsBot, DSisyphBot, FrescoBot, Jonesey95, Overjive, Reaper Eternal, Look2See1, Dewritech, Moswento, ChuispastonBot, EdoBot, ClueBot NG, Alara69, Snotbot, Angelobellomo, Onehstrybuff, Frannyd22, BG19bot, Dee McPherson, BattyBot, Netherzone, Fettlemap, Armanjafari, Mogism, Ueutyi, Wuerzele, Bazzy11, Monkbot, JackArmstrong1463, Chatsworthhistory, Bender the Bot, Jamierakers, LewFamAdv and Anonymous: 66

- **Soviet submarine K-19** *Source:* https://en.wikipedia.org/wiki/Soviet_submarine_K-19?oldid=766408447 *Contributors:* The Epopt, Dze27, Andre Engels, Tox~enwiki, Lightning~enwiki, Александър, Jll, Schneelocke, David Newton, Kbk, Kierant, Averell23, Topbanana, JonathanDP81, Dale Arnett, Moriori, Lupo, Erics, Ezhiki, Matt Crypto, Bobblewik, Btphelps, Karl Dickman, Davidstrauss, N328KF, Jkl, Rich Farmbrough, Alistair1978, Bender235, Aranel, Zscout370, 3mta3, AndromedaRoach, Titanium Dragon, Axeman89, Nightstallion, Walshga, Woohookitty, Bellhalla, Zealander, Jeff3000, Ardfern, Hbdragon88, Deltabeignet, BD2412, Wachholder0, Ketiltrout, Biederman, Mark Sublette, Quuxplusone, Murphyscat, Cloviscassiani, YurikBot, Traianus, Paulschreiber, RussBot, Ytrottier, Gaius Cornelius, Shaddack, Alex Bakharev, Lavenderbunny, Stassats, Schrepfler, Michalis Famelis, Julienlecomte, Robyvecchio, David Underdown, VodkaJazz, John-moore7e7400, Curpsbot-unicodify, Sailboatd2, Trekster, Attilios, SmackBot, Sam8, IstvanWolf, Brianski, Captain scarlet, Valley2city, Chris the speller, Bluebot, Philosopher, Modest Genius, Jmlk17, Fitzhugh, DMacks, Sjester, Acdx, Sambot, John, Gobonobo, AEMoreira042281, Irides-cent, Clarityfiend, Oorgh, Neil Evans, Winston Spencer, HDCase, Soccerdudeet, Cydebot, Jackyd101, Tec15, Myscrnnm, Aldis90, Slo186, CynicalMe, SGGH, Dawkeye, Nick Number, Uruiamme, AntiVandalBot, Luckz, Youknowthatoneguy, Caroldermoid, Bernd vdB~enwiki, Pawl Kennedy, LorenzoB, MartinBot, Limbo@MX~enwiki, Nono64, Sindresolberg, Khathi, Zipzipzip, ElectricValkyrie, Mufka, Tanaats, Kei-thF0064, Idioma-bot, Law Lord, Xandell, Rei-bot, Yegor Chernyshev, Ahm2307, Plutonium27, Prinz.W, SieBot, VVVBot, Lightmouse, Void-Point, Martin Velek, Wuhwuzdat, Wikiman999224, Matrek, All Hallow's Wraith, UKoch, Rockfang, Alexbot, 7&6=thirteen, Thewellman, Antediluvian67, DumZiBoT, Magus732, M.nelson, LaaknorBot, SCSInet, Mdnavman, Lightbot, Anyname21, Miden, Rave, Legobot, TaBOT-zerem, Zarabelda, AnomieBOT, Rubinbot, Xqbot, RicHard-59, Hébus, Backspacekey, Kobrabones, I dream of horses, Kjnelan, Mr68000, NicoScPo, Trappist the monk, Tubbs662, Dlambe3, EmausBot, John of Reading, Markwpowell64, A2soup, Ὁ οἶστρος, Schleppnik, Targaryen, Dragon623, Widr, Sephalon1, Tom soldier, BG19bot, Illyukhina, Neøn, Kendall-K1, Mdy66, Dimmizer, Chalim Kenabru, RoxyFlox, ÄDA - DÄP, Dexbot, Henrygg98, Monkbot, Cornersss, Vinegarymass911, Zachpoo, Kontakr, I2padams, TitaniumIron and Anonymous: 180

- **SL-1** *Source:* https://en.wikipedia.org/wiki/SL-1?oldid=765039172 *Contributors:* Ewen, Gabbe, Wwwwolf, Mac, Andrewa, Azazello, GCarty, JonathanDP81, Fvw, Jamesday, Chris Roy, Rfc1394, Rhombus, Catbar, Matt Gies, DocWatson42, Lproven, Inkling, Prosfilaes, Plasma east, Abdull, D6, Rich Farmbrough, Alistair1978, Bender235, Neko-chan, Sietse Snel, Aaron D. Ball, Mizchalmers, R. S. Shaw, Kjkolb, Tom Yates, Radical Mallard, Dhartung, DV8 2XL, Gene Nygaard, Dan100, Tafinucane, Woohookitty, Ardfern, Rjwilmsi, Lockley, Vegaswikian, The wub, Ground Zero, NekoDaemon, Jared Preston, Simesa, Wavelength, Aekolman, Limulus, Christy747, Shaddack, Baldur~enwiki, Howcheng, Ke4djt, LegalBeagle, Ospalh, John Sheu, Georgewilliamherbert, Bayerischermann, Ringler, CmdrFirewalker, Petri Krohn, DasBub, Groyolo, Victor falk, KnightRider~enwiki, SmackBot, Hux, Incnis Mrsi, Hmains, Ottawakismet, Kasyapa, Croquant, Colonies Chris, Backspace, Elendil's Heir, Markmit, Glacier109, Calvados~enwiki, John, Vgy7ujm, Gobonobo, Langhorner, Dicklyon, Meckser~enwiki, DELutz, Highspeed, Carl-willis, CmdrObot, Baiji, Benwildeboer, Cydebot, A876, Zgystardst, Zginder, Cuhlik, Thijs!bot, Dtgriscom, Ekashp, Uruiamme, Armist~enwiki, AniRaptor2001, MER-C, Magioladitis, Wowolaf, LorenzoB, Jseliger, Hbent, Vytal, Wikianon, Wittyname, Read-write-services, Padillah, Third-dright, Cdamama, Trumpet marietta 45750, DadaNeem, Underjack, Teewinot, Makrisj, Speciate, Hugo999, Johnfos, Yilloslime, Cootiequits, Joseph A. Spadaro, Bporopat, AHMartin, 4wajzkd02, OKBot, Pauljoffe, FieldMarine, Rift621, XLinkBot, Fujimuji, Addbot, Jacopo Werther, AkhtaBot, Neweb, Download, Protonk, Windward1, Fraggle81, AnomieBOT, Götz, Captain Quirk, Eilandtje, Materialscientist, Citation bot, Beeline23, Shirik, Rb88guy, Rainald62, Full-date unlinking bot, Lotje, Gaffershellofishyboy, Markos Strofyllas, WikitanvirBot, Angrytoast, Ida Shaw, Josve05a, A2soup, JoeSperrazza, Jlsicard, Rocketrod1960, Frietjes, Meltdown627, Helpful Pixie Bot, Tomhung357, Johnny Squeaky, Asauers, ChrisGualtieri, Jamesallain85, Nouniquenames, Loganfalco, Jloughry, GRK Astronomer, Kennethaw88, American Money, Monkbot, Tigercompanion25, Fightallignorance, Yoshams, EERob, Bender the Bot, Lsun42256 and Anonymous: 103

- **Johnston Atoll** *Source:* https://en.wikipedia.org/wiki/Johnston_Atoll?oldid=763754973 *Contributors:* The Epopt, Bryan Derksen, Koyaanis Qatsi, Rmhermen, Maury Markowitz, Hoshie, Docu, TUF-KAT, Jiang, Jonathan Griffitts, Vanished user 5zariu3jisj0j4irj, Marshman, Anon-Moos, Olathe, Jerzy, Denelson83, PuzzletChung, Branddobbe, Robbot, Pigsonthewing, Sbisolo, Postdlf, Filemon, Alan Liefting, DocWatson42, Whitti, Folks at 137, Fastfission, Bkonrad, Wyss, Gilgamesh~enwiki, Bobblewik, Golbez, Stevietheman, Beland, Känsterle~enwiki, Kaldari, Urhixidur, Acad Ronin, D6, Dufekin, ChrisRuvolo, A-giau, Twinxor, Vsmith, Calion, Xezbeth, Sarrica, Risacher, Bender235, Evice, Can-isRufus, *drew, El C, Kwamikagami, Aude, Whosyourjudas, Bollar, Viriditas, .:Ajvol:., Giraffedata, Sparkgap, Pharos, Jumbuck, Ua747sp, Sl, Djlayton4, Hadlock, Henry W. Schmitt, Axeman89, Richard Weil, Deror avi, Ratzer, Miss Madeline, Kelisi, Isnow, BD2412, Election-world, Snarfhound, Nightscream, Koavf, Erebus555, Ground Zero, RexNL, Gheorghe Zamfir, Taichi, Chobot, Bgwhite, YurikBot, Wavelength,

Cgblaine2, Aquillion, Kjkolb, Nk, Rajah, Bobbis, Tcp-ip, Pschemp, TACD, Helix84, Sam Korn, Krellis, Pearle, RazorChicken, Jonathunder, Wonglijie, Espoo, Storm Rider, Alansohn, Gary, PS-2507, Udo Altmann, QVanillaQ, Civvi~enwiki, Arthena, Rd232, Keenan Pepper, Trainik, Jeltz, Supine, Lord Pistachio, Riana, Ashley Pomeroy, Yamamoto114, Lightdarkness, Kurieeto, Mrmiscellanious~enwiki, Redfarmer, Neilmckillop, Mysdaao, Ruleke, Rwendland, Bart133, Scott5114, Snowolf, GeorgeStepanek, Miltonhowe, Jannev~enwiki, Wtmitchell, Velella, Tauwasser, BRW, Uucp, Mark Bergsma, Uffish, Evil Monkey, RJFJR, Lapinmies, RainbowOfLight, CloudNine, Dave.Dunford, Versageek, Gene Nygaard, Redvers, Axeman89, Karderio, Dan100, Richard Weil, BadSeed, SmthManly, Tafinucane, RyanGerbil10, Falcorian, Madame, Tristessa de St Ange, Itinerant, Bobrayner, R5gordini, Jkt, Angr, Richard Arthur Norton (1958-), Kelly Martin, Distantbody, Reinoutr, Woohookitty, Sinanozel, Milen~enwiki, Ataru, Camw, LOL, Swamp Ig, PoccilScript, Jftsang, Oliphaunt, Carcharoth, BillC, Ekem, KrisK, MrWhipple, Bratsche, Astator, Ruud Koot, Urod, WadeSimMiser, Ardfern, Duncan.france, Hdante, Jwanders, Tabletop, Chris Buckey, Sicooke, GregorB, Male1979, J M Rice, Jagvar, Kralizec!, Ryan Reich, Wayward, NightOnEarth, Prashanthns, Gimboid13, ShadowHunter, Palica, Pfalstad, Gallaghp, Holek, MrSomeone, Youngamerican, Mandarax, RedBLACKandBURN, 0pera, Graham87, TaivoLinguist, Stromcarlson, Deltabeignet, Magister Mathematicae, Ajcomeau, BD2412, Qwertyus, Li-sung, Monk, FreplySpang, JIP, BorgHunter, Chickenhead, Phoenix-forgotten, Ciroa, Crzrussian, Rjwilmsi, Tim!, Nightscream, Dr.Gonzo, .digamma, Koavf, Ghoat, Zincomog, Joffan, Jivecat, Urbane Legend, Vary, Loudenvier, Darthsco, Bruce1ee, Collard, Alban, Smithfarm, Vegaswikian, Nneonneo, Ligulem, Gene Wood, StephanieM, Czalex, Docether, TheIncredibleEdibleOompaLoompa, Williamborg, Fred Bradstadt, Ev, MJGR, JPat, KaiMartin, Tommy Kronkvist, Titoxd, CCRoxtar, Reve100, Gringo300, SchuminWeb, G Clark, Ground Zero, Old Moonraker, Winhunter, Ysangkok, Nihiltres, Nivix, Chanting Fox, Itinerant1, John Z, Mark Sublette, Rune.welsh, Pathoschild, RexNL, Gurch, Kolbasz, Intgr, Atrix20, Preslethe, Wikipedia Admin, Priyanksharma, Alphachimp, Okto8, Srleffler, Jswanhart, Russavia, Coolhawks88, King of Hearts, Chobot, Jdhowens90, Benlisquare, Jared Preston, Benjamin Gatti, Bgwhite, Ahpook, Simesa, Yacoob, WriterHound, Gwernol, Tone, Whosasking, IHUB.org Founder, UkPaolo, RedGreenInBlue, The Rambling Man, YurikBot, Wavelength, Eirik, Angus Lepper, Themepark, Tommyt, Freerick, Sceptre, Xcrivener, Jimp, Brandmeister (old), NMS, Midgley, Petiatil, Armistej, Sillybilly, Muchness, Brandre, Lexi Marie, Chris Capoccia, Limulus, Splette, IByte, Cpc464, Akamad, Stephenb, Barefootguru, Njh~enwiki, Shell Kinney, Gaius Cornelius, Sandpiper, Shaddack, Ihope127, Neilbeach, Wimt, Brooza, Daveswagon, Anomalocaris, Axel-Berger, Evstafiev, Shanel, NawlinWiki, Alex Whittaker, Alzhaid~enwiki, Mipadi, Janke, Grafen, Joel7687, Mmccalpin, Berend de Boer, RazorICE, Howcheng, BlackAndy, Seegoon, Jeff Carr, Irishguy, Nick, Matnkat, Mosquitopsu, Johantheghost, Anetode, Andygui, Cholmes75, Diotti, Matticus78, Cruise, RL0919, Snagglepuss, Tony1, Zwobot, Ospalh, MakeChooChooGoNow, Egh0st, Dbfirs, BOT-Superzerocool, Gadget850, CorbieVreccan, Klutzy, Jhinman, Caspian, Fenian Swine, Black Falcon, User27091, MaxDZ8, Volantares, Natmaka, Jesusjonez, Noosfractal, Syrae, Fallout boy, Roger Gianni~enwiki, FF2010, Georgewilliamherbert, Johndrinkwater, Superdude99, 21655, Entropy238, 2over0, SilentC, Ninly, Ageekgal, Nikkimaria, Theda, Closedmouth, Lorus77, Th1rt3en, Reyk, Jmackaerospace, BrownHornet21, JQF, Sean Whitton, BorgQueen, Petri Krohn, JoanneB, Ipstenu, SPTimoshenko, CWenger, Hurricane Devon, HereToHelp, ArielGold, ChipperGuy, Curpsbot-unicodify, Rwh86, Garion96, Nixer, Jack Upland, RunOrDie, Paul D. Anderson, Aryah, Junglecat, DasBub, CeeKay, Auroranorth, Elliskev, Ben.c, Groyolo, DVD R W, Tom Morris, Bibliomaniac15, Dandelions, Mark76uk, Itub, IslandHopper973, FlashFM, Attilios, Biddlesby, Iorek85, AtomCrusher, SmackBot, Illuminattile, Smadge1, Dweller, Kuban kazak, Unschool, Haymaker, DocS, Hux, Oxford Comma, Dookie611, Arniebuteft, InverseHypercube, KnowledgeOfSelf, Hydrogen Iodide, Bjelleklang, Unyoyega, Pgk, CyclePat, C.Fred, Mr link, Wegesrand, Blue520, Kilo-Lima, Btm, Chairman S., Nickst, Jedikaiti, Okinasevych, Pandion auk, VxP, Jrockley, Renesis, Eskimbot, Hardyplants, Mdd4696, Alsandro, Yauhin, Gaff, PeterSymonds, Peter Isotalo, Gilliam, 9591353082, Hmains, Oscarthecat, TRosenbaum, Grassynoel, Durova, Dinosnake, Poulsen, Chris the speller, Bluebot, Ottawakismet, JonRidinger, SynergyBlades, Cadmium, 91.josh, Sept298901, Tree Biting Conspiracy, Carbonrodney, KolyaFrankovich, SchfiftyThree, RayAYang, Joel.Gilmore, JoeBlogsDord, Heikoh, Oni Ookami Alfador, Exitr, Whispering, Baa, Robth, DHN-bot~enwiki, Cassivs, A51Abductee, Darth Panda, Verrai, MaxSem, Scwlong, Modest Genius, Veggies, Zsinj, Badger151, Can't sleep, clown will eat me, Bigfoot hunter, Dbiagioli, Kronn, Koubiak, Vankrugermeer, Scarletsmith, OrphanBot, Sephiroth BCR, Onorem, Jennica, MJCdetroit, Gurps npc, Ukrained, Britmax, Hippo43, Darwin's Bulldog, Aces lead, CorbinSimpson, Addshore, Sholom, Thrane, Magmagirl, The tooth, Stevenmitchell, Mandrak, Jmlk17, Theanphibian, KathyRyan, Ghiraddje, Rgrant, Khukri, Nakon, Axel-berger, Bob92, VegaDark, Eroyce, Got Milked, RandomP, Fuzzypeg, S-man, Gump Stump, Kotjze, Copysan, Daniel.Cardenas, Roger.lee, Riurik, DDima, Rodeosmurf, Wikipedical, Tesseran, Alþykkr, Ohconfucius, Will Beback, Thejerm, Angela26, Bige1977, Snowgrouse, SashatoBot, Spinolio, Lambiam, Esrever, Severisth, Krashlandon, Wamini, Swatjester, Ser Amantio di Nicolao, Nareek, Molerat, John, Skittlesjc, Euchiasmus, Zapptastic, Lapaz, Ulner, Vgy7ujm, J 1982, Calum MacÙisdean, Simongraham, Kingfisherswift, Lazylaces, Evenios, JoshuaZ, Minna Sora no Shita, Scetoaux, IronGargoyle, Keber, Speedboy Salesman, A. Parrot, Remigiu, JHunterJ, BillFlis, Illythr, Slakr, Joeylawn, Beetstra, NcSchu, Avs5221, Dblecros, Larrymcp, Waggers, TastyPoutine, IceHunter, TPIRFanSteve, Peyre, LaMenta3, RHB, Crossbyname, Wwagner, Sifaka, Valkian, MBob, Iridescent, Dakart, Clarityfiend, Mikiman, BobDBilde, Shoeofdeath, Newone, Carlwillis, Kadams1970, DonL, Cls14, Octane, CapitalR, Domitori, Witchyrose, Levg79, Courcelles, IanOfNorwich, Thebigone45, I state facts675, LegalSwoop, Tawkerbot2, CodemauL, Filelakeshoe, Chris55, Greverod, Lahiru k, MightyWarrior, Fritz28408, Rabidchipmunk666, Carroy~enwiki, Croctotheface, Alexander Iwaschkin, JForget, ScottW, Unreal128, Sleeping123, CmdrObot, Tanthalas39, Raysonho, Mattbr, PorthosBot, Zarex, Unionhawk, Van helsing, Cyrus XIII, Pikul~enwiki, Seattledude, BeenAroundAWhile, Rawling, W guice, Slkong, Picaroon, Styler 13, Page Up, TheMightyOrb, Dub8lad1, Mr Echo, Ruslik0, GHe, Kylu, SofieAriana, RTNM, Stuart Drewer, Argon233, Duality Rules, Birdhurst, SEJohnston, ArmenG, Outriggr (2006- 2009), Finn bell, MapLover, Moreschi, Stebulus, Ken Gallager, DonalCahill, Ptrsmth7, Tex, Karenjc, Skybon, Myasuda, A Chougle, Xilog, Whoody, Seejyb, Elyscape, CMG, Cydebot, Honk squeak, KPbIC, Kanags, Whichone, Reywas92, Righttovanish1, MC10, Steel, Mato, Mortus Est, UncleBubba, TAz69x, Gogo Dodo, Vinyanov, JFreeman, D9qhd8, Corpx, Tec15, Harpwolf, Lugnuts, Cuhlik, SquareWave, LMAnthony, Sloth monkey, Shotmenot, AlexMS, BCap8021, Murileemartin, Yummy mummy, Carlroller, JEdlund, SETh of MONROVIA, Msnicki, JByrd, DumbBOT, RottweilerCS, Szandor~enwiki, Kozuch, ShaneBaker, ErrantX, Duhon~enwiki, Helvetica, TheJC, BRT01, Btharper1221, Vanished User jdksfajlasd, Ubikvist, Gimmetrow, Pipatron, Lyverbe, Corlen, Killer Swath, Insanimaniac2005, Thijs!bot, Epbr123, Wikid77, Willworkforicecream, Jwt015, Kablammo, N5iln, Andyjsmith, Smiklas, J.Ring, Gralo, Mereda, PierceG, Simeon H, Jacksav, Dtgriscom, Marek69, Drmemory, John254, Frank, A3RO, SGGH, Peace01234, Patthedog, James086, Java13690, X201, JustAGal, Grahamdubya, Spirit Of Truth, Einsidler, Grayshi, Greg L, Nick Number, Gierszep, Dawnseeker2000, MattyDienhoff, SparhawkWiki, Tidy~enwiki, Dzubint, Sbandrews, Mentifisto, Gossamers, AntiVandalBot, Majorly, Nutshack1, Abu-Fool Danyal ibn Amir al-Makhiri, Format, Luna Santin, Chubbles, Seaphoto, Opelio, QuiteUnusual, DarkAudit, Userboy87, Cchhrriiss, TimVickers, Madbehemoth, Timrog@gmail.com, Yoosq, Vanjagenije, Roothog, Echo5Joker, Aspensti, Alphachimpbot, SkoreKeep, Tkirton, Lantios, Myanw, Lklundin, Kaini, Gökhan, Hayesgm, Obeattie, Darrenhusted, Golgofrinchian, ClassicSC, Kcowolf, Ioeth, JAnDbot, Demonkey36, Maj0r Tom, Wbarnhill, Harryzilber, Tony Myers, MER-C, Planetary, Mfrphoto, Arch dude, Owenozier, Jgaray, Dave rabbit, Hut 8.5, GurchBot, NSR77, Savant13, Delius1967, Kirrages, LittleOldMe, Globalhealth, Acroterion, Usws, Moni3, Gtation, Magioladitis, WolfmanSF, Mattb112885, Bongwarrior, VoABot II, Djkeddie, Transcendence, Dekimasu,

Simmons, ThinkEnemies, JV Smithy, Reach Out to the Truth, Webirsn~enwiki, Minimac, Stephen Donnan1988, Hess108, Chasmosaurus13, DARTH SIDIOUS 2, MelonEllen, AXRL, Obsidian Soul, The Utahraptor, Truthful Scientist, RjwilmsiBot, Undescribed, Bento00, NameIs-Ron, Beyond My Ken, Kotopoto, Forenti, 1947enkidu, Balph Eubank, Aircorn, Kiko4564, Samdacruel, Grondemar, Bowei Huang, Jpatros, Steve03Mills, Mukogodo, EmausBot, John of Reading, WikitanvirBot, Immunize, Karyn Devlin, Botb75, Mordgier, TheGbomb, Year7, Haon 2.0, Matt1611, Chris9lfc, Fly by Night, Boundarylayer, Marzenka101, Japs 88, GoingBatty, Alexb163, Ordoo, Active Banana, MW2-throwing-knife-god, PanthonGan, DominicConnor, Ebe123, Qrsdogg, Tommy2010, Jordyjordy, Winner 42, Rockdevourer, Dcirovic, TeleComNasSpr-Ven, Arian1314, 2007littledude, Bravo Foxtrot, JSquish, Ronk01, Mattyjunior, Archer888, John Cline, Ida Shaw, Bongoramsey, Johnmlindsay, Josve05a, Lani1234, A2soup, Neun-x, Hycesar, Westley Turner, Wackywace, Theonewiththeone, 07joe2k7, Speakerman454, Druzhnik, Midas02, BredoteauU2, The Lord of the Allosaurs, Brazmyth, Nick4-26-86, Bobby987654, Mofo ho, H3llBot, Erobern1994, SporkBot, AlexH555, Troneatem, Bluebrigand, Eyeshot, QEDK, Tangoman.fr, Iharper16, Kindzmarauli, Staszek Lem, Jsayre64, Rcsprinter123, Jbnla, The Writer 2.0, IGeMiNix, Brandmeister, L Kensington, Jguy, Morgankevinj, MonoAV, Xiaoyu of Yuxi, Theshipscook, Ego White Tray, Orange Suede Sofa, Ecomiles, Ipsign, ChuispastonBot, Aliencore, L'ecrivant, EdoBot, JanetteDoe, Vlad21263, Ace of Raves, Anandteke, Ladnadruk, Jill Orly, TitaniumCarbide, Flores,Alberto, Ivolocy, Queen083, E. Fokker, Will Beback Auto, ClueBot NG, Dromend, Crazyman121, Mansmokingacigar, Jordanhey1993, Khushwant singh best, RaptorHunter, Somedifferentstuff, Theaitetos, Chicken59, Kayz911, Lordweiner27, Hon-3s-T, Bped1985, Georgepauljohnringo, Baseball Watcher, Apulmullen, Amorgan103, Physics is all gnomes, Hoygiv, Cjweber, Cj005257, Frietjes, Cntras, Hazhk, Mesoderm, Asif.al.noor, Jamoringstut, Legendre2k, Antiqueight, Cognate247, NEISORG, MarcusBritish, 100eme, Cheezey banana 101, NiclasCage, XXD4YBR3AKERXX, Datheisen, Adgard1, Jofgoodwood, Mohasin1, Pread13, Helpful Pixie Bot, Bowiechen, Goatcheeze98, Charles mcquoi, Jessica&Asia, AzureAnt, Hhhggg, Newyork1501, Barkvark27, Exeb75, Miguel0327, Anuclanus, Sokavik, Kinaro, Luigi.scorzato, Yash4g, Bminor7th, BG19bot, Amtkhogger94, Wzrd1, PhnomPencil, Marcocapelle, ERJANIK, Not A Superhero, Harizotoh9, Asauers, Zedshort, Marky777, Polmandc, 007a83, Kimelea, WebTV3, Energy4All, BattyBot, Regicide1649, Netherzone, Haymouse, NWRGeek, Cyberbot II, Aliwal2012, Popopo8776, Khazar2, Azure94, IjonTichyIjonTichy, Soni, Dexbot, Boiler Room 4, Ukrained2012, Andrux, Tony Mach, Stephen David Williams, PeterWesco, Remus Octavian Mocanu~enwiki, Advanceddeepspacepropeller, Epicgenius, Happymarmotte, Datdyat, EvergreenFir, Cebr1979, ElHef, ArmbrustBot, DDear99, Itc editor2, Le Grand Bleu, Somchai Sun, Jarash, Two kinds of pork, SeventeenTurtle, Poppy Dickson, Mfb, Stamptrader, Man of Steel 85, Andrew J.Kurbiko, Uthorr, Angelgreat, Monkbot, WikiOriginal-9, KBH96, Brightnsalty, Cynulliad, Kaciemonster, Pishcal, Hobbitschuster, Francjñg, Bweilz, Titaniumman23, Amccann421, Nøkkenbuer, Isaacjohnson1, My Chemistry romantic, Dorpater, Electrosharkskin, Buerish, Prevalence, InternetArchiveBot, The Voidwalker, GreenC bot, Bender the Bot, Heart of Destruction, Acopyeditor, Mr14159, JosVan, Medhat moussa and Anonymous: 2681

- **List of Chernobyl-related articles** *Source:* https://en.wikipedia.org/wiki/List_of_Chernobyl-related_articles?oldid=750680251 *Contributors:* Altenmann, Tweenk, Mzajac, 朝彦, Herzen, Wongm, Simesa, Shaddack, Nickst, Scwlong, DDima, BrownHairedGirl, FairuseBot, CmdrObot, Chrisahn, Jacrio, Dtgriscom, Frank, Mchl, Cooper-42, D.h, CommonsDelinker, DadaNeem, Johnfos, Bentley4, GirasoleDE, Danimations, Baclark, DerBorg, Good Olfactory, Komischn, Materialscientist, Jonkerz, Nordmensch, Ladnadruk, ClueBot NG, BG19bot, MeanMotherJr, Andrux, Finnusertop, Wikiwikiwikiwikiwikiwikiwikiwiki, Libertarian12111971, Andrew76-77 and Anonymous: 15

- **Effects of the Chernobyl disaster** *Source:* https://en.wikipedia.org/wiki/Effects_of_the_Chernobyl_disaster?oldid=765586618 *Contributors:* Lquilter, Yann, Kierant, Altenmann, Carnildo, Beefman, Wwoods, Mboverload, Grant65, Sonjaaa, Eregli bob, OwenBlacker, Imjustmatthew, Irpen, Klemen Kocjancic, 朝彦, D6, Discospinster, Rich Farmbrough, Bender235, ESkog, Zscout370, Tom, O18, Stesmo, Eritain, Towel401, Schnolle, Alansohn, Arthena, Wtmitchell, Velella, Deacon of Pndapetzim, Antony Ivanoff, Gene Nygaard, Ron Ritzman, Bobrayner, Angr, Woohookitty, PoccilScript, Stefanomione, Rjwilmsi, Tim!, Koavf, Joffan, Amire80, Nick R, Eubot, Ground Zero, Crazym108, Old Moonraker, Ysangkok, Gurch, Banazir, Kolbasz, Lightsup55, Simesa, WriterHound, Wavelength, Drdisque, Chris Capoccia, Limulus, Shaddack, -OOPSIE-, Tony1, Nikkimaria, Fourohfour, DasBub, Biddlesby, SmackBot, DocS, Tonyr68uk, 0x6adb015, Mdd4696, Kjaergaard, Gilliam, Ohnoitsjamie, Hmains, Cadmium, Carbonrodney, Ikiroid, JRPG, OrphanBot, Ukrained, Wen D House, Axel-berger, Jgrahamc, Fuzzypeg, Riurik, Ohconfucius, Ajnosek, John, Lapaz, Vgy7ujm, Calum MacÙisdean, HonestTom, Even248, Noahspurrier, RomanSpa, Infidus, Dl2000, Hu12, Ewulp, Clay, Zarex, Nadyes, IntrigueBlue, Cydebot, Mr.weedle, Tec15, NorthernThunder, Helvetica, Gimmetrow, DavidBeoulve, Headbomb, Ruber chiken, Marek69, Frank, Josh.karli, Aspensti, Simon Burchell, Magioladitis, Bongwarrior, VoABot II, -Kerplunk-, Cgingold, Bellemichelle, Cooper-42, Talon Artaine, Frye0031, Philg88, D.h, Gandydancer, CommonsDelinker, Captain panda, Uncle Dick, Maurice Carbonaro, Acalamari, Enuja, Apostle12, Katharineamy, Grosscha, DadaNeem, Master shepherd, Boothferry, Joshua Issac, Dr John Wells, ACSE, Vranak, Johnfos, Philip Trueman, Deleet, Flyte35, Marjolijn, MatthewHaywood, DanteHicks79, Leafyplant, PDFbot, Brockle, Maxim, Dirkbb, SylviaStanley, GirasoleDE, LovelyLillith, Swliv, Dhgwren, Flyer22 Reborn, Bc789, The posp, Cyfal, ImageRemovalBot, ClueBot, Multi-AC, Lwnf360, Niceguyedc, Hagen Scherb, Piledhigheranddeeper, Jusdafax, Hscherb, Sun Creator, CAVincent, SchreiberBike, Chaosdruid, WooteleF, Tdslk, DumZiBoT, Penguinlover95, Dthomsen8, Noctibus, Addbot, Some jerk on the Internet, Guoguo12, Ethanpet113, Fluffernutter, MrOllie, Download, Favonian, Isobelrosepa, Soccer-holic, Verbal, Lightbot, Legobot, Yobot, Fraggle81, AnomieBOT, DemocraticLuntz, 1exec1, Jim1138, OpenFuture, Citation bot, RevelationDirect, G6cid, Xqbot, Kyeller, Nagleman, Bangabandhu, Buttons0603, Leor klier, PascalMichelSI, FrescoBot, Decwrites, Ag97, Mattlox, Citation bot 1, Wojtek.miller, Pinethicket, I dream of horses, Urae, Trappist the monk, Robertiki, Lenaraz2, Johnny Sand, Anhn, RjwilmsiBot, Salvio giuliano, JohnThomasTucker, John of Reading, Boundarylayer, Wiki-asd-97, Uploadvirus, Dcirovic, Number10a, ZéroBot, John Cline, H3llBot, AManWithNoPlan, Tolly4bolly, Noohoo, Donner60, Vismatarchivist, ClueBot NG, Silicaslip, Bped1985, Toddrmcallister, Shanehpatterson, Helpful Pixie Bot, AzureAnt, Gob Lofa, Sokavik, BG19bot, MusikAnimal, Klilidiplomus, Rytyho usa, Klas2k, BattyBot, Rohanmanocha, Jcarver6, Russiabuff, Haymouse, Cyberbot II, Nelsonwill, Mediran, Popopo8776, Khazar2, Editfromwithout, Webclient101, Clayber1234, Andrux, Frosty, Haileyturner, Stratoprutser, Faizan, Epicgenius, Smarty9108, Redd Foxx 1991, DavidLeighEllis, Finnusertop, RainCity471, Kombinezon, Racer Omega, Fixuture, Monkbot, AKS.9955, MarshmallowMan3, Demoniccathandler, Graemem56, Ceosad, Sshenk, UglowT, Andywass123, Editor abcdef, Avery Miller, Nikelilman13, Liechtenstein96, InternetArchiveBot, DaisyMeRolling, VladyslavGubin, GreenC bot, Fatengineer, Bender the Bot, JueLinLi, DrStrauss, IamWill and Anonymous: 256

- **Goiânia accident** *Source:* https://en.wikipedia.org/wiki/Goi%C3%A2nia_accident?oldid=762093078 *Contributors:* AxelBoldt, Bryan Derksen, The Anome, Rmhermen, Shii, TomCerul, Leandrod, Edward, Delirium, Eric119, Aarchiba, Ineuw, Ehn, Mulad, RodC, Warmfuzzygrrl, WhisperToMe, LMB, JonathanDP81, AnonMoos, Scott Sanchez, Wetman, Hajor, Mjmcb1, Robbot, Tlogmer, Korath, Sanders muc, Cecropia, Mattflaschen, Carnildo, Alan Liefting, Inkling, Wwoods, Mboverload, Solipsist, Pne, Dvavasour, LucasVB, DragonflySixtyseven, Clarknova, Quota, Deglr6328, Abdull, Adambondy, Eyrian, Brianhe, Pjacobi, Adam850, MeltBanana, Sam Derbyshire, Pavel Vozenilek, Quietly, Mashford, Neko-chan, MisterSheik, Alereon, Orlady, Jpgordon, Reinyday, Vortexrealm, Giraffedata, Helix84, Caeruleancentaur, Hooperbloob, Blahma,

Keenan Pepper, Axl, Hu, Dental, Dhartung, Velella, Evil Monkey, Vuo, Dave.Dunford, ShawnVW, Gene Nygaard, Dan100, Falcorian, Feezo, Firsfron, Ylem, Oliphaunt, Crackerbelly, Ikescs, Ardfern, Cbustapeck, Jacj, Stevey7788, Graham87, Mucky Duck, Tim!, Urbane Legend, XP1, Rillian, BlueMoonlet, SLi, Patrickr, Kolbasz, Wgfcrafty, JonathanFreed, Pstevens, Dylan Thurston, EamonnPKeane, YurikBot, Wavelength, NTBot~enwiki, Huw Powell, Kiscica, Brandmeister (old), Guslacerda, Epolk, Shawn81, Shaddack, Lusanaherandraton, G®iffen~enwiki, Howcheng, FreelanceWizard, Mgcsinc, Supten, Chris S, Tetracube, Leviramsey, EdX20, Nippoo, Dsyzdek, A bit iffy, SmackBot, Yellow-Monkey, Fireworks, Franny Wentzel, Septegram, Yellowbounder, AxelHarvey, Alias777, Cadmium, Moshe Constantine Hassan Al-Silverburg, Rolypolyman, Scwlong, Tsca.bot, Racklever, Thatnewguy, BrianTung, Fuhghettaboutit, Jumping cheese, BWDuncan, Dogosaurus, Acdx, Captainbeefart, John, Microchip08, Minna Sora no Shita, Beta34, Tonsa, The-Pope, TJ Spyke, Lapinbleu, Joseph Solis in Australia, Drlegendre, Rocketman768, Nixxonvaldez, CmdrObot, Rwflammang, Gyopi, Mateus Hidalgo, Lentower, Deusnoctum, Location, Rifleman 82, Headbomb, Sijarvis, Mdotley, Ingolfson, Serpent's Choice, Darlosity, WolfmanSF, JamesBWatson, Schoowru, Swpb, The Anomebot2, Nick Cooper, Jim-jamjak, ZackTheJack, Diotime, Lionfish0, Welshleprechaun, Cfrydj, Akronym, RockMFR, Pharaoh of the Wizards, McSly, Trumpet marietta 45750, Woodega, Nwbeeson, Tvbrichmond, DeFaultRyan, Widders, GrahamHardy, Hugo999, VolkovBot, Johnfos, Lear's Fool, Mazarin07, Truthanado, Ian Glenn, Gorpik, Rob.bastholm, Afernand74, Johnnywiggle, Anyeverybody, ClueBot, Badger Drink, Palantir-palantir, Bbalasub, Roxport, Redrocketred, Aaroncorey, Ktr101, AssegaiAli, PixelBot, Shinkolobwe, NuclearWarfare, InternetMeme, FellGleaming, Ltmboy, Eleman, Cabayi, Addbot, Jacopo Werther, CanadianLinuxUser, Ashanda, LaaknorBot, Doniago, 84user, Bwrs, Verbal, Lightbot, The Bushranger, Luckas-bot, Yobot, Yngvadottir, Amy Baily, AnomieBOT, KDS4444, Archon 2488, Materialscientist, Aff123a, Citation bot, Xqbot, Anonymous from the 21st century, Johndcurry, Tibidibtibo, Intrepid-NY, Kgrad, RjwilmsiBot, EmausBot, Trofobi, ElTorbe, Dadaist6174, Dazman83, A2soup, SporkBot, Demiurge1000, Prandr, Whoop whoop pull up, ClueBot NG, Jnorton7558, DieSwartzPunkt, LucaNevski, CopperSquare, Reify-tech, Avecmonami, Gabriel Yuji, MeanMotherJr, Cyberbot II, Zbjornson, Zerabat, Tigraan, Agnichuck, Opencooper, Fer48, InternetArchiveBot, GreenC bot and Anonymous: 171

- **1990 Clinic of Zaragoza radiotherapy accident** *Source:* https://en.wikipedia.org/wiki/1990_Clinic_of_Zaragoza_radiotherapy_accident? oldid=758386345 *Contributors:* Chirlu, Bender235, Reinyday, Foobaz, Keenan Pepper, Tim!, FAR, Shaddack, BOT-Superzerocool, Airconswitch, Oneac, John, K7aay, Fayenatic london, Jetstreamer, The Anomebot2, Read-write-services, Kojozone, Hugo999, Johnfos, Ttonyb1, Doctorfluffy, Alexbot, Skbkekas, Nettings, The Bushranger, AnomieBOT, Hairhorn, Xqbot, Erik9bot, Bigweeboy, Rubisco85, Diannaa, EmausBot, Werieth, Lawlar, SporkBot, BattyBot, Fer48, InternetArchiveBot, GreenC bot and Anonymous: 8

- **1996 San Juan de Dios radiotherapy accident** *Source:* https://en.wikipedia.org/wiki/1996_San_Juan_de_Dios_radiotherapy_accident?oldid= 756649849 *Contributors:* Bender235, Wtmitchell, Jedikaiti, Scwlong, Rosarinagazo, Mr pand, Fayenatic london, Waacstats, The Anomebot2, R'n'B, Hugo999, Johnfos, Karmos, Afernand74, Mumiemonstret, Addbot, The Bushranger, Hazard-Bot, ClueBot NG, Netherzone, Fer48, Julietdeltalima, B2ubried and Anonymous: 3

- **Tokaimura nuclear accident** *Source:* https://en.wikipedia.org/wiki/Tokaimura_nuclear_accident?oldid=759727269 *Contributors:* Sebastian Wallroth, Achurch, Andycjp, Kusunose, Mathias Schindler, Bender235, Alansohn, Rwendland, RJFJR, Axeman89, Oliphaunt, Ardfern, Rjwilmsi, Mr.Unknown, Nneonneo, DVdm, Wrightchr, DAJF, Tony1, Ospalh, Pr1268, Cojoco, Tim R, SmackBot, Nihonjoe, CSZero, Theanphibian, John, Cydebot, Greg L, Alexduric, Ironiridis, Wasell, Jetstreamer, The Anomebot2, Destynova, D.h, CommonsDelinker, Thirdright, Paris1127, ACSE, Hugo999, Johnfos, Nburden, Satani, Pmoir, 4wajzkd02, Breawycker, Grapeofdeath, Pac72, Danimations, ClueBot, Eiland, Philarete~enwiki, Boneyard90, Wnt, Addbot, Zup326, CarsracBot, Leucius, C933103, Gail, Snaily, Vectoor, PMLawrence, AnomieBOT, Archon 2488, Pontificalibus, FrescoBot, Szaszicska, Trkiehl, Maher27777, Brett R. Stone, EmausBot, WikitanvirBot, ZéroBot, Redhanker, ClueBot NG, Diowerefox, EvilResident, Wzrd1, Armchairgangsta, AlexHeylin, YiFeiBot, Thibaut120094, BrayLockBoy, Uamaol, Charlotte Aryanne, Chrisleese, GreenC bot and Anonymous: 53

- **Instituto Oncologico Nacional** *Source:* https://en.wikipedia.org/wiki/Instituto_Oncologico_Nacional?oldid=677476987 *Contributors:* Mboverload, D6, Rich Farmbrough, RJFJR, Mindmatrix, Koavf, Vegaswikian, Kolbasz, Bgwhite, FF2010, SmackBot, Radioheadhst, Wizardman, Skapur, Jesse Viviano, Cydebot, Road Wizard, Fayenatic london, Magioladitis, Jllm06, The Anomebot2, Schmloof, Johnpacklambert, Clerks, Johnfos, Cyphine, Good Olfactory, Addbot, Leszek Jańczuk, LilHelpa, Lamppis, Debpty, John of Reading, Whoop whoop pull up, AdventurousSquirrel, Wiki CRUK John and Anonymous: 3

- **Fukushima Daiichi nuclear disaster** *Source:* https://en.wikipedia.org/wiki/Fukushima_Daiichi_nuclear_disaster?oldid=766361325 *Contributors:* Rmhermen, Deb, Miguel~enwiki, Shii, Maury Markowitz, Heron, Leandrod, Michael Hardy, Ixfd64, Komap, Yann, Mcarling, Mkweise, Jpatokal, Darrell Greenwood, Nahum, Bueller 007, Kragen, Julesd, Glenn, Cherkash, Rl, Mxn, Hashar, IceKarma, Tpbradbury, Dragons flight, Furrykef, AnonMoos, Chris Rodgers, Frank A, Ke4roh, Kizor, Mrickard, Nurg, Sekicho, Zidane2k1, Trevor Johns, Cyberia23, Xanzzibar, Alan Liefting, Ancheta Wis, Dinomite, Sj, Fukumoto, BenFrantzDale, Karn, Xerxes314, Wwoods, Alison, Markus Kuhn, Ketil, AJim, Cloud200, Macrakis, Tweenk, Btphelps, Andycjp, Geni, Beland, Piotrus, IGEL, Anythingyouwant, Bumm13, Gscshoyru, Qiq~enwiki, Sam, Quota, Klemen Kocjancic, Frau Holle, Dbaron, Arcataroger, Discospinster, Hydrox, Vsmith, Smyth, Alistair1978, Bender235, Cphoenix, Diamonddavej, PatrikR, Nigelj, Touriste, Stesmo, Viriditas, R. S. Shaw, Cavrdg, Deryck Chan, Wikinaut, Elrey, Ardric47, Ikester8, Jjron, Espoo, Spitzl, Eric Kvaalen, Joshbaumgartner, Rodw, AzaToth, Axl, Sligocki, Rwendland, Radical Mallard, Burwellian, Ksnow, Fivetrees, Wtmitchell, Velella, Unconventional, Capzloc, Geraldshields11, Cmprince, Dan East, Coolgamer, Tariqabjotu, Hijiri88, Brycen, Bobrayner, TheIguana, Henrik, Bryan986, Daira Hopwood, Barrylb, JFG, Pol098, Brentdax, Trödel, MONGO, Miss Madeline, BartBenjamin, GregorB, BlaiseFEgan, Naomechateies, Rotten, Darkoneko, Kralizec!, Stefanomione, Theo F, Mandarax, BD2412, SDalley, DePiep, Edison, Ciroa, Sjö, Ilyak, Rjwilmsi, Koavf, Gryffindor, Strait, XP1, BlueMoonlet, Mike Peel, Vegaswikian, Guinness2702, Mbutts, LjL, Yug, Titoxd, Wragge, Ground Zero, Project42, Nihiltres, Thunderbird~enwiki, Survivor, Itinerant1, Kolbasz, Jrtayloriv, KFP, Intgr, Nicapicella, Psantora, Chobot, Costas Skarlatos, Benlisquare, DVdm, Bgwhite, WriterHound, Wavelength, TexasAndroid, Todd Vierling, MikeFoss18, Gregconquest, Bhny, Limulus, Jengelh, Groogle, Chaser, Shawn81, Hydrargyrum, Akamad, Gaius Cornelius, Sandpiper, Eleassar, Rsrikanth05, Akhristov, TheMandarin, Petmal, Nowa, Joel7687, CarlFink, Lexicon, Robert McClenon, Aaron Brenneman, DAJF, CecilWard, Juanpdp, Voidxor, Ospalh, Adicarlo, Mtu, Strolls, Bruce Hall, SamuelRiv, Zunaid, Silverchemist, Intershark, Bdell555, BMT, Maphisto86, Miblo, OtherDave, Y23, MarsJenkar, Arthur Rubin, Petri Krohn, AnimeJanai, Macropode, Superp, Xil, Garion96, Mardus, Cmglee, Groyolo, Finell, Kyaa the Catlord, Tallard, Shenhemu, AndrewWTaylor, MaeseLeon, MartinGugino, SmackBot, Prebys, Elonka, Twerges, F, Rtc, Prodego, Mjspe1, Rouenpucelle, Cla68, Kintetsubuffalo, Nil Einne, Yamaguchi 先生, Jim Casper, Gilliam, Brianski, Tomathy, Skizzik, Ephraim33, Robin Whittle, Chris the speller, Ottawakismet, Snori, Exprexxo, Yozi66, IIXII, Kurtm, Sbharris, Arsonal, Jdthood, A. B., Scwlong, Audriusa, Rogermw, Tharikrish, Vanis314, Leoboudv, Squigish, Greenshed, Zirconscot, Joshua 70448, Theanphibian, Cybercobra, Mytwocents, Tank-en-mate, Salt Yeung, MBCF, Btm1, Henning Makholm, Gossg, Salamurai, Daniel.Cardenas, Mion, Brainfood, Ohconfucius, Will Beback, L337p4wn, Tburke261, Esrever, Djmitche,

Arnoutf, PSeibert~enwiki, John, Rousse, Ptroxler, JorisvS, L41n, Mgiganteus1, Scetoaux, Jaywubba1887, Vamoose, Johan Jönsson, Hvn0413, Stwalkerster, Rock4arolla, TPIRFanSteve, Hiroe, Dl2000, 08-15, Hu12, Mdanziger, Haitani~enwiki, Iridescent, Majorbonkers, 293.xx.xxx.xx, Lathrop1885, UncleDouggie, Bobamnertiopsis, Frank Lofaro Jr., Luminaux, TreyGreene, Stifynsemons, Vaughan Pratt, 850 C, Lamiot, Vejet, Edward Vielmetti, Stevo1000, Jesse Viviano, CuriousEric, N2e, AshLin, MarsRover, Simply south, Rowellcf, Johnlogic, SKS2K6, Cydebot, Raamin, Reywas92, A876, Wannesvdh, Grahamec, UncleBubba, Kslotte, Adrian Glamorgan, Boardhead, Blaisorblade, DavidRF, Quibik, Ssilvers, Dpu, Rspeed, Xanthis, Thijs!bot, Epbr123, Cimbalom, Hervegirod, Badams5115, Pietrodn, N5iln, Dougsim, Hirnbeiss~enwiki, Speedyboy, Peace01234, Second Quantization, Grayshi, Dawnseeker2000, Nemilar, BokicaK, Rprout520, Seaphoto, Carolmooredc, Robzz, Yellowdesk, SkoreKeep, John Moss, Lfstevens, Ossiman, Kaini, Ingolfson, Daytona2, Golgofrinchian, Mikenorton, Hthalljr, NapoliRoma, Barek, Josephmarty, Skomorokh, Ericoides, Arch dude, Hello32020, Ccrrccrr, Greensburger, Acroterion, Mchl, TransControl, Nicransby, Magioladitis, Bongwarrior, Yakushima, Rami R, Cgingold, LorenzoB, Beagel, Rif Winfield, Markus451, Coffeepusher, Nankai, Read-write-services, Gandydancer, Andre.holzner, Numero4, Keith D, NikonMike, CommonsDelinker, CalvinTy, Lilac Soul, FourTildes, Eplack, Peter Chastain, Hans Dunkelberg, Maurice Carbonaro, Jesant13, Theeurocrat, Maproom, Mathglot, Dispenser, Snokite, Eric in SF, Naniwako, Mikael Häggström, Rehrenberg, Molly-in-md, C1010, Jaysarvage, Yanni576, DadaNeem, Facorread, Student7, Shadow Android, Topamo, Reebzz, RVJ, Squids and Chips, GrahamHardy, Funandtrvl, Jmcdon10, ACSE, Vranak, Carpetmaster101, VolkovBot, Error9312, Johnfos, Timmonsgray, Iwavns, Jeff G., Lear's Fool, Ryan032, Martinevans123, TXiKiBoT, Oshwah, Deleet, Jkstark, Mark v1.0, Ozdawn, Anonymous Dissident, Piperh, Imasleepviking, Joren, Rdfox 76, Pmoir, Kmhkmh, GDuwen, Lamro, Typ932, Vchimpanzee, Galuboo, Elriana, Lkiller123, Truthanado, Monty845, Rep07, Doc James, Logan, SylviaStanley, Pmarshal, Adaviel, Fanatix, Wjl2, Enkyo2, TCO, Sam Brightman, Swliv, WereSpielChequers, Cwkmail, Yintan, Minxlj, Crash Underride, Bleeisme, France3470, PookeyMaster, Flyer22 Reborn, Musicandnintendo, A. Carty, MaynardClark, Oda Mari, Colfer2, Snideology, Paolo.dL, Hzh, Avidallred, Brokendata, Devangel77b, Steven Crossin, Danimations, Goofball222, Eugen Simion 14, Autumn Wind, Sunrise, Lynntoniolondon, Rooboy715, Spitfire19, Rkarlsba, Mojoworker, Chrisrus, SuperSaiyaMan, Henk Poley, Pre1mjr, Asocall, Nergaal, Hadseys, ImageRemovalBot, RobinHood70, Vanished user qkqknjitkcse45u3, Martarius, GorillaWarfare, Nielspeterqm, Rustic, Kennvido, Abhinav, Elmao, Heracletus, Unbuttered Parsnip, MichaelVernonDavis, Porksoda1978, Cochonfou, Der Golem, Marethyu316, VQuakr, Mild Bill Hiccup, Eiland, Watti Renew, Doseiai2, Boing! said Zebedee, Jwihbey, Mezigue, JTSchreiber, U5K0, Niceguyedc, Daniel D.L., Pross356, Ottawahitech, Namazu-tron, Counteraction, Deselliers, Ktr101, Excirial, Jtmonone~enwiki, Jusdafax, John Nevard, Rhatsa26X, Shinkolobwe, Megiddo1013, Sun Creator, Holden yo, Coinmanj, NuclearWarfare, L.tak, Aurora2698, Arjayay, Wrin, Peter.C, LarryMorseDCOhio, Salty1984, Hunanuk, Yohananw, Iohannes Animosus, 7&6=thirteen, Chaosdruid, Buckethed, The Wicked Twisted Road, JPLeRouzic, L.smithfield, GFHandel, HarrivBOT, Johnuniq, Freelion, Vigilius, Humanengr, Alastair Carnegie, Ginbot86, Semitransgenic, InternetMeme, Jax 0677, MWadwell, XLinkBot, MarmotteNZ, Nathan Johnson, Wikiuser100, Mr vanhorn, R Stillwater, Pgallert, Starviking, David.Boettcher, WikHead, Badgernet, Vitaltrust, MystBot, JCDenton2052, Dfoxvog, Light theworld, PYX-340, Addbot, Razr Nation, The Sage of Stamford, Roentgenium111, RPHv, Fancy-cats-are-happy-cats, Ocdnctx, Wda, MartinezMD, Scientus, MrOllie, Download, Mjr162006, Lihaas, Alandeus, Sun Ladder, Chzz, Colt9033, EvelynFl, Elen of the Roads, 84user, 102orion, Lightbot, OlEnglish, C933103, Ettrig, Mps, Rchoetzlein, Ben Ben, Legobot, Luckas-bot, Yobot, Boardersparadise, A.k.a., Fraggle81, Amirobot, DanniDK, Angel ivanov angelov, Claverhouse, Shinkansen Fan, LemonMonday, Dmarquard, AnomieBOT, Climatedragon, John Holmes II, 1exec1, Götz, Jim1138, Shock Brigade Harvester Boris, Bsea, Plastdroid, TParis, Knowledgekid87, Flinders Petrie, AnAnthro, Materialscientist, Ahumphr, Danno uk, Citation bot, A333, Eunomiac, Suit, MegaPedant, LovesMacs, Quebec99, LilHelpa, Xqbot, J G Campbell, TracyMcClark, Nrpf22pr, Pontificalibus, Sylwia Ufnalska, Tagryn, Akio Hasegawa, Jeffwang, Sklapro, Anna Frodesiak, Srich32977, Martijnd, Delerium2k, Champlax, Porttimies, Quiqsilver, Mark Schierbecker, Freja Beha Erichsen, Amaury, Argos42, Yann322, Sqgl, Moxy, J. Sbg, Geronimo355, JayJay, Dylandn, 老陳, Biem, CES1596, GliderMaven, Dailycare, Jilkmarine, FrescoBot, Surv1v4l1st, Pepper, Filippo83, RicHard-59, Ironboy11, YOKOTA Kuniteru, Astronomyinertia, Juze~enwiki, Mistakefinder, Amuchmoreexotic, Goodbye Galaxy, Ravendrop, Cs32en, Berny68, Armigo~enwiki, Ahmer Jamil Khan, Hugo75, Stuthulhu, Diwas, Gautier lebon, Kenrick95, Abductive, Jonesey95, Andynct, Supreme Deliciousness, Varkstuff, Geogene, Lancrey, Jusses2, Delboy1978uk, Serols, Σ, MKFI, Just a guy from the KP, Bashokinenkan, Pbsouthwood, Bgpaulus, Jauhienij, Double sharp, Trappist the monk, KevinRachel2010, Samlikeswiki, Flodded, Callanecc, Now wiki, Comet Tuttle, RoadTrain, Oracleofottawa, Vrenator, MarkForeman, Robertiki, StevenCSimmons, Mabeenot, Jeffrd10, Chronulator, Cstrosser, Suffusion of Yellow, PleaseStand, Tbhotch, Reach Out to the Truth, Gcsjapan, Marie Poise, Obankston, AXRL, Hullernuc, RjwilmsiBot, Azure777, Ripchip Bot, Jonathan Levy, Fzz85, Highcc, TheFallenCrowd, 1947enkidu, RAN1, Nihola, TGCP, Prosopon, Whywhenwhohow, EmausBot, John of Reading, Nima1024, WikitanvirBot, Trofobi, Ghostofnemo, Boundarylayer, Dewritech, Racerx11, NotAnonymous0, Solarra, Dwalin, The Mysterious El Willstro, Jim Michael, Winner 42, Carbo1200, Stormchaser89, Wikipelli, Dcirovic, K6ka, The Blade of the Northern Lights, Anirudh Emani, Enyby, Tanner Swett, Miki08, Mz7, Werieth, PBS-AWB, JLGD, Endicott65, Fæ, Josve05a, A2soup, Neun-x, August571, Rppeabody, KaliumPropane, Kalin.KOZHUHAROV, Mrmatiko, Redhanker, RaymondSutanto, Jonfkessler, Falcon2700, Gavbadger, Radical Edward2, Varus2319, Hanjifi, Aeonx, Qqchose2sucre, Rexprimoris, Dagashin, Stageivsupporter, AlexH555, Francois Boulogne, Gatyonrew, Bob drobbs, Wingman4l7, Tolly4bolly, Wabbott9, L1A1 FAL, Ocdncntx, Fletch the Mighty, Cwill151, Wiggles007, Brandmeister, John KB, Sahimrobot, Jguy, Awjrichards, Thatmonk, Asdfsfs, Wipsenade, Shigeru23, ChuispastonBot, Unga Khan, Brad78, HandsomeFella, Matewis1, Grenade J, BennyJ, SSDGFCTCT9, LikeLakers2, Teapeat, Sven Manguard, 28bot, Whoop whoop pull up, TitaniumCarbide, Mjbmrbot, Ivolocy, Mikhail Ryazanov, ジャコウネズミ, Will Beback Auto, Rememberway, ClueBot NG, PatchesTheCaveman, Mansmokingacigar, Dager2345, Gareth Griffith-Jones, RaptorHunter, Adam Majer, Theaitetos, Abc-mn-xyz, Trigonomie, Catlemur, Satellizer, Pacific813, Joefromrandb, PaleCloudedWhite, Physics is all gnomes, Wonderdwarf, Theimmaculatechemist, Jcgoble3, The Master of Mayhem, Cntras, NRC OPA, Mr. D. E. Mophon, Hahahafr, Tinytrev, Subsider34, Masssly, Axis of eran, AritaMoonlight81, Zenez, Bzee91, Cj005257-public, Toddrmcallister, Suresh 5, Widr, Geofferybard, NuclearBWR, Chaosprimus, DrMandarin, Mouramoor, Iusethis, Reference Desker, Dougmcdonell, Lonewolf9196, Penyulap, Rootover, Cjdobber01, Blelbach, Noidh4xor, Avecmonami, Mayem~enwiki, Nicercat, Serazahr, Roger.nkata, Pasrich, Sexandlove, Frandomar, Slum125, SimonEdward, Chingchingchinching1234, Katharina giesler, AlexTheBarbarian, Oso revólver, Singer987987, NuclearEnergy, You are really focking ghey, Ini uyo, Bhermann, RobertSegal, Alexmuscat, Zyon788, Onewheeler, N7 Steph, YellowTurban, Kaisersushi, Helpful Pixie Bot, Twittofon, Jtmonone2, Hagoth, Marktine 23, Hopelessgleek, EarthquakeDisassemble, TSUBAME2, Joshi.sameer, N.leblanc88, Datamedic, IgglesPickles, Leopd, Faridhusain, Popcornduff, Drnovog, Tipthejust, Anentiresleeve, Madoc.1, Nyra1963, Hhhggg, Tourletour, MauchoEagle, 乾隆帝, FatTrebla, BulborbHunter, Wbm1058, Slotmad70, Spine001, Bibcode Bot, ⊠寄⊠像, Jdogburger, WNYY98, BG19bot, Brukner, Makedonija, NewsAndEventsGuy, Mikckeymouse, L48g, Lisamccabe, EvilResident, Andol, PM3, Jim Sukwutput, Rockchevy, Elinkwest, MusikAnimal, Frze, AvocatoBot, Kendall-K1, Calvin Marquess, MrsEcoGreen, Mark Arsten, Compfreak7, Ninney, Kay Uwe Böhm, Yowanvista, AdventurousSquirrel, Puneet nanda, E11107, Watson system, ZinniaJones, Blaspie55, Corinna39, Polmandc, Ices2Csharp, NoelyNoel, Wendie 1970, ConradMayhew, Alexjohnj, EdwardH, Smarandi, WebTV3, Cmoney1998, Caseyjdblue, Comfr, Señorsnazzypants, MeanMotherJr, Tititotone,

BattyBot, Ajeff003, Soccerfan98, StarryGrandma, Mdann52, Zwitt.the.twitt, Stausifr, Cyberbot II, ChrisGualtieri, Shikharsingh01, Arcandam, Khazar2, Aaascj, CallumA, TestBanNow, Iamozy, Winkelvi, Peppernox, P3Y229, BrightStarSky, Siphon06, HAmin86, Dexbot, Validpoint, Mysterious Whisper, Mogism, Kephir, Ptrw08, Cerabot~enwiki, Yogwi21, NevErSn0w, Lugia2453, AldezD, Stewartx97, Tony Mach, SFK2, Jamesx12345, Jo-Jo Eumerus, Lsquare40, Lindeman4m, Hbr001, Haileyturner, Fuku-2012, Ariealso, The Sickest One, Epicgenius, Haraldmmueller, NJoyitt, Ruby Murray, Abishai 300, Tina Gerhardt, Morg00, Melonkelon, EvergreenFir, Froglich, DavidLeighEllis, Crispulop, Jop2~enwiki, Baptx, JustBerry, Mandruss, Finnusertop, JohnLewen, Haikurambler, Internationalpoliticaleconomy, Balljust, Bouthaley-II, MrScorch6200, Juliusz Gonera, Gubino, Curtiscarnathan, AstroFizzy, Tistscien, 加集友規, SpiritedMichelle, Hidehicampers, Fixuture, FrB.TG, JonathanN.IFCP, RubberBroke, Suelru, SaintAviator, Suzuke87, Alexbelyakov, Nomikaniro, GuineaPigC77, Monkbot, Ozmol, Ricardo.alvarez12, Vieque, Mokhisultanova, Aliciamontoyadelaaldea, Graemem56, TheSATMD, Wikiwizard1212, Trackteur, Rarapas, Crazypdj, Halal Capone, KBH96, Dukon, Ghostsax, Kambiz12, Karsmars, StewdioMACK, Nuve307, AlphaBetaGamma01, Thecyne, Hadron137, CaptainPiggles, VexorAbVikipædia, Cewbot, Helper201, Sc50682, Pardonsnowden, Srednuas Lenoroc, Misinformation68, Marcus1093, イマジン, Modulus12, James Hare (NIOSH), New Media Theorist, A.K. Ferrara, LG88JH, InternetArchiveBot, Gunnar2701, Tyobos, Ethanbas, Marianna251, GreenC bot, Future fm Fukushima, Chrimas1, SiJayGreen, Hardtanker, Bender the Bot, Acopyeditor, Kodman1111, Wikishovel, Bepismaster123, Kurtvillagracia, Srednaus Lenoroc and Anonymous: 1095

- **Russian submarine Ekaterinburg (K-84)** *Source:* https://en.wikipedia.org/wiki/Russian_submarine_Ekaterinburg_(K-84)?oldid=724826759 *Contributors:* Ezhiki, Btphelps, Dziban303, Tabletop, Kolbasz, Wavelength, Whoisjohngalt, Petri Krohn, Ohconfucius, Dl2000, The ed17, Cydebot, Brad101, Buckshot06, GroveGuy, Truthanado, Pijuvwy, VargaA, Arjayay, The Wicked Twisted Road, Rave, Wikieditoroftoday, Truth or consequences-2, Tupsumato, Σ, EmausBot, Sp33dyphil, ZéroBot, August571, The rakish fellow, BG19bot, JustSomePics, Khazar2, ÄDA - DÄP, Monkbot, XANTHO GENOS and Anonymous: 11

- **International Nuclear Event Scale** *Source:* https://en.wikipedia.org/wiki/International_Nuclear_Event_Scale?oldid=761708085 *Contributors:* Michael Hardy, Ixfd64, Pipian, Doradus, Katana0182, Kierant, Grendelkhan, LMB, Omegatron, AnonMoos, Northgrove, Twang, Marnanel, Jonathan O'Donnell, Tweenk, Mzajac, FlohEinstein, Discospinster, Regebro, ESkog, JJJJust, Theshowmecanuck, Hooperbloob, Jumbuck, Nuke-Operator, Rd232, Rwendland, LOL, GregorB, MushroomCloud, Mandarax, BD2412, Rjwilmsi, Rillian, Nneonneo, Keimzelle, Tbone, Mahlon, Srleffler, Bgwhite, Simesa, Whosasking, YurikBot, Wavelength, Huw Powell, RadioFan, Gaius Cornelius, Dilaudid~enwiki, Gillis, Yoninah, NormanGray, Amakuha, Ospalh, Bota47, Oliverdl, Petri Krohn, Stuinzuri, Cmglee, SmackBot, Akranzel, Rtc, Unyoyega, Brianski, Squiddy, Ottawakismet, Cadmium, Jonathon Bolster, Egsan Bacon, Zirconscot, Theanphibian, Wen D House, Bigturtle, DMacks, John, Tomhubbard, Gobonobo, Jaywubba1887, Cowbert, Prunk, Plebeian, JoeBot, Twas Now, Pmyteh, ShelfSkewed, Cydebot, Quibik, Teratornis, Kablammo, Seaphoto, Roothog, -m-i-k-e-y-, Yellowdesk, JAnDbot, NapoliRoma, Andonic, Bongwarrior, Soulbot, Cgingold, CommonsDelinker, Nono64, Pomte, Dinkytown, Maurice Carbonaro, Rod57, Trumpet marietta 45750, Fongyun, Olegwiki, Cometstyles, Bogdan~enwiki, ACSE, X!, Hammersoft, VolkovBot, Larryisgood, Johnfos, RingtailedFox, ArbitraryConstant, Kyle the bot, A4bot, Rei-bot, PDFbot, Truthanado, Silver Spoon, Flyer22 Reborn, Colfer2, Lightmouse, Nonukes, RW Marloe, Rkarlsba, Mtaylor848, Pre1mjr, Tjayh913, ClueBot, Nielspeterqm, Nailedtooth, VQuakr, Eiland, Kancat, Laurapalmersdead, Excirial, Alexbot, PixelBot, Megiddo1013, L.tak, Jroovers, Protectthehuman, Addbot, Jonbryce, Roentgenium111, DOI bot, Betterusername, Ronhjones, Cuaxdon, Fluffernutter, Download, LaaknorBot, Morning277, CarsracBot, Glane23, Sun Ladder, Alexandre 12, KitemanSA, Luckas-bot, Yobot, Fomalhaut76, Dmarquard, AnomieBOT, Ciphers, SwiftlyTilt, Rubinbot, Flinders Petrie, Danno uk, Xqbot, Anna Frodesiak, Shadowjams, Rcmaniac25, Biem, Rünno, FrescoBot, Prisonermonkeys, Blubbaloo, Citation bot 1, Javert, Watchpup, RedBot, Nora lives, FoxBot, Rivkid007, NeonSalad, Flodded, Connelly90, Brambleclawx, Techplex.Engineer, RjwilmsiBot, Ripchip Bot, Bossanoven, Noommos, Jowa fan, 1947enkidu, Boundarylayer, Oyoyoy, The Mysterious El Willstro, Dcirovic, Fæ, Redhanker, S trinitrotoluene, SporkBot, Cymru.lass, Jsayre64, SBaker43, ChuispastonBot, HandsomeFella, Mjbmrbot, ClueBot NG, RaptorHunter, Tim PF, O.Koslowski, Mr. D. E. Mophon, Widr, Dougmcdonell, Majesty of the Commons, Barbicanboy, Ndencker, Ondarribia67, Hpelgrift, Iliaarpad, Compfreak7, WebTV3, Cyberbot II, Fbetti9, Raymond1922A, Joeinwiki, ImVeryAwesome, Limnalid, Tistscien, Stamptrader, Sukuluoko, Monkbot, Cccp3, Ceosad, Julietdeltalima, JJMC89, Primetime637, Sam Brown, GreenC bot, Bender the Bot and Anonymous: 247

- **Nuclear safety and security** *Source:* https://en.wikipedia.org/wiki/Nuclear_safety_and_security?oldid=765547099 *Contributors:* Edward, Furrykef, Pstudier, Alan Liefting, Wwoods, Jason Quinn, Andycjp, ConradPino, Antandrus, Beland, Gscshoyru, Gerrit, FT2, Bender235, Kjkolb, Snowolf, Pauli133, DV8 2XL, Woohookitty, Mandarax, Rjwilmsi, Joffan, Vegaswikian, Gurch, Kolbasz, Benjamin Gatti, Bgwhite, Simesa, TexasAndroid, Limulus, Gillis, FF2010, Pb30, Petri Krohn, Tobixen, SmackBot, Gilliam, Chris the speller, Audriusa, Royboycrashfan, Theanphibian, DMacks, Daniel.Cardenas, Mion, Qmwne235, Gobonobo, IronGargoyle, Rkmlai, Cydebot, Gogo Dodo, Hebrides, Clovis Sangrail, DumbBOT, Jelliott4, Dougsim, Headbomb, Dtgriscom, Nick Number, Dreadengineer, AlmostReadytoFly, Magioladitis, Bongwarrior, Canonymous, Twsx, Zatoichi26, Catgut, Cgingold, ArmadilloFromHell, DerHexer, NAHID, Leyo, J.delanoy, ACSE, Johnfos, Alexandria, Philip Trueman, Steve3849, MADe, AlleborgoBot, Fanra, T oldberg, Alpineglider, Buday.csaba, Flyer22 Reborn, Vykk, Wuhwuzdat, ClueBot, Eiland, Niceguyedc, AndyFielding, Layla27, Togokill, Westinghouse 31, Thingg, Silkwood 64, Gotmebegginplease, RadioactiveKiller, Prisonisland 720, DumZiBoT, FellGleaming, Badgernet, Addbot, Some jerk on the Internet, MrOllie, HatlessAtlas, Ben Ben, Luckas-bot, Yobot, Ptbotgourou, Daphonmaster, AnomieBOT, Ulric1313, Gemtpm, Asittig, Nasnema, Jack B108, FrescoBot, Ionium Dope, Tutorboi124, Gaba p, Pinethicket, Notedgrant, Jonesey95, Rotblats09, Cnwilliams, Gerda Arendt, Robertiki, Diannaa, Hullernuc, RjwilmsiBot, AndyHe829, WikitanvirBot, Trofobi, Boundarylayer, Dewritech, Dcirovic, ZéroBot, Daonguyen95, Midas02, Tkreh, Ivolocy, Miguel.baillon, Rememberway, ClueBot NG, Satellizer, SunCountryGuy01, Dougmcdonell, Helpful Pixie Bot, Ledjazz, Neøn, Trefohm, Asauers, Tgiesler, WanoComms, Energy4All, BattyBot, Netherzone, Cyberbot II, Fbetti9, Khazar2, Darryl from Mars, Cerabot~enwiki, Frosty, Jodosma, JamesMoose, Akatian 8912, Limnalid, Internationalpoliticaleconomy, Fixuture, Monkbot, Filedelinkerbot, Demoniccathandler, Fer48, Graemem56, Vivekkush1983, Yufeiyzchn, CaptainCarlosdeCorona, InternetArchiveBot, GreenC bot, Bender the Bot and Anonymous: 100

- **Atomic spies** *Source:* https://en.wikipedia.org/wiki/Atomic_spies?oldid=763814686 *Contributors:* Stevertigo, Fastfission, Ehusman, Discospinster, Bender235, Richard Arthur Norton (1958-), Woohookitty, Stefanomione, Ashmoo, Rjwilmsi, Koavf, Jivecat, Ryk, Vegaswikian, MarnetteD, Ground Zero, Midgley, Nobs01, Hawkeye7, RG2, SmackBot, RedSpruce, Hmains, Moshe Constantine Hassan Al-Silverburg, Colonies Chris, Mwinog2777, Percommode, Stevenmitchell, Drummondt, Atomspytroy, Nishkid64, John, Mike1901, MTSbot~enwiki, Cydebot, Tec15, Difluoroethene, Ronny8, Keith D, Maurice Carbonaro, Theeurocrat, FrummerThanThou, Apostle12, Wikiwonky, Student7, Pphaneuf, Diletante, Synthebot, RHodnett, Hertz1888, IdreamofJeanie, Francvs, Newzild, TomNativeNewYorker, Richard-of-Earth, HexaChord, Addbot, Aboctok, Ederiel, Htews, AnomieBOT, LilHelpa, MeDrewNotYou, Chaheel Riens, LucienBOT, LittleWink, Jandalhandler, Full-date unlinking bot, Mr.98, C4K3, RjwilmsiBot, DASHBot, EmausBot, John of Reading, Italia2006, Ridoking, Jay-Sebastos, Choobashek, Will Beback

43.6.2 Images

- **File:NVZateyev.jpg** *Source:* https://upload.wikimedia.org/wikipedia/en/f/f2/NVZateyev.jpg *License:* Fair use *Contributors:* ? *Original artist:* ?

- **File:NagasakiHypocenter.jpg** *Source:* https://upload.wikimedia.org/wikipedia/commons/6/61/NagasakiHypocenter.jpg *License:* CC BY-SA 2.5 *Contributors:* Transferred from en.wikipedia to Commons. *Original artist:* The original uploader was Unit3000-21 at English Wikipedia

- **File:Nagasaki_City_view_from_Hamahira01s3.jpg** *Source:* https://upload.wikimedia.org/wikipedia/commons/c/ce/Nagasaki_City_view_from_Hamahira01s3.jpg *License:* CC BY 2.5 *Contributors:* Own work *Original artist:* 663highland

- **File:Nagasaki_Lantern_Festival_-_01.jpg** *Source:* https://upload.wikimedia.org/wikipedia/commons/3/39/Nagasaki_Lantern_Festival_-_01.jpg *License:* CC BY 3.0 *Contributors:* Taken by JKT-c. *Original artist:* JKT-c

- **File:Nagasaki_One_Legged_Torii_C1946.jpg** *Source:* https://upload.wikimedia.org/wikipedia/commons/e/ed/Nagasaki_One_Legged_Torii_C1946.jpg *License:* Public domain *Contributors:* Own work *Original artist:* Fg2

- **File:Nagasaki_Trolley_M5199.jpg** *Source:* https://upload.wikimedia.org/wikipedia/commons/7/71/Nagasaki_Trolley_M5199.jpg *License:* Public domain *Contributors:* Own work *Original artist:* Fg2

- **File:Nagasaki_illustration2.jpeg** *Source:* https://upload.wikimedia.org/wikipedia/commons/8/8a/Nagasaki_illustration2.jpeg *License:* Public domain *Contributors:* UBC Library Digital Collections *Original artist:* Ōhata, Bunjiemon

- **File:Nagasaki_in_Nagasaki_Prefecture_Ja.svg** *Source:* https://upload.wikimedia.org/wikipedia/commons/b/b9/Nagasaki_in_Nagasaki_Prefecture_Ja.svg *License:* CC-BY-SA-3.0 *Contributors:* Japanese wiki [1] *Original artist:* ja: 利用者:Lincun

- **File:Nagasaki_panorama.jpg** *Source:* https://upload.wikimedia.org/wikipedia/commons/c/c4/Nagasaki_panorama.jpg *License:* CC BY-SA 3.0 *Contributors:* I took it. *Original artist:* Jason7825

- **File:Nagasaki_peace_memorial_hall.jpg** *Source:* https://upload.wikimedia.org/wikipedia/commons/4/49/Nagasaki_peace_memorial_hall.jpg *License:* CC-BY-SA-3.0 *Contributors:* Own work *Original artist:* Aude

- **File:Nagasakibomb.jpg** *Source:* https://upload.wikimedia.org/wikipedia/commons/e/e0/Nagasakibomb.jpg *License:* Public domain *Contributors:* http://www.archives.gov/research/military/ww2/photos/images/ww2-163.jpg National Archives image (208-N-43888) *Original artist:* Charles Levy from one of the B-29 Superfortresses used in the attack.

- **File:NanbanCarrack.jpg** *Source:* https://upload.wikimedia.org/wikipedia/commons/0/00/NanbanCarrack.jpg *License:* Public domain *Contributors:* Kobe City Museum *Original artist:* Kano Naizen

- **File:Naval_Ensign_of_Russia.svg** *Source:* https://upload.wikimedia.org/wikipedia/commons/4/4c/Naval_Ensign_of_Russia.svg *License:* Public domain *Contributors:* N 162-ФЗ - О знамени Вооруженных Сил Российской Федерации, знамени Военно-Морского Флота, знаменах иных видов Вооруженных Сил Российской Федерации и знаменах других войск *Original artist:* Zscout370, SeNeKa

- **File:Naval_Ensign_of_the_Soviet_Union.svg** *Source:* https://upload.wikimedia.org/wikipedia/commons/6/64/Naval_Ensign_of_the_Soviet_Union.svg *License:* Public domain *Contributors:* http://flagspot.net/flags/su~{}ru.html *Original artist:* User:Zscout370

- **File:Nuclear_Power_History.png** *Source:* https://upload.wikimedia.org/wikipedia/commons/5/58/Nuclear_Power_History.png *License:* CC-BY-SA-3.0 *Contributors:* ? *Original artist:* ?

- **File:Nuclear_power_history.svg** *Source:* https://upload.wikimedia.org/wikipedia/commons/c/c7/Nuclear_power_history.svg *License:* CC0 *Contributors:* Own work *Original artist:* Delphi234

- **File:Nuclear_power_plant_construction.jpg** *Source:* https://upload.wikimedia.org/wikipedia/commons/2/26/Nuclear_power_plant_construction.jpg *License:* CC BY-SA 3.0 *Contributors:* Own work *Original artist:* Ypna

- **File:Nuvola_apps_kaboodle.svg** *Source:* https://upload.wikimedia.org/wikipedia/commons/1/1b/Nuvola_apps_kaboodle.svg *License:* LGPL *Contributors:* http://ftp.gnome.org/pub/GNOME/sources/gnome-themes-extras/0.9/gnome-themes-extras-0.9.0.tar.gz *Original artist:* David Vignoni / ICON KING

- **File:Okonomiyaki_2.jpg** *Source:* https://upload.wikimedia.org/wikipedia/commons/5/5c/Okonomiyaki_2.jpg *License:* CC BY-SA 2.5 *Contributors:* ? *Original artist:* ?

- **File:Operation_Crossroads_-_Moving_of_Bikinians_to_Rongerik,_March_7,_1946.jpg** *Source:* https://upload.wikimedia.org/wikipedia/commons/b/b0/Operation_Crossroads_-_Moving_of_Bikinians_to_Rongerik%2C_March_7%2C_1946.jpg *License:* Public domain *Contributors:* Operation Crossroads - The official pictorial record, page 21 *Original artist:* Joint Task Force 1

- **File:Operation_Emery_-_Baneberry.jpg** *Source:* https://upload.wikimedia.org/wikipedia/commons/b/b2/Operation_Emery_-_Baneberry.jpg *License:* Public domain *Contributors:* This image is available from the National Nuclear Security Administration Nevada Site Office Photo Library under number P-2746. *Original artist:* Photo courtesy of National Nuclear Security Administration / Nevada Site Office

- **File:Ostural-Spur.png** *Source:* https://upload.wikimedia.org/wikipedia/commons/7/79/Ostural-Spur.png *License:* CC BY-SA 3.0 *Contributors:* Own work, background image from maps-for-free.com, File:Russia conic location map.svg *Original artist:* Jan Rieke, maps-for-free.com; Minimap: NordNordWest, Historicair, Bourrichon, Insider, Kneiphof

- **File:Oura_Church_Nagasaki.JPG** *Source:* https://upload.wikimedia.org/wikipedia/commons/f/fa/Oura_Church_Nagasaki.JPG *License:* CC BY-SA 3.0 *Contributors:* ? *Original artist:* ?

- **File:Ozersk_Broadway.jpg** *Source:* https://upload.wikimedia.org/wikipedia/commons/4/41/Ozersk_Broadway.jpg *License:* CC BY-SA 3.0 *Contributors:* Own work *Original artist:* Sergey Nemanov

- **File:Pacific_Fishbowl.png** *Source:* https://upload.wikimedia.org/wikipedia/commons/c/c0/Pacific_Fishbowl.png *License:* Public domain *Contributors:* http://commons.wikimedia.org/wiki/File:Pacific_Proving_Grounds.png (with my own modifications). *Original artist:* U.S. Federal Government produced map with my own additions.

- **File:Pacific_Ocean_laea_location_map.svg** *Source:* https://upload.wikimedia.org/wikipedia/commons/9/96/Pacific_Ocean_laea_location_map.svg *License:* CC BY-SA 3.0 *Contributors:* Own work *Original artist:* Tentotwo

- **File:Richland_Washington.jpg** *Source:* https://upload.wikimedia.org/wikipedia/commons/9/90/Richland_Washington.jpg *License:* Public domain *Contributors:* Image #N1D0060216 at http://www5.hanford.gov/ddrs/index.cfm *Original artist:* USGov-DOE (work of the U.S. federal government)

- **File:Rocky_Flats_Plant_Historic_District.jpg** *Source:* https://upload.wikimedia.org/wikipedia/commons/7/71/Rocky_Flats_Plant_Historic_District.jpg *License:* Public domain *Contributors:* Transferred from en.wikipedia to Commons by FastilyClone using MTC!. *Original artist:* ?

- **File:Room_damaged_by_1969_Rocky_Flats_Fire.jpg** *Source:* https://upload.wikimedia.org/wikipedia/commons/6/63/Room_damaged_by_1969_Rocky_Flats_Fire.jpg *License:* Public domain *Contributors:* http://www.lm.doe.gov/land/sites/co/rocky_flats/haer/base/776.htm *Original artist:* United States Department of Energy

- **File:Rubik'{}s_cube_v3.svg** *Source:* https://upload.wikimedia.org/wikipedia/commons/b/b6/Rubik%27s_cube_v3.svg *License:* CC-BY-SA-3.0 *Contributors:* Image:Rubik'{}s cube v2.svg *Original artist:* User:Booyabazooka, User:Meph666 modified by User:Niabot

- **File:SL-1Burial.jpg** *Source:* https://upload.wikimedia.org/wikipedia/commons/9/96/SL-1Burial.jpg *License:* Public domain *Contributors:* http://ar.inel.gov/images/pdf/200311/2003110400237GSJ.pdf *Original artist:* John R. Giles, EPA

- **File:SL-1_The_Accident_Phases_I_and_II_Animated.gif** *Source:* https://upload.wikimedia.org/wikipedia/commons/4/48/SL-1_The_Accident_Phases_I_and_II_Animated.gif *License:* Public Domain *Contributors:* http://www.archive.org/download/gov.ntis.A13886VNB1/gov.ntis.a13886vnb1.gif *Original artist:* Internet Archive, derivative of a U.S. Government work

- **File:SSFL_Aerial_2005.jpg** *Source:* https://upload.wikimedia.org/wikipedia/commons/9/9e/SSFL_Aerial_2005.jpg *License:* Public domain *Contributors:* http://etec.energy.gov/library/June-2005-Meeting/Presentation.pdf *Original artist:* U.S Department of Energy

- **File:SSFL_Burnpit_illegal_waste.jpg** *Source:* https://upload.wikimedia.org/wikipedia/commons/7/78/SSFL_Burnpit_illegal_waste.jpg *License:* Public domain *Contributors:* US Federal Government documents *Original artist:* ?

- **File:SSFL_barrels_Shooter.jpg** *Source:* https://upload.wikimedia.org/wikipedia/commons/7/71/SSFL_barrels_Shooter.jpg *License:* Public domain *Contributors:* US Federal government source. *Original artist:* ?

- **File:Salmon_at_Hanford_Site.jpg** *Source:* https://upload.wikimedia.org/wikipedia/commons/2/2f/Salmon_at_Hanford_Site.jpg *License:* Public domain *Contributors:* US Department of Energy. (Image 2014455 at http://www.doedigitalarchive.doe.gov/) *Original artist:* US Department of Energy

- **File:SanFranHouses06.JPG** *Source:* https://upload.wikimedia.org/wikipedia/commons/e/ec/SanFranHouses06.JPG *License:* Public domain *Contributors:*

  Scanned from the personal collection of en:User:Infrogmation Originally from en.wikipedia; description page is/was here.

  Original artist: ?

- **File:Sanno_torii_boxed_in_red.jpg** *Source:* https://upload.wikimedia.org/wikipedia/commons/4/4e/Sanno_torii_boxed_in_red.jpg *License:* Public domain *Contributors:* digitized image with red box; Originally photographed by the U.S. Strategic Bombing Survey, 1945; Committee for Research of Photographs and Materials of the Atomic Bombing, Nagasaki Foundation for Promotion of Peace -- Nagasaki Atomic Bomb Museum *Original artist:* Lieutenant R.J. Battersby

- **File:Satellite_image_map_of_Mayak.jpg** *Source:* https://upload.wikimedia.org/wikipedia/commons/b/b0/Satellite_image_map_of_Mayak.jpg *License:* Public domain *Contributors:* NASA World Wind screenshot (Landsat Global Mosaic visual layer) *Original artist:* NASA, Jan Rieke (color correction, borders and labels)

- **File:Seal_of_the_Marshall_Islands.svg** *Source:* https://upload.wikimedia.org/wikipedia/commons/8/86/Seal_of_the_Marshall_Islands.svg *License:* Public domain *Contributors:* Own work *Original artist:* Ericmetro

- **File:Setokazenooka-park01.jpg** *Source:* https://upload.wikimedia.org/wikipedia/commons/9/9d/Setokazenooka-park01.jpg *License:* CC-BY-SA-3.0 *Contributors:* File:Setokazenooka-park01.jpg from the Japanese Wikipedia *Original artist:* Muramasa(talk / Contributions) at the Japanese Wikipedia

- **File:Shadow_picture_of_Nagasaki_prefecture.png** *Source:* https://upload.wikimedia.org/wikipedia/commons/4/45/Shadow_picture_of_Nagasaki_prefecture.png *License:* Copyrighted free use *Contributors:* Data is from here *Original artist:* Created by LERK

- **File:Sikorsky_SH-3G_Sea_King_from_Helicopter_Combat_Support_Squadron_1_in_flight_during_an_aerial_radiation_survey_over_Bikini_Atoll_in_November_1978.jpg** *Source:* https://upload.wikimedia.org/wikipedia/commons/4/4c/Sikorsky_SH-3G_Sea_King_from_Helicopter_Combat_Support_Squadron_1_in_flight_during_an_aerial_radiation_survey_over_Bikini_Atoll_in_November_1978.jpg *License:* Public domain *Contributors:* HD.10.141 *Original artist:* ENERGY.GOV

- **File:Sl-1-ineel61-667.jpg** *Source:* https://upload.wikimedia.org/wikipedia/commons/d/dc/Sl-1-ineel61-667.jpg *License:* Public domain *Contributors:* Proving the Principle. U.S. Department of Energy, Idaho Operations Office. ISBN 0-16-059185-6. http://www.inl.gov/proving-the-principle/chapter_15.pdf *Original artist:* Idaho National Engineering and Environmental Laboratory, INEEL 61-667

- **File:Sl-1-ineel61-9.jpg** *Source:* https://upload.wikimedia.org/wikipedia/commons/5/51/Sl-1-ineel61-9.jpg *License:* Public domain *Contributors:* As found in Proving the Principle by Stacy, Susan M., U.S. Department of Energy, Idaho Operations Office. ISBN 0-16-059185-6. http://www.inl.gov/proving-the-principle/chapter_15.pdf *Original artist:* Idaho National Engineering and Environmental Laboratory, INEEL 61-9

- **File:Sl-1-ineel81-3966.jpg** *Source:* https://upload.wikimedia.org/wikipedia/commons/4/40/Sl-1-ineel81-3966.jpg *License:* Public domain *Contributors:* As found in Proving the Principle by Stacy, Susan M., U.S. Department of Energy, Idaho Operations Office. ISBN 0-16-059185-6. http://www.inl.gov/proving-the-principle/chapter_16.pdf *Original artist:* Idaho National Engineering and Environmental Laboratory, INEEL 81-3966

- **File:Tobol_basin.png** *Source:* https://upload.wikimedia.org/wikipedia/commons/9/9d/Tobol_basin.png *License:* CC BY-SA 3.0 *Contributors:* Own work *Original artist:* СафроновАВ

- **File:Totalexternaldoseratecher.png** *Source:* https://upload.wikimedia.org/wikipedia/commons/0/01/Totalexternaldoseratecher.png *License:* Public domain *Contributors:* Transferred from en.wikipedia to Commons. *Original artist:* Mrmiscellanious at English Wikipedia

- **File:TowerCity-Nagasaki.jpg** *Source:* https://upload.wikimedia.org/wikipedia/commons/8/80/TowerCity-Nagasaki.jpg *License:* CC BY-SA 3.0 *Contributors:* Own work *Original artist:* user:Atsasebo

- **File:Towns_evacuated_around_Fukushima_on_April_11th,_2011.png** *Source:* https://upload.wikimedia.org/wikipedia/commons/b/b0/Towns_evacuated_around_Fukushima_on_April_11th%2C_2011.png *License:* CC BY-SA 3.0 *Contributors:*

- File:Iitate_vs_Fukushima_evacuation_zones_large.svg *Original artist:*

- User:Mayhew

- **File:Twenty-six_Martyrs_of_Japan_2011.jpg** *Source:* https://upload.wikimedia.org/wikipedia/commons/c/cf/Twenty-six_Martyrs_of_Japan_2011.jpg *License:* CC BY-SA 3.0 *Contributors:* 投稿者自身による撮影 *Original artist:* ぱちょぴ

- **File:UCHIDA_KUICHI_Nagasaki.png** *Source:* https://upload.wikimedia.org/wikipedia/commons/9/97/UCHIDA_KUICHI_Nagasaki.png *License:* Public domain *Contributors:* Ebay.com *Original artist:* Uchida Kuichi

- **File:USA_Colorado_location_map.svg** *Source:* https://upload.wikimedia.org/wikipedia/commons/1/19/USA_Colorado_location_map.svg *License:* CC BY 3.0 *Contributors:* own work, using

  - World Data Base II data
  - U.S. Geological Survey (USGS) data

  Original artist: NordNordWest

- **File:USA_Idaho_location_map.svg** *Source:* https://upload.wikimedia.org/wikipedia/commons/7/79/USA_Idaho_location_map.svg *License:* CC BY 3.0 *Contributors:* Own work *Original artist:* Alexrk2

- **File:USA_New_Mexico_location_map.svg** *Source:* https://upload.wikimedia.org/wikipedia/commons/1/10/USA_New_Mexico_location_map.svg *License:* CC BY 3.0 *Contributors:* Own work *Original artist:* Alexrk2

- **File:USS_Tanager_(AM-5).jpg** *Source:* https://upload.wikimedia.org/wikipedia/commons/4/46/USS_Tanager_%28AM-5%29.jpg *License:* Public domain *Contributors:* http://coris.noaa.gov/about/eco_essays/nwhi/surveys.html *Original artist:* Possibly Donald Ryder Dickey, official photographer of the Tanager Expedition. Current photo hosted by NOAA, Marine Field Surveys and Data Collection.

- **File:US_AEC_SL-1.JPG** *Source:* https://upload.wikimedia.org/wikipedia/commons/b/bb/US_AEC_SL-1.JPG *License:* Public domain *Contributors:* English Wikipedia *Original artist:* US Atomic Energy Commission

- **File:Unbalanced_scales.svg** *Source:* https://upload.wikimedia.org/wikipedia/commons/f/fe/Unbalanced_scales.svg *License:* Public domain *Contributors:* ? *Original artist:* ?

- **File:United_Nuclear_Corporation_Church_Rock_Uranium_Mill.jpeg** *Source:* https://upload.wikimedia.org/wikipedia/commons/d/d5/United_Nuclear_Corporation_Church_Rock_Uranium_Mill.jpeg *License:* Public domain *Contributors:* http://theenergylibrary.com/node/11489 *Original artist:* EPA

- **File:Usa_edcp_location_map.svg** *Source:* https://upload.wikimedia.org/wikipedia/commons/2/20/Usa_edcp_location_map.svg *License:* CC BY-SA 3.0 *Contributors:* Own work *Original artist:* Uwe Dedering

- **File:Usa_edcp_relief_location_map.png** *Source:* https://upload.wikimedia.org/wikipedia/commons/f/f3/Usa_edcp_relief_location_map.png *License:* CC BY-SA 3.0 *Contributors:* Own work *Original artist:* Uwe Dedering

- **File:VOA_Markosian_-_Chernobyl02.jpg** *Source:* https://upload.wikimedia.org/wikipedia/commons/1/16/VOA_Markosian_-_Chernobyl02.jpg *License:* Public domain *Contributors:* D. Markosian: *One Day in the Life of Chernobyl*, VOA News, photo gallery. *Original artist:* Diana Markosian

- **File:View_of_Chernobyl_taken_from_Pripyat.JPG** *Source:* https://upload.wikimedia.org/wikipedia/commons/6/6e/View_of_Chernobyl_taken_from_Pripyat.JPG *License:* Public domain *Contributors:* This photo is the author's own work *Original artist:* Jason Minshull

- **File:View_of_Chernobyl_taken_from_Pripyat_zoomed.JPG** *Source:* https://upload.wikimedia.org/wikipedia/commons/e/e0/View_of_Chernobyl_taken_from_Pripyat_zoomed.JPG *License:* Public domain *Contributors:* Image:View of Chernobyl taken from Pripyat.JPG *Original artist:* Jason Minshull

- **File:WIPP_DoE_2014-05-15_5_15_Image_lrg.jpg** *Source:* https://upload.wikimedia.org/wikipedia/commons/a/af/WIPP_DoE_2014-05-15_5_15_Image_lrg.jpg *License:* Public domain *Contributors:* http://www.wipp.energy.gov/wipprecovery/photo_video.html *Original artist:* DoE photographer

- **File:Western_Style_Houses_at_Higashiyamate_Nagasaki_Japan05s3.jpg** *Source:* https://upload.wikimedia.org/wikipedia/commons/f/f5/Western_Style_Houses_at_Higashiyamate_Nagasaki_Japan05s3.jpg *License:* CC BY 2.5 *Contributors:* Own work *Original artist:* 663highland

- **File:Wfm_sts_overview.png** *Source:* https://upload.wikimedia.org/wikipedia/commons/c/cb/Wfm_sts_overview.png *License:* CC-BY-SA-3.0 *Contributors:* ? *Original artist:* ?

- **File:Whippoorwill_(AT-O-$-$169).jpg** *Source:* https://upload.wikimedia.org/wikipedia/commons/b/b1/Whippoorwill_%28AT-O-$-$169%29.jpg *License:* Public domain *Contributors:* http://www.navsource.org/archives/11/02035.htm *Original artist:* National Archives photo

43.6.3   Content license